Environmental Issues of Deep-Sea Mining

Rahul Sharma
Editor

Environmental Issues of Deep-Sea Mining

Impacts, Consequences and Policy Perspectives

Springer

Editor
Rahul Sharma
CSIR-National Institute of Oceanography
Dona Paula, Goa, India

ISBN 978-3-030-12695-7 ISBN 978-3-030-12696-4 (eBook)
https://doi.org/10.1007/978-3-030-12696-4

Cover Image: A schematic showing the processes involved in deep-sea mining for the three main types of mineral deposits. (Left to Right: hydrothermal sulphides, polymetallic nodules, ferromanganese crusts - Not to scale) (Adopted from: Kathryn A. Miller, Kirsten Thompson, Paul Johnston, David Santillo, 2018. An Overview of Seabed Mining Including the Current State of Development, Environmental Impacts, and Knowledge Gaps. Front. Mar. Sci., volume 4, https://doi.org/10.3389/fmars.2017.00418).

This Springer imprint is published by the registered company Springer Nature Switzerland AG
The registered company address is: Gewerbestrasse 11, 6330 Cham, Switzerland

Below: Seabed photographs of benthic organisms associated with deep-sea minerals in different oceans:

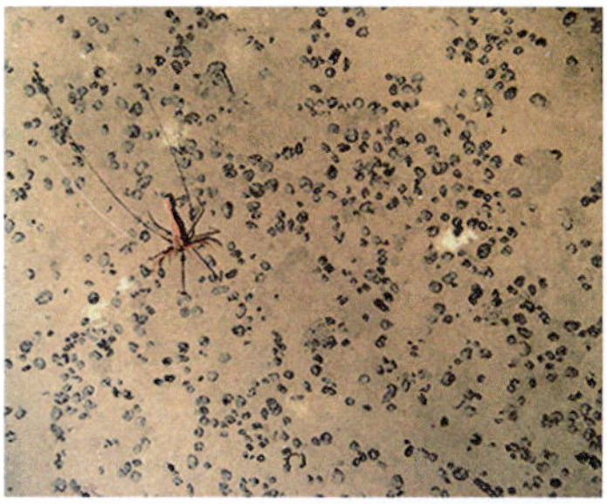

Image of the seafloor in the abyssal Pacific showing manganese nodules and large deep-water prawn (Bathystylodactylus sp.). Image shows an area of seafloor approximately 50cm across. (Credit: Image courtesy Dr Daniel Jones, National Oceanography Centre, Southampton)

Typical area of rocky seabed away from the ridge axis with the crinoid Anachalypsicrinus nefertiti and some large sponges. Mid Atlantic Ridge, depth c. 2400m. (Credit: Image courtesy Dr Daniel Jones, National Oceanography Centre, Southampton, UK. ECOMAR Project)

Abundant Chrysomallon squamiferum and Gigantopelta aegis, with Kiwa n. sp. “SWIR”, Bathymodiolus marisindicus, and Mirocaris fortunata on platform of “Tiamat” vent chimney, Southwest Indian Ridge, depth 2778m (Credit: Image courtesy NERC University of Southampton, SWIR_2011-11-27_10-24-08_James Cook_JC67_2_ROV01)

Foreword

INTERNATIONAL SEABED AUTHORITY

Foreword

It is a great pleasure once again to write a foreword to a second volume by Dr. Rahul Sharma on the subject of deep sea mining. Whilst his previous volume focused on issues related to resource potential and technology development, the present volume turns its attention to the environmental impacts and consequences of deep sea mining and the management responses that will be needed to manage, reduce and mitigate negative environmental impacts.

Part XI of the United Nations Convention on the Law of the Sea and its 1994 Implementing Agreement, together with the rules, regulations and procedures that have been developed by the International Seabed Authority since 1994, contain a comprehensive and complete body of international law governing environmental management of deep sea mining. These include detailed and substantive provisions for the assessment of possible environmental impacts arising from exploration for marine minerals in the Area, as well as guidance on baseline studies, monitoring and reporting.

Currently, the Authority is engaged in the development of a comprehensive code for exploitation of marine mineral resources, which includes detailed provisions relating to the protection and preservation of the marine environment. Under the current draft, contractors would be obligated to submit an environmental impact statement to document and report the results of an environmental impact assessment process which must be designed to identify, predict, evaluate and mitigate the biophysical, social and other relevant effects of proposed mining operations.

Proper environmental management relies upon the availability of the best scientific advice and information. The papers in the present volume not only provide a comprehensive overview of the state of scientific knowledge regarding the impacts of marine mining but also review some of the policy options available to managers for managing those impacts.

Thus, the first section of the volume provides an overview of the environmental issues associated with deep sea mining from geological, biological and geochemical perspectives. The second part of the volume provides a more detailed review of the status of the scientific studies and modelling work that has been done over the past 30 years to estimate the impacts of mining activities. The book then goes on to review the available policy responses, including issues of data management, environmental impact assessment, mitigation technology and environmental management practices.

Overall, this book makes an important contribution to the extensive and rapidly growing body of scholarly literature addressing the potential effects on the marine environment of deep-sea mining. It must be emphasized that deep-sea mining, considered both on its own and in the context of the overall global stressors on the health of this planet, has much to offer that is environmentally constructive. The present volume does a good job of putting the discussion of the environmental impacts of deep sea mining into context and presents the various management options and responses available in a balanced way. I congratulate Dr. Sharma and the experts that have contributed to this volume on their achievement.

Michael Lodge
Secretary General
International Seabed Authority
Kingston, Jamaica

14-20 Port Royal Street, Kingston, Jamaica • Phone: (876) 922-9105-9 • Fax: (876) 922-0195 • E-mail: postmaster@isa.org.jm

OFFICE OF THE PERMANENT OBSERVER FOR THE INTERNATIONAL SEABED AUTHORITY TO THE UNITED NATIONS
One United Nations Plaza, Room 1140, New York, N.Y. 10017, United States of America
Phone: (212) 963-6470, 963-6411 • Fax: (212) 963-0908

Foreword

सत्यमेव जयते
डॉ. एम. राजीवन
DR. M. RAJEEVAN

सचिव
भारत सरकार
पृथ्वी विज्ञान मंत्रालय
पृथ्वी भवन, लोदी रोड़, नई दिल्ली-110003
SECRETARY
GOVERNMENT OF INDIA
MINISTRY OF EARTH SCIENCES
PRITHVI BHAVAN, LODHI ROAD, NEW DELHI-110003

FOREWORD

Mining of deep-sea minerals has generated significant interest in recent years owing to their potential as alternative source of metals to the depleting land resources. Whereas deep-sea mining could be a possible option to meet our demands for critical metals in future, maintaining an environmental balance of these hitherto undisturbed marine environs is a challenge for the mankind.

All the stakeholders such as the contractors, researchers, technocrats as well as the policy makers and financial institutions have a responsibility to ensure a sustainable exploitation of these mineral resources. In this endeavor, the International Seabed Authority is playing a key role in developing guidelines for regulating the exploitation of these minerals that form a common heritage of mankind.

I commend the efforts of all the authors who have contributed chapters towards this book, the editor for initiating and putting together this volume and the publishers for providing a platform for this book at this opportune time.

I urge the world community to come forward and make deep-sea mining a successful venture while ensuring environmental balance in the marine ecosystem.

(M. Rajeevan)

Tel. : +91-11-24629771, 24629772 ❑ Fax : +91-11-24629777 ❑ E-mail : secretary@moes.gov.in

Preface

Deep-sea mining is currently in a transitional phase between exploration and exploitation of deep-sea mineral deposits that are projected as alternative source of metals to depleting land resources in future. On one hand, long-term prospecting and resource evaluation has led to the identification of potential mining areas on the deep seafloor. On the other, the development of mining and processing technology is gaining momentum, with a few entities planning their sea trials in the near future. However, the commencement of mining of deep-sea minerals on a commercial scale depends on metal prices and their availability in the world market.

In view of the concerns over potential disturbances in the marine environment due to various offshore and onshore activities, the world community is focusing its attention to the environmental issues of deep-sea mining. This is more so because many of the deep-sea minerals occur in the "Area", that is, areas that lie in international waters beyond the national jurisdiction of any state. As the mining operations could be expected to commence in the coming decades, pertinent questions that need to be answered include what are the possible environmental impacts, who is responsible for it, how do we regulate the activities in this area, what if the concerned party does not (or cannot) do anything about it, what are the mitigation measures, and how do we restore or conserve the marine environment.

This book brings forth various issues with contributions from leading experts under different themes such as the environmental issues of deep-sea mining, its potential impacts, environmental data standardization and applications, environmental management, and economic considerations. The contributions from all the authors are highly acknowledged with a hope that this book will serve as a comprehensive reference material for addressing various environmental issues of deep-sea mining.

As deep-sea mining is an activity of the future, with increasing environmental awareness, it is incumbent on all stakeholders, including the potential contractors, the sponsoring states, the international regulating agencies, and the environmental groups, to devise strategies for economically and environmentally sustainable deep-sea mining ventures to meet the future demand for metals in the world and preserve the marine environment within acceptable limits.

It is important to realize that just as it is our responsibility to give a healthy environment to the next generation, it is equally incumbent on us to ensure the availability of adequate resources for their future.

Dona Paula, Goa, India Rahul Sharma

Acknowledgments

This book on *Environmental issues of Deep-Sea Mining – Impacts, Consequences and Policy Perspectives* is a sequel to *Deep-Sea Mining – Resource Potential, Technical and Environmental Considerations* published by Springer in 2017. Both the publications have been possible due to the confidence entrusted by the publishers in the topics addressed in these books. I acknowledge the support extended by them in this endeavor, in particular Dr. Sherestha Saini, Mr. Aaron Schiller, and Ms. Susan Westendorf from the Springer New York Office, as well as the staff of SPi Global, particularly Ms. M. K. Chandhini and Ms. S. Kanimozhi for production of the book.

All the authors of the chapters deserve a special mention for their outstanding contributions, despite having multiple commitments, that has made this publication possible. Each chapter is unique in its content, and the ideas presented give the book a broad perspective. This shows the rich expertise that the authors have and their willingness to share the same is highly appreciated.

The Foreword by Mr. Michael Lodge, Secretary General, International Seabed Authority, Jamaica, gives a comprehensive overview of the issues related to the subject of deep-sea mining and environment and sets the tone for this book. Also the Foreword by Prof. M. Rajeevan, Secretary, Ministry of Earth Sciences (Government of India), New Delhi, provides a way forward in the field of deep-sea mining and environmental conservation. The encouragement and support received from Mr. Lodge and Prof. Rajeevan are sincerely acknowledged.

This book is the result of a suggestion from Dr. T. R. P. Singh, Ex-General Manager, Engineers India Limited, New Delhi, to bring together a large volume of information on the subject in one place, including the experimental data, regulations, and management of deep-sea mining from an environmental perspective. Discussions with officials of the Ministry of Earth Sciences, Government of India, as well as the inputs of Prof. PK Sen, IIT Kharagpur, were very helpful during this project and in writing my chapters.

CSIR-National Institute, Goa, where I have worked for almost 36 years, holds a very special place in shaping my career and developing my understanding of the subject that led me to take up the challenge of putting this book together.

Special thanks are due to my colleagues for their inputs as well as the directors of the Institute for their support during the compilation of this book.

And finally, the members of my immediate as well as extended families have been the source of constant encouragement through this endeavor, and their support is highly appreciated.

May God bless us all.

Rahul Sharma
CSIR-National Institute of Oceanography
Dona Paula, Goa, India

Contents

Part I
Environmental Issues

Deep-Sea Mining and the Environment: An Introduction

Rahul Sharma and Samantha Smith

Abstract Seafloor minerals, many of which occur in the deep ocean in international waters, have attracted significant attention due to the discovery of deposits with high metal grades and large volumes, in addition to the growth in global demand for strategic metals such as copper, nickel, cobalt, and rare earths. Furthermore, much of the world is recognizing the need to transition to a clean energy, low-carbon economy, and to do so requires metals used in clean energy infrastructure and technologies, metals such as manganese, nickel, copper, and cobalt (World Bank 2017), the same metals found in, for example, polymetallic nodule deposits. This has led to several entities obtaining exploration contracts for areas of the seafloor governed under international regulations and developing technologies for their extraction. At the same time, environmental groups have raised concerns over the possible environmental impacts of deep-sea mining on seafloor and deep-sea ecosystems. This chapter provides an overview of the general environmental issues and concerns being raised in relation to deep-sea mining, introduces some of the mechanisms being put in place to ensure the effective protection of the marine environment, and raises pertinent questions that are being or will need to be addressed as the deep-sea minerals industry moves forward into reality.

Keywords Deep-sea mining · Environmental issues · Sustainable development

R. Sharma (✉)
CSIR-National Institute of Oceanography, Dona Paula, Goa, India

S. Smith
Blue Globe Solutions, Toronto, ON, Canada

Nauru Ocean Resources Inc., Aiwo, Republic of Nauru
e-mail: samantha@blueglobesolutions.com

R. Sharma (ed.), *Environmental Issues of Deep-Sea Mining*,
https://doi.org/10.1007/978-3-030-12696-4_1

1 Background

It is well known that any human interference with nature perturbs natural conditions. Seas and open oceans are often thought of by the general public as pristine parts of the earth's surface that have regularly symbolized relatively undisturbed, well-balanced ecosystems that humankind would like to preserve eternally, especially after having seen the ill effects of anthropogenic activities on land. However, the oceans, including the deep sea, are not entirely pristine with a number of activities already occurring such as shipping, waste disposal (including nuclear, plastics, and mine tailings), fishing including bottom trawling, and others. Increasing demand for resources in order to satisfy the growing requirements of humankind have also pushed the boundaries of exploring and exploiting marine resources in the last few decades, in shallow waters, and in the deep sea.

One such marine resource entails seafloor mineral deposits such as polymetallic nodules, polymetallic/hydrothermal/seafloor massive sulfides and ferromanganese/cobalt-rich crusts (Table 1). These deposits are considered alternatives to depleting land resources of strategic metals such as copper, nickel, cobalt, lead, zinc, molybdenum, platinum (Cronan 1980; Rona 2003), and rare earths (Takaya et al. 2018) that are required for various industrial as well as domestic purposes (Lenoble 2000; Glumov et al. 2000; Kotlinski 2001) (Table 2).The largest known deposits are located in the international seabed area, called "The Area," and all activities in relation to these seabed resources are regulated by the International Seabed Authority (ISA) established in 1994 under the 1982 United Nations Convention on the Law of the Sea and the 1994 Agreement relating to the Implementation of Part XI of the United Nations Convention on the Law of the Sea (www.isa.org.jm). Currently ISA has signed 17 exploration contracts for polymetallic nodules, 7 for polymetallic sulfides, and 5 for ferromanganese/cobalt-rich crusts (Tables 3a, 3b, and 3c) in different oceans (Fig. 1a–e).

Table 1 Salient features of deep-sea minerals

Type	Description	Volume	Metals and their mean concentration[a]	Principal deposits
Polymetallic nodules	Concretions of layered iron and manganese oxides with associated metals from the water column or sediment	Nodules: average 5–10 cm; deposits: up to thousands of km^2	*Mn* (28,4%), *Ni* (13 ppm), *Cu* (10,7 ppm), *Co*(2098 ppm), *Mo* (590 ppm), *Zn* (1366 ppm), *Zr* (307 ppm), *Li*(131 ppm), *Pt* (128 ppm), *Ti* (199 ppm), *Y* (96 ppm), *REEs* (813 ppm) (CCZ)	Clarion-Clipperton Zone, Peru Basin, Central Indian Ocean and Penrhyn Basin

(continued)

Table 2 (continued)

Type	Description	Volume	Metals and their mean concentration[a]	Principal deposits
Seafloor massive sulfides (SMS)	Concentrated deposits of sulfidic minerals (>50–60%) resulting from hydrothermal activity on the seabed	Up to several km^2; up to tens of meters thick	*Cu* (0.8–17.9%), *Au* (0.4–13.2 ppm), *Ag* (64–1260 ppm), *Zn* (2.7–17.5%), *Pb* (0.02–9.7%), *Co, As, Al, Si, REEs*	Red Sea, back-arc basins, mid-oceanic ridges, and other plate boundaries, oceanic hotspots (intraplate volcanoes)
Ferromanganese crusts	Layered manganese and iron oxides with associated metals on hard substrate rock of subsea mountains and ridges	Up to several km2; <0.3 m thick	*Mn* (21%), *Co* (6647 ppm), *Ni* (4326 ppm), *Cu* (573 ppm), *Te* (34 ppm), *Mo* (431 ppm), *Zr* (423 ppm), *Ti* (TiO_2–1.4%), *Pt* (0.273 ppm), *W* (68 ppm), *REEs* (1628 ppm)	Equatorial Pacific Ocean and Central Atlantic Ocean

Modified from Cuyvers et al. (2018)
[a]Concentrations for sulfides from Cherkashov (2017), nodules from Hein et al. (2013), and crusts from Halbach et al. (2017)

Table 2 Uses and status of key metals found in deep-sea minerals

Metal	Main uses	World reserves on land in 2018 (https://www.usgs.gov)	Production rate in 2016 (https://www.usgs.gov)	Increase in production rate per year, %
Cu	Electric energy transmission (26%), electric motors (12%), traction motor (9%), household heating appliances (8%), data transfer/communication (5%), architecture and consumer goods (10%), water supply (13%), mechanical components (6%), electronic contact/heat conduction (3%), car wiring (5%), others (3%) (Zepf et al. 2014)	790 million t	20,100 thousand t	3.1
Ni	Stainless/alloy steel (66%), nonferrous alloys and super alloys (18%), electroplating (8%), others (8%) (Zepf et al. 2014), increasingly used in energy storage units (e.g., Li-ion batteries)	74 million t	2,090,000 t	3.7
Co	Batteries (27%), super alloys and magnets (26%), hard metals (14%), pigments (10%), catalysts (9%), others (14%) (Zepf et al. 2014)	7,100,000 t	111,000 t	8.3

(continued)

Table 2 (continued)

Metal	Main uses	World reserves on land in 2018 (https://www.usgs.gov)	Production rate in 2016 (https://www.usgs.gov)	Increase in production rate per year, %
Mn	Metallurgy, aluminum alloys, reagent in organic chemistry, batteries, coinage (https://en.wikipedia.org)	680 million t (manganese content in ore)	15,700 thousand t	4.3
Fe	Metallurgy, industry, alloys, automobiles, machines, trains, ships, buildings, glass (https://en.wikipedia.org)	83,000 million t (iron content in ore)	1450 million t (iron content of usable ore)	5.1
Pb	Lead bullets, protective sheath for underwater cables, construction industry, brass and bronze, lead-acid batteries, oxidizing agent in organic chemistry, lead-based semiconductors (https://en.wikipedia.org)	88 million t	4710 thousand t	2.6
Zn	Galvanizing, alloys, anode material for batteries, manufacture of chemicals, daily vitamin and mineral supplement, cosmetics (https://en.wikipedia.org)	230 million t	12,600 thousand t	2.9
Cd	Rechargeable batteries, photovoltaic cells, pigment in paints, stabilizers in plastics, corrosion-resistant coatings and plating (Zepf et al. 2014)	500,000 t (information of 2014)	23,900 t	1.1
Mo	Carbon steel (35%), chemicals and catalysts (14%), stainless steel (25%), tool steel (9%), cast iron (6%), molybdenum metal (6%), others (5%) (Zepf et al. 2014)	17 million t	279 million t	4.2
Pt	Autocatalyst (40%), jewelry (35%), investment (6%), medical and biomedical (3%), glass (2%), chemicals (6%), electrical (2%), petroleum (2%), others (4%) (Zepf et al. 2014)	69,000,000 kg (PGM)	191,000 kg	1.7
Au	Coinage, jewelry, industry (10%), electrical contacts, alloys (https://en.wikipedia.org)	54,000 t	3110 t	1.4
Ag	Jewelry (34%), electronics (24%), photography/mirrors (20%), catalysts (6%), others (16%) (Zepf et al. 2014)	530,000 t	25,700 t	2.5
REE	Magnets (25%), catalysts (24%), batteries (15%), polishing (11%), glass (6%), steel (9%), others (10%) (Zepf et al. 2014)	120 million t	129,000 t	2.9

Table 3a Contractors for exploration of polymetallic nodules

Contractor	Sponsoring State	General location of the exploration area under contract	Contract start date
InterOceanMetal Joint Organization	Bulgaria, Cuba, Czech, Poland, Russia, Slovakia	Clarion-Clipperton Fracture Zone (CCFZ), Pacific Ocean	29 March 2001
JSC Yuzhmorgeologiya	Russia	CCFZ, Pacific Ocean	29 March 2001
Government of the Republic of Korea	Republic of Korea	CCFZ, Pacific Ocean	27 April 2001
China Ocean Mineral Resources Research and Development Association	China	CCFZ, Pacific Ocean	22 May 2001
Deep Ocean Resources Development Co.	Japan	CCFZ, Pacific Ocean	20 June 2001
Institut français de recherché pour l'exploitation de lamer	France	CCFZ, Pacific Ocean	20 June 2001
Government of India	India	Indian Ocean	25 March 2002
Federal Institute for Geosciences and Natural Resources of Germany	Germany	CCFZ, Pacific Ocean	19 July 2006
Nauru Ocean Resources Inc.	Nauru	CCFZ, Pacific Ocean	22 July 2011
Tonga Offshore Mining Limited	Tonga	CCFZ, Pacific Ocean	11 January 2012
Global Sea Mineral Resources NV	Belgium	CCFZ, Pacific Ocean	14 January 2013
UK Seabed Resources Ltd. – I	UK and Northern Ireland	CCFZ, Pacific Ocean	8 February 2013
Marawa Research and Exploration Ltd.	Kiribati	CCFZ, Pacific Ocean	19 January 2015
Ocean Mineral Singapore Pte Ltd	Singapore	CCFZ, Pacific Ocean	22 January 2015
UK Seabed Resources Ltd. – II	UK and Northern Ireland	CCFZ, Pacific Ocean	29 March 2016
Cook Islands Investment Corporation	Cook Islands	CCFZ, Pacific Ocean	15 July 2016
China Minmetals Corporation	China	CCFZ, Pacific Ocean	12 May 2017

Table 3b Contractors for exploration of ferromanganese crusts

Contractor	Sponsoring State	General location of the exploration area under contract	Contract start date
Japan oil, Gas and Metals National Corporation	Japan	Pacific Ocean	27 January 2014
China Ocean Mineral Resources Research and Development Association	China	Western Pacific Ocean	29 April 2014
Ministry of Natural Resources and Environment of the Russian Federation	Russia	Pacific Ocean	10 March 2015
Companhia De Pesquisa de Recursos Minerais	Brazil	South Atlantic Ocean	9 November 2015
Republic of Korea	Republic of Korea	Western Pacific Ocean	27 March 2018

Table 3c Contractors for exploration of hydrothermal sulfides

Contractor	Sponsoring State	General location of the exploration area under contract	Contract start date
China Ocean Mineral Resources Research and Development Association	China	Southwest Indian Ridge	18 November 2011
Government of the Russian Federation	Russia	Mid-Atlantic Ridge	29 October 2012
Government of the Republic of Korea	Republic of Korea	Central Indian Ridge	24 June 2014
Institut français de recherche pour l'exploitation de la mer	France	Mid-Atlantic Ridge	18 November 2014
Federal Institute for Geosciences and Natural Resources of Germany	Germany	Southeast and Central Indian Ridge	6 May 2015
Government of India	India	Central Indian Ocean	26 September 2016
Government of Republic of Poland	Poland	Mid-Atlantic Ridge	12 February 2018

Source: www.isa.org accessed on 2 December 2018

a

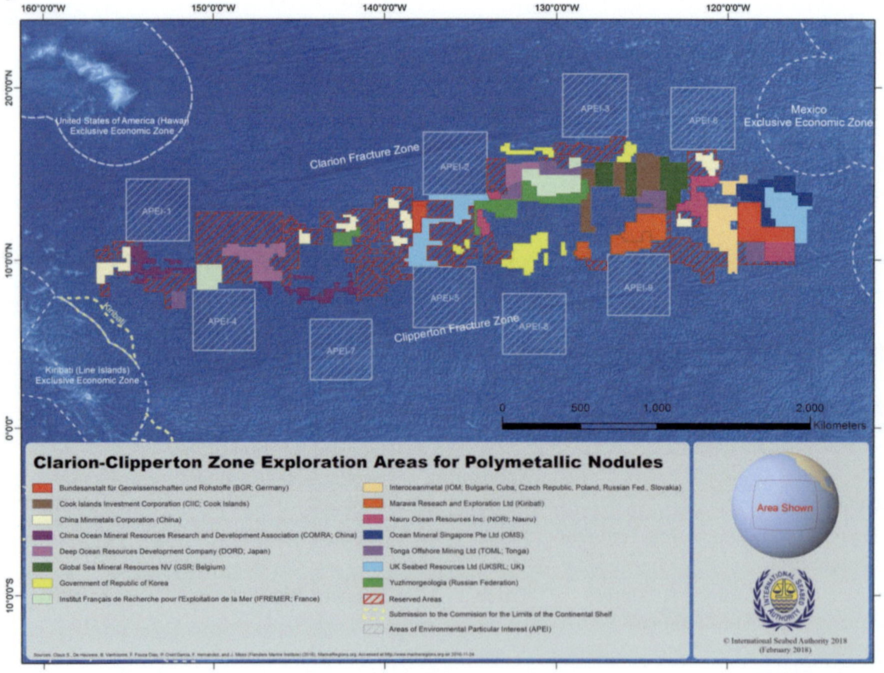

Fig. 1 (**a**) Exploration areas for polymetallic nodules in Clarion-Clipperton Zone, Pacific Ocean. (Courtesy: International Seabed Authority, Jamaica). (**b**) Exploration areas for polymetallic nodules and sulfides, Indian Ocean. (Courtesy: International Seabed Authority, Jamaica). (**c**): Exploration areas for polymetallic sulfides on the Mid-Atlantic Ridge. (Courtesy: International Seabed Authority, Jamaica). (**d**) Exploration areas for Cobalt-rich ferromanganese crusts in the Pacific Ocean. (Courtesy: International Seabed Authority, Jamaica). (**e**) Exploration areas for cobalt-rich ferromanganese crusts on South Atlantic seamounts. (Courtesy: International Seabed Authority, Jamaica)

b

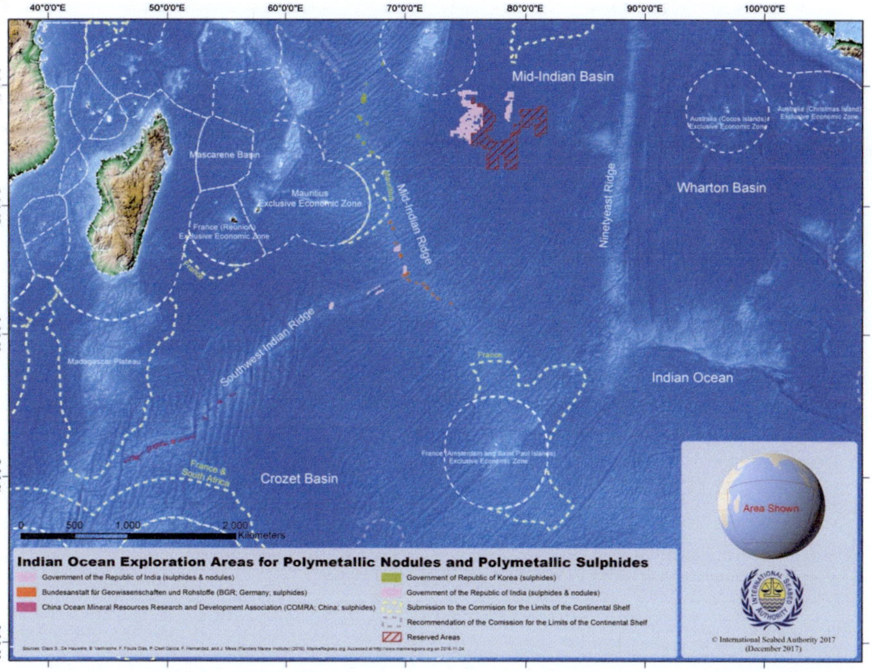

c

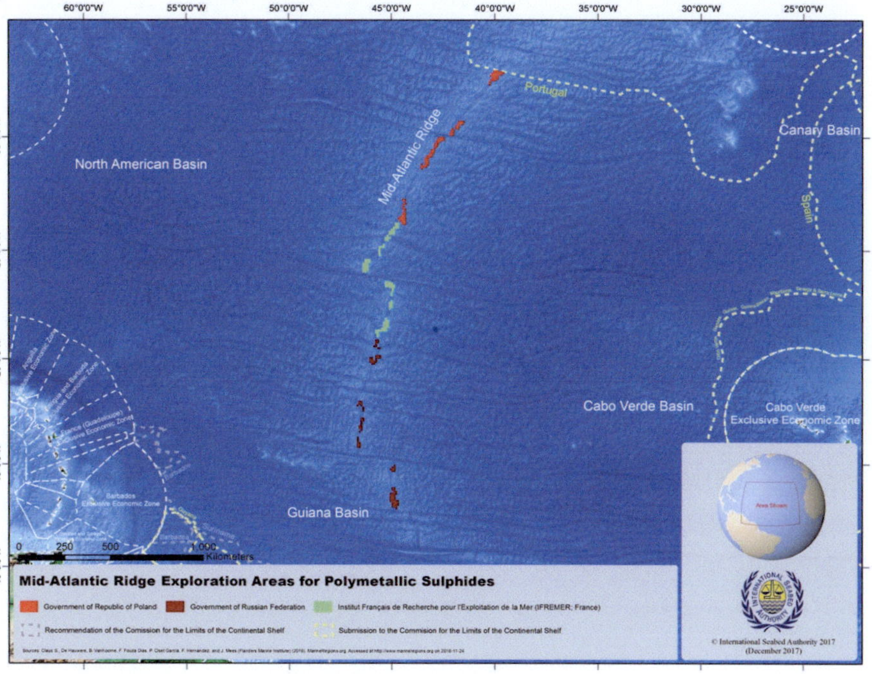

Fig. 1 (continued)

d

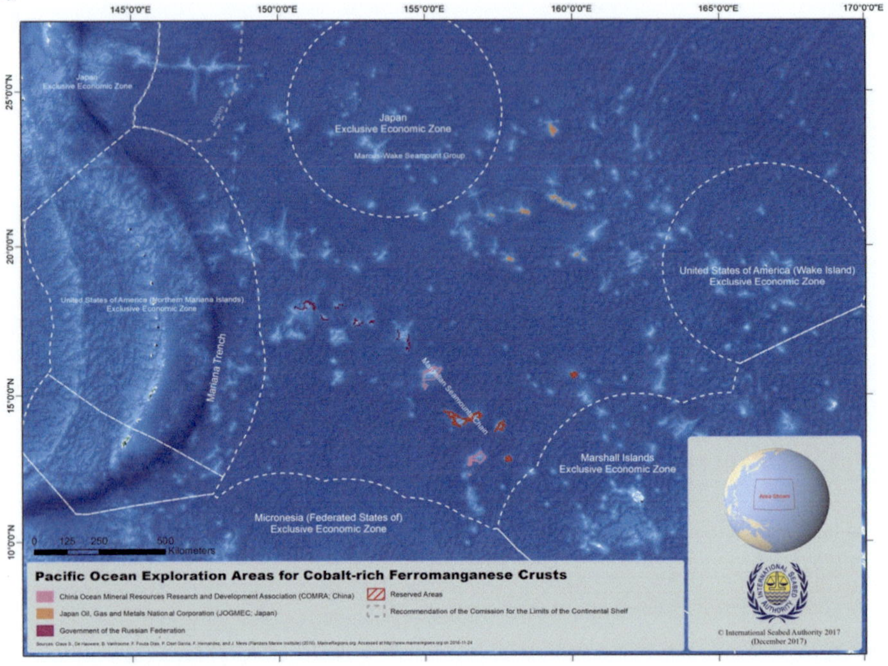

e

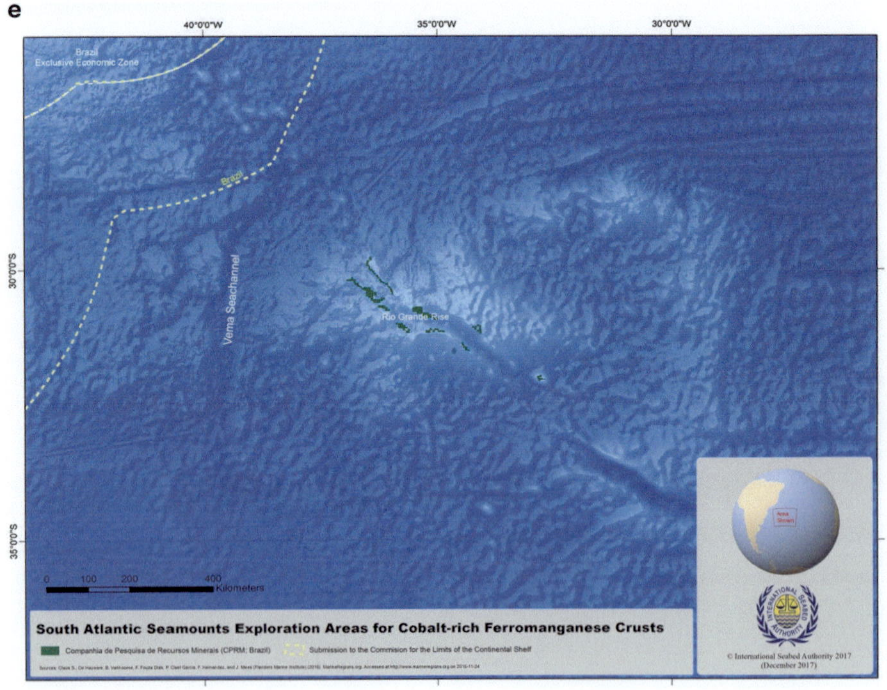

Fig. 1 (continued)

Seafloor mineral exploration in the deep ocean is sometimes considered akin to exploring beyond the normal limits of human endeavors because of the extreme conditions associated with the deep-sea environment such as:

(i) Most deep-sea mineral deposits are located in the international seabed area, at least 1000 km from the nearest landmass or habitation.
(ii) The deposits are associated with geological features such as deep abyssal plains (nodules), mid-oceanic ridges and back-arc regions (sulfides), and seamounts (crusts) that generally occur at the depths of 1.5–6 km below the ocean's surface.
(iii) The deposits occur under extreme environmental conditions such as high pressure (150–600 bars), complete darkness, and, sometimes, complex current regimes.

ISA has put in place regulations for prospecting and exploration for polymetallic sulfides (ISA 2010), ferromanganese crusts (ISA 2012), and polymetallic nodules (ISA 2013a) and, at the time of writing, exploitation regulations are being developed, with an expected completion date around 2020 (ISA 2018).

2 Key Issues of Deep-Sea Mining

Concerns of possible damages to the marine environment have been raised through several articles in the scientific literature (e.g., Van Dover et al. 2017). Several benthic impact experiments conducted to understand the biological responses to disturbances associated with nodule extraction have reported variable results due to different means adopted for conducting the studies as well as different time scales of monitoring on the restoration process (summarized by Jones et al. 2017). Most of these experiments entailed plowing or suction mechanisms to mimic nodule collection and disturbing the seafloor conditions leading to vertical mixing and lateral migration of sediments, alteration in physicochemical conditions, and reduction in biomass (Sharma et al. 2001). Not only was the scale of disturbance caused by these experiments much smaller than what is expected from a large-scale mining operation (Yamazaki and Sharma 2001), most of these studies were restricted to studying the impacts on the seafloor from where the minerals will be picked up and did not include the study of secondary effects such as sediment redistribution (i.e., sediment plumes). The concern is that the sediment plume could smother benthic organisms (Thiel et al. 1997). It is expected that knowledge around sediment plumes will soon be greatly advanced (e.g., through programs such as JPI Oceans II) as projects move closer to production and collect long-term physical oceanography data, allowing for plume modeling and then validation testing of the anticipated plumes through, for example, component testing of prototype mineral harvesting vehicles offshore.

In the water column, the main effects from full-scale mining operations are expected to occur as a result of the presence of a lifting system designed to take the minerals from the seafloor to the sea surface (most if not all Contractors are designing these to be fully enclosed) and the occasional passage of the mineral harvesting tools and remotely operated vehicles.

Additionally, on board a surface vessel, the mineral will be separated from seawater in a dewatering plant, and the seawater (and any remaining sediment) will be discharged back to the ocean, at some depth below the euphotic (light) zone and possibly back at depth near the seafloor (this discharge is called "return water").

It is also possible that there could be accidental discharges, for example, if the lifting system were to break and all contents were lost. In this case, the impact is expected to be short lived given the minerals should sink back to the seafloor.

Any discharge (through normal operations or accidental events) could locally increase the turbidity in the water column at the depth of discharge, and some spreading is likely to occur and could possibly affect productivity (Pearson 1975; ISA 1998), although how real or large an issue this is likely depends primarily on the depth of discharge and may not be a major issue if Contractors do as currently expected and design their mining systems to avoid surface and shallow water mining discharges.

Transportation of several thousand tons of mineral ore to land for onshore processing would require ore carriers adding somewhat to maritime traffic in the associated region and would also increase the possibility of oil spills, accidental losses of ship or large equipment at sea, and, also possible, although unlikely for modern and reputable maritime operators, unintentional or intentional dumping of garbage that cannot be monitored easily in open seas (Pearson 1975).

On land, the minerals obtained from the seafloor will be processed to recover the metals they contain. Following mineral processing onshore, any waste or tailings left behind after extraction of metal from the ores will need to be disposed suitably so as to avoid the risk of serious impacts on land. Due to the often high-grade and multi-metal nature of seafloor mineral deposits, it is anticipated that the waste generated from seafloor mining operations will be significantly less than the industry's land-based counterparts.

A detailed description of the likely environmental impacts of nodule, crust, and sulfide mining is provided by Weaver and Billett (2019), Chap. 3, this volume.

3 Major Concerns Raised Around Deep-Sea Mining

In light of incomplete knowledge as well as perceived threats, several concerns are being raised in relation to deep-sea mining that need to be addressed. Some of these are discussed below:

3.1 *Concern Raised: Large Areas of Seafloor Beneath All Oceans Will Be Mined*

Seafloor areas up to 75,000 km^2 for nodules, 2500 km^2 for sulfides, and 1000 km^2 for crusts could be allotted to each Contractor for exploitation through a contract with the ISA in the international seabed area (ISA 2018), and, although extremely unlikely, there seems to be a concern that all of the areas currently under exploration contract could be converted to exploitation contracts at the same time. Considering that currently there are 17 contracts for nodules, 7 for sulfides, and 5 for crusts (Tables 3a, 3b, and 3c), the total area of the seafloor under contract could be 1,297,500 km^2. A seemingly common fear is that the entire ~1.3 million km^2 would be mined simultaneously at the time when mining commences. Some of the facts (with examples specific to polymetallic nodule deposits) that need consideration are:

(i) An area of 75,000 km^2 with a minimum abundance of 5 kg/m^2 (which is estimated as an example cutoff abundance for commercial viability) would contain a resource of 375 million tons (wet) or 280 million tons of (dry) nodules that can provide resources for 187 years of mining at an annual mining rate of 1.5 million tons containing 2.8 million tons of nickel and copper each year (at 1% concentration) and 0.28 million tons of cobalt (at 0.1%) and 61 million tons of manganese (at 24%) (Sharma 2017). This means that even if a few mines are operational in different oceans, they can cater to the world's demand of copper, nickel, and cobalt based on current production rates (www.statista.com).

(ii) The above calculation assumes that the entire 75,000 km^2 exploration area contains commercially viable nodule abundances and that the entire area is mineable. However, it is unlikely that the entire 75,000 km^2 would be mined, due to nodule abundance and seafloor slope restrictions, with some Contractors stating that only 18–50% of the ground they are exploring is expected to be mined depending on the type of mineral.

(iii) It is important to mention here that a 75,000 km^2 area is just 0.044% of the total area of the Pacific Ocean, 0.088% of the Atlantic Ocean, and 0.10% of the Indian Ocean. It is quite likely that mining might commence at a couple of mine sites located in different oceans; thus the mining areas could be far apart with smaller areas of influence.

(iv) It is expected that the average abundance of nodules in the First Generation Mine Sites (FGMs) will be much higher (Singh and Sudhakar 2015) than the cutoff abundance considered here, and the actual area being mined to achieve the targeted mining rate may be further reduced.

(v) The current exploration contracts are at various stages of development, and, based on current knowledge, it is extremely unlikely that there would be more than 3 to 4 mines operating worldwide within the first 20 years of the deep seafloor minerals industry commencing.

3.2 Concern Raised: Seafloor Environments Over Millions of Square Kilometers Will Be Destroyed

It has been estimated that for a cutoff abundance of 5 kg/m^2 and a mining rate of 1.5 MT/year with 300 days of operation, with an overall efficiency of the mining system of at least 25%, an area of 6400 km^2 will be actually mined over a period of 20 years (lifetime of a mine site as per UNOET 1987), which is only 8.5% of the contract area (75,000 km^2), and the actual area mined will only be 300 km^2/year (i.e., 1 km^2 per day) (Sharma et al. 2019, Chap. 19, this volume). These are conservative estimates, and the actual area impacted could be much smaller due to higher nodule abundances and mining system efficiencies.

3.3 Concern Raised: Sediment Plumes Will Impact the Marine Environment

Sediment plumes at and above the seafloor will occur due to movement of the nodule collector and the separation and collection of nodules from the surrounding sediment. According to one estimate with every 1 ton of nodules recovered, 2.5–5.5 tons of sediment will be resuspended (Amos and Roels 1977). It is expected that the sediment that is resuspended in the near bottom waters would mainly contain very fine clayey particles that may either remain in suspension for a long period of time or get transported to adjacent areas by bottom currents creating a layer of resedimented particles on the seafloor in an area larger than the area directly being mined (Pearson 1975). Many of the collector designs are proposing screening of the associated sediments close to the seafloor, so as to minimize the amount of unwanted material being lifted to the surface. Contactors are also looking to minimize impacts to the water column through the use of fully enclosed lifting systems and by designing systems which will put the return water from the dewatering plant at the deepest depths possible or at the most appropriate depth as determined through the Environmental Impact Assessment process. Contractors are also considering engineering solutions along with styles and patterns of mining operations in order to concentrate plumes within smaller areas (e.g., Hong et al. 2019, Chap. 5, this volume).

3.4 Concern Raised: Deep-Sea Biota Will Be Destroyed

Although experimental results have shown that the biota, both sessile and mobile, will be affected not only within the active mining area but also in the adjacent areas due to compaction, lifting, screening, and redistribution of seafloor sediments (Ozturgut et al. 1980; Foell et al. 1990; Fukushima 1995; Tkatchenko et al. 1996;

Trueblood et al. 1997; Thiel et al. 1997; Shirayama 1999; Radziejewska 1997; Ingole et al. 1999), the following points should also be considered:

(i) Faunal diversity over these areas is highly variable over small distances within a nodule field, meaning some areas are more diverse than others.
(ii) Mining of seafloor massive sulfides may have footprints of a few square kilometers only, and hydrothermal vent fauna have an ability to recover quickly from disturbance (Gollner et al. 2017) and that population of benthic animals on fast spreading ridges can recover in a few years (Van Dover 2011).
(iii) In case of sulfides and crusts, the threat of any species or groups becoming extinct is rare as it is impossible to mine 100% of the seafloor, and several other areas with similar habitats would not be mined due to various reasons.
(iv) ISA has established and is establishing networks of Areas of Particular Environmental Interest [APEIs] which are to remain unimpacted by mining where representative groups of biota will remain, allowing for the maintenance of ecosystem health and function.

3.5 Concern Raised: Not Enough Is Being Done for the Marine Environment

This concern probably stems from the general experience of mining terrestrial deposits, and in certain parts of the world, environmental protection has not always been given its due consideration, and this has led to serious damage to ecosystems. However, it is expected that in case of deep-sea mining, these could be avoided or at least minimized due to several stipulations that are being put in place in advance of the industry's commencement, such as:

(i) Under the Regional Environmental Management Plan for Clarion-Clipperton Fracture Zone in the Pacific Ocean, nine 160,000 m^2 Areas of Particular Environmental Interest (APEIs) have been established for environmental monitoring (ISA 2011), and these are to remain untouched by mining.
(ii) Several EIA studies have been conducted since the 1970s including benthic impact experiments by some of the Contractors to gain an understanding of the possible impacts (Jones et al. 2017) and to inform environmental management decisions.
(iii) ISA has put in place environmental guidelines for all Contractors to follow for data collection, impact assessment, and monitoring of the environment (ISA 2013b, which is being updated).
(iv) Draft regulations on Exploitation of Mineral Resources in the Area (ISA 2018) propose that each Contractor will be required to submit the following before commencement of mining to ensure the details of the project; its likely impact and proposed management strategies are well understood prior to the commencement of mining:

- Mining work plan
- Financing plan
- Environmental Impact Statement (following the completion of an Environmental Impact Assessment)
- Emergency response and contingency plan
- Health, safety, and maritime security plan
- Environmental management and monitoring plan
- Closure plan

All of the above will be reviewed and assessed prior to the signing of the exploitation contract between the ISA and Contractor.

4 Mechanisms for Responsible Environmental Management

Efforts are being made to propose several mechanisms, technical as well as regulatory, so as to ensure avoidance or reduction in environmental impacts as follows:

(i) Several measures have been suggested for consideration during design and operation of the mining system (Sharma 2015; Hong et al. 2019, Chap. 5, this volume; Billett et al. 2019, Chap. 15, this volume) that include:

- Employing methodologies to minimize sediment penetration and redistribution
- Separation of minerals from the associated substrates near the seafloor
- Lifting of minimum possible sediment to the surface
- Discharge of return water (dewatering plant discharge) below the oxygen minimum zone
- The use of biodegradable fluids in all subsea equipment
- Efficient mineral processing that removes as many of the metals as possible
- "Constructive" use of unwanted material after extraction of metals

(ii) Each Contractor is expected to comply with the proposed "Regulations on Exploitation of Mineral Resources in the Area" (ISA 2018), which, as stated above, include the following:

- Completion of an Environmental Impact Assessment and submission of an Environmental Impact Statement as per the EIS Template issued by the ISA
- Environmental management and monitoring plan submission and implementation
- Mine closure plan submission and implementation when appropriate

5 Considerations for Sustainable Deep-Sea Mining

Thinking globally and holistically about our planet and its resources, deep-sea mining may have many environmental and social advantages compared to its land-based counterparts. For example, seafloor deposits are being found with higher grades of metal than what is currently being mined on land. Seafloor deposits often contain a number of metals, for example, seafloor polymetallic nodules often contain high amounts of nickel, copper, cobalt, and manganese. As a result, one seafloor nodule mine could potentially replace the future need to develop three additional mines on land (i.e., a nickel, copper, and manganese mine). Additionally, there are no human communities living in, or even near, the environments associated with deep seafloor mineral deposits – meaning, there is no need to relocate human communities and land-use conflicts are avoided. Due to the high-grade, multi-metal nature of seafloor mineral deposits, they are anticipated to create much less waste than their land-based counterparts.

As the concepts, policies, and technologies are gradually evolving, deep-sea mining has the advantage of not only learning from experiences but also employing best available technologies. Besides offering access to critical metals, deep-sea mining contributes to marine scientific research (in hitherto unexplored oceanic regions), capacity building (in new research fields), as well as developing technological spinoffs (for extreme conditions) (Van Nijen et al. 2018). While these advantages exist, it is generally well accepted by the relevant stakeholders that it is important that deep-sea mining is developed in a sustainable, responsible way for which the following questions need to be considered:

(i) Are there alternatives that could be considered (e.g., artificial substitutes for the required metals)?
(ii) What are the engineering solutions that can be employed to minimize sediment plumes?
(iii) Are there mining patterns/styles which can be adopted to minimize impacts (including strip mining, discharge depths of sediment plumes?
(iv) Is there a limit to the number of mine sites needed to meet the anticipated demand?
(v) What environmental commitments can the industry make (e.g., biodegradable fluids in all subsea equipment)?
(vi) Should regulators be considering incentives to encourage continual environmental performance improvement?
(vii) What are the potential mitigation and management strategies, and of these which are feasible?
(viii) Is deep-sea ecosystem restoration realistic and should it be considered?
(ix) How do we ensure both exploitation and conservation needs are met?

6 Conclusions

The demand for minerals and metals is rising with population growth and urbanization. Also, as a global society, we are seeking a clean energy, low-carbon future, which also requires significant amounts of metal. Land resources have become stretched, while significant deposits of the metals needed for clean energy solutions have been identified on the seafloor. While there appear to be many environmental and social responsibility advantages to going to the sea for minerals and metals, concerns remain about the future impact on deep-sea ecosystems. Unless alternatives are found (such as synthetic substitutes), it is looking more and more likely that we will need seafloor minerals to meet societal needs and goals. As we move forward, we need to find solutions that are both environmentally sound and economically viable (Lodge et al. 2017).

Acknowledgment The maps showing exploration areas for minerals in different oceans are from the website (www.isa.org.jm) of the International Seabed Authority (ISA), Jamaica. The permission granted by ISA to reproduce these maps is gratefully acknowledged. Tables 1 and 2 have been compiled by Ms. Farida Mustafina, student of POMOR program at St. Petersburg State University, Russia, during her internship at the National Institute of Oceanography, Goa, India.

References

Amos, A. F., & Roels, O. A. (1977). Environmental aspects of manganese nodule mining. Marine Policy, 1, pp. 160–162.

Billett, D. S. M., Jones, D. O. B., & Weaver, P. P. E. (2019). Improving environmental management practices in deep-sea mining. In R. Sharma (Ed.), *Deep-sea mining and environment – Impacts, consequences and management*. Cham: Springer International Publishers AG.

Cherkashov, G. (2017). Seafloor massive sulfide deposits: Distribution and prospecting. In R. Sharma (Ed.), *Deep-sea mining – Resource potential, technical and environmental considerations* (pp. 143–164). Cham: Springer International Publishing AG.

Cronan, D. S. (1980). *Underwater minerals* (p. 362). London: Academic Press.

Cuyvers, L., Berry, W., Gjerde, K., Thiele, T., & Wilhem, C. (2018). *Deep seabed mining: a rising environmental challenge*. Gland: IUCN and Gallifrey Foundation. x + 74 pp.

Foell, E. J., Thiel, H., & Schriever, G. (1990). DISCOL: A long term largescale disturbance – Recolonisation experiment in the abyssal eastern tropical Pacific Ocean. *Proceedings of Offshore Technology Conference* (pp. 497–503). Houston, USA, Paper No. 6328.

Fukushima, T. (1995). Overview Japan deep-sea impact experiment = JET. *Proceedings of 1st ISOPE Ocean Mining Symposium* (pp. 47–53). Tsukuba, Japan, ISOPE.

Glumov, I. F., Kuzneicov, K. M., & Prokazova, M. S. (2000). Ocenkaznaczzenija mineralych resursov meidunarodnogorajonamorskogodna w mineral nosyriewompotenciale Rossijskoj Federacii (in Russian). In *Proceedings of Geological Congress* (pp. 27–29). St. Petersburg.

Gollner, S., Kaiser, S., Menzel, L., Jones, D. O. B., Brown, A., Mestre, N. C., van Oevelen, D., Menot, L., Colaço, A., Canals, M., Cuvelier, D., Durden, J. M., Gebruk, A., Egho, G. A., Haeckel, M., Marcon, Y., Mevenkamp, L., Morato, T., Pham, C. K., Purser, A., Sanchez-Vidal, A., Vanreusel, A., Vink, A., & Martinez Arbizu, P. (2017). Resilience of benthic deep-sea fauna to mining activities. *Marine Environmental Research, 129*(Suppl. C), 76–101.

Halbach, P., Jahn, A., & Cherkashov, G. (2017). Marine co-rich ferromanganese crust deposits: Description and formation, occurrences and distribution, estimated world-wide resources. In R. Sharma (Ed.), *Deep-sea mining – Resource potential, technical and environmental considerations* (pp. 65–141). Cham: Springer International Publishing AG.

Hein, J. R., Mizell, K., Koschinsky, A., & Conrad, T. A. (2013). Deep-ocean mineral deposits as a source of critical metals for high- and green-technology applications: Comparison with land-based resources. *Ore Geology Reviews, 51*, 1–14.

Hong, S., Kim, H.-W., Yeu, T., Choi, J.-S., Lee, T. H., Bae, D. S., & Lee, J.-G. (2019). Technologies for safe and sustainable mining of deep-seabed minerals. In R. Sharma (Ed.), *Deep-sea mining and environment – impacts, consequences and management*. Cham: Springer International Publishers AG.

Jones, D. O. B., Kaiser, S., Sweetman, A. K., Smith, C. R., Menot, L., Vink, A., Trueblood, D., Greinert, J., Billett, D. S. M., Arbizu, P. M., Radzeijewska, T., Singh, R., Ingole, B., Stratman, T., Simon-Lledo, S., Durden, J. M., & Clark, M. R. (2017). Biological responses to disturbance from simulated deep-sea polymetallic nodule mining. *PLoS One, 12*(2), e0171750.

Ingole, B. S., Ansari, Z. A., Matondkar, G. P., & Rodrigues, N. (1999). Immediate response of meio and macrobenthos to disturbance caused by a benthic disturber. *The Proceedings of the Third ISOPE Ocean Mining Symposium* (pp. 191–197). Goa, India.

ISA. (1998). Draft guidelines for the assessment of the possible environmental impacts arising from exploration of deep-seabed polymetallic nodules from the area. *Proceedings of Workshop by the International Seabed Authority* (pp. 289). Sanya, China.

ISA. (2010). Decision of the assembly of the International Seabed Authority relating to the regulations on prospecting and exploration for polymetallic sulphides in the area. *International Seabed Authority ISBA/16/A/12/Rev.1* (pp. 49).

ISA. (2011). Environmental management plan for the Clarion-Clipperton zone. *International Seabed Authority ISBA/17/LTC/7* (pp. 18).

ISA. (2012). Decision of the Assembly of the International Seabed Authority relating to the regulations on prospecting and exploration for cobalt-rich ferromanganese crusts in the area. *International Seabed Authority ISBA/18/A/11* (pp. 49).

ISA. (2013a). Decision of the Council of the International Seabed Authority relating to amendments to the regulations on prospecting and exploration for polymetallic nodules in the area and related matters. *International Seabed Authority ISBA/19/C/17* (pp. 49).

ISA. (2013b). Recommendations for the guidance of contractors for the assessment of the possible environmental impacts arising from exploration for marine minerals in the area. *International Seabed Authority ISBA/19/LTC/8* (pp. 32).

ISA. (2018). Draft regulations on the exploitation of mineral resources in the area. *International Seabed Authority, Jamaica, ISBA/24/LTC/WP.1* (pp. 97). April 2018.

Kotlinski, R. (2001). Mineral resources of the world's ocean – Their importance for global economy in the 21st century. In *Proceedings of 4th ISOPE ocean mining symposium* (pp. 1–7). Szczecin, Poland: International Society for Offshore and Polar Engineers.

Lenoble, J. P. (2000). A comparison of possible economic returns from mining deep-sea polymetallic nodules, polymetallic massive sulphides and cobalt-rich ferromanganese crusts. In *Proceedings of workshop on mineral resources* (pp. 1–22). Jamaica: International Seabed Authority.

Lodge, M. W., Segerson, K., & Squires, D. (2017). Sharing and preserving the resources in the deep-sea: Challenges for the International Seabed Authority. *International Journal of Marine and Coastal Law, 32*, 427–457.

Ozturgut, E., Lavelle, J. W., Steffin, O., & Swift, S. A. (1980). Environmental investigation during manganese nodule mining tests in the north equatorial Pacific. In *November 1978, NOAA Technical Memorandum ERL MESA-48*. Colorado, USA: National Oceanic and Atmospheric Administration, 50.

Pearson, J. S. (1975). *Ocean floor mining* (p. 201). Park Ridge: Noyes Data Corporation.

Radziejewska, T. (1997). Immediate responses of benthic meio- and megafauna to disturbance caused by polymetallic nodule miner simulator. In *Proceedings international symposium environmental studies for deep-sea mining* (pp. 223–236). Tokyo, Japan: Metal Mining Agency of Japan.

Rona, P. A. (2003). Resources of the ocean floor. *Science, 299*, 673–674.

Sharma, R., Nath, B. N., Parthiban, G., & Sankar, S. J. (2001). Sediment redistribution during simulated benthic disturbance and its implications on deep seabed mining. *Deep-Sea Research II, 48*, 3363–3380.

Sharma, R. (2015). Environmental issues of deep-sea mining. *Procedia Earth and Planetary Science, 11*, 204–211.

Sharma, R. (2017). Deep-sea mining: Current status and future considerations. In R. Sharma (Ed.), *Deep-sea mining* (pp. 3–21). Cham: Springer International Publishers AG.

Sharma, R., Mustafina, F., & Cherkashov, C. (2019). Review of mining rates, environmental impacts, metal values and investments for polymetallic nodules mining. In R. Sharma (Ed.), *Deep-sea mining and environment – Impacts, consequences and management*. Cham: Springer International Publishers AG.

Shirayama, Y. (1999). Biological results of JET project: An overview. In *Proceedings 3rd ISOPE ocean mining symposium* (pp. 185–190). Goa, India: ISOPE.

Singh, T. R. P., & Sudhakar, M. (2015). Estimating potential of additional mine-sites for polymetallic nodules in Pacific and Indian oceans. *International Journal of Earth Sciences and Engineering, 8*, 1938–1941.

Takaya, Y., Yasukawa, K., Kawasaki, T., Fujinaga, K., Ohta, J., Usui, Y., Nakamura, K., Kimura, J.-I., Chang, Q., Hamada, M., Dodbiba, G., Nozaki, T., Iijima, K., Morisawa, T., Kuwahara, T., Ishida, Y., Ichimura, T., Kitazume, M., Fujita, T., & Kato, Y. (2018). The tremendous potential of deep-sea mud as a source of rare-earth elements. *Scientific Reports, 8*, 5763. https://doi.org/10.1038/s41598-018-23948-5.

Thiel, H., Angel, M. V., Foell, E. J., Rice, A. L., & Schriever, G. (1997). Environmental risks from large-scale ecological research in the deep sea: A desk study. Marine Science and Technology, pp. 1–209.

Tkatchenko, G., Radziejewska, T., Stoyanova, V., Modlitba, I., & Parizek, A. (1996). Benthic impact experiment in the IOM pioneer area: Testing for effects of deep-sea disturbance. In *International seminar on deep sea-bed mining technology*. Beijing, China: China Ocean Mineral Resources R&D Association (C55-C68).

Trueblood, D. D., Ozturgut, E., Pilipchuk, M., & Gloumov, I. F. (1997). The ecological impacts of the joint U.S.-Russian benthic impact experiment. In *Proceedings international symposium environmental studies for deep-sea mining* (pp. 237–243). Tokyo, Japan: Metal Mining Agency of Japan.

UNOET. (1987). *Delineation of mine sites and potential in different sea areas* (p. 27). London: UN Ocean Economics and Technology Branch and Graham & Trotman Limited.

Van Dover, C. L. (2011). Mining seafloor massive sulphides and biodiversity: What is at risk? *ICES Journal of Marine Science, 68*, 341–348.

Van Dover, C. L., Ardron, J. A., Escobar, E., Gianni, M., Gjerde, K. M., Jeckel, A., Jones, D. O. B., Levin, L. A., Niner, H. J., Pendleton, L., Smith, C. R., Thiele, T., Turner, P. J., Watling, L., & Weaver, P. P. E. (2017). Biodiversity loss from deep-sea mining. *Nature Geoscience, 10*(7), 464.

Van Nijen, K., Passel, S. V., & Squires, D. (2018). A stochastic techno-economic assessment of seabed mining of polymetallic nodules in the clarion Clipperton fracture zone. *Marine Policy, 95*, 133–141.

Weaver, P., & Billett, D. (2019). Environmental impacts of nodule, crust and sulphide mining. In R. Sharma (Ed.), *Deep-sea mining and environment – Impacts, consequences and management*. Cham: Springer International Publishers AG.

World Bank. (2017). *The growing role of minerals and metals for a low carbon future*. Washington, DC: World Bank Publications. (http://documents.worldbank.org/curated/en/207371500386458722/pdf/117581-WP-P159838-PUBLIC-ClimateSmartMiningJuly.pdf).

Yamazaki, T., & Sharma, R. (2001). Estimation of sediment properties during benthic impact experiments. *Marine Georesources and Geotechnology, 19*, 269–289.

Zepf, V., Simmons, J., Reller, A., Ashfield, M., & Rennie, C. (2014). *BP 2014 Materials critical to the energy industry. An introduction* (2nd ed., p. 94). London: British Petroleum.

en.wikipedia.org. Wikipedia, the free encyclopedia. Accessed Sept 2018.

www.isa.org.jm. Accessed 12 June 2018.

www.statista.com. Accessed 12 June 2018.

www.usgs.gov. Accessed Sept 2018.

Dr. Rahul Sharma (rsharma@nio.org, rsharmagoa@gmail.com) retired as Chief Scientist from the CSIR-National Institute of Oceanography in Goa, India, with a career spanning 35 years in the field of exploration and exploitation of marine minerals. He has led a multidisciplinary group on "environmental studies for marine mining." He has a master's degree in Geology and a doctorate in Marine Science. His professional interests include application of exploration and environmental data to deep-sea mining. He has edited 3 special issues of journals, published 37 scientific papers, authored 22 articles and 41 technical reports, and presented more than 50 papers at national and international conferences. He has also edited a book Deep-Sea Mining: Resource Potential, Technical and Environmental Considerations published by Springer International Publishers in 2017 that has chapters contributed by experts from around the world.

His international assignments include Visiting Scientist to Japan; Visiting Professor to Saudi Arabia; member of the UNIDO mission "to assess the status of Deep-sea mining technologies" in Europe, the USA, and Japan; and invited speaker and consultant for the International Seabed Authority, Jamaica. He has contributed to the "World Ocean Assessment report I" of the United Nations and has also been invited to contribute a chapter on "Potential impacts of deep-sea mining on marine ecosystem" for the Oxford Encyclopedia for environmental science. In addition to his research career, he has been involved with several activities relating to science communication and outreach as well as training programs for international participants, professionals and students.

Dr. Samantha Smith (samantha@blueglobesolutions.com; samantha@nauruoceanresources.com) has 20 years' experience conducting environmental assessments in a number of different countries around the world, over five continents, and has 13 years' experience working with the deep seafloor minerals sector.

Samantha has a PhD in Environmental Biology/Biogeochemistry from the University of Bristol (UK) and is a Director and Former President of the International Marine Minerals Society and a Fellow of AusIMM. She is also the former Vice President Corporate Social Responsibility for Nautilus Minerals Inc. During her 9-year tenure with Nautilus, Samantha led an international, multidisciplinary team to complete the Solwara 1 Environmental Impact Statement and deliver the world's first Environment Permit for seafloor massive sulfide extraction.

Since 2014, Samantha has run the environmental consultancy Blue Globe Solutions, based in Canada, consulting to various entities in the marine minerals space, including industry and government, for projects both within national jurisdiction and beyond. Much of Samantha's time is spent advising and consulting to Nauru Ocean Resources Inc., an ISA Contractor and fully owned subsidiary of DeepGreen Metals Inc.

Samantha is a coauthor of several peer-reviewed manuscripts related to deep-sea environmental management.

Environmental Issues of Deep-Sea Mining: A Law of the Sea Perspective

Philomène A. Verlaan

Abstract Addressing the environmental issues raised by deep-sea mining may provide an example for the international community on how to implement correctly the unqualified requirement in the United Nations Convention on the Law of the Sea (LOSC) that "States have the obligation to protect and preserve the marine environment". This chapter offers an overview of how this could work.

Keywords Deep-sea mining · Marine environmental protection · States' obligation · LOSC

Addressing the environmental issues raised by deep-sea mining may provide an example for the international community on how to implement correctly the unqualified requirement in the United Nations Convention on the Law of the Sea[1] (LOSC) that "States have the obligation to protect and preserve the marine environment".[2] Correct implementation entails considering the marine environment as a whole, as the LOSC does. Jurisdictional, sectoral and resource divisions in the LOSC (which, alas, retains more of these divisions than would be expected in an instrument whose Preamble states that "the problems of ocean space are closely interrelated and need to be considered as a whole")[3] cannot be invoked to justify, qualify or otherwise create an exception to the LOSC's fundamental marine environmental protection obligation. This obligation not only applies throughout "ocean space", but it also

[1] United Nations Convention on the Law of the Sea (Montego Bay, 10 December 1982, in force 16 November 1994) 1833 *UNTS* 3 (LOSC). The LOSC is our world's "Constitution for the Oceans" (Koh, 1983). TTB Koh (1983) 'A Constitution for the Oceans. Remarks by Tommy T. B. Koh of Singapore, President of the Third United Nations Conference on the Law of the Sea.' In: United Nations Convention on the Law of the Sea, with Index and Final Act of the Third United Nations Conference on the Law of the Sea (United Nations Publication No. E.83.V.5, New York, NY) pp. xxxiii-xxxvii.

[2] LOSC Article 192.

[3] LOSC Preamble, 3rd paragraph.

P. A. Verlaan (✉)
University of Hawai'i at Manoa, Honolulu, HI, USA

R. Sharma (ed.), *Environmental Issues of Deep-Sea Mining*,
https://doi.org/10.1007/978-3-030-12696-4_2

applies to the rest of our planet, both the land[4] and the atmosphere,[5] when activities conducted there either "result or are likely to result"[6] in adverse effects on the marine environment. Even the likelihood of adverse effects triggers the obligation to act.

Unfortunately, so far the international community has neither adequately considered the environmental consequences of its activities in terms of their likely and actual adverse effects on the marine environment as a whole, nor implemented the clear and unequivocal requirements set out in the LOSC to "prevent, reduce and control"[7] these effects accordingly.

For example, the scientific consensus on the demonstrably harmful effects of greenhouse gas emissions on the environment in general and on the marine environment in particular (e.g. ocean acidification, warming, deoxygenation) has still not yet triggered the mandatory actions unequivocally required by the LOSC. From the feeble international instruments promulgated so far under the auspices of the United Nations Framework Convention on Climate Change (UNFCCC),[8] including the UNFCCC itself, it would appear that States continue to assume that they have a legal option on whether or not to prevent, reduce and control the production and emission of greenhouse gases. At least for the 167 States, and the European Union, that are party to the LOSC (as of 31.08.2018), this assumption is incorrect. The same erroneous assumption applies to the growing plague of plastics infesting the oceans.

Efforts at achieving legally binding marine environmental protection do exist and are growing, but they have also largely been characterized by fragmentation rather than integration. It is ever more starkly evident that the marine environment has no natural boundaries that correspond to any anthropogenic ones. Nevertheless, jurisdictional, sectoral, resource and other forms of partitioning approaches to addressing adverse effects on the marine environment from our activities persist. The most recent example, involving two partitions of the marine environment itself, is the decision by the United Nations General Assembly to develop an international legally binding instrument (ILBI) under the LOSC on the conservation and sustainable use of marine biological diversity of areas beyond national jurisdiction (ABNJ) (hereinafter as the BBNJ negotiations).[9] Setting a human-devised (ABNJ) and a biological (marine biodiversity) partition as the focus for the ILBI disregards the

[4] LOSC Articles 194, 207, 213.

[5] LOSC Articles 194, 212, 222.

[6] Note the precautionary language.

[7] See, e.g. LOSC Articles 194–196, 207–212, 213–222.

[8] United Nations Framework Convention on Climate Change (Rio de Janeiro, 9 May 1992, in force 21 March 1994) 31 *ILM* 849 (UNFCCC).

[9] UN General Assembly Resolution A/RES/72/249: International legally binding instrument under the United Nations Convention on the Law of the Sea on the conservation and sustainable use of marine biological diversity of areas beyond national jurisdiction (hereinafter: BBNJ negotiations), available at http://www.un.org/depts/los/general_assembly/general_assembly_resolutions.htm; accessed 6 July 2018. The first round of BBNJ negotiations took place in September 2018.

stark physical realities of the marine environment and the pervasive nature of the increasingly mounting threats it faces. Whether this legally and scientifically flawed fragmented approach by the BBNJ negotiations to marine environmental protection will result in an ILBI that meets the LOSC's marine environmental protection requirements remains to be seen.[10]

Deep-sea mining, by contrast, is an emerging industry whose stakeholders have accepted the undeniably daunting challenge of developing an integrated approach to marine environmental protection. These stakeholders include the States Parties to the LOSC: the latter are all *ipso facto* members of the International Seabed Authority (ISA), the organization set up under the LOSC "through which States Parties shall, in accordance with this Part [LOSC Part XI], organize and control activities in the Area, particularly with a view to administering the resources[11] of the Area".[12] Minerals (i.e. resources recovered from the Area[13]) are, so far, the only example of a global resource under global intergovernmental management by a global intergovernmental organization (the ISA) established exclusively for this purpose. The ISA's member states emphasize the need for a global, multiregional approach to development and implementation of better environmental policy and operational frameworks for site-specific deep-sea mining and related activities.[14]

Unfortunately, the LOSC's own fragmented approach to the Area (defined as "the seabed and ocean floor and subsoil thereof, beyond the limits of national jurisdiction")[15] does not facilitate the task of the ISA, because the LOSC does not limit its marine environmental protection requirements[16] to the Area. For example, in the context of deep-sea mining, the scope of the ISA's marine environmental responsibilities extends to "the coastline", i.e. well beyond the Area and far into waters within national jurisdiction, and must include "prevention, reduction and control of interference with the ecological balance of the marine environment".[17] Political will can resolve the issues raised by the former obligation, but scientific information remains inadequate to offer confident guidance on how to achieve the latter at the level of operational sophistication required.

---

[10] A detailed elaboration of these arguments is set out in Verlaan, P. (2018). The interface of science and law: A challenge to the privileging of 'marine biodiversity' over 'marine environment'. In R. A. Barnes, & R. Long (Eds.), *Frontiers in international environmental law: Oceans and climate challenges* (Brill, Leiden) in press.

[11] For purposes of LOSC Part XI, these are defined as "all solid, liquid or gaseous mineral resources in situ in the Area at or beneath the seabed". LOSC Article 133(a).

[12] LOSC Article 157. It is ironic that these same state parties are also participating in the BBNJ negotiations, which are being conducted on the opposite premise.

[13] LOSC Article 133(b).

[14] Lodge, M., & Verlaan, P. (2018). Deep-sea mining: International regulatory challenges and responses. *Elements* (in press).

[15] LOSC Article 1(1)(1).

[16] See LOSC Article 145, which is the governing article applicable specifically to "activities in the Area"; other marine environmental protection requirements for these activities are found elsewhere in the LOSC, including in Part XII, which is dedicated to the marine environment.

[17] LOSC Article 145(b).

Despite this uncertainty, the ISA must establish a comprehensive framework for sustainable – i.e. environmentally and commercially responsible – management of the emerging deep-sea mining industry. The present book, for which it is a signal honour and privilege to add these brief reflections, will make an invaluable contribution to assist the ISA, as the representative of the global community of stakeholders in sustainable deep-sea mining, in achieving this compelling mandate.

Dr. Philomène A. Verlaan is an oceanographer specialized in the biogeochemistry and ecology of deep-sea ferromanganese nodules and crusts (Ph.D., Imperial College London) with extensive sea-going experience (23 – so far – oceanographic research cruises and 9 submersible dives). She is also an attorney-at-law specialized in international law of the sea (J.D., Florida State University; Member of the Florida Bar). She assists international public and private organizations in negotiating the complex interface between marine science and law of the sea to achieve environmentally and commercially responsible uses of marine resources compatible with the Law of the Sea Convention, the world's "Constitution for the Oceans". Author of over 50 (so far) refereed publications, she is a Visiting Colleague at the Department of Oceanography, University of Hawai'i, Senior Technical Adviser, Advisory Committee on Protection of the Sea, Coordinator of the International Marine Minerals Society's Code of Environmental Management of Marine Mining, on the editorial board for Marine Georesources and Geotechnology and the International Journal of Marine and Coastal Law, Member, World Commission on Environmental Law, and Fellow of the Institute of Marine Engineering, Science and Technology.

Environmental Impacts of Nodule, Crust and Sulphide Mining: An Overview

Philip P. E. Weaver and David Billett

Abstract The new industry of deep-sea mining (DSM) potentially offers abundant supplies of several metals from the deep ocean, but the ores will need to be recovered from pristine environments in which the ecosystems are often poorly known. Information that is available for some of these environments suggests that organisms may struggle to recover from the impacts of DSM, whilst in other areas the impacts may be somewhat less.

Deep-sea mining is focussed on three distinct resources – manganese nodules (also known as polymetallic nodules), cobalt crusts and seafloor massive sulphides (SMS) (sometimes called polymetallic sulphides). These occur in different seafloor settings, each hosting very different ecosystems and each with its own set of environmental issues.

Manganese nodules occur in the deep basins of the ocean where lack of sediment supply results in very slow sediment accumulation – rates that can be as low as 1 mm per thousand years – thus allowing nodules to form from slow precipitation of metals. Interest in mining manganese nodules is focussed mainly on the Clarion Clipperton Zone in the eastern equatorial Pacific and Central Indian Basin in the Indian Ocean. Here the seabed faunas are sparsely distributed but are very varied in composition. Many different species live in the upper few centimetres of the sediment or attached to the nodules. The mining process will disrupt this surface sediment layer and remove the nodules. Experiments have shown that species are very slow to return to the disrupted areas. Combined with the large areas that will need to be mined for manganese nodules, this gives rise to potentially a high environmental and ecological impact.

Cobalt crusts occur as layers up to 26 cm thick coating the rocky tops and upper flanks of seamounts, with the most promising deposits occurring between 800 and 2500 m water depth. The absence of sedimentation due to currents in these areas

P. P. E. Weaver (✉)
Seascape Consultants Ltd, Romsey, UK
e-mail: phil.weaver@seascapeconsultants.co.uk

D. Billett
Deep Seas Environmental Solutions Ltd., Ashurst, Hampshire, UK
e-mail: david.billett@deepseasolutions.co.uk

R. Sharma (ed.), *Environmental Issues of Deep-Sea Mining*,
https://doi.org/10.1007/978-3-030-12696-4_3

allows the slow growth of the crust via the precipitation of minerals from seawater. Seamount faunas are not well studied but they include a large number of species, many of which are slow-growing, long-lived and slow to reproduce. This makes it difficult for the ecosystem to recover from disruption. Large areas will need to be mined because the ore occurs in a very thin layer and whole seamounts may be affected.

The third resource – seafloor massive sulphides – differs from the previous two, being formed from precipitation of metals from hydrothermal fluids at oceanic plate boundaries. This process creates three-dimensional ore bodies extending metres into the seabed which are similar to some ore bodies that occur on land. Ecosystems comprising specialist organisms that can tolerate and make use of the harsh biochemical conditions are often found at active hydrothermal vents. These vent sites are probably too hot to ever be mined, so ore bodies are being sought some distance away from the active ridge axis in areas where venting is weaker or has stopped. The species occurring 'off axis' are more akin to those from the surrounding rocky slopes and possibly on the continental slopes in the same ocean basin. The species may occur over wide areas, and the impact of localised mining may be relatively small.

In all types of deep-sea mining, the generation of plumes of sediment-laden water, both by the mining process and the transport of ores to a support ship, will have an impact on benthic and mid-water ecosystems away from the mining site. If uncontrolled, such impacts could be comparable to or of greater scale to impacts in the mined areas.

Keywords Deep-sea mining · Polymetallic sulphides · Manganese nodules · Cobalt crusts · Environmental impact · Impact of plumes · Habitat loss

1 Introduction

Deep-sea manganese nodules and cobalt crusts were first discovered during the Challenger expeditions of 1873–1876 (Murray 1876; Murray and Renard 1891), and the idea that metals could be mined from the seabed in the deep remote parts of the ocean was first mooted in the 1960s (Mero 1965). However, after some initial efforts to assess the potential of commercial-scale mining of manganese nodules in the late 1960s and 1970s, the concept was largely abandoned. Following the establishment of the International Seabed Authority (ISA) in 1994 under the United Nations Convention on the Law of the Sea (UNCLOS), rules were established to enable contracts to be signed for exploration for seabed minerals. Six 15-year contracts were signed in 2001 and one in 2002 for exploration for manganese nodules. These early contracts were awarded to States, in one case a group of States working together. Many more exploration contracts have been signed subsequently, including for cobalt crusts and seafloor massive sulphides (SMS), but no mining has taken place to date. Indeed regulations to govern exploitation (the act of mining) are currently under development by the ISA.

In the development of its exploitation regulations, the ISA needs to take into account its dual remit of regulating deep-sea mining and ensuring that the marine environment is protected from any harmful effects of mining. This is seen by many as a difficult task because the effects of the mining equipment are not known; many of the subcomponents required for mining are only now under development and have not been tested in a mining situation. Another complicating factor is the lack of information on many aspects of deep-sea environments including species distributions, interactions between different taxa and between the species and their environment. Evidence, though scant in many cases, seems to suggest that many seabed locations subject to deep-sea mining will not be able to recover for tens or even hundreds of years (Gollner et al. 2017). The precautionary approach has therefore been promoted as a sensible way to develop this new industry (Ellis et al. 2017), but how this will be accommodated in the emerging exploitation regulations is not clear.

One of the major obstacles to generating scientific knowledge of deep-sea ecosystems in areas of potential mining interest is their remoteness. The logistics and costs of mounting independent scientific studies have meant that only a few studies have been undertaken. In addition, it is only relatively recently that advanced technologies such as autonomous underwater vehicles (AUVs) and remotely operated vehicles (ROVs) have become more widely available. In recent years two projects have produced valuable information. The MIDAS project (www.eu-midas.net/) investigated the environmental impacts of deep-sea mining and the JPI Oceans 'Mining Impact' project investigated the "Ecological aspects of deep-sea mining" specifically for manganese nodules (www.jpi-oceans.eu/ecological-aspects-deep-sea-mining). Both of these projects focused on seabed ecosystems; there has been no concerted effort to examine potential impacts on organisms in the water column.

The continuing uncertainties of the environmental impacts of deep-sea mining have the potential to be a show-stopper unless greater knowledge is generated on the environmental impacts of mining and on plans to mitigate the expected impacts (Miller et al. 2018). Uncertainties about the environmental impact of seafloor mining led a decision-making committee to reject a proposal to mine seabed phosphorite deposits in waters off New Zealand (Ellis et al. 2017). In Japan test mining was carried out in 2017 for deep-sea minerals in the Okinawa Trough, and whilst the test mining technologies were reported to be successful, no information on assessing the environmental impacts has been released (www.meti.go.jp/english/press/2017/0926_004.html).

2 Deep-Sea Minerals and Their Occurrence

The ocean floor varies considerably from place to place across the world. New crust is formed at the mid-ocean ridges where oceanic plates spread apart, and volcanic activity is common (Fig. 1). These areas are rocky, but as the plates move away from the ridge axis through seafloor spreading, thin layers of sediment begin to

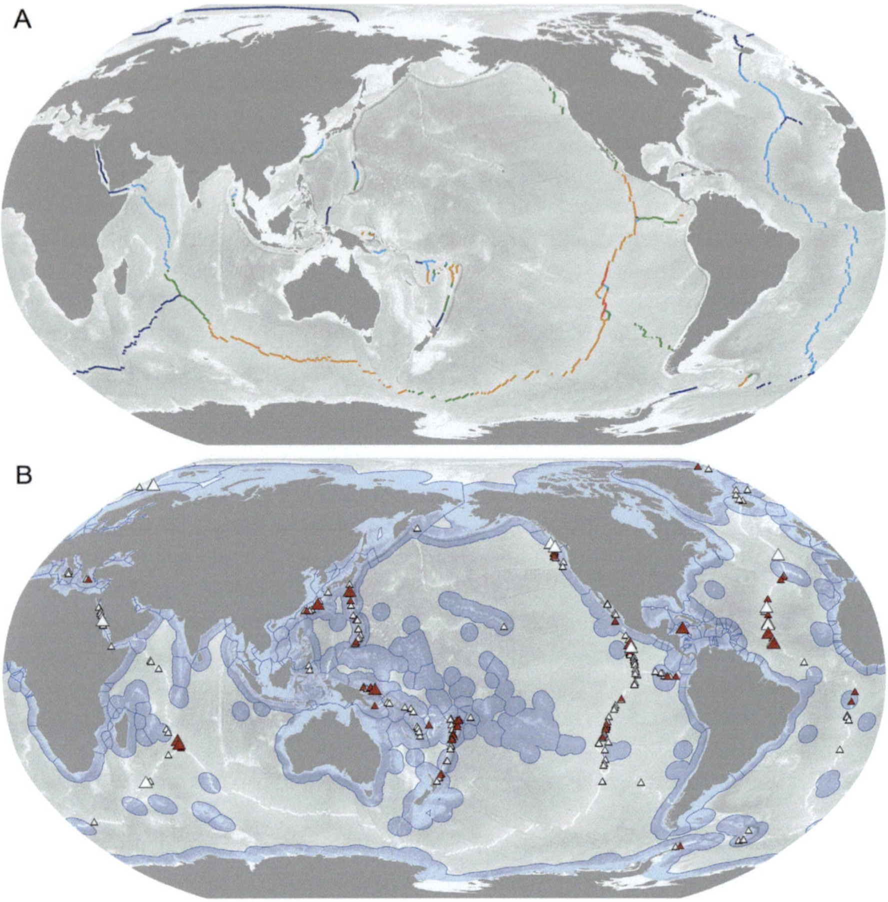

Fig. 1 (**a**) Locations of mid-ocean ridges and back-arc spreading centres important for the formation of seafloor massive sulphides. Colours denote the spreading rate of each segment. Dark blue = ultraslow spreading (140 mm/year). (**b**) Location of high-temperature seafloor hydrothermal systems and associated seafloor mineralization, where red colour indicates occurrences with economically interesting metal concentrations (average grade of the deposit is either >5 wt% Cu, >15 wt% Zn or >5 ppm Au) and large symbols indicate occurrences with size estimates above 1 million tonnes. Using these criteria, only a few occurrences of economic interest have been identified. Note that geochemical analyses are commonly only available for surface samples that are not representative for the entire occurrence. A quantitative resource assessment for seafloor massive sulphides is only available for two occurrences (Solwara 1 and Solwara 12, both within the EEZ of Papua New Guinea; [56]). Light blue areas delineate the Exclusive Economic Zones. (Figure from Petersen et al. 2016 Marine Policy)

accumulate, and the rock is eventually covered by thick layers of mud. These layers of mud can accumulate relatively quickly (10 s of cm per 1000 years) in areas with strong sediment supply to very slowly (~1 mm per 1000 years) in the middle of the large ocean basins such as the Pacific (http://www.boscorf.org/repository/

curatorial-reference/accumulation -rates). Exceptions to this are seamounts where rocky edifices may tower above the general seafloor reaching high into the water column. These seamounts are often swept clean of sediment by currents acting on their steep slopes.

Metalliferous deposits can accumulate in the top layers of the oceanic crust at mid-ocean ridges, on the seabed in areas with very low sediment accumulation ratcs and on the rocky flanks of seamounts.

On volcanically active ocean ridges (mid-ocean ridges, volcanic arcs and back-arc basins: Hannington et al. 2011), high-temperature hydrothermal fluids associated with the volcanism can dissolve metals in the upper crust, and in some circumstances, these metals can be deposited when the fluids are released at the seabed (Fig. 2). These are known as seafloor massive sulphide (SMS) deposits and can be rich in zinc, lead, copper, gold and silver (Fig. 3). They form three-dimensional ore bodies that can be metres to tens of metres thick. Ancient equivalents are mined on land today. There is a relationship between the rate of seafloor spreading with the occurrence and longevity of hydrothermal vents. On slow-spreading ridges, such as those in the Atlantic Ocean, vents can be hundreds of kilometres apart with individual vents lasting for millennia, whilst on fast-spreading ridges, such as those on the East Pacific Rise, vents can be just tens of kilometres apart with individual vents

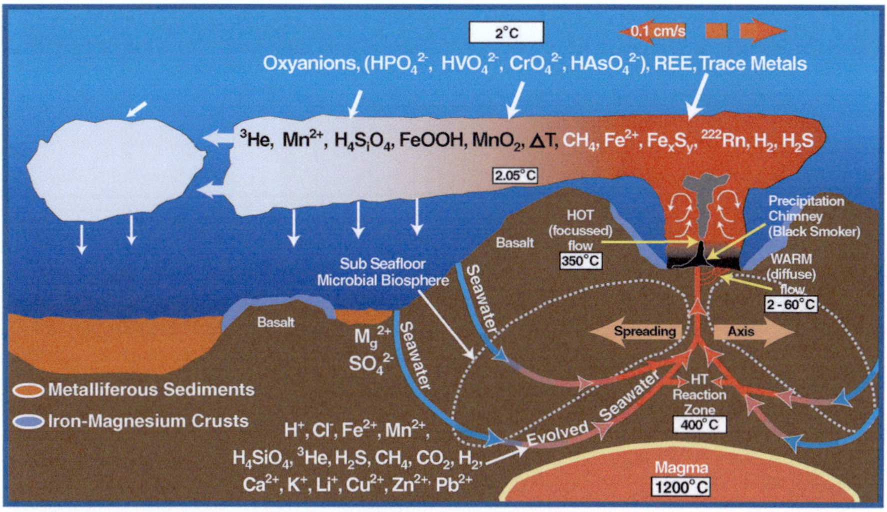

Fig. 2 Schematic diagram of hydrothermal vent system. Volcanic heat at the mid-ocean ridge axis drives hydrothermal circulation and chemical exchange between the ocean crust and seawater. A mid-ocean ridge hydrothermal system, plume, and resulting deposits and precipitates are featured. Some metals represented: Mn = Manganese, Mg = Magnesium, Co = Copper, Zn = Zinc, Fe = Iron. Some gases represented: 3He = Helium from the Earth's mantle, H2 = Hydrogen, CH4 = Methane, H2S = Hydrogen Sulphide, CO2 = Carbon Dioxide. The boiling point of water at sea level is 100 degrees C (Celsius) or 200 degrees F (Fahrenheit). 350 °C = 690 °F. (Image courtesy of Submarine Ring of Fire 2002, NOAA/OER)

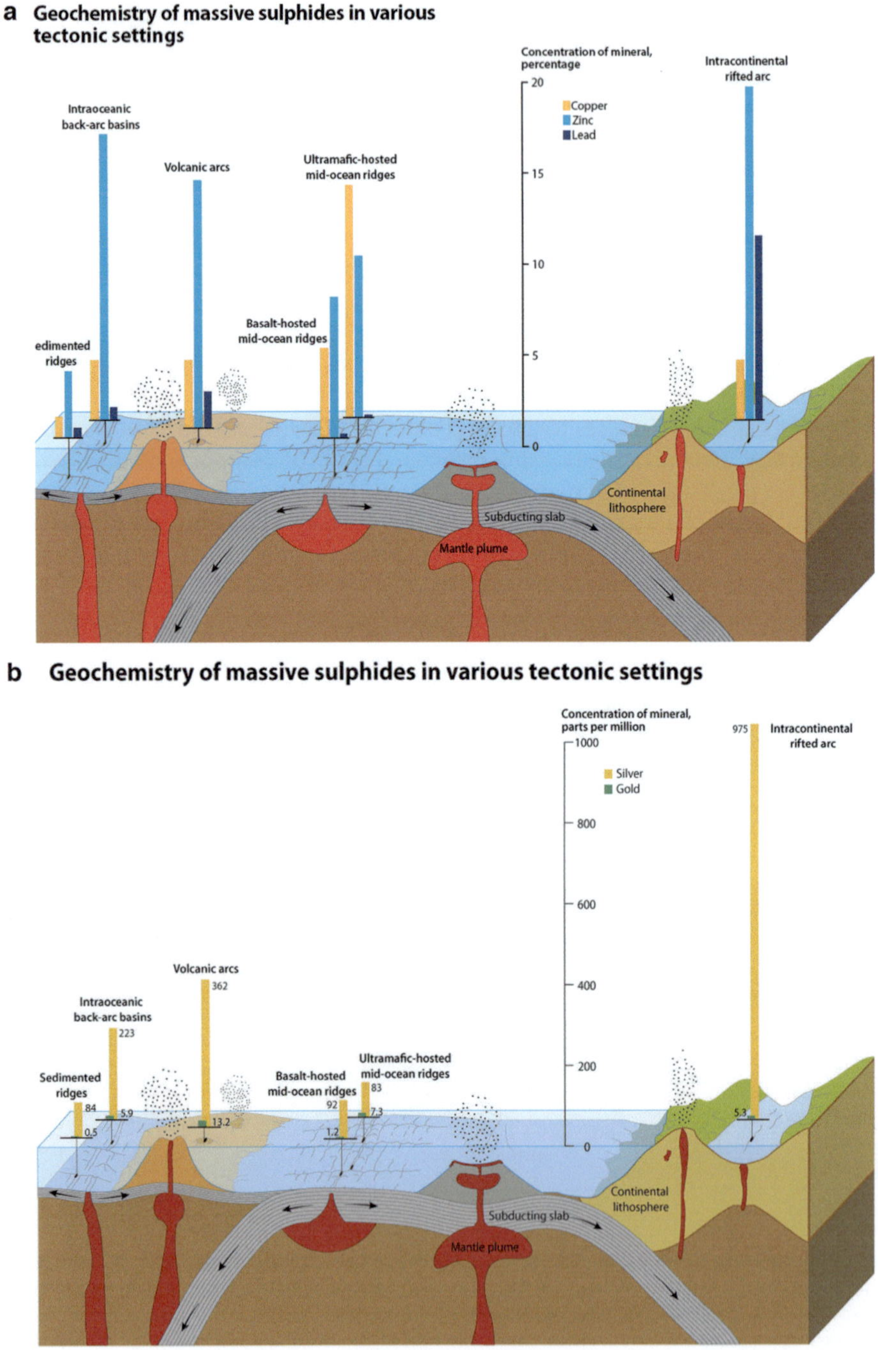

Fig. 3 Potential concentrations of metals in sulphide deposits in various plate boundary settings (**a**) copper, zinc and lead (**b**) gold and silver. Note metals do not necessarily precipitate in all locations depending on local conditions. (From SPC 2013, www.grida.no/resources/8159)

lasting for decades (Copley et al. 2016). Slow-spreading ridges therefore have the time to develop larger and more commercially attractive metal deposits.

Metalliferous deposits in the form of cobalt crusts on seamounts and manganese nodules on abyssal plains are both formed from the very slow precipitation of metals from seawater – hence the requirement for very slow or no sediment accumulation that would prevent the growth of significant deposits (Maribus 2014). On the flat ocean floor such deposits take the form of nodules – golf ball or potato-sized concretions – that lie strewn across the seabed in water depths between 3500 and 6500 m (Fig. 4). These are formed both from precipitation of minerals directly from overlying seawater (hydrogenously) and from seawater held in the sediments beneath the nodules (diagenetically). Although they take millions of years to form, very few of them are found completely buried in the sediment, perhaps because they are overturned periodically by animals searching for food. Manganese nodules contain a large range of metals. But the most commercially attractive are nickel, copper, cobalt, manganese and molybdenum. The Pacific Ocean is particularly rich in manganese nodules with large deposits in the Clarion Clipperton Zone (CCZ), Peru Basin and Penrhyn Basin. Additional deposits have been found in the Indian Ocean

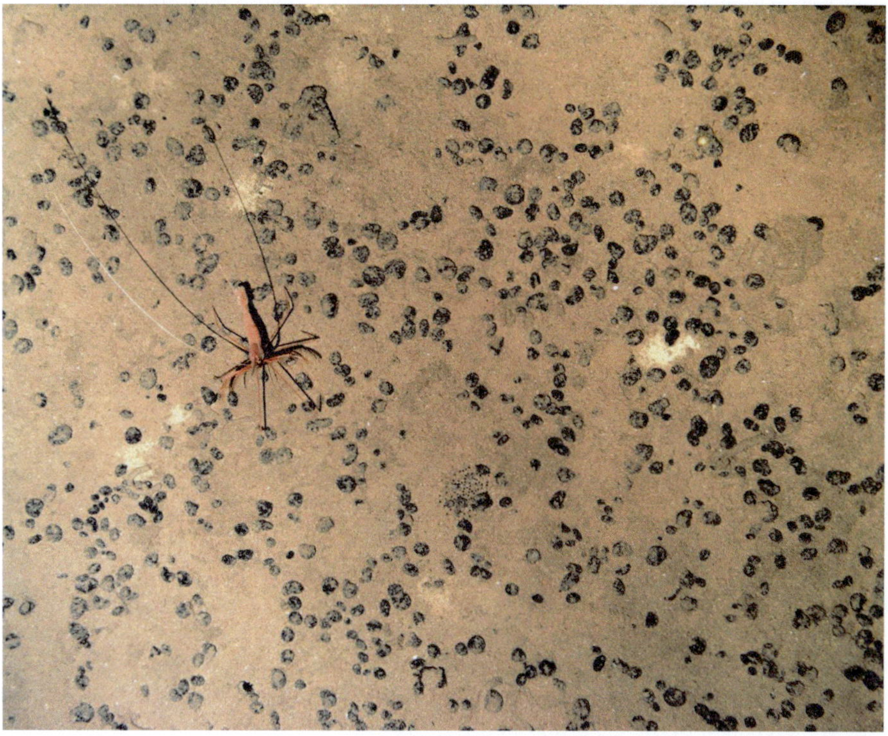

Fig. 4 Image of the seafloor in the abyssal Pacific showing manganese nodules and large deepwater prawn (*Bathystylodactylus* sp.). Image shows an area of seafloor approximately 50 cm across. (Credit: Image courtesy of Dr Daniel Jones, National Oceanography Centre, Southampton (NOC))

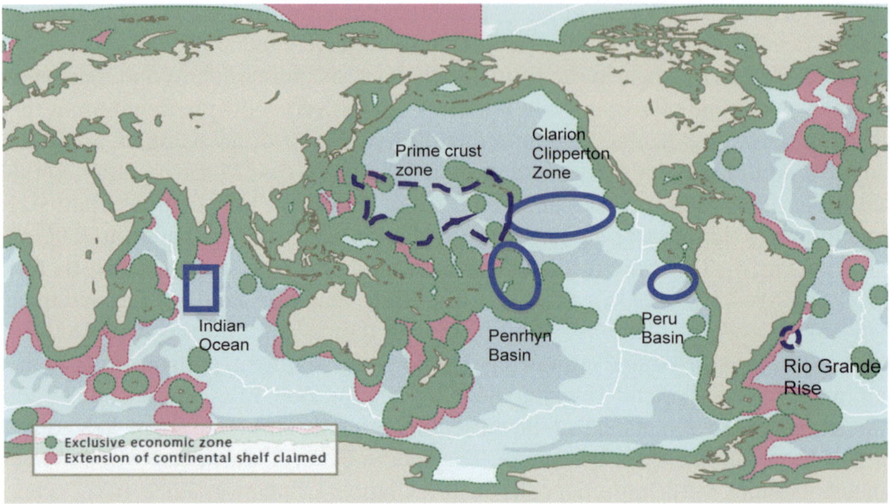

Fig. 5 Areas of the world ocean where manganese nodules (solid blue lines) and cobalt crusts (dashed blue lines) are known to occur in high enough quantities and in sufficient thickness in relatively benign geomorphology to be economically interesting. (From Maribus 2014, 2017)

Fig. 6 Cobalt-rich crust on a hyaloclastic substrate from the flank of a seamount in the Central Pacific; thickness of crust ca. 8 cm. (Source: Prof. Halbach, FU Berlin)

(Fig. 5). The CCZ, Peru Basin and Indian Ocean deposits all lie in international waters, whereas the most promising deposits in the Penrhyn Basin lie within the territory of the Cook Islands (McCormack 2016). To be commercially attractive, nodules need to be abundant on the seafloor: a density of more than 15 kg/m^2 over areas of more than several tenths of a square kilometre is generally accepted as a minimum (www.isa.org.jm/files/documents/EN/Brochures/ENG7.pdf). This is frequently exceeded in the CCZ and Penrhyn Basin.

Cobalt crusts form by precipitation of minerals from seawater onto bare rock outcrops in the ocean (Fig. 6). The most economically interesting crusts are found at water depths between 800 and 2200 m on the flanks and summits of seamounts and plateaux where currents are sufficiently strong to prevent sediment from depositing (Hein 2002). In these areas crusts can build up to thickness of 26 cm, with growth rates of 1–6 mm per million years, so the oldest seamounts tend to have the thickest crusts. The most promising areas for exploitation lie in the Pacific Ocean, in an area known as the prime crust zone, although other areas such as the Rio Grande Rise in the South Atlantic also have crust formations (Fig. 5). Cobalt crusts are composed predominantly of iron and manganese oxides and hydroxides but can contain significant quantities of other metals such as cobalt, nickel and platinum, alongside rarer metals such as tellurium, bismuth, tungsten, niobium and some rare earth elements. The main metals of commercial interest are cobalt, nickel and platinum, though rare earth elements may also be recovered.

3 Progress Towards Deep-Sea Mining

Interest in mining metals in the deep sea began in the 1960s, when it appeared there would be a bountiful supply of cobalt, nickel and other metals in the manganese nodules that could be found in the deep parts of the oceans (Mero 1965). Initial interest was shown by a number of companies (Jones et al. 2017) including a consortium led by Lockheed in the USA, called the Ocean Minerals Company (OMCO), who built the ship Glomar Explorer and fitted it with a test mining system for manganese nodules. This consisted of a seabed-mining vehicle connected to the ship via a 5000-m-long steel pipe (Chung 2009). The mining system was successfully tested in the mid-1970s but never went into commercial production. In fact, all companies involved in this early phase gave up due to a combination of factors, including relatively low metal prices, the engineering challenges and the lack of a legal framework in the "Area." The term "Area" is defined in UNCLOS as "The seabed and ocean floor and subsoil thereof, beyond the limits of national jurisdiction" and relates to the entire ocean floor that is not owned or governed by individual countries (Fig. 7) (for a description of UNCLOS and its history, see http://www.un.org/depts/los/conventionagreements/conventionoverviewconvention.htm).

The legal framework for controlling deep-sea mining in the Area was established as part of the 1982 United Nations Convention on the Law of the Sea (UNCLOS) and the 1994 Agreement relating to the implementation of Part XI of UNCLOS (known as the 1994 Implementation Agreement). The International Seabed Authority is an autonomous international organisation created by UNCLOS to organise, carry out and control activities in the Area. This role includes both setting the detailed standards that govern the operations of deep-sea mining companies and supervising the implementation of these standards and the general provisions of Part XI. The Authority also has the responsibility for the protection of the marine environment (Article 145), the promotion of marine scientific research (Article 143) and the protection of underwater cultural heritage in the Area (Article 149). The Authority became fully operational in June 1996 and has its headquarters in Kingston, Jamaica (Article 156). All States parties to the 1982 Convention are

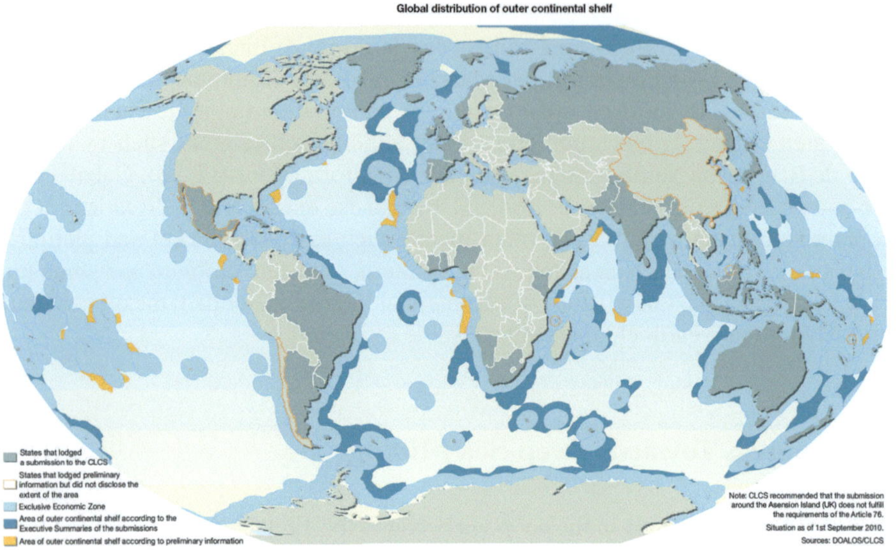

Fig. 7 Exclusive economic zones, extended continental shelf areas and claims for extended continental shelf areas under UNCLOS. Within these areas states maintain control of activities on the seabed such as mining. The pale blue regions on the map are collectively known as the Area, and deep-sea mining here is managed by the International Seabed Authority. (Courtesy: Riccardo Pravettoni GRID Arendal www.grida.no/resources/7922)

members of the Authority. On 15 January 2015, there were 167 members of the Authority plus the European Union, with the notable exception of the USA.

By the mid-1990s with the legal framework established, and major advances in offshore engineering brought about by the offshore hydrocarbon industry, the deep-sea mining industry was poised to take off. The ISA approved six contracts for exploration for manganese nodules in 2001, one more in 2002 (Fig. 8) and a further contract in 2006. However, the sharp rise in metal prices in the early 2000s (Fig. 9) seems to have been responsible for a surge in applications for contracts beginning in 2011, such that by the end of 2018, a total of 29 exploration contracts had been awarded (ISA 2019). Of these 17 were for manganese nodules, 7 for SMS and 5 for cobalt crusts (Fig. 8). As of the end of 2018, however, the ISA had not issued any contracts for exploitation of minerals (actual mining). This is because the contractors are not yet ready to begin mining because they are still engaged in equipment development and because the rules that will govern exploitation remain under discussion and development at the ISA, with an expectation that they will be approved by 2020. Until these exploitation regulations are in place, it is difficult for the companies to estimate costs and profit margins.

Whilst progress towards mining may be proceeding slowly in international waters, the mining of SMS deposits within the exclusive economic zones (EEZs) of some countries is close to becoming a reality. The Canadian company Nautilus Minerals claim to be almost ready to mine SMS deposits in the Bismarck Sea off Papua New Guinea (www.nautilusminerals.com). The Japanese corporation JOGMEC also began test mining of SMS deposits in the Okinawa Trough off Japan

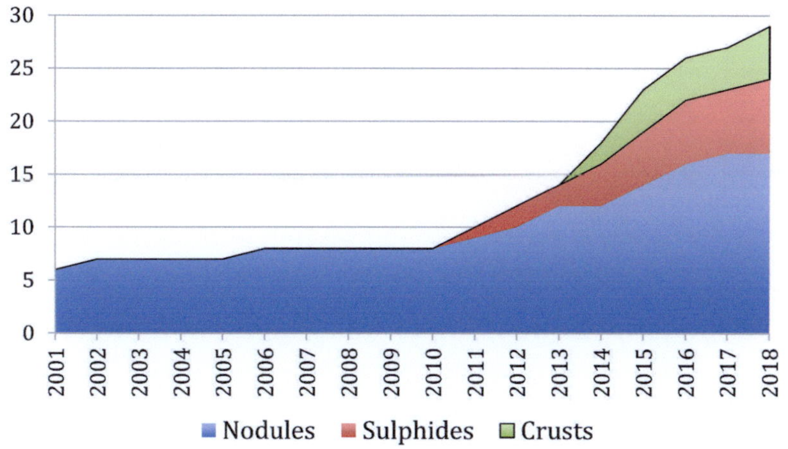

Fig. 8 Cumulative number of exploration contracts awarded by the International Seabed Authority (ISA 2019)

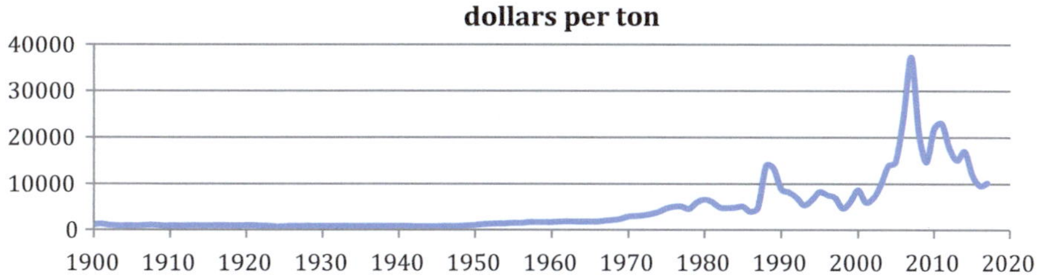

Fig. 9 Nickel prices since 1900. (Data combined from USGS 1999 and www.indexmundi.com/commodities/?commodity=nickel&months=360, 29th November, 2019)

in 2017 (www.meti.go.jp/english/press/2017/0926_004.html). The mining of SMS deposits has advanced more quickly because the deposits can be found at relatively shallow water depths, and they are similar to those mined on land, so existing mining technology can be adapted to the deep sea rather than needing to be invented.

Nodule and crust mining presents more acute technological challenges. Exploitation of manganese nodules requires the development of riser pipes that can transfer the ore from the seabed to the sea surface over 5000–6000 m water depth and the ability to control mining vehicles on the seabed at these remote depths (Fig. 10). The mining of cobalt crusts requires the development of sophisticated equipment that can remotely measure the thickness of the crust over large areas to facilitate resource assessment. Equipment needs to be developed that will remove only the metal-rich crust and not the underlying rocks. This will require systems that constantly vary the depth of the cutting blades as the mining machine progresses over rugged terrain.

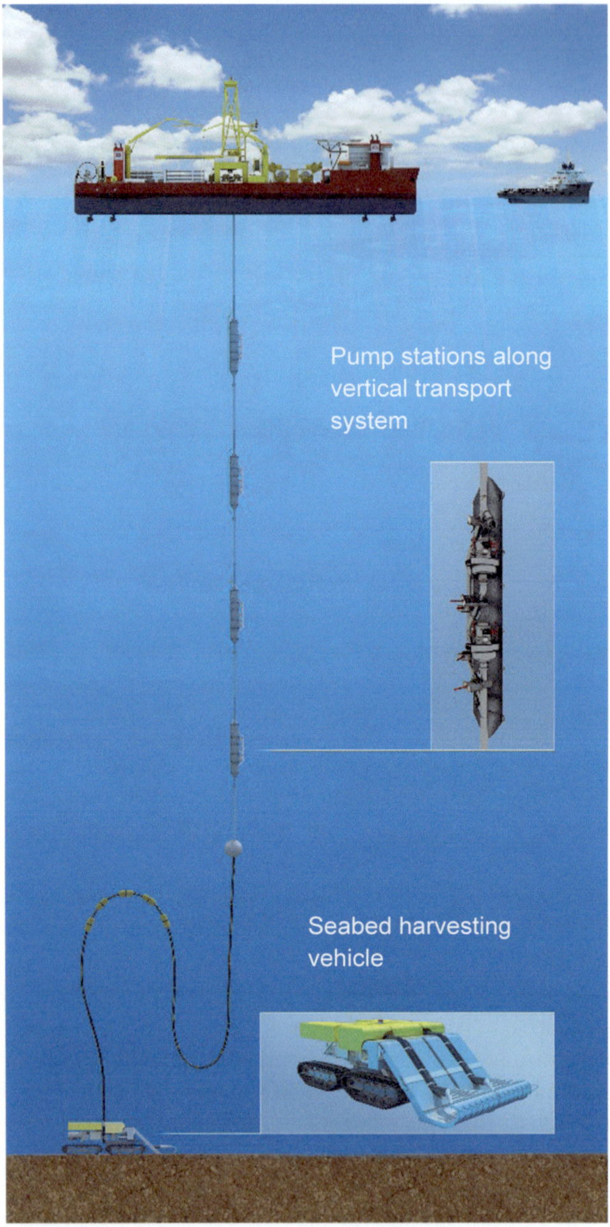

Fig. 10 Schematic diagram showing the main components of a deep-sea mining system for manganese nodules (not to scale). (Slide courtesy of IHC Mining, the Netherlands)

4 Environmental Issues Related to Deep-Sea Mining

The potential environmental impacts of deep-sea mining were recognised at an early stage (Thiel et al. 1991). There are many concerns relating to physical impacts of the mining systems on the seafloor, the creation of sediment plumes by seabed operations, the integrity of the riser pipes and the release of wastes following pre-processing of the minerals at the sea surface. A number of concerns are specific to certain deposit types, for example, specialist hydrothermal vent communities are at risk from exploitation of SMS deposits at mid-ocean ridges, whilst the ecosystems in areas of nodule fields are diverse with sparsely distributed individuals that may be impacted by the very large areas (Weaver et al. 2018). Other issues are common to all types of deep-sea mining such as the impact of noise and of mid-water plumes of sediment-laden water that may affect animals living in the water column.

The biggest issues relating to the impact of deep-sea mining revolve around:

- Lack of scientific knowledge of deep-sea ecosystems
- Lack of knowledge about the performance of the technology that is still under development
- Lack of a regulatory framework that will set limits on environmental impact

Gaining more scientific knowledge in the available timeframe before mining begins may be difficult in regions such as the CCZ, some seamount chains and some ocean ridges because they are so remote from land. Nevertheless there is a pressing need for resources to be dedicated to such activities (Boetius and Haeckel 2018). In the Area the contractors themselves may generate the most information as they are mandated to collect environmental data within their contract areas. However, these data need to be generated as exploration progresses and to be reported early because Regional Environmental Assessments (REAs) are likely to be required before mining begins (Jones and Weaver 2017). Standardisation of data collection methods and plans for environmental surveys to achieve statistically significant results are two critical challenges so that data from large regions of the seabed incorporating multiple contract areas will provide a comprehensive assessment of cumulative and regional impacts. It is not clear who will collect information from outside of the contract areas to complete this work.

The technology issue relates to the large monetary investment that is needed before the equipment can be field-tested and hence its full environmental impact assessed. There will consequently be reticence to significantly modify the equipment designs if the environmental impacts are greater than predicted. Cumulative impacts – where the effect increases as time progresses – will be difficult to identify in the first few years of mining. A strategy for dealing with these unknowns is to use adaptive management to deal with problems as they arise (Jaeckel 2016). The new exploitation regulations under consideration (ISA 2018) will have to ensure that technologies to mitigate for environmental impacts are considered in detail at the equipment design stage to reduce the need for costly modifications at a later date. In addition the Regulations should be formulated in such a way that will compel new technologies to be introduced as we learn more about environmental aspects. This might mirror the way in which measures introduced by the International Maritime Organisation are now being introduced to regulate toxic emissions from shipping.

Development of legislation brings both reassurance and new challenges: on one hand, setting a regulatory framework for a new industry should ensure that activities meet modern standards and have minimal impact on the environment. On the other hand, the regulations must deal with uncertainties and unknowns and anticipate issues that can only be solved by greater investment. Such regulation must adhere to the law as it was laid down under UNCLOS and be enforceable across all the mining nations. One way forward would be to apply the precautionary approach, which was adopted by the Rio Conference, or Earth Summit, in 1992 (UNCED 1992). Principle 15 of the Rio Declaration states "*where there are threats of serious or irreversible damage, lack of full scientific certainty shall not be used as a reason for postponing cost-effective measures to prevent environmental degradation.*" A brochure explaining how the Precautionary Principle might be applied to deep-sea mining has been produced by the Secretariat to the Pacific Community (SPC) (see http://dsm.gsd.spc.int/public/files/resources/ Deep_Sea_Minerals_in_the_Pacific_Islands_Region_Brochure_13_Precautionary_Principle.pdf). This type of approach has been advocated by Ellis et al. (2017).

5 Deep-Sea Ecosystems

The impacts of deep-sea mining on deep-sea ecosystems can be divided into habitat destruction and/or modification at the mine site, impacts in the area surrounding the mine site related to the dispersal of sediment-laden plumes generated by the mining process, impacts related to plumes released in the water column after dewatering of the ores on the ship and transport barge and impacts related to other factors such as sound and light.

The three ores – seafloor massive sulphides, manganese nodules and cobalt crusts – are each associated with very different ecosystems, each posing different environmental issues during their exploitation. Each will therefore be treated separately below, followed by a section on impacts that are common to all deep-sea mining activities.

For all three ores, the application of the 'mitigation hierarchy' moving in sequence from 'avoid', to 'minimise', to 'restore' and finally, if relevant, to 'offset' should be considered. The rate at which deep-sea ecosystems recover from impacts in many cases will present particular challenges for the restoration of deep-sea ecosystems, and offsetting impacts by improving degraded similar deep-sea habitats is unlikely to be appropriate (Van Dover et al. 2017).

5.1 Potential Environmental Impacts from Mining SMS Deposits

SMS deposits have higher metal contents than the surrounding rocks, for example, at the Solwara 1 mine site in the Bismarck Sea off Papua New Guinea, the ore contains just over 7% copper and 5.0 g/t gold (Lipton 2012). Rocks containing metals with these percentages are difficult to determine by geophysical prospecting, and prospectors currently often use geochemical prospecting methods such as the occurrence of hydrothermal venting (expulsion of hot fluids at the seabed) as an indication of where to look in detail (Holz et al. 2015). The vent fluids can be identified through physical and chemical anomalies in the water column (temperature, chemical variations of elements such as Mn, Fe, redox potential and/or particle concentrations). The vent fluids can themselves sustain unique biological communities of bacteria, shrimp, tubeworms and other organisms (Figs. 11, 12, and 13) that are adapted to this hot and inhospitable environment (Fisher et al. 2013). Such chemosynthetic organisms have limited distributions because they are confined to the hot springs which are themselves limited to small areas distributed along the ocean ridges.

Different ridge systems also host different organisms, for example, giant tubeworms dominate many vent habitats in the eastern Pacific but are not known from the Atlantic, Indian or Southwest Pacific ridges. Vent shrimp are very common in the Atlantic, whilst shrimp, snails and anemones are common in the Indian Ocean (Fisher et al. 2013; Clark and Smith 2013). At least 11 different biogeographical

Fig. 11 White shrimp (including *Rimicaris* sp.) on hydrothermal chimney at Von Damm vent site on Mount Dent Caribbean Sea and smaller reddish shrimp with green head. (NOAA Okeanos Explorer Program, Mid-Cayman Rise Expedition 2011 www.photolib.noaa.gov/htmls/expl6952.htm accessed December 18th 2017)

Fig. 12 Vent mussels, shrimp and crabs from the Mid-Atlantic Ridge © MISS√O SEHAMA, 2002 (funded by FCT, PDCTM 1999/MAR/15281). (Photographs taken by VICTOR6000/IFREMER)

zones have been proposed for vent fauna (Rogers et al. 2012; Chown 2012) (Fig. 14). The geographic distributions of particular species may be limited, but in some cases, notably polychaete worms, geographic distributions appear to be extensive (Copley et al. 2016).

Loss of habitat at the seabed due to SMS mining will be relatively small scale because SMS deposits are three-dimensional and have a small footprint on the seabed. The Solwara 1 mine off Papua New Guinea will directly impact 1.4 km^2 of seabed over a projected 3-year mine life (Batker and Schmidt 2015). This is a very small deposit, but even large SMS deposits may only have footprints of a few square kilometres. However, as hydrothermal vent chemosynthetic communities are also potentially restricted to a small footprint, it may be necessary to take particular regard of these communities in environmental impact assessments and regional environmental management plans. If large numbers of hydrothermally active sites are mined or impacted by mining, then these unique ecosystems, which occur in relatively few locations, could suffer loss of connectivity between populations (Van Dover 2014). SMS mining in areas without present-day hydrothermal vent communities may have little impact in terms of loss of habitat because deep-sea taxa associated with them will be more typical of rocky slope fauna, which have wide

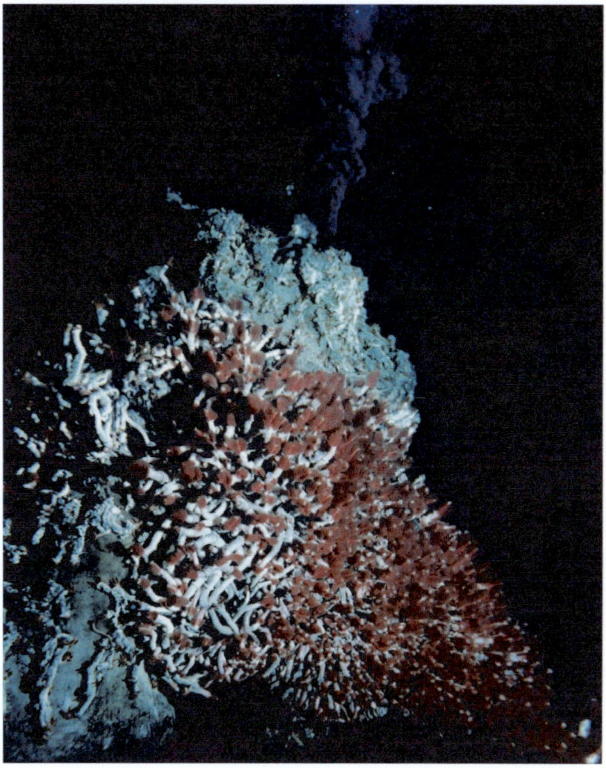

Fig. 13 Hydrothermal vent surrounded by giant red tubeworms and squat lobsters. From the Main Endeavour hydrothermal field on the Juan de Fuca Ridge. Vibrant colonies of tubeworms with red gills thrive on this vent, which is predominantly composed of iron- and sulphur-bearing minerals. (Credit: University of Washington; NOAA/OAR/OER.www.flickr.com/photos/51647007@N08/5277263409/)

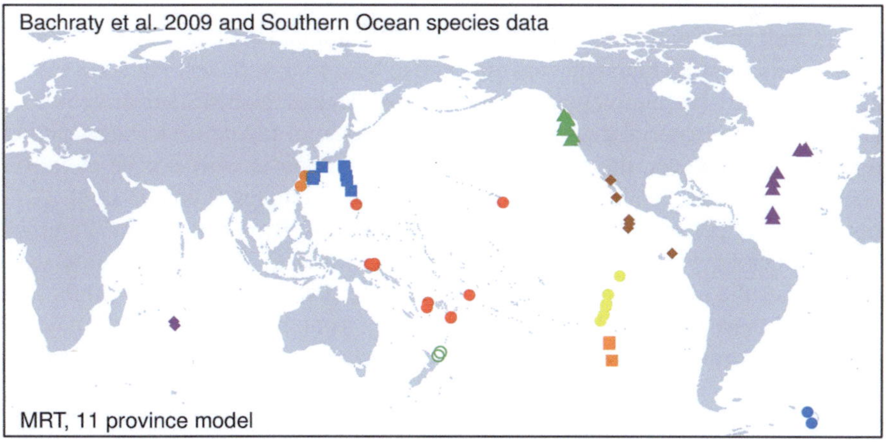

Fig. 14 Biogeographic provinces of hydrothermal vent faunas. The data shows 11 provinces identified in different colours each of which has distinct groups of species. (From Rogers et al. 2012)

geographic distributions. Such low impact will however depend on the spread and impact of plumes. These communities are described further below.

Plans have been developed to temporarily relocate some organisms and to create new substrates to replace those that have been mined (http://nusc.live.irmau.com/IRM/Company/ShowPage.aspx?CategoryId=190&CPID=1176&EID=83153213). However, the success of these plans has not been proven, and much greater sustained research on the restoration of hydrothermal vent communities is required. Closely spaced areas of active hydrothermal venting can show many very different habitats and biodiversity indicating that subtle variations in environmental factors and biological interactions may control distribution of organisms (Boschen et al. 2015) making it very difficult, if not impossible, to temporarily relocate organisms to other sites.

One characteristic of hydrothermal vent faunas is their ability to recover relatively quickly from disturbance, especially in comparison to other deep-sea faunas (Gollner et al. 2017). This is because hydrothermal venting does not necessarily occur at constant rates and may even cease for some periods, thus organisms need to be adapted to colonise other areas. Populations of animals on fast-spreading ridges can recover in just a few years (Van Dover 2011; Gollner et al. 2017), whilst those on slower-spreading ridges may recover more slowly (Boschen et al. 2013). It is not certain what the impacts of mining might be on the long-term health of vent communities in the different setting of fast-, slow- and ultraslow-spreading ridges – especially the cumulative impacts if mining is intense along any particular interval of ocean ridge.

Apart from the seabed area that is directly impacted by the mining, adjacent areas may be impacted by plumes of sediment-laden water generated by the mining machines. These plumes will spread away from the mine site and may bury or smother seafloor organisms and habitats and prevent recolonisation (Levin et al. 2016; Clark and Smith 2013). Such organisms could include corals, sponges and a wide variety of other fauna that are attached to hard rock surfaces. For SMS mining the plumes may be relatively small in volume because the SMS substrate is hard, and most of the material generated may be too large or too dense to be suspended in the water. However, the act of mining sulphide deposits will expose metal surfaces to oxidation, which will produce dissolved sulphides (Fuchida et al. 2017; Simpson and Spadaro 2016; Brown and Hauton 2018; Brown et al. 2017) and will also produce very small sulphide particles (Fuchida et al. 2017; Hauton et al. 2017). Thus the plumes are also likely to be toxic (Fallon et al. 2018; Knight et al. 2018), and these toxins will be transported away from the mine site. This may lead to very significant releases of toxic metals, such as cadmium, arsenic, antimony and copper, especially if the crushed ore is stockpiled on the seabed for any length of time. Increases in acidity following oxidation may lead to runaway reactions within ores stored on the seabed. Consequently, care should be taken to minimise any leakage of toxic material from stockpiled sulphides either on the seabed or onshore (Steiner 2009).

Gwyther (2008) suggested that plumes will spread only for 1 km at the Solwara 1 site, but this remains to be tested. Plumes may also be generated if sediment overlying the ore body has to be removed from the mine site. This sediment, which is likely to be pumped downslope from the mine site, will be easily disaggregated and less dense than the ores (Gwyther 2008) and hence may spread further. It is likely that the volume of these near-seabed plumes can be reduced through good engineering design and mining practice. One possibility could be to encase the mining vehicles in a shroud that would contain the plume and force all of it up the riser pipe (Van Dover et al. 2017).

When the ores are received by the mining vessel, they will be dewatered, and the waste water will be returned to the ocean. The dewatering process cannot remove all the smallest particles and those smaller than a few tens of microns will be discharged in the returned water. Gwyther (2008) suggested that only particles smaller than about 8 μm will be released at Solwara 1, and the plume will be released close to the seabed. In many cases toxic metals attach preferentially to fine particles. Even though the discharge plume may be diluted very quickly within the environment (Gwyther 2008), it may still have biological effects, especially through bioaccumulation. Discharge plumes therefore have the potential for wide-ranging effects at scales larger than those evident directly at the mine site and in both the pelagic and benthic environment.

Mining at active hydrothermal sites may be limited because these locations are still in the process of precipitating metals and the deposits may be small (German et al. 2016: Van Dover et al. 2018). Furthermore, active vents are very hot and can pose serious challenges to the development and operation of equipment (Clark and Smith 2013). It is more likely therefore that SMS deposits will be sought at some distance from the ridge axes at sites where the ore bodies have had chance to fully form and to cool (Petersen et al. 2017). These ore bodies are currently difficult to locate because their geochemical and geophysical signatures become progressively weaker away from the ridge as the minerals become buried by the gradually thickening sediment layers and/or by lava flows. Work is underway to develop new high-resolution prospecting equipment including magnetic and self-potential sensors (measuring weak electrical currents produced by chemical reactions occurring between the sulphide deposit and seawater) as well as mineralogical and geochemical techniques (Petersen et al. 2017). It is hoped that potential ore bodies will eventually be located up to a few tens of kilometres from the ridge axes and potentially beneath several metres of sediment.

The ecosystems present on ore bodies away from the ridge axes have not been extensively studied but are likely to be more representative of the regional benthic fauna with few, if any, of the chemosynthetic organisms (Gollner et al. 2017: Boschen et al. 2015) (Fig. 15). Exceptions may occur if low-level hydrothermal activity is still in progress, and research needs to be conducted to ensure that specialist faunas have not developed to use weathered sulphide deposits as a substrate (Erickson et al. 2009). The regional fauna will be dependent on a number of factors such as substrate type, water depth, water mass characteristics (temperature and salinity) and particulate organic flux to the seafloor. These characteristics were used by Watling et al. (2013) to develop a scheme to describe the biogeographic distribution of species across the seabed of the world in which a number of different benthic

Fig. 15 Typical area of rocky seabed away from the ridge axis with the crinoid *Anachalypsicrinus nefertiti* and some large sponges. Mid-Atlantic Ridge, depth c. 2400 m. (Credit: Image courtesy of Dr. Daniel Jones, National Oceanography Centre, Southampton, UK. ECOMAR Project)

provinces were proposed. Substrate is important because hard rocks allow species that attach to the seabed, and soft sediment allows burrowing animals. In many areas both hard and soft seabeds are intermixed (Priede et al. 2013).

The siting of an SMS mine away from the ridge axis in one of the faunal provinces could have a small footprint that may not make a significant impact. Nevertheless, these areas remain poorly known, and a precautionary approach is prudent until the impacts of actual mining can be determined. The International Seabed Authority is likely to adopt environmental management plans for the Atlantic and Indian Oceans in the next few years, and one element of these may be networks of Areas of Particular Environmental Interest (APEIs) that will establish areas for the conservation of biodiversity and across the ridge axis where representative ecosystems at different depths can be preserved. Such a network has been suggested for part of the North and Equatorial Atlantic Ocean (Dunn et al. 2018).

5.2 *Potential Environmental Impacts from Mining Manganese Nodules*

Manganese nodules are located in some of the remotest places on the planet. Consequently these areas have received less attention from the scientific community than sedimented areas closer to continental margins. It has actually been the prospect of mining in areas such as the Clarion Clipperton Zone (CCZ) in the equatorial eastern Pacific that has stimulated research. Ecosystems with manganese nodules have been found to have a great diversity of organisms and with many species appearing apparently at only one sample location. This gives the impression that individuals from each species are very rare, but this may reflect merely the very small number of samples that have been collected and examined. With greater sampling effort, many species may eventually be found to have wide distributions. There is not enough information to understand why such high diversity of species occurs and how different species interact. There is little information on connectivity between populations of all taxa (Glover et al. 2016a). What is known is that the seabed of the CCZ has a very stable environment with very low sedimentation rates and generally low input of food. Organic matter inputs come from particulate organic flux from the sea surface, and inputs are slightly greater towards the east and south of the CCZ owing to slightly higher primary productivity in these regions. These variations in food supply are reflected in variations in fauna across this vast region (Smith 2013), which extends for about 6 million km^2. In this environment it is expected that many organisms will be long-lived and slow to reproduce making it difficult for them to recover quickly from mining disturbances (Gollner et al. 2017).

Animals in the CCZ, and likewise all manganese nodule areas, occur within the sediment, attached to the nodules and in the water column above the seabed. Specialist benthopelagic organisms occur in the benthic boundary layer (BBL) immediately above the seabed (Billett et al. 1985). In addition, the BBL is an important region for the dispersal of larvae from benthic organisms.

The fauna within and on the sediment generally occur within the upper few centimetres of the seabed. The dominant meiofaunal groups (organisms that can pass through a 1 mm mesh sieve but will be retained by a 45 μm mesh) are nematodes and harpacticoid copepods, whilst polychaetes and isopod crustaceans are the most important macrofaunal taxa (larger than 1 mm). Typical megafauna (animals visible in sea bottom photographs and trawls) include holothurians, fish and giant protists. The latter provide shelter for numerous invertebrate groups such as nematodes and isopods. A specialist fauna attaches to the nodules, which provides a hard substrate supporting a wide variety of encrusting fauna such as corals, bryozoans, xenophyophores, komokiaceans and sponges (Fig. 16) (Vanreusel et al. 2016), whilst smaller animals such as foraminiferans and nematodes colonise crevices in the nodule surfaces. A detailed account of the biology of manganese nodule areas is given by Smith (2013).

A study of megafauna in the CCZ by Vanreusel et al. (2016) found that the epifauna (animals living on top of the sediment and on nodules) in areas associated with polymetallic nodules are more abundant and diverse than in areas without nodules or with only low nodule numbers (Fig. 17). The loss of nodules will reduce the substrate upon which a wide variety of fauna depend. As the nodules take millennia to grow, the loss of substrate will have permanent impacts on biodiversity unless spatial management plans safeguard large areas where nodules occur, including areas with dense nodule cover.

Fig. 16 Images of megafauna associated with nodules from the CCZ (**a–c**) actiniarians; (**d–f**) alcyonacean corals; (**g**) antipatharian coral; (**h–l**) hexactinellid sponges. Copyright: ROV Kiel 6000 Team/ GEOMAR Kiel. (From Vanreusel et al. 2016)

The main environmental concerns relating to the mining of manganese nodules are:

1. The slow rate at which abyssal fauna recolonize the environment following mining disturbances
2. The large aerial scale of the impacts
3. Lack of knowledge of the organisms due to vastness of the area and limited scientific study

Recolonisation in the deep sea is much more complex than on land because many of the species are long-lived and reproduce very slowly, and they are not adapted to environments where disturbances occur. This is particularly true for the nodule areas where mining will scrape the top ~10 to 15 cm of the seabed, disaggregate the sediment and sift out the nodules, dumping the sediment particles behind the collector machine (Fig. 18). This dumped material will have different properties to the original seabed, containing much more water, lacking cohesion and will be devoid of life. Experiments in the MIDAS project have shown that even the bacteria are lost in this process.

Jones et al. (2017) carried out a comprehensive review of the effects of nodule mining on sediment communities based on small-scale commercial test mining and

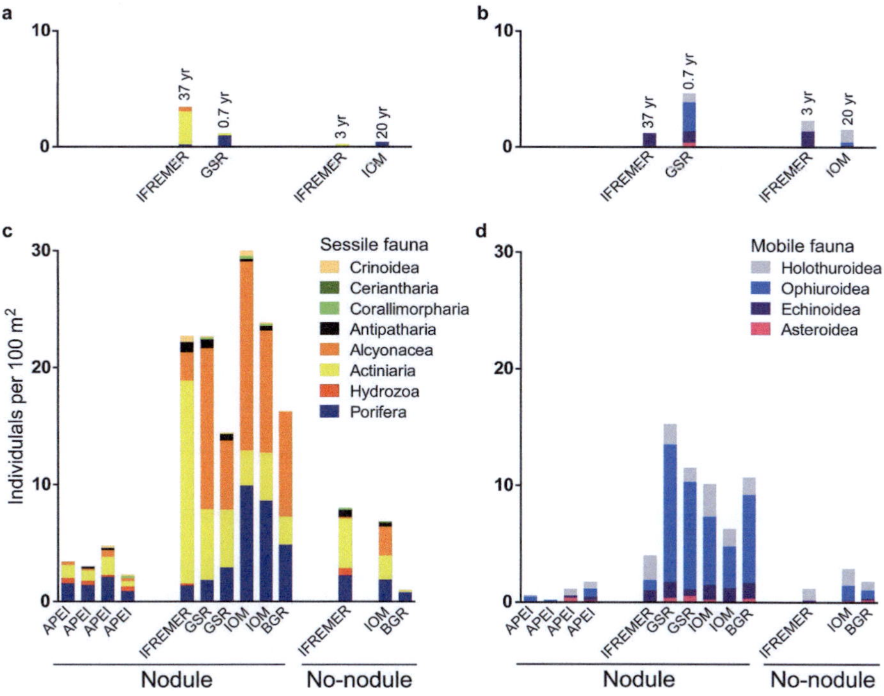

Fig. 17 Densities (individuals/100 m^2) of sessile (**a**, **c**) and mobile (**b**, **d**) epifauna for separate ROV video transects from areas of the CCZ rich in nodules and nodule-free areas. (**a**, **b**) Densities from ROV video transects in experimentally disturbed areas of various age; (**c**, **d**) Densities from undisturbed areas. (From Vanreusel et al. 2016)

scientific disturbance experiments. They found that whilst severe changes in abundance and diversity of benthic communities occurred immediately after the simulated mining disturbance, some recovery occurred in faunal density and species numbers for meiofauna and mobile megafauna over a number of years. Even after two decades, however, the communities had not returned to the pretest baseline condition. In addition, where the sediment surface is compacted with no evidence of significant resedimentation and restructuring of a surface sediment layer, meiofaunal communities showed no recovery even after 26 years (Miljutin et al. 2011).

In another study the seabed of the DISCOL area in the Peru Basin has been repeatedly revisited since it was deliberately disturbed in 1989 (Thiel and Schreiver 1990). While faunal densities of most taxa were almost back to pre-disturbance levels after 7 years, diversity and faunal composition had not recovered over a longer 26-year period (MIDAS 2016; Gollner et al. 2017). The MIDAS project also showed the importance of ultrahigh-resolution bathymetric mapping to set the disturbance experiment in the context of the local sub-metre scale geomorphology, and the need for precision sampling by ROV-placed cores for accurate determination of spatial variability in recovery rates. Jones et al. (2017) also highlight the limitations

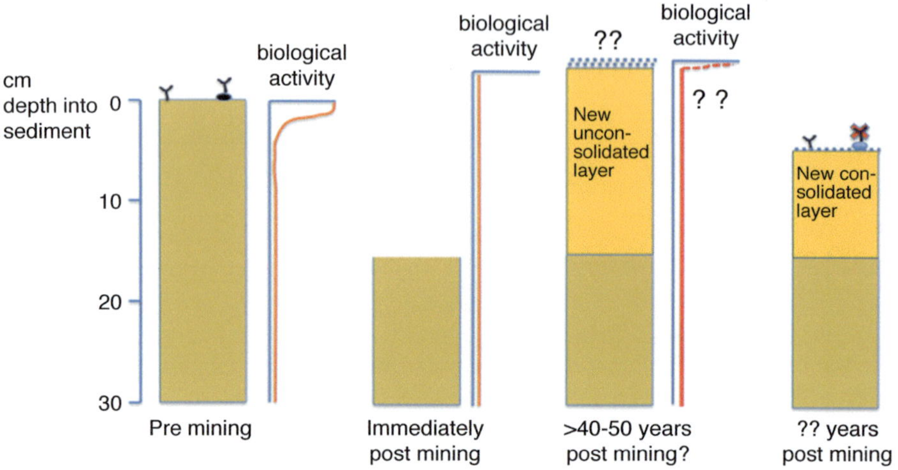

Fig. 18 Diagram showing the process of mining may inhibit recolonisation of organisms. Mining is assumed to remove the upper 15 cm of the seabed including the nodules from the surface. The nodules will be sifted out and taken to the ship in a riser pipe, and the remaining sediment will be dumped behind the mining vehicle. Organisms and any available food material were originally concentrated in the upper few cm but are now lost into the water column or dispersed in the new unconsolidated layer. This layer may be too soft to support recolonisation even if and when food becomes available. Animals that attach to nodules are not able to return because the hard substrate is no longer available

of the data of some studies, most notably owing to small sample sizes in meiofaunal and macrofaunal studies.

The mining vehicles will alter the physical structure of the seabed by destruction of the bioactive surface layer and through the deposition of an unconsolidated sediment layer at least 10 cm thick. This disaggregated layer will be more readily resuspended by near-seabed currents even though these currents are relatively sluggish. It is not known how long the redeposited layer will take to return to its original semi-consolidated and more cohesive state, but other deep abyssal sediments, such as on the Madeira Abyssal Plain in the Atlantic show layers of redeposited sediments, introduced by turbidity currents from the continental margins, have remained soupy for 1000 years (Thomson and Weaver 1994). Re-establishing an appropriate supply of organic material from the nutrient-poor surface waters overlying nodule areas may be required to stimulate the consolidation of sediments by key organisms, notably microbes and meiofauna, which bind the sediment and create microstructure for subsequent recolonisation by other fauna.

The clouds of suspended sediment particles created by the mining systems at the seabed may be a particular problem in nodule mining because the sediments consist mainly of small clay particles (Weaver et al. 2018; Jones et al. 2017). The plumes will be denser than the surrounding water and will flow across the seabed depositing sediment as they progress. Plume deposits may have an immediate effect if the seabed organisms are overwhelmed by particles or their filter-feeding mechanisms are blocked. In addition, plumes may have a chronic effect if the organisms are subject to low levels of deposition over many years or decades.

Plumes will spread for long distances but will have little impact in their far range. The area over which plumes will damage or kill animals is not known and may vary according to the composition of the plume and the organisms impacted. Early estimates suggested impacts could reach as far as 100 km from the mine site (Rolinski et al. 2001), but estimates produced by modelling plume behaviour by the MIDAS project (see http://www.eu-midas.net/sites/default/files/downloads/MIDAS_recommendations_ for_policy_lowres.pdf) suggested 50 km may be more realistic. For areas of manganese nodules, the problem is that background sedimentation rates are extremely low, and thus the faunas are not adapted to cope with even small amounts of fallout from plumes.

Plumes are one area where good technology can reduce the environmental impact. For example, a mechanical nodule collector may generate a much smaller plume than removing the nodules with a suction hopper dredge (www.isa.org.jm/files/documents/EN/Brochures/ENG7.pdf) that uses water jets to loosen the nodules.

For manganese nodule mining to be commercially viable, a nodule abundance of over 13.7 kg/m^2 is required to provide an annual yield of 1.5–2 million tons of nodules per year. This means that each mining operation must cover a total area of 120–180 km^2 per year (Volkmann and Lehnen 2017). Owing to the slow rates of recovery indicated above contractors involved in deep-sea mining may be required to continue monitoring the mine site for considerable time after mining activities have ceased. There is an economic case therefore for minimising mining impacts through engineering solutions and for considering ecosystem restoration actions that might be undertaken to speed up ecosystem recovery processes. These actions would reduce the costs and commitments of continued monitoring as part of a closure plan apart from demonstrating good environmental management.

Our knowledge of biology of the abyssal seabed and the overlying water column in terms of (1) species diversity and distribution, breeding patterns and connectivity between communities and (2) ecological functions of the various components of the ecosystem is poor and cannot be addressed quickly. This is particularly true for the assessment of the efficacy of ecosystem restoration measures which may take 10 years or so to generate statistically significant results. Work is urgently needed here because the results could feed into better equipment design and better working practices.

New techniques in using Autonomous Underwater Vehicles and environmental-DNA (e-DNA) may provide fast, accurate and cost-effective approaches to monitoring of mining impacts. Molecular taxonomic approaches have the potential of making information more robust by linking DNA sequencing with traditional morphological techniques (Glover et al. 2016b). For instance, Janssen et al. (2015) used morphological and DNA barcoding to study connectivity between populations of isopods and of polychaetes between two locations spaced 1300 km apart in the CCZ. Their results showed a higher proportion of wide-ranging species in polychaetes in contrast to apparently more restricted distributions in isopods. Many more similar studies are needed for a good scientific understanding of ecosystems in nodule areas.

5.3 *Potential Environmental Impacts from Mining Cobalt Crusts*

The most economically important cobalt crust deposits are found on the tops and upper flanks of seamounts in a water depth range of 800–2500 m (Hein and Koschinsky 2014). Since the crusts form over millions of years, the oldest seafloor tends to have the thickest deposits, and for this reason the western Pacific is an area that is being actively explored. In addition, crusts formed on top of flat-topped guyots will make mining operations easier. As of the end of 2017, the International Seabed Authority had signed three contracts for the exploration of cobalt crusts in the Western Pacific, and a further one was awaiting signature. In addition, there is also a contract for exploration of cobalt crusts in the South Atlantic on the Rio Grande Rise (ISA 2019).

Areas with cobalt crusts are characterised by having reasonably swift currents, which sweep away sediments leaving rocky substrates on which the crust can develop. These hard surfaces provide anchorage points for sessile filter-feeding animals (Fig. 19), and some of these, such as corals and sponges, provide structural habitat for many other species (Nalesso et al. 1995). In a study off New Zealand, Rowden et al. (2010) showed that biomass was four times greater on seamounts than on adjacent continental margin slopes at equivalent depths due to high densities of stony corals. The flanks of the seamounts occupy a broad depth range with different species occupying different depth ranges, and thus many different species can be found on single seamounts (Clark et al. 2010). Significant questions remain about seamount faunas such as how many species are endemic to particular seamounts or

Fig. 19 Fauna including deep-sea corals attached to manganese crust on the Vogt Seamount in the western Pacific. (Images courtesy of the NOAA Office of Ocean Exploration and Research, 2016 Deepwater Exploration of the Marianas)

chains of seamounts. In addition, it is not certain whether cobalt crusts support different species assemblages to seamounts without cobalt crusts, i.e. whether there are species adapted to the higher metal contents of the crusts. One study on Hawaiian seamounts has shown no significant differences (Clark et al. 2011), whilst a more recent study by Schlacher et al. (2014) showed that the faunal composition was different with lower abundance of certain species in areas with cobalt crusts.

Apart from the corals and sponges, a wide range of species has been recorded from seamounts including squids and echinoderms (sea stars, sea cucumbers, feather stars, crabs and sea squirts). Large foraminiferans including xenophyophores may be common on some crust areas (Mullineaux1987, Fukushima 2007), and high densities of suspension feeders, especially feather stars and sea pens, have also been recorded along the edges of crusts (Fukushima 2007).

The mining of cobalt-rich crust will involve grinding or scraping the surface of the seamounts or other topographic features to a depth of about 25 cm (see https://www.isa.org.jm/sites/default/files/files/documents/eng9.pdf). As with polymetallic nodules, crusts are essentially a two-dimensional feature, and therefore large areas would need to be mined per year. Assuming a crust thickness of 30–60 mm, Hein et al. (2009) estimated that approximately 9–17 km^2 will be mined per million tons of ore extracted. He et al. (2011), using a different mining model, estimated that up to 60 km^2 would need to be mined per year for the same yield. Each operator is likely to mine between 1 and 2 million tons of ore per year over a 20-year mining operation.

Apart from the loss of this area of habitat, benthic plumes created by the mining process will likely travel downslope on the flanks of the seamount or other feature. As the habitats and organisms change with depth, a range of different habitats may be impacted. The extent of the impact of the benthic plume will depend on its volume and composition, which will be dependent on the mining process and equipment design. Shrouds or other mechanisms could be used to trap the plume and force it all up the riser pipe (Van Dover et al. 2017). The plume will affect organisms by clogging the filter-feeding mechanisms of animals that rely upon clean current flow containing zooplankton and particulate organic material. In addition, deposits from plumes may overwhelm organisms and prevent juveniles from settling (Rogers 1999).

Many of the species that live on seamounts are slow-growing, long-lived and slow to reproduce. Some cold-water corals are known to live for hundreds to thousands of years (Roark et al. 2006; Rogers et al. 2007; Carreiro-Silva et al. 2013). Schlacher et al. (2014) considered that the life history characteristics of the cobalt crust fauna would make recovery from the mechanical impacts of mining very slow. Indeed studies of faunal recovery on seamounts off New Zealand following intense bottom trawling showed few signs of recolonisation after 10 years (Williams et al. 2010).

Hence the environmental impacts from mining cobalt crusts will be similar to the mining of nodules – large areas impacted that could be considerable larger if plumes are not controlled – and poor recovery potential of the ecosystems. The organisms impacted however will be represented by very different species to the nodule areas.

In terms of environmental management, regional spatial management plans will be important and may require whole seamounts to be set aside in their entirety for the conservation of biodiversity. This may require a high level of planning and agreement by different contractors working in the same region. In addition, restoration measures for impacted ecosystems may be important. Experimentation is already being conducted in the restoration of deep-sea corals (https://www.bbva.com/en/hospital-corals-mediterranean/; http://www.merces-project.eu/sites/default/files/MERCES_D1.1_0.pdf) although the costs of large-scale restoration may prove to be very high (Van Dover et al. 2014).

6 Common Issues Related to Environmental Impact Affecting all Resource Types

Apart from the resource-specific issues discussed above, there are several issues that are common to all resource types. The most important of these are mid-water plumes, noise and light.

6.1 Mid-Water Plumes

Once mined the ores from all three resources will be transferred to the surface support ship as a slurry with a high-water content, and this water will need to be removed from the slurry so that the ores can be stored in a dry state, ready for transfer to shore. The recovered water will be returned to the ocean but will contain small particles that are impossible to remove (Weaver et al. 2018). This returned water is known as the mid-water plume, and no research to date has been carried out on its potential impacts. However, introducing large volumes of small particles into the ocean could impact mid-water organisms by clogging their filter-feeding mechanisms if the concentrations are high. In addition there may be issues of toxicity especially from sulphide mining. The returned water may have different properties to the water it is added to, e.g. temperature and pH, which are critical factors since organisms are very sensitive to these properties. In addition, changes in oxygen content or the presence of elements such as iron could cause changes in plankton and or microbial populations. In shallower mining operations, it may be possible to pump this returned water back to the seabed, e.g. Nautilus Minerals has proposed to drive seabed pumps with the returned water from Solwara 1 where the mine will be at a depth of 1600 m (Gwyther 2008). Returning the mid-water plume to the seabed at 4000–5000 m in nodule mines may be technically difficult.

The supply ship will periodically offload its ores to barges that will transport the ores to shore. This transhipment may require the ores to be rewetted and then dewatered on the barge (see http://www.motorship.com/news101/industry-news/digging-deep-the-new-seafloor-industry). This will create an additional plume similar to the mid-water plume.

Returned water could be added back to the ocean at the deepest feasible depth thereby impacting less layers in the water column. It may even be possible to release the returned water at the base of the riser pipe where it will mix with the volumetrically much larger seabed plume thus converting the plume problem to a single hazard.

6.2 Noise and Light

The seafloor production tools, the riser pipe and the ship will all generate noise and vibration. Noise and vibration together can affect the auditory senses and systems of some animals and may cause discomfort, e.g. by causing avoidance reactions or even damage. Background noise can also interfere with communication between animals or limit their ability to detect prey (Popper et al. 2003). Evidence is beginning to emerge that many deep-sea fishes depend on underwater sound for communication during mating and possibly also for navigation, e.g. https://news.agu.org/press-release/new-research-reveals-sound-of-deep-water-animal-migration/. Sound may have an impact at all water depths but could be a particular problem in depths down to 2000 m where it may interfere with marine mammals. Any noise generated at the base of the thermocline in the "deep-sound channel" has the potential to travel long distances (see https://dosits.org/wp-content/uploads/2017/07/DOSITS-Booklet-2015-web.pdf), and since the depths between 800 and 1000 m may coincide with mine sites for SMS and cobalt crust deposits, particular attention may need to be given to controlling sound generation in this layer. The positioning of pumps along the vertical transport or riser pipe may need to avoid the deep-sound channel as well as the deep scattering layer where large concentrations of fish and invertebrates are found, as well as deep-diving marine mammals that feed on them.

Seafloor mining operations may require lights. They will also be used on the support ship, which, unlike ships in transit, will move very slowly as the mining proceeds. Although light does not penetrate more than a few hundred metres in the ocean, most deep-sea fishes have well-developed eyes, which are sensitive to very dim light such as that produced by bioluminescence. Depending on the species, deep-sea fishes either avoid or are attracted by artificial light, which can destroy or temporarily hamper their vision. Most deep-sea invertebrates have light receptors too, and artificial lights can severely damage eyes of invertebrates, or even blind them permanently. Bright artificial light can also obscure or completely block the function of bioluminescence, which is fundamental to deep-sea organisms for, e.g. orientation (especially in rough terrain), communication, finding food, mating and defence against predation.

Light emitted from the support ship may have an impact on birds as well as on near surface marine life. The slow speed of the ship will make it a permanent target, being in a nearby position every night. Birds can be affected by disruption of their migrations, by collisions with the ship and by disruption of their feeding habits, though the use of lighting with specially adapted wavelengths (Marquenie et al. 2014) can reduce the hazard. Marine life in the upper ocean (e.g. squid and other predators, plus animals in the deep scattering layer that migrate up and down daily) is also attracted by light, and locally this can disrupt normal community behaviour and structure and predator-prey relationships.

7 Conclusion

When considering the potential environmental impacts from deep-sea mining, it is important to distinguish between the three types of deep-sea mining for metalliferous ores – manganese nodules found on deep abyssal plains, cobalt crusts found on rocky surfaces such as seamounts and seafloor massive sulphides found at and near to oceanic tectonic plate boundaries. Each of these ores is associated with its own unique ecosystem, but these ecosystems are poorly understood due to their remoteness from land and the consequent cost of carrying out research. The impacts of new activities, from a whole new industry, on poorly understood ecosystems are therefore difficult to predict. Research has shown that many animals in nodule and crust environments are long-lived and slow to reproduce. This combined with alteration of the seabed structure, particularly in nodule areas, means that it will be difficult for the ecosystem to re-establish except over long timescales perhaps taking tens to hundreds of years in the case of manganese nodules and cobalt crusts.

Following initial enthusiasm for mining metalliferous ores from the deep sea in the 1980s, only little activity took place until 2002. Since that date many contracts for exploration have been signed, and effort has been put into technology development. The International Seabed Authority is developing its regulations for exploitation, which are expected to be in place by 2020, and these may herald the start of this new industry. Mining for manganese nodules and cobalt crusts will be quite different from any metal mining on land since large areas will be mined per year as the ores only exist as thin layers at the seabed. On the other hand, mining for seafloor massive sulphides will be analogous to some on-land mines and have a much smaller footprint.

Research into all aspects of seafloor ecosystems in areas with potential for DSM is urgently required, with particular emphasis needed on community structure, ecosystem function and connectivity between populations. In the water column, very little is known about mid-water ecosystems and how these might be affected by deep-sea mining. Information is also needed on how the mining machines will operate, particularly on how successful they will be in limiting plume development and spread, and how they will restrict noise and light pollution.

Acknowledgements The research leading to these results has received funding from the European Community's Horizon 2020 Research and Innovation Programme for Blue Growth under grant agreement no. 688975 (Blue Nodules) for PPEW and grant agreement no. 689518 (Marine Ecosystem Restoration in Changing European Seas, MERCES) for DSMB.

References

Batker, D., & Schmidt, R. (2015). Environmental and social benchmarking analysis of the Nautilus Minerals Inc. Solwara 1 Project. Nautilus Document Reference SL01-NMN-XEE-RPT-0180-001. http://www.nautilusminerals.com/irm/content/pdf/eartheconomics-reports/earth-economics-may-2015.pdf.

Billett, D.S.M., Hansen, B., & Huggett, Q.J. (1985). Pelagic Holothurioidea (Echinodermata) of the northeast Atlantic. In: Keegan, B.F. & B.D.S. O'Connor (Ed.), *Echinodermata: Proceedings of the 5th International Echinoderms Conference*, Galway, 399–411.

Boetius, A., & Haeckel, M. (2018). Mind the seafloor. *Science, 359*, 34–36.

Boschen, R. E., Rowden, A. A., Clark, M. R., & Gardner, J. P. A. (2013). Mining of deepsea seafloor massive sulfides: A review of the deposits, their benthic communities, impacts from mining, regulatory frameworks and management strategies. *Ocean and Coastal Management, 84*, 54–67. https://doi.org/10.1016/j.ocecoaman.2013.07.005.

Boschen, R. E., et al. (2015). Megabenthic community structure on three New Zealand seamounts: Implications for seafloor massive sulfide mining. *Marine Ecology Progress Series, 523*, 1–14.

Brown, A., & Hauton, C. (2018). Ecotoxicological responses to chalcopyrite exposure in a proxy for deep-sea hydrothermal vent shrimp: Implications for seafloor massive sulphide mining. *Chemistry and Ecology, 34*(4), 391–396.

Brown, A., Thatje, S., & Hauton, C. (2017). The effects of temperature and hydrostatic pressure on metal toxicity: Insights into toxicity in the deep sea. *Environmental Science & Technology, 51*, 10222–10231.

Carreiro-Silva, M., Andrews, A. H., Braha-Henriques, A., de Matos, A., Porteiro, F. M., & Santos, R. S. (2013). Variability in growth rates of long-lived black coral *Leiopathes* sp. from the Azores. *Marine Ecology Progress Series, 473*, 189–199.

Chown, S. L. (2012). Antarctic marine biodiversity and deep-sea hydrothermal vents. *PLoS Biology, 10*(1), e1001232. https://doi.org/10.1371/journal.pbio.1001232.

Chung J. S. (2009) Deep-ocean mining technology III: Developments. Proceedings of The Eighth (2009) ISOPE Ocean Mining Symposium September 20±24, Chennai.

Clark, M., & Smith S. (2013) Environmental management considerations. In E. Baker, Y. Beaudoin (Eds.), *(2013) Deep sea minerals: Cobalt-rich ferromanganese crusts, a physical, biological, environmental and technical review*. Vol. 1C (pp. 23–40). Secretariat of the Pacific Community.

Clark, M. R., Rowden, A. A., Schlacher, T., Williams, A., Consalvey, M., Stocks, K. I., Rogers, A. D., O'Hara, T. D., White, M., Shank, T. M., & Hall-Spencer, J. (2010). The ecology of seamounts: Structure, function, and human impacts. *Annual Review of Marine Science, 2*, 253–278.

Clark, M.R., Kelley, C., Baco, A., & Rowden, A. (2011). Fauna of cobalt-rich ferromanganese crust seamounts. International Seabed Authority Tech. Study No. 8. 83 pp.

Copley, J. T., Marsh, L., Glover, A. G., Hühnerbach, V., Nye, V. E., Reid, W., Sweeting, C., Wigham, B. D., & Wiklund, H. (2016). Ecology and biogeography of megafauna and macrofauna at the first known deep-sea hydrothermal vents on the ultraslow-spreading Southwest Indian Ridge. *Scientific Reports, 6*, 39158. https://doi.org/10.1038/srep39158.

Dunn, D. C., Van Dover, C. L., Etter, R. J., Smith, C. R., Levin, L. A., Morato, T., Colaço, A., Dale, A. C., Gebruk, A. V., Gjerde, K. M., Halpin, P. N., Howell, K. L., Johnson, D., Perez, J. A. A., Ribeiro, M. C., Stuckas, H., Weaver, P. P. E., & the SEMPIA Workshop Participants. (2018). A strategy for the conservation of biodiversity on mid-ocean ridges from deep-sea mining. *Science Advances, 4*(7), EAAR4313.

Ellis, J. I., Clark, M. R., Rouse, H. L., & Lamarche, G. (2017). Environmental management frameworks for offshore mining: The New Zealand approach. *Marine Policy, 84*(Supplement C), 178–192. https://doi.org/10.1016/j.marpol.2017.07.004.

Erickson, K. L., Macko, S. A., & Van Dover, C. L. (2009). Evidence for a chemoautotrophically-based food web at inactive hydrothermal vents (Manus Basin). *Deep-Sea Research Part II, 56*, 1577–1585.

Fallon, E. K., Niehorster, E., Brooker, R. A., & Scott, T. B. (2018). Experimental leaching of massive sulphide from TAG active hydrothermal mound and implications for seafloor mining. *Marine Pollution Bulletin, 126*, 501–515.

Fisher, C. R., Rowden, A. A., Clark, M. R., & Desbruyères, D. (2013). Biology associated with sea-floor massive sulphide deposits. In E. Baker, Y. Beaudoin (Eds.), *(2013) Deep sea minerals: Sea-floor massive sulphides, a physical, biological, environmental and technical review*. Vol. 1A (pp. 19–26). Secretariat of the Pacific Community.

Fuchida, S., Yokoyama, A., Fukuchi, R., Ishibashi, J., Kawagucci, S., Kawachi, M., & andKoshikawa, H. (2017). Leaching of metals and metalloids from hydrothermal ore particulates and their effects on marine phytoplankton. *ACS Omega, 2*(7), 3175–3182. https://doi.org/10.1021/acsomega.7b00081.

Fukushima, T. (2007). Amounts of megabenthic organisms in areas of manganese nodules, cobalt-rich crusts and polymetallic sulphides occurrences. Proceedings of the International Seabed Authority's (ISA) Workshop, September 2004: Polymetallic Sulphides and Cobalt-Rich Ferromanganese Crust Deposits: Establishment of Environmental Baselines and an Associated Monitoring Programme During Exploration (ed. by ISA), pp. 356–368. International Seabed Authority, Kingston, Jamaica. Available at: http://www.isa.org.jm/en/documents/publications.

German, C. R., Petersen, S., & Hannington, M. D. (2016). Hydrothermal exploration of mid ocean ridges: Where might the largest sulfide deposits be forming? *Chemical Geology*, 420114–420126. https://doi.org/10.1016/j.chemgeo.2015.11.006.

Glover, A., Dahlgren, T., Taboada, S., Paterson, G., Wiklund, H., Waeschenbach, A., Cobley, A., Martínez, P., Kaiser, S., Schnurr, S., Khodami, S., Raschka, U., Kersken, D., Stuckas, H., Menot, L., Bonifacio, P., Vanreusel, A., Macheriotou, L., Cunha, M., Hilário, A., Rodrigues, C., Colaço, A., Ribeiro, P., Błażewicz, M., Gooday, A., Jones, D., Billett, D., Goineau, A., Amon, D., Smith, C., Patel, T., McQuaid, K., Spickermann, R., & andBrager, S. (2016a). The London workshop on the biogeography and connectivity of the Clarion-Clipperton Zone. *Research Ideas and Outcomes, 2*, e10528. https://doi.org/10.3897/rio.2.e10528.

Glover, A. G., Dahlgren, T. G., Wiklund, H., Mohrbeck, I., & Smith, C. R. (2016b). An end-to-end DNA taxonomy methodology for biodiversity survey in the central Pacific abyssal plain. *Journal of Marine Science and Engineering, 4*. https://doi.org/10.3390/jmse4010002.

Gollner, S., Kaiser, S., Menzel, L., Jones, D. O. B., Brown, A., Mestre, N. C., van Oevelen, D., Menot, L., Colaço, A., Canals, M., Cuvelier, D., Durden, J. M., Gebruk, A., Egho, G. A., Haeckel, M., Marcon, Y., Mevenkamp, L., Morato, T., Pham, C. K., Purser, A., Sanchez-Vidal, A., Vanreusel, A., Vink, A., & Martinez Arbizu, P. (2017). Resilience of benthic deep-sea fauna to mining activities. *Marine Environmental Research, 129*(Supplement C), 76–101.

Gwyther, D. (2008). Environmental impact statement: Nautilus minerals Niugini Limited, Solwara 1 Project, Vol. A. Coffey Natural Systems, Queensland, Australia. 226 pp. Available at http://www.cares.nautilusminerals.com/assets/documents/main%20document%20text.pdf. Accessed 27th Nov 2017.

Hannington, M., Jamieson, J., Monecke, T., Petersen, S., & Beaulieu, S. (2011). The abundance of seafloor massive sulfide deposits. *Geology, 39*, 1155e1158.

Hauton, C., Brown, A., Thatje, S., Mestre, N. C., Bebianno, M. J., Martins, I., Bettencourt, R., Canals, M., Sanchez-Vidal, A., Shillito, B., Ravaux, J., Zbinden, M., Duperron, S., Mevenkamp, L., Vanreusel, A., Gambi, C., Dell'Anno, A., Danovaro, R., Gunn, V., & Weaver, P. (2017). Identifying toxic impacts of metals potentially released during Deep-Sea mining – A synthesis of the challenges to quantifying risk. *Frontiers in Marine Science, 4*, 368. https://doi.org/10.3389/fmars.2017.00368.

He, G., Ma, W., Song, C., Yang, S., Zhu, B., Yao, H., Jiang, X., & Cheng, Y. (2011). Distribution characteristics of seamount cobalt-rich ferromanganese crusts and the determination of the size of areas for exploration and exploitation. *ActaOceanologicaSinica, 30*(3), 63–75.

Hein, J. R. (2002). Cobalt-rich ferromanganese crusts: Global distribution, composition, origin and research activities. International seabed authority, Technical Study, 2, 36–89.

Hein, J. R., & Koschinsky, A. (2014). Deep-ocean ferromanganese crusts and nodules, treatise on geochemistry: Elsevier (pp. 273–291), URL: http://linkinghub.elsevier.com/retrieve/pii/B9780080959757011116.

Hein, J. R., Conrad, T. A., & Dunham, R. E. (2009). Seamount characteristics and mine-site model applied to exploration and mining lease block selection for cobalt-rich ferromanganese crusts. *Marine Georesources and Geotechnology, 27*, 160–176.

Hölz, S., Jegen, M., Petersen, S., & Hannington, M. (2015). How to find buried and inactive seafloor massive sulfides using transient electromagnetics. Nearshore underwater mining: Critical commodities for the future 44th underwater mining conference 2015 Tampa Bay, FL.

International Seabed Authority (2018). Draft regulations on exploitation of mineral resources in the Area. ISBA/24/LTC/WP.1/Rev.1, 113pp

International Seabed Authority (2019) Status of contracts for exploration and related matters, including information on the periodic review of the implementation of approved plans of work for exploration. ISBA/25/C/9 International Seabed Authority, Kingston, Jamaica, pp. 1–5

Jaeckel, A. (2016). Deep seabed mining and adaptive management: The procedural challenges for the International Seabed Authority. *Marine Policy, 70*, 205–211. https://doi.org/10.1016/j.marpol.2016.03.008.

Janssen, A., Kaiser, S., Meißner, K., Brenke, N., Menot, L., & Martínez Arbizu, P. (2015). A reverse taxonomic approach to assess macrofaunaldistribution patterns in abyssal Pacific polymetallicnodule fields. *PLoS One, 10*(2), e0117790. https://doi.org/10.1371/journal.pone.0117790.

Jones, D. O. B., & Weaver, P. P. E. (2017) Overarching issues around regional governance of deep seabed mining. In: Towards an ISA Environmental Management Strategy for the Area: Report of an International Workshop convened by the German Environment Agency (UBA), the German Federal Institute for Geosciences and Natural Resources (BGR) and the Secretariat of the International Seabed Authority (ISA) in Berlin, Germany, 20–24 March 2017. Pages 74–77. (ISA Technical Study; 17).

Jones, D. O. B., Kaiser, S., Sweetman, A. K., Smith, C. R., Menot, L., Vink, A., et al. (2017). Biological responses to disturbance from simulated deep-sea polymetallic nodule mining. PLoS One, 12(2), e0171750.

Knight, R. D., Roberts, S., & Cooper, M. J. (2018). Investigating monomineralic and polymineralic reactions during the oxidation of sulphide minerals in seawater: Implications for mining seafloor massive sulphide deposits. *Applied Geochemistry, 90*(August 2017), 63–74.

Levin, L. A., Baco, A. R., Bowden, D. A., Colaco, A., Cordes, E. E., Cunha, M. R., Demopoulos, A. W. J., Gobin, J., Grupe, B. M., Le, J., Metaxas, A., Netburn, A. N., Rouse, G. W., Thurber, A. R., Tunnicliffe, V., Van Dover, C. L., Vanreusel, A., & Watling, L. (2016). Hydrothermal vents and methane seeps: Rethinking the sphere of influence. *Frontiers in Marine Science, 3*, 72. https://doi.org/10.3389/fmars.2016.00072.

Lipton, I. (2012). Mineral resource estimate: Solwara Project, Bismarck Sea, PNG. Technical Report compiled under NI43–101. Golder Associates, for Nautilus Minerals Nuigini. 218 pp.

Maribus. (2014). World ocean review 3. Marine resources – Opportunities and risks. 163 pp maribusgGmbH, Pickhuben 2, 20457 Hamburg http://worldoceanreview.com/wp-content/downloads/wor3/WOR3_english.pdf.

Maribus. (2017). World ocean review 5. Coasts – A vital habitat under pressure. 207 pp. maribusgGmbH, Pickhuben 2, 20457 Hamburg http://worldoceanreview.com/wp-content/downloads/wor5/WOR5_en.pdf.

Marquenie, J. M., Wagner, J., Stephenson, M. T., & Lucas, L. (2014). Green lighting the way: Managing impacts from offshore platform lighting on migratory birds. *Society of Petroleum Engineers*. https://doi.org/10.2118/168350-MS.

McCormack, G. (2016). *Cook Islands seabed minerals – A precautionary approach to mining*. Rarotonga: Cook Islands Natural Heritage Trust. 34pp.

Mero, J. L. (1965). *The mineral resources of the sea* (Oceanography series). New York: Elsevier. 1. 312 pp.

MIDAS. (2016). Managing impacts of deep-sea resource exploitation: Research highlights. www.eu-midas.net.
Miljutin, D. M., Miljutina, M. A., Martinez Arbizu, P., & Galéron, J. (2011). Deep-sea nematode assemblage has not recovered 26 years after experimental mining for polymetallic nodules (Clarion-Clipperton Fracture Zone, Tropical Eastern Pacific). *Deep-Sea Research Part I, 58*, 885–897.
Miller, K. A., Thompson, K. F., Johnston, P., & Santillo, D. (2018). An overview of seabed mining including the current state of development, environmental impacts, and knowledge gaps. *Frontiers in Marine Science, 4*, 418.
Mullineaux, L. S. (1987). Organisms living on manganese nodules and crusts: Distribution and abundance at three North Pacific sites. *Deep-Sea Research, 34*, 165–184.
Murray, J. (1876). Preliminary report on specimens of the sea bottom obtained in soundings, dredgings, and trawlings of H.M.S. Challenger in the years 1873–1875 between England and Valparaiso. *Proceedings of the Royal Society, 24*, 471–547.
Murray, J., & Renard, A.F. (1891). Report on deep sea deposits, Report on the Scientific Results of the Voyage of H.M.S. Challenger, Deep Sea Deposits, 525 pp.
Nalesso, R. C., Duarte, L. F. L., Rierozzi, I., & Enumo, E. F. (1995). Tube epifauna of the polychaete *Phyllo chaetopterus socialis* Claparede. *Estuarine, Coastal and Shelf Science, 41*, 91–100.
Petersen, S., Krätschell, A., Augustin, N., Jamieson, J., Hein, J. R., & Hannington, M. D. (2016). News from the seabed – Geological characteristics and resource potential of deep-sea mineral resources. *Marine Policy, 70*, 175–187.
Petersen, S., Hannington, M., & Kratschell A. (2017). Technology developments in the exploration and evaluation of deep-sea mineral resources. Responsabilité & Environnement – January 2017 – N°85, p 14–18.
Popper, A. R., Fewtrell, J., Smith, M. E., & McCauley, R. D. (2003). Anthropogenic sound: Effects on the behaviour and physiology of fishes. *Marine Technology Society Journal, 37*(4), 35–40.
Priede, I. G., Billett, D. S. M., Brierley, A. S., Hoelzel, A. R., Inall, M., Miller, P. I., Cousins, N. J., Shields, M. A., & Fujii, T. (2013). The ecosystem of the Mid-Atlantic Ridge at the sub-polar front and Charlie–Gibbs Fracture Zone; ECO-MAR project strategy and description of the sampling programme 2007–2010. *Deep Sea Research, Part II, 98*, 220–230. https://doi.org/10.1016/j.dsr2.2013.06.012.
Roark, E. B., Guilderson, T. P., Dunbar, R. B., & Ingram, B. L. (2006). Radiocarbon- based ages and growth rates of Hawaiian deep-sea corals. *Marine Ecology Progress Series, 327*, 1–14.
Rogers, A. D. (1999). The biology of *Lophelia pertusa* (Linnaeus 1758) and other deep-water reef-forming corals and impacts from human activities. *International Review of Hydrobiology, 84*, 315–406.
Rogers, A. D., Baco, A., Griffiths, H., Hart, T., & Hall-Spencer, J. M. (2007). Corals on seamounts. In T. J. Pitcher, T. Morato, P. J. B. Hart, M. R. Clark, N. Haggan, & R. S. Santos (Eds.), *Seamounts: Ecology, fisheries and conservation* (Blackwell Fisheries and Aquatic Resources Series 12) (pp. 141–169). Oxford: Blackwell Publishing.
Rogers, A. D., Tyler, P. A., Connelly, D. P., Copley, J. T., James, R., et al. (2012). The discovery of new deep-sea hydrothermal vent communities in the Southern Ocean and implications for biogeography. *PLoS Biology, 10*(1), e1001234. https://doi.org/10.1371/journal.pbio.1001234.
Rolinski, S., Segschneider, J., & Sundermann, J. (2001). Long-term propagation of tailings from deep-sea mining under variable conditions by means of numerical simulations. *Deep-Sea Research Part II, 48*, 3469–3485. https://doi.org/10.1016/S0967-0645(01)00053-4.
Rowden, A. A., Schlacher, T. A., Williams, A., Clark, M. R., Stewart, R., Althaus, F., Bowden, D. A., Consalvey, M., Robinson, W., & Dowdney, J. (2010). A test of the seamount oasis hypothesis: Seamounts support higher epibenthic mega faunal biomass than adjacent slopes. *Marine Ecology, 31*(suppl. 1), 95–106.
Schlacher, T. A., Baco, A. R., Rowden, A. A., O'Hara, T. D., Clark, M. R., Kelley, C., & Dower, J. F. (2014). Seamount benthos in a cobalt-rich crust region of the central Pacific: Conservation

challenges for future seabed mining. *Diversity and Distributions, 20*, 491–502. https://doi.org/10.1111/ddi.12142.

Secretariat of the Pacific Community. (2013). In E. Baker, Y. Beaudoin (Eds.) Deep sea minerals: Sea-floor massive sulphides, a physical, biological, environmental and technical review. Vol. 1A. Secretariat of the Pacific Community.

Simpson, S. L., & Spadaro, D. A. (2016). Bioavailability and chronic toxicity of metal sulfide minerals to benthic marine invertebrates: Implications for deep sea exploration, mining and tailings disposal. *Environmental Science & Technology, 50*, 4061–4070.

Smith, C. (2013). Biology associated with manganese nodules. In E. Baker, & Y. Beaudoin (Eds.) *(2013) Deep sea minerals: Manganese nodules, a physical, biological, environmental and technical review*. Vol. 1B (pp. 19–26). Secretariat of the Pacific Community.

Steiner, R. (2009). Independent review of the environmental impact statement for the proposed Nautilus Minerals Solwara 1 Seabed Mining Project, Papua New Guinea. http://www.deepseaminingoutofourdepth.org/wp-content/uploads/Steiner-independent -review-DSM1.pdf.

Thiel, H., & Schriever, G. (1990). Deep-sea mining, environmental impact and the DISCOL project. *Ambio, 19*, 245–252.

Thiel, H., Foell, E. J., & Schriever, G. (1991). Potential environmental effects of deep seabed mining. Berichte des ZentrumsfürMeeres- und Klimaforschung der Universität Hamburg, Vol. 26, 243 pp.

Thomson, J., & Weaver, P. P. E. (1994). An AMS radiocarbon method to determine the emplacement time of recent deep-sea turbidites. *Sedimentary Geology, 89*, 1–7.

UNCED. (1992). Rio declaration on environment and development. United Nations Conference on Environment and Development, Rio de Janeiro, 3–14 June 1992. http://www.unesco.org/education/pdf/RIO_E.PDF.

United States Geological Survey. (1999). Metal prices in the United States through 1998 compiled by Patricia A. Plunkert and Thomas S. Jones. 179 pp.

Van Dover, C. L. (2011). Mining seafloor massive sulphides and biodiversity: What is at risk? *ICES Journal of Marine Science, 68*, 341–348.

Van Dover, C. L. (2014). Impacts of anthropogenic disturbances at deep-sea hydrothermal vent ecosystems: A review. *Marine Environmental Research, 102*, 59–72.

Van Dover, C. L., Aronson, J., Pendleton, L., Smith, S., Arnaud-Hoand, S., Mareno-Mateos, D., Barbier, E., Billett, D., Bowers, K., Danovaro, R., Edwards, A., Kellert, S., Morato, T., Pollard, E., Rogers, A., & Warner, R. (2014). Ecological restoration in the deep sea: Desiderata. *Marine Policy, 44*, 98–106.

Van Dover, C. L., Ardron, J. A., Escobar, E., Gianni, M., Gjerde, K. M., Jaeckel, A., Jones, D. O. B., Levin, L. A., Niner, H. J., Pendleton, L., Smith, C. R., Thiele, T., Turner, P. J., Watling, L., & Weaver, P. P. E. (2017). Biodiversity loss from deep-sea mining. *Nature Geoscience, 10*, 464–465. 2017/06/26/online10, 464.

Van Dover, C. L., Arnaud-Haond, S., Gianni, M., Helmreich, S., Huber, J. A., Jaeckel, A. L., Metaxasg, A., Pendletonh, L. H., Peterseni, S., Ramirez-Llodraj, E., Steinbergk, P. E., Tunnicliffel, V., & Yamamoto, H. (2018). Scientific rationale and international obligations for protection of active hydrothermal vent ecosystems from deep-sea mining. *Marine Policy, 90*(January), 20–28. https://doi.org/10.1016/j.marpol.2018.01.020.

Vanreusel, A., Hilario, A., Ribeiro, P. A., Menot, L., & Arbizu, P. M. (2016). Threatened by mining, polymetallic nodules are required to preserve abyssal epifauna. *Scientific Reports, 6*, 26808. https://doi.org/10.1038/srep26808.

Volkmann, S. E., & Lehnen, F. (2017). Production key figures for planning the mining of manganese nodules. *Marine Georesources & Geotechnology, 36*(3), 360–375, https://doi.org/10.1080/1064119X.2017.1319448

Watling, L., Guinotte, J., Clark, M. R., & Smith, C. R. (2013). A proposed biogeography of the deep ocean floor. *Progress in Oceanography, 111*, 91e112.

Weaver, P. P. E., Billett, D. S. M., & Van Dover, C. L. (2018). Environmental risks of deep-sea mining. In M. Salomon & T. Markus (Eds.), *Handbook on marine environment protection (Science, impacts and sustainable management)* (pp. 215–245). Cham: Springer.

Williams, A., Schlacher, T. A., Rowden, A. A., Althaus, F., Clark, M. R., Bowden, D. A., Stewart, R., Bax, N. J., Consalvey, M., & Kloser, R. J. (2010). Seamount megabenthic assemblages fail to recover from trawling impacts. *Marine Ecology, 31*(Suppl 1), 183–199.

Philip P. E. Weaver is a marine scientist with research background in seabed environments including geological and ecological as well as human aspects. He spent the early part of his career looking at sedimentary processes in the NE Atlantic before broadening his expertise to look at human impacts especially on deep-sea ecosystems. He spent most of his career at the National Oceanography Centre, Southampton and its predecessor the Institute of Oceanographic Sciences, becoming Deputy Director in 2006. In 2010 he left to set up his own company – Seascape Consultants that specialises in environmental marine work. Phil has coordinated a series of EC funded projects that investigated deep-sea ecosystems from geological and biological perspectives. He coordinated the MIDAS project (Managing Impacts of Deep-sea Resource Exploitation) funded by the European Commission from 2013 to 2016. This project examined the environmental impacts of deep-sea mining. Phil has carried out a number of assignments for the International Seabed Authority.

David Billett is a deep-sea biologist with over 40 years' experience of the deep-sea environment, including mid-ocean ridges, abyssal sediments, seamounts, coral mounds, submarine canyons and continental slopes. He has a particular interest in (1) the potential effects of climate change on deep-sea ecosystems; (2) distinguishing between natural and man-made impacts; (3) the environmental management of offshore deepwater fisheries, oil and gas production and mining; and (4) the spatial and temporal management of man's activities in the deep ocean. David was a member of the ISA Legal and Technical Commission (LTC) between 2007 and 2017 specialising in environmental matters. David is now a Visiting Research Fellow at the National Oceanography Centre, Southampton, UK, and is coordinating putting industry at the heart of the marine ecosystem restoration agenda as part of a large multinational research project 'Marine Ecosystem Restoration in Changing European Seas' (MERCES) funded by the European Commission (http://www.merces-project.eu/).

Towards an Ecosystem Approach to Environmental Impact Assessment for Deep-Sea Mining

Kate J. Thornborough, S. Kim Juniper, Samantha Smith, and Lynn-Wei Wong

Abstract There is growing recognition that a clean energy, low carbon future will require additional metals to be inserted into the world economy. This has led to increased interest in obtaining these metals from the seafloor of the deep ocean, with many proponents of the seafloor mineral industry claiming environmental and social advantages such as minimal waste, no disruption of indigenous populations, no need for deforestation including large areas of rainforest, and multi-metal and/or high-grade deposits. With interest in mining the deep seabed on the rise, an increasing number of exploration licences and contracts granted, and one mining project expected to be soon ready to commence operations, the deep seabed mineral industry is emerging, bringing with it a recognised need for thoughtful environmental assessment and management. This chapter examines the current state of knowledge of the services provided by ecosystems associated with deep-sea mineral deposits and how this knowledge can support the future inclusion of ecosystem services in environmental impact assessments (EIAs) for individual mining operations and in the regional-scale planning of resource extraction and conservation measures.

Faced with an incomplete understanding of deep-sea environments and the management strategies that could be deployed to minimise ecological losses from mining operations, scientists have expressed concern about the potential for related species extinctions, changes in ecosystem structure and function, and a loss of deep-sea ecosystem services (McGeoch et al. *Diversity and Distributions, 16*(1), 95–108, 2010; Van Dover *Marine Environmental Research, 102*, 59–72, 2014;

K. J. Thornborough (✉)
Independent Researcher, Manly, NSW, Australia
e-mail: kate@thornborough.com.au

S. K. Juniper
Ocean Networks Canada, University of Victoria, Victoria, BC, Canada

S. Smith
Blue Globe Solutions, Toronto, ON, Canada

Nauru Ocean Resources Inc., Aiwo, Republic of Nauru

L.-W. Wong
Asian School of the Environment, Nanyang Technological University, Singapore, Singapore

R. Sharma (ed.), *Environmental Issues of Deep-Sea Mining*,
https://doi.org/10.1007/978-3-030-12696-4_4

Van Dover et al. *Nature Geoscience, 10*(7), 464–465, 2017; Folkersen et al. *Marine Policy, 94*, 71–80, 2018). The environmental costs of extracting deep-sea mineral resources have been the subject of an increasing number of studies, yet remain difficult to quantify (Thurber et al. *Biogeosciences, 10*(11), 18193–18240, 2014; Jobstvogt et al. *Ecological Economics, 97*, 10–19, 2014; Folkersen et al. *Marine Policy, 94*, 71–80, 2018). Regulators of future deep-sea mining activities have been developing rules, regulations, and guidelines aimed at the responsible use of seabed mineral resources and that conservation goals are met. Key players in the seabed mineral industry have committed to a precautionary and ecosystem-based approach to environmental management that aims to mitigate the potential impacts of mining. Part of this approach includes employing best environmental practice and tools used in environmental management, including robust EIAs and adaptive management.

Keywords Environmental impact assessment · Ecosystem services · Deep-sea biodiversity · Adaptive management

1 Environmental Impact Assessment and Deep-Sea Mining

Environmental impact assessments (EIAs) are the most widely accepted tool for assessing the potential environmental impacts of commercial and industrial projects (Morgan 2012; Lallier and Maes 2016). The EIA process has been adopted by governments worldwide and incorporated into international law and lending institution standards since the 1970s in response to growing environmental awareness. EIAs focus on the impacts of a project on the environment and related social, economic, and cultural values, bringing input from ecologists, economists, and social scientists into a single assessment framework. EIAs also include impact mitigation strategies and the expected residual impacts. Over time, EIAs have evolved in response to the changing needs of decision-makers, the experience of practice, and the changing values and priorities of society (Morgan 2012; Jha-Thakur and Fischer 2016). Global trends in EIA developments include mandatory requirements for EIAs; comprehensive assessments that go beyond the biophysical to include social, economic, and cultural values; the inclusion of EIAs earlier in the project development process; increasing scrutiny of assessments including public access to EIAs; more ambitious assessments with regard to sustainability objectives; and a growing recognition of scientific uncertainties and the need to take a precautionary approach to decision-making (Gibson 2002; Ellis et al. 2017).

The EIA process, in principle, fosters communication between the proponent, the regulator, and stakeholders, ensuring that uniform and consistently high environmental standards are applied to all project proponents (Durden et al. 2018) and that environmental decisions are based on sound scientific grounds (Lallier and Maes 2016). The EIA process for the seabed mineral industry is recognised as a key component to ensuring effective protection of marine ecosystems, and this was explicitly mentioned in the leaders' declaration from the G7 Summit in Germany in 2015 along with scientific research as a priority issue for sustainable deep-sea mining. However, deep-sea mining as an industry does not yet exist, and to date, there have been fewer uses of EIA as a decision-making tool in the deep sea relative to terrestrial and coastal environments (e.g. Petts 1999; Glasson et al. 2013). EIAs conducted in the deep sea have so far focused on offshore hydrocarbon resources (e.g. Husky Oil 2001), commercial fisheries (e.g. CCAMLR 2008; Government of New Zealand 2008; SPRFMO 2008), and seabed minerals, including the exploration and extraction of minerals (e.g. NOAA 1981; Coffey Natural Systems/Nautilus Minerals 2008).

The United Nations Convention on the Law of the Sea (UNCLOS) sets the legal framework for seabed mining and became effective in 1994. The UNCLOS established the International Seabed Authority (ISA) as the regulatory body governing mineral resources in the area of the seabed beyond national jurisdiction (referred to as "the Area"). Under UNCLOS, states are given rights to exploit the resources of their continental shelves and seek rights from the ISA to undertake or sponsor deep-sea mining exploration and exploitation within the area and are under a general obligation to protect and preserve the marine environment and must be "no less effective than international rules, regulations and procedures" (UNCLOS Article 192). Several national jurisdictions have or are in the process of developing their own rules and regulations to govern the nascent seafloor mineral industry within their own exclusive economic zones (EEZs).

The ISA requires an EIA if any one exploratory sampling activity of the seafloor or "test" mining activity in the area exceeds 10,000 m^2 (ISBA/19/LTC/8). The ISA developed a provisional Environmental Impact Statement (EIS) template for deep-sea mining activities (the document prepared to present the environmental effects of mining), partly based on an EIS developed for Nautilus Minerals' Solwara 1 Project proposal to mine polymetallic seafloor massive sulphide deposits in Papua New Guinea (Coffey Natural Systems/Nautilus Minerals Ltd 2008). The Solwara 1 EIS is based on extensive studies of the deep-sea benthic communities at the site. Because it is the first seafloor massive sulphide mining operation to receive a mining licence, it is regarded as an engineering experiment, an environmental experiment, an economic experiment, and a policy experiment (Filer and Gabriel 2016). The EIS template is intended to guide contractors to achieve consistency in EIA information and has subsequently evolved since 2012. A workshop convened in Germany in 2017 considered several models for EIA procedures in national legislation (e.g. Clark et al. 2017) and international law, identified procedural and technical issues with EIA processes and offered key recommendations and options to address these issues (ISA 2017). The ISA issued an updated EIS template in 2018 (ISBA/24/LTC/WP.1/Add.1) as an Annex to the draft exploitation regulations.

By design, EIAs are project-scale assessments and are often limited in spatial and temporal scale to the project being assessed. If the disturbance footprint of any activity, land or sea-based, occurs at the scale of an entire ecosystem, there is a potential that it will bring risks of long-term, ecosystem-scale losses of biodiversity, ecosystem function, and ecosystem services that need to be considered during the EIA process. Assessment of the services provided by deep-sea ecosystems is a relatively recent field whose development is hampered by the so far limited knowledge of deep-sea biodiversity and ecosystem function. Given the spatial extent of potential disturbances resulting from deep-sea mining operations and the likely extended recovery times, it is necessary to compile existing information into a framework that will enable meaningful consideration of ecosystem services in the EIA process in the future (Karjalainen et al. 2013). The ISA's EIS template accounts for biodiversity and ecosystem functions, which underpin deep-sea biodiversity, and has added consideration of ecosystem services to the scope of EIAs for future mining contracts (ISBA/24/LTC/WP.1 Annex IV), so far without specific guidelines for identifying and quantifying services provided by deep-sea ecosystems. As environmental assessment of deep-sea mining continues to develop, it is an appropriate time to evaluate what is required for a robust EIA process and the nature and scope of EIAs for the major types of deep-sea mineral resources and their environments. Some of these approaches, as well as linkages between different components of ecosystem management have been covered in a separate chapter in this book (Cormier 2019, Chap. 14, this volume).

2 Ecosystem Services Provided by the Deep Sea

Mineral deposits that are the current focus of regulatory efforts by the ISA include polymetallic sulphide deposits at active and dormant hydrothermal vents, cobalt-rich ferromanganese crusts found primarily on seamounts, and polymetallic nodules on abyssal plains. The spatial extent of deposits is a function of deposit type: polymetallic sulphide deposits typically occupy hectare-scale areas of seafloor, while crusts occur at square kilometer scales on individual seamounts, and manganese nodule fields/provinces extend over thousands of square kilometers. These scale differences will be reflected in the environmental footprints of individual mining operations and the environmental management strategies that will need to be considered.

Deep-Sea Mineral Deposits of Economic Interest
The International Seabed Authority has issued exploration contracts for three common types of deep-sea mineral deposits, each occurring within distinct geological settings. Associated ecosystems operate at different spatial and temporal scales (Levin et al. 2016a, b) and have varying degrees of resilience to mining activities (Gollner et al. 2017; Suzuki et al. 2018).

Polymetallic sulphide deposits are mainly restricted to mid-ocean ridges, back-arc basins, and submerged volcanic arcs. Most deposits cover only a few hectares. To date, Nautilus Minerals Niugini Ltd's Solwara 1 Project in Papua New Guinea holds the only licence to mine polymetallic sulphide deposits. The Solwara 1 deposit is high grade and relatively small in scale (10 ha or 0.1 km^2). Elsewhere, as of October 2018, the ISA has entered into 15-year contracts with seven contractors for the exploration (not exploitation or mining) of polymetallic sulphides in the international seabed area. Each exploration contract covers an area of 10,000 km^2 (ISA Exploration Areas). Within the exploration areas, it is envisaged that individual future mine sites will encompass a much smaller area with direct impacts covering up to a few square kilometres. However, a series of mine sites may be required to create a profitable enterprise within a single mining contract area (Petersen et al. 2016).

Cobalt-rich ferromanganese crusts accrete onto nearly all rock surfaces in the deep ocean and can also coat rock pebbles and cobbles. They are particularly abundant on current-swept flanks of seamounts, knolls, ridges, and plateaux that are free of sediments over long periods (Hein 2004) at depths of 400–5,000 m, with the thickest and most cobalt-rich crusts found at depths of ~800 to 2,500 m. While no exploitation contracts for ferromanganese crusts on seamounts have been issued, the ISA has entered into 15-year contracts with five contractors for exploration of cobalt-rich ferromanganese crusts with exploration areas for each contractor set at 3,000 km^2, encompassing several seamounts (ISA Exploration Areas). Areas of this size are needed to provide enough ore for a 20–30 year mining operation. Individual mining blocks on seamounts are anticipated to be in the order of 20 km^2.

Polymetallic nodules are found spread over the abyssal plain where they may cover 10–30% of the seafloor at any given location (Collins et al. 2013). While no contracts to mine polymetallic nodules have been granted, as of October 2018, the ISA has entered into 15-year exploration contracts for polymetallic nodules with 17 contractors. Exploration areas for each contractor are in the order of 75,000 km^2, areas roughly equivalent to the size of Panama (ISA Exploration Areas). However, some contractors have stated that the maximum area they will be able to mine (for economic and/or technical reasons) is 18–30% of the contract area (ISBA/19/LTC/8), and it is expected that ISA contractors will be required to set aside a preservation reference zone (PRZ).

Ecosystem services are goods and services derived from ecosystem processes ("ecosystem functions") that contribute to human well-being. The term "ecosystem services" has been evolving since the 1970s when it first gained traction in the scientific literature (e.g. SCEP 1970; Holdren and Ehrlich 1974; Ehrlich et al. 1977; Westman 1977). Underpinning the many facets of ecosystem functioning and the ecosystem services provided by the deep sea is biodiversity (Thurber et al. 2014; Niner et al. 2018). Biodiversity, from local to global scales, can affect multiple ecosystem functions including maintenance of ecosystem productivity (Tilman et al. 1997; Worm et al. 2006; Foley et al. 2010; Thurber et al. 2014), stabilisation of food webs (Myers et al. 2007), and the ability to recover from disturbances (Leslie and Kinzig 2009; Palumbi et al. 2009). Consideration of ecosystem services is implied

Table 1 Overview of deep-sea ecosystem services in ecosystems currently associated with mineral deposits

Ecosystem service category	Deep-sea ecosystem service	Deep-sea ecosystem/habitat		
		Seamounts	Hydrothermal vents	Abyssal plains
Supporting services	Nutrient cycling	nc	nc	+
	Habitat	++	++	+
	Primary production	+	+	nc
	Water circulation and exchange	+	+	–
Regulating services	Gas and climate regulation	nc	nc	nc
	Waste absorption and detoxification	–	+	++
	Bioregulation	–	–	–
	Carbon capture and storage	nc	–	+
Provisioning services	Finfish, shellfish	++	+	–
	Oil, gas, minerals	+, ++	++	+
	Chemical compounds for industrial and pharma use (biotic materials)	+	++	nc
Cultural services	Educational	+	+	nc
	Scientific	+	++	nc
	Aesthetic	nc	nc	–
	Existence/bequest	–	–	–
Option-use value	Future unknown and speculative benefits of the whole ecosystem	+	+	+

Adapted from Fletcher et al. (2011) and Thurber et al. (2014)
"++" = widespread or abundant; "+" = present or assumed; "–" = unknown, not present; and "nc" = no consensus in the literature

in the current ISA EIA template by the inclusion of biodiversity and ecosystem functions, but it is not explicitly articulated (ISBA/19/LTC/8). Below we provide an overview of the major categories of ecosystem services provided by the deep sea and examine some of the services that are specific to nodules, crusts, and sulphides (Table 1) and how these relate to biodiversity and ecosystem function.

Ecosystem services are generally characterised as supporting services, provisioning services, regulating services, and cultural services, with an additional category of "option use value" to capture future unknown and speculative benefits of individual ecosystem components or the whole ecosystem (Walsh et al. 1984; Hanemann 1989; Beaumont et al. 2007; Armstrong et al. 2012; Jobstvogt et al. 2014). Ecosystem service frameworks tend to focus on biotic services, those provided by living components of the ecosystem (e.g. MA 2005; Beaumont et al. 2007; Mace et al. 2009; Bateman et al. 2011). However, abiotic goods and services, such as mineral resources, are recognised in some frameworks (e.g. Haines-Young et al. 2009) either directly as a provisioning service or indirectly through their roles in supporting and regulating services (Armstrong et al. 2012). Supporting services in the deep sea are functions (processes) that support the production and maintenance of other ecosystem services (Armstrong et al. 2012), including habitat, nutrient cycling, water circulation and exchange, and chemosynthetic primary production. Regulating services are the benefits obtained through the natural regulation of habitats and ecosystem processes (MA 2005), including gas and climate regulation, waste absorption and detoxification, biological regulation, and carbon dioxide (CO_2) capture and storage. Provisioning services include the products used by humans that are obtained directly from habitats and ecosystems (MA 2005) and tend to be the most tangible ecosystem services to conceptualise. Cultural services include the non-material benefits people obtain from habitats and ecosystems (MA 2005), and because the deep sea is not accessible to the majority of the population, the foci of cultural services are scientific, educational, and bequest/existence services. Option-use value includes future unknown and speculative benefits (Beaumont et al. 2007), related to future discoveries of provisioning, regulating, and cultural service values. In the context of deep-sea mining, the most discernable ecosystem service that the deep sea offers is the provision of minerals, which are abiotic ecosystem components. However, in accessing a single provisioning service such as seabed mineral deposits, it is crucial to consider the potential impact of removing these mineral deposits on other ecosystem services and the biodiversity and ecological functions underpinning them. Assessing the impact of the potential loss of ecosystem services as a result of deep-sea mineral extraction first requires identification of the relevant services and determination of their value (Thurber et al. 2014).

Ecosystem Services Provided by Hydrothermal Vents

Hydrothermal vents support highly localised faunal communities (Smith et al. 2008b), whose dependency on chemosynthetic processes, together with the physiochemical severity of the vent environment, have resulted in a high level of endemism, where the majority of species are found only at vents, and many are confined to vents within distinct biogeographic provinces (Van Dover et al. 2002; Thurber et al. 2014).

Vents are hotspots of microbial productivity and microbial genetic diversity (Van Dover 2000, 2011; Levin et al. 2016a). The chemosynthetic production of new organic matter by vent microbial communities sustains high-biomass faunal communities that have low species diversity but exhibits considerable genetic novelty that includes sophisticated symbioses, extreme thermotolerance, and resistance to hydrogen sulphide and heavy metals. Microbial chemosynthetic processes remove sulphide and dissolved metals from discharging vent fluids and transform climate-relevant substances such as carbon and methane (Hinrichs and Boetius 2002; Cochonat et al. 2007; Jørgensen and Boetius 2007; Dekas et al. 2009; Boetius and Wenzhofer 2013).

Hydrothermal vents hold high scientific value and have been the source of continual discoveries of novel organisms with adaptations to extreme environments (Deming 1998). The same technological advances that have enabled the exploration of these environments (e.g. developments in submersibles) have also led to scientific breakthroughs. Repeat submersible surveys, autonomous instruments, and cabled seafloor observatories monitor hydrothermal vent biology, geology, and chemistry, and promote community and educator engagement.

One option-use value of hydrothermal vents is likely to be found in the genetic novelty of vent organisms and the potential future discovery of as yet unidentified enzymes and chemical compounds that may be relevant to human health and environmental remediation (Deming 1998) or have industrial applications. For example, the high, and so far unquantified, diversity of microorganisms at hydrothermal vents has led to growing interest in potential biotechnology applications, including bioenergy and biofuel production (Chaudhuri and Lovley 2003). A more specific example involves hyperthermostable forms of carbonic anhydrase, the enzyme that rapidly catalyses the reversible hydration of CO_2 in aqueous solutions. Hyperthermostable carbonic anhydrase is widespread among the chemoautotrophs of hydrothermal vents (Goffredi et al. 1999; Jo et al. 2014; Fredslund et al. 2018) and has attracted research interest for its medical relevance and biotechnical potential for carbon sequestration (Fredslund et al. 2018).

Ecosystem Services Provided by Seamounts

Seamounts rise up in stark contrast to the surrounding seafloor and occur either singularly or as a chain. They are comprised of a mixture of soft and hard substrata and often harbour fragile, vulnerable, and long-lived, habitat-forming epifauna that create areas of high biodiversity and support rich fishing grounds (Clark et al. 2010; Chivers et al. 2013; Thurber et al. 2014). In the Pacific Ocean, there are more than 11,000 seamounts (57% of the global total) (Yesson et al. 2011), and many more might exist in uncharted waters (Kim and Wessel 2011). The Atlantic Ocean has fewer seamounts, and their cobalt-rich ferromanganese crusts are mostly associated with hydrothermal activity at seafloor-spreading centres, with the exception of seamounts in the northeast and northwest continental margin areas.

Seamounts and the water column above them often serve as important habitats, aggregation areas, feeding grounds, and reproduction sites for many open-ocean and deep-sea species of fish, sharks, marine mammals, seabirds, and benthic organisms (Genin 2004; Baco 2007; Danovaro et al. 2008; Morato et al. 2009; Clark et al. 2010). Seamounts can be biological hotspots with distinct, abundant, and diverse fauna and are often a source of previously unidentified species (Worm et al. 2003; Morato et al. 2008, 2010). Studies of megabenthic fauna (large invertebrates larger than 1–2 cm) have been conducted as part of the Japan-SOPAC surveys (1985–2005) in ferromanganese crust areas within the EEZs of Kiribati, Tuvalu, Samoa, the Marshall Islands, and the Federated States of Micronesia.

Deep-water corals and sponges are often the most conspicuous megafaunal organisms on seamounts (Genin et al. 1986; Lundsten et al. 2009; Rowden et al. 2010). These suspension-feeding organisms concentrate particulate organic matter from the surrounding water column and provide physical habitat for other species. Coral diversity and community composition along the Hawaiian-Emperor seamount chain have been extensively investigated in video imagery and samples obtained from submersible surveys (Baco 2007; Sinniger et al. 2013).

The high productivity on seamounts provides provisioning services of fisheries and by attracting mobile organisms. To date, fisheries represent the most important provisioning service derived from seamount ecosystems. The main fisheries occurring on seamounts use trawls, longlines, gillnets, and pots and traps, but other methods including hook and line are also present in some small-scale seamount fisheries (Clark and Koslow 2007; Da Silva and Pinho 2007).

As yet undiscovered benthic biodiversity is the most obvious option-use value of seamounts. The seamount benthos is poorly sampled worldwide, and expeditions frequently report species new to science (Smith et al. 2008b).This genetic resource reservoir includes sponges, the richest known marine source of novel compounds (Witherell and Coon 2001; Grehan et al. 2003; Maxwell et al. 2005; Armstrong et al. 2012).

While most seamounts are remote and difficult to access, several cultural services related to seamounts have been identified. Compared to the surrounding seafloor, seamounts are often biodiversity hotspots of considerable scientific interest, particularly in relation to endemism and genetic connectivity (De Forges et al. 2000; Sautya et al. 2011). Those seamounts that are easily accessible (close to land) support tourism industry growth through marine mammal watching (Armstrong et al. 2012). A number of seamounts within EEZs have cultural value through their designation as marine protected areas (e.g. SGaan Kinghlas-Bowie Seamount Marine Protected Area in Canadian Pacific waters, Marine Protected Area of the Mid-Atlantic Ridge North of the Azores in Portugal, Papahānaumokuākea Marine National Monument in Hawaiian waters, Huon Marine Park in Australian waters).

Ecosystem Services Provided by the Areas of Abyssal Plains that Host Polymetallic Nodules

The abyssal zone of the oceans, from 3,000 to 6,000 m below sea level, covers 54% of the geographic surface of the planet (Gage and Tyler 1992). The areas of the abyss where polymetallic nodules form are relatively restricted by comparison but still represent vast expanses of seafloor. For example, the Clarion-Clipperton Zone (CCZ), the primary area of commercial interest for polymetallic nodule mining (Wiklund et al. 2017), covers 5 million km^2 (~1.4% of the total ocean area), ~65% of the size of Australia's 7.7 million km^2 land area.

The abyssal plains that host polymetallic nodules enhance the biodiversity of the deep-sea benthos (Smith et al. 2008a; De Smet et al. 2017). The hard substratum provided by manganese nodules supports faunal assemblages distinct from those in adjacent sediments (Mullineaux 1987; Bussau et al. 1995; Veillette et al. 2007; Smith et al. 2008a, b). The presence of nodules has been shown to influence the abundance, community composition, and distribution of meiofauna, macrofauna, and megafauna at regional scales (Mullineaux 1987; Glover et al. 2015; Amon et al. 2016; De Smet et al. 2017). These faunal assemblages play a key role in supporting and regulating ecological and biochemical processes (Jahnke and Jackson 1992; MA 2005; Danovaro et al. 2008).

Abyssal plain ecosystems provide services in the form of remineralization and detoxification that are poorly quantified but possibly of global significance given the huge area of the abyss (MA 2005; Armstrong et al. 2012). Deep-sea ecological processes regulate the burial of carbon in seabed sediments, which ultimately feeds back to climate regulation. The so-called *conveyor belt* of global thermohaline circulation moves cold deep-sea water masses move around the globe, providing buffering capacity for nutrient and carbon cycles (Thurber et al. 2014). Through this large-scale circulation system, the remineralization of nutrients in the deep sea eventually fuels surface ocean productivity and fisheries and supports the production of oxygen (Beaumont et al. 2007; Armstrong et al. 2012; Thurber et al. 2014; Niner et al. 2018).

Nodule ecosystems have high scientific value as a source of continued discoveries of previously undescribed species and genetic diversity. Few published data on the macrofaunal biodiversity and community structure are available (De Smet et al. 2017), but studies have recorded large numbers of species from relatively small areas, noting a long list of rare species (i.e. where only small numbers, or even singletons, have been found) and fauna largely undescribed by taxonomists (Grassle and Maciolek 1992; Glover et al. 2002; Smith et al. 2006). For example, 183 species of polychaetes were discovered in 19.25 m^3 of sediment and nodules in the CCZ (Glover et al. 2002), with estimates that >90% of polychaete species collected in the abyss currently remain undescribed by taxonomists (Grassle and Maciolek 1992). In addition to the scientific discovery potential of abyssal seabed biodiversity, the sedimentary strata beneath the deep seafloor contain extensive and scientifically important records of global climate, ocean productivity, and deep-sea conditions over long-time scales (Nellemann et al. 2008).

3 State of Knowledge of Deep-Sea Ecosystem Services

The deep sea represents a huge knowledge challenge. Only 5% of the deep sea has been explored, and less than 0.01% (an area the size of a few football fields) has been sampled and studied in detail (Ramirez-Llodra et al. 2010), although this is set to increase, particularly in the areas of commercial interest. There are only 200 or so scientists worldwide working to fill knowledge and data gaps around deep-sea ecosystems, and researchers have so far identified only a fraction of the deep-sea faunal species (Higgs and Attrill 2015; Sinniger et al. 2016; Shulse et al. 2017; Niner et al. 2018). There is not a single deep-sea species for which we understand its entire life history (Armstrong et al. 2012). Environmental studies conducted by contractors with exploration licences or contracts are helping to fill these knowledge gaps, often through collaborations with the scientific community. This will contribute to inform decision-making about potential impacts and how to minimise them. The knowledge and publications that have resulted from seabed mineral industry EIA processes represent a major contribution to deep-sea science. For example, the Solwara 1 EIS has resulted in more than 40 publications and presentations and directly involved ~14 academic institutions, and ~21 were consulted during the process and/or reviewed the EIS. The Abyssal Baseline Project (ABYSSLINE) undertook biological baseline studies within the UK Seabed Resources Ltd. Exploration contract area in the Clarion-Clipperton Zone in the Pacific and produced a database of deep-sea biodiversity. More than 60 publications have resulted from this work. DISCOL/ATESEPP (disturbance and recolonization experiment; Auswirkungen technischer Eingriffe in das Ökosystem der Tiefsee im Süd-Ost-Pazifik vor Peru (German programme)), the first experiment to investigate the potential impacts from mining in a manganese nodule area in the deep South East Pacific Ocean resulted in 124 publications, including a special issue of Deep-Sea Research II (Volume 48) in 2001. The Indian Deep-sea Environment Experiment (INDEX) has resulted in 32 publications on the deep-sea environment from the Central Indian Ocean in three special issues (Sharma and Nath 2000; Sharma 2001, 2005). A survey of peer-reviewed publications from biological studies in the area published between July 2017 and July 2018 found 32 new publications (Glover 2018), with the most focus on polymetallic nodules (17 publications), followed by 10

Table 2 State of knowledge of deep-sea ecosystem services in ecosystems currently associated with mineral deposits

Ecosystem service category	Deep-sea ecosystem service	Deep-sea ecosystem / habitat		
		Seamounts	Hydrothermal vents	Abyssal plains
Supporting services	Nutrient cycling	1–4	38–43, 53, 54	71–76, 87
	Habitat	5–15	41, 44–48, 51	77–85
	Primary production	15–18	39, 49–53	86
	Water circulation and exchange	16,19	51	–
Regulating services	Gas and climate regulation	1–4	39, 50, 54	73, 86–89
	Waste absorption and detoxification	1–4	55–56, 61	73, 86–90
	Bioregulation	18	–	–
	Carbon capture and storage	–	54	72, 86–89, 91–92
Provisioning services	Finfish, shellfish	19–23, 37	46–48	91–92
	Oil, gas, minerals	24–32	53, 57–61	80
	Chemical compounds for industrial and pharma use (biotic materials)	4	42, 54, 62–66	–
Cultural services	Educational	33	39	95–98
	Scientific	3, 34–35, 37	39, 42, 67–68	95–98
	Aesthetic	36–37	69–70	–
	Existence/bequest			
Option-use value	Future unknown and speculative benefits of the whole ecosystem	33	69,70	97

Green shading = some knowledge; yellow shading = little knowledge; blue shading = no knowledge; no shading = irrelevant. Numbering represents references presented in Appendix A

publications on polymetallic sulphides and five on cobalt-rich crusts. There are also several recent and existing programmes outside of the seabed mining industry that explore and catalogue deep-sea biodiversity and habitats, such as the Census of the Diversity of Abyssal Marine Life (CeDAMar) a field project of the Census of Marine Life that catalogued the biodiversity of abyssal plains and the Ocean Biogeographic Information System (OBIS), a global database documenting marine biodiversity from observations around the world.

However, the quantification and valuation of deep-sea ecosystem services have developed slowly, primarily because of the challenges associated with conducting ecosystem-level observations and experiments in remote environments where ecosystem processes unfold across vast spatial and temporal sales (Niner et al. 2018). The focus has been twofold, with economists and social scientists examining the economic value of deep-sea ecosystem services (Costanza et al. 1997; Ressurreição et al. 2011; Wattage et al. 2011; Jobstvogt et al. 2014; Binney and Fleming 2016; Folkersen et al. 2018) and deep-sea researchers studying biodiversity and ecological functions and interactions that support these ecosystem services (e.g. Smith et al. 2008a; Beaumont et al. 2007; MA 2005; Rex and Etter 2010; Danovaro et al. 2008; Thurber et al. 2014). Beyond the commercial value of deep-sea mineral deposits, studies of the economic value of the deep sea are so varied that it is impossible with any confidence to estimate value in monetary terms, let alone determine the value of individual ecosystem services that the deep sea provides (Thurber et al. 2014; Jobstvogt et al. 2014; Folkersen et al. 2018). Yet, broad cost-benefit analyses of deep-sea mining are needed for informed decision-making (Thurber et al. 2014) and are an important component of the EIA process.

Several frameworks have been proposed for assigning monetary values to the ecosystem functions and services of the deep sea (van den Hove and Moreau 2007; Armstrong et al. 2012); however, there are major impediments to accurate monetary assessments of marine ecosystem services and a general lack of knowledge of the economic value of marine ecosystem services that managers and society require for quantitative evaluations (Thurber et al. 2014). Scientific understanding of the ecological value of biodiversity and ecosystem functions is growing but remains very limited. Consequently, there is currently high uncertainty about the environmental trade-offs that will be considered during evaluations of deep-sea mining proposals (Folkersen et al. 2018). Yet, as interest in exploiting deep-sea minerals grows and knowledge is gained by contractors fulfilling their environmental study obligations, it is important to develop knowledge-based, decision-making frameworks to weigh losses of biodiversity and ecosystem services against the economic returns from mining operations.

The need to identify and better understand the ecosystem services provided by the deep sea has gained momentum as the seabed mining industry has taken form, and several recent studies have attempted to identify the services present (e.g. Armstrong et al. 2010, 2012; Fletcher et al. 2011; Thurber et al. 2014) and quantify their value (e.g. Beaumont et al. 2007; Armstrong et al. 2012; Folkersen et al. 2018). These studies highlighted the limited state of knowledge of some ecosystem services and where future investigation could be directed. Building on this work, Table 2 presents a summary of the state of knowledge of deep-sea ecosystem services in ecosystems currently associated with mineral deposits. Ecosystem services that can be linked to simple quantifiable metrics, such as connections between provisioning services and fishery catch rates, or cultural services and numbers of scientific publications are easier to identify and quantify than more subjective services, such as regulating services. On the other hand, the temporal and spatial characteristics of deep-sea ecosystem functions and services will be very difficult to incorporate into decision-making frameworks, based on current knowledge. For example, deep-sea ecosystem functions such as faunal-microbial interactions that accelerate nutrient cycling in sediments operate at scales of micrometres to metres over periods of years, yet their extent and derived services can be widespread but only become apparent after centuries of integrated activity (Thurber et al. 2014). Clearly, there will be a need to incorporate an assessment of uncertainty and knowledge gaps in decision-making frameworks. The industry, ISA, and relevant stakeholders are seeking an effective approach to assessing environmental effects including explicit consideration of managing uncertainty and risk and approaches that provide opportunities for adaptive management (Ellis et al. 2017).

The seabed mineral industry is moving towards an ecosystem-scale assessment and management approach. An ecosystem-based approach to EIA will require drawing links between biodiversity and ecosystem function, commonly included in these assessments, and as far as possible and practicable, ecosystem services with the aim of increasing understanding and reducing risk of harm. The seabed mineral industry has repurposed a number of approaches to provide regulators and managers with tools to begin including ecosystem services in EIAs and ultimately in management.

4 Evaluation Criteria

Criteria for identifying areas of high ecological value are a fundamental tool for balancing ecological and socioeconomic objectives and focusing limited resources for conservation in marine spatial planning (Gilman et al. 2011). Effective criteria need to appropriately represent the ecosystem and the functions and services it provides and are used in EIAs, risk assessments, vulnerability assessments, and marine spatial planning. Criteria can focus on taxa individually or jointly, on specific habitats or communities, or regionally to identify networks of sites important to the maintenance of ecosystems at a global scale (e.g. IMO 2006a, b; Langhammer et al. 2007; OSPAR Commission 2007; FAO 2009; O et al. 2015). The use of evaluation criteria can result in the identification and prioritisation of species, communities, habitats, or areas that can act as reference points for baseline assessments or those that require monitoring in relation to human activities. Additionally, criteria may be used to select simple indicators of some of the more tangible ecosystem services such as provisioning services (finfish, shellfish, oil, gas, minerals), cultural services (educational, research), and supporting services (habitat). As knowledge increases from deep-sea research and the first mining contractors, it may be possible to use criteria to identify indicators for some of the more intangible ecosystem services that are currently difficult to incorporate in assessments, such as regulating services.

Several United Nations agencies have proposed suites of criteria for identifying areas of high ecological value. These include Ecologically or Biologically Significant Areas (EBSAs) (Convention for Biological Diversity (CBD); CBD 2008), Vulnerable Marine Ecosystems (VMEs) (Food and Agriculture Organisation; FAO 2009), and Particularly Sensitive Sea Areas (PSSAs) (International Maritime Organization; IMO 2006a, b). A 2009 CBD workshop that considered the application of EBSA criteria to areas beyond national jurisdiction found no inherent incompatibilities between the criteria that have been applied by intergovernmental organisations such as the ISA and other national and non-governmental organisations (CBD 2010). Other commonly used criteria include Key Biodiversity Areas for identifying globally significant sites for biodiversity conservation (Langhammer et al. 2007) and Valued Ecosystem Components to identify species, habitats, and communities crucial to the health and functioning of an ecosystem (O et al. 2015). Gilman et al. (2011) provide an in-depth review of available criteria.

While the purposes for which these criteria were developed often differ, each suite employs similar indices to assess ecosystems, with the aim of ensuring species conservation and maintenance of environmental quality. Criteria include but are not limited to rare, unique, endemic, habitat forming, sensitive, nutrient importer/exporter, keystone role in the food web, ecological processes critical for ecosystem functioning, functional groups that play a critical role in ecosystem functioning, functional significance of the habitat (importance for vulnerable life history stages), and structural complexity of the habitat. The concepts and protocols for evaluation can be applied in screening and scoping processes related to deep-sea EIAs, for site screening and scoping for strategic environmental assessments, and providing knowledge to formulate conservation plans. The use of such criteria in spatial planning allows for more informed and transparent decision-making and can help establish a framework for monitoring and managing ecosystem services and the biodiversity and function supporting them. These criteria are needed to overcome data gaps and address limitations in knowledge of factors responsible for maintaining ecosystem integrity (Gilman et al. 2011). Including ecological value considerations in EIAs facilitates the process of EIS review and public consultation and adds transparency to the assessment, which can be a significant factor in obtaining social licence (Lallier and Maes 2016). In the case of deep-sea mining, internationally recognised value criteria can provide a basis for using limited knowledge as a starting point for developing an adaptive management framework.

5 Adaptive Management Approach

Adaptive management has been identified as a practical tool to address the development of environmentally responsible and cost-effective observation and monitoring systems that can be used in the deep sea to feed into conservation measures. The industry is not expected to grow exponentially at this point but will expand gradually, allowing time to incorporate feedback from the work of the first contractors to improve EIA methods and management. An adaptive management approach used in this way allows for management to scale up as the industry scales up. The ISA has been proactive in developing rules, regulations, and procedures that incorporate standards for the protection and preservation of the marine environment during exploration and mining. Part of the adaptive management toolbox for deep-sea mining is the development of regional environmental management plans (REMPs) and marine spatial planning to protect biodiversity, ecosystem functions and services, and the sustainability of resources.

The first REMP for abyssal polymetallic nodule fields in the CCZ was approved in 2012 (ISBA/17/LTC/7) with the goal of facilitating exploitation and cooperative research, monitoring the environment, fostering area-based management, and applying an ecosystem-based approach to management and broad stakeholder participation. A key component of a REMP is a network of areas of particular environmental interest (APEIs) for the preservation of representative ecosystems for the protection of biodiversity and ecosystem structure and function (Dunn et al. 2018). APEI networks contribute to a precautionary approach to environmental management of deep-sea mining by ensuring that representative benthic habitats and associated ecosystems are protected from serious harm on regional scales. The CCZ was divided into nine subregions, each containing a 400 × 400 km APEI, representing a total of ~24% of the CCZ management area. Two additional APEIs were proposed in 2016 (22nd Session of the ISA), which, if implemented, would increase the total protected area to ~29% (Dunn et al. 2018). The location of these APEIs within each subregion is currently flexible, allowing for adaptive management based on new knowledge and the location of mining sites.

The ISA has called for REMPs in other areas where there are currently exploration contracts (ISBA/20/C/1; ISBA/21/C/20; ISBA/22/C/28) in response to recommendations from the United Nations General Assembly (UNGA; UNGA 69/245; UNGA 70/235). The ISA is yet to approve a REMP for any region of polymetallic sulphide or ferromanganese crusts but has encouraged the scientific community to support the development of REMPs for key areas of commercial interest (Dunn et al. 2018). In response, guidelines were proposed for the establishment of networks of chemosynthetic ecosystem reserves to protect environments associated with polymetallic sulphide deposits (Dinard Guidelines for Chemosynthetic Ecological Reserves; Van Dover et al. 2011). Building on these guidelines, ISA workshops have discussed the design of a network of protected areas for the Mid-Atlantic Ridge, resulting in a framework for the design and assessment of APEI network scenarios. Dunn et al. (2018) subsequently developed a network design framework for the Mid-Atlantic Ridge, based on a suite of criteria for evaluating potential network performance against conservation targets and CBD network design criteria (EBSA criteria). APEIs for ferromanganese-encrusted seamounts are currently in the initial stages of development. However, some of these seamounts are included in existing marine protected areas. A key example is the Pacific Remote Islands Marine National Monument (PRIMNM), one of the world's largest MPAs that covers an area of ~1.3 million km^2.

While areas protected from the impacts of mining are important reference sites, it is possible that the APEIs may not be well studied or understood prior to the commencement of mining. However, it is an ISA requirement as part of the EIA to designate impact reference zones (IRZs) and preservation reference zones (PRZs) for the primary purposes of ensuring the preservation of biota representative of those impacted and facilitating monitoring of biota impacted by mining activities (ISBA/19/LTC/8 Annex I). Both IRZs and PRZs are included in the CCZ REMP. IRZs are designated within the claim area and are representative of the site to be mined and the recovery potential. PRZs are reference sites and should include some of the mineral deposit to be as ecologically similar as possible to the impact zone and be removed from potential mining impacts (ISBA/17/LTC/7). PRZs hold high scientific value as locations where meaningful baselines may be established for comparison with IRZs and a key component of the EIA. Continued efforts to map and monitor the deep sea will provide a strong foundation to support the increasing efforts to spatially manage these areas (Wedding et al. 2013).

Encouragement of scientific research and monitoring, both related to and independent from deep-sea mining impacts in APEIs and PRZs, is an important step towards understanding deep-sea biodiversity and ecosystem functions that underpin ecosystem services. Simple indicators, selected using available criteria, could be used to measure and monitor not only these key factors but some tangible ecosystem services and the potential impacts on them for incorporation in an EIA. The ISA actively encourages data sharing and collaboration to ensure there is access to environmental data for EIAs and the formulation by the ISA of rules, regulations, and procedures while encouraging the scientific community to publish their findings. A culture of data sharing and collaboration will be essential for accelerating the rate at which knowledge is gained, identifying knowledge gaps, prioritising future research, and increasing the accuracy and speed at which EIAs can be conducted, leading to informed decision-making. An inventory of the data holdings from each contractor should be accessible (ISBA/19/LTC/8), photographs should be archived (ISBA/19/LTC/8, Annex I), and a list of relevant publications in peer-reviewed journals published during the reporting year (ISBA/21/LTC/15) should be shared by the contractor. Long-term, data sharing and collaboration could result in building databases of environmental information (such as population connectivity, genetics, and biodiversity at a regional level) over several decades that will be an invaluable resource for future EIAs.

The information gained from EIAs and the scientific research conducted at reference sites can feed into adaptive management to continually improve the EIA process and management. By building on knowledge, new priorities for research can be set and, consequently, certainty will be increased.

6 Conclusions

The seabed mineral industry is moving towards an ecosystem-based approach to EIAs, which would include assessing the impacts of mining on ecosystem services. At this time, knowledge of deep-sea ecosystem services is limited to tangible ecosystem services, such as provisional and cultural services, and even less is known of the less tangible services, such as regulatory and supporting services. However, by using the available tools and taking an adaptive management approach to environmental assessments, allowing management to scale up as the industry does, the more tangible ecosystem services could be included in future EIAs, building knowledge with the aim of eventually being able to assess ecosystem services more completely.

Appendix A: Studies Included in Table 2

1	Meyers et al. (2014)	26	Hein et al. (2010)	51	McCollom and Shock (1997)	76	Jørgensen and Boetius (2007)
2	Edwards et al. (2011)	27	Wang et al. (2011)	52	Petersen et al. (2011)	77	Beaulieu (2001)
3	Huo et al. (2015)	28	Edwards et al. (2004)	53	Kilias et al. (2013)	78	Smith et al. (2006)
4	Emerson and Moyer (2010)	29	Verlaan (1992)	54	Minic and Thongbam (2011)	79	Gage and Tyler (1992)
5	Clark et al. (2010)	30	Halbach (1986)	55	Zierenberg et al. (2000)	80	Mullineaux (1987)
6	Koslow et al. (2015)	31	Wang and Müller (2009)	56	Minic et al. (2006)	81	Linley et al. (2017)
7	Schlacher et al. (2014)	32	Jiang et al. (2017)	57	Ehrlich (1983)	82	Vanreusel et al. (2016)
8	Morgan et al. (2015)	33	Koschinsky et al. (1996)	58	Juniper and Sarrazin (1995)	83	Bussau et al. (1995)
9	Tittensor et al. (2009)	34	Baco et al. (2016)	59	Hiriart et al. (2010)	84	Rex et al. (2004)
10	Worm et al. (2003)	35	Sogin et al. (2006)	60	Juniper et al. (1992)	85	Kuhnz et al. (2014)
11	Morato et al. (2010)	36	Sewell 2017	61	Holden and Adams (2003)	86	Smith et al. (2008b)
12	Tracey et al. (2011)	37	Ressurreição and Giacomello (2013)	62	Wery et al. (2002)	87	Weslawski et al. (2004)
13	Boehlert and Sasaki (1988)	38	Campbell et al. (2006)	63	Jiang et al. (2014)	88	Smith et al. (2009)
14	Pereira-Filho et al. (2011)	39	Levin et al. (2016b)	64	Chen et al. (2015)	89	Ruhl et al. (2008)
15	Genin et al. (1986)	40	Holden et al. (2012)	65	Thornburg et al. (2010)	90	Hannides and Smith (2003)
16	White et al. (2007)	41	Reysenbach et al. (2006)	66	Cripps et al. (2009)	91	Mayor et al. (2014)
17	Morato et al. (2009)	42	Poli et al. (2017)	67	Leary (2004)	92	Lampitt et al. (2001)
18	Boehlert and Genin (1987)	43	Mehta et al. (2003)	68	Van Dover et al. (2018)	93	Pearcy et al. (1982)
19	Genin (2004)	44	Lonsdale (1977)	69	Van Dover (2014).	94	Priede et al. (2010)
20	Clark (2017)	45	Mullineaux et al. (2003)	70	Baross and Hoffman (1985)	95	Brandt et al. (2005)
21	Clark and Dunn (2012)	46	Salinas-de-León et al. (2018)	71	Brunnegård et al. (2004)	96	Snelgrove (1999)
22	Watson et al. (2007)	47	Biscoito et al. (2002)	72	Smith (1992)	97	Webb et al. (2010)
23	Koslow (1997)	48	Lutz and Kennish (1993)	73	Smith and Druffel (1998)	98	Ramirez-Llodra et al. (2010)
24	Krishnan et al. (2006)	49	Van Dover and Lutz (2004)	74	Deming and Colwell (1982)		
25	Wang et al. (2009)	50	Nakagawa and Takai (2008)	75	Scheckenbach et al. (2010)		

References

Amon, D. J., Ziegler, A. F., Dahlgren, T. G., Glover, A. G., Goineau, A., Gooday, A. J., Wiklund, H., & Smith, C. R. (2016). Insights into the abundance and diversity of abyssal megafauna in a polymetallic-nodule region in the eastern Clarion-Clipperton Zone. *Scientific Reports, 6*, 30492.

Armstrong, C.W., Foley, N., Tinch, R., & van den Hove, S. (2010). *Ecosystem goods and services of the deep sea*. Deliverable D6.2HERMIONE Project, 68 p.

Armstrong, C. W., Foley, N. S., Tinch, R., & van den Hove, S. (2012). Services from the deep: Steps towards valuation of deep sea goods and services. *Ecosystem Services, 2*, 2–13.

Baco, A. R. (2007). Exploration for deep-sea corals on North Pacific seamounts and islands. *Oceanography, 20*(4), 108–117.

Baco, A. R., Etter, R. J., Ribeiro, P. A., Von der Heyden, S., Beerli, P., & Kinlan, B. P. (2016). A synthesis of genetic connectivity in deep-sea fauna and implications for marine reserve design. *Molecular Ecology, 25*(14), 3276–3298.

Baross, J. A., & Hoffman, S. E. (1985). Submarine hydrothermal vents and associated gradient environments as sites for the origin and evolution of life. *Origins of Life and Evolution of the Biosphere, 15*(4), 327–345.

Bateman, I. J., Mace, G. M., Fezzi, C., Atkinson, G., & Turner, K. (2011). Economic analysis for ecosystem service assessments. *Environmental and Resource Economics, 48*(2), 177–218.

Beaulieu, S. E. (2001). Life on glass houses: Sponge stalk communities in the deep sea. *Marine Biology, 138*(4), 803–817.

Beaumont, N. J., Austen, M. C., Atkins, J. P., Burdon, D., Degraer, S., Dentinho, T. P., Derous, S., Holm, P., Horton, T., van Ierland, E., Marboe, A. H., Starkey, D. J., Townsend, M., & Zarzycki, T. (2007). Identification, definition and quantification of goods and services provided by marine biodiversity: Implications for the ecosystem approach. *Marine Pollution Bulletin, 54*(3), 253–265.

Binney, J., & Fleming, C. (2016). Counting the potential coast of deep sea-bed mining to Fiji. Brisbane, Australia: Report for WWF International, Mainstream Economics and Policy.

Biscoito, M., Segonzac, M., Almeida, A., Desbruyeres, D., Geistdoerfer, P., Turnipseed, M., & Van Dover, C. (2002). Fishes from the hydrothermal vents and cold seeps – an update. *CBM-Cahiers de Biologie Marine, 43*(3–4), 359–362.

Boehlert, G. W., & Genin, A. (1987). A review of the effects of seamounts on biological processes. *Seamounts, Islands and Atolls, 43*, 319–334.

Boehlert, G. W., & Sasaki, T. (1988). Pelagic biogeography of the armorhead, pseudopentaceros wheeleri, and recruitment to isolated seamounts in the North Pacific Ocean. *Fishery Bulletin, 86*(3), 453–466.

Boetius, A., & Wenzhöfer, F. (2013). Seafloor oxygen consumption fuelled by methane from cold seeps. *Nature Geoscience, 6*(9), 725–734.

Brandt, A., Brenke, N., Andres, H. G., Brix, S., Guerrero-Kommritz, J., Mühlenhardt-Siegel, U., & Wägele, J. W. (2005). Diversity of peracarid crustaceans (Malacostraca) from the abyssal plain of the Angola Basin. *Organisms Diversity & Evolution, 5*, 105–112.

Brunnegård, J., Grandel, S., Ståhl, H., Tengberg, A., & Hall, P. O. (2004). Nitrogen cycling in deep-sea sediments of the porcupine abyssal plain, NE Atlantic. *Progress in Oceanography, 63*(4), 159–181.

Bussau, C., Schriever, G., & Thiel, H. (1995). Evaluation of abyssal metazoan meiofauna from a manganese nodule area of the eastern South Pacific. *Oceanographic Literature Review, 10*(42), 883.

Campbell, B. J., Engel, A. S., Porter, M. L., & Takai, K. (2006). The versatile ε-proteobacteria: Key players in sulphidic habitats. *Nature Reviews Microbiology, 4*(6), 458.

CBD. (2008). IX/20 Marine and Coastal Biodiversity. Decision Adopted by the Conference of the parties to the Convention on Biological Diversity at its Ninth Meeting, Bonn, 19–30 May 2008. Annex I. Scientific Criteria for Identifying Ecologically or Biologically Significant Marine Areas in Need of Protection in Open-Ocean Waters and Deep-Sea Habitats. UNEP/CBD/COP/DEC/IX/20. Convention on Biological Diversity Secretariat, Montreal, Canada.

CBD. (2010). Report of the expert workshop on scientific and technical aspects relevant to environmental impact assessment in marine areas beyond national jurisdiction. 14th meeting of the Subsidiary Body on Scientific, Technical & Technological Advice. Nairobi, 10–21 May 2010.

CCAMLR. (2008). Preliminary assessments of known and anticipated impacts of proposed bottom fishing activities on vulnerable marine ecosystems. Collated by the Secretariat. CCAMLR XXVII/26*, 24 September, 2008, Agenda Item No. 5, SC Agenda Item No. 4, 107pp.

Chaudhuri, S. K., & Lovley, D. R. (2003). Electricity generation by direct oxidation of glucose in mediatorless microbial fuel cells. *Nature Biotechnology, 21*(10), 1229–1232.

Chen, J., Zhang, Z., Zhang, C., & Yu, B. (2015). Genome sequence of *Geobacillus thermoglucosidasius* DSM2542, a platform hosts for biotechnological applications with industrial potential. *Journal of Biotechnology, 216*, 98–99.

Chivers, A. J., Narayanaswamy, B. E., Lamont, P. A., Dale, A., & Turnewitsch, R. (2013). Changes in polychaete standing stock and diversity on the northern side of Senghor Seamount (NE Atlantic). *Biogeosciences, 10*(6), 3535–3546.

Clark, M. R. (2017). Deep-sea seamount fisheries: A review of global status and future prospects. *Submission article platform-Latin American Journal of Aquatic Research, 37*(3), 501–512.

Clark, M. R., & Dunn, M. R. (2012). Spatial management of deep-sea seamount fisheries: Balancing sustainable exploitation and habitat conservation. *Environmental Conservation, 39*(3), 204–214.

Clark, M. R., & Koslow, J. A. (2007). Impacts of fisheries on seamounts. In T. J. Pitcher, T. Morato, P. J. B. Hart, M. R. Clark, N. Haggan, & R. S. Santos (Eds.), *Seamounts: Ecology, fisheries and conservation* (Blackwell Fisheries and Aquatic Resources Series 12) (pp. 413–441). Oxford: Blackwell Publishing.

Clark, M. R., Rowden, A. A., Schlacher, T., Williams, A., Consalvey, M., Stocks, K. I., Rogers, A. D., O'Hara, T. D., White, M., Shank, T. M., & Hall-Spencer, J. M. (2010). The ecology of seamounts: Structure, function, and human impacts. *Annual Review of Marine Science, 2*, 253–278.

Clark, M. R., Rouse, H. L., Lamarche, G., Ellis, J. I., & Hickey, C. (2017). *Preparation of environmental impact assessments: General guidelines for offshore mining and drilling with particular reference to New Zealand.* NIWA Science and Technology Series No. 81. 110 pp.

Cochonat, P., Durr, S., Gunn, V., Herzig, P., Mevel, C., Mienert, J., Schneider, R., Weaver, P., & Winkler, A. (2007). *The deep-sea frontier: Science challenges for a sustainable future.* Luxembourg: Office for Official Publications of the European Communities. 53 pp.

Coffey Natural Systems/Nautilus Minerals Ltd. (2008) Environmental impact statement. Solwara 1 project. Volume A. Main report. 226 p.

Collins, P. C., Croot, P., Carlsson, J., Colaço, A., Grehan, A., Hyeong, K., Kennedy, R., Mohn, C., Smith, S., Yamamoto, H., & Rowden, A. (2013). A primer for the environmental impact assessment of mining at seafloor massive sulfide deposits. *Marine Policy, 42*, 198–209.

Cormier, R. (2019). Ecosystem approach for management of deep-sea mining activities. In R. Sharma (Ed.), *Environmental issues of deep-sea mining – impacts, consequences and policy perspectives*. Cham: Springer International publishing AG.

Costanza, R., d'Arge, R., De Groot, R., Farber, S., Grasso, M., Hannon, B., Limburg, K., Naeem, S., O'neill, R. V., Paruelo, J., & Raskin, R. G. (1997). The value of the world's ecosystem services and natural capital. *Nature, 387*, 253–260.

Cripps, R. E., Eley, K., Leak, D. J., Rudd, B., Taylor, M., Todd, M., Boakes, S., Martin, S., & Atkinson, T. (2009). Metabolic engineering of *Geobacillus thermoglucosidasius* for high yield ethanol production. *Metabolic Engineering, 11*(6), 398–408.

Da Silva, P. M., & Pinho, M. R. (2007). Small-scale fishing on seamounts. In T. J. Pitcher, T. Morato, P. J. B. Hart, M. R. Clark, N. Haggan, & R. S. Santos (Eds.), *Seamounts: Ecology fisheries and conservation* (Fisheries and aquatic resource series) (pp. 335–360). Oxford: Blackwell Scientific.

Danovaro, R., Gambi, C., Dell'Anno, A., Corinaldesi, C., Fraschetti, S., Vanreusel, A., Vincx, M., & Gooday, A. J. (2008). Exponential decline of deep-sea ecosystem functioning linked to benthic biodiversity loss. *Current Biology, 18*(1), 1–8.

De Forges, B. R., Koslow, J. A., & Poore, G. C. B. (2000). Diversity and endemism of the benthic seamount fauna in the southwest Pacific. *Nature, 405*, 944–947.

De Smet, B., Pape, E., Riehl, T., Bonifácio, P., Colson, L., & Vanreusel, A. (2017). The community structure of deep-sea macrofauna associated with polymetallic nodules in the eastern part of the Clarion-Clipperton Fracture Zone. *Frontiers in Marine Science, 4*, 1–14.

Dekas, A. E., Poretsky, R. S., & Orphan, V. J. (2009). Deep-sea archaea fix and share nitrogen in methane-consuming microbial consortia. *Science, 326*, 422–426.

Deming, J. W. (1998). Deep ocean environmental biotechnology. *Current Opinion in Biotechnology, 9*(3), 283–287.

Deming, J. W., & Colwell, R. R. (1982). Barophilic bacteria associated with digestive tracts of abyssal holothurians. *Applied and Environmental Microbiology, 44*(5), 1222–1230.

van den Hove, S., & Moreau, V. (2007). *Ecosystems and biodiversity in deep waters and high seas: A scoping report on their socio-economy, management and governance* (p. 84). Switzerland: UNEP-WCMC.

Dunn, D. C., Van Dover, C. L., Etter, R. J., Smith, C. R., Levin, L. A., Morato, T., Colaço, A., Dale, A. C., Gebruk, A. V., Gjerde, K. M., Halpin, P. N., Howell, K. L., Johnson, D., Perez, J. A. A., Ribeiro, M. C., Stuckas, H., & Weaver, P. (2018). A strategy for the conservation of biodiversity on mid-ocean ridges from deep-sea mining. *Science Advances, 4*(7), 1–16.

Durden, J. M., Lallier, L. E., Murphy, K., Jaeckel, A., Gjerde, K., & Jones, D. O. B. (2018). Environmental impact assessment process for deep-sea mining in "the Area". *Marine Policy, 87*, 194–202.

Edwards, K. J., Bach, W., McCollom, T. M., & Rogers, D. R. (2004). Neutrophilic iron-oxidizing bacteria in the ocean: Their habitats, diversity, and roles in mineral deposition, rock alteration, and biomass production in the deep-sea. *Geomicrobiology Journal, 21*(6), 393–404.

Edwards, K. J., Wheat, C. G., & Sylvan, J. B. (2011). Under the sea: Microbial life in volcanic oceanic crust. *Nature Reviews Microbiology, 9*(10), 703.

Ehrlich, H. L. (1983). Manganese-oxidizing bacteria from a hydrothermally active area on the Galapagos Rift. *Ecological Bulletins, 35*, 357–366.

Ehrlich, P. R., Ehrlich, A., & Holden, J. (1977). *Ecoscience: Population, resources, environment*. San Francisco: W.H. Freeman. xi, 1051 p.

Ellis, J. I. I., Clark, M. R. R., Lamarche, G., Rouse, H. L. L., & Lamarche, G. (2017). Environmental management frameworks for offshore mining: The New Zealand approach. *Marine Policy, 84*, 178–192.

Emerson, D., & Moyer, C. L. (2010). Microbiology of seamounts: Common patterns observed in community structure. *Oceanography, 23*(1), 148–163.

FAO. (2009). *International guidelines for the management of deep-sea fisheries in the High Seas*. Rome: Food and Agriculture Organization of the United Nations.

Filer, C., & Gabriel, J. (2016). How could nautilus minerals get a social licence to operate the world's first deep sea mine? *Marine Policy, 95*, 394–400.

Fletcher, S., Saunders, J., & Herbert, R. J. H. (2011). A review of the ecosystem services provided by broad-scale marine habitats in England's MPA network. *Journal of Coastal Research, 64*(1), 378–383.

Foley, M. M., Halpern, B. S., Micheli, F., Armsby, M. H., Caldwell, M. R., Crain, C. M., Prahler, E., Rohr, N., Sivas, D., Beck, M. W., Carr, M. H., Crowder, L. B., Emmett Duffy, J., Hacker, S. D., McLeod, K. L., Palumbi, S. R., Peterson, C. H., Regan, H. M., Ruckelshaus, M. H., Sandifer, P. A., & Steneck, R. S. (2010). Guiding ecological principles for marine spatial planning. *Marine Policy, 34*(5), 955–966.

Folkersen, M. V., Fleming, C. M., & Hasan, S. (2018). The economic value of the deep sea: A systematic review and meta-analysis. *Marine Policy, 94*, 71–80.

Fredslund, F., Borchert, M. S., Poulsen, J. C. N., Mortensen, S. B., Perner, M., Streit, W. R., & Leggio, L. L. (2018). Structure of a hyperthermostable carbonic anhydrase identified from an active hydrothermal vent chimney. *Enzyme and Microbial Technology, 114*, 48–54.

Gage, J. D., & Tyler, P. A. (1992). *Deep-sea biology: A natural history of organisms at the deep-sea floor*. Cambridge, UK: Cambridge University Press.

Genin, A. (2004). Bio-physical coupling in the formation of zooplankton and fish aggregations over abrupt topographies. *Journal of Marine Systems, 50*(1–2), 3–20.
Genin, A., Dayton, P. K., Lonsdale, P. F., & Spiess, F. N. (1986). Corals on seamount peaks provide evidence of current acceleration over deep-sea topography. *Nature, 322*, 59–61.
Gibson, R. B. (2002). From Wreck Cove to Voisey's Bay: The evolution of federal environmental assessment in Canada. *Impact Assessment and Project Appraisal, 20*(3), 151–159.
Gilman, E., Dunn, D., Read, A., Hyrenbach, K. D., & Warner, R. (2011). Designing criteria suites to identify discrete and networked sites of high value across manifestations of biodiversity. *Biodiversity and Conservation, 20*(14), 3363–3383.
Glasson, J., Therivel, R., & Chadwick, A. (2013). *Introduction to environmental impact assessment*. London: UCL Press Ltd, University College. 416 pp.
Glover, A. (2018). New scientific data from exploration contract areas relevant to the management of deep-sea mining. Presentation provided at a side event at the 24th Session of the International Seabed Authority, August 2018.
Glover, A. G., Smith, C. R., Paterson, G. L. J., Wilson, G. D. F., Hawkins, L., & Sheader, M. (2002). Polychaete species diversity in the central Pacific abyss: Local and regional patterns, and relationships with productivity. *Marine Ecology Progress Series, 240*, 157–170.
Glover, A. G., Dahlgren, T. G., Wiklund, H., Mohrbeck, I., & Smith, C. R. (2015). An end-to-end DNA taxonomy methodology for benthic biodiversity survey in the Clarion-Clipperton Zone, central Pacific abyss. *Journal of Marine Science and Engineering, 4*(1), 2.
Goffredi, S. K., Girguis, P. R., Childress, J. J., & Desaulniers, N. T. (1999). Physiological functioning of carbonic anhydrase in the hydrothermal vent tubeworm *Riftia pachyptila*. *Biological Bulletin, 196*(3), 257–264.
Gollner, S., Kaiser, S., Menzel, L., Jones, D. O. B., Brown, A., Mestre, N. C., van Oevelen, D., Menot, L., Colaço, A., Canals, M., Cuvelier, D., Durden, J. M., Gebruk, A., Egho, G. A., Haeckel, M., Marcon, Y., Mevenkamp, L., Morato, T., Pham, C. K., Purser, A., Sanchez-Vidal, A., Vanreusel, A., Vink, A., & Martinez Arbizu, P. (2017). Resilience of benthic deep-sea fauna to mining activities. *Marine Environmental Research, 129*, 76–101.
Government of New Zealand. (2008). Preliminary assessment of bottom fishing activities in 2008–2009 for New Zealand: Implementing conservation measure 22-06 (Bottom Fishing in the Convention Area). Impact Assessment submitted to CCAML R, CCAML R-XX VII/26*, 24 September 2008. Original: English Agenda Item No. 5, SC Agenda Item No. 4, 70pp.
Grassle, J. F., & Maciolek, N. J. (1992). Deep-sea species richness: Regional and local diversity estimate from quantitative bottom samples. *The American Naturalist, 139*(2), 313–341.
Grehan, A., Long, R., Deegan, B., & Ó Cinneide, M. (2003). The Irish coral task force and Atlantic coral ecosystem study: Report on two deep-water coral conservation stakeholder workshops held in Galway in 2000 and 2002. *Marine Environment and Health Series, 11*, 117.
Haines-Young, R., Potschin, M., de Groot, D., Kienast, F., & Bolliger, J., (2009). *Towards a common international classification of ecosystem services (cices) for integrated environmental and economic accounting (draft v1). Report to the European Environment Agency for Contract no.: EEA/BSS/07/007*. Centre for Environmental Management, School of Geography, University of Nottingham.
Halbach, P. (1986). Processes controlling the heavy metal distribution in Pacific ferromanganese nodules and crusts. *Geologische Rundschau, 75*(1), 235–247.
Hanemann, W. M. (1989). Information and the concept of option value. *Journal of Environmental Economics and Management, 16*, 23–37.
Hannides, A., & Smith, C. R. (2003). The northeast abyssal Pacific plain. In *Biogeochemistry of marine systems* (pp. 208–237). Oxford: Blackwell.
Hein, J. R. (2004). Cobalt-rich ferromanganese crusts: Global distribution, composition, origin and research activities. In: Minerals other than polymetallic nodules of the International Seabed Area. Proceedings of a workshop help on 26-30 June, 2000 (Vol. 1, pp 188–256). Kingston, Jamaica: International Seabed Authority.
Hein, J. R., Conrad, T. A., & Staudigel, H. (2010). Seamount mineral deposits: A source of rare metals for high-technology industries. *Oceanography, 23*(1), 184–189.

Higgs, N. D., & Attrill, M. (2015). Biases in biodiversity: Wide-ranging species are discovered first in the deep sea. *Frontiers in Marine Science, 2*, 61.
Hinrichs, K., & Boetius, A. (2002). The anaerobic oxidation of methane: New insights in microbial ecology and biogeochemistry. In G. Wefer, D. Billett, D. Hebbeln, B. Jorgensen, M. Schulter, & T. Van Weering (Eds.), *Ocean margin systems* (pp. 457–477). Berlin: Springer-Verlag.
Hiriart, G., Prol-Ledesma, R.M., Alcocer, S., & Espíndola, S. (2010). Submarine geothermics: Hydrothermal vents and electricity generation. Proceedings World Geothermal Congress, 1–6.
Holden, J. F., & Adams, M. W. (2003). Microbe–metal interactions in marine hydrothermal environments. *Current Opinion in Chemical Biology, 7*(2), 160–165.
Holden, J. F., Breier, J. A., Rogers, K. L., Schulte, M. D., & Toner, B. M. (2012). Biogeochemical processes at hydrothermal vents: Microbes and minerals, bioenergetics, and carbon fluxes. *Oceanography, 25*(1), 196–208.
Holdren, J. P., & Ehrlich, P. R. (1974). Human population and the global environment: Population growth, rising per capita material consumption, and disruptive technologies have made civilization a global ecological force. *American Scientist, 62*(3), 282–292.
Huo, Y., Cheng, H., Post, A. F., Wang, C., Jiang, X., Pan, J., Wu, M., & Xu, X. (2015). Ecological functions of uncultured microorganisms in the cobalt-rich ferromanganese crust of a seamount in the Central Pacific are elucidated by fosmid sequencing. *Acta Oceanologica Sinica, 34*(4), 92–113.
Husky-Oil. (2001). *White rose oilfield comprehensive study report* (p. 94). Canada: Husky Oil Operations Limited.
IMO. (2006a). Revised guidelines for the identification and designation of particularly sensitive areas. Resolution A.982(24). 24th session, International Maritime Organization. 6 February 2006.
IMO. (2006b). *Revised guidelines for the identification and designation of Particularly Sensitive Sea Areas (PSSAs)* (Resolution A.982(24)). London: International Maritime Organization.
ISA. (2017). Towards an ISA environmental management strategy for the area: Report of an international workshop convened by the German Environment Agency (UBA), the German Federal Institute for Geosciences and Natural Resources (BGR) and the Secretariat of the International Seabed Authority (ISA) in Berlin, Germany, 20–24 March 2017. ISA Technical study No. 17.
ISA (International Seabed Authority). (n.d.) Exploration Areas. https://www.isa.org.jm/contractors/exploration-areas. Accessed 15 Aug 2018.
ISBA/17/LTC/7. (2011). Environmental management plan for the Clarion-Clipperton Zone, 17th session, Legal and Technical Commission, Kingston, Jamaica, 11–22 July 2011.
ISBA/19/LTC/8. (2013). Recommendations for the guidance of the contractors for the assessment of possible environmental impacts arising from exploration for marine minerals in the Area. 19th session. Legal and Technical Commission, Kingston, Jamaica, 15–26 July 2013.
ISBA/20/C/1. (2014). Status of prospecting and of contracts for exploration for polymetallic nodules, polymetallic sulphides and cobalt-rich ferromanganese crusts in the Area. 20th session, Legal and Technical Commission. Kingston, Jamaica, 14–25 July 2014.
ISBA/21/C/20. (2015). Decision of the Council of the International Seabed Authority relating to the summary report of the Chair of the Legal and Technical Commission. 21st session, Kingston, Jamaica. 13–24 July 2015.
ISBA/21/LTC/15. (2015). Recommendations for the guidance of contractors on the content, format and structure of annual reports. Legal and Technical Commission, August 2015.
ISBA/22/C/28. (2016). Decision of the Council of the International Seabed Authority relating to the summary report of the Chair of the Legal and Technical Commission. 22nd session, Kingston, Jamaica, 11–22 July 2016.
ISBA/24/LTC/WP.1. (2018). Draft regulations on exploitation of mineral resources in the Area. 24th session, Legal and Technical Commission, part II, Kingston, Jamaica, 2–13 July 2018.
ISBA/24/LTC/WP.1/Add.1. (2018). Draft Environmental Impact Statement template. 24th session, Legal and Technical Commission, part II. Kingston, Jamaica, 2–13 July 2018.
Jahnke, R. A., & Jackson, G. A. (1992). The spatial distribution of sea floor oxygen consumption in the Atlantic and Pacific Oceans. In *Deep-sea food chains and the global carbon cycle* (pp. 295–307). Dordrecht: Springer.

Jha-Thakur, U., & Fischer, T. B. (2016). 25 years of the UK EIA system: Strengths, weaknesses, opportunities and threats. *Environmental Impact Assessment Review, 61*, 19–26.

Jiang, L., Long, C., Wu, X., Xu, H., Shao, Z., & Long, M. (2014). Optimization of thermophilic fermentative hydrogen production by the newly isolated *Caloranaerobacter azorensis* H53214 from deep-sea hydrothermal vent environment. *International Journal of Hydrogen Energy, 39*(26), 14154–14160.

Jiang, X. D., Sun, X. M., Guan, Y., Gong, J. L., Lu, Y., Lu, R. F., & Wang, C. (2017). Biomineralisation of the ferromanganese crusts in the Western Pacific Ocean. *Journal of Asian Earth Sciences, 136*, 58–67.

Jo, B. H., Seo, J. H., & Cha, H. J. (2014). Bacterial extremo-α-carbonic anhydrases from deep-sea hydrothermal vents as potential biocatalysts for CO_2sequestration. *Journal of Molecular Catalysis B: Enzymatic, 109*, 31–39.

Jobstvogt, N., Hanley, N., Hynes, S., Kenter, J., & Witte, U. (2014). Twenty thousand sterling under the sea: Estimating the value of protecting deep-sea biodiversity. *Ecological Economics, 97*, 10–19.

Jørgensen, B. B., & Boetius, A. (2007). Feast and famine—microbial life in the deep-sea bed. *Nature Reviews Microbiology, 5*(10), 770–781.

Juniper, S. K., & Sarrazin, J. (1995). Interaction of vent biota and hydrothermal deposits: present evidence and future experimentation. In *Seafloor hydrothermal systems: Physical, chemical, biological, and geological interactions* (pp. 178–193). Washington, DC: American Geophysical Union.

Juniper, S. K., Jonasson, I. R., Tunnicliffe, V., & Southward, A. J. (1992). Influence of a tube-building polychaete on hydrothermal chimney mineralization. *Geology, 20*(10), 895–898.

Karjalainen, T. P., Marttunen, M., Sarkki, S., & Rytkönen, A. M. (2013). Integrating ecosystem services into environmental impact assessment: An analytic–deliberative approach. *Environmental Impact Assessment Review, 40*, 54–64.

Kilias, S. P., Nomikou, P., Papanikolaou, D., Polymenakou, P. N., Godelitsas, A., Argyraki, A., Carey, S., Gamaletsos, P., Mertzimekis, T. J., Stathopoulou, E., & Goettlicher, J. (2013). New insights into hydrothermal vent processes in the unique shallow-submarine arc-volcano, Kolumbo (Santorini), Greece. *Scientific Reports, 3*, 2421.

Kim, S. S., & Wessel, P. (2011). New global seamount census from altimetry-derived gravity data. *Geophysical Journal International, 186*(2), 615–631.

Koschinsky, A., Halbach, P., Hein, J. R., & Mangini, A. (1996). Ferromanganese crusts as indicators for paleoceanographic events in the NE Atlantic. *Geologische Rundschau, 85*(3), 67–576.

Koslow, J. A. (1997). Seamounts and the ecology of deep-sea fisheries: The firm-bodied fishes that feed around seamounts are biologically distinct from their deepwater neighbors—and may be especially vulnerable to overfishing. *American Scientist, 85*(2), 168–176.

Koslow, J.A., Auster, P., Bergstad, O.A., Roberts, J.M., Rogers, A., Vecchione, M., Harris, P.T., Rice, J., & Bernal, P. (2015). Biological communities on seamounts and other submarine features potentially threatened by disturbance. United Nations World Ocean Assessment.

Krishnan, K. P., Fernandes, C. E., Fernandes, S. O., & Bharathi, P. L. (2006). Tolerance and immobilization of cobalt by some bacteria from ferromanganese crusts of the Afanasy Nikitin Seamounts. *Geomicrobiology Journal, 23*(1), 31–36.

Kuhnz, L. A., Ruhl, H. A., Huffard, C. L., & Smith, K. L., Jr. (2014). Rapid changes and long-term cycles in the benthic megafaunal community observed over 24 years in the abyssal Northeast Pacific. *Progress in Oceanography, 124*, 1–11.

Lallier, L. E., & Maes, F. (2016). Environmental impact assessment procedure for deep seabed mining in the area: Independent expert review and public participation. *Marine Policy, 70*, 212–219.

Lampitt, R. S., Bett, B. J., Kiriakoulakis, K., Popova, E. E., Ragueneau, O., Vangriesheim, A., & Wolff, G. A. (2001). Material supply to the abyssal seafloor in the Northeast Atlantic. *Progress in Oceanography, 50*(1–4), 27–63.

Langhammer, P.F., Bakarr, M.I., Bennun, L., & Brooks, T.M. (2007). Identification and gap analysis of key biodiversity areas: targets for comprehensive protected area systems (No. 15). IUCN.

Leary, D. K. (2004). Bioprospecting and the genetic resources of hydrothermal vents on the high seas: What is the existing legal position, where are we heading and what are our options. *Macquarie Journal of International and Comparative Environmental Law, 1*, 137.

Leslie, H. M., & Kinzig, A. P. (2009). Resilience science. In K. L. McLeod & H. Leslie (Eds.), *Ecosystem-based management for the oceans* (pp. 55–73). Washington, DC: Island Press.

Levin, L. A., Baco, A. R., Bowden, D. A., Colaco, A., Cordes, E. E., Cunha, M. R., Demopoulos, A. W. J., Gobin, J., Grupe, B. M., Le, J., Metaxas, A., Netburn, A. N., Rouse, G. W., Thurber, A. R., Tunnicliffe, V., Van Dover, C. L., Vanreusel, A., & Watling, L. (2016a). Hydrothermal vents and methane seeps: Rethinking the sphere of influence. *Frontiers in Marine Science, 3*(72), 1–23.

Levin, L. A., Mengerink, K., Gjerde, K. M., Rowden, A. A., Van Dover, C. L., Clark, M. R., Ramirez-Llodra, E., Currie, B., Smith, C. R., Sato, K. N., Gallo, N., Sweetman, A. K., Lily, H., Armstrong, C. W., & Brider, J. (2016b). Defining "serious harm" to the marine environment in the context of deep-seabed mining. *Marine Policy, 74*, 245–259.

Linley, T. D., Stewart, A. L., McMillan, P. J., Clark, M. R., Gerringer, M. E., Drazen, J. C., Fujii, T., & Jamieson, A. J. (2017). Bait attending fishes of the abyssal zone and hadal boundary: Community structure, functional groups and species distribution in the Kermadec, New Hebrides and Mariana trenches. *Deep Sea Research Part I: Oceanographic Research Papers, 121*, 38–53.

Lonsdale, P. (1977). Clustering of suspension-feeding macrobenthos near abyssal hydrothermal vents at oceanic spreading centers. *Deep-Sea Research, 24*(9), 857–863.

Lundsten, L., Barry, J. P., Cailliet, G. M., Clague, D. A., DeVogelaere, A. P., & Geller, J. B. (2009). Benthic invertebrate communities on three seamounts off southern and central California, USA. *Marine Ecology Progress Series, 374*, 23–32.

Lutz, R. A., & Kennish, M. J. (1993). Ecology of deep-sea hydrothermal vent communities: A review. *Reviews of Geophysics, 31*(3), 211–242.

MA (Millennium Ecosystem Assessment). (2005). *Ecosystems and human well-being: Synthesis*. Washington, DC: Island Press.

Mace, G., Bateman, I.J., Albon, S., Balmford, A., Church, A., & Winn, J. (2009). Conceptual framework and methodology. Report to the UK National Ecosystem Assessment.

Maxwell, S., Ehrlich, H., Speer, L., & Chandler, W. (2005). Medicines from the deep: The importance of protecting the high seas from bottom trawling. Natural Resources Defense Council issue paper, (pp. 1–14).

Mayor, D. J., Sanders, R., Giering, S. L., & Anderson, T. R. (2014). Microbial gardening in the ocean's twilight zone: Detritivorous metazoans benefit from fragmenting, rather than ingesting, sinking detritus: Fragmentation of refractory detritus by zooplankton beneath the euphotic zone stimulates the harvestable production of labile and nutritious microbial biomass. *BioEssays, 36*(12), 1132–1137.

McCollom, T. M., & Shock, E. L. (1997). Geochemical constraints on chemolithoautotrophic metabolism by microorganisms in seafloor hydrothermal systems. *Geochimica et Cosmochimica Acta, 61*(20), 4375–4391.

McGeoch, M. A., Butchart, S. H. M., Spear, D., Marais, E., Kleynhans, E. J., Symes, A., Chanson, J., & Hoffmann, M. (2010). Global indicators of biological invasion: Species numbers, biodiversity impact and policy responses. *Diversity and Distributions, 16*(1), 95–108.

Mehta, M. P., Butterfield, D. A., & Baross, J. A. (2003). Phylogenetic diversity of nitrogenase (nifH) genes in deep-sea and hydrothermal vent environments of the Juan de Fuca Ridge. *Applied and Environmental Microbiology, 69*(2), 960–970.

Meyers, M. E. J., Sylvan, J. B., & Edwards, K. J. (2014). Extracellular enzyme activity and microbial diversity measured on seafloor exposed basalts from Loihi Seamount indicate importance of basalts to global biogeochemical cycling. *Applied and Environmental Microbiology, 80*(16), 4854–4864.

Minic, Z., & Thongbam, P. D. (2011). The biological deep sea hydrothermal vent as a model to study carbon dioxide capturing enzymes. *Marine Drugs, 9*(5), 719–738.

Minic, Z., Serre, V., & Hervé, G. (2006). Adaptation of organisms to extreme conditions of deep-sea hydrothermal vents. *Comptes Rendus Biologies, 329*(7), 527–540.

Morato, T., Varkey, D. A., Damaso, C., Machete, M., Santos, M., Prieto, R., Santos, R. S., & Pitcher, T. J. (2008). Evidence of a seamount effect on aggregating visitors. *Marine Ecology Progress Series, 357*, 23–32.

Morato, T., Bulman, C., & Pitcher, T. J. (2009). Modelled effects of primary and secondary production enhancement by seamounts on local fish stocks. *Deep Sea Research Part II: Topical Studies in Oceanography, 56*(25), 2713–2719.

Morato, T., Hoyle, S. D., Allain, V., & Nicol, S. J. (2010). Seamounts are hotspots of pelagic biodiversity in the open ocean. *Proceedings of the National Academy of Sciences, 107*, 9707–9711.

Morgan, R. K. (2012). Environmental impact assessment: The state of the art. *Impact Assessment and Project Appraisal, 30*(1), 5–14.

Morgan, N. B., Cairns, S., Reiswig, H., & Baco, A. R. (2015). Benthic megafaunal community structure of cobalt-rich manganese crusts on Necker Ridge. *Deep Sea Research Part I: Oceanographic Research Papers, 104*, 92–105.

Mullineaux, L. S. (1987). Organisms living on manganese nodules and crusts: Distribution and abundance at three North Pacific sites. *Deep Sea Research Part A. Oceanographic Research Papers, 34*(2), 165–184.

Mullineaux, L. S., Peterson, C. H., Micheli, F., & Mills, S. W. (2003). Successional mechanism varies along a gradient in hydrothermal fluid flux at deep-sea vents. *Ecological Monographs, 73*(4), 523–542.

Myers, R. A., Baum, J. K., Shepherd, T. D., Powers, S. P., & Peterson, C. H. (2007). Cascading effects of the loss of apex predatory sharks from a coastal ocean. *Science, 315*(5820), 1846–1850.

Nakagawa, S., & Takai, K. (2008). Deep-sea vent chemoautotrophs: Diversity, biochemistry and ecological significance. *FEMS Microbiology Ecology, 65*(1), 1–14.

Nellemann, C., Hain, S., & Alder, J. E. (2008). *In dead water – merging of climate change with pollution, over-harvest and infestations in the world's fishing grounds. United Nations Environment Programme*. Norway: GRID-Arendal.

Niner, H. J., Ardron, J. A., Escobar, E. G., Gianni, M., Jaeckel, A., Jones, D. O. B., Levin, L. A., Smith, C. R., Thiele, T., Turner, P. J., Van Dover, C. L., Watling, L., & Gjerde, K. M. (2018). Corrigendum: Deep-sea mining with no net loss of biodiversity – an impossible aim. *Frontiers in Marine Science, 5*(53), 1–12.

NOAA. (1981). *Deep seabed mining. Draft programmatic environmental impact statement*. Washington, DC: Office of Ocean Minerals and Energy.

O, M., Martone, R., Hannah, L., Greig, L., Boutillier, J., Patton, S. (2015). An Ecological Risk Assessment Framework (ERAF) for Ecosystem-based Oceans Management in the Pacific Region. DFO Can. Sci. Advis. Sec. Res. Doc. 2014/072.vii + 59 p.

OSPAR Commission. (2007). Guidelines for the identification and selection of marine protected areas in the OSPAR maritime area. *Update*, pp. 2003–17.

Palumbi, S. R., Sandifer, P. A., Allan, J. D., Beck, M. W., Fautin, D. G., Fogarty, M. J., Halpern, B. S., Incze, L. S., Leong, J. A., Norse, E., & Stachowicz, J. J. (2009). Managing for ocean biodiversity to sustain marine ecosystem services. *Frontiers in Ecology and the Environment, 7*(4), 204–211.

Pearcy, W. G., Stein, D. L., & Carney, R. S. (1982). The deep-sea benthic fish fauna of the northeastern Pacific Ocean on Cascadia and Tufts Abyssal Plains and adjoining continental slopes. *Biological Oceanography, 1*(4), 375–428.

Pereira-Filho, G. H., Amado-Filho, G. M., de Moura, R. L., Bastos, A. C., Guimarães, S. M., Salgado, L. T., Francini-Filho, R. B., Bahia, R. G., Abrantes, D. P., Guth, A. Z., & Brasileiro, P. S. (2011). Extensive rhodolith beds cover the summits of southwestern Atlantic Ocean seamounts. *Journal of Coastal Research, 28*(1), 261–269.

Petersen, J. M., Zielinski, F. U., Pape, T., Seifert, R., Moraru, C., Amann, R., Hourdez, S., Girguis, P. R., Wankel, S. D., Barbe, V., & Pelletier, E. (2011). Hydrogen is an energy source for hydrothermal vent symbioses. *Nature, 476*(7359), 176.

Petersen, S., Krätschell, A., Augustin, N., Jamieson, J., Hein, J. R., & Hannington, M. D. (2016). News from the seabed – geological characteristics and resource potential of deep-sea mineral resources. *Marine Policy, 70*, 175–187.

Petts, J. (1999). *Handbook of environmental impact assessment: Volume 2: Impact and limitations* (Vol. 2). Chichester: Wiley.

Poli, A., Finore, I., Romano, I., Gioiello, A., Lama, L., & Nicolaus, B. (2017). Microbial diversity in extreme marine habitats and their biomolecules. *Microorganisms, 5*(2), 25.

Priede, I. G., Godbold, J. A., Niedzielski, T., Collins, M. A., Bailey, D. M., Gordon, J. D., & Zuur, A. F. (2010). A review of the spatial extent of fishery effects and species vulnerability of the deep-sea demersal fish assemblage of the Porcupine Seabight, Northeast Atlantic Ocean (ICES Subarea VII). *ICES Journal of Marine Science, 68*(2), 281–289.

Ramirez-Llodra, E., Brandt, A., Danovaro, R., De Mol, B., Escobar, E., German, C. R., Levin, L. A., Martinez Arbizu, P., Menot, L., Buhl-Mortensen, P., Narayanaswamy, B. E., Smith, C. R., Tittensor, D. P., Tyler, P. A., Vanreusel, A., & Vecchione, M. (2010). Deep, diverse and definitely different: Unique attributes of the world's largest ecosystem. *Biogeosciences, 7*(9), 2851–2899.

Ressurreição, A., & Giacomello, E. (2013). Quantifying the direct use value of Condor seamount. *Deep Sea Research Part II: Topical Studies in Oceanography, 98*, 209–217.

Ressurreição, A., Gibbons, J., Ponce, T., Kaiser, M., Santos, R. S., & Edwards-Jones, G. (2011). Economic valuation of species loss in the open sea. *Ecological Economics, 70*(4), 729–739.

Rex, M. A., & Etter, R. J. (2010). *Deep-sea biodiversity: Pattern and scale*. Cambridge, MA: Harvard University Press.

Rex, M. A., McClain, C. R., Johnson, N. A., Etter, R. J., Allen, J. A., Bouchet, P., & Warén, A. (2004). A source-sink hypothesis for abyssal biodiversity. *The American Naturalist, 165*(2), 163–178.

Reysenbach, A. L., Liu, Y., Banta, A. B., Beveridge, T. J., Kirshtein, J. D., Schouten, S., Tivey, M. K., Von Damm, K. L., & Voytek, M. A. (2006). A ubiquitous thermoacidophilic archaeon from deep-sea hydrothermal vents. *Nature, 442*(7101), 444.

Rowden, A. A., Schlacher, T. A., Williams, A., Clark, M. R., Stewart, R., Althaus, F., Bowden, D. A., Consalvey, M., Robinson, W., & Dowdney, J. (2010). A test of the seamount oasis hypothesis: Seamounts support higher epibenthic megafaunal biomass than adjacent slopes. *Marine Ecology, 31*, 95–106.

Ruhl, H. A., Ellena, J. A., & Smith, K. L. (2008). Connections between climate, food limitation, and carbon cycling in abyssal sediment communities. *Proceedings of the National Academy of Sciences, 105*(44), 6–11.

Salinas-de-León, P., Phillips, B., Ebert, D., Shivji, M., Cerutti-Pereyra, F., Ruck, C., Fisher, C. R., & Marsh, L. (2018). Deep-sea hydrothermal vents as natural egg-case incubators at the Galapagos Rift. *Scientific Reports, 8*(1), 1788.

Sautya, S., Ingole, B., Ray, D., Stöhr, S., Samudrala, K., Raju, K. K., & Mudholkar, A. (2011). Megafaunal community structure of Andaman seamounts including the Back-arc Basin – a quantitative exploration from the Indian Ocean. *PLoS One, 6*(1), e16162.

SCEP (Study of Critical Environmental Problems). (1970). *Man's impact on the global environment*. Cambridge, MA: MIT Press.

Scheckenbach, F., Hausmann, K., Wylezich, C., Weitere, M., & Arndt, H. (2010). Large-scale patterns in biodiversity of microbial eukaryotes from the abyssal sea floor. *Proceedings of the National Academy of Sciences, 107*(1), 115–120.

Schlacher, T. A., Baco, A. R., Rowden, A. A., O'Hara, T. D., Clark, M. R., Kelley, C., & Dower, J. F. (2014). Seamount benthos in a cobalt-rich crust region of the central Pacific: Conservation challenges for future seabed mining. *Diversity and Distributions, 20*(5), 491–502.

Sewell, B. (2017). America's deep sea treasures: The northeast canyons and seamounts marine national monument. NRDC. Fact sheet. https://www.nrdc.org/sites/default/files/northeast-canyons-and-seamounts-marine-national-monument-fs.pdf. Accessed 28 Oct 2018.

Sharma, R. (2001). Indian deep-sea environment experiment (INDEX): A study for environmental impact of deep seabed mining in Central Indian Ocean. *Deep Sea Research Part II: Topical Studies in Oceanography, 48*(16), 3295–3426.

Sharma, R. (2005). Indian deep-sea environment experiment (INDEX): Monitoring the restoration of marine environment after artificial disturbance to simulate deep-sea mining in central Indian Basin. *Marine Georesources and Geotechnology, 23*, 253–427.

Sharma, R., & Nath, B. N. (2000). Indian BIE: Indian deep-sea environment experiment (INDEX). *Marine Georesources and Geotechnology, 18*, 177–294.

Shulse, C. N., Maillot, B., Smith, C. R., & Church, M. J. (2017). Polymetallic nodules, sediments, and deep waters in the equatorial North Pacific exhibit highly diverse and distinct bacterial, archaeal, and microeukaryotic communities. *Microbiologyopen, 6*, 1–16.

Sinniger, F., Ocaña, O. V., & Baco, A. R. (2013). Diversity of zoanthids (Anthozoa: Hexacorallia) on Hawaiian seamounts: Description of the Hawaiian Gold Coral and additional zoanthids. *PLoS One, 8*(1), e52607.

Sinniger, F., Pawlowski, J., Harii, S., Gooday, A. J., Yamamoto, H., Chevaldonné, P., Cedhagen, T., Carvalho, G., & Creer, S. (2016). Worldwide analysis of sedimentary DNA reveals major gaps in taxonomic knowledge of deep-sea benthos. *Frontiers in Marine Science, 3*, 92.

Smith, K. L. (1992). Benthic boundary layer communities and carbon cycling at abyssal depths in the central North Pacific. *Limnology and Oceanography, 37*(5), 1034–1056.

Smith, K. L., & Druffel, E. R. M. (1998). Long time-series monitoring of an abyssal site in the NE Pacific: An introduction. *Deep-Sea Research II, 45*, 573–586.

Smith, C.R., Drazen, J., & Mincks, S.L. (2006). Deep-sea biodiversity and biogeography: Perspectives from the abyss. International Seabed Authority Seamount Biodiversity Symposium, March 1–13.

Smith, C. R., De Leo, F. C., Bernardino, A. F., Sweetman, A. K., & Arbizu, P. M. (2008a). Abyssal food limitation, ecosystem structure and climate change. *Trends in Ecology and Evolution, 23*(9), 518–528.

Smith, C. R., Levin, L. A., Koslow, A., Tyler, P. A., & Glover, A. G. (2008b). The near future of the deep-sea floor ecosystems. In N. V. Polunin (Ed.), *Aquatic ecosystems: Trends and global prospects* (pp. 334–350). Cambridge, UK: Cambridge University Press.

Smith, K. L., Ruhl, H. A., Bett, B. J., Billett, D. S. M., Lampitt, R. S., & Kaufmann, R. S. (2009). Climate, carbon cycling, and deep-ocean ecosystems. *Proceedings of the National Academy of Sciences, 106*(46), 19211–19218.

Snelgrove, P. V. (1999). Getting to the bottom of marine biodiversity: Sedimentary habitats: Ocean bottoms are the most widespread habitat on earth and support high biodiversity and key ecosystem services. *BioScience, 49*(2), 129–138.

Sogin, M. L., Morrison, H. G., Huber, J. A., Welch, D. M., Huse, S. M., Neal, P. R., Arrieta, J. M., & Herndl, G. J. (2006). Microbial diversity in the deep sea and the underexplored "rare biosphere". *Proceedings of the National Academy of Sciences, 103*(32), 2115–12120.

SPRFMO. (2008). Revised draft bottom fishery impact assessment standard. Eighth International Meeting: Science Working Group, SP-08-SWG-DW-01, 39pp.

Suzuki, K., Yoshida, K., Watanabe, H., & Yamamoto, H. (2018). Mapping the resilience of chemosynthetic communities in hydrothermal vent fields. *Scientific Reports, 8*(1), 9364.

Thornburg, C. C., Zabriskie, T. M., & McPhail, K. L. (2010). Deep-sea hydrothermal vents: Potential hot spots for natural products discovery? *Journal of Natural Products, 73*(3), 489–499.

Thurber, A. R., Sweetman, A. K., Narayanaswamy, B. E., Jones, D. O. B., Ingels, J., & Hansman, R. L. (2014). Ecosystem function and services provided by the deep sea. *Biogeosciences, 10*(11), 18193–18240.

Tilman, D., Knops, J., Wedin, D., Reich, P., Ritchie, M., & Siemann, E. (1997). The influence of functional diversity and composition on ecosystem processes. *Science, 277*(5330), 1300–1302.

Tittensor, D. P., Baco, A. R., Brewin, P. E., Clark, M. R., Consalvey, M., Hall-Spencer, J., Rowden, A. A., Schlacher, T., Stocks, K. I., & Rogers, A. D. (2009). Predicting global habitat suitability for stony corals on seamounts. *Journal of Biogeography, 36*(6), 1111–1128.

Tracey, D. M., Rowden, A. A., Mackay, K. A., & Compton, T. (2011). Habitat-forming cold-water corals show affinity for seamounts in the New Zealand region. *Marine Ecology Progress Series, 430*, 1–22.

UNGA 69/245. (2015). Resolution adopted by the General Assembly on 29 December 2014. 69/245. Oceans and law of the sea. 69th session. Agenda Item 74(a). 24 February 2015.

UNGA 70/235. (2016). Resolution adopted by the General Assembly on 23 December 2015. 70/235. 70th session. Agenda Item 79 (a). 15 March 2016.

Van Dover, C. (2000). *The ecology of deep-sea hydrothermal vents*. Princeton: Princeton University Press.
Van Dover, C. L. (2011). Mining seafloor massive sulphides and biodiversity: What is at risk? *ICES Journal of Marine Science, 68*(2), 341–348.
Van Dover, C. L. (2014). Impacts of anthropogenic disturbances at deep-sea hydrothermal vent ecosystems: A review. *Marine Environmental Research, 102*, 59–72.
Van Dover, C. L., & Lutz, R. A. (2004). Experimental ecology at deep-sea hydrothermal vents: A perspective. *Journal of Experimental Marine Biology and Ecology, 300*(1–2), 273–307.
Van Dover, C. L., German, C. R., Speer, K. G., Parson, L. M., & Vrijenhoek, R. C. (2002). Evolution and biogeography of deep-sea vent and seep invertebrates. *Science, 295*(5558), 1253–1257.
Van Dover, C. L., Smith, C. R., Ardron, J., Arnaud, S., Beaudoin, Y., Bezaury, J., Boland, G., Billet, D., Carr, M., Cherkashov, G., Cook, A., De Leo, F., Dunn, D., Fisher, C. R., Godet, L., Gjerde, K., Halpin, P., Levin, L., Lodge, M., Menot, L., Miller, K., Milton, D., Naudts, L., Nugent, C., Pendleton, L., Plouviez, S., Rowden, A., Santos, R., Shank, T., Smith, S., Tao, C., Tawake, A., Thurnherr, A., & Treude, T. (2011). *Environmental management of deep-sea chemosynthetic ecosystems: Justifications of and considerations for a spatially based approach* (ISA Technical study; no. 9). Kingston: International Seabed Authority. 90pp.
Van Dover, C. L., Ardron, J. A., Escobar, E., Gianni, M., Gjerde, K. M., Jaeckel, A., Jones, D. O. B., Levin, L. A., Niner, H. J., Pendleton, L., Smith, C. R., Thiele, T., Turner, P. J., Watling, L., & Weaver, P. P. E. (2017). Biodiversity loss from deep-sea mining. *Nature Geoscience, 10*(7), 464–465.
Van Dover, C. L., Arnaud-Haond, S., Gianni, M., Helmreich, S., Huber, J. A., Jaeckel, A. L., Metaxas, A., Pendleton, L. H., Petersen, S., Ramirez-Llodra, E., & Steinberg, P. E. (2018). Scientific rationale and international obligations for protection of active hydrothermal vent ecosystems from deep-sea mining. *Marine Policy, 90*, 20–28.
Vanreusel, A., Hilario, A., Ribeiro, P. A., Menot, L., & Arbizu, P. M. (2016). Threatened by mining, polymetallic nodules are required to preserve abyssal epifauna. *Scientific Reports, 6*, 26808.
Veillette, J., Sarrazin, J., Gooday, A. J., Galéron, J., Caprais, J. C., Vangriesheim, A., Étoubleau, J., Christian, J. R., & Juniper, S. K. (2007). Ferromanganese nodule fauna in the tropical North Pacific Ocean: Species richness, faunal cover and spatial distribution. *Deep Sea Research Part I: Oceanographic Research Papers, 54*(11), 1912–1935.
Verlaan, P. A. (1992). Benthic recruitment and manganese crust formation on seamounts. *Marine Biology, 113*(1), 171–174.
Walsh, R. G., Loomis, J. B., & Gillman, R. A. (1984). Valuing option, existence, and bequest demands for wilderness. *Land Economics, 60*(1), 14–29.
Wang, X., & Müller, W. E. (2009). Marine biominerals: Perspectives and challenges for polymetallic nodules and crusts. *Trends in Biotechnology, 27*(6), 375–383.
Wang, X. H., Schloßmacher, U., Natalio, F., Schröder, H. C., Wolf, S. E., Tremel, W., & Müller, W. E. (2009). Evidence for biogenic processes during formation of ferromanganese crusts from the Pacific Ocean: Implications of biologically induced mineralization. *Micron, 40*(5–6), 526–535.
Wang, X., Wiens, M., Schröder, H. C., Schloßmacher, U., & Müller, W. E. (2011). *Molecular biomineralization: Toward an understanding of the biogenic origin of polymetallic nodules, seamount crusts, and hydrothermal vents. Molecular biomineralization* (pp. 77–110). Berlin: Springer.
Watson, R., Kitchingman, A., & Cheung, W. (2007). *Catches from world seamount fisheries. Seamounts: Ecology, fisheries and conservation* (pp. 400–412). Oxford: Blackwell.
Wattage, P., Glenn, H., Mardle, S., Van Rensburg, T., Grehan, A., & Foley, N. (2011). Economic value of conserving deep-sea corals in Irish waters: A choice experiment study on marine protected areas. *Fisheries Research, 107*(1–3), 59–67.
Webb, T. J., Berghe, E. V., & O'Dor, R. (2010). Biodiversity's big wet secret: The global distribution of marine biological records reveals chronic under-exploration of the deep pelagic ocean. *PLoS One, 5*(8), e10223.

Wedding, L. M. M., Friedlander, A. M. M., Kittinger, J. N. N., Watling, L., Gaines, S. D. D., Bennett, M., Hardy, S. M. M., & Smith, C. R. R. (2013). From principles to practice: A spatial approach to systematic conservation planning in the deep sea. *Proceedings of the Royal Society B: Biological Sciences, 280*(1773), 1–37.

Wery, N., Cambon-Bonavita, M. A., Lesongeur, F., & Barbier, G. (2002). Diversity of anaerobic heterotrophic thermophiles isolated from deep-sea hydrothermal vents of the mid-Atlantic ridge. *FEMS Microbiology Ecology, 41*(2), 105–114.

Weslawski, J. M., Snelgrove, P. V., Levin, L. A., Austen, M. C., Kneib, R. T., Iliffe, T. M., Carey, J. R., Hawkins, S. J., & Whitlatch, R. B. (2004). *Marine sedimentary biota as providers of ecosystem goods and services. Sustaining biodiversity and ecosystem services in soils and sediments* (pp. 73–98). Washington, DC: Island Press.

Westman, W. E. (1977). How much are nature's services worth? *Science, 197*(4307), 960–964.

White, M., Bashmachnikov, I., Arístegui, J., & Martins, A. (2007). Physical processes and seamount productivity. In *Seamounts: ecology, fisheries and conservation* (pp.65–84).

Wiklund, H., Taylor, J. D., Dahlgren, T. G., Todt, C., Ikebe, C., Rabone, M., & Glover, A. G. (2017). Abyssal fauna of the UK-1 polymetallic nodule exploration area, Clarion-Clipperton Zone, central Pacific Ocean: Mollusca. *ZooKeys, 2017*(707), 1–46.

Witherell, D., & Coon, C. (2001). Protecting gorgonian corals off Alaska from fishing impacts. In J. H. M. Willison, J. Hall, S. Gass, E. Kenchington, M. Butler, & P. Doherty (Eds.), *First international symposium on deep-sea corals* (pp. 117–125). Halifax: Ecology Action Centre.

Worm, B., Lotze, H. K., & Myers, R. A. (2003). Predator diversity hotspots in the blue ocean. *Proceedings of the National Academy of Sciences, 100*(17), 9884–9888.

Worm, B., Barbier, E. B., Beaumont, N., Duffy, J. E., Folke, C., Halpern, B. S., Jackson, J. B., Lotze, H. K., Micheli, F., Palumbi, S. R., & Sala, E. (2006). Impacts of biodiversity loss on ocean ecosystem services. *Science, 314*, 787–790.

Yesson, C., Clark, M. R., Taylor, M. L., & Rogers, A. D. (2011). The global distribution of seamounts based on 30 arc seconds bathymetry data. *Deep Sea Research Part I: Oceanographic Research Papers, 58*(4), 442–453.

Zierenberg, R. A., Adams, M. W., & Arp, A. J. (2000). Life in extreme environments: Hydrothermal vents. *Proceedings of the National Academy of Sciences, 97*(24), 12961–12962.

Dr. Kate J. Thornborough is a marine scientist with experience working for federal and state government agencies (Fisheries and Oceans Canada and NSW Department of Primary Industries – Fisheries), a university research initiative (Ocean Networks Canada, University of Victoria), and a university (University of Sydney). Her work focuses on developing and implementing programmes and tools to provide the best science advice to managers, including ecological risk assessments, indicator selection frameworks, ecological vulnerability assessments, monitoring strategies and plans, and georeferenced databases for spatial management. She has worked on coral reefs in the Great Barrier Reef and the South China Sea, hydrothermal vents, seamounts, glass sponge reefs, and the coastal waters of Canada and Australia. Kate is currently an independent researcher and also works for NSW Department of Primary Industries – Fisheries in Australia on a National Environmental Science Program project to develop guidelines for the analysis of cumulative effects and risks.

Dr. S. Kim Juniper is Chief Scientist with Ocean Networks Canada, a University of Victoria-based organisation that operates cabled ocean observatories in the Pacific and Arctic Oceans. He is also a professor in UVic's School of Earth and Ocean Sciences and holder of the British Columbia Leadership Chair in Ocean Ecosystems and Global Change. He has authored more than 120 peer-reviewed publications on the microbiology, biogeochemistry and ecology of deep-sea hydrothermal vents and other marine habitats. He has provided occasional scientific expertise to the International Seabed Authority since 2000 and co-authored the Marine Genetic Resources chapter in the recent UN World Ocean Assessment. He is a member of the Scientific Advisory Committee for the Canadian Healthy Oceans Network, a government-academia partnership that is researching the ecological importance of biodiversity in Canada's three oceans.

Dr. Samantha Smith (samantha@blueglobesolutions.com) has 20 years experience conducting environmental assessments in a number of different countries around the world, over five continents, and has 13 years experience working with the deep seafloor minerals sector. Samantha has a PhD in Environmental Biology/Biogeochemistry from the University of Bristol (UK) and is a Director and Former President of the International Marine Minerals Society and an International Fellow of AusIMM. She is also the former Vice President of Corporate Social Responsibility for Nautilus Minerals Inc. During her 9-year tenure with Nautilus, Samantha led an international, multidisciplinary team to complete the Solwara 1 Environmental Impact Statement and deliver the world's first environmental permit for seafloor massive sulphide extraction. Since 2014, Samantha has run the environmental consultancy Blue Globe Solutions, based in Canada, consulting to various entities in the marine mineral space, including industry and government, for projects both within national jurisdiction and beyond. Much of Samantha's time is spent advising and consulting to Nauru Ocean Resources Inc., an ISA Contractor and fully owned subsidiary of DeepGreen Metals Inc. Samantha is a co-author of several peer-reviewed manuscripts related to deep-sea environmental management.

Lynn-Wei Wong received her BSc in Biology with a Marine Concentration from the University of Victoria and BSc (Hons) in Accounting and Finance from Warwick Business School at the University of Warwick. She currently assists in research relating to the quantification of tropical urban coastal ecosystem services at the Asian School of the Environment (ASE) at Nanyang Technological University of Singapore. Prior to joining ASE, Lynn-Wei compiled The State of the Oceans and Coasts Report with a blue economy theme for Singapore's submission to the Partnership in Environmental Management for the Seas of East Asia (PEMSEA) under the Coastal and Marine Branch at the National Biodiversity Centre, National Parks Board.

Technologies for Safe and Sustainable Mining of Deep-Seabed Minerals

Sup Hong, Hyung-Woo Kim, Taekyung Yeu, Jong-Su Choi, Tae Hee Lee, and Jong-Kap Lee

Abstract Safety and sustainability are both critical for the sake of profitability in deep-seabed mining. Environmental impacts and potential harms to ecosystems would be caused basically by benthic intervention and materials transportation. A commercial mining has to be based on minimizing of the environmental impacts. For safe and sustainable mining, the entire mining system should be designed with the objectives of the production efficiency of minerals and the feasibility of treatment of by-products (seawater, sediment) and operated in an integrated and smart fashion. Utilization of techniques of simulation-based design (SBD) and multidisciplinary design optimization (MDO) and also implementation of underwater robot technology are required to ensure operability and sustainability as well as to reduce development risks as well.

Keywords Benthic intervention · By-products · Simulation-based design (SBD) · Multidisciplinary design optimization (MDO) · Automation · Tailings treatment

1 Introduction

Polymetallic nodules (PMN) are the most explored resource among deep-seabed minerals. Their deposits are found at upper sediment layers of abyssal basins (see Fig. 1). The "Nodule Belt" in Clarion-Clipperton Zone of Northeast Pacific is the biggest area with high abundance. The metals and minerals contained in PMN are being highlighted as so-called energy metals, which are widely utilized in energy harvesting and generation, energy transportation, and energy saving. Those are mostly trace and rare earth metals and include elements such as *manganese*, *nickel*,

S. Hong (✉) · H.-W. Kim · T. Yeu · J.-S. Choi · J.-K. Lee
Korea Research Institute of Ships and Ocean Engineering (KRISO), Deajeon, Republic of Korea
e-mail: suphong@kriso.re.kr

T. H. Lee
Hanyang University, Seoul, Republic of Korea

R. Sharma (ed.), *Environmental Issues of Deep-Sea Mining*,
https://doi.org/10.1007/978-3-030-12696-4_5

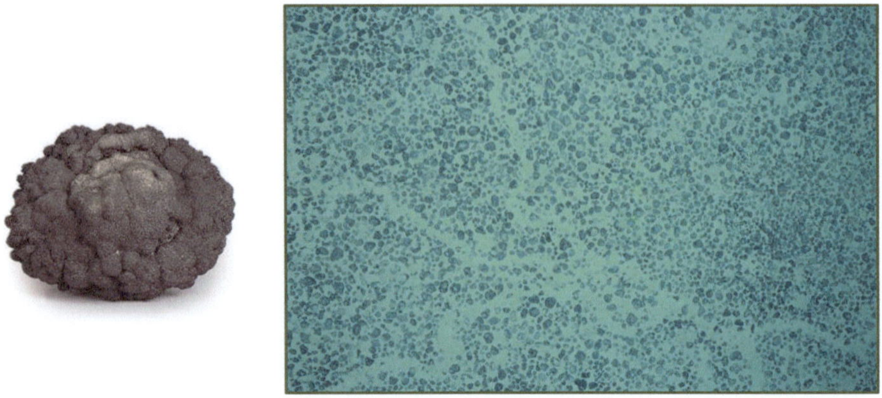

Fig. 1 Polymetallic nodules sample and as seen in seabed photograph. (Courtesy: KIOST)

cobalt, *copper*, *molybdenum*, *tellurium*, *selenium*, *platinum*, *lanthanide*, *neodymium*, *praseodymium*, *cerium*, etc. Toward the low carbon society, for global sustainability, the role of metals and minerals is growing in the industries of energy and transportation (World Bank Group 2017). In this context, the approaches for the energy metals from deep-seabed have to be based on the global sustainability as well.

Deep-seabed mining (DSM) of polymetallic nodules will be a quite complex and challenging operation. For technological feasibility of these operations, the commonly asked questions are the following: How to dredge the ores from seafloor? How to transport them up to the sea surface? What kinds of equipment should be applied? How to operate these equipment? How to reduce and manage the undesired by-products? How to minimize the sediment plume and control the dispersion? How serious will be the environmental impacts and potential harms?

What is also critical is that equipment and materials of entire subsystems will be exposed all the time to wear and tear due to corrosion, collision, friction, and fatigue. Regular inspection, maintenance, and repair are highly mandatory. Some core parts of equipment and subsystems could be resolved by applying the industry standards. Nonetheless, the underwater mining equipment and systems have not been fully tested and proven through field operations yet anywhere in the world.

The first primary factor in selection of concept and scenario of DSM is the ultra-deep water depth of 5000 m. It causes a highly nonlinear and coupled dynamic behavior of underwater systems, which is critical for control of relative positions between subsystems. The second one is the geological setting that polymetallic nodules are deposited on the soft and cohesive sediment on seafloor. Collection of nodules from the top of this fluffy sediment layer with high water content is a huge challenge to seafloor dredging operations.

Simultaneous accomplishment of profitability and sustainability in DSM with that complexity will be an extraordinary challenging mission. It should be treated based on multi- and interdisciplinary approaches. Occurrence of downtimes in mining operation has to be minimized. With the aims of safety, sustainability, and profitability, it is proposed to operate the entire mining procedure by integrated means and in a smart fashion.

A concept of continuous mining should take account of the safe and reliable operations of underwater system including launching-and-retrieval (L&R) and maintenance-and-repair (M&R) and of the seawater treatment and tailing process on board. The safety and reliability contains a wide spectrum ranging from design and operation to risk management.

For sustainability, requirement of continuous mining should take into account safe and reliable operations of all underwater systems, including launching-and-retrieval (L&R) and maintenance-and-repair (M&R), as well as of seawater treatment and tailing discharge on board. The safety and reliability of such systems encompasses a wide spectrum ranging from design and operation to risk management.

In this chapter, initially, the functional requirements that are mandatory and prerequisite for the sake of safe, profitable, and sustainable DSM are reviewed. Subsequently, it is described about the concepts of continuous mining as well as design and engineering technologies. Thereafter, some operations that have been conducted recently are discussed in regard to safety and sustainability in deep-sea mining.

2 Functional Requirements for Deep-Sea Mining

Functional requirements describe the necessary aspects required to achieve the comprehensive performances of DSM and will provide the basis for making decision of mining concept. The most important requirements are *annual production*, *safety*, and *sustainability*, to be considered by the contractor(s) based on the concerned regulations. Functional requirements shall be based on operational feasibility targeting fulfillment of these requirements.

2.1 Production of Functional Requirements

The *Functional Requirements* (FR) are expressed in form of a hierarchical structure with multiple aspects having multiple levels (see Fig. 2). Production of FR is the utmost critical step influencing the success of subsequent procedures, and so it has to be considered meticulously.

Function domain is built up by *Functional Parameters* (FPs) corresponding to FR. The FPs define the factors to accomplish the goals of sustainable commercial mining. Design domain will correspond to the function domain, where *Design Parameters* (DPs) are defined to fulfil the desired performance values of the FP of each level. This relationship can be schematically shown as Fig. 3.

The functional requirements for deep-seabed mining could be described as in Table 1, where FR are limited within three levels and the third level is expressed by representative feature. The FR of each level are parameterized in Table 2 in order to

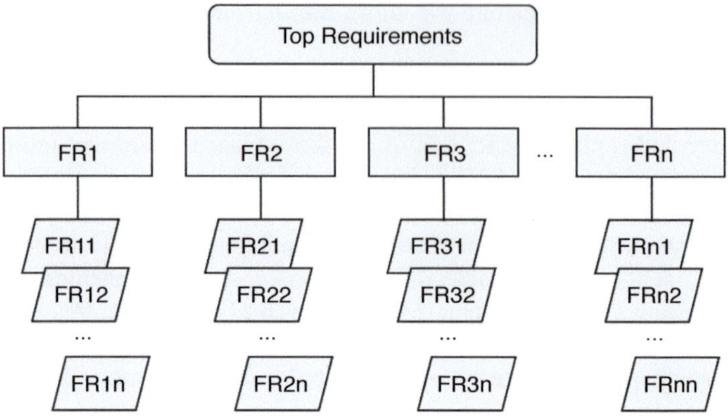

Fig. 2 Hierarchy of functional requirements

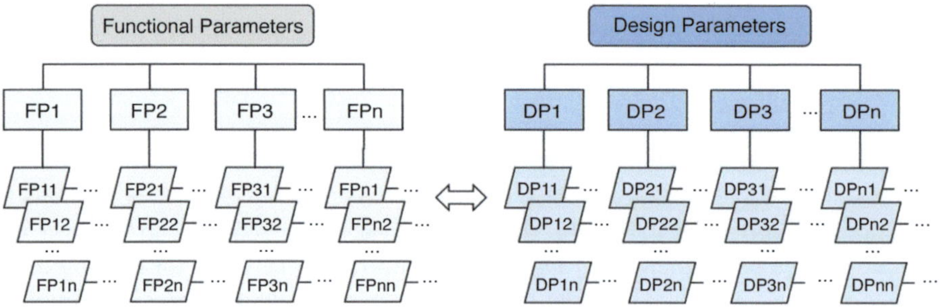

Fig. 3 Corresponding hierarchies between FPs (left) and DPs (right)

define the design domain. The DPs are determined through participation of experts and engineers of various disciplines. *Design Variables* would be defined corresponding to the DP in turn.

2.2 *Hazards and Risks in Deep-Seabed Mining*

The intrinsic hazards have to be resolved by means of an appropriate operational concept, development of technologies, utilization of proven technologies, and application of industrial standards. Risks anticipated in DSM are attributed primarily to scenario of mining operation, which stems fundamentally from investigation of technological viability.

Hazard identification and risk assessment are needed prior to the development of not existing (or unproven) technologies. Participation and role of experts group are utmost critical in this phase. Various kinds of design and analysis schemes and concerned field experiences are required in order to find appropriate means and

Table 1 Functional requirements for deep-seabed mining

Level 1	Level 2	Level 3	Remarks
To be capable of eco-friendly collecting ores from seafloors into buffering station	Moving on seafloors along designed paths	Automatic or remote control	Dependent on geological setting
	Dredging the ores from seafloors	Automatic or remote control	Dependent on geological setting
	Fragmentizing the dredged ores	Optimum fragmentizing of ores	Materials durability
	Discharging the fragmented ores to buffering station	Prevention of congestion/clogging	Coping with inconstant dredging
	Reducing the disturbances of benthic zones	Prevent excessive disturbances	As low as possible
To be capable of safe and effective vertical transportation of the collected ores to sea surface	Conveying the discharged ores into buffering station	Preventing clogging	Optimum flexible/conduit design
	Optimum operation of vertical lifting system	Optimum design of lifting system	Slurry flow
	Maximizing the mass flowrate of ore transportation	Feeding control of ores	Max. volume concentration
	Avoiding failures in vertical transportation	Preventing overloadings	Anti-clogging, bypass
To be capable of ore-seawater separation and of eco-friendly tailing	Separation of seawater matching with pump flowrate	Multi-stage separations	Prevention of overflow
	Treatment of separated seawater to allowable criteria	Satisfaction of eco-criteria	To be defined
	Eco-friendly tailing of treated seawater	Injection below thermocline depth	To be determined
To be capable of storage and offloading of the separated ores	Conveying the separated ores into cargo holds	Loading procedure	Weight-and-ballast calculation
	Storage management of the ores in cargo holds	Prevention from agglomeration	Drying or dehumidifying
	Offloading ores to ore-carriers	Offloading with mining operation	Tandem positioning
To be capable of supporting the underwater mining operations	Launching and retrieval of underwater systems	Heave compensation, automation	Safety and contingency plan
	Power generation, supply and control	Maximizing efficiency	Contingency plan
	Dynamic positioning relative to underwater systems	DP2-class	Safety and contingency plan
	Maintenance and repairs	Regular inspection procedures	Redundancy and contingency
	Underwater navigation and tele-communication	High-precision, real-time comm.	LBL, USBL
	Provision of accommodation, foods, and fresh water	Two-shift system	Emergency plan

Table 2 Functional parameters for deep-seabed mining referred to Table 1

Level 1	Level 2	Level 3	Remarks
Collecting capacity of ores	Capacity of moving on seafloors	Mobility and controllability; speed, steering, stability	Sustainability criteria
	Capacity of seafloor dredging	Ore dredging efficiency, amount of plume generation	Sustainability criteria
	Capacity of ore crushing	Optimum size of fragmented ores	Material durability
	Capacity and efficiency of discharge of ore fragments	Max. ore flowrate, max. volume concentration	Robust performance
	Capacity of minimizing benthic intervention, max. overall efficiency of ore collecting	Area sweep efficiency, penetration into substrata, spreading range of plume, co-transportation of sediments, noise levels, etc.	Sustainability criteria
Lifting capacity of ores	Capacity of ore influx into buffering station	Capacity of buffering and sediment plume discharge	Sediment plume extraction
	Optimum performance of vertical lifting system	Head, flowrate, pipe diameter, ore passage, working depth	Material durability
	Max. volume concentration of ores	Capacity and controllability of ore feeding	Flow assurance
	Capacity of overload monitoring	Capability of overloading prevention; bypass	Safety criteria
Treatment capacity of ore-seawater mixture	Capacity of seawater separation	Seawater separation flowrate via multistage	Matching with pump flowrate
	Capacity of separated seawater treatment	Seawater treatment flowrate to allowable criteria	Eco-criteria to be defined
	Capacity of eco-friendly seawater tailing	Flowrate, pressure loss, injection water depth	Below thermocline depth
Storage and offloading capacity of ores	Capacity of conveying separated ores	Automatic control of ore conveying into cargo holds	Weight-and-ballast calculation
	Capacity of management of ores in cargo holds	Power to dry (or dehumidify) ores in cargo holds	Design of cargo holds
	Capacity of ore offloading	Simultaneous offloading with mining operation	Offloading operation sea-state
Support capacity for underwater operations	Capacity of launch-and-retrieval equipment	Sea-state limit, automation ratio	Safety and minimizing downtime
	Capacity of power generation, supply, and control	Efficiency and optimization	Design of contingency backup
	Capacity of dynamic positioning	Sea-state limit	Safety and contingency plan
	Capacity of maintenance-and-repair on board	Inspection standards/manual, spares	Minimizing downtime
	Capacity of UW navigation and telecommunication	Precision, latency, robustness, bandwidth	Sensor fusion algorithm
	Capacity of accommodation, foods, and fresh water	Capability of two-shift system	Health, safety, environment

measures for reduction of risks and damage. Qualitative description about hazards and risks is followed by risk assessment identifying levels and scopes of risks and damages. Table 3 shows an example of qualitative descriptions of hazards and risks of a continuous mining scenario using hydraulic lifting.

The functional requirements, the hazard identification and risk assessment, as well as the subsequent concept design are interrelated so that *iterative procedure* will be necessary until all kinds of risks are estimated as acceptable. Various kinds of countermeasures will be required to reduce the risks below certain levels.

Among the functional requirements, those ones related with risks and damages to ecosystems can be categorized in *Sustainability Requirements* and are described in the following section.

2.3 *Sustainability Requirements*

In deep-seabed mining, the sustainability focuses on the impacts by mining operations and the resulting potential harms to ecosystems in benthic zones and water column. The environmental impacts are caused basically by intervention in benthic zones and transportation of materials (Hong and Yang 2018).

Benthic Intervention

The collecting process of polymetallic nodules from seafloor sediment layer requires a collecting machine, crawling and dredging the nodules from seafloor. This process will cause inevitable impacts on benthic ecosystem. The upper layers of sediment, swept by a collecting machine, will be destroyed to a certain depth. An ideal condition of picking-up only nodules from sediments would not be feasible in respect of commercial mining.

The primary concern about seabed sustainability is the size of mine site, which is calculated by Eq. (1):

$$A_s = \frac{A_r \cdot D}{A_n \cdot E \cdot M} \quad (1)$$

where:

A_s is the size of mine site (km^2)
A_r is annual nodule production rate (dry tons/year)
D is duration of mining operation (years)
A_n is average nodule abundance in mineable area (kg/m^2)
E is overall efficiency of collector (%)
M is mineable area (km^2) (UNOET 1987)

Table 3 A general description about hazards and risks in deep-seabed mining

Activity	Hazards	Risks	Damages
On-board operation	Handling heavy weights: moving, hanging	Collision and/or drop accidents Damage(s) of equipment Mortality and/or injury accidents	Loss of costs and time
	Repeated launch-and-retrievals of underwater system	Connection/disconnection failures of pipe units and cables Loss of underwater system(s) Damage(s) of equipment Mortality and/or injury accidents	Loss of costs and time
	Use of high power capacity equipment and systems	Power interruption or overload Damage(s) of equipment and/or system Fire and/or explosion accidents Mortality and/or injury accidents	Loss of costs and time Environmental impact
	Necessity of a large number of various signals for communication and control	Loss of signal(s) Stop of mining operation Damage(s) of equipment Loss of system	Loss of costs and time
Collecting	Moving on soft and sticky sediment seafloor	Excessive digging and penetration into sediments Reduction and/or loss of mobility and production rate Damage or loss of machine Total stop of mining operation	Loss of costs Increase of downtime Environmental impact
	Dredging nodules half-buried in sediments	Loss of dredging capability Over-dredging Reduction of production rate Excessive generation of plumes	Loss of costs Environmental impact
	Necessity of crushing nodules	Crushing failure Clogging in collector Damage of crusher/collector Total stop of mining operation	Loss of costs Increase of downtime
	Necessity of discharging dredged nodules to prevent overweight of collector	Discharging failure Clogging in collector Clogging in flexible conduit Total stop of mining operation	Loss of costs Increase of downtime
Lifting	Necessity to support the total weight and tension force of free-hanging lifting pipeline	Failure of lifting pipeline Loss of underwater system(s) Damage of mining vessel Total stop of mining operation Mortality or injury accidents	Loss of costs and time Environmental impact
	Continuous vertical transportation of slurry through lifting pipeline	Pipe and/or pump clogging Damage of pump station(s) Total stop of mining operation	Loss of costs and time
	Use of high power underwater pump stations	Electric shortage Occurring of water hammering Damage and failure of pump station(s) Total stop of mining operation	Loss of costs and time Environmental impact

Table 3 (continued)

Activity	Hazards	Risks	Damages
Positioning	Keeping positions in harsh environmental conditions	Failure in DPS Total stop of mining operation Damage and/or loss of underwater system Damage of mining vessel (derrick)	Loss of costs and time Environmental impact
	Dependent on GPS and underwater localization system, and signal delay	Irrelevant positioning Decrease of production rate Damage of flexible conduit and umbilical Damage and/or loss of collector	Loss of costs and time Environmental impact
On-board treatments	Massive transportation of ores-seawater-sediment mixture	Clogging or overflow in ores-seawater separation Flooding of mining vessel Total stop of mining operation Damage of on-board facility Accident of lives	Loss of costs and time Environmental impact
	Massive seawater amount to be treated	Overflow in seawater treatment Flooding of mining vessel Total stop of mining operation Damage of on-board facility Accident of lives	Loss of costs and time Environmental impact
	Necessity of sediments separation from seawater	Overload in sediment caking [optional] Delay of mining operation Damage of on-board facility	Loss of costs and time Environmental impact
	Massive tailing amount to treat	Overload and/or failure of tailing treatment Flooding of mining vessel Total stop of mining operation Damage of on-board facility Accident of lives	Loss of costs and time Environmental impact
	Dumping out tailing below thermocline water depth	Loss of power for tailing treatment Failure of tailing pipeline Total stop of mining operation	Loss of costs and time Environmental impact
	Repeats of loading and offloading operations of huge amount of nodules in wet-and-dirty condition	Congestion in ore storage Agglomeration of ores in cargo holds Delay and/or failure of offloading Stop of mining operation	Loss of costs and time

Higher is the overall efficiency (E), smaller will be the size of mine site (A_s). Of course, it will make profitability higher. The overall efficiency (E) of a collecting system is estimated as:

$$E = e_d \times e_s \tag{2}$$

where:

e_d is nodule dredging efficiency
e_s is area sweep efficiency

Dredging Efficiency

The nodule dredging efficiency (e_d) is the ratio of minerals gathered by dredge head versus the minerals on the seafloor initially. There are various kinds of dredge head designs showing e_d over 80%: principally classified in hydraulic, mechanical, and hybrid designs (Oebius 1993). However, those were mostly tested in laboratory conditions, since it is not easy to define the nodule coverage in real conditions.

The dredging efficiency wouldn't be sufficient for sustainability. An optimum operation condition has to be found in order to satisfy the required efficiency and to minimize the benthic impacts, e.g., generation and dispersion of sediment plumes. As an example, a hydraulic dredge head was tested by Hong et al. (1999), where the term "pick-up" was used instead of dredge. It was shown that a flowrate ratio (q_r) defined as Eq. (3) has correlation with the dredging efficiency.

$$q_r = \frac{q_w}{v \cdot h} \tag{3}$$

where:

v is the forward speed of dredge head
q_w is the water-jet flowrate per unit breadth
h is the gap between dredge head and seafloor

An optimum control of h and q_w with respect to the forward speed (v) can reduce generation of plume and energy consumption and secure productivity.

Sweep Efficiency

The area sweep efficiency (e_s) is the area percentage of the seafloor swept by collecting machine with respect to the designed area of mine site. The area sweep efficiency will be governed by mobility and controllability of collector vehicle, which is utmost important for delineation of mine sectors and design of mining paths in each mine sector. For the sake of high area sweep efficiency, an automatic path tracking of collector vehicle, based on high-precision seafloor navigation system, will be mandatory. Mobility of collector vehicle, represented by *maximum speed*, *minimum turning radius*, *maximum upslope*, and *relevant sediment penetration*, and precision of tracking control algorithm are the most critical points for development of collector vehicle.

If e_s is around 60%, the overall efficiency (E) will be about 50%. On the other hand, in case e_s is 80%, the E of collecting system will rise up to 64%. This makes a big difference in the aspect of sustainability: reduction of 25% in the size of mine site (A_s).

Because the sediment on the seafloor is very sensitive and the remolded strength of sediments is quite low, an overlapping of mining tracks should be prevented for securing seafloor mobility. In order to maximize the area sweep efficiency (e_s) and the overall efficiency (E), an optimum design of mining paths is utmost important together with the performance of collector vehicle.

In order to reduce the scope of benthic intervention, it is necessary to have a collecting system of high overall efficiency. The major factors for enhancement of the overall efficiency (E) are summarized as follows (Hong and Yang 2018):

- Optimum design of dredge head and collector vehicle
- Optimum operation of collector
- Robust algorithm for path tracking control
- Optimum design of mining paths based on mobility and controllability

Moreover, the collecting system needs to be designed as a robot system capable of automatic path tracking, since a remote control will be totally paralyzed due to sediment plumes.

Materials Transportation

The materials transported in DSM will be typically nodules, seawater, and sediments. Whereas nodules are the target material, the others are undesired by-products. Sediments and seawater when mixed together form sediment plumes. One part of the sediment plumes is spread out horizontally in benthic zone, and the other part enters to lifting line together with nodules.

The horizontal dispersion of sediment plume will resettle back to seafloor within certain distances. The vertical transportation of sediments has to be reduced as much as possible before reaching to water surface. The functional parameters of collecting and lifting processes regarding with discharging of sediment plume as in Table 2 are based on this aspect. The sediments lifted up to the mining vessel have to be treated to satisfy an (*to be defined*) environmental regulation.

Vertical Ore Lifting

Vertical lifting of nodules requires the maximum energy in the entire mining operation. Hydraulic lifting, air-lifting, and mechanical lifting have been studied by a large number of academicians, researchers, and experts. In the aspects of durability, reliability, and safety, hydraulic lifting and air-lifting are front-runners for commercial application. Hydraulic lifting has comparatively higher efficiency but has disadvantages in durability and maintenance-and-repair of underwater pump stations. Air-lifting has advantages of maintenance-and-repair and with relatively less volume of water but has disadvantages in efficiency and noise. Also, it is difficult to foresee other problems of durability and safety due to drastic expansion of injected air bubbles and consequent impacts and collisions.

The efficiency of hydraulic lifting, typically represented by maximum *volume concentration* of ores in pipe, is related with energy consumption and by-products (seawater and sediments) required for ore transportation.

In the aspects of profitability and sustainability, the consequences of increasing efficiency are as follows:

- Cost reduction (CAPEX, OPEX) for on-board treatment facilities and mining vessel
- Reduction of by-products and tailing

It is highly recommended for safety and efficiency to operate the lifting pumps constantly in optimum design condition. On the other hand, the collecting rate of nodules (ton/hour) will not be constant because the nodule abundance (kg/m^2) is not uniform. Moreover, due to non-homogeneity of sediment properties and undulation of seafloor topography, the speed of collector cannot be kept constant, and the dredging efficiency will not be constant as well.

Therefore, a buffering mechanism between collecting and lifting operations plays a significant role for safety and efficiency (flow assurance) and sustainability (sediments extraction). Feeding control, i.e., an optimum ore charging into lifting line, is of great importance for the sake of preventing pipe clogging and keeping optimum efficiency of lifting pumps. A further function of discharging sediment, collected during mining operation, will reduce the sediment transportation to the mining vessel (see below Fig. 4).

By-products and Tailing

The inevitable by-products in nodule lifting are a great burden for sustainable mining. Figure 5 shows a schematic diagram of material flows in on-board treatment facilities (Hong and Yang 2018). The treatment objects are nodules, seawater, and sediments. In order to avoid flooding, the on-board treatment facilities of the mining vessel should be designed as per suitable types and capacities of sub-units in Fig. 5. Moreover, the on-board treatment processes have to be integrated in capacities to avoid any congestion and clogging to secure smooth material flows.

In view of sustainability, the design of deep-seabed mining system should not focus only on the achievement of annual nodule production but also consider the feasibility of treating the by-products.

Figure 6 represents the correlations between materials transportation and treatments. As of current approaches, the treatment of pumped seawater-sediment mixture would be a challenge for commercial mining. The criteria of seawater

Fig. 4 Concept of sediment plume discharges at collector and buffer

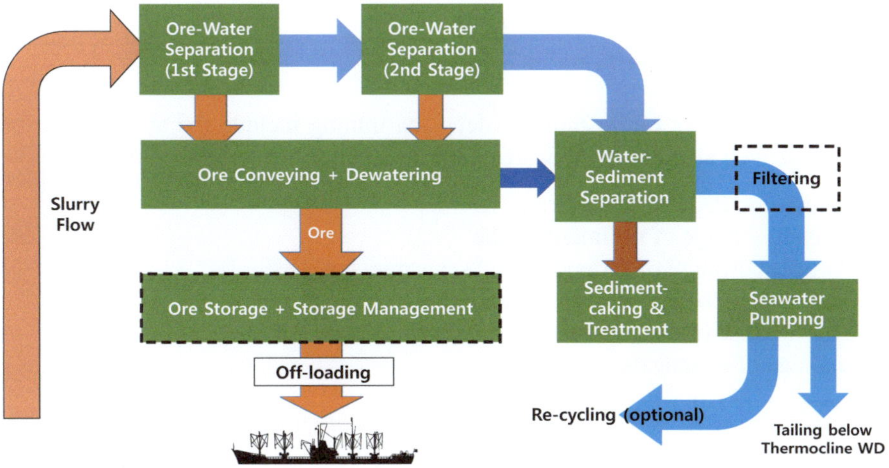

Fig. 5 A schematic diagram of on-board treatment processes

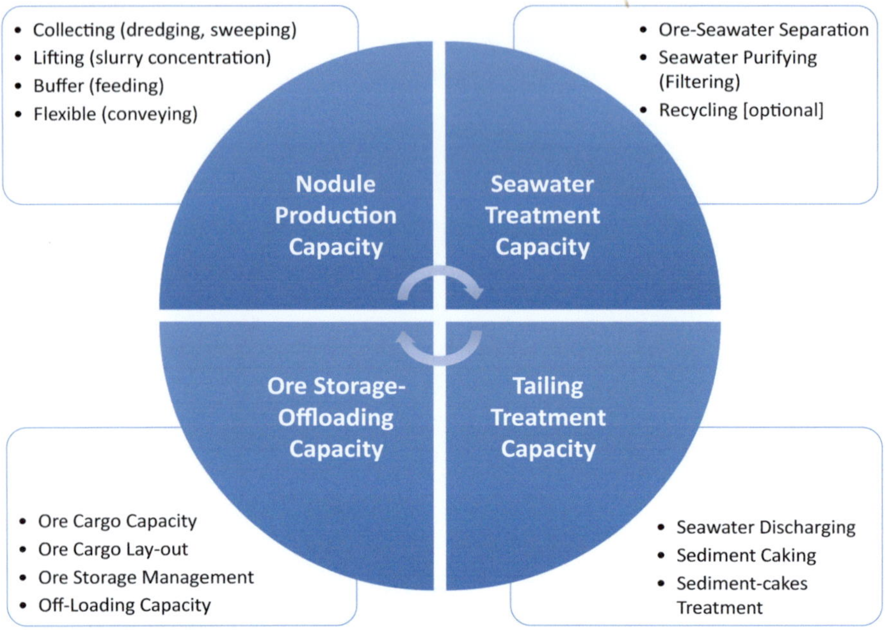

Fig. 6 Major aspects of material flows in deep-seabed mining system

tailing (*to be defined*) will be another significant factor affecting economic and technological feasibility. A relevant *recycling* (or utilization) of treated seawater for a certain purpose could be one of keys for the future of deep-sea mining.

Key Parameters for Sustainable Mining

The key parameters for sustainable deep-sea mining include the activities on the seafloor as well as those in the water column which are described below:

- Penetration depth into substrata
- Spreading range of sediment plume
- Co-transportation of sediments and seawater
- Underwater noise levels (sound and light)
- Criteria of seawater treatment and tailing
- Leakage of chemicals
- Energy efficiency

Seafloor Penetration Depth

The seafloor penetration depth (PD) is defined as the depth of penetration by the collector in to the seafloor sediment, either because of the removal of top layer of sediment or due to sinking of the collector under its own weight and during locomotion. The removal of top layer means the removal of benthic habitats and could result in compaction, removal, transportation, and dispersion of sediments. Thus, PD needs *to be minimized* as much as possible.

Spreading Range of Sediment Plume

The sediment plumes will be inevitably generated by seafloor mining operations. It will be made by crawling, harvesting, and excavating on seafloors. The spreading and resettlement of sediment plumes cause changes of benthic ecosystem (Nakata et al. 1997; Yamazaki et al. 2000). Visibility degradation hinders safe and efficient mining operations as high turbidity blocks video monitoring and disturbs acoustic signals. Hence, the spreading range should be controlled and managed for the sake of sustainability as well as operational safety.

Co-transportation of Sediments and Seawater

The co-transportation (CT) of sediments and seawater up to mining vessel would be unavoidable in DSM. The decision and management about the amounts of co-transported by-products (sediments and seawater) is one of the critical factors for the design of mining platform. Topside of mining platform will be heavily occupied by the facilities for seawater and sediment separation and treatment (filtering option) and discharge of tailings. This will strongly affect the FEED (Front-End Engineering and Design) of the mining platform based on the concept and scenario of DSM. The CT of sediment and seawater has to be minimized as much as possible for the feasibility and sustainability of DSM.

Underwater Noises

The underwater noises have not been treated enough in deep-sea mining. It seems that benthic faunas in the silent and dark environment are rather less habituated to external lights and sounds than marine mammals. It has been observed that the underwater sound noises have harmful effects on marine mammals like whales and dolphins. Underwater illumination would have rather local effects. However, underwater sounds of low frequencies propagate long distance and disturb the marine mammals. Main sources of underwater sounds will be the operations of engines, thrusters, generators, and on-board facilities of mining platform, making structure-borne noises radiating in wide range. IMO (2014) provided guidelines for the reduction of underwater noise from commercial shipping. And the concept of eco-ship would have to be adapted for the design of mining platform (vessel). The underwater mining system (nodule collector) will also generate sound noises that include machine noises and hydrodynamic noises. Eventually, many of hydraulic actuators will have to be replaced by electric ones in order to reduce the noise level. Moreover, countermeasures may have to be figured out through shielding design, utilizing anechoic materials and shape, based on characteristics of sound sources.

Capacity of Seawater Treatment and Tailing

The capacity of seawater treatment and tailing is also one of the critical points in decision of mining concept and design of mining platform. It would not be only concerned with sustainability but also related with feasibility of mining scenario. The capacity of seawater treatment facility has to meet the pumped volume of seawater per unit time. The tailing treatment is required to be conducted below sufficient water depth, e.g., thermocline layer, to exclude re-upwelling. The design and engineering of the seawater treatment facilities will be coupled with general arrangement of topside, hull strength, ballasting, and so on.

Leakage of Chemicals

The leakage of chemicals (LC) is forbidden by London Convention on prevention of marine pollution by dumping of wastes and other matters (IMO 1972). Use of any chemicals in treatment of sediment-seawater mixture to enhance the resettling speed and capacity should be banned. Leakages of hydraulic oils for actuation and pressure compensation have to be meticulously monitored and prevented. Sealing of hydraulic actuators and pressure compensation devices is a critical issue. Standards and routines of monitoring, inspection, and maintenance have to consider this problem as well.

Energy Efficiency

The enhancement of energy efficiency (EF) is one of critical issues in design and engineering of DSM system. EF is to lower the costs of construction and operation and to reduce CO_2 emission. It is related with sustainability concerns in all kinds of resource development.

As a consequence, it has to be kept in mind that the eco-friendliness will play a dominant role for ensuring profitability as well as safety. Minimum disturbance of benthic zones and less transportation of sediments and seawater up to sea surface will bring a drastic reduction of the costs in CAPEX and OPEX.

The concerned regulations and standards on safety and environment protection in maritime and offshore activities such as shipping, navigation, offshore platforms, subsea equipment, and sea operations can be found in IMO (1974, 1978), ISO (2010), ISO/IEC (2014), API (2006, 2008), IEC (2009), and even in the rule books of classification societies.

The regulations on sustainability of deep-seabed are being developed by the initiative of International Seabed Authority (ISA). The DSM technologies, in particular, the seafloor dredging technology, should meet the sustainability requirements. It is supposed that these regulations that are expected to be implemented soon will provide reasonable guidelines between *permitted*, *non-permitted*, and *accidental impacts* in deep-seabed mining activities. Safety design will have to focus on the prevention of structural and operational failure of mining system and avoiding the accidental impacts in turn. The technologies for DSM have to be based on the limitation of environmental impacts within the permitted ranges satisfying the sustainability requirements.

3 Deep-Seabed Mining Technology

Polymetallic nodules (PMN) are found mostly in half-buried condition in the upper layer of deep-sea sediments of abyssal basins. So, the mining of PMN has to be performed based on the concept of moving on seafloors, maximizing the nodule collecting efficiency and minimizing the by-products.

The profitability of deep-seabed mining can be secured through mass production of PMN in magnitude of *millions tons per year*, which would be feasible by means of continuous mining. The concept of continuous mining is realized on the basis that the mass flowrate of ores can be kept constant as designed through the entire system (see Fig. 6). Congestion and clogging in materials transportation and treatment at any locations might cause shutdown or total failure of the entire mining operation.

The technologies for continuous mining can be grouped in two categories:

- Design and engineering technology
- Operation technology

The design and engineering is required to figure out the equipment, facilities, and systems for the purpose of commercial mining based on safety, operability, sustainability, and productivity. It will govern CAPEX and OPEX. The operation technology will have to realize the functional performances of total mining system. The DSM operation is executed on the basis of system integration of equipment, powers, and signals. The performance of operation technology can be verified and validated only through large-scale sea trials. It is necessary to operate deep-seabed mining system by integrated means and in a smart fashion.

3.1 Concept of Continuous Mining System

In continuous mining concept, polymetallic nodules are harvested from seafloor to sea surface in the following sequence: *dredging-and-collecting* → *buffering* → *lifting* → *separation* → *treatment-and-tailing*.

A continuous mining system would consist of the following major subsystems:

- Seafloor dredging robot
- Buffer station
- Lifting system

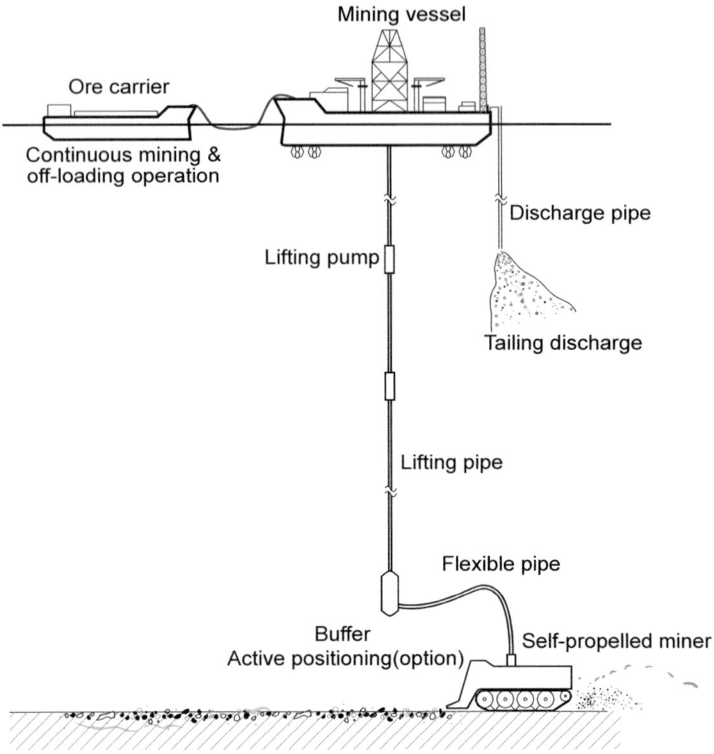

Fig. 7 A concept of continuous mining of polymetallic nodules

- Seawater separation and treatment facility
- Nodule handling and offloading system
- Tailing treatment facility
- Mining platform

Figure 7 represents a schematic view of continuous mining system using hydraulic lifting. The collector mechanism moving on the seafloor is assisted by positioning of the relevant distances between subsystems, such as seafloor collector, buffer station, and surface mining vessel. That increases the complexity of design and operation of the deep-seabed mining system. The condition of moving-end on bottom (seafloor dredging machine or robot) of the production line would give rise to significant problems of safety and control, not experienced in conventional offshore industry.

The vertical lifting line is almost freely hanging. The top tension forces magnified by axial vibrations of lifting line will play a predominant role in structural safety design. The swinging motions of the long vertical line will get long natural periods, and it makes the motion control extremely difficult.

The following components are the prerequisite core technologies for design and operation of continuous mining system:

- Estimation of highly nonlinear coupled dynamics of total mining system
- Dynamic positioning of subsystems with relevant distances
- Understanding of slurry flow characteristics and flow assurance measures
- Mobility of seafloor mining machine/robot covering productivity
- On-board batch processing of seawater treatment and trailing in a row
- High-precision underwater navigation
- Multidisciplinary design of seafloor mining machines and robots
- Real-time remote control of underwater systems
- Maximizing of automation processes in mining operation
- Security of equipment/system reliability and materials durability

On the basis of the systematic accomplishments of the above components, the following issues important for profitability, sustainability, and safety could be achieved:

- Minimizing of downtime
- Risk management
- Safety and security measures

It is important for a vertical lifting pipe that natural period of axial vibration modes is determined only by length (L_p), elastic modulus (E), and mass density (ρ) as Eq. (4), independent from pipe cross section. The period of first mode is about 4 s for a pipe of 5000 m length.

$$T_n = \frac{4}{n} \cdot \frac{L_p}{c_o} \left(n : \text{odd numbers} \right) \tag{4}$$

where $c_o = \sqrt{E / \rho}$ (phase speed) is about 5164 m/s in case of steel.

A lifting system with pumps and buffer will have a natural period of around 5.0 s. It is in the range of high wave energy density. Therefore, it is mandatory to compensate the heave motions in order to avoid a resonant vibration.

3.2 Design and Engineering Technology

Design and engineering of a continuous mining system, like other system designs, will be executed in the sequence of *conceptual design*, *preliminary design*, *basic design*, and *detail design*. The design sequence is repeated until requirements of safety, operability, profitability, and sustainability are satisfied in allowance limits.

The conceptual design is to examine the feasibility of mining scenarios for the annual production and to show candidate(s). In the preliminary design, the safety, operability, and power requirements of mining system are investigated in priority, and the main dimensions of total mining system are determined. The basic design

Table 4 Scope of work and technology requirements in design sequence

Design and engineering of DMS and technology requirements		
Design phase	Goals	Technology requirements
Conceptual design	**Selection of mining scenario** Annual production plan System configuration Operational concept Types and capacities Sustainability	**Rules and codes** Modeling and simulation: dynamic behaviors Modeling and estimation of environmental impacts
Preliminary design	**Fundamental investigations** Safety Operability Power requirements Main dimensions Risk assessment Environment assessment	**Rules and codes** Modeling and simulation: dynamic behaviors Experiment and analysis on fundamental phenomena Multidisciplinary design optimization Real-time remote control Model-based control Modeling and estimation of environmental impacts
Basic design	**Calculations and drawings** Collecting system Lifting system Launch and retrieval system Seawater treatment system Tailing process and system Power system Communication system Mining vessel Costs assessment Environment assessment (review/revision)	**Rules and codes** Engineering design VR-based design Cost evaluation Modeling and estimation of environmental impacts
Detail design	**Documentations and planning** Drawings and procedures: manufacture, fabrication, construction, assemblage Procurement plan Health/safety/environment (HSE) procedure Commissioning plan	**Rules and codes** Manufacture/fabrication/construction technology Quality control Production management HSE management

will result in the compilation of numerous and diverse drawings based on the main dimensions up to the level of subsystem and provide cost assessment. The detail design will show a great deal of drawings and procedures about manufacture, fabrication, construction, and assemblage of components of mining system. The scope of work in design sequence and technology requirements are listed as Table 4.

The design and engineering are to accomplish the following four "core" subjects on the basis of safety:

- *C*ost effectiveness
- *O*perability

- *R*eliability
- *E*nvironment-friendliness

Fundamental Tests and Experiments

There are some needs of tests and experiments, in particular, concerned with dredging and lifting of nodules. The terramechanics of collector vehicle on deep-sea sediments, the performance of dredge head, and the characteristics of slurry pipe flows are the representative examples.

Terramechanics and Experiments

The geotechnical properties of deep-sea sediments are a tough hurdle for the movement of the nodule collector on the seafloor. The deep-sea sediments are very cohesive and adhesive so that traction force is obtained mainly from the shear resistance (Yamazaki et al. 1989, 1995). Test results on the maximum shear resistance, that is, *cohesion*, are highly variable between sites. The average cohesion is about 3.5 kPa in surface layer and increases with penetration depth. The sensitivity, defined by the ratio of the maximum shear strength versus the residual strength, is scattered also in range of 2–6. Figure 8 shows an example of shear characteristics of cohesive soil according to a mathematical formulation of Wong (1993).

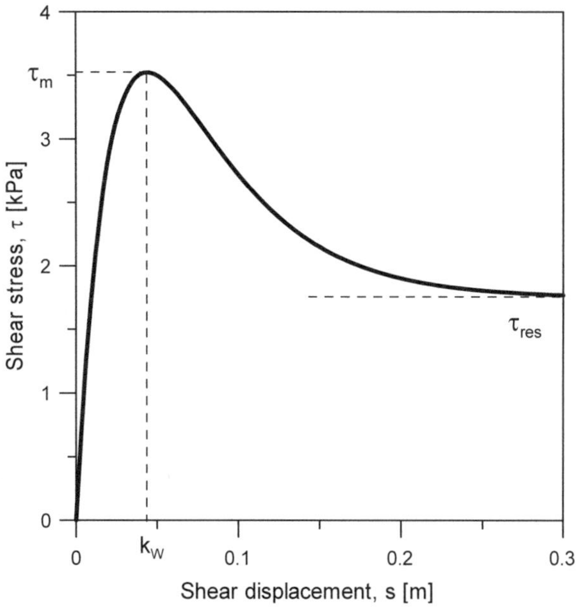

Fig. 8 Shear characteristics of cohesive soil (τ_m, maximum shear stress; τ_{res}, residual shear stress; and k_W, shear deformation of τ_m)

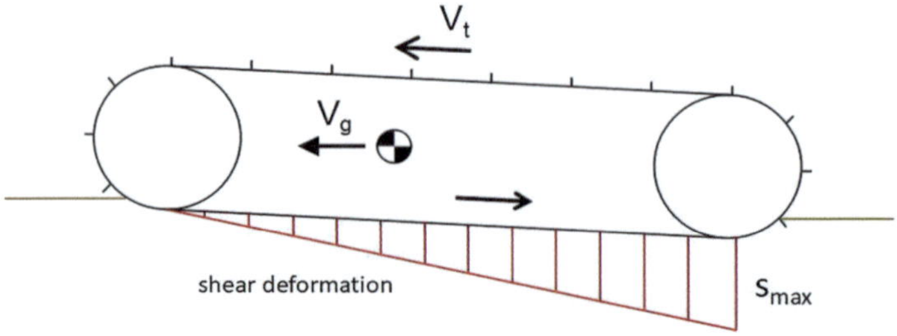

Fig. 9 Development of shear deformation below tracked vehicle

A collector vehicle should satisfy floatation, trafficability, and steerability, where floatation concerns about static sinkage of vehicle, trafficability is related with speed and dynamic sinkage, and steerability decides controllability. In the aspect that an excessive sinkage must be prevented, track-type vehicle is preferred to an Archimedean-screw type.

For driving of a tracked vehicle on cohesive sediment, the decision and control of an optimum *slip* is the principal issue. The slip velocity (V_s), defined as Eq. 5, determines the distribution of shear deformation of sediment below the tracks and the shear resistance from sediment in turn.

$$V_s = V_g - V_t \tag{5}$$

where V_g is the ground speed of vehicle and V_t is the speeds of tracks. Integration of V_s during the time span from start to end of bottom contact results in the development of shear deformation along the contact surface of tracks (Fig. 9).

Because of scaling problem of grain size, porosity, and cohesion, a track segment is tested, instead of a scaled vehicle model, to investigate relationships between bottom pressure, static sinkage, shear stress, shear deformation, and dynamic (or slip) sinkage, where sediments are imitated by cohesion (for instance, 3.5 kPa).

By means of track segment tests, it was known that an excessive sinkage occurs at bottom pressure over 5~6 times of the residual shear strength, and the average of traction strength is about 1.2–2.0 times the residual shear stress depending on grouser shapes, and an additional dynamic (slip) sinkage occurs proportionately to the shear deformation (Hong and Choi 2001). It is shown in Figs. 10 and 11.

Performance Test of Dredge Head

The principles of dredge heads are classified in three categories of hydraulic, mechanical, and hybrid ones, though a hydraulic principle is preferred in respect of simplicity and robustness requirements.

Investigation of dredging performance should be based on large-scale model tests because of the same problems as in terramechanics experiments. The efficiency of

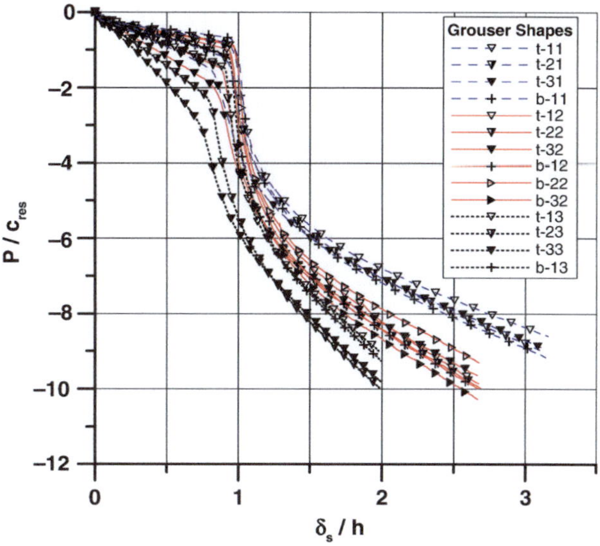

Fig. 10 Experimental results of bottom pressure (P) and static sinkage (δ_s) depending on grouser shapes (c_{res}, residual stress; h, grouser height)

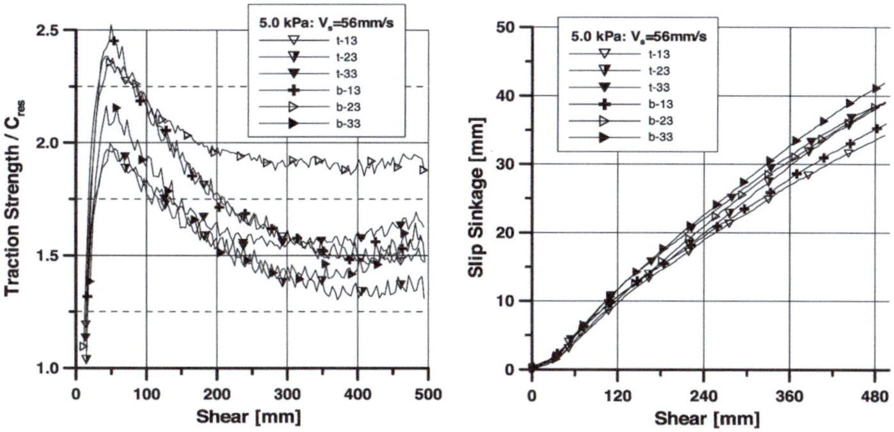

Fig. 11 Examples of experimental results of traction strength (left) and slip sinkage (right) versus shear deformation (c_{res}, residual stress; bottom pressure, 5.0 kPa; slip velocity, 56 mm/s)

nodule dredging will be the main objective of model tests. For instance, various kinds of hydraulic dredge head showing a high dredging efficiency over 80% can be found in Oebius (1993).

In addition to the nodule dredging efficiency, particularly for hydraulic dredging, an optimum operational condition has to be found out, and thereby the sediment plume can be reduced and controlled. Figure 12 shows experimental results of dredging (pick-up) efficiency versus flowrate ratio (Eq. 3) of a hydraulic dredge head (Hong et al. 2001).

Fig. 12 Dredging (pick-up) efficiency of a hydraulic dredge head

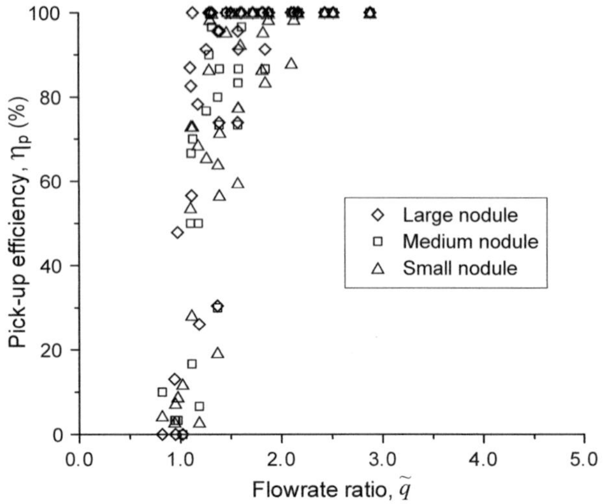

Slurry Flow Characteristics and Experiments

The characteristics of slurry flows (solid-liquid, solid-liquid-gas) are represented by *volume concentration*, *solid velocity*, and *hydraulic gradient*. The important factors are ratio of solid density (ρ_s) to water density (ρ_w), ratio of solid diameter (d) to pipe diameter (Δ), load ratio (μ_s), and Reynolds number.

The volume concentration (C_v) is related with the load ratio (μ_s) by the following equation (Engelmann 1978):

$$C_v = \mu_s \cdot \frac{\overline{v}_{wo}}{\overline{v}_s} \cdot \frac{1}{S} \tag{6}$$

where:

$\overline{v}_s$ is the mean velocity of solids
$\overline{v}_{wo}$ is the mean velocity of water only
$S = \rho_s/\rho_w$ is the density ratio

The load ratio (μ_s) is defined as

$$\mu_s = \frac{\dot{m}_s}{\dot{m}_w} \tag{7}$$

In condition that the load ratio (μ_s) is given, it is needed to measure $\overline{v}_s$ in order to know about the volume concentration (C_v), and vice versa.

The total hydraulic gradient consists of the static gradient of mixture and the dynamic gradient. The static gradient corresponds to the holdup of the solids in pipe. The dynamic gradient is due to the friction between water and the pipe inner wall. The static hydraulic gradient (i_s) is obtained by the volume concentration (C_v) as follows:

$$i_s = 1 + C_v \left(S - 1 \right) \tag{8}$$

The mean water velocity can be obtained from the volume concentration as well.

$$\overline{v}_w = \frac{\overline{v}_{wo}}{\left(1 - C_v \right)} \tag{9}$$

Figures 13 and 14 show the experiment results of the volume concentration and the hydraulic gradients based on the measurement of the solids velocities by means of PTV (Particle Tracking Velocimetry) methodology (Hong et al. 2002b).

Some important results from the fundamental experiments on solid-liquid slurry flows in vertical pipe are summarized as follows:

- Reynolds number, based on the mean water velocity of solid-water mixture flow, has significant effects on volume concentration and hydraulic gradient.
- Load ratio of pump has a dominant effect on the volume concentration of solids in pipe, which determines the mixture density of slurry flow and the static head loss in turn. At a constant load ratio, the volume concentration increases gradually by reduction of Reynolds number.

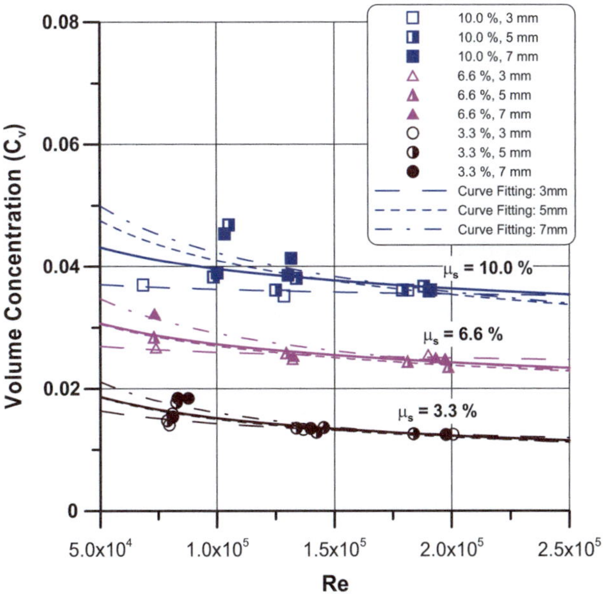

Fig. 13 Experimental results of volume concentration as a function of Reynolds number, load ratio, and particle size

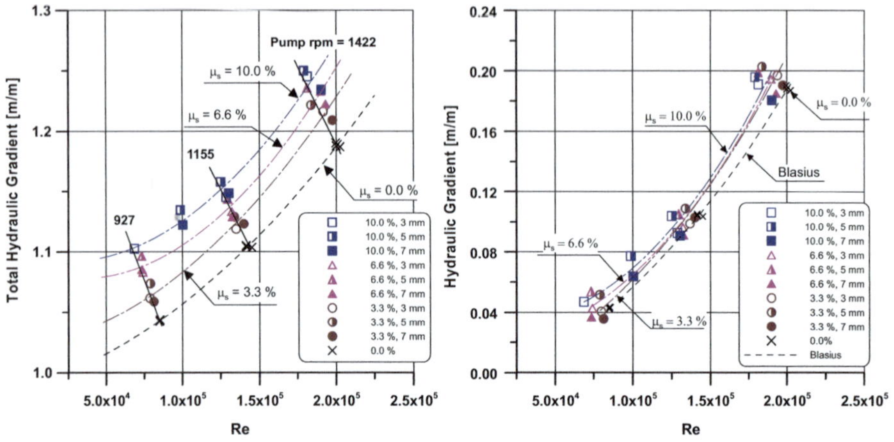

Fig. 14 Total hydraulic gradient (left) and hydraulic gradient (right) as a function of Reynolds number, load ratio, and particle size

- Load ratio has a minor effect on the hydraulic gradient of slurry flow, which is evaluated by subtraction of the static head loss from the total hydraulic gradient. The difference of hydraulic gradient of slurry flow from that of water only flow becomes larger at higher range of Reynolds number, which is caused by the change of water velocity profile near the pipe wall due to the presence of solid particles.
- Particle size has significant effects on volume concentration and hydraulic gradients. The effect on volume concentration is restrained relatively in lower range of Reynolds number, whereas the effects on hydraulic gradients are in the whole ranges of Reynolds number. There might be a certain diameter ratio of d/D producing largest hydraulic gradient. The effects of particle size, however, will need more thorough experimental investigation.

In practice, it is possible to adjust the load ratio (μ_s) by means of a feeder at buffering station (Hong et al. 2014). Maximizing of the volume concentration (C_v) and minimizing of the water flowrate is a trade-off problem in the aspects of safety, productivity, and sustainability.

It is expected that lifting pipe will oscillate and vibrate in waves and currents. Experimental data on the characteristics of slurry flows in oscillating and vibrating pipes are still limited: as examples, Xia and Huang (2000) and Xia et al. (2004). Also, it is necessary to investigate more about slurry transportation through a flexible (dynamic) pipe. In curved pipes, where the *Magnus Effect* is no longer expected, it is likely that a bed of solid particles will be formed on the pipe inner bottom. This phenomenon will drop transportation efficiency and could result in fatal failure due to "pipe clogging" and bring about reduction of lifetime of conduits.

Modeling and Simulation

Modeling and simulation (M&S) techniques are used mostly in the stages of conceptual design and preliminary design. M&S is particularly effective for development of a new system, which has no precedents. The lack of experiences in operational practice can be compensated by means of modeling and simulation. Furthermore, the integration of design and operation in early stage could decrease the development risks.

The main subjects of M&S for the development of a continuous mining system are several and diverse as follows:

Nonlinear coupled dynamics of underwater systems
Nonlinear dynamics/terramechanics of seafloor crawling vehicle
Dynamic positioning of surface mining platform
Multibody dynamics of seafloor dredging machine/robot
Structural responses of subsea equipment/device
Launching and retrieval operation of underwater system
Heave compensating mechanism
Signal characteristics of underwater localization
Hydrodynamic behaviors of nodules and/or fragments
Slurry flows (solid-liquid and/or solid-liquid-gas)
Separation and filtering of ores and fine particles
Spreading of sediment plumes
Discharging and spreading of tailing

Numerical Analysis Models

Computer simulations, based on mathematical modelings, are a powerful tool in science, technology, and engineering. In engineering applications, it covers a broad spectrum, typically, structural stress and deformation responses, hydrodynamics/fluid mechanics, multiphase flows, terramechanics, rigid-body dynamics, multibody dynamics, and fluid-structure interactions. This enables simulation-based design (SBD) and multidisciplinary design optimization (MDO) for deep-seabed mining system.

Scopes and contents of modeling and simulations for the design of DSM system can be summarized as in Table 5.

To begin it is most important to understand global dynamics of underwater mining system. An analysis on coupled dynamics of underwater systems is to provide critical information for the following purposes:

- Understanding of motions of underwater systems
- Estimation of mechanical interactions between subsystems
- Evaluation of safety and operability
- Feedbacks to functional requirements
- Global design of underwater mining system

Table 5 Scope and contents of modeling and simulation techniques for design of deep-seabed mining system

Field	Theory/methodology	Topics	Applications
Structure mechanics	Finite element method (FEM)	Structural static analysis Structural dynamic analysis Linear analysis Nonlinear analysis	Structure design: mining platform hull, topside facility frame structure, topside equipment/device frame, underwater equipment/system frame, pipe/riser and connector/flange, pressure vessel Basic design MDO
	Finite difference method (FDM)	Structural dynamic analysis Nonlinear analysis	Global dynamic analysis: coupled dynamic behavior of underwater mining system Fast (real-time) simulations Preliminary design SBD
Hydrodynamics/ fluid mechanics	Potential theory (inviscid flow) FEM Boundary element method (BEM)	Estimation of external environmental forces (wave loads) Coupled dynamic analysis of multiple floating bodies	Structure design of mining platform Seakeeping analysis of mining platform Design of DPS (dynamic positioning system) Dynamic analysis of floating bodies Preliminary design Basic design
	Navier-stoke theory (viscous flow) Computational fluid dynamics (CFD)	Analysis of viscous flows around bodies Analysis of density flows	Analysis of local flow patterns in ducts Analysis of flow patterns around lifting riser Analysis of dispersion of sediment plumes Preliminary design Basic design
Multiphase fluid mechanics	Fluid particle dynamics (FPD)	Analysis of particles dynamics in ducts and vessel Analysis of dynamics of sediment plume	Analysis of slurry flows: particles dynamics in ducts, buffer/feeder and pumps Analysis of particles dynamics in nodule separator Analysis of dispersion and resettlement of sediment plume Preliminary design Basic design

Terramechanics	Empirical modeling	Characteristics of soil-machine interactions	Modeling: pressure-sinkage, shear-resistance, shear-sinkage Concept design Preliminary design
	FEM	Characteristics of global soil mechanics	Analysis of soil response behavior
Rigid-body dynamics	Classical mechanics on rigid body	Dynamic analysis of underwater systems Analysis of coupled nonlinear dynamics of underwater systems	Dynamic analysis of collector vehicle Dynamic analysis of buffer Analysis of coupled dynamics of underwater mining system Concept design Preliminary design BD MDO (by means of meta-modeling)
Multi-body dynamics	Theory of multi-body dynamics	Analysis of dynamic systems with multiple bodies with joints	Dynamic analysis of collector and buffer Dynamic analysis of heave motion compensator Analysis of launch and retrieval operation Preliminary design Basic design
Fluid-structure interaction	FEM/FDM+CFD	Analysis of interactions between fluids and structure responses	Dynamic analysis of lifting pipe with respect to external and internal flows Analysis of vortex-induced vibration (VIV) of pipe Fatigue analysis of lifting pipe Preliminary design Basic design
	FEM/FDM+ Empirical modeling	Analysis of interactions between fluids and structure responses	Dynamic analysis of lifting pipe with respect to external and internal flows Fatigue analysis of lifting pipe Fast (real-time) simulation Preliminary design Basic design

A simulation model for global dynamics of underwater mining system is to be achieved by means of integration of the following simulation models on:

- 3D nonlinear dynamics of lifting pipe and flexible hose
- 3D nonlinear dynamics of seafloor vehicle
- 3D dynamics of joints between subsystems

There are a number of references on dynamic analysis of lifting system: for instance, Brink and Chung (1981), Chung and Whitney (1981, 1983), Aso et al. (1992, 1994), Hong (1995, 1997), and Hong et al. (2003a, b). Typical results of axial vibrations and towing simulations are shown in Figs. 15 and 16.

An example of simulation model on nonlinear dynamics of tracked vehicle crawling on seafloor sediments can be found in Hong et al. (2002a, b), where tracked

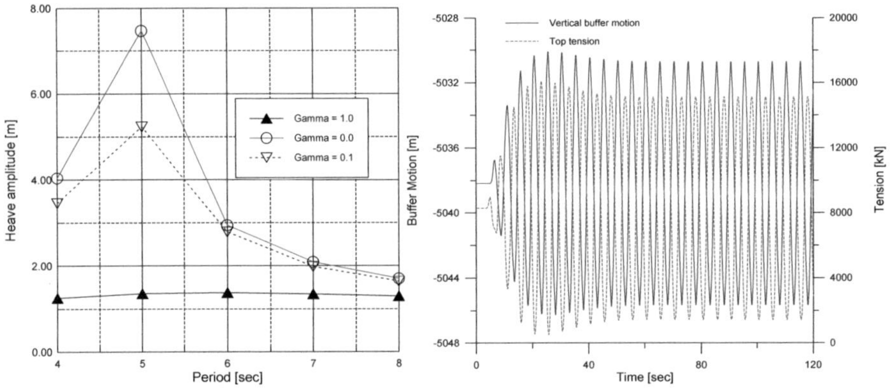

Fig. 15 Heave amplitudes of buffer (left) and resonant responses of top tension and buffer motion (right): total length 5030 m, total weight in water 848 ton, top excitation amplitude of 1 m, forced excitation period of 5 s (Hong 1997)

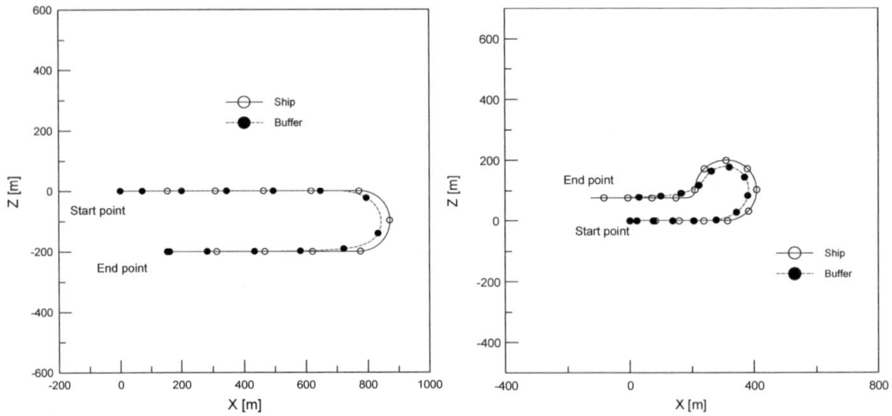

Fig. 16 Trajectories of ship and buffer at U-turn (left) and at keyhole turn (right): total length 5030 m, total weight in water 848 ton, towing speed 1 knot (Hong 1997)

vehicle is modeled as a single body and interacts with soft and cohesive sediments based on the characteristics of terramechanics of track segments (Figs. 10 and 11). The single-body simulation model of tracked vehicle was verified through comparative study with multi-body dynamics model (Kim et al. 2003). The single-body simulation model (TRACIM) realizes "real-time" simulations and can be utilized to produce a vast data set about driving performance of vehicle with respect to variation of design parameters. Figures 17 and 18 show the design variables of a single-body vehicle model and examples of simulation results of turning trajectory.

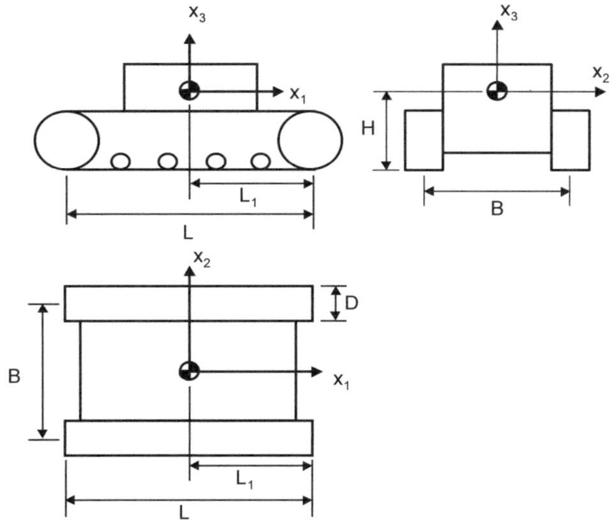

Fig. 17 Design variables of single-body model of self-propelled collector

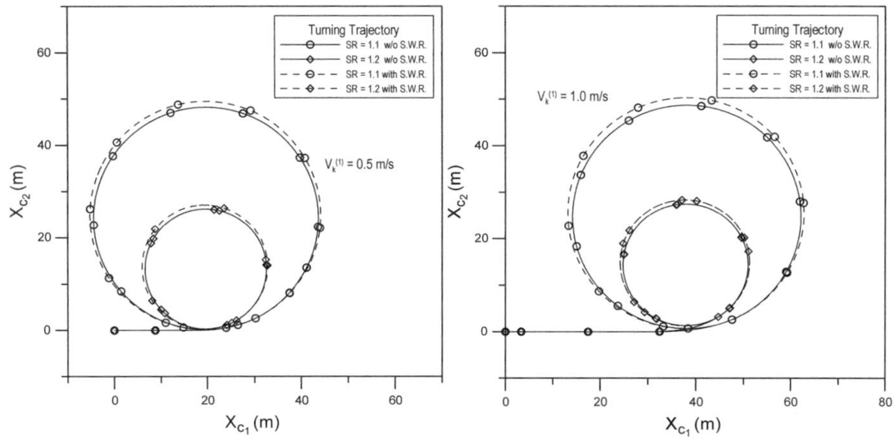

Fig. 18 Examples of simulation results: turning trajectories according to variation of track speeds and steering ratios (Hong et al. 2002b)

Examples of global dynamic analysis of total integrated mining system are shown in Fig. 19 (Hong et al. 2006) and in Fig. 20 (Kim et al. 2006). This kind of fast "real-time" simulation provides a powerful tool for SBD of deep-seabed mining system. It is available to investigate the global safety and operability of underwater mining system with respect to the design variations of subsystems, such as

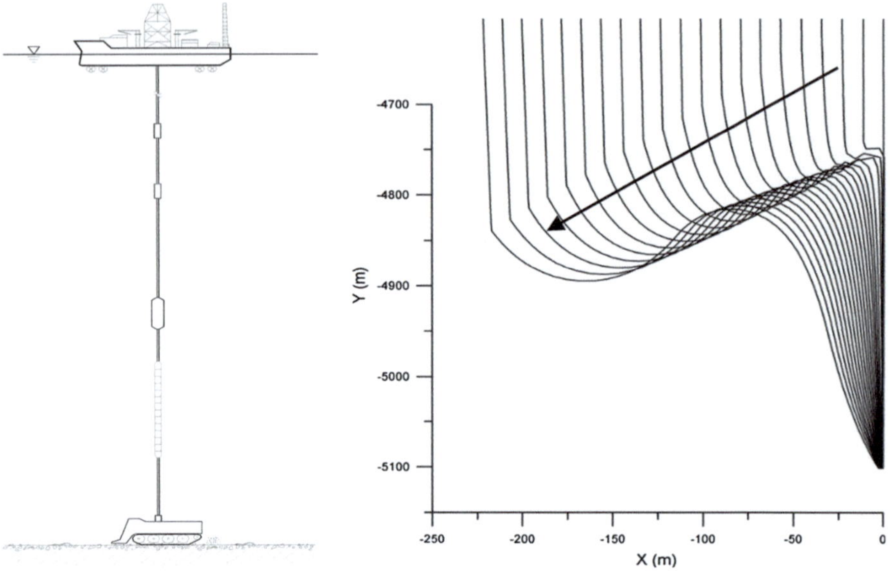

Fig. 19 An example of global dynamic simulation: landing (Hong et al. 2006)

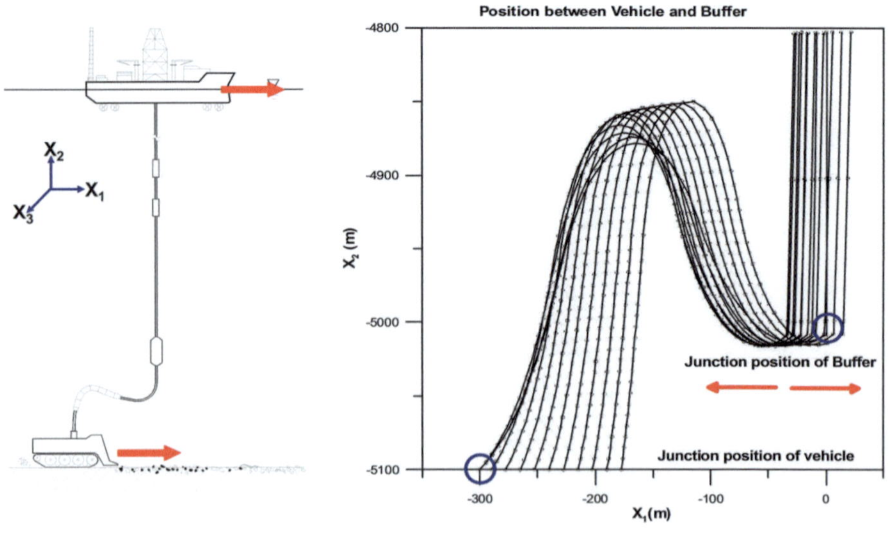

Fig. 20 An example of global dynamic simulation: mining (Kim et al. 2006)

configuration of collector vehicle (bottom pressure, center of mass, center of buoyancy, track shape), flexible conduit (length, buoyancy distribution), buffer (weight, joints, size, damper) and required buoyancy for lifting pipe, and so on. Moreover, based on the real-time simulation, it would be possible to implement a diagnostic and prognostic measure in dynamic positioning of mining platform.

Meta-models

Meta-modeling is to build up a surrogate model, which is a model of the model. A meta-model of a performance response, achieved from an experimental results or by means of a theoretical (mathematical) model with respect to variation of design parameters, can be effectively utilized for multidisciplinary design optimization.

For an instance, the efficiency of dredge head and the characteristics of slurry pipe flows can be transformed in response surface models (RSM). On the other hand, the driving performance response of collector vehicle (driving speeds, turning radii, static and dynamic sinkages, slips, pitch and roll) can be obtained by using a numerical simulation model and can be meta-modeled by kriging (simplified as *kriging model*).

1. *RSM of dredging efficiency*: Experimental results of dredging efficiency (η) such as in Fig. 12 can be curve-fitted as a polynomial quadratic function with respect to the independent design variables, which are driving speed (v), gap above seafloor also influenced by vehicle pitching (h) and water-jet flowrate (q), as follows:

$$\eta = \beta_0 + \beta_1 v + \beta_2 h + \beta_3 q + \beta_4 \mathrm{vh} + \beta_5 \mathrm{vq} + \beta_6 \mathrm{hq} + \beta_7 v^2 + \beta_8 h^2 + \beta_9 q^2 \qquad (10)$$

   where the coefficients β_0, …, β_9 are determined by means of the least square regression about the set of experimental results.

2. *RSM of characteristics of slurry pipe flow*: Experimental results of volume concentration and hydraulic gradients of Figs. 13 and 14 can be also meta-modeled as a function of Reynolds number with respect to variation of design parameters, i.e., load ratio and particle size.
3. *Kriging model of vehicle performance*: Kriging originally comes from the field of geostatistics as a model to predict spatial correlated data, such as thickness of ore layers, from a limited number of exploration data. A kriging model for prediction of driving performance of tracked vehicle crawling on soft cohesive soil was developed by Lee et al. (2007). The kriging model was constructed based on the "limited" data from 108 simulation cases of vehicle designs, which were generated by means of numerical simulation model (TRACSIM) of Hong et al. (2002a, b).

Utilization of the meta-models for DSM system aims at the achievement of multidisciplinary design optimization (MDO) of collector robot for satisfaction of functional requirements and sustainability constraints. It is expected that MDO could reduce the development risks and accomplish the seabed sustainability and achieve the optimum trade-offs of performances.

Multidisciplinary Design Optimization

As in Fig. 6, all the sub-units of DMS system are fully coupled in capacity, and each operational principle is various and diverse. Thus, in order to figure out appropriate design solution(s), a number of different physical disciplines need to be applied, such as fluid mechanics, hydrodynamics, material and structural mechanics, hydraulics, geotechnique, rock mechanics, terramechanics, slurry flow dynamics, electrics and electronics, and so on.

In particular, the designs of underwater systems have to be approached through multidisciplinary design optimization (MDO). Utilizing MDO, structural safety, power supply, equipment purchase, maintenance and repair, hydrostatic and hydrodynamic resistance, weight limit and balance, functions, and performances are to be secured together with satisfaction of sustainability requirements but all within the limits of the supporting capacities of mining platform (vessel).

Figure 21 shows, in case of a self-propelled collecting machine (or robot), the functions of sub-units are strongly interrelated and a design optimization must be done. Also, the flexible conduit in Fig. 21 has to satisfy two conflicting functions: one is a sufficient structural flexibility assuring the mobility of collector vehicle, and the other is a flow assurance preventing pipe clogging. This contradictory situation is schematically expressed in Fig. 22.

An example of quality function deployment (QFD) for PMN collector (Fig. 23), conducted in early development stage (Hong et al. 2011b), visualizes the correlations between design requirements and design variables. The relationships between design variables were evaluated, and the relative importance for the design requirements were quantified with respect to 28 design variables. At that time, a hybrid pickup device (dredge head) was adopted.

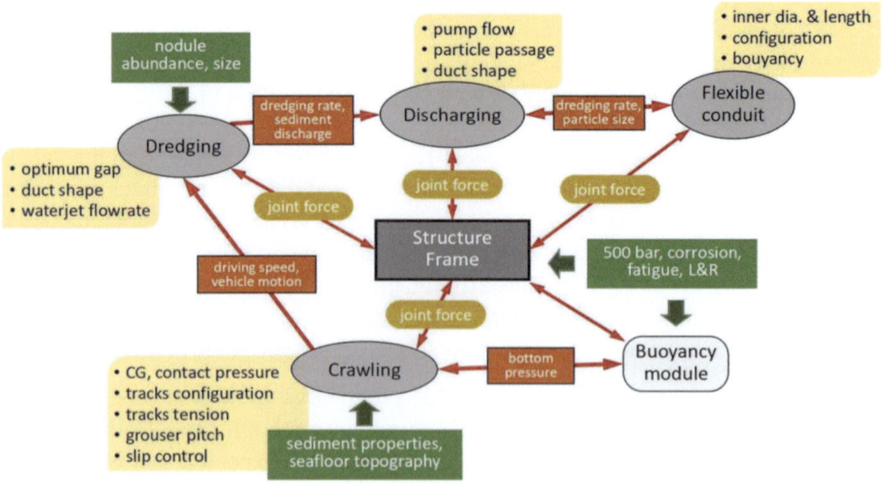

Fig. 21 Interrelationships among design requirements and design variables of a self-propelled collector. (Courtesy: Sup Hong)

Fig. 22 Design conflict of flexible conduit: design for vehicle (left) versus design for discharge pump (right)

MDO of Nodule Collector

Figure 24 shows a schematic view of the framework for MDO of self-propelled collector. The dredging efficiency and the discharging efficiency are meta-modeled by RSM based on experimental results. The vehicle performance is formulated by kriging model based on the numerical simulation results using TRACIM. The performance (specific strength) of chassis frame structure is (high-fidelity) modeled by FEM.

Lee et al. (2012) demonstrated the convergence of MDO iteration procedure, performed for a test collector, where the optimization problem was formulated to minimize the total power (objective function) with satisfaction of the design constraints (productivity, structure safety, trafficability limits of vehicle). It was resulted that the total power was reduced by 14.1% compared to the initial (given) design after 25 runs of iteration.

The correlations between subsystems and the data flows of design variables are represented as block diagram in Fig. 25. The meanings of symbols are as follows:

- h, Q_{jet}, b_w: gap, water-jet flowrate, and width of dredge head
- ΔP, C_v, Q_{pump}: pressure loss, volume concentration, and pump flowrate
- d_f, L_f, κ_f: inner diameter, length, and curvature of flexible conduit
- A_m, A_l, A_d: cross sections of main frame, loading frame, dredge head frame
- W, H, L, L_1, B, D: design variables of single-body model (Fig. 17)
- δ_z, θ_{pitch}: sinkage and pitch angle of single-body collector model
- V_g, V_t: ground velocity and track velocity
- $\dot{M}$, σ_{max}: production rate (kg/s) and maximum stress of chassis frame

3.3 Practical Implementations and Sea Trials

In this section, two cases of practical implementation are introduced, and the results from sea trials, which have been performed and conducted in recent years, are discussed. One is development of a pilot mining robot, and the other one is development of pilot lifting system.

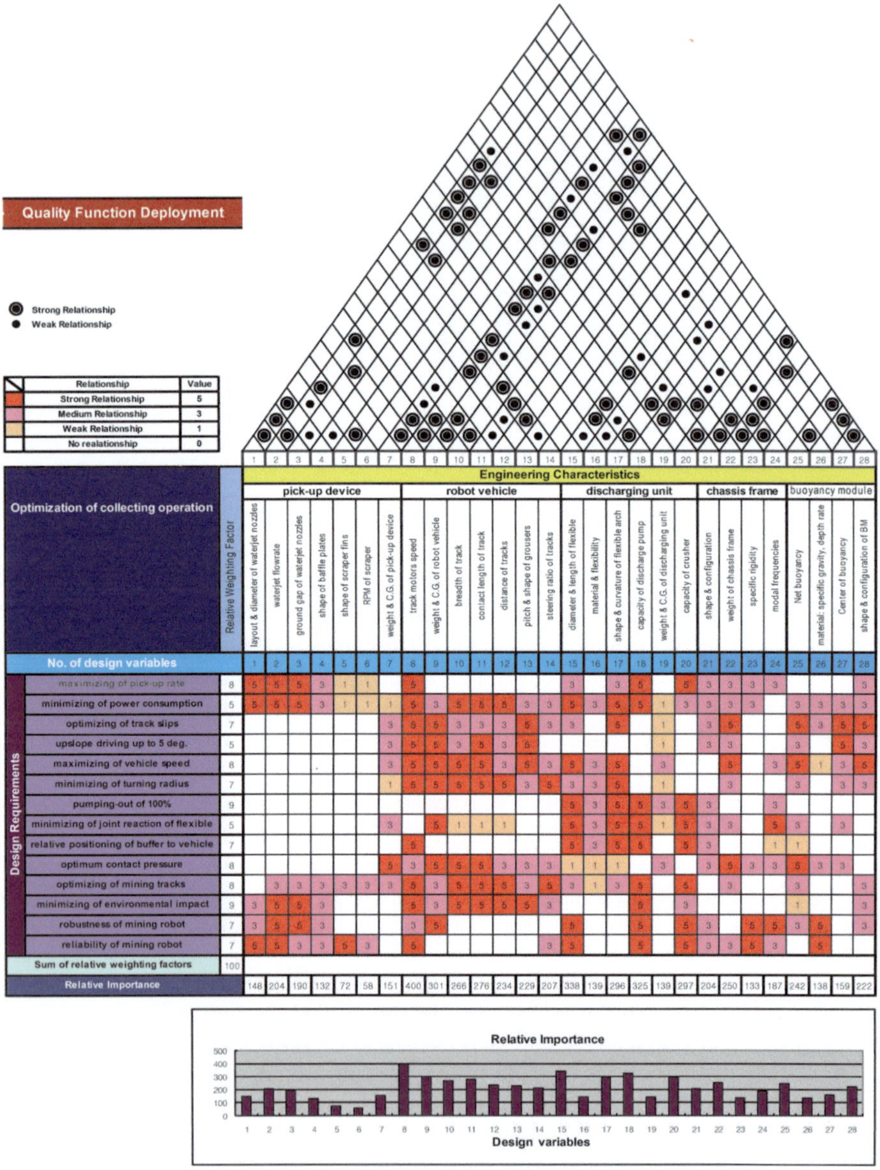

Fig. 23 An example of QFD for collecting system of polymetallic nodules

A Pilot Mining Robot

Figure 26 explains a pilot mining robot (MineRo II), developed by using SBD and MDO during 2011–2013 (Hong et al. 2013). It was aimed at pilot mining tests in 5000 m water depth. The main specifications are as follows:

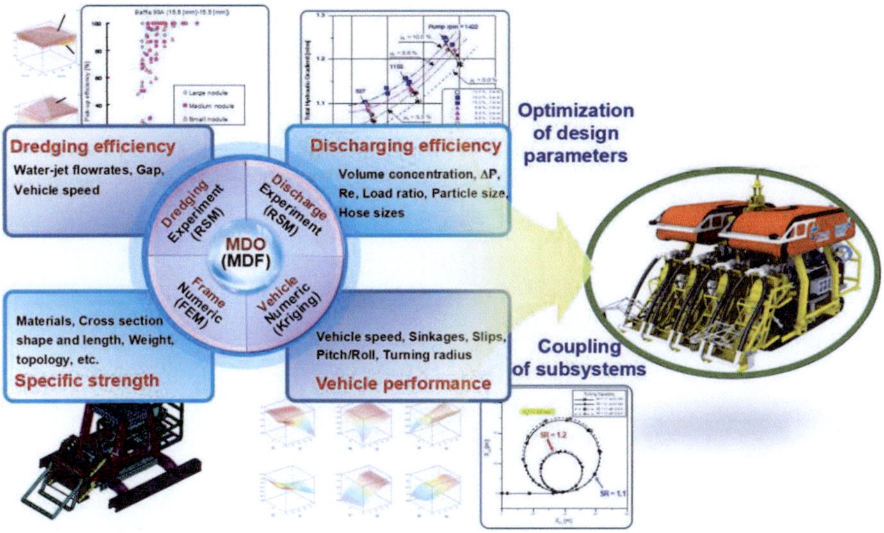

Fig. 24 Schematic view of MDO framework for a self-propelled collector

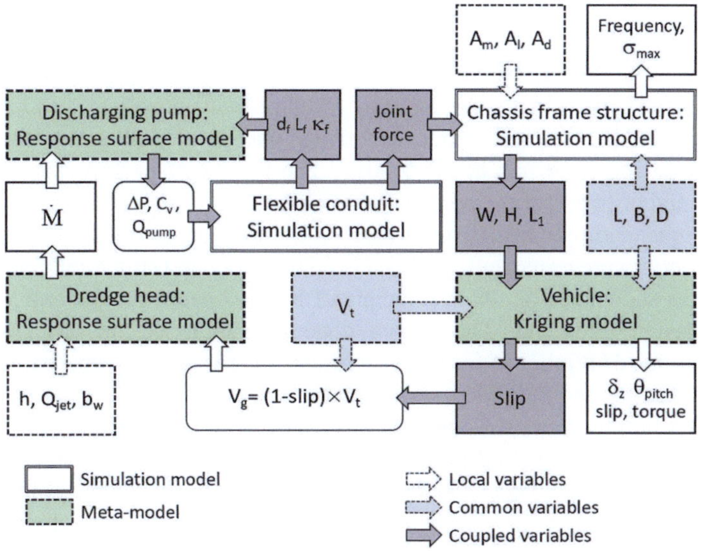

Fig. 25 Correlations between subsystems and flows of design variables in MDO of self-propelled collector. (Revised from Lee et al. 2012)

- Dredging capacity: 300,000 tons (dry) per year
- Weight: 28 tons (9 ton under water)
- Size: 6.5 m (*L*) × 4.7 m (*W*) × 3.6 m (*H*)
- Power: 530 kW (500 kW hydraulic, 30 kW electric) by 4000VAC
- Operating depth: 5000 m

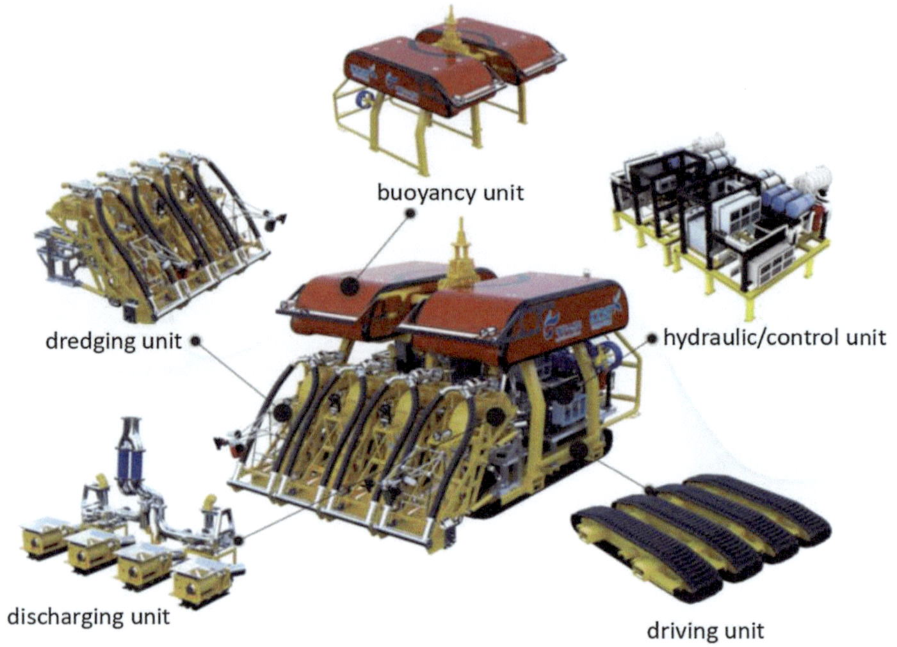

Fig. 26 A pilot mining robot (MineRo II) and its main units

Prior to MineRo II, a test miner (MineRo I) was developed, and it was operated two times during sea trials in 2009/2010. The results of performance tests were evaluated and utilized for the concept of MineRo II.

The unique feature of MineRo II is that it is a combination robot of two unit robots. The unit robots were tested and validated in a laboratory and structurally integrated as one robot as shown in Fig. 26. A principle of KISS (*K*eep *I*t *S*imple and *S*mart) was kept as design guideline. The design concept was based on 3M (*M*odularity, *M*aintainability, *M*ultiplicity) for the sake of expandable mining capacity, simple and easy maintenance, and multiple backup solutions.

The pilot mining robot (MineRo II) was tested in June of 2013 (Fig. 27). Nodule (artificial) dredging tests and path tracking control tests were conducted, and the performances were validated. The configuration of H/W for remote control of mining robot and the data flow of remote control S/W are shown schematically in Figs. 28 and 29, respectively (Yeu et al. 2013).

The algorithm of path tracking control is referred to the numerical simulation results of single-body model (Kim et al. 2009a, b). Figures 30 and 31 are the test results of underwater localization algorithm (sensor fusion by Integrated Kalman Filter) and path tracking control algorithm, respectively (KRISO 2013).

The results of path tracking control of pilot mining robot (MineRo II) show the possibilities of automation of collecting operation. It is prerequisite to automate the collecting operation, because just a remote control of collector will be totally paralyzed in the sediment plumes.

Fig. 27 Sea trial of a pilot mining robot (MineRo II)

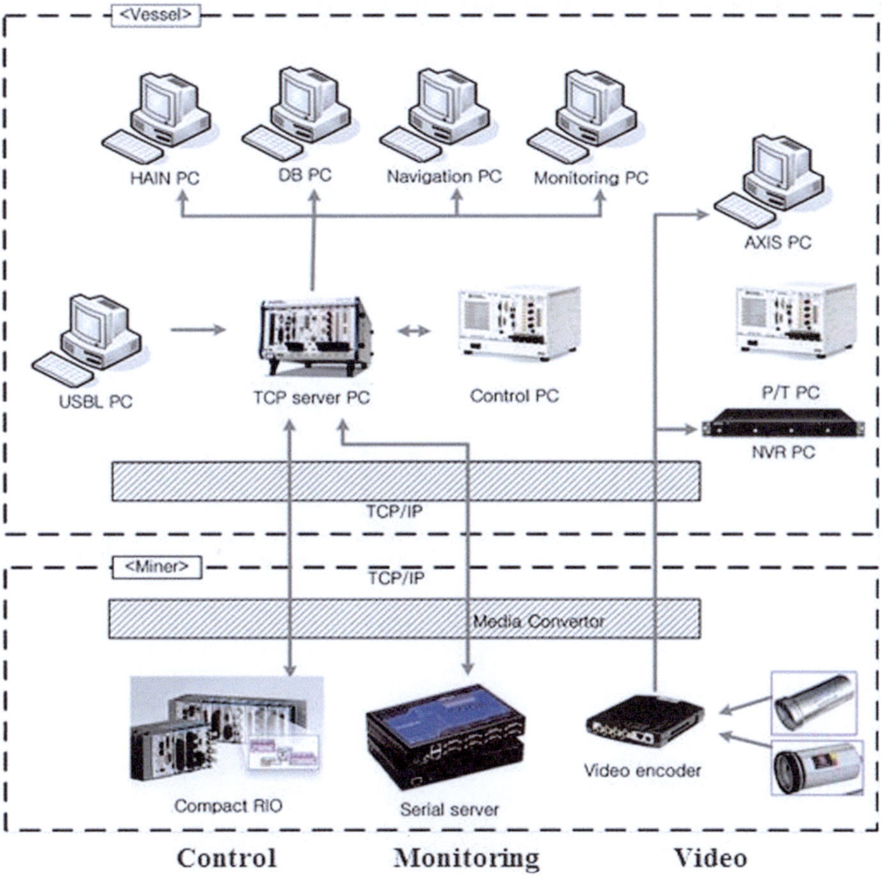

Fig. 28 Configuration of H/W for remote control of mining robot (Yeu et al. 2013)

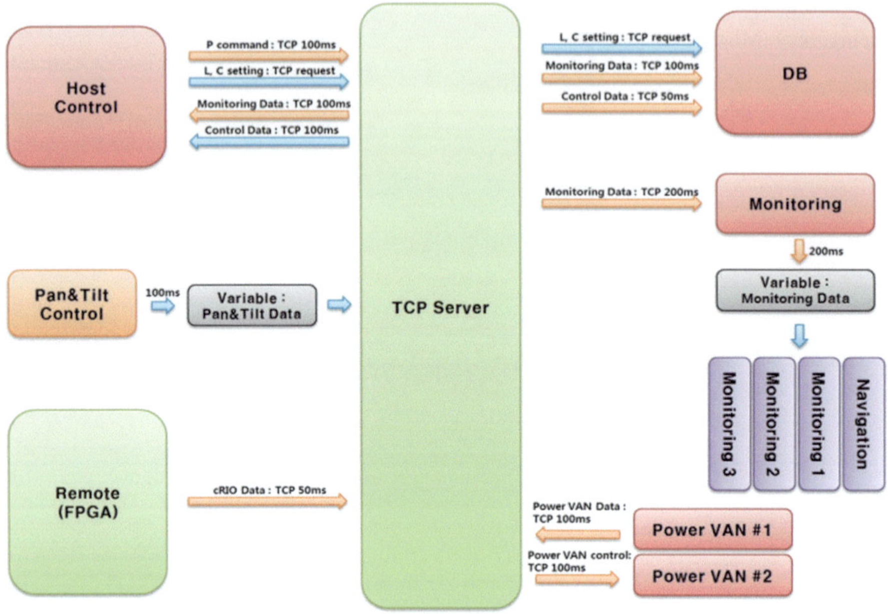

Fig. 29 Concept of data flow of remote control S/W (Yeu et al. 2013)

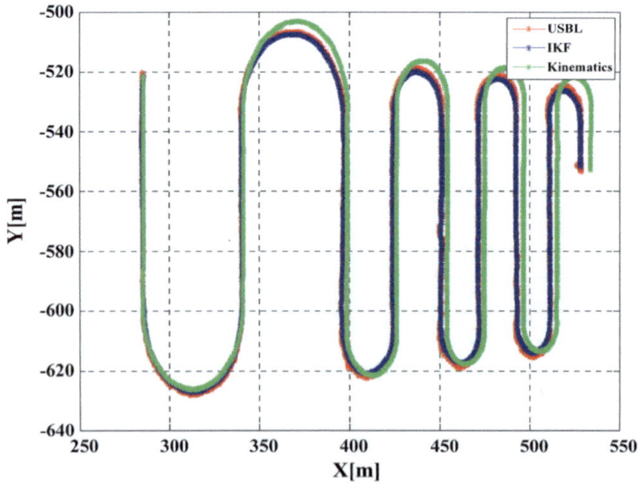

Fig. 30 Localization algorithm test through turning driving

An automatic control of seafloor mining robot for tracking optimum designed mining paths will enhance the seafloor sweeping efficiency and the overall collecting efficiency together. It will reduce the benthic intervention and provide a sound basis for deep-seabed sustainability (see Sect. 2.2.3).

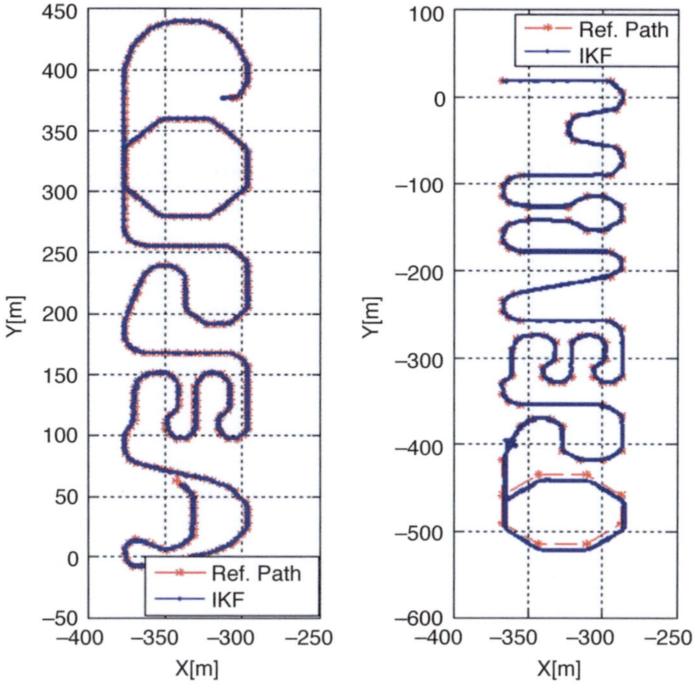

Fig. 31 Results of path tracking controls: COREA (left) and MineRo (right)

A Pilot Lifting System

A pilot lifting system was developed and tested during 2014–2015 (Hong et al. 2014, 2016). It consists of buffer station, lifting pump, and lifting pipe. The main specifications of three subsystems are in Table 6.

A large-scale sea trial was conducted in November to December of 2015. The performances of lifting pump and buffer station were tested with a truncated lifting pipe of 500 m, where the rest of pressure drop was simulated on board. The facilities for launch-and-retrieval through moon pool, nodule separation, and slurry measurement were designed and constructed and integrated with test support vessel (Fig. 32).

It was verified that the lifting pump and the buffer station satisfy the design requirements, integrated operation of pump and buffer is available, and an on-board measurement of volume concentration is feasible.

The main checkpoints in the sea trial were as follows:

- Heading control of buffer in sea currents
- Feeding control for flow assurance in slurry lifting
- Remote operation of underwater lifting pump
- Launch and retrieval of lifting system
- On-board treatment and measurement of liquid-solid slurry flow (Figs. 33 and 34)

Table 6 Main specifications of pilot lifting system

Subsystem	Component	Specifications
Lifting pump	6-stage slurry pump	Flowrate: 500 m^3/h
		Head: 250 m
		Passage: 40 mm (max)
	Submersible electric motor	Power: 800 kW, 6000VAC
		Working depth: 500 m
	Shroud	Function: shield and slurry ducts
		Flange: connection with lifting pipes
Lifting pipe	Pipe unit	Inner diameter: 202 mm
		Length: 12.5 m
		Material: API 5L X80 (SAW)
		Quantity: 38ea
	Flange	API 6A 6B 5000 psi, ASTM A182 F22 CLASS 3
Buffer station	Hopper	Size: 6 m^3 (storage for 15 min)
		Screen: sediment plume discharge
	Feeder	Capacity: max. 10 kg/s (controllable)
	Thruster	Power: 400 kgf
		Quantity: 2ea
	Sensors	Pressures, flowrates, temperature, gyro heading, yaw-rate, cameras, transducer, hydrophone, etc.

Fig. 32 Photos at sea trial of lifting system: lifting pump launching, buffer station, and on-board separation and measurement facility (from left clockwise)

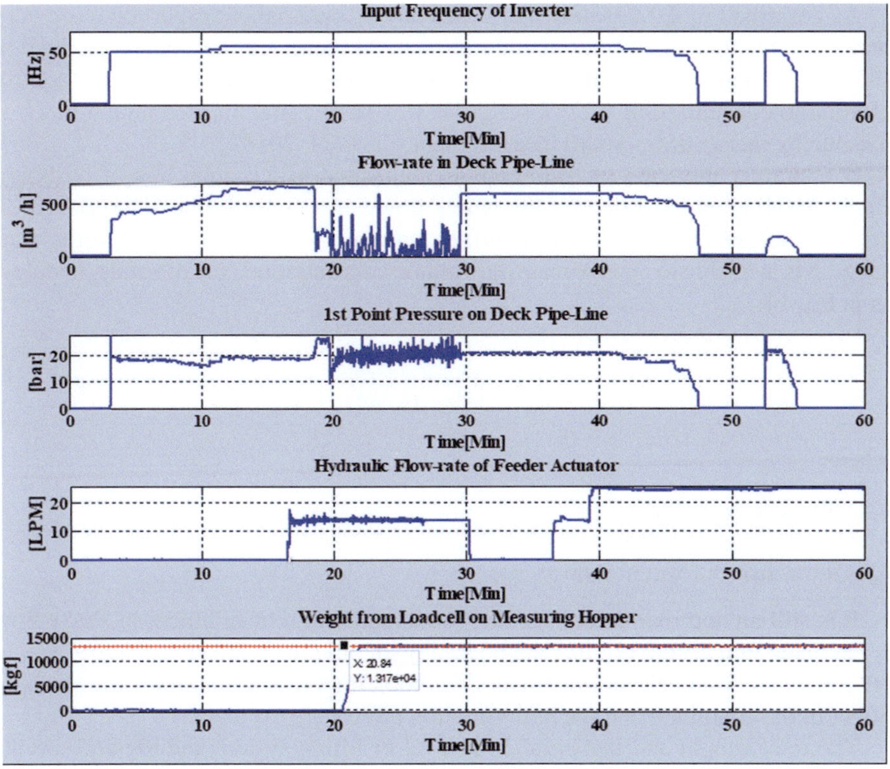

Fig. 33 An example of measurements during sea trial (KRISO 2015)

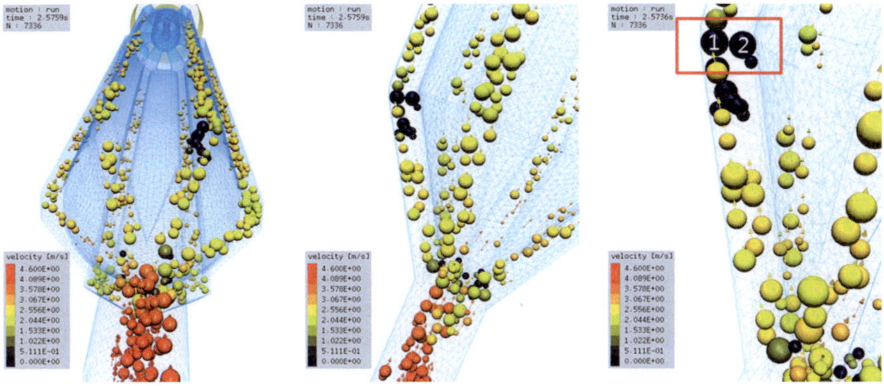

Fig. 34 A simulation result of particles dynamics through duct channels of pump shroud (KRISO 2015)

As described in 2.2.2, safety and efficiency of materials transportation are very important for the sake of sustainable mining. Maximizing the nodule production and minimizing the by-products at the same time are closely related with the control of volume concentration (C_v) of solids. It was shown that approximation of C_v is feasible by means of on-board measurement (Oh et al. 2018).

A container mounted on load cells is connected to a branch line of on-board slurry flowline and is filled by the slurry flow for a certain time span. Based on measurements of the weight (W_s) and the volume (V_s) of slurry filled in the container, it is available to approximate the volume concentration (C_v) of solids (nodules) as in Eq. 11:

$$C_v = \frac{W_s}{(\rho_n - \rho_l) g V_s} \tag{11}$$

where:

ρ_s is the density of slurry
ρ_l is the density of seawater
ρ_n is the density of wet nodule

It is still an approximation, because the content ratio of sediment in slurry flow is not taken into account in this methodology. In order to evaluate the slurry weight (W_s) from the load cell measurement data, the effect of ship motions, the sloshing effect in the container, and the hull vibration effects are excluded.

The total weight and launch-and-retrieval of lifting system and the undesirable by-products (sediment and seawater) are a great concern for design and operation of mining platform.

Concerned with sustainability issue, as illustrated in Fig. 5, the huge amount of seawater pumped up to surface has to be separated fast from nodule fragments and also sediment particles as much as possible, and the separated seawater should be treated according to environment regulation. The capacities of treatment facilities have to be matched with the lifting capacity.

With regard to safety issues, the lifting pipe is exposed to lasting loads from inside and outside: total weight, tension and bending, abrasion and collision, corrosion, forced excitation by ship and waves, fluid-structure interactions such as vortex-induced vibrations (VIV), and internal flow effects as well (Stanton and Yu 2010). The weight and top tension is the most critical parameter for design of derrick structure and of course for mining platform. Next, the on-board treatment facilities for separation, filtering (optional), and tailing process will be laid out on deck aiming for a smooth flow. Then, the design engineers would face conflicting problems in deck space, hull strength, and stability of mining platform.

Small-sized nodule particles will have advantage for slurry lifting, because free fall (terminal) velocity is lower than larger ones and higher volume concentration of solids might be expected. However, a higher crushing ratio, defined as crushed size versus original size, would raise design complexity of collector and lower material

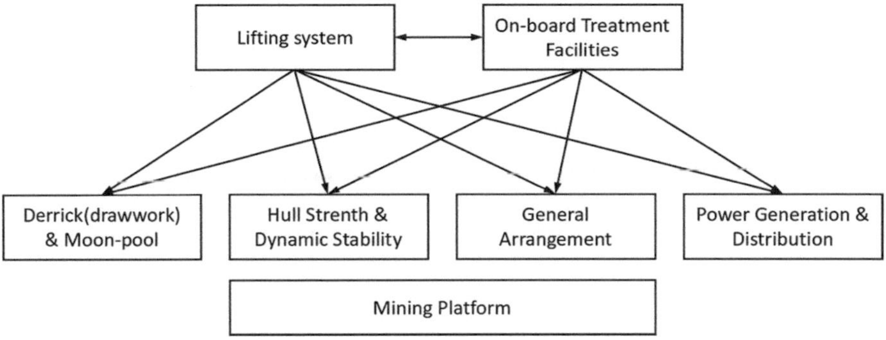

Fig. 35 Relationships of lifting system and on-board treatment system for design of mining platform

durability. Fine (powderized) particles generated by crushing and even in lifting pumps (Yamazaki and Sharma 2001) could increase the wear damage of materials by abrasion and further would make the on-board separation and treatment process more difficult.

The coupled relationships between lifting system, on-board treatment facilities, and design of mining platform are schematically shown in Fig. 35.

In functional aspects, a mining platform will get a combined hull form of drill-ship and ore carrier, where drilling equipment and mud system are replaced with the on-board treatment facilities. However, the seawater pumping in DSM is not same with the mud circulation in drilling operation. It is expected that the water depth of 5000 m and the amount of seawater pumping in DSM would pose severe hindrance to recycling of seawater.

The deep-seabed mining platform is the most important unit of total DSM system in the aspects of operation and investment. The design of mining platform has to meet the functional and sustainability requirements (see Table 1). It is necessary to integrate the design of on-board treatment system with the hull structure design of mining platform. It will require to change the approach of design and engineering of mining platform, that is, not to see the on-board treatment facility as a topside facility to be simply mounted on the deck of mining platform. In this way, general desk arrangement, arrangement of ballasting tanks, tailing treatment process, hull strength design, and dynamic stability of mining platform could be influenced in a positive sense.

4 Conclusions

Sustainability is a substantial subject dealing with safety and profitability in deep-seabed mining. Environmental impacts and potential harms to ecosystem are mainly caused by benthic intervention and materials transportation. Safety of deep-seabed mining can be ensured by minimizing and control of environmental impacts.

The functional requirements for safe and sustainable mining should be based on the operability of mining system. The design and engineering has to be integrated with investigation on operational feasibility of deep-seabed mining system.

Simulation-based design (SBD) and multidisciplinary design optimization (MDO) are powerful tools for the design and engineering of DSM system, and it can reduce development risks and fulfill safety and sustainability.

A seafloor mining robot, capable of tracking mining paths covering most area of mine site, is prerequisite for enhancement of overall collecting efficiency and for minimizing the benthic intervention. Integrated design of lifting system, on-board treatment system, and mining platform is very critical for the sake of a sustainable mining system. New approach for an integrated design of hull structure with on-board treatment facilities would give rise to more benefits.

References

American Petroleum Institute (API). (2006). Recommended practice for design of risers for Floating Production and Storage Systems (FPSOs) and Tension Leg Platforms (TLPs), API 2RD, June 1998 edition reaffirmed May, 2006.

American Petroleum Institute (API). (2008). Recommended practice for development of safety and environmental management program for offshore operations and facilities, API RP 75 (The 3rd edition).

Aso, K., Kan, K., Doki, H., & Iwato, K. (1992). Effect of vibration absorbers on the longitudinal vibration of a pipe string in the deep sea—Part 1: In case of mining cobalt crusts. *International Journal of Offshore and Polar Engineering, 2*(4), 309–317.

Aso, K., Kan, K., Doki, H., & Ohkoshi, T. (1994). The effects of vibration absorbers on the longitudinal vibration of a pipe string in the deep sea—Part 2: A case for mining manganese nodules. *International Journal of Offshore and Polar Engineering, 4*(1), 62–67.

Brink, A. W., & Chung, J. S. (1981). Automatic position control of a 300,000-ton ship ocean mining system. *OTC, 4091*, 205–224.

Chung, J. S., & Whitney, A. K. (1981, 1983). Axial stretching oscillation of an 18,000-ft vertical pipe in the ocean. *Journal Energy Resources Technology, ASME 105*, 195–200, OTC(1981).

Engelmann, H. E. (1978). Vertical hydraulic lifting of large-size particles – a contribution to marine mining. In: *Proceedings of offshore technology conference, OTC 3137* (pp 731–740).

Hong, S. (1995). Three dimensional analysis on behaviour of mining pipe of deep seabed mineral resources. In: *Proceedings of 1st ISOPE ocean mining symposium*, Tsukuba, Japan (pp 69–74).

Hong, S. (1997). 3-D dynamic analyses of lifting pipe systems in deep seabed mining. In: *Proceedings of 2nd ISOPE ocean mining symposium*, Seoul.

Hong, S., & Choi, J. S. (2001). Experimental study on grouser shape effects on traffic ability of extremely soft seabed. In: *Proceedings of 4th ISOPE ocean mining symposium*, Szczecin (pp. 115–118).

Hong, S., & Yang, N. (2018). Sustainability aspects in deep-seabed mining of polymetallic nodule (to be presented). In: *Proceedings of annual underwater mining conference*, Bergen, Norway, September 10–14.

Hong, S., Choi, J. S., Kim, J. H., & Yang, C. K. (1999). Experimental study on hydraulic performance of hybrid pick-up device of manganese nodule collector. In: *Proceedings of 3rd ISOPE ocean mining symposium*, Goa (pp 69–77).

Hong, S., Choi, J. S., Kim, J. H., & Yang, C. K. (2001). A note on design and operation of water-jet nodule lifter of manganese nodule collector. *International Journal of Offshore and Polar Engineering, 11*(3), 231–233.
Hong, S., Choi, J. S., & Yang, C. K. (2002a). Experimental study on solid-water slurry flow in vertical pipe by using PTV method. In: *Proceedings of annual conference ISOPE.*
Hong, S., Kim, H. W., & Choi, J. S. (2002b). Transient dynamic analysis of tracked vehicles on extremely soft cohesive soil. In: *Proceedings of 5th ISOPE Pacific/Asia offshore mechanics symposium*, Seoul (pp. 100–107).
Hong, S., Choi, J. S., & Kim, H. W. (2003a). Effects of internal flow on dynamics of underwater flexible pipes. In: *Proceedings of 5th ISOPE ocean mining symposium*, Tsukuba (pp. 91–98).
Hong, S., Kim, H. W., & Choi, J. S. (2003b). A new method using Euler parameters for 3D nonlinear analysis of marine risers/pipelines. In: *Proceedings of 5th ISOPE ocean mining symposium*, Tsukuba (pp. 83–90).
Hong, S., Kim, H. W., Choi, J. S., & Yeu, T. K. (2006). Launching simulation of deep-seabed integrated mining system. In: *Proceedings of annual autumn conference of KSOE* (pp. 311–314) (in Korean with English abstract).
Hong, S., Choi, J. S., Kim, H. W., Yeu, T. K., Kim, J. H., Kim, Y. S., Kang, S. G., & Rheem, C. K. (2011a). Experimental study on vortex-induced vibration of a long flexible pipe in sheared flows. In: *Proceedings of the 9th ISOPE ocean mining symposium*, Maui (pp. 70–77).
Hong, S., Kim, H. W., Choi, J. S., Yeu, T. K., Chi, S. B., & Lee, K. Y. (2011b). Comparative study on mining robots design for polymetallic nodules and seabed massive sulfides. Poster Presentation at Annual Conference of Underwater Mining Institute, Hawaii, Sept 2011.
Hong, S., Kim, H. W., Yeu, T. K., Choi, J. S., Yoon, S. M., Kim, J. H., Lee, C. H., Min, C. H., Lee, M. U., Sung, K. Y., Kim, S. S., & Oh, J. W. (2013). Pilot mining robot for polymetallic nodules and pre-pilot mining tests. In: *Proceedings of annual conference of Underwater Mining Institute*, Rio de Janeiro.
Hong, S., Choi, J. S., Yeu, T. K., Kim, H. W., Yoon, S. M., Min, C. H., Sung, K. Y., Lee, C. H., & Oh, J. W. (2014). Development of buffer station for safe and eco-friendly mining of seabed mineral resources. In: *Proceedings of annual conference of Underwater Mining Institute*, Lisbon.
Hong, S., Yeu, T. K., Kim, H. W., Jung, J. Y., Min, C. H., Cho, S. G., Yoon, S. M., Kim, S. S., Kim, J. H., Lee, C. H., Sung, K. Y., Oh, J. W., & Park, S. H. (2016). Technology achievements by means of 3rd pre-pilot mining test: Pilot lifting system. In: *Proceedings of annual conference of Underwater Mining Institute*, Singdo.
International Electrotechnical Commission (IEC). (2009). Functional safety of electrical/electronic/programmable electronic safety-related systems, IEC 61508.
International Maritime Organization (IMO). (1972). Convention on the prevention of marine pollution by dumping of wastes and other matters.
International Maritime Organization (IMO). (1974). International convention for the safety of life at sea, IMO SOLAS 74.
International Maritime Organization (IMO). (1978). International convention for the prevention of pollution from ships, IMO MARPOL 73/78.
International Maritime Organization (IMO). (2014). Guidelines for the reduction of underwater noise from commercial shipping to address adverse impact on marine life. IMO MEPC.1/Circ.833.
International Organization for Standardization/International Electrotechnical Commission (ISO/IEC). (2014). Safety aspects – guidelines for their use in standards, ISO/IEC Guide 51:2014.
International Organization for Standards (ISO). (2010). Safety of machinery – General principles for design – Risk assessment and risk reduction, ISO 12100:2010(en).
Kim, H. W., Hong, S., & Choi, J. S. (2003). Comparative study on tracked vehicle dynamics on soft soil, single-body dynamics vs. multi-body dynamics. In: *Proceedings of 5th ISOPE ocean mining symposium*, Tsukuba (pp. 141–148).

Kim, H. W., Hong, S., Choi, J. S., & Yeu, T. K. (2006). Total dynamic analysis of deep-seabed integrated mining system. In: *Proceedings of annual autumn conference of KSOE* (pp. 311–314) (in Korean with English abstract).

Kim, H. W., Hong, S., Lee, C. H., Choi, J. S., & Yeu, T. K. (2009a). A study on steering performance of four column tracked vehicle on extremely cohesive soft soil. In: *Proceedings of 8th ISOPE ocean mining symposium*, Chennai. September 20–24.

Kim, H. W., Lee, C. H., Hong, S., Yeu, T. K., & Choi, J. S. (2009b). A study on steering characteristics of pilot Mining Robot (MinerRo II) on extremely cohesive soft soil. In: *Proceedings of annual conference of Underwater Mining Institute*, Rio de Janeiro.

KRISO (2013). Annual report on development of integrated mining technology for deep-seabed mineral resources for Korea. (in Korean) March, 2014.

KRISO (2015). Annual report on development of integrated mining technology for deep-seabed mineral resources for Korea. (in Korean) March, 2016.

Lee, T. H., Jung, J. J., Hong, S., Kim, H. W., & Choi, J. S. (2007). Prediction for motion of tracked vehicle traveling on soft soil using kriging metamodel. *International Journal of Offshore and Polar Engineering, 17*(2), 132–138.

Lee, M. U., Cho, S. G., Choi, J. S., Kim, H. W., Hong, S., & Lee, T. H. (2012). Metamodel-based multidisciplinary design optimization for development of a deep-sea manganese nodules test miner. *Journal of Applied Mathematics, Vol. 1*, pp. 1–18.

Nakata, K., Kubota, M., Aoki, S., & Taguchi, K. (1997). Dispersion of resuspended sediments by ocean mining activity-modeling study. In: *Proceedings of the international symposium on environmental studies for deep-sea mining*, Metal Mining Agency of Japan, Tokyo (pp. 169–186).

Oebius, H. (1993). Entwicklung eines umweltschonenden Manganknollenabbau- und -gewinnungsverfahrens. Report 1228/94, VWS, BMFT Forschungsvorhaben.

Oh, J. W., Jung, J. Y., & Hong, S. (2018). On-board measurement methodology for the liquid-solid slurry production of deep-seabed mining. *Ocean Engineering, 149*, 170–182.

Stanton, P., & Yu, A. (2010). Interim use of API codes for the design of dynamic riser systems for the deep sea mining industry. In: *Proceedings of the 29th international conference on ocean, offshore and arctic engineering*, Paper No OMAE2010–20189.

UNOET. (1987). *Delineation of mine sites and potential in different sea area*. London: UN Ocean Economics and Technology Branch and Graham & Trotman Limited, 79 pp.

World Bank Group. (2017). *The growing role of minerals and metals for a low carbon future*. Report of World Bank Group and Extractives Global Programmatic Support, June 2017.

Xia, J. X., & Huang, J. Z. (2000). Additional pressure losses of solid-liquid flow in an oscillating pipeline. *Journal of Hydrodynamics, B., 1*, 75–80.

Xia, J. X., Ni, J. R., & Mendoza, C. (2004). Upward flow of large size particles-water mixtures through swaying pipe. *Journal of Transportation Engineering ASCE, 130*(4), 535–543.

Yamazaki, T., & Sharma, R. (2001) Preliminary experiment on powderization of deep-sea sedimentduring hydraulic transportation. In: *Proceedings of the 4th ISOPE ocean mining symposium*, Szczecin (pp. 44–49).

Yamazaki, T., Tomishima, Y., Handa, K., & Tsurusaki, K. (1989). Experimental study of adhesion appearing between plate and clay. In: *Proceedings of the 8th international conference on offshore mechanics and arctic engineering, vol. 1* (pp 573–579).

Yamazaki, T., Tsurusaki, K., Handa, K., & Inagaki, T. (1995). Geotechnical properties of deep ocean sediment layer. *Journal Mining Material Process Instituute Japan, 111*, 309–315. (in Japanese with English abstract).

Yamazaki, T., Kuboki, E., Yoshida, H., & Suzuki, T. (2000). A consideration on size distribution of resuspended deep-sea sediments. In: *Proceedings of the 10th international offshore and polar engineering conference, vol. 1* (pp 507–514).

Yeu, T. K., Yoon, S. M., Hong, S., Kim, J. H., Kim, H. W., Choi, J. S., & Min, C. H. (2013). Operating system of KIOST pilot mining robot in inshore test. In: *Proceedings of the 10th ISOPE ocean mining symposium*, Szczecin (pp. 265–268).

Dr. Sup Hong is a principal researcher at Korea Research Institute of Ships and Ocean Engineering (KRISO). He majored in Naval Architecture and Offshore Engineering at Seoul National University receiving Bachelors (1983) and Masters (1985) degrees. He holds a Doctorate for offshore engineering from Technical University of Aachen (1992). Since 1993, he has worked in the research field of deep-sea engineering. The main research focus is on development of deep-seabed mining technology. Based on the experiences of sea trials with test collector (2009–2010), pilot mining robot "MineRo" (2012–2013), and pilot lifting system (2015), his interest is focused on sustainable mining of deep-seabed minerals.

Part II
Environmental Impact Assessment

Assessment of Deep-Sea Faunal Communities-Indicators of Environmental Impact

Virginie Tilot

Abstract Our assessment of deep-sea faunal communities is based on the results of a comprehensive UNESCO/IOC baseline study of the megafaunal assemblages of the metallic nodule ecosystem of five areas within the Clarion Clipperton Zone (CCZ) of the eastern Pacific Ocean. This study serves as benchmark to interpret the structure of megafaunal populations associated with benthic biotopes in areas targeted for mining. It identifies on a large scale the variability of nodule and sediment facies and their associations with specific megafaunal communities. An appropriate set of management tools and options have been developed, in particular indicators of sensitivity to environmental changes anthropogenically or naturally induced. The general characteristics of the nodule ecosystem in the CCZ and its sensitivity to deep-sea mining are discussed from the surface to the seabed in relation to recent research on the description of water masses and dynamics and an assessment of their vulnerability. A tridimensional multiparametric rapid environmental assessment (REA) has been applied on one pilot site of the French contract area using GIS zoning, ecohydrodynamics, and sensitivity indexes.

Keywords Polymetallic nodules · Deep-sea mining · Clarion Clipperton Zone · Deep-sea benthic communities · Sensitivity indicators · Environmental impact

1 Introduction

Knowledge of the structure of its megafaunal assemblages is essential to understanding the functioning of any deep-sea ecosystem (Rex and Etter 2010). Typically this faunal component includes a significant fraction (17–50%) of benthic abyssal biomass (Haedrich and Rowe 1977). Since the megafauna (Fig. 1) is also one of the principal agents of bioturbation at the depositional interface of the deep-sea benthos (Mauviel and Sibuet 1985; Levin et al. 1986), it can influence many other biological

V. Tilot (✉)
Muséum National d'Histoire Naturelle, Patrimoine Naturel (PATRINAT), Paris, France

Instituto Español de Oceanografía, Malaga, Spain

R. Sharma (ed.), *Environmental Issues of Deep-Sea Mining*,
https://doi.org/10.1007/978-3-030-12696-4_6

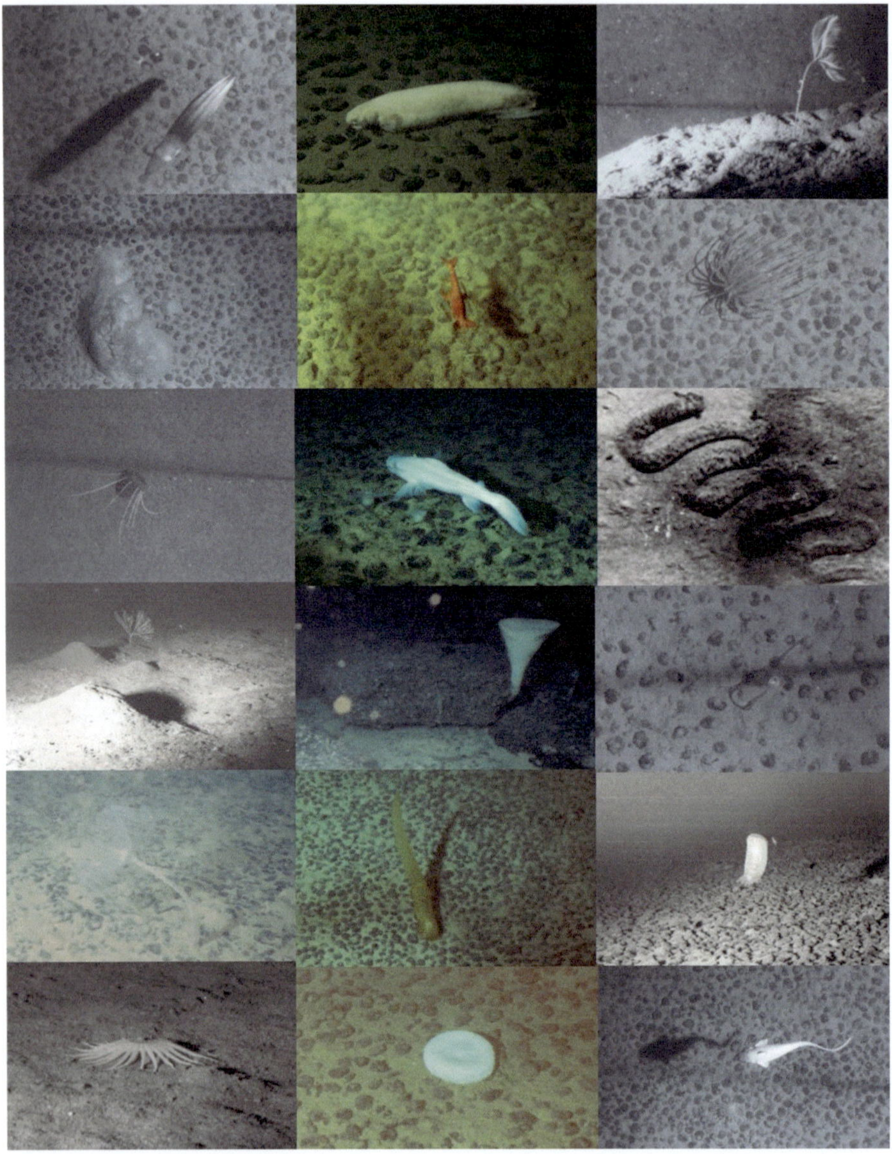

Fig. 1 Photos of megafauna in the nodule ecosystem in the CCZ taken by the platforms "Epaulard, RAIE, TROIKA and the manned submersible Nautile" from the three French study sites. From top left to bottom right: 1 Cephalopod *Cirrata Cirroteuthis* sp. over nodule facies C 40% ©IFREMER, 2 Fish Ophidiid *Barathrites* sp. over facies C15% ©IFREMER, 3 Crinoid *Hyocrinus bethellianus* on an escarpment © IFREMER, 4 Mound and burrow of an echiurian worm Bonellidae *Jacobia birsteini* on facies C 40% ©IFREMER, 5 Crustacean Decapoda *Plesiopenaeus* sp. or *Aristaeomorpha* sp. on facies B 50% ©IFREMER, 6 Ceriantharian Aliciidae, *Cerianthus* sp. on facies B 50% ©IFREMER, 7 Scyphomedusa *Periphylla periphylla* over facies B ©IFREMER, 8 Fish *Bathysaurus mollis* over facies C 30% ©IFREMER, 9 Enteropneusta Spengelitae *Glandiceps abyssicola* on facies O ©IFREMER, 10 Octocorallia Gorgonacea Primnoidae, Bioturbation and

and geochemical components of the deep ocean (Sharma and Rao 1992). In particular, the benthic fauna plays an important role in carbon cycling and mineralization within the epibenthos. Interestingly, it also contributes to the genesis of polymetallic nodules, bioturbation along with bottom currents playing a role in allowing nodules to remain on the seafloor (Friedrich and Plüger 1974; Du Castel 1985; Halbach et al. 1988).

Megafaunal assemblages also serve as good indicators of the status of a habitat in the face of natural and anthropogenic impacts and may be used to identify areas, environmental parameters, faunal communities, and associations on a large scale through video/photo imagery (Tilot et al. 2018), to measure the variation in flux of particulate organic carbon (Smith et al. 1997), to identify critical environmental parameters, to characterize selected habitats and associated nodule facies (Tilot 2006a; Tilot et al. 2018), and to measure rates of recolonization under natural or impacted conditions (Tilot 1988, 1989, 1990a, b, 1991, Bluhm 1997).

Echinoderms and cnidarians, the two dominant phyla in the CCZ, are good indicators of nodule facies and edaphic parameters as developed further. On the photo/video imagery of the CCZ, echinoderms, in particular holothurians, were studied in detail (Tilot 2006b). Comparisons have been made with other sites outside the CCZ in particular the nodule ecosystem of the Peru basin (Tilot 1989) where echinoderms in particular have served as good indicators for impact studies (Thiel et al. 1993; Bluhm et al. 1995; Bluhm and Gebruk 1999; Glover et al. 2016).

Fixed organisms, generally suspension feeders, and swimming holothurians (Elasipods), such as *Enypniastes eximia* and *Peniagone leander*, are benthopelagical indicators of bottom currents. Some swimming holothurians can be seen at several hundreds of meters above the seabed (Rogacheva et al. 2012).

The large equatorial polymetallic nodule belt of the CCZ (118–157° W/9–16° N), wherein lie the most economically important deposits, covers about 2 million km^2 of abyssal hills and escarpments in the eastern Pacific (Halbach et al. 1988; Kotlinski 1998). Considerable regional-scale variation in the geological environment (topography, erosion by deep ocean currents, and regional deposition of sediments) has led to a classification of nodule deposits and the recognition of a series of distinct *nodule facies* (Hoffert 2008; Hoffert and Saget 2004), differentiation of which has been based on a combination of photographic study and of collection of samples for morphological and geochemical assessment (Table 1).

Fig. 1 (continued) mounds on facies O ©IFREMER, 11 Sponge Hexactinellida *Bathydorus* sp. on rocky facies ©IFREMER, 12 Ascidian Sorberacea *Phlebobranchiata* sp. on facies C15% ©IFREMER, 13 Hydrozoan *Branchiocerianthus imperator* on facies C 40% ©IFREMER, 14 Holothurian *Psychropotes longicauda* on facies B 40% ©IFREMER, 15 Holothurian *Peniagone leander* on facies B 40% ©IFREMER, 16 Actinian *Ophiodiscus* sp. or *Bolocera* sp. on facies O ©IFREMER, 17 Actinian *Liponema* sp. on facies C 20% ©IFREMER, and 18 Fish Macrouridae *Coryphaenoides armatus* or *C. yaquinae* on facies B 50% ©IFREMER. (Source: Tilot (2010b))

Table 1 Nodules-facies classification based on visual identification from samples and imagery (Voisset and Hein1978; Hoffert and Saget 2004)

	Nodule facies				
Criteria	A	BP	B	C	C+
Degree of burial (%)	0	0–30	30–60	60–100	30–60
Est. average diameter in cm	<2 to 10	2.5 to 4.5	2.5 to 10	2.5 to >15	2.5 to 7.5
Morphology	Spherical, polylobate	Heterogen fragments	Disk, ellipsoid polylobate, fragments	Spherical, ellipsoid	Spherical, ellipsoid fragments
Surface texture	Very smooth	Smooth	Smooth, some granular	Very hummocky equatorial thickening	Hummocky granular some smooth
Sediment on upper surface	No	Some	Some	A lot	Yes
Aggregated nodules	Very frequent	Very frequent	Never	Never	Never
Density on the seabed	High (30–50%)	High (35–50%)	High (40–50%)	Low (5–20%)	Medium (20–40%)

The comprehensive study of the biodiversity and distribution of the epibenthic megafauna of the CCZ presented here is based on data collected from five study sites (Fig. 2). The research was originally undertaken at IFREMER with funding from EU, IFREMER, and NOAA (Tilot 1988, 1989, 1990a, b, 1991, 1992, 1995, ESCO CNRS IFREMER 2014) and then updated and expanded with the additional BIE site as part of a project supported by the Intergovernmental Oceanographic Commission (IOC) of UNESCO (published in 3 vol. see: http://unesdoc.unesco.org/images/0014/001495/149556e.pdf#223; Tilot 2006a, b, c, 2010a, b) in order to establish a UNESCO/IOC baseline. This research led to the development of options for the management and conservation of the nodule ecosystem, taking into account relevant scientific, legal, and institutional issues (UNESCO/IOC vol 3, Tilot 2010a). Similar options have been proposed for other deep-sea habitats and discussed at a panel meeting of experts held at UNESCO (ASOM 2010). This provided an international platform for consideration of these issues (Tilot 2011) and initiated the development of a number of management and conservation tools (Tilot 2013, 2014). In particular, the application of marine spatial planning was considered (Ardron et al. 2008; Tilot 2004). In this context, a multilayer Rapid Ecological Assessment (REA) (Tilot 2014), sensitivity indexes (Tilot 2003, 2016; Tilot et al. 2008a, b), GIS, zoning, and ecohydrodynamics have been applied to a pilot site in a French contract area, NIXO 45 (Tilot et al. 2018).

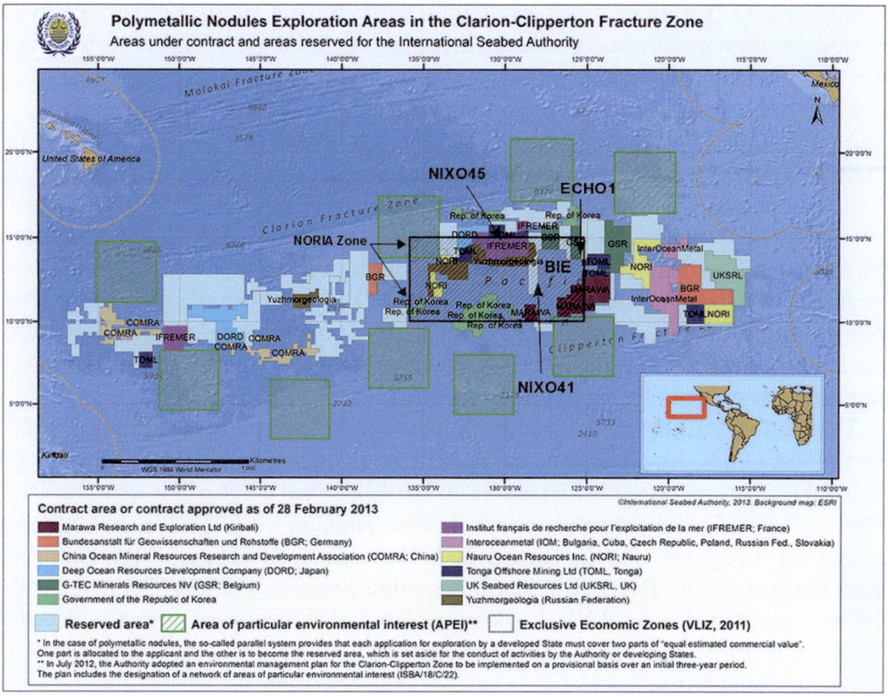

Fig. 2 Updated map adapted from https://www.isa.org.jm of the CCZ showing the areas under contract and the areas reserved for the International Seabed Authority, as well as the Areas of Particular Environmental Interest (APEI) and the UNESCO/IOC Baseline Study areas (NORIA, NIXO 45, NIXO 41, Echo 1, BIE)

2 Methods and Study Areas

The study areas used during the UNESCO/IOC baseline study (Tilot 2006a, b, 2010a, b) were the NORIA site, which include NIXO 45, NIXO 41, the ECHO 1, and the Benthic Impact Experiment (BIE) sites.

The NORIA (AFERNOD) area which covers about 450,000 km^2 delimited between 125°W/135°W and 11°N/16°N was surveyed from 1974 on a grid pattern with a station every 2.4 km, in order to determine the structural, bathymetric, and sedimentary environment of the nodules. The NIXO 45 site lies within 130°00′W/130°10′W, 13°56′N/14°08′N, at a mean depth of 4950 m. The NIXO 41 site lies within 127°W/130°W, 12°10′N/13°35′W at a depth of 4700–5000 m and extends eastward to NIXO 45. The ECHO I site is located at 14°40′N-125°25′W at a depth of 4500 m (Fig. 1). The BIE site is located at 12°59′N/128°22′W at a depth of 4800–5000 m.

Photographic data have been collected by the Deep Tow System during two cruises on ECHO I site, during the ECHO I cruise in 1983 (Spiess et al. 1987) and during the SIO Quagmire II in 1990 (Tilot 1990a; Wilson 1990). Both cruises were in situ environmental impact studies investigating biological impacts for mining operations conducted during 1978.

The BIE was sponsored by the National Oceanic and Atmospheric Administration's (NOAA) Ocean Minerals and Energy Division on board the YUZHMORGEOLOGIYA in collaboration with Russian, Japanese, German, and French scientists during July 1991 (Smith and Trueblood 1991). The BIE objectives were to assess the potential environmental impacts of deep-sea mining on the deep-sea benthos by blanketing a large area of the deep-sea floor with sediment to simulate deep-sea mining and monitor the response of the benthic community both spatially and temporally. The assessment of megafaunal communities has been performed (Tilot 1991) prior to the blanketing phase and represents a baseline assessment of megafauna in the area.

The platforms used for video and still photography and physical sampling within the UNESCO/IOC baseline study areas were the French towed device "Remorquage Abyssal d'Instruments pour l'Exploration" (RAIE), the suprabenthic camera sleigh "Troika," the French camera coupled free sampler "ED1," the autonomous unmanned "Epaulard," the French manned submersible, "the Nautile" for the French sites, the American "Deep Tow Instrumentation System" for ECHO I and for the BIE site, and the Russian MIR-1 system which includes a series of boxes mounted on the same frame with photo and video cameras GOU453, USNEL-type box cores, and multi-cores Barnett-type.

The assessment of megafauna presented here is based on the analysis of more than 200,000 photos of the seafloor and more than 55 h of videos. So much imagery permitted the compilation of an exhaustive database. Comparisons were made with the fauna from the Peruvian basin analyzed during DISCOL cruise (Tilot 1989). The description of megafauna was facilitated by establishing a taxonomic atlas annotated with observations on the morphology, ethology, and feeding behavior of each taxon identified on imagery. Information on the geographic and bathymetric distribution of each taxon was compiled from data from abyssal regions, mainly in Pacific and Atlantic, outside the CCZ (UNESCO/IOC baseline vol 1, Appendix IV) (Tilot 2006a). Hypotheses of identification for each taxa were collated on the advice of international specialists and the literature of the region and other deep oceans in the world. A photographic atlas was assembled for all morphotypes identified as displayed in Appendix I, II of UNESCO/IOC baseline vol 1 (Tilot 2006a), in particular for echinoderms (UNESCO/IOC baseline vol 2, Tilot 2006b).

3 Results

The main results obtained from survey work in the five different areas of the UNESCO/IOC baseline may be summarized as follows:

3.1 NORIA

– A total of 159 taxa were recorded representing 13 phyla. These were mainly cnidarians (59 taxa), echinoderms (50 taxa including 32 holothurians), chordates (23 taxa including 17 fish), and sponges (21 taxa) (Fig. 3a).

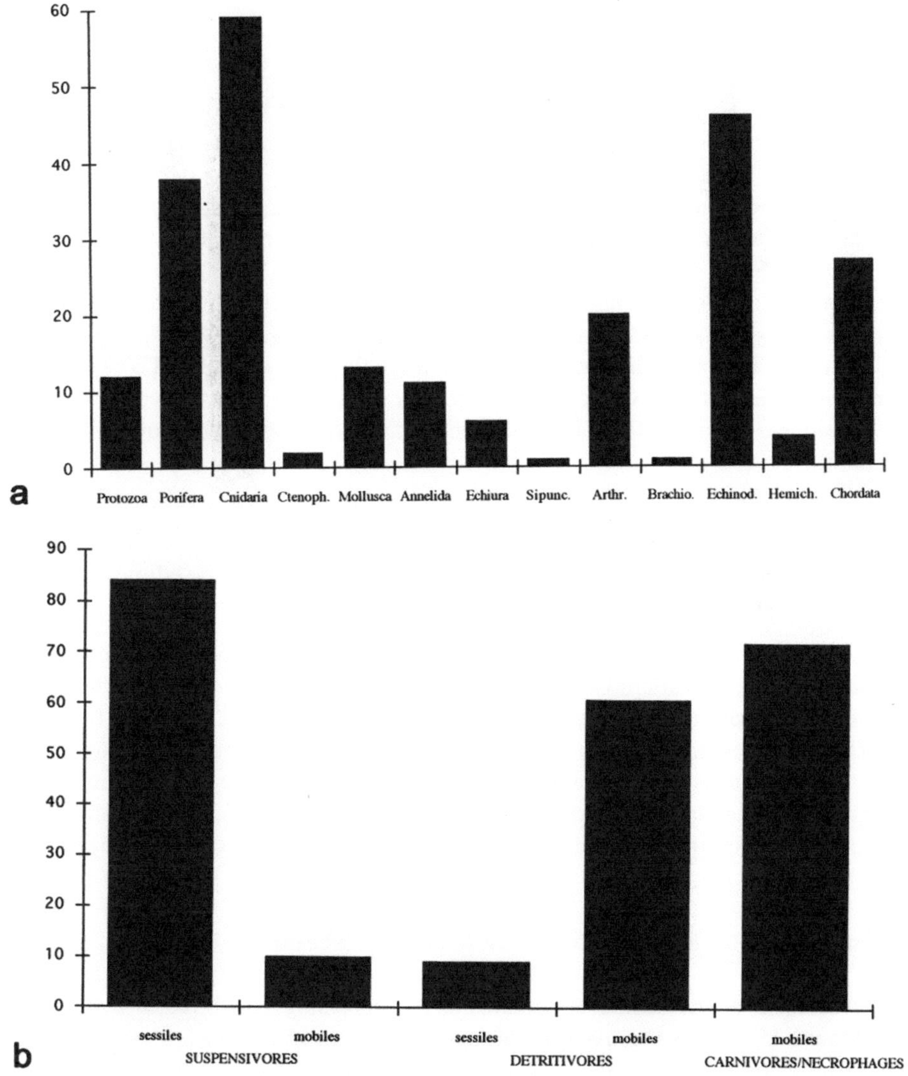

Fig. 3 Histograms for (**a**) the number of taxa in each phylum for the megafauna in CCZ and for (**b**) the taxonomic richness considering trophic and functional groups. (http://unesdoc.unesco.org/images/0014/001495/149556f.pdf, volume 1 Figs. 35 and 37)

- There was a prevalence of 84 taxa sessile, of which 10 taxa were mobile suspension feeders.
- Suspension feeders (68 sessile and 10 mobile taxa), which were mainly cnidarians and sponges, displayed a higher taxonomic richness than deposit feeders, which were mainly echinoderms (60 taxa, among which 50 taxa are mobile) and carnivores/scavengers (45 taxa) (Fig. 3b).

3.2 NIXO 45

- A total of 134 taxa were identified from the analysis of 70,000 m^2 of photographed surface; cnidarians outnumbered echinoderms (as in NORIA) and sponges (Fig. 4a).
- Nodule coverage and abiotic factors such as slope, topography, and bathymetry appeared to largely determine the abundance and composition of the faunal assemblage.
- Overall a higher taxonomic richness was recorded on all nodule facies (Fig. 5a), with a maximum of 48 taxa recorded on facies C 2–15%. However, 36 taxa were found on facies O with recent sediments, more than on facies B and BP (34 taxa), likely because facies O recent is located on the same valleys as facies C and is considered as a transition facies between patches of facies C with low to medium nodule coverage.

More specifically the habitats, in decreasing order of species richness, were facies C 10, facies O with recent sediments, facies C 15%, facies C 20%, facies BP 35%, facies C 2–5%, facies B 50%, facies BP 50%, facies O old sediments, facies C 0%, facies C 30%, and facies B 40%.

- The overall faunal abundance of 553 ind/ha encompassed mainly, in decreasing order, 258 ind/ha of cnidarians (mainly actinians and octocorallians), 180 ind/ha of echinoderms (holothurians and crinoids), and 56 ind/ha of sponges (Fig. 5).
- There were overall more suspension feeders than deposit feeders and carnivores/scavengers, irrespective of nodule facies (Fig. 5b). More specifically, there were more sessile suspension feeders than mobile deposit feeders, mobile carnivores/scavengers, motile suspension feeders, or sessile deposit feeders (Fig. 5).

Faunal Communities Identified as Indicators of Nodule Facies

- Overall, taxa were more abundant on, in decreasing order, facies C 10% on slopes >15°, facies O in recent sediments, facies C 15%, facies C 20%, facies BP 35%, facies C 2–5%, facies B 50%, facies BP 50%, facies O in old sediments, facies C 40%, facies C 30%, and facies B 40% (Fig. 5).
- Preferential habitats for suspension feeders were facies O with recent sediments and facies C 10% on slopes. Among motile suspension feeders, a predominance

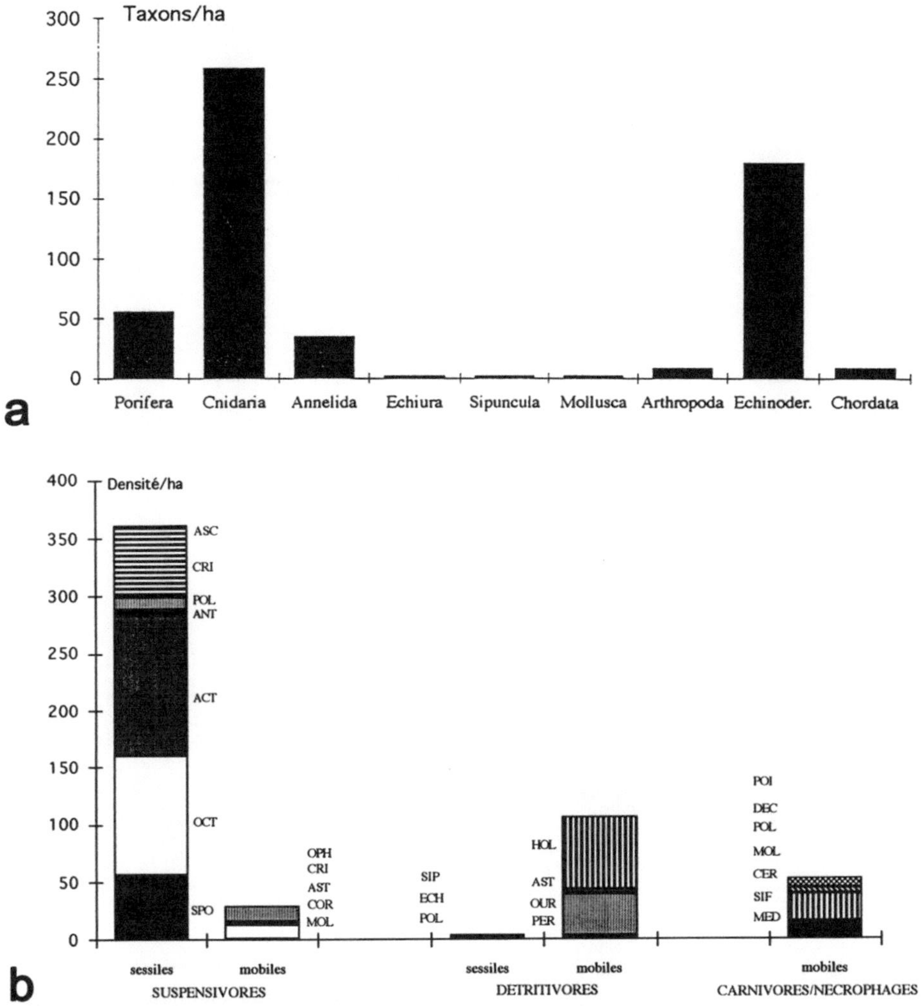

Fig. 4 Histograms for (**a**) abundance for each phylum and of (**b**) partitioned faunal abundance within trophic and functional groups of the megafauna of NIXO 45 site. (http://unesdoc.unesco.org/images/0014/001495/149556f.pdf, volume 1 Figs. 39 and 42)

of actinians was recorded, in decreasing order, on facies BP 35%, facies C 30%, facies B 50%, facies C 15%, and on facies O (Fig. 5b).

- The preferential habitat for deposit feeders, mainly echinoids and holothurians, was facies C 10%.
- Some phyla prevailed on specific facies, e.g., cnidarians, echinoderms, and sponges, on facies C 10% and facies C 30% on slopes.

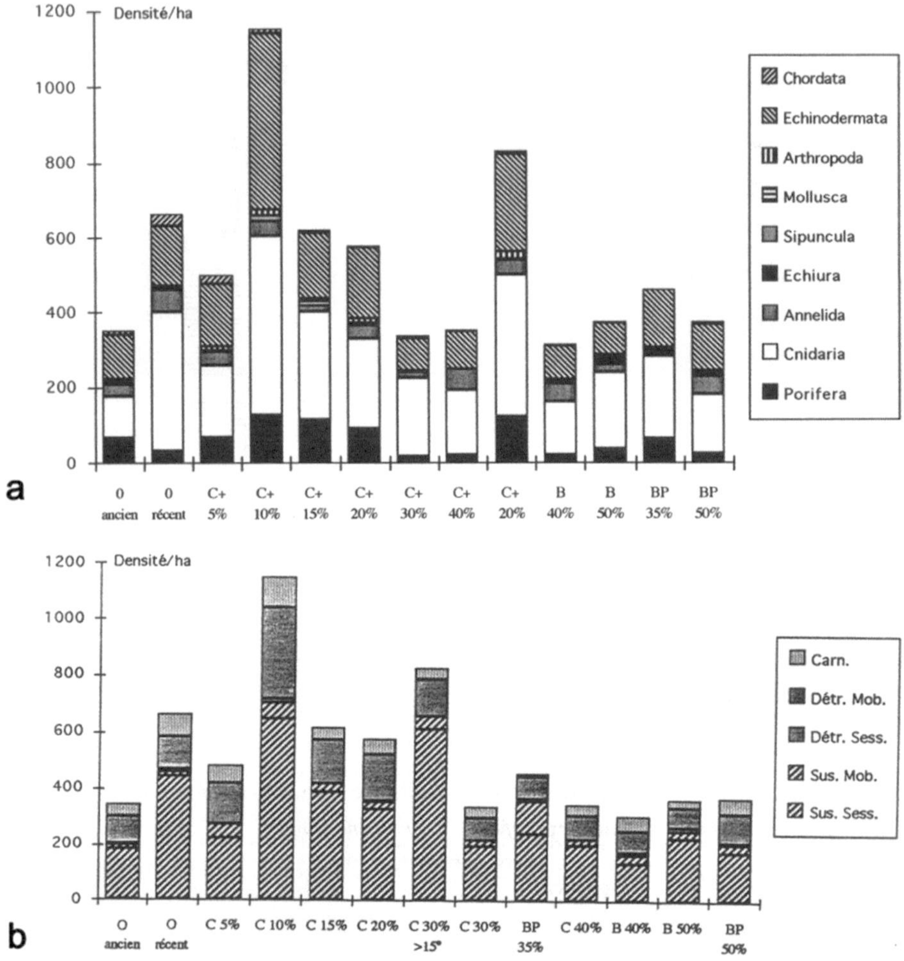

Fig. 5 Histograms for (**a**) the partitioned faunal abundance per phylum and for (**b**) trophic and functional groups within each nodule facies on NIXO 45 site. (http://unesdoc.unesco.org/images/0014/001495/149556f.pdf, volume 1 Figs. 41 and 43)

A Singular and Conspicuous Taxa as Nodule Facies Indicator

- The echiurian, *Jacobia birsteini* (Fig. 1 photo 4), a bonnelid worm, is known to construct large mounds (more than 2 m long, 80 cm wide and 50 cm high) (Tilot 1992, 1995, 2005, 2006a) which increase turbulence, induce an exchange with the interstitial water in galleries, and influence the depth of nodule cover. These bonnelid worms were present, at densities of up to 16 ind/ha, mainly on facies B40%, facies B50%, facies BP35%, and facies BP35%. They could be good indicators of these facies. By the mounds they build, they generate habitat heterogeneity and thus influence the dynamics of epifaunal assemblages; they also have a salient role in sediment irrigation and mixing and so an action on nodule diagenesis (Tilot 1995, 2006a, b, 2010b).

Other organisms are good indicators of edaphic conditions (slope, topography, heterogeneity of substrate) and currents as seen previously and developed further for NIXO 45 with the identification of preferential habitats.

As evidenced by the multi-criteria statistical analysis (Tilot 1992, 2006a) shown in Fig. 6, there were **13 preferential habitats** which could be ranked according to nodule coverage, habitat heterogeneity, and slope >15%, as follows:

- **Facies O on ancient sediments** was characterized by a majority of suspension feeders, notably sponges *Pheronema* sp., ringed hexactinellids, *Cladorhiza* sp. (most abundant on this facies), octocorallians (Primnoidae and Isididae), fixed crinoids Hyocrinidae, and actinians Hormathiidae and Actinoscyphiidae and Amphianthidae (*Amphianthus bathybium* sp. prevailed on this facies). Detritus feeders recorded were mainly asterids (Pterasteridae), and holothurians (*Synallactes aenigma, Benthodytes lingua).* Taxa unique to this facies were sponges Cladorhizidae, sedentary polychaetes with characteristic rounded mounds, asteroids *Hymenaster violaceus,* holothurians *Benthodytes lingua*, gastropods Pterotracheidae and Liparid fish.
- **Facies O on recent sediments** was characterized mainly by octocorallians among suspension feeders, by isopod Munnopsidae and by asteroids Porcellenasteridae among detritus feeders, and by Ophidioid and Ipnopid fishes among carnivores. A particular form of bioturbation, the "witch ring" (Heezen and Hollister 1971), associated with polychaete worms, was abundant. Unique to this facies were the holothurians *Psychropotes longicauda,* siphonophores *Physonectes*, and sedentary polychaetes of the family Cirratulidae.
- **Facies C 5–10%** was characterized by a peak in abundance of suspension feeders including octocoralliarids of the families Isididae, Primnoidae, and Umbellulidae, corallimorpharids of the family Sideractiidae (*Nectatis singularis* displayed high abundance only on this facies), and hexactinellid sponges of the family Hyalonematidae (*Hyalonema* sp., *Poecillastra* sp., *Phakellia* sp., and *Esperiopsis* sp. were most abundant on this facies). Detritus feeders were mainly the holothurians *Peniagone gracilis*, *Mesothuria murrayi, Paelopatides* sp., *Pannychia moseleyi*, *Enypiastes eximia*, the echinoids *Plesiodiadema globulosum*, the isopods of the family Munnopsidae and the sipunculids *Nephasoma elisae*. Carnivores were well represented by the Ophidioid fishes, jellyfishes of the family Trachynemidae and the decapods *Plesiopenaeus* sp. Unique to this facies were alveolate hexactinellid sponges, the demosponges *Phakellia* sp., bivalves of the family Vesicomyidae, and the decapods *Plesiopenaeus* sp.
- **Facies C 15–25% including slopes >15%** was characterized by a high diversity dominated by the suspension feeding actinids *Liponema, Actinernus verrill, Bolocera* sp. *Actinoscyphia* sp., the sponges *Caulophacus* sp. (which were most abundant on this facies) and *Poecillastra* sp., ring- or dish-shaped hexactinellids of the family Rosselidae, Euretidae, and Cladorhizidae and *Cornucopia* sp. (on slopes), crinoids of the family Antedonidae and polychaetes of the family

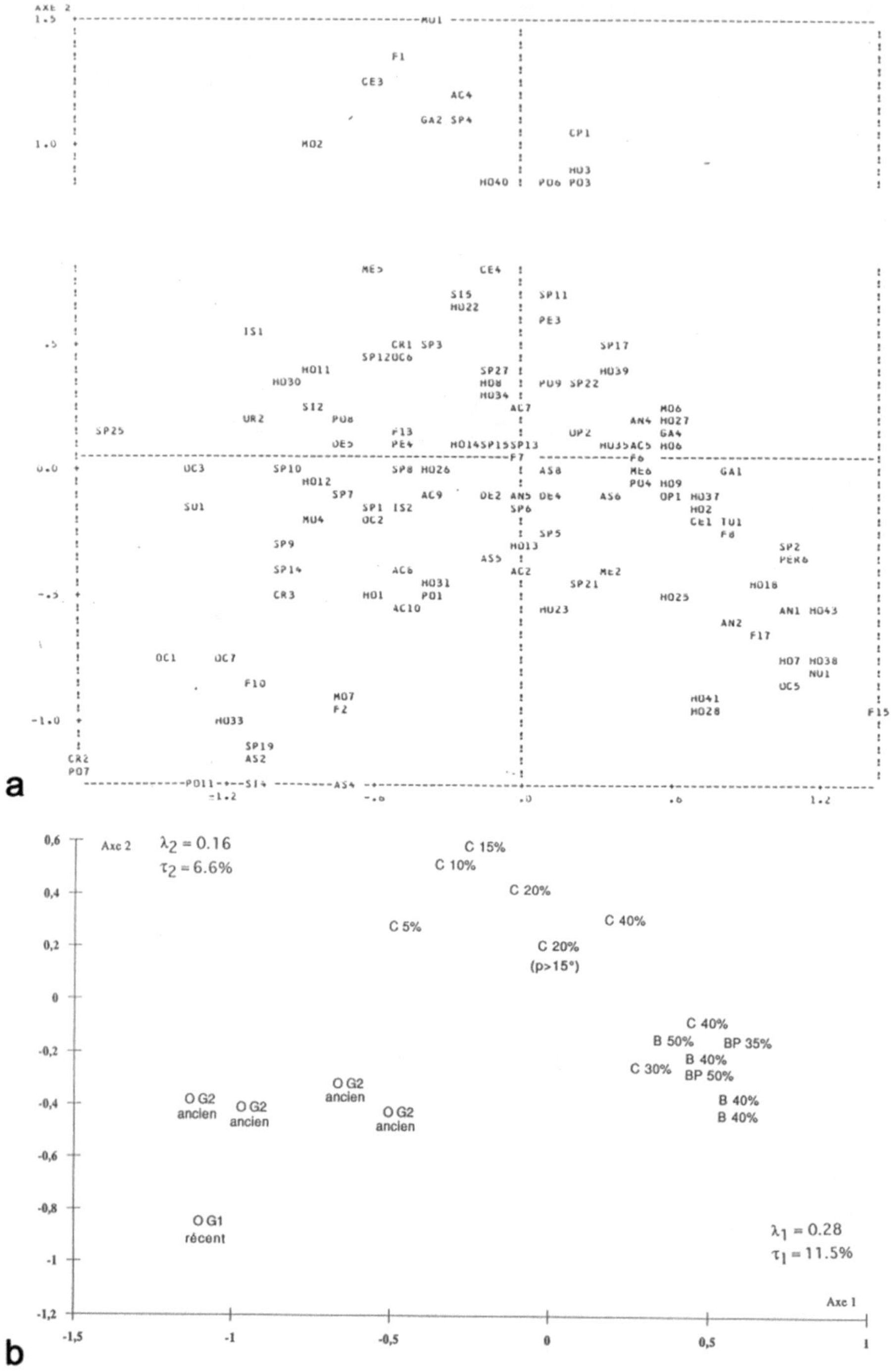

Fig. 6 Multidimensional scaling (MDS) plot to show (**a**) the degrees of similarity and difference in distribution patterns of the principal taxa and (b) in distribution patterns of the nodule facies within NIXO 45. (http://unesdoc.unesco.org/images/0014/001495/149556f.pdf, volume 1 Figs. 46 and 47)

Sabellidae. Detritus feeders were mostly echinoids of the family Aeropsidae, Holothurians *Peniagone vitrea*, *Meseres macdonaldi*, *Benthodytes* sp., Brisingidae with ten arms, and Peracarida Cumaceans. Carnivores were mostly decapods *Nematocarcinus* sp., archaeogastropods, siphonophores of the Rhodaliidae and fishes Bithitidae (*Typhlonus* sp.). Unique to this facies were sponges *Hyalonema* sp., members of the family Caulophacidac, siphonophores *Physonectes*, Chiroteuthid cephalopods, Galatheids hermit crabs with a rounded rostrum, neogastropods of the family Turridae, polychaetes of the family Polynoidae or Aphroditidae, and fishes *Coryphaenoides yaquinae*.

- **Facies C 30–40%** was characterized mainly by the suspension feeding sponges *Euplectella* sp. and the dark ophiuroids *Ophiomusium*. Common detritus feeders were the holothurians *Peniagone intermedia* and, in particular, the swimming holothurians *Enypniastes eximia*. Carnivore polychaetes of the family Hesionidae and Aphroditidae were present. Unique to this facies were the cephalopods *Benthesicymus* sp., jellyfishes of the family Nausithoidae, the holothurians *Orphnurgus* sp., and *Amperima naresi* and an unknown apparently free-living ascidian about 30 cm diameter.
- **Facies B 40–50%** was characterized by a dominance of suspension feeders over detritus feeders. The suspension feeders were mostly the antipatharids *Bathypates patula, Bathypates lyra* and the brisingid *Freyella* sp. The detritus feeders mostly recorded were the holothurians *Psychronaetes hanseni, Benthodytes typica* and the asteroids *Hymenaster* sp. There were also aggregations of the actinids *Sincyonis tuberculata* and the swimming aphroditid polychaetes. Unique to this facies were octocoralliarids of the family Umbellulidae, the antipatharids *Schizpathes crassa*, and the holothurians of the family Deimatidae *Deima validum*.
- **Facies BP 35–50%** was characterized by a prevalence of fixed suspension feeders including vase-shaped sponges, *Poecillastra sp*., which were most abundant on this facies as well as the actinids *Bolocera* sp., *Sincyonis tuberculata, Actinoscyphia* and the ophiuroids *Ophiomusium armatum*. Detritus feeders were mainly the holothurians *Synallactes aenigma, Synallactes profundi, Peniagone leander* and *Benthodytes* sp. (e.g., *B. incerta),* and carnivores mostly polynoid polychaetes. Unique to this facies were two-horned-shaped Hexactinellid sponges.
- **Facies A 30%**, mainly present in NIXO 41, was characterized by a dominance of suspension feeders over detritus feeders, not due to cnidarians as on all other facies but to ophiuroids. The echinoderms present included, in decreasing order of abundance, ophiuroids, *Ophiomusium armatum*, holothuriods, echinoids *Plesiodiadema globulosum*, and asteroids. Were recorded the holothurians *Synallactes profundi, Deima validum, Benthodytes* sp., *Mesothuria murrayi, Peniagone leander*, *Psychronaetes hanseni* and the crinoids *Fariometra parvula*. Cnidarians were mainly represented by actinians *Actinernus verrill*, *Sincyonis tuberculata,* sponges by *Pheronema* sp. and other hexactinellids. The bonellid worms *Jacobia birsteini* associated to specific elongated mounds were

particularly abundant on this facies. Unique to this facies were the holothurians *Peniagone vitrea* and *Amperima rosea*.

- **Facies RCVO** (Rocky, with Carbonate bars and Volcanic Outcrops) has much smaller numbers of megafauna, mostly suspension feeding octocorallians, and, in decreasing order of abundance, sponges (*Cornucopia* sp.), actinians, and crinoids. Very few detritus feeders or carnivores were present.

Megafaunal Taxa Distribution

Habitat heterogeneity was the main factor structuring the distribution of megafaunal assemblages at different scales.

There is a marked patchiness in the distribution of megafauna, at a taxon level, as observed in the ratios of Morisita indexes for quadrats within the four main nodule facies. Concerning the most abundant taxa:

- Actinids *Sincyonis tuberculata* aggregated above 1600 m^2 on facies C and 800 m^2 on facies O.
- Holothurians *Mesothuria murrayi* aggregated with a peak at 800 m^2(and a lesser one at 200 m^2) on nodule-facies BP, at 200 m^2 and above 800 m^2 on facies B, at less than 400 m^2 on facies C. No *Mesothuria murrayi* have been recorded on facies O.

3.3 NIXO 41

- NIXO 45 and NIXO 41 appeared very similar in terms of taxonomic richness and faunal abundance, despite the different methods and different platforms employed (different towed cameras and the manned submersible Nautile). The two areas showed the same order of dominance and proportionality of suspension feeders over deposit feeders and carnivores, whatever the facies, but tended to be characterized by a prevalence of ophiuroids instead of cnidarians.
- Overall, the order of abundance of trophic groups may be characterized as sessile deposit feeders > mobile deposit feeders > sessile suspension feeders > mobile carnivores/scavengers > motile suspension feeders.
- The relative abundances of megafauna were, in decreasing order of abundance, sipunculids, echinoderms (mainly echinoids and holothurians) > echiurians, > sponges, and cnidarians (mainly actinians).
- No taxon was found to be exclusive to NIXO 41, i.e., all the taxa recorded had already been identified in NIXO 45.
- Overall the relative taxonomic richness of facies was, in decreasing order, facies A 30% >C 30% >B 35%.

- The taxonomic richness, faunal composition, and levels of abundance of individual taxa on facies B 35% were very similar to those recorded on the same facies in NIXO 45.
- Facies C+ 30% in NIXO 41 had a greater faunal abundance than the same facies in NIXO 45.
- The preferential habitat of suspension and deposit feeders was facies C+ 30% with recent sediments.
- Facies A 30% resembled facies B 40% in NIXO 45 in its population of cnidarians and resembled sloping facies C 20–40% in NIXO 45 in having a majority of echinoderms, and a relatively high density of sessile deposit feeders with the bonnelid echiurian worms present.

3.4 *ECHO I*

- ECHO1 displayed an overall taxonomic richness of 61 taxa recorded over around 25,200 m^2, sampled over 3 nodule facies and a higher faunal richness (36 taxa) on facies C 40% than on facies B 45% (23 taxa).
- Dredge tracks, produced by pilot-scale mining tests (OMA) in 1978, were still undisturbed and partly recolonized although with a minimum taxonomic richness (8 taxa) on facies O.
- There were a greater number of taxa on facies C 40% (densely covered with big nodules).
- There was an overall faunal dominance of deposit feeders over suspension feeders and carnivores/scavengers, and more specifically an overall relative dominance of sessile deposit feeders > mobile deposit feeders > sessile suspension feeders > mobile carnivores/scavengers > motile suspension feeders.
- Results that differed from NIXO 45 and NIXO 41 were that suspension feeders prevailed on facies B 45% and facies C 40% on old sediments, with the most actinians (167 ind/ha) occurring on facies C40%.
- Deposit feeders prevailed on facies O, mostly echinoderms, mainly echinoids and holothurians.
- Overall the order of abundance of the main taxa was sipunculids (50 ind/ha) >echinoderms (mainly echinoids and holothurians) >echiurians > sponges and cnidarians (mainly actinians).
- The overall predominance of echinoderms, mainly holothurians and echinoids, on photographs from ECHO I, taken with a 50 mm lens, differed from that recorded in NIXO 45 but resembled that recorded in NIXO 41 for facies A 30% and facies C 30%, although echinoderms were mainly represented by ophiuroids instead of holothurians as in NIXO 45.
- Deposit feeders, sipunculids, echinoids, and holothurians were particularly abundant (332 ind/ha) on facies O with old sediments.

3.5 *BIE*

- There was a greater faunal abundance and richness on facies B 45% and facies C 40% than on facies O.
- There were overall about the same proportions of deposit feeders to suspension feeders to carnivores/scavengers. However suspension feeders prevailed on the nodule-bearing facies, while deposit feeders were slightly more abundant on facies O as in ECHO I.
- The predominance of echinoderms (mainly holothurians and ophiuroids) over sponges and cnidarians (hydrozoans) was less marked on all substrates compared to that on other sites.
- As in ECHOI, sponges had a marked preference for nodule facies C 40%, where echiurians were also particularly abundant.
- The highest diversities were recorded on gently undulating plains and horst slopes, intermediate diversities on horst tops and trough axles, and low diversities on trough slopes, with the lowest diversity being recorded on volcanic slopes.
- Megafauna appeared more abundant on nodules that were diagenetically grown, D-type, and D1 subtype (Cu > Ni > 1.2%).

4 Discussion

The associations between different functional assemblages (Steneck 2001) and different nodule facies, as presented here, reflect the critical factors that drive the species population dynamics; these will likely include environmental heterogeneity, sediment loads, and current variability and strength, as well as biotic factors, on/in the benthos, through the whole water column to the surface and above. This dependence of the biological communities on precise physical environment needs to be borne in mind when considering the management of such physically driven biotopes (Tortell 1992), as does the fact that these environments are one of the world's few that remain relatively pristine (Tilot 2010a; Ramirez-Llodra et al. 2011). The changes to the seabed and overlying water column brought about by the mining activity will inevitably impact the different benthic communities present. Furthermore one must consider cumulative impacts with both natural impacts (natural climate variation, benthic storms, El Niño events, etc.) and anthropogenic disturbances (pollution, fishing, seabed mining, oil and gas extraction, disposal of wastes, etc.) in the same area, generally resulting in degradation and homogenization of habitats across broad areas (Glover and Smith 2003; Smith et al. 2008a; Thiel 2003).

Thus the indicators of sensitivity to environmental changes should be identified at different levels in the general characteristics and processes involved in the nodule ecosystem functioning in the CCZ from the surface (and above) to the seabed in a tridimensional perspective. The main findings discussed in the next sections lead to the designation of indicators of sensitivity to environmental changes anthropogenically or naturally induced as outlined further in Table 2.

Table 2 Summary of the general ecological characteristics of the water column and the water-sediment interface and processes involved in the nodule ecosystem functioning in the CCZ and indicators of sensitivity to environmental changes anthropogenically or naturally induced (adapted from Tilot et al. 2018)

General ecological characteristics and processes involved in the nodule ecosystem functioning in the CCZ	Indicators of sensitivity to environmental changes anthropogenically or naturally induced
Primary productivity, POCF, trophic input	The area is more sensitive; the flux of trophic input can be variable in space in time, generally very low on the benthos except events and horizontal currents. There will be a cumulative impact of natural and anthropogenic activities at the surface, above and in the water column, a recovery rate to assess for each stage and when cumulated
Biological features of abyssal fauna	The area is more sensitive, slower recovery rate if survival. Recent research evidences rapid adaptation to variable trophic input, opportunistic behavior; one would have to test the response of abyssal fauna to deep-sea mining impact and identify thresholds of impact for survivors et reorganization of food chains
The structure of faunal communities: Taxonomic richness, abundance and distribution at different temporal and spatial scales	Highly sensitive. Many species are new to science and need to be protected from extinction. There is a variability of biotic and abiotic parameters that conditions the structure of megafaunal populations at a regional scale and intra-site scale. Biodiversity and abundance of megafauna are enhanced on nodule areas vs nodule-free areas, in particular facies C 5–20% nodules + slopes >15%. Among Cnidarians and Echinoderms, prevailing components of megafauna in CCZ, some taxonomic populations evidence a non-random distribution. Information is lacking on sensitivity to spatial-scale dependence of recolonization of benthic communities
Temperature 1–4 °C, pressure increasing with depth (1 bar/10 m; 500 bar/5000 m), absence of light, climate change	Sensitive. Characteristics of an extreme environment may become harsher to fauna to survive with alternative state of the environment after impact. The response of megafauna to noise and lights during DSM should be monitored on a long term. Abyssal fauna is strongly affected by climate change. Primary production and carbon export to the deep sea will be reduced
Deep ocean circulation, water masses, bottom currents	Unknown impact but most probably long lasting with multiple effects due to complexity. The low oxygen values of the thermohaline and intermediate waters would hinder the organic material decomposition, thus increase the sinking rates to the ocean floor. From Catala et al. (2015b), two water masses are highly sensitive: at 500–1500 m a unique layer with great molecular diversity and the oldest water masses of the oceans at 1800–3500 m. Bottom currents are variable, up to 25 cm/s, but there can be benthic storms and other events impinging on the sediment surface and fauna associated
Sedimentation rates and mixing	Highly sensitive. There is a lack of information on the sensitivity to sediment burial at the scale of the CCZ. DSM would probably have a considerable negative biological impact on a long term and at a regional scale, over an area of more than 3 million km^2 of the deep-sea floor, overall the water column and surface of the ocean (and layer of air over the ocean)

(continued)

Table 2 (continued)

General ecological characteristics and processes involved in the nodule ecosystem functioning in the CCZ	Indicators of sensitivity to environmental changes anthropogenically or naturally induced
Deep seafloor, global geochemical cycles, topography	More sensitive as heterogeneity of the substrate and topography play an important role in the structure of faunal assemblages
Polymetallic nodules, growth, and processes to maintain nodules on the seabed	Highly sensitive. The degree of exposure to currents and terrain slope are factors that determine the growth of nodules (von Stackelberg and Beiersdorf 1991; Wang et al. 2001; Hoffert 2008). The rate of growth of nodules can be variable in space and time. It has been estimated by radiochronological methods at several millimeters (4–9 mm) per million years (Harada and Nishida 1979; Heye 1988) and the time for reconstitution of nodules at one million years for 1–2 mm size nodules (McMurtry 2001); thus nodule mining cannot be considered as a sustainable activity. The role of bioturbation, in particular by deposit feeders, and the role of currents, which are sufficient to transport fine particles, alter chemical properties and reduce the real rate of sedimentation. Both factors have been proposed as important in maintaining the position of the nodules at the sediment surface (Hartmann 1979; Schneider 1981; Hoffert 2008). Foraminifera are systematically associated with nodules (Saguez 1985)

Adapted from Tilot et al. (2018)

4.1 The Structure of Faunal Communities in the CCZ

The UNESCO/IOC baseline (Tilot 2006a, 2010a) provided evidence of both intra-regional and within-area variability of the structure of epibenthic faunal assemblages in the CCZ in relation to variation in biotic and abiotic parameters at different scales of space and time. The data revealed a relatively high taxonomic diversity for all faunal categories, with many species new to science. These findings have been confirmed in other studies (Smith 1999; ISBA 2008; Levin et al., 2001; Glover et al. 2002, 2016; Janssen et al. 2015; Paterson et al. 2015; Dahlgren et al. 2016; Amon et al. 2016b, 2017). Key faunal groups within the CCZ were the cnidarians, echinoderms and sponges among the megafauna, and polychaete worms, nematode worms, and protozoan foraminifera among the macrofauna and meiofauna; these taxa represented >50% of faunal abundance and species richness in abyssal sediments and displayed a broad range of ecological and life history types. The faunal assemblages encompassed true abyssal species, notably among them Isopoda, Nematoda and Foraminifera, and Echinodermata.

All comparable studies have provided evidence that habitat heterogeneity ("nodule facies," micro-heterogeneity at nodule level, patches of detritus, biogenic structures, and bioturbation), in combination with other factors (bottom currents, sediment chemistry, varying trophic input, and sedimentation rate), was responsible for driving the structure of the epibenthic faunal assemblages. There was clearly a markedly greater abundance and taxonomic richness of megafauna and macrofauna in nodule-bearing areas, while meiofauna preferred nodule-free areas. This general pattern has been observed in the CCZ (Amon et al. 2016b; Galeron et al. 2006; ISBA 2008; Morgan 1991; Mullineaux 1987; Radziejewska 1997; Radziejewska and Stoyanova 2000; Stoyanova 2008; Durden et al. 2015; Vanreusel et al. 2016) and also within the eastern Pacific in the Peru basin, where the nodule crevice fauna has been found to be distinctly different from that living in the sediment in the proximity of the nodules (Tilot 1989; Thiel et al. 1993).

Polymetallic nodules clearly provided a distinct habitat for both infaunal communities and for the sessile suspensivore fauna, e.g., sponges, actinids, stalked crinoids, octocorallians, sedentary polychaetes, antipatharids, and tunicates that settle on the nodules. In addition, it appeared that the nature of sediments differed from nodule areas vs nodule-free areas where sediments were softer. The UNESCO/IOC baseline (Tilot 2006a, 2010a) also provided evidence that megafaunal suspension feeders prevailed on nodule-covered areas, while sponges and deposit feeders, particularly holothurians, dominated in nodule-free areas.

The fact that nodule abundance and abiotic factors, such as slope, topography, and bathymetry, largely determined the abundance and composition of faunal assemblages has been supported in UNESCO/IOC baseline (Tilot 2006a, 2010a) in which different megafaunal assemblages were found associated with 13 distinctive habitats. However, at a regional scale, the work also provided evidence that (a) there tended to be more suspension feeders at western sites and more deposit feeders at eastern sites, mirroring a gradient in oceanic primary productivity from west to east, and (b) there was an increase in taxonomic richness of megafaunal and macrofaunal assemblages moving from east to west, reflecting a trend of greater substrate heterogeneity at both nodule and facies levels in the western area, in particular for facies C 2–15% and nodule-facies C 30%.

The community structure of the faunal assemblages varied substantially, with a significant turnover in species being evident over a latitudinal range. For foraminifera and polychaetes, the turnover occurred at scales of 1000–3000 km across the CCZ (ISBA/Kaplan/Nodinaut). For megafauna, the turnover occurred at a larger scale (UNESCO/IOC baseline). Herein, the high values of the standard deviations of overall abundances of taxa indicated a high within-site variability, likely the result of the nonrandom dispersion of fauna widely reported in the literature (Schneider et al. 1987; Kaufman and Smith 1997). This phenomenon was particularly evident at the taxon level with, e.g., the actinid *Sincyonis tuberculata* aggregating at densities greater than 1600 m^2 on facies C but occurring at random at 400–800 m^2 on facies O, as judged by applying the Levis, David, and Moore indexes and Fisher's coefficient (UNESCO/IOC baseline vol 1, Tilot 2006a).

4.2 Primary Productivity, Particulate Organic Carbon Fluxes (POCF), and Trophic Input

The CCZ is defined as mesotrophic abyss (Primary productivity NECC range from 210 to 327 mg C m^3 d^{-1}, POCF range from 0.5 to 1.6 g C m^{-2} y^{-1}) (Smith and Demopoulos 2003; Hannides and Smith 2003; Pennington et al. 2006) with however seasonal patterns, upwelling and El Niño events (Dymond and Collier 1988; Ruhl and Smith 2004).

There is a variation in flux of deposition of phytodetritus, often in patches, on the sediment surface (Karl et al.1996; Smith et al. 1997, 1999, 2008b; Baldwin et al. 1998; Radziejewska and Stoyanova 2000; Radziejewska 2002; Karl 2002; Smith and Rabouille 2002; Lutz et al. 2007; Stoyanova 2008; Amon et al. 2016a, b), and the dynamic benthic-pelagic coupling (Rowe 1971; Pfannkuche and Lochte 1993) induces significant changes in the structure of the epibenthic megafaunal assemblages and measurable effects on the ecosystem function (Smith et al. 1993; Lauerman et al. 1997; Beaulieu and Smith 1998; Drazen et al. 1998; Thurston et al. 1998; Danovaro 2008; Loreau 2008).

There is a horizontal contribution to nutritive sediment particles in suspension as a result of currents originating to the west in zones of high primary production (Smith and Demopoulos 2003). Particular edaphic and hydrological conditions above the bottom layer, perhaps related to the topography of the ocean floor, can favor the collection of nutritive sediment particles (Beaulieu and Smith 1998).

In addition, the carcasses of larger pelagic organisms provide a significant input of carbon to benthic food webs (Dayton and Hessler 1972; Haedrich and Rowe 1977; Smith and Baco 2003; Tyler 2003).

4.3 Biological Features of Abyssal Megafauna, Community Dynamics, and Diversity

Deep-sea benthic megafauna are generally small in body size (with some exceptions, e.g., bonnelid worms), show low biomass, and are extremely delicate, feeding on a thin layer of organic matter at the water-sediment interface or in the water column above the benthos (Gage and Tyler 1991).

The deep-sea megafauna is characterized by slow metabolic growth, slow maturation, low reproductive potential, and low rates of colonization, although compensated for to some extent by greater longevity (Gage and Tyler 1991; Gray 1997; Smith and Demopoulos 2003), all adaptations to extreme environmental conditions (Rex 1983).

Temporal and spatial environmental variability are driving factors within the deep-sea ecosystem (Smith et al. 2001). As a result, community dynamics would follow a patch mosaic model with a dynamic mosaic of microhabitats on which species might specialize (Grassle and Sanders 1973). Thus, contrary to possible expectation, the deep-sea benthos is surprisingly species-rich (Snelgrove and Smith 2002).This variability induces fluctuations in biological processes (Tyler 1988; Gage and Tyler 1991; Smith and Kaufmann 1999; Gooday 2002) including opportunistic feeding (Billett et al. 1988; Jumars et al. 1990) and rapid community responses to variation at the deep-sea floor (Smith et al. 1986, 2001).

4.4 Water Masses and Major Currents

When comparing the three major ocean basins, the Pacific Ocean has the lowest salinity. As a result, the northern North Pacific shows no deep water formation and only a weakened intermediate water formation (Fiedler and Talley, 2006). As the deep global circulation ends in the northern North Pacific, this zone is considered as the oldest deep waters of the world (Talley et al.2011).

This is an important factor to be born in mind in relation to deep-sea mining, especially given the associated operations (including sediment dispersal, see further) anticipated to take place in mid-water.

The mid-latitude North Pacific surface circulation is characterized by a two-gyre circulation driven by westerly and trade winds. The CCZ is located in the anticyclonic subtropical gyre and is dominated by the North Equatorial Current (NEC), which is the broad westward flow of the subtropical gyre between 8 and 20°N (Fiedler et al. 2013). This current is associated with equatorial upwelling in the east that brings high nutrients and biological productivity. According to Longhurst's biogeographic provinces, this area is covered by the North Pacific Tropical Gyre province (NPTG) (Longhurst 2007).

Below the upper layer (>200 m), based on the analysis of samples collected by the 2010 Spanish Malaspina Expedition within the area 7–20°N and 99–167°W, seven separate water masses can be identified at depths between 200 m and 4000 m (Catalá et al. 2015a, b) (Fig. 7; for more information about the water masses, see Table S1 of Catalá et al. 2015b).

In the central domain (200–500 m), the central waters make up the thermocline/pycnocline in the subtropics (Talley et al. 2011). In the CCZ, the dominant central water is the 13 degrees Equatorial Water (13EqPac) (Tsuchiya 1981), which represents 18.1% of the total water volume of the defined area. It is advected eastward across the Pacific from the low-latitude western boundary currents and is associated with the subsurface countercurrents (Talley et al. 2011). At a lesser degree, the North Pacific Central Mode Water (CMW_{NP}) and the North Pacific Subtropical Mode Water ($STMW_{NP}$) are also present with 14.1% and 1%, respectively. These water masses are found in the North Pacific subtropical gyre (Suga et al. 1997).

The intermediate domain (500–1500 m) is mostly covered by the North Pacific Intermediate Water (NPIW) with a proportion of 22.3%. This water mass is formed by the mixing of different waters in the NW Pacific, in the Okhotsk Sea, and at the Oyashio Front (Johnson 2008; Bostock et al. 2010) and is characterized by a salinity minimum with low oxygen and low density (Talley et al. 2011). For the case of the

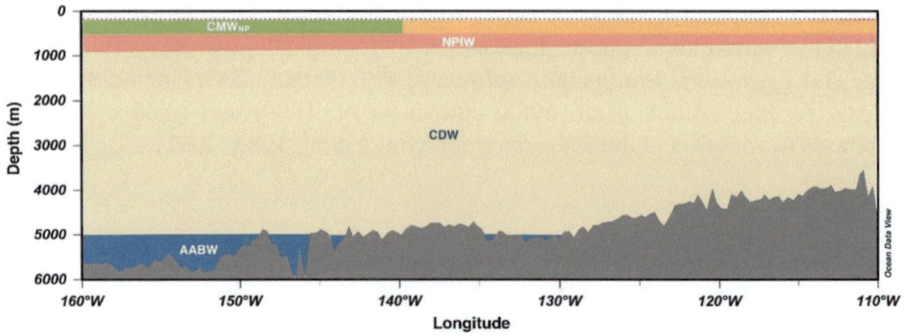

Fig. 7 Water masses in the CCFZ region. The results are based on the water mass analysis performed in the Malaspina 2010 Circumnavigation (Catalá et al. 2015a, b). Only the most abundant water mass for each site was plotted. The dashed red line represents the upper limit of the water mass analysis (200 m). The lower limit of the Malaspina water mass analysis was at 4000 m. Below this depth, the information has been extracted from Talley et al. (2011). AABW, Antarctic Bottom Water; CDW, Circumpolar Deep Water; NPIW, North Pacific Intermediate Water; CMWNP, North Pacific Central Mode Water; 13EpPac, Equatorial Pacific Central Water (13 °C)

Malaspina Expedition, the NPIW showed a water mass weighted-average salinity of 34.53 ± 0.02 and an apparent oxygen utilization of 255 ± 6 μmol kg^{-1} (see Table 1 from Catalá et al. 2015a). The coldest branch of the Antarctic Intermediate Water ($AAIW_{3.1}$) (McCartney 1982; Talley 1996) was slightly present at the CCFZ with 0.9%.

The thermohaline and intermediate waters (500–1500 m) in the CCZ present unique features as they undergo excessive aging (Catala et al. 2015b). This anomaly is associated with (1) the high biological productivity of the area, which exports organic material from the surface to the deep ocean (Feely et al. 2004), and (2) the sluggish circulation of the North Pacific that facilitates organic matter sinking. The high productivity of this area together with the weak sub-thermocline currents results in a pronounced oxygen minimum layer (OML) (Kessler 2006). We hypothesize that the low oxygen values hinder the organic material decomposition, thus increasing the sinking rates to the ocean floor (Catala, pers. comm.).

The abyssal flow of the CCFZ consists of an anticyclonic circulation (Talley et al. 2011). The abyssal domain (>1500 m) is covered mostly by the Circumpolar Deep Water (CDW), also named "common water," formed by mixing in the Antarctic Circumpolar current of mid-depth Indian, Pacific, and Atlantic deep water with Antarctic Bottom Water (AABW) and the North Atlantic Deep Water (NADW) (Montgomery 1958; Georgi 1981; Broecker et al. 1985). The AABW, at its purity state, is also detected in the area at lower proportions (i.e., 4.4%). This water mass is formed in the Weddle and Ross Sea (Onken 1995; Arhan et al. 1999) and is recognized by lower temperatures and higher salinity than the overlying CDW (Talley et al. 2011).

A potentially significant management issue is the desirability of preserving the oldest water masses located in the North East Pacific between 1800 and 3500 m according to Talley et al. (2011), because of their unique properties, which stem from their perpetual reworking in the deep ocean. They are characterized by a distinct biogeochemistry and greater molecular diversity, heterogeneity, and complexity of dissolved organic matter (Dittmar 2014). Given the water masses age, the microbial communities are expected to be adapted to these hypoxic and inhospitable conditions, resulting in a different deep-sea microbial biodiversity. This singular biogeochemistry and microbial flora are presumed to have an impact on the nodule ecosystem via the water column food chain and pelagic-benthic transfer to the sediments; it may also influence nodule genesis. These oldest water masses are thus very vulnerable, but the impact of deep-sea mining on them remains difficult to quantify.

4.5 Eddy Kinetic Energy and Bottom Currents

The maximum sea-height variability denoting excess eddy kinetic energy at the surface is transmitted to the seafloor in water depths exceeding 3000 m (Hollister and Nowell 1991; Kontar and Sokov 1994). Many regions of the world ocean show evidence of benthic storm activity in which cold currents in the deep ocean accelerate into powerful sediment-transporting events (e.g., Gardner and Sullivan 1981; Hollister and McCave 1984; Klein 1987; Nowell et al. 1982; Quirchmayr 2015). Bottom currents in the northeastern tropical Pacific are predominantly weak (Shor 1959), between 2 and 25 cm/s (Amos et al. 1977), but the flow is characterized by alternation of periods of stronger, quasi-unidirectional currents, and periods of slower water movement (Kontar and Sokov 1994, 1997). Nevertheless, several hydrographic disturbances, such as changes in deep ocean current or periodic "benthic storms," have been recorded during long-term in situ monitoring (Aller 1997; Kontar and Sokov 1994); such changes will in turn impinge on the sediment surface itself (Tkatchenko and Radziejewska, 1998).

4.6 Climate Variation and Global Warming

Deep-sea communities are strongly affected by climate variation in the Pacific Ocean basin (Barange et al. 2010; Smith et al. 2009), as are those of the upper ocean (Barnett et al. 2005; Levitus et al. 2000). Global warming is anticipated to increase stratification while reducing vertical mixing and nutrient exchange from deeper depths (Behrenfeld et al. 2006, Sarmiento et al. 2004); this in turn is expected to reduce primary production and carbon export to the deep-sea (Huisman et al. 2006), thus impacting the deep ocean ecosystem (Glover et al. 2010), including in particular that of the northeast Pacific (McGowan et al. 1998).

4.7 Sedimentation Rates and Mixing: The Sedimentary Environment of Nodules

In the Eastern Equatorial Pacific Ocean, sediment coverage is about 20% thicker within morphological depressions than on the ridges, and, under the influence of sedimentary processes, the contours tend to soften (Shor 1959). The sediment cover in the CCZ is mainly biogenic but has been subjected to diagenetic processes to varying extent (Hannides and Smith 2003; Hoffert 2008). The surface layer is highly hydrated and can be several hundred meters thick, depending on the age of the crust and prevailing sedimentation rates (Kotlinski 1999; Menzies et al. 1975).

Sedimentation rates in the CCZ are assessed to be 0.1–10 cm per thousand years (Gage and Tyler 1991; Smith and Demopoulos 2003). Although the currents are weak (2–25 cm/s), they are sufficient to transport fine particles (in some places for a thickness of about 400 m) and thus, in places, to reduce the rates of sedimentation. Yet they are too slow to cause the erosion of even lightly consolidated sediments (Hoffert 2008). This explains why in ECHOI dredge tracks produced in 1978 are still visible over 30 years later (ISBA 2008; Tilot 1991, 2006a).

According to Hoffert et al. (1992), the presence of nodule deposits on the seabed is due entirely to the existence of present-day erosion (less than 210,000 years) which once interrupted, causes rapid nodule burial. Lonsdale and Southard (1974) also showed that in identical conditions, nodule-bearing sediments are more sensitive to erosion created by lower current speeds than are sediments without nodules.

The mechanical resistance of the ocean floor at the NIXO 45 site has been estimated from cores of recent clayey silt and analysis by a field vane tester (NIXONAUT-GEMONOD cruise, 1988). The cohesion is weak for recent silt; it rises from 1 to 2 kPa in the interface layer to 3–5 kPa at several tens of centimeters depth. In contrast, it is clearly stronger (upto 20–80 kPa) for older sediments.

The ecological characteristics of the nodule ecosystem within the CCZ are summarized in Table 2. As outlined, a series of issues are relevant to the management and conservation of biodiversity in this environment.

5 Options for Management of the Impacts of Deep-Sea Mining

Unfortunately, unlike fisheries, for example, polymetallic nodule mining cannot be managed as a sustainable activity, since the reconstitution of nodules following removal, and the restoration of the nodule ecosystem, if ever possible, will take several million years, based on measured rates of nodule formation and growth (Ghosh and Mukhopadhyay 2000; McMurtry 2001). Further, commercial nodule mining is envisioned to be the largest-scale human activity ever to affect the deep sea, most probably taking place at a regional scale over an area of more than

3 million km^2 of the deep-sea floor. The processes will impact not only the benthos but also the water column and the surface of the ocean (and even the layer of air over the ocean), at a global regional scale. Thus, faunal communities particular to the nodule ecosystem may even be threatened with extinction.

The temporal and spatial scales of disturbances will naturally determine whether these habitats are able to sustain themselves or change to the extent that the original pristine state can never be recovered (Berkes et al. 2003; Glover and Smith 2003; Gundersen and Pritchard 2002; Ramirez-Llodra et al. 2011; Barbier et al. 2014). The extent of impacts will have crucial consequences for management, since it is easier to repair a damaged ecosystem than restore it once destroyed. During impact and post-impact phases, changes in species composition of the benthos almost invariably occur, often favoring short-lived species that can quickly colonize after the disturbances (Hughes et al. 2003, 2005; Khripounoff et al. 2006). Subsequently alternate ecosystem states may be maintained through density-dependent mortality effects (e.g., owing to altered predator-prey ratios) or as a result of populations failing to achieve density thresholds required for reproductive success (Cury and Shannon 2004).

The environmental impact studies that have been conducted on nodule ecosystems in the CCZ, the Indian Ocean, and the Peru basin (e.g., Amos et al. 1977; Bluhm 2001; Chung et al. 2001; Foell et al. 1990; Fukushima 1995; Fukushimi et al. 2000; Oebius et al. 2001; Ozturgut et al. 1980; Radziejewska 1997; Schriever et al. 1997; Sharma et al. 2001; Shirayama 1999; Thiel et al. 1992, 2001; Tilot 1989; Tkatchenko et al. 1996; Trueblood et al.1997; Thiel 2001, 2003) mostly focused on alterations to the sediment-nodule interface and impact to epibenthic faunal communities over very localized areas, compared to the size and the number of areas to be mined. Several reviews (e.g., GESAMP 2016; Jones et al. 2017; Sharma 2015; Smith 1999) concluded that there is a lack of information on the potential impact of sediment burial at the scale of the CCZ and that there would be considerable long-term negative effects on the ecosystem.

These arise because, according to Sharma (2015), for every ton of manganese nodule mined, 2.5–0.5 tonnes of sediment will be resuspended. Adjacent areas will obviously experience the highest sedimentation rates, but sediment plumes will remain in suspension over long periods and also travel laterally (Jankowski and Zielke 2001). It is considered that significant sediment loads will clog the filter feeding apparatus of most benthic fauna. Hannides and Smith (2003) estimate that the mining proposed would severely affect benthic communities over 20,000–45,000 km^2 of seabed, with resettlement of up to 95% of the sediment particles released within a period of 3–14 years, depending on the depth of release (Rolinski et al. 2001). That this is so can be appreciated if one considers that even with a current as low as 1 cm per second, waterborne sediment would travel more than 3000 km within a period of 10 years. Even while still suspended in the water column, the sediment plume is anticipated to have various effects depending on the particular water masses involved.

In conformity with the principle of ecosystem-based management and the commitments enshrined in the Convention on Biological Diversity, marine spatial planning in the high seas/deep sea needs to pursue an integrated approach that both considers all stakeholders and sectors of activity and fully appreciates the tridimensional status of the oceans. The necessary marine spatial planning will require cumulative impact studies that monitor the extent of impact at all scales (Ardron et al. 2008; Gianni et al. 2016; Foley et al. 2010).

In the pelagic domain, numerous species are using the CCZ as feeding grounds near the surface (0–200 m for Bluefin tuna according to Block et al. (2003, 2011), see map). Some species extend their range to depth of 1000 m such as for yellowfin tuna, bigeye tuna, and swordfishes to 1500 m for bottlenose whale or approximately 3000 m for sperm whale (FAO Fisheries and Aquaculture 2000; Karleskint et al. 2013; Schor et al. 2014, see maps), and recently at 4258 m, evidence of contact on the nodule seabed of Cuvier's beaked whales has been hypothesized (Marsh et al. 2018).

5.1 Monitoring and Rapid Ecological Assessment (REA)

A wide variety of long-term monitoring methodologies are used to collect time series data in shallow-water ecosystems for research and management purposes (Davies et al. 2001; Price 2004), many incorporating the use of still or video imagery (see, e.g., Tilot 2003; Tilot et al. 2008a, b). However, as with baseline assessments, a major issue is the limited time and resources that can be allocated to sampling at any one site, to which may be added the extremely large areas over which competent monitoring is often desired. Such issues are especially acute when the need to monitor the benthic environment of the CCZ is considered.

Monitoring of deep-sea benthos using the methods described is comparable in that only reasonably conspicuous megafauna can be recorded and that key environmental factors such as sediment type, nodule density, and seabed slope must be assessed visually on a predefined scale.

Adoption of REA principles does not however exclude the necessity for monitoring of key water column parameters, such as turbidity, current speed, dissolved oxygen concentration, production (chlorophyll), alkalinity, etc. It is essential to establish a permanent or semipermanent tridimensional monitoring system recording key environmental variables related to the different mining activities that will take place not only on the seabed but also in the water column and at the surface. Most of these variables can be measured with sensors, for example, current velocity using Lowered Acoustic Doppler Current Profilers (LADCP).

Judging by the recent results of Catalá et al. (2015a, b) and research on hydrodynamics close to the seabed (Henry et al. 2013; Moreno Navas et al. 2014) and on nepheloid layers (Gardner and Sullivan 1981; Gardner et al. 1984) in particular for nodule fields (Hoffert 2008), attention will need to be paid to the selected environmental parameters within five particular depth zones: (i) from the surface to 200 m,

(ii) on the central domain (200–500 m), (iii) within the intermediate domain (500–1500 m), with a pronounced high productivity and pronounced oxygen minimum layer, (iv) in the abyssal domain (>1500–up to 200 m above the seabed) where are located the oldest water masses of the oceans (between 1800 and 3500 m), and (v) close to the seabed (200 m above to the seabed including the bottom nepheloid layer which may extend >200 m above the seabed into the abyssal domain), a most interesting layer with direct implications for the functioning of the nodule ecosystem and its genesis. The data should enable the management body to set up guidelines that can be adapted over time and space to reflect expected impacts (Tilot 2004, 2010a, 2014) and consider pelagic protected areas (Game et al. 2009), such as in Tilot (2004) and Ardron et al. (2008).

5.2 Management Tools Applied to a Pilot Study: NIXO 45

A multilayer Rapid Ecological Assessment (REA), sensitivity indices and the study of bottom currents have been applied to a pilot site, NIXO 45 (Tilot et al. 2018).

Description of the Site from Imagery

The geomorphological landscape of the nodule area of NIXO 45 is a central depression of maximum depth 5150 m, bordered by two hills to the west and east, which rise to a depth of about 4750 m. On basis of photographic data and multibeam sounder, it has been interpreted as composed of basins, valleys, hills, and plateaux all generally cut into more or less regular successive steps with subvertical walls. In addition, there are slopes and cones of scree as well as outcrops of the basaltic platform. The slopes rarely exceed 7° though can reach 15° locally, particular on the sides of the hills (du Castel 1985). The whole morphological structure has a principal north-south orientation, parallel to the ancient axis of accretion. Two types of sediments are present "red clays" and "contemporary siliceous silts." The latter forms most of the substratum, several meters thick, on which the nodules grow. On steep slopes and gravitational landslides (slope 15–20%), nodules are generally smaller (nodule-facies BP) or absent.

The main results of the REA, illustrated in Fig. 8, show that:

- There is an overall dominance of suspension feeders > detritus feeders > carnivores and scavengers.
- Megafauna is more abundant on big nodules C+10% and C+30% on slopes (>15°).
- The values and sensitivity indexes outline that overall facies C may be considered the most vulnerable, in particular facies C 15–20% on a slope of >15°. Thus special focus will have to be given to manage impacts on facies C as it is the most interesting commercially with facies B and Bp (Fig. 8).

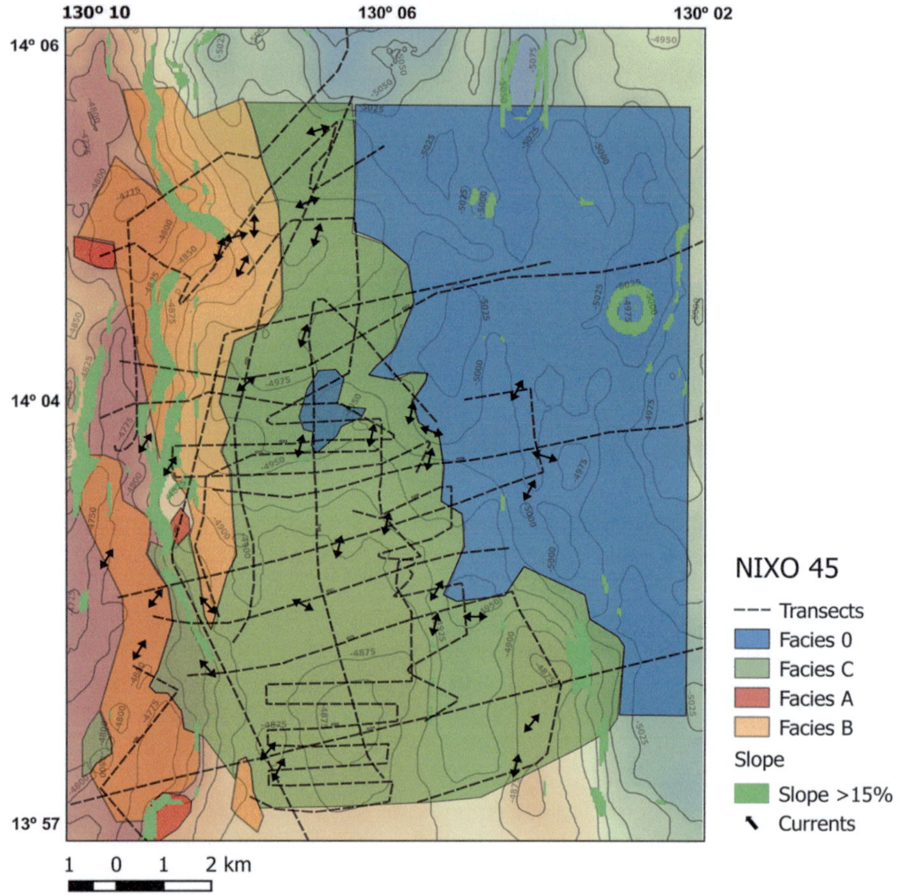

Fig. 8 Pilot Site NIXO 45 where REA and management indices have been applied. GIS Bathymetry, currents and the 4 main facies represented. The areas with slope bigger than 15% are highlighted in green. Bottom current directions derived from the analysis of imagery of the seabed, in particular of the direction of suspension feeders

Fixed Organisms as Indicators of Bottom Currents

As evidenced in Heezen and Hollister (1971) and Kennett (1982), fixed organisms are good indicators of bottom currents. Seabed morphology in NIXO 45 comprises horsts and grabens (linear ripples and furrows) that tend to generate stronger currents on slopes that are favored by sessile suspension feeders (UNESCO/IOC baseline vol 1, Tilot 2006a). The varying near-bed velocity field plays a role in the supply of edible particles to benthic suspension feeders since the horizontal flux of organic seston is a function of the interaction of particles of varying densities with the near-bed velocity field (Muschenheim 1987).

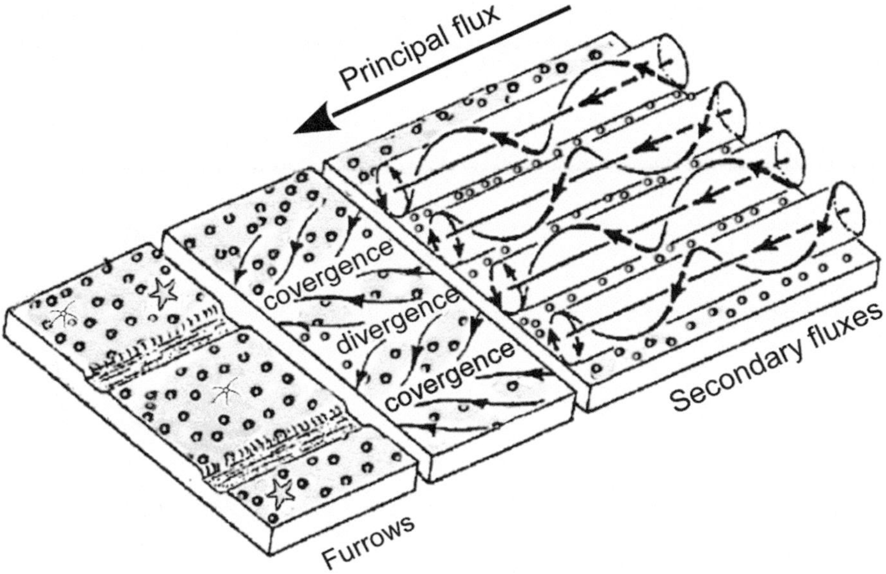

Fig. 9 Diagram representing the possible local bottom current pattern and the setting up of structures in "furrows" as observed in NIXO 45. (Adapted from Flood 1983)

The analysis of the imagery from NIXO 45 found that most suspension feeding assemblages occurred on facies B and C in the eastern and western hills, in particular on facies C 5–10%, C 15–20% with slope >15° (Tilot 2006a). Their distribution indicated the general direction of N40-N50 bottom currents (Tilot et al. 2018) (Fig. 9).

Value and Sensitivity Indices for NIXO 45

Based on data gathered during the UNESCO/IOC baseline survey of NIXO 45, we have developed a provisional sensitivity index, based on the values of a series of ecological and environmental indicators. We have also assessed the typical (modal) values of each indicator and hence the value of the index, for each facies, arriving at the values shown in Tables 3a and 3b. These values support the view that overall facies C, followed by facies C 25–40% and facies BP 35–50% may be considered the most vulnerable, with facies C 5–10% and facies C 15–20% on a slope of >15% obtaining very similar scores. These facies are hosting the most diverse and abundant megafauna according to the multivariate analysis. Facies O with recent sediments, adjacent to facies C (see results for NIXO 45), have the second highest richness and abundance values. To note that miners are most interested on facies C and facies BP with this range of nodule coverage (Tilot 2006a).

Table 3a "Value Indexes" for NIXO 45 based on results of UNESCO/COI baseline survey (Tilot 2006a)

Value index	O	O	A	B	C	C	C	BP	RCVO
Facies/criteria (top number highest value)	Old Sed	Recent Sed	30–50%	40–50%	5–10%	15–20% With slope >15°	25–40%	35–50%	Rock-outcrop
% nod cover (1–6)	–	–	3	6	1	2	4	5	–
Interest for miners (1–3)	–	–	1	3	2	3	3	3	–
Slope/escarpments (0–3)	0	0	1	2	0	3	0	1	3
Hills (1–3)	1	3	2	3	1	2	2	1	1
Plateaux, valleys (0–3)	0	2	1	3	3	3	3	2	1
Deeper basins (0–1)	1	0	0	0	0	0	0	0	0
Taxon abundance (1–9)	3	7	5	2	9	8	4	6	1
Taxon diversity (1–9)	5	8	4	2	9	7	3	6	1
Susp feeders (1–9)	2	7	3	5	9	8	4	6	1
Detr feeders (1–9)	5	6	4	3	9	8	7	2	1
Carnivores (1–9)	7	5	1	9	6	3	4	8	2
Echinoderms (1–9)	3	6	4	2	9	8	7	5	1
Cnidarians (1–9)	2	6	2	3	9	8	7	4	1
Sponges (1–9)	5	3	2	4	7	9	8	6	1
Echiurian (1–3)	2	3	3	3	2	2	3	3	–
Bioturbation (1–3) superficial to deep	3	3	2	3	(1), 3	(1), 3	1	2	–
Total value index	**39**	**59**	**38**	**53**	**79**	**77**	**60**	**60**	**14**
Value index/10	**4**	**6**	**4**	**5**	**8**	**8**	**6**	**6**	**2**

[a]Bioturbation (1–3) was scored as: 1, surface tracks (between nodules, circular furrows); 2, intermediary bioturbation (furrows, lines, imprints); 3, deep bioturbation (several cm to ms): tumuli, burrows, and rosette. The simplified Value Index Score was obtained by summing all values and dividing by 10 to give values on a 10-point scale (Tilot et al. 2018)

Table 3b "Sensitivity Index" values for NIXO 45 based on results of UNESCO/COI baseline survey (Tilot 2006a)

Facies/criteria	O	O	A	B	C	C	C	BP	RCVO
(Top number highest value)	Old Sed	Recent Sed	30–50%	40–50%	5–10%	15–20% With slope >15°	25–40%	35–50%	Rock-outcrop
% nod cover (1–6) weight 1	–	–	3	6	1	2	4	5	–
Interest for miners (1–3) weight 0	–	–	0	0	0	0	0	0	–
Slope/escarpments (0–3) weight 2	0	0	2	4	0	6	0	2	6
Hills (1–3) weight 2	2	6	4	6	2	2	4	2	2
Plateaux, valleys (0–3) weight 1	0	2	1	3	3	3	3	2	1
Deeper basins (0–1) weight 0	0	0	0	0	0	0	0	0	0
Taxon abundance (1–9) weight 1	3	7	5	2	9	8	4	6	1
Taxon diversity(1–9) weight 2	10	16	8	4	18	14	6	12	2
Susp feeders (1–9) weight 2	4	14	6	10	18	16	8	12	2
Detr feeders (1–9) weight 0	0	0	0	0	0	0	0	0	0
Carnivores (1–9) weight 1	7	5	1	9	6	3	4	8	2
Echinoderms (1–9) weight 1	3	6	4	2	9	8	7	5	1
Cnidarians (1–9) weight 2	4	12	4	6	18	16	14	8	2
Sponges(1–9) weight 2	10	6	4	8	14	18	16	12	2
Echiurian(1–3) weight 2	4	6	6	6	4	4	6	6	0
Bioturbation (1–3) superficial to deep weight 2	6	6	4	6	4	4	2	4	0
Total	**53**	**86**	**52**	**72**	**106**	**104**	**78**	**84**	**21**
Vulnerability index/12	**5**	**9**	**5**	**7**	**11**	**10**	**8**	**8**	**2**

The values in the table were obtained by weighting the values from Table 3a (by a factor 0–2) according to their likely sensitivity to environmental impacts such as nodule mining, as judged from knowledge of the ecosystem and information contained in the literature. The Simplified Sensitivity Index Score was obtained by summing all values and dividing by 10 to give values on a 12-point scale (Tilot et al. 2018)

Acknowledgments The research presented in this paper would not have been possible without the support of the "Institut Français de Recherche pour l'Exploitation de la Mer" (IFREMER), which made available its facilities, cruises, data, and expertise, through teams from the departments of Deep Ocean Ecosystems and Marine Geosciences, with the funding from the European Commission, DG Enterprise and Industry, the "Institut Océanographique de Paris" (France) and the National Oceanic and Atmospheric Administration (NOAA), US Department of Commerce (USA). Later support was provided by the Intergovernmental Oceanographic Commission of UNESCO (IOC-UNESCO) and the Government of Flanders to update and expand the work with BIE site. Cruises and data were also been made available by SCRIPPS (Dr. George Wilson) for the ECHO I site and for Quagmire cruise, by NOAA for BIE site (Dr. Craig Smith and D. Trueblood), by the University of Hamburg, (Prof H. Thiel) for the DISCOL cruise in the Peru Basin, South Pacific. Data for the water mass analysis was available thanks to the Malaspina expedition (grant number CSD2008-00077). The author thanks the department of PATRINAT of the National Museum of Natural History (MNHN), Paris (France) and the "Instituto Español de Oceanografía," Malaga (Spain), for their support.

References

Aller, R. (1997). Benthic community response to temporal and spatial gradients in physical disturbance within a deep-sea western boundary region. *Deep-Sea Research Part I, 44*, 39–69.

Amon, D. J., Hilario, A., Martinez Arbizu, P., & Smith, C. R. (2016a). Observations of organic falls in the abyssal Clarion Clipperton zone, tropical Eastern Pacific Ocean. *Marine Biodiversity*. https://doi.org/10.1007/s12526-016-0572-4.

Amon, D., Ziegler, A., Dahlgren, T., Glover, A., Goineau, A., Gooday, A., Wiklund, H., & Smith, C. (2016b). First insights into the abundance and diversity of abyssal megafauna in a polymetallic-nodule regíón in the eastern Clarion Clipperton zone. *Nature Scientific Reports, 6*, 30492. https://doi.org/10.1038/srep30492.

Amon, D. J., Ziegler, A., Kremenetskaia, A., Mah, C., Mooi, R., O'Hara, T., Pawson, D., Roux, M., & Smith, C. (2017). Megafauna of the UKSRL exploration contract area and eastern Clarion Clipperton zone in the Pacific ocean: Echinodermata. *Biodiversity Data Journal, 5*, e11794. https://doi.org/10.3897/BDJ.5.e11794.

Amos, A. F., Roels, O. A., Mos, A. F., Roels, O. A., Garside, C., Malone, T. C., et al. (1977). Environmental aspects of nodule mining. In G. P. Glasby (Ed.), *Marine manganese deposits* (pp. 391–437). Amsterdam: Elsevier Publication Company.

Ardron, J., Gjerde, K., Pullen, S., & Tilot, V. (2008). Marine spatial planning in the high seas. *Marine Policy The International Journal of Ocean Affairs, 32*(5), 832–839.

Arhan, M., Mercier, H., & Lutjeharms, J. R. E. (1999). The disparate evolution of three Agulhas rings in the south Atlantic Ocean. *Journal of Geophysical Research, 104*, 20,987–21,005. https://doi.org/10.1029/1998JC900047.

ASOM (Académie des Sciences d'Outre-Mer). (2010). Recommendations on the conservation and management of biodiversity of three deep-sea ecosystems targeted by mining (polymetallic nodules, cobalt rich ferro-manganese crusts and hydrothermal sulfurs). *Proceedings of the international workshop organized at UNESCO*, 15 December 2010.

Baldwin, R. J., Glatts, R. C., & Smith, K. L. (1998). Particulate matter fluxes into the benthic boundary layer at a long time-series station in the abyssal NE Pacific: Composition and fluxes. *Deep-Sea Research II, 45*, 643–665.

Barange, M., Field, J. G., Harris, R. P., Hofmann, E. H., Perry, R. I., & Werner, F. E. (Eds.). (2010). *Marine ecosystems and global change*. Oxford: Oxford University Press, 436 pp.

Barbier, E. B., Moreno-Mateos, D., Rogers, A. D., Aronson, J., Pendleton, L., Danovaro, R., Henry, L. A., Morato, T., Ardron, J., & Van Dover, C. L. (2014). Ecology: Protect the deep sea. *Nature, 505*, 7484.

Barnett, T. P., Adam, J. C., & Lettenmaier, D. P. (2005). Potential impacts of a warming climate on water availability in snow-dominated regions. *Nature, 438*, 303–309. https://doi.org/10.1038/nature04141.

Beaulieu, S. E., & Smith, K. L., Jr. (1998). Phytodetritus entering the benthic boundary layer and aggregated on the sea floor in the abyssal NE Pacific: Macro- and microscopic composition. *Deep-Sea Research II, 45*, 781–815.

Behrenfeld, M. J., O'Malley, R. T., Siegel, D. A., McClain, C. R., Sarmiento, J. L., Feldman, G. C., et al. (2006). Climate-driven trends in contemporary ocean productivity. *Nature, 444*, 752–755. https://doi.org/10.1038/nature05317.

Berkes, F., Colding, J. C., & Folke, C. (2003). Navigating socio-ecological systems: Building resilience for complexity and change, ecology and society. *Ecology and Society, 9*(1), 1.

Billett, D. S. M., Llewellyn, C., & Watson, J. (1988). Are deep-sea holothurian selective feeders? In R. D. Burke et al. (Eds.), *Echinoderm biology* (pp. 421–429). Rotterdam: A. Balkima Publishers.

Block, B. A., Costa, D. P., Boehlert, G. W., & Kochevar, R. E. (2003). Revealing pelagic habitat use: The tagging of Pacific pelagics program. *Oceanologica Acta, 25*, 255–266.

Block, B. A., Jonsen, S. J., Winship, A. J., Schaefer, S. A., Bograd, S. J., et al. (2011). Tracking apex marine predator movements in a dynamic ocean. *Nature, 475*(7354), 86–90. https://doi.org/10.1038/nature10082.

Bluhm, H. (1997). Megafauna as indicators for the recolonization of abyssal areas impacted by physical disturbances. *International symposium on environmental studies for deep-sea mining* (Tokyo, Japan), 49–63.

Bluhm, H. (2001). Re-establishment of an abyssal megabenthic community after experimental physical disturbance of the seafloor. *Deep-Sea Research II, 48*, 3841–3868.

Bluhm, H., & Gebruk, A. (1999). Holothuroidea (Echinodermata) of the Peru basin. Ecological and taxonomical remarks based on underwater images. *Marine Ecology, 20*(2), 167–195.

Bluhm, H., Schriever, G., & Thiel, H. (1995). Megabenthic recolonization in an experimentally disturbed abyssal manganese nodule area. *Marine Georesources and Geotechnology, 13*, 393–416.

Bostock, H. C., Opdyke, B. N., & Williams, M. J. M. (2010). Characterising the intermediate depth waters of the Pacific Ocean using δ13C and other geochemical tracers. *Deep-Sea Research Part I: Oceanographic Research Papers, 57*(7), 847–859.

Broecker, W. S., Takahashi, T., & Takahashi, T. (1985). Sources and flow patterns of deep-ocean waters as deduced from potential temperature, salinity and initial phosphate concentration. *Journal of Geophysical Research, 90*, 6925–6939.

Catalá, T. S., Reche, I., Fuentes-Lema, A., Romera-Castillo, C., Nieto-Cid, M., Ortega-Retuerta, E., et al. (2015a). Turnover time of fluorescent dissolved organic matter in the dark global ocean. *Nature Communications, 6*, 5986. https://doi.org/10.1038/ncomms6986.

Catalá, T. S., Reche, I., Álvarez, M., Khatiwala, S., Guallart, E. F., Benítez-Barrios, V. M., et al. (2015b). Water mass age and ageing driving chromophoric dissolved organic matter in the dark global ocean. *Global Biogeochemical Cycles, 29*, 1–18. https://doi.org/10.1002/2014GB005048.

Chung, J. S., Schriever, G., Sharma, R., & Yamazaki, T. (2001). Deep seabed mining environment: engineering and environment assessment. *Proceedings of ISOPE-ocean mining symposium*, Sczczcin, Poland (pp. 8–14).

Cury, P., & Shannon, L. (2004). Regime shifts in upwelling ecosystems: Observed changes and possible mechanisms in the northern and southern Benguela. *Progress Oceanography, 60*, 223–243.

Dahlgren, T. G., Wiklund, H., Rabone, M., Amon, D. J., Ikebe, C., Watling, L., Smith, C. R., & Glover, A. G. (2016). Abyssal fauna of the UK-1 polymetallic nodule exploration claim, Clarion-Clipperton zone, Central Pacific Ocean: Cnidaria. *Biodiversity Data Journal, 4*, e9277. https://doi.org/10.3897/BDJ.4.e9277.

Danovaro, R. (2008). Exponential decline of deep-sea ecosystem functioning linked to benthic biodiversity loss. *Current Biology, 18*, 1–8.

Davies, J., Baxter, J., Bradley, M., Connor, D., Khan, J., Murray, E., et al. (2001). *Marine monitoring handbook*. Peterborough: Joint Nature Conservation Council, 405 pp. (ISBN 1 85716 550 0).

Dayton, P. K., & Hessler, R. R. (1972). Role of biological disturbance in maintaining diversity in the deep-sea. *Deep Sea Research, 19*, 199–208.

Dittmar, T. (2014). The biogeochemistry of marine dissolved organic matter. In D. A. Hansell & C. A. Carlson (Eds.), Academic.

Drazen, J. C., Baldwin, R. J., & Smith, K. L., Jr. (1998). Sediment community response to a temporally varying food supply at an abyssal station in the NE Pacific. *Deep-Sea Research Pt II, 45*, 893–913.

Du Castel, V. (1985). Etablissement d'une carte géologique au 1/20 000e d'un domaine océanique profond dans une zone riche en nodules polymétalliques du Pacifique Nord (zone Clarion-Clipperton). Thèse de Doctorat, Université de Bretagne Occidentale, Brest, France.

Durden, J. M., Bette, B. J., Jones, D. O. B., Huvenne, V. A., & Ruhl, H. A. (2015). Abyssal hills-hidden source of increased habitat heterogeneity, benthic megafaunal biomass and diversity in the deep sea. *Progress in Oceanography, 137*(Part A), 209–218.

Dymond, J., & Collier, R. (1988). Biogenic particle fluxes in the equatorial Pacific: Evidence for both high and low productivity during the 1982–1983 El Niño. *Global Biogeochemical Cycles, 2*, 129–137.

ESCO CNRS IFREMER. (2014). Impacts environnementaux de l'exploitation des ressources minerales marines profondes. *Rapport d'expertise scientifique collective*. 939 pp.

FAO Fisheries and Aquaculture. (2000). *Biological characteristics of tuna*. Michel Goujon and Jacek Majkowski. http://www.fao.org/fishery/topic/16082/en.

Feely, R. A., Sabinelli, C. L., Schlitzer, R., Bullister, J. L., Mecking, S., & Greeley, D. (2004). Remineralization in the upper water column of the Pacific Ocean. *Journal of Oceanography, 60*, 45–52.

Fiedler, P., & Talley, L. D. (2006). Hydrography of the eastern tropical Pacific: A review [Hidrografía del Pacífico tropical oriental: Una revisión]. *Progress in Oceanography, 69*, 143–180.

Fiedler, P. C., Mendelssohn, R., Palacios, D. M., & Bograd, S. J. (2013). Pycnocline variations in the eastern tropical and north Pacific, 1958–2008. *Journal of Climate, 26*, 583–599. https://doi.org/10.1175/JCLI-D-11-00728.1

Flood, R. D. (1983). Classification of sedimentary furrows and a model for furrows initiation and evolution. *Geological Society of America Bulletin*, 630–639.

Foell, E., Thiel, H., & Schriever, G. (1990). DISCOL: A long-term large scale disturbance-recolonisation experiment in the Abyssal Eastern Tropical Pacific Ocean. *Proceedings of offshore technology conference*, Houston. (pp. 497–503).

Foley, M. M., Halpern, B. S., Micheli, F., Armsby, M. H., Caldwell, M. R., Crain, C. M., et al. (2010). Guiding ecological principles for marine spatial planning. *Marine Policy, 34*(5), 955–966. https://doi.org/10.1016/j.marpol.2010.02.001.

Friedrich, G., & Plüger, W. (1974). The distribution of manganese, iron, cobalt, nickel, copper and zinc in manganese nodules in various fields. *Meerestechnik, 6*, 203–206.

Fukushima, T. (1995). Overview "Japan Deep-Sea impact experiment = JET". *Proceedings of ISOPE ocean mining symposium*, Japan (pp. 47–53).

Fukushima, T., Shirayama, Y., & Kuboki, E. (2000). The characteristics of deep-sea epifaunal megabenthos community two years after an artificial rapid deposition event. *Publications of the SetoMarine Laboratory, 39*, 17–27.

Gage, J. D., & Tyler, P. A. (1991). *Deep-Sea biology: A natural history of organisms at the Deep-sea floor*. Cambridge, UK: Cambridge University Press.

Galeron, J., Scolan P., Fifis, A., & Sibuet, M. (2006). Spatial variability of megafaunal assemblages in deep-sea polymetallic nodule fields in the North-East Pacific. Oral presentation, *11th deep-sea biology symposium* (10–14 July 2006), Southampton, UK.

Game, E. T., Grantham, H. S., Hobday, A. J., Pressey, R. L., Lombard, A. T., Beckley, L., et al. (2009). Pelagic protected areas: The missing dimension in ocean conservation. *Trends in Ecology and Evolution, 24*, 360–369.

Gardner, W. D., & Sullivan, L. G. (1981). Benthic storms: Temporal variability in a deep-ocean nepheloid layer. *Science, 213*, 3.

Gardner, W. D., Sullivan, L. G., & Thorndike, E. M. (1984). Long-term photographic, current, and nephelometer observations of manganese nodule environments in the Pacific. *Earth and Planetary Science Letters, 70*, 95–109.

Georgi, D. T. (1981). On the relationships between the large-scale property variations and fine structures of the Antarctic circumpolar current. *Journal of Geophysical Research, 86*, 6556–6566.

GESAMP. (2016). Proceedings of the GESAMP International Workshop on the Impacts of Mine Tailings in the Marine Environment. (IMO/FAO/UNESCO-IOC/UNIDO/WMO/IAEA/UN/UNEP/UNDP Joint Group of Experts on the Scientific Aspects of Marine Environmental Protection). *Rep. Stud. GESAMP*, 94, 84 pp.

Ghosh, A. K., & Mukhopadhyay, R. (2000). *Mineral wealth of the ocean*. Rotterdam: A.A. Balkema.

Gianni, M., Fuller, S., Currie, D. E. J., Schleit, K., Pike, B., Goldsworthy, L., et al. (2016) How much longer will it take? *A ten-year review of the implementation of United Nations General Assembly resolutions 61/105, 64/72 and 66/68 on the management of bottom fisheries in areas beyond national jurisdiction*. Deep Sea Conservation Coalition.

Glover, A. G., & Smith, C. R. (2003). The deep-sea floor ecosystem: Current status and prospects of anthropogenic change by the year 2025. *Environmental Conservation, 30*(3), 219–241.

Glover, A. G., Smith, C. R., Paterson, G. I. J., Wilson, G. D. F., Hawkins, L., & Sheader, M. (2002). Polychaete species diversity in the Central Pacific abyss: Local and regional patterns and relationships with productivity. *Marine Ecology Progress Series, 2002*, 157–170. https://doi.org/10.3354/meps240157.

Glover, A. G., Gooday, A. J., Bailey, D. M., Billett, D. S. M., Chevaldonné, P., Colaço, A., et al. (2010). Temporal change in deep-sea benthic ecosystems: A review of the evidence from recent time-series studies, Chapter 1. In *Advances in marine biology* (Vol. 58, pp. 1–95). Burlington: Elsevier Press.

Glover, A. G., Wiklund, H., Rabone, M., Amon, D. J., Smith, C. R., O'Hara, T., Mah, C. L., & Dahlgren, T. G. (2016). Abyssal fauna of the UK-1 polymetallic nodule exploration claim, Clarion Clipperton zone, Central Pacific Ocean: Echinodermata. *Biodiversity Data Journal, 4*, e7251. https://doi.org/10.3897/BDJ.4.e7251.

Gooday, A. J. (2002). Biological responses to seasonally varying fluxes of organic matter to the ocean floor: A review. *Journal of Oceanography, 58*, 305–332.

Grassle, J. F., & Sanders, H. L. (1973). Life histories and the role of disturbance. *Deep-Sea Research, 20*, 643–659.

Gray, J. S. (1997). Marine biodiversity: Patterns, threats and conservation needs. *Biodiversity and Conservation, 6*, 153–175.

Gundersen, L. H., & Pritchard, L. (2002). *Resilience and the behaviour of large-scale systems*. In L. H. Gundersen, & L. Pritchard (Eds.), Island Press. Washington, USA

Haedrich, R. L., & Rowe, G. T. (1977). Megafaunal biomass in the deep-sea. *Nature, 269*, 141–142. https://doi.org/10.1038/269121a0.

Halbach, P., Friedrich, G., & von Stackelberg, U. (1988). *The manganese Nodule Belt of the Pacific Ocean: Geological environment, nodule formation, and mining aspects*. Stuttgart: F. Enke Verlag.Halbach, P. and Manheim, F.T.

Hannides, A. K., & Smith, C. R. (2003). The northeastern Pacific abyssal plain. In K. D. Black & G. B. Shimmield (Eds.), *Biogeochemistry of marine systems, chapter 10*. Oxford: Blackwell Science.

Harada, K. & Nishida, S. (1979). Biochronology of some Pacific manganese nodules and their growth mechanism. *Colloques Internationaux du CNRS*, Gif-sur-Yvette: 25–30.

Hartmann, M. (1979). Evidence for early diagenetic mobilization of trace metals from dislocation of pelagic sediments. *Chemical Geology, 26*, 277–293.

Heezen, B., & Hollister, C. (1971). *The face of the deep* (pp. 1–659). New York: Oxford University Press.

Henry, L.-A., Moreno Navas, J., & Roberts, J. M. (2013). Multi-scale interactions between local hydrography, seabed topography, and community assembly on cold-water coral reefs. *Biogeosciences, 10*, 2737–2746.

Heye, D. (1988). The internal microstructure of manganese nodules and their relationship to the growth rate. *Marine Geology, 26*, 59–66.

Hoffert, M., Le Suave, R., Pautot, G., Schaaf, A., Cochonat, P., Morel, Y., & Larque, P. (1992). *Caractérisation, par observations directes à l'aide du submersible « Nautile », de structures érosives actuelles des zones à nodules polymétalliques du Pacifique Nord équatorial: les "taches grises"*. Compte rendu à l'Académie des Sciences.

Hoffert, M. (2008). Les nodules polymétalliques dans les grands fonds océaniques, une extraordinaire aventure minière et scientifique sous-marine. *Société géologique de France. Editions VUIBERT*. ISBN 978-2-7117-7166-0. 430 pp.

Hoffert, M. & Saget, P. (2004). Manuel d'identification des "facies-nodules" pour la zone de plongées NIXO45. *Rapport Interne IFREMER*, Plouzané, France.

Hollister, C. D., & McCave, I. N. (1984). Sedimentation under deep-sea storms. *Nature, 309*, 220–225.

Hollister, C. D., & Nowell, A. R. M. (1991). HEBBLE epilogue. *Marine Geology, 99*, 445–460.

Hughes, T. P., Baird, A. H., Bellwood, D. R., Card, M., Connolly, S. R., Folke, C., et al. (2003). Climate change, human impacts and the resilience of coral reefs. *Science, 301*(5635), 929–933.

Hughes, T. P., Bellwood, D. R., Folke, C., Steneck, R., & Wilson, J. (2005). New paradigms for supporting the resilience of marine ecosystems. *Trends in Ecology and Evolution, 20*, 380–385.

Huisman, J., Thi, N. N. P., Karl, D. M., & Sommeijer, B. (2006). Reduced mixing generates oscillations and chaos in the oceanic deep chlorophyll maximum. *Nature, 439*, 322–325.

International Sea Bed Authority (I.S.B.A.). (2008). Technical study no. 3: Biodiversity, species ranges, and gene flow in the abyssal Pacific nodule province: predicting and managing the impacts of deep seabed mining. In C. R. Smith, J. Galeron, A. Gooday, A. Glover, H. Kitazato, L. Menot, et al. (Ed.), *Edited for the International Seabed Authority of the "Kaplan Project"*, 45 pp. ISBN: 978-976-95217-2-8.

Jankowski, J., & Zielke, W. (2001). The mesoscale sediment transport due to technical activities in the deep sea. *Deep-Sea Research II, 48*, 3487–3521.

Janssen, A., Kaiser, S., Meißner, K., Brenke, N., Menot, L., & Martínez Arbizu, P. (2015). A reverse taxonomic approach to assess macrofaunal distribution patterns in abyssal pacific polymetallic nodule fields. *PLoS One, 10*, e0117790. https://doi.org/10.1371/journal.pone.0117790.

Jones, D., Kaiser, S., Sweetman, A., Smith, C., Menot, L., Vink, et al. (2017). Biological responses to disturbance from simulated deep-sea polymetallic nodule mining. *Plos ONE.*, Ed by K. Vopel, *12*(2), e0171750. https://doi.org/10.1371/journal.pone.0171750.

Johnson, G. C. (2008). Quantifying Antarctic bottom water and north Atlantic deep water volumes. *Journal of Geophysical Research, 113*, C05027. https://doi.org/10.1029/2007JC004477.

Jumars, P. A., Mayer, L. M., Deming, J. W., Baross, J. A., & Wheatcroft, R. A. (1990). Deep-sea deposit feeding strategies suggested by environmental and feeding constraints. *Philosophical Transactions of the RoyalSociety of London. A, 331*, 85–101.

Karl, D. (2002). Nutrient dynamics in the deep blue sea. *Trends in Microbiology, 10*, 410–418.

Karl, D. M., Christian, J. R., Dore, J. E., Hebel, D. V., Letelier, R. M., Tupas, L. M., et al. (1996). Seasonal and interannual variability in primary production and particle flux at Station ALOHA. *Deep-Sea Research, II*(43), 539–568.

Karleskint, G., Turner, R., & Small, J. (2013). *Introduction to marine biology*. 4th Edition. Edition Brooks Cole. pp 529.

Kaufman, R. S., & Smith, K. L. (1997). Activity patterns of mobile epibenthic megafauna at an abyssal site in the eastern North Pacific; results from a 17 month time-lapse photography study. *Deep sea Research (Part I), 44*(4), 559–579.

Kessler, W. S. (2006). The circulation of the eastern tropical Pacific: A review. *Progress in Oceanography, 69*, 181–217.
Kennett, J. P. (1982). *Marine geology* (pp. 1–815). Englewood Cliffs: Prentice Hall.
Khripounoff, A., Caprais, J. C., & Crassous, P. (2006). Geochemical and biological recovery of the disturbed seafloor in polymetallic nodule fields of the Clipperton-Clarion Fracture Zone (CCFZ) at 5000m depth. *Limnol Oceanogr, 51*, 2033–2041.
Klein, H. (1987). Benthic storms, vortices and particle dispersion in the Deep West European Basin. *Dtsch Hydrol Z., 49*, 87–102.
Kontar, E. A., & Sokov, A. V. (1994). A benthic storm in the northeastern tropical Pacific over the fields of manganese nodules. *Deep-Sea Research 1, 41*(7), 1069–1089.
Kontar, E. A., & Sokov, A. V. (1997). On the benthic boundary layer's dynamics. *Journal of Marine Systems, 11*, 369–385.
Kotlinski, R. (1998). The present state of knowledge on oceanic polymetallic ores as exemplified by Interoceanmetal Joint Organization's activity. *Mineralogia Polonica, 29*, 77–89.
Kotlinski, R. (1999). Metallogenesis of the world's ocean against the background of oceanic crust evolution. *Polish Geological Institute Special Papers, 4*, 1–59.
Lauerman, L. M., Smoak, J. M., Shaw, T. J., Moore, W. S., & Smith, K. L. (1997). 234 Th and 210 Pb evidence from rapid ingestion of settling particles by mobile epibenthic megafauna in the abyssal NE Pacific. *Limnology and Oceanography, 42*(3), 589–595.
Levin, L. A., Demaster, D. J., Mccann, L. D., & Thomas, C. L. (1986). Effects of giant protozoans (class Xenophyophorea) on deep-seamount benthos. *Marine Ecology Progress Series, 29*, 99–104.
Levin, L. A., Etter, R. J., Rex, M. A., Gooday, A. J., Smith, C. R., Pineda, J., et al. (2001). Environmental influences on regional deep-sea species diversity. *Annual Review of Ecology and Systematics, 32*, 51–93.
Levitus, S., Antonov, J. I., Boyer, T. P., & Stephens, C. (2000). Warming of the world ocean. *Science, 287*, 2225–2229.
Longhurst, A. R. (2007). *Ecological geography of the sea* (2nd ed.). Burlington: Academic. https://doi.org/10.1016/B978-012455521-1/50000-0.
Lonsdale, P., & Southard, J. B. (1974). Experimental erosion of north Pacific red clay. *Marine Geology, 17*, M51–M60.
Loreau, M. (2008). Biodiversity and ecosystem functioning: The mystery of the deep sea. *Current Biology, 18*, R126–R128.
Lutz, M. J., Caldeira, K., Dunbar, R. B., & Behrenfeld, M. J. (2007). Seasonal rhythms of net primary production and particulate organic carbon flux to depth describe the efficiency of biological pump in the global ocean. *Journal of Geophysical Research, 112*, 1–26.
Marsh, L., Huvenne, V. A., & Bones, D. (2018). Geomorphological evidence of large vertebrates interacting with the seafloor at abyssal depths in a region designated for deep-sea mining. *Royal Society Open Science*. https://doi.org/10.1098/rsos.180286.
Mauviel, A., & Sibuet, M. (1985). Répartition des traces animales and importance de la bioturbation. In: L. Laubier & C. Monniot (Eds), *Peuplements profonds du Golfe de Gascogne. Campagne Biogas*. (Ifremer publication).
McCartney, M. S. (1982). The subtropical recirculation of mode waters. *Journal of Marine Research, 40*, 427–464.
McGowan, J. A., Cayan, D. R., & Dorman, L. M. (1998). Climate–ocean variability and ecosystem response in the northeast Pacific. *Science, 281*, 210–217.
McMurtry, G. (2001). Authigenic deposits. In S. A. Thorpe & K. K. Turekian (Eds.), *Encyclopedia of ocean sciences* (pp. 201–220). London: Academic.
Menzies, R. J., George, R., & Rowe, G. T. (1975). Abyssal environment and ecology of the world oceans. New York: Wiley-Interscience. *44*(1), 346.
Montgomery, R. B. (1958). Water characteristics of Atlantic Ocean and World Ocean. *Deep Sea Research, 5*, 134–148.

Moreno Navas, J., Henry, L.-A., Miller, P., & Roberts, J. M. (2014). Ecohydrodynamics of cold-water coral reefs: A case study of the Mingulay reef complex (Western Scotland). *PLoS One, 9*(8), e106208. https://doi.org/10.1371/journal.pone.0098218.

Morgan, C. (1991). Geographical distributions of benthic megafauna in the Clarion-Clipperton zone. *NOAA report*.

Mullineaux, L. S. (1987). Organisms living on manganese nodules and crusts: Distribution and abundance at three North Pacific sites. *Deep-Sea Research, 34*, 165–184.

Muschenheim, D. K. (1987). The dynamics of near-bed seston flux and suspension-feeding benthos. *Journal of Marine Research, 45*(2), 473–496.

Nowell, A. R. M., Hollister, C. D., & Jumars, P. A. (1982). High energy benthic boundary layer experiment: HEBBLE. *Eos Transactions American Geophysical Union, 63*(31), 594–595.

Oebius, H. U., Becker, H. J., Rolinski, S., & Jankowski, J. (2001). Parametrization and evaluation of marine environmental impacts produced by deep-sea manganese nodule mining. *Deep-Sea Research II, 48*, 3453–3467.

Onken, R. (1995). The spreading of lower circumpolar deep water into the Atlantic Ocean. *Journal of Physical Oceanography, 25*, 3051–3063.

Ozturgut, E., Lavelle, J. W., Steffin, O., & Swift, S. A. (1980). Environmental investigation during manganese nodule mining tests in the North Equatorial Pacific, in November 1978. *NOAA Tech. Memorandum ERL MESA-48*, NOAA, USA, 50.

Paterson, G. L. J., Menot, L., Colaço, A., Glover, A. G., Gollner, S., Kaiser, S., et al. (2015). Biogeography and connectivity in deep-sea habitats with mineral resource potential gap analysis. *MIDAS Delivery, 4*, 45. Available online at: http://www.vliz.be/imisdocs/publications/41/268441.pdf.

Pennington, T. J., Mahoney, K. L., Kuwahara, V. S., Kolber, D. D., Calienes, R., et al. (2006). Primary production in the eastern tropical Pacific: A review. *Progress in Oceanography, 69*, 285–317.

Pfannkuche, O., & Lochte, K. (1993). Open ocean pelago-benthic coupling: Cyanobacteria as tracers of sedimenting salp feces. *Deep-Sea Research Part I, 40*, 727–737.

Price, A. R. G. (2004). Rapid coastal environmental assessment. In: *Standard survey methods for the red sea and Gulf of Aden*. (PERSGA/GEF). *PERSGA Technical Series* 10. PERSGA, Jeddah, 1–2.

Quirchmayr, R. (2015). On the existence of benthic storms. *Journal of Nonlinear Mathematical Physics, 22*(4), 540–544. https://doi.org/10.1080/14029251.2015.1113053.

Radziejewska, T. (1997). Immediate responses of benthic meio- and megafauna to disturbance caused by polymetallic nodule miner simulator. *Proceedings of the international symposium on environmental studies for deep-sea mining*, Tokyo, Japan (November 20–21), 223–235.

Radziejewska, T. (2002). Responses of deep-sea meiobenthic communities to sediment disturbance simulating effects of polymetallic nodule mining. *International Review of Hydrobiology, 87*, 459–479.

Radziejewska, T., & Stoyanova, V. (2000). Abyssal epibenthic megafauna of the Clarion-Clipperton area (NE Pacific): changes in time and space versus anthropogenic environmental disturbance. *Oceanological Studies, XXIX*(2), 83–101.

Ramirez-Llodra, E., Tyler, P. A., Baker, M. C., Bergstad, O. A., Clark, M. R., Escobar, E., et al. (2011). Man and the last great wilderness: Human impact on the deep sea. *PLoS One, 6*, e22588. https://doi.org/10.1371/journal.pone.0022588.

Rex, M. A. (1983). Geographic patterns of species diversity in the deep-sea benthos. In G. T. Rowe (Ed.), *Deep sea biology* (pp. 453–472). New York: Wiley.

Rex, M., & Etter, R. (2010). *Deep-sea biodiversity; pattern and scale*. Cambridge, MA: Harvard University Press, 355 pp.

Rogacheva, A., Gebruk, A., & Alt, C. (2012). Swimming deep-sea holothurians (Echinodermata: Holothuroidea) on the northern mid-atlantic ridge. *Zoosymposia, 7*, 213–224.

Rolinski, S., Segschneider, J., & Sündermann, J. (2001). Long-term propagation of tailings from deep-sea mining under variable conditions by means of numerical simulations. *Deep-Sea Research II, 48*, 3469–3485.

Rowe, G. T. (1971). Benthic biomass and surface productivity. In J. D. Costlow (Ed.), *Fertility of the sea* (pp. 97–122). New York: Gordon and Breach.
Ruhl, H. A., & Smith, K. L., Jr. (2004). Shifts in Deep-Sea community structure linked to climate and food supply. *Science, 305*, 513–515.
Saguez, G. (1985). Etude de la morphologie, de la structure interne and de la lithologie des nodules polymétalliques de la zone Clarion-Clipperton. Relations avec l'environnement. Thèse de Doctorat, Université de Bretagne Occidentale, Brest: 1–228.
Sarmiento, J. L., Slater, R., Barber, R., Bopp, L., Doney, S. C., Hirst, A. C., et al. (2004). Response of ocean ecosystems to climate warming. *Global Biogeochemical Cycles*. https://doi.org/10.1029/2003GB002134.
Schneider, J. (1981). Manganese nodules in the deep sea. Formation, ocurrence and ecological consequencesof mining. *Natur und Museum, III*, 114–124.
Schneider, D. C., Gagnon, J. M., & Gilkinson, K. D. (1987). Patchiness of epibenthic megafauna on the outer Grand Banks of Newfoundland. *Marine Ecology Progress Series, 39*, 1–13.
Schor, G., Falcone, E., Moretti, D., & Andrews, D. (2014). First Long-Term Behavioral Records from Cuvier's Beaked Whales (*Ziphius cavirostris*) reveal record-breaking dives. *PLOS one, 9*(3), e92633.
Schriever, G., Ahnert, A., Borowski, C., & Thiel, H. (1997). Results of the large scale deep-sea impact study DISCOL during 8 years of investigation. In *Proceedings international symposium environmental studies for deep-sea mining* (pp. 237–243). Japan: Metal Mining Agency of Japan.
Sharma, R. (2015). Environmental issues of deep-sea mining. Global challenges, policy framework and sustainable development for mining of mineral and fossil energy resources (GCPF2015). *Procedia Earth and Planetary Science, 11*(2015), 204–211.
Sharma, R., & Rao, A. (1992). Geological factors associated with megabenthic activity in the Central Indian Basin. *Deep-Sea Research, 39*(3/4), 705–713.
Sharma, R., Nagender Nath, B., Parthiban, S., & Jai Sankar, S. (2001). Sediment redistribution during simulated benthic disturbance and its implications on deep seabed mining. *Deep-Sea Research Part II, 48*, 3363–3380.
Shirayama, Y. (1999). Biological results of JET project: An overview. *Proceedings of ISOPE ocean mining symposium,* Goa, India, 1999 (pp. 185–190).
Shor, G. (1959). Reflexion studies in the eastern equatorial Pacific. *Deep-Sea Research, 5*, 283–285.
Smith, C. R. (1999). The biological environment in the nodule provinces of the deep sea. In N. Odunton (Ed.), *Deep-seabed polymetallic Nodule Exploration: Development of environmental guidelines* (pp. 41–68). Kingston: International Seabed Authority.
Smith, C. R., & Baco, A. R (2003). Ecology of whale falls at the deep-sea floor. *In Oceanography and Marine Biology: An Annual Review*. 41, 311–354
Smith, C. R., & Demopoulos, A. (2003). Ecology of the deep. Pacific Ocean floor. In P. A. Tyler (Ed.), *Ecosystems of the world, 28: Ecosystems of the deep. Ocean*. Amsterdam: Elsevier.
Smith, K. L., Jr., Kaufmann, R. S., & Wakefield, W. W. (1993). Mobile megafaunal activity monitored with a time-lapse camera in the abyssal North Pacific. *Deep Sea Research Part I: Oceaonographic Research Papers, 40*(11–12), 2307–2324. https://doi.org/10.1016/0967-0637(93)90106-D.
Smith, K. L., & Kaufmann, R. S. (1999). Long-term discrepancy between food supply and demand in the deep eastern north Pacific. *Science, 284*, 1174–1177.
Smith, C. R., & Rabouille, C. (2002). What controls the mixed-layer depth in deep-sea sediments? The importance of POC flux. *Limnology and Oceanography, 47*, 418–426.
Smith, C. R., & Trueblood, D. D. (1991). BIE I cruise report. R/V Yuzhmorgeologiya. Funded by NOAA.
Smith, C. R., Jumars, P. A., & DeMaster, D. J. (1986). In situ studies of megafaunal mounds indicate rapid sediment turnover and community response at the deep-sea floor. *Nature, 323*, 251–252.
Smith, C. R., Berelson, W., Demaster, D. J., Dobbs, F. C., Hammond, D., Hoover, D. J., et al. (1997). Latitudinal variations in benthic processes in the abyssal equatorial Pacific: Control by biogenic particle flux. *Deep-Sea Research, II*(44), 2295–2317.

Smith, K. L., Kaufmann, R. S., Baldwin, R. J., & Carlucci, A. F. (2001). Pelagic-benthic coupling in the abyssal eastern North Pacific: An 8-year time-series study of food supply and demand. *Limnology and Oceanography, 46*, 543–556.

Smith, C. R., Levin, L., Koslow, A., Tyler, P., & Glover, A. (2008a). The near future of the deep sea floor ecosystems. In N. Polunin (Ed.), *Aquatic ecosystems: trends and global prospects* (pp. 334–349). Cambridge: Cambridge University Press.

Smith, C., Gaines, S., Watling, L., Friedlander, A., Morgan, C., Thurnherr, A. et al. (2008b). Rationale and recommendations for the establishment of preservation reference areas for nodule mining in the Clarion-Clipperton Zone. *Fourteenth session*. Kingston, Jamaica, 26 May–6 June 2008. http://www.isa.org.jm/en/sessions/2008/documents. Legal and Technical Commission, International Seabed Authority. Technical document no. ISBA/14/LTC/2.

Smith, K. L., Ruhl, H., Bett, B., Billett, D., Lampitt, R., & Kaufmann, R. (2009). Climate, carbon cycling and deep-ocean ecosystems. *Proceedings of the National Academy of Sciences of the United States of America, 106*(46), 19211–19218.

Snelgrove, P. V. R., & Smith, C. R. (2002). A riot of species in an environmental calm: The paradox of the species-rich deep sea. *Oceanography and Marine Biology Annual Review, 40*, 311–342.

Spiess, F. N., Hessler, R., Wilson, G., & Weydert, M. (1987). Environmental effects of deep sea dredging, NOAA report, 87–5: 1–85.

Steneck, R. S. (2001). Functional groups. In S. A. Levin (Ed.), *Encyclopedia of biodiversity* (Vol. 3, pp. 121–139). San Diego: Academic Press.

Stoyanova, V. (2008). Some peculiarities of distribution patterns of polymetallic nodules associated megabenthic community". Interoceanmetal Joint Organization. Oral Communication at the panel discussion *"A proposition of strategy of conservation for a nodule ecosystem in the Clarion-Clipperton Fracture Zone, NE Pacific Ocean"* organized by UNESCO/IOC on 12 Dec 2008.

Suga, T., Takei, Y., & Hanawa, K. (1997). Thermostad distribution in the North Pacific subtropical gyre: The central mode water and the subtropical mode water. *Journal of Physical Oceanography, 27*, 140–152.

Talley, L. D. (1996). Antarctic intermediate water in the South Atlantic. In G. Wefer et al. (Eds.), *The South Atlantic Present and Past Circulation* (pp. 219–238). Berlin/Heidelberg: Springer Verlag.

Talley, L. D., Pickard, G., Emery, W., & Swift, J. (2011). *Descriptive physical oceanography: An introduction* (6th ed.). Boston: Elsevier Academic Press, 560 p.

Thiel, H., Bluhm, H., Borowski, C., Bussau, C., Gooday, A., Maybury, C., & Schriever, G. (1992). The impact of mining on deep sea organisms. The DISCOL-Project. *Ocean Challenge, 3*(1), 40–46.

Thiel, H. (2001). Evaluation of the environmental consequences of polymetallic nodule mining based on the results of the TUSCH Research Association. *Deep-Sea Research Part II, 48*, 3433–3452.

Thiel, H. (2003). Anthropogenic impacts on the deep-sea. In P. A. Tyler (Ed.), *Ecosystems of the world: The Deep Sea*. Amsterdam: Elsevier.

Thiel, H., Schriever, G., Bussau, C., & Borowski, C. (1993). Manganese nodule crevice fauna. *Deep-Sea Research, 40*(2), 419–423.

Thiel, H., Schriever, G., Ahnert, A., Bluhm, H., Borowski, C., & Vopel, K. (2001). The large scale environmental impact experiment DISCOL – Reflection and foresight. *Deep-Sea Research II, II48*, 3869–3882.

Thurston, M. H., Rice, A. L., & Bett, B. J. (1998). Latitudinal variation in invertebrate megafaunal abundance and biomass in the North Atlantic Ocean abyss. *Deep-Sea Research II, 45*, 203–224.

Tilot, V. (1988). Review of Environmental Impact studies on deep-sea polymetallic mining in the Eastern Central Pacific Ocean. Contract CEE CDC/88/7730/IN/01: 1–59. Report for the Commission of the European Communities Directorate-General for Industry.

Tilot, V. (1989). Report to EU of the german oceanographic on – A DIS-turbance and re-COL-onization experiment DISCOL II in the Peru Basin, South Pacific. Contract CEE CDC/88/7730/IN/01: 1–59. Report for the Commission of the European Communities Directorate-General for Industry.

Tilot, V. (1990a). Report to the Scripps Institution of Oceanography on the description of megafaunal assemblages in ECHO I on basis of the analysis of photographs of the seafloor by Deep Tow cameras on pilot study site of environmental impact.
Tilot, V. (1990b). Report on U.S. Quagmire II cruise. R/V New Horizon-RUM III cruise, 23 April–17May 1990. Funded by EU, DG for industry and NOAA. *IFREMER, Brest, France*.
Tilot, V. (1991). Report on the description of megafaunal assemblages and manganese nodule coverage along transects taken during the Benthic Impact Experiment (NOAA-BIE) cruise by NOAA and SCRIPPS, USA. June 24–July 1991. Contract CEE CDC/90/7730/IN/01: 1–59. Report for the Commission of the European Communities Directorate-General for Industry.
Tilot, V. (1992). La structure des assemblages mégabenthiques d'une province à nodules polymétalliques de l'océan Pacifique tropical Est. [PhD dissertation thesis]. [Brest (France)]: University of Bretagne Occidentale.
Tilot, V. (1995). An ecologically important type of biological activity in a deep-sea ferro-manganese nodule environment of the tropical north east Pacific. *Mesogée, 54*, 101–120.
Tilot, V. (2003). Final report on the Gulf of Aqaba Monitoring Programme (GAMP). *Egyptian Environmental Agency Affairs (EEAA) and European Commission, National Parks of Egypt Protectorates Development Programmes project*. SEM 04/220/027 A.
Tilot, V. (2004). Management plan for the marine mammals sanctuary in the Mediterranean sea «Pelagos» adopted by the governments of France, Italy and Monaco in October 2004, signatories to the Agreement on the creation of a sanctuary for marine mammals concluded on 25 November 1999. 107 pp.
Tilot, V. (2005). Eritrea marine, coastal and islands biodiversity. Lectures on marine ecology, oceanography, marine organisms, seagrass ecology, mangrove ecology, fisheries, reef fish ecology, survey technics (coral, fish and invertebrate monitoring). ECMIB-GEF/UNDP project.
Tilot, V. (2006a). Biodiversity and distribution of the megafauna Vol 1 The polymetallic nodule ecosystem of the Eastern Equatorial Pacific Ocean. Intergovernmental Oceanographic Commission. Technical Series 69, Project Unesco COI/Min Vlanderen, Belgium. http://unesdoc.unesco.org/images/0014/001495/149556f.pdf.
Tilot, V. (2006b). Biodiversity and distribution of the megafauna Vol 2 Annotated photographic Atlas of the echinoderms of the Clarion-Clipperton fracture zone. Intergovernmental Oceanographic Commission. Technical Series 69, Project Unesco/COI/MinVlanderen, Belgium. http://unesdoc.unesco.org/images/0014/001495/149556f.pdf.
Tilot, V. (2006c). Megabenthic assemblages of polymetallic nodule ecosystems in the Clarion-Clipperton Fracture Zone within the abyssal tropical East Pacific. In: *Proceedings of workshop on prospects for international collaboration in marine environmental research to enhance understanding of the deep-sea environment, international seabed authority*, Kingston, Jamaica, Appendix A, pp. 1–14.
Tilot, V., Leujak, W., & Ormond, R. (2008a). Monitoring of south Sinai coral reefs: Influence of natural and anthropogenic factors. *Aquatic Conservation: Marine and Freshwater ecosystems, 18*(7), 1109–1126.
Tilot, V., Veron, C., & Jeudy de Grissac, A. (2008b). The coral reefs of Eritrea: Little known gems. In C. Wilkinson (Ed.), *Status of coral reefs of the world: 2008* (p. 78). Townsville: Global Coral Reef Monitoring Network and Reef and Rainforest Research Centre.
Tilot, V. (2010a). Biodiversity and distribution of faunal assemblages Vol 3 Options for the management and conservation of the nodule ecosystem in the Clarion Clipperton Fracture Zone (NE Pacific Ocean): scientific, legal and institutional aspects. *Technical Series 69, Project Unesco COI/Min*, Vlaanderen, Belgium. http://unesdoc.unesco.org/images/0014/001495/149556e.pdf#223.
Tilot, V. (2010b). Panorama of epibenthic megafaunal communities in a deep sea nodule ecosystem of the CCZ, NE Pacific Ocean. In UNESCO/IOC, Tech. Series 69, vol 3. http://unesdoc.unesco.org/images/0014/001495/149556e.pdf#223.
Tilot, V. (2011). Recommendations for the conservation and management of the biodiversity of deep sea ecosystems targeted by seabed mining (Polymetallic Nodule Deposits, Cobalt-Rich

Ferromanganese Crusts and Hydrothermal Sulfides). *Proceedings 40th annual conference of the Underwater Mining Institute*, Hilo, Hawaii, 14–18 September, 2011.

Tilot, V. (2013). Tools for marine spatial management in deep seas (A guide for Research, Conservation and Development Practitioners). *Proceedings of the International Marine Protected Areas Congress (IMPAC3)*, IUCN/Agence des Aires Marines Protégées (Fr), Marseille, France, 21–27 October 2013.

Tilot, V. (2014). An innovative multilayer rapid ecological assessment adapted to monitoring and management of Deep Sea Ecosystems. *Proceedings deep sea mining summit 2014*, London, UK. 17–19 March 2014.

Tilot, V. (2016). General features of the coral reef communities of the Daymaniyat islands, Oman sultanate and designation of management indexes for Environmental Sensitivity mapping. *IUCN/ROWA report.*

Tilot, V., Ormond, R., Moreno-Navas, J., & Catala, T. (2018). The benthic megafaunal assemblages of the CCZ (Eastern Pacific) and an approach to their management in the face of threatened anthropogenic impacts. *Frontiers in Marine Science, 5*, 7. https://doi.org/10.3389/fmars.2018.00007.

Tkatchenko, G., & Radziejewska T. (1998). Recovery and recolonization processes in the area disturbed by a polymetallic nodule collector simulator. In: J. S. Chung, M. Olagnon, C. H. Kim, & W. Koterayama (Eds.), *Proceedings of the 8th ISOPE conference*, Montreal, Canada, vol. 2, pp. 282–286.

Tkatchenko, G., Radziejewska, T., Stoyanova, V., Modlitba, I., Parizek, A., et al. (1996). Benthic impact experiment in the IOM pioneer area: Testing for effects of deep-sea disturbance. In *International seminar on deep sea mining, 1997* (pp. 223–236). Tokyo, Japan: Metal Mining Agency of Japan.

Tortell, P. (1992). Coastal zone sensitivity mapping and its role in marine environmental management. *Marine Polution Bulletin, 25*, 88–93.

Trueblood, D. D., Ozturgut, E., Philipchuk, M., & Gloumov, I. F. (1997). The ecological impacts of the Joint US-Russian Benthic Impact Experiment. In *Proceedings of the international symposium on environmental studies for deep-sea mining* (pp. 237–243). Tokyo, Japan: Metal Mining Agency of Japan.

Tsuchiya, M. (1981). The origin of the Pacific equatorial 13°C water. *Journal of Physical Oceanography, 11*(6), 794–812.

Tyler, P. A. (1988). Seasonality in the deep sea. *Oceanography and Marine Biology. Annual Review, 26*, 227–258.

Tyler, P. A. (2003). Ecosystems of the world. In *Ecosystems of the deep oceans* (Vol. 28). Amsterdam/Boston: Elsevier Press. ISBN 0 444 82619 X. 569 pp.

Vanreusel, A., Hilario, A., Ribiero, P., Menot, L., & Martinez Arbizu, P. (2016). Threatened by mining polymetallic nodules are required to preserve abyssal epifauna. *Scientific Reports, 6*, 26808. https://doi.org/10.1038/srep26808(2016).

Voisset, M., & Hein, P. (1978). Géochimie des nodules du Pacifique Nord-Est. *Rapports scientifiques et techniques du CNEXO*. Ifremer, France.

Von Stackelberg, U., & Beiersdorf, H. (1991). The formation of manganese nodules between the Clarion and Clipperton fracture zones southeast of Hawaii. In *Marine geology* (Vol. 98, pp. 411–423). Englewood Cliffs: Prentice-Hall.

Wang, Y., Li, J., Han, X., & WU, Z. (2001). The influence of terrain slope on the distribution of polymetallic nodules. *Acta Oceanologica Sinica, 23*(1), 65–23.

Wilson, G. D. F. (1990). Cruise report on RUM 3-R/V New Horizon Cruise "Quagmire II": 23 April–17 May 1990. *SIO reference series, number 90–25, August 1990.*

Dr. Virginie Tilot has a PhD in Biological Oceanography and an Environmental Engineering degree, with specialization in Conservation, Aquaculture and Fisheries. She is "attaché honoraire" at the UMS (AFB CNRS MNHN) "Patrimoine Naturel" (PATRINAT) of the Museum National d'Histoire Naturelle, Paris, and is a visiting research scholar at the marine station in Malaga of the Instituto Español de Oceanografía. She has 35 years of experience in the areas of biological oceanography, marine (high seas/coastal/island) ecology, deep-sea ecology, coral reef ecology, biodiversity assessment, conservation, marine protected areas, marine spatial management and planning, strategic environmental impact assessment, and governance. She has worked for diverse national and international organizations and institutions, universities in particular for agencies of the United Nations (UNEP, UNDP, FAO, UNESCO/IOC. etc.), the European Commission, and non governmental organizations. Her activities took place in different areas of the Atlantic, Pacific, Indian Ocean, Red Sea, Oman Sea, Mediterranean Sea, and in more than 50 countries. She produced a total of 110 publications, among which 91 at an international level. She has participated at different levels in 35 regional projects in the world. She teaches in an European master program Erasmus Mundus "Marine Biodiversity and Conservation," in the doctoral school of the Museum National of Natural History, Paris, and the virtual Academy of the University of Bremen within the framework of EU COST Ocean Governance. She produced 60 communications at international meetings, 5 scientific films, and 10 annotated guides/ photographic atlas of the marine fauna and participated in 4 cartographic atlas/georeferenced databases of natural resources and human activities, 6 inventories and collections of the faunal specimens, and 6 scientific exhibits.

Long-Term Monitoring of Environmental Conditions of Benthic Impact Experiment

Tomohiko Fukushima and Akira Tsune

Abstract The Japan Deep-Sea Impact Experiment (JET) is an in situ experiment aimed at evaluating the environmental impacts and recovery process arising from associated manganese nodule mining. The experiment was initiated in 1994 and ended in 1996. However, owing to the short monitoring period, it was not possible to gain adequate understanding of the recovery process. Subsequently, the Deep Ocean Resources Development Co., Ltd., revisited the site 17–18 years later and surveyed the environmental conditions (long-term monitoring survey). This survey found that the environmental impacts that were recognizable immediately after and 1 year after benthic disturbance were not visible after 17–18 years. This chapter presents the results of the long-term monitoring survey and discusses future plans of environmental impact evaluations.

Keywords Deep-sea mineral resources · Manganese nodules · Impact experiment · Environmental impact assessment (EIA) · Meiobenthos · Clarion-Clipperton Fracture Zone (CCZ)

1 Benthic Impact Experiment

1.1 Historical Background

In view of the shortage of metal resources in the near future, the development of seabed mineral resources has received increased attention. At the same time, it has become necessary for environmental impact assessments (EIAs) to be more rigorous than ever before (Fukushima and Nishijima 2017). However, since there is no

T. Fukushima (✉)
Japan Agency for Marine-Earth Science and Technology (JAMSTEC), Yokosuka, Kanagawa, Japan
e-mail: fukushimat@jamstec.go.jp

A. Tsune
Deep Ocean Resources Development Co., Ltd. (DORD), Tokyo, Japan

R. Sharma (ed.), *Environmental Issues of Deep-Sea Mining*,
https://doi.org/10.1007/978-3-030-12696-4_7

active mining being carried out at present, it is not easy to correctly assess the environmental impact. In 1972, the Lamont Doherty Geological Survey undertook a desk study on the environmental impact of deep-sea mining (Ozturgut et al. 1997). Subsequently, the National Oceanic and Atmospheric Administration (NOAA) conducted a large-scale baseline study DOMES (Deep Ocean Mining Environmental Study: 1976–1980) (Ozturgut et al. 1978; Ozturgut 1981). The results of that survey were used to list the factors that were thought to potentially adversely impact an environment in Programmatic Environmental Impact Statement (PEIS) (NOAA 1981). The following year, NOAA narrowed the list down to two activities that were thought to be particularly serious (NOAA 1982). The first possibility was that mining devices or collectors could impact benthic organisms directly (physically), and the second was that the resuspension and resettlement of marine sediment accompanying the operation of those machines could indirectly impact benthic organisms in surrounding areas. Since then, the NOAA has been conducting verification experiments, such as the Acute Mortality Experiment (1987) (Smith et al. 1988) and QUAGMIRE Expedition II (1990) (Wilson 1990), which were specific to sediments resuspension and resettlement (hereinafter called benthic disturbance). In 1992, a new device (hereinafter, called the disturber) was developed to create benthic disturbance, which was successfully operated in the Clarion-Clipperton Fracture Zone (CCZ) (Trueblood 1993). This experiment was known as the Benthic Impact Experiment (BIE) (1992). Subsequently, the same device was used to carry out the Japan Deep-Sea Impact Experiment (JET) (1994) and IOM'BIE (1995) in the CCZ, as well as the Indian Deep-Sea Environment Experiment (INDEX) (1997) in the Indian Ocean (hereinafter, referred to as the BIE-type experiment) (Fukushima 1995; Kotlinski 1995; Sharma et al. 2000).

Prior to these, researchers from Hamburg University, Germany, conducted a Disturbance and Recolonization (DISCOL) experiment using a plow harrow in Peru Basin (Thiel 1991). Although there are differences in methods between the German experiment and the BIEs, both were field experiments for deep-sea mineral resource development. The four BIE-type experiments revealed that benthic disturbance reduced benthic organisms, at least temporarily immediately after the experiments. However, as the monitoring was discontinued after 1 year, 2 years, 5 years, and 3 years (44 months) for BIE, JET, IOM'BIE, and INDEX, respectively, the duration required for complete restoration was not ascertained. Similarly, DISCOL monitoring was discontinued within 7 years, so the recovery process has remained unknown.

In this context, Deep Ocean Resources Development Co., Ltd. (DORD), revisited the JET site in 2011 and 2012 and attempted to survey the present status.[1] Similarly, JPI-Oceans, which is a project launched by the EU, revisited the IOM'BIE site in 2015 and conducted environmental monitoring activity 20 years later (JPI Oceans 2016). Moreover, the same research vessel revisited the DISCOL site and carried out a survey equivalent to monitoring 26 years later and reported the long-term effects on megafauna (Vanreusel et al. 2016).

[1] JET (1994–1996) was conducted by the Metal Mining Agency of Japan (MMAJ; now called Japan Oil, Gas and Metals National Corporation (JOGMEC).

1.2 Short-Term Impact of Benthic Disturbance

Briefly, the BIE-type experiment involved understanding the natural environmental conditions, followed by an artificial impact, and comparing environmental conditions before and after the impact. In this process, understanding the natural environmental conditions is referred to as the baseline study, conducting artificial impact is referred to as the disturbance, and evaluating the changes in the seabed conditions is referred to as monitoring. Before discussing the long-term monitoring results, the results of short-term monitoring revealed by JET are introduced.

JET was carried out in the western part of the CCZ, which is located in the exploration area of DORD (hereinafter, the JET site; Fig. 1). The water depth at the experimental site is 5200 m and it exhibits a valley type of topography with relatively small amounts of manganese nodules. A baseline survey (JET1) was carried out in August 1994, and the seabed was disturbed by towing a disturber in September

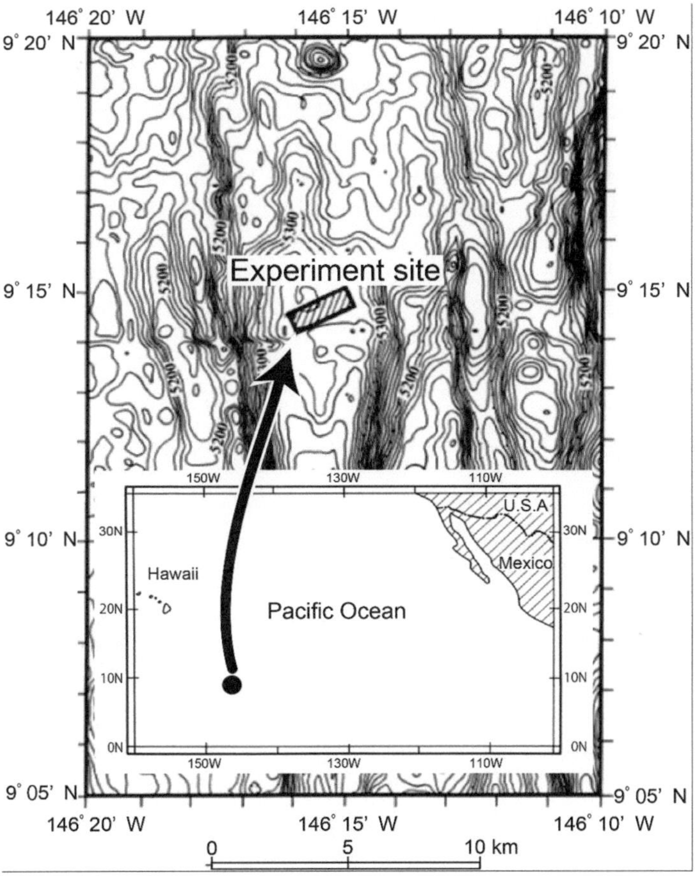

Fig. 1 Location and topography of JET site. (After Fukushima 2004)

1994 (Fig. 2). A monitoring survey was carried out immediately after benthic disturbance (JET2), 1 year later (JET3), and 2 years later (JET4). In this paper, the area inside the track where the disturber was towed is called the "disturbance" area, and the area within 300 m of the track is called the "disturbed area" (Fig. 3). Also, in order to differentiate from the survey conducted in 2011–2012 (i.e., JET5), the JET1, JET2, JET3, and JET4 surveys are collectively referred to as the previous JET surveys. In JET1, it was confirmed through analysis of variance (ANOVA) (all $p < 0.01$) that the concentration of total organic carbon (TOC), total nitrogen (TN), and calcium carbonate ($CaCO_3$) in the sediment decreased significantly with depth (Fukushima et al. 2001). Moreover, it was confirmed that the vertical distribution of meiobenthos is concentrated within 1 cm of the surface (Fukushima 2004). In other words, under natural conditions, there is a vertical gradient distribution of sediment and meiobenthic components.

In order to create the disturbance, the benthic disturber was towed for 20 h and 27 min. Using data and analysis of sediment traps, seabed video observations, and numerical models, the amount of sediment discharge was calculated to be about 352 tonnes (converted to dry weight) (Fukushima 1995), and the sediment was estimated to have spread along a distance of at least 500 m (Barnett and Yamauchi 1995). In addition, from sediment trap samples, it was found that the released sediment consisted mainly of inorganic material (Fukushima et al. 2002).

JET2 results did not show vertical reduction in TOC, TN, and $CaCO_3$ concentrations in the disturbance adjacent area, as was seen in JET1 (Fukushima 2004). Furthermore, an integrated indicator was calculated from the first principal component (PZ1) of principal component analysis (PCA) using these three components. The relationship between the value and the distance from the disturbance area was then compared, and statistically significant slopes were found in the 0–0.25 cm and 0.25–0.5 cm layers (both $p < 0.05$) (Fukushima et al. 2001). In other words,

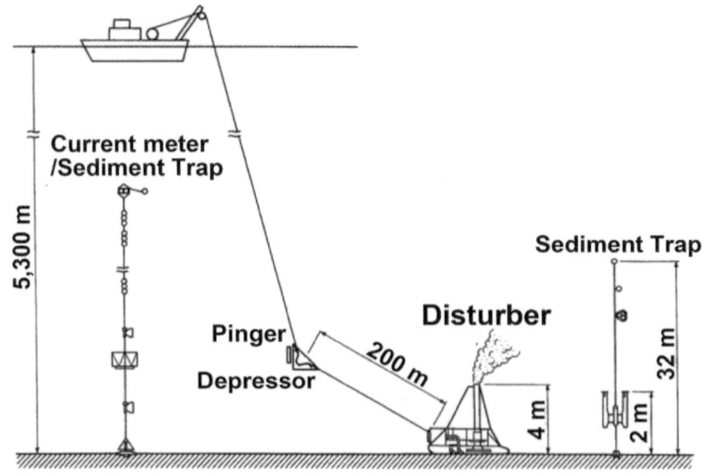

Fig. 2 Schematic arrangement for the disturber towing. (After Fukushima 2004)

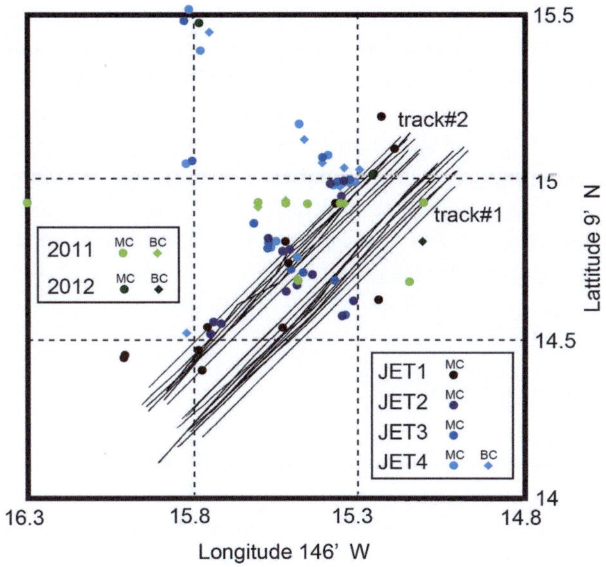

Fig. 3 Disturber tow tracks and sediment sampling positions. (After Fukushima and Tsune 2018)

redeposition is considered to have changed the chemical composition of the surface layer of the sediment. This means that the food environment of benthic organisms (especially infauna that depend on sediments for nutrients, deposit feeders, etc.) deteriorated. Subsequently, as compared to the results from JET1, the abundance of meiobenthos was statistically found to be significantly smaller (*t-test*; each $p < 0.01$). From these results, it can be estimated that benthic disturbance changes the chemical composition of the surface sediments and the distribution and abundance of meiobenthos (Fukushima 2004).

A comparison of the results of JET3 and JET4 with those of JET1 reveals no remarkable difference in the chemical compositions of sediment and the abundance of meiobenthos, as was the case with JET2. In JET3, the correlation between the distance and PZ1 was significant in the 0–0.25 cm layer ($p < 0.05$), but not significant in the 0.5 cm layer. Furthermore, no significant relationship was found in any layer in JET4. As for the abundance of meiobenthos, no statistical significance was found between JET1 and JET3 and JET4. Thus, while the chemical composition of sediment and biomass of benthic organisms seemed to return to their original state, there were changes in species diversity. Comparing the species diversity of JET1 and JET4, although it was limited to nematodes, it was seen that JET4 had a higher species number and diversity index than JET1(expected species number (Es (n)) (Fukushima 2004). Besides, in the case of macrobenthoss, it was found that the influence on infaunal species is more serious than that on epifaunal species (Fukushima and Imajima 1997). Further, in the case of megabenthos, it was inferred that the influence on the deposit feeders was larger than that on the suspension feeders (Fukushima et al. 2000).

To summarize the above, the chemical composition of the sediment temporarily changed as a result of benthic disturbance, but the change was limited vertically to the surface layer up to 0.5 cm, and decreased with the passage of time. Even in the case of meiobenthos, as with changes in chemical components in sediments, there was a change in the vertical distribution and abundance immediately after benthic disturbance, but it approached its original level with time. Although limited results, species diversity of nematode were different between JET1 and JET4. The above is the information of previous JET survey, and these observations were compared during long-term monitoring carried out in JET5.

2 Methods and Materials

JET5 was carried out for long-term monitoring at the JET site in 2011 and 2012, i.e., 17–18 years after benthic disturbance during which data was collected on current direction and velocity, water quality, sediment properties, and benthic organisms (meio-, macro-, and megabenthos). This study describes the results of analyses for sediment properties, such as total organic carbon (TOC), total nitrogen (TN), and calcium carbonate ($CaCO_3$), as well as the habitat condition, distribution, abundance, and diversity of meiobenthos. A total of eight samples (seven from 2017 and one from 2018) were obtained from the disturbance area and the disturbance adjacent area (Fig. 4). The distances from the disturbance area for each sampling point are shown in Table 1.

A multiple corer (MC) was used for sampling, and tubes 28 mm in diameter were used for subsampling from the MC corer. Slices were made at 0–0.25, 0.25–0.5, 0.5–0.75, 0.75–1.0, 1.0–2.0, and 2.0–3.0 cm intervals and were split into groups for analysis of meiobenthos and chemical composition (Fig. 4). Based on ISA guidelines, the meiobenthos was set to be between 32 and 250 μm, which were grouped into nematodes, crustaceans, and others. Furthermore, foraminifera are generally included as a main component of meiobenthos, but this site is below the carbonate compensation depth (CCD); therefore, species that were difficult to count have been excluded from the quantitative data (Fig. 5) (Goineau and Goody 2017). Therefore, the meiobenthos within the scope of this survey is metazoan meiobenthos.[2] As per the previous JET survey, the TOC and TN of the sediments were analyzed using the CHN analyzer, and $CaCO_3$ was analyzed using a coulometer.

Statistical analysis was basically the same as that used in the previous JET survey. The vertical profile of the sediment components was tested using analysis of variance (ANOVA), and the horizontal environmental gradient was evaluated using regression analysis of the distance from the disturbance area and the first principal component (PZ1) obtained in the principal component analysis (PCA). The abundance of meiobenthos was compared using a *t-test*. As previously mentioned, based

[2] In this paper, meiobenthos excluding protozoa, such as foraminifera, are called metazoan meiobenthos.

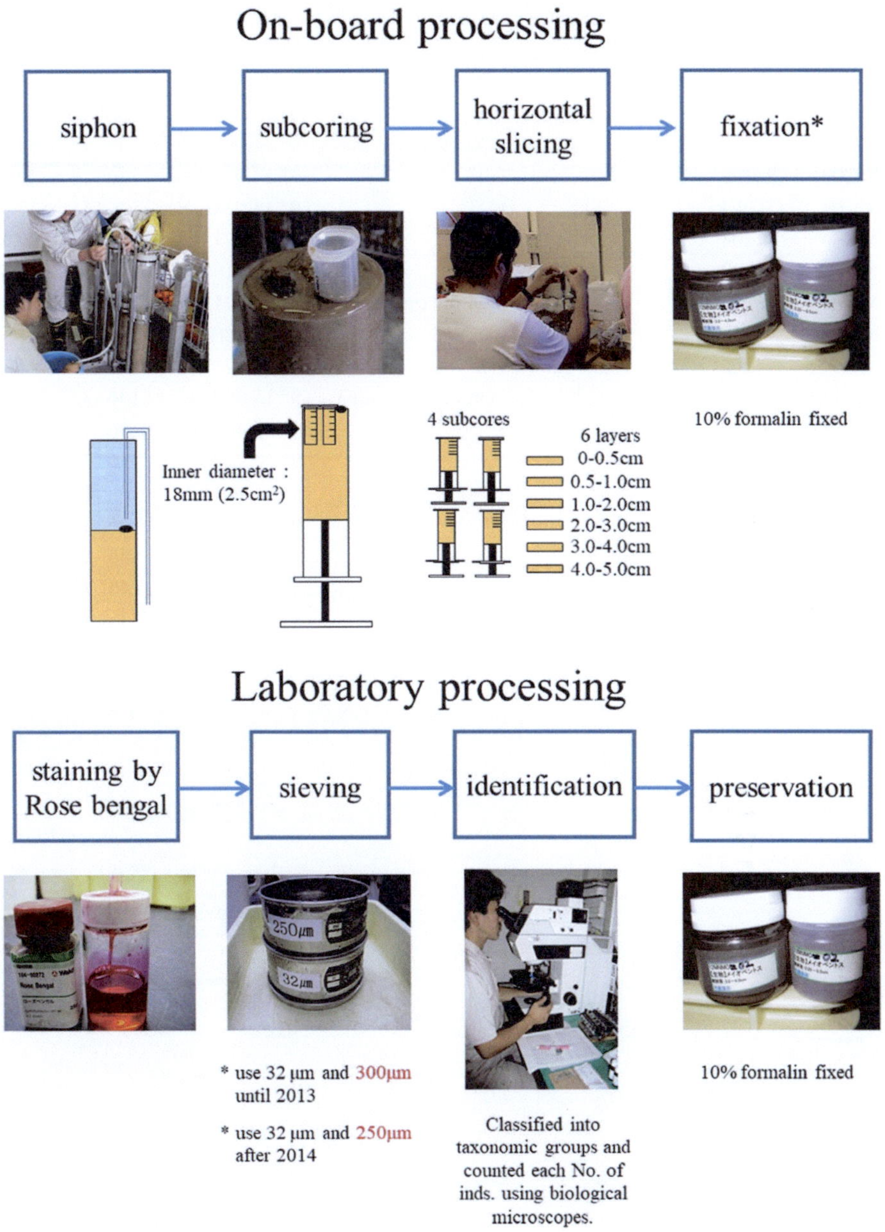

Fig. 4 Meiobenthos sample processing flow

Table 1 Sampling positions of multiple corer (MC) and spade corer (SC)

Sample name	Distance from the disturbance track area (km)	Remark	Latitude	Longitude
MC01	1.352		9° 14.9191′N	146° 16.4118′W
MC02	0	On the track area	9° 14.9194′N	146° 15.3514′W
MC03	0.295		9° 14.9227′N	146° 15.6020′W
MC04	0.094		9° 14.9186′N	146° 15.4533′W
MC05	1.208		9° 14.9212′N	146° 16.2997′W
MC06	0.179		9° 14.9190′N	146° 15.5172′W
MC07	0	On the track area	9° 14.9211′N	146° 15.1021′W
MC08	0.033		9° 14.6804′N	146° 15.4804′W
MC09	0.203		9° 14.6758′N	146° 15.1450′W
MC13	0	On the track area	9° 15.008′N	146° 15.257′W
MC18	1.259		9° 15.481′N	146° 15.779′W
SC01	1.342		9° 14.9108′N	146° 16.4121′W
SC02	0	On the track area	9° 14.9187′N	146° 15.3404′W
SC03	0.275		9° 14.9090′N	146° 15.6008′W
SC05	0.189		9° 14.9264′N	146° 15.5178′W
SC06	0.025		9° 14.6829′N	146° 15.4847′W
SC09	0.116		9° 14.802′N	146° 15.106′W

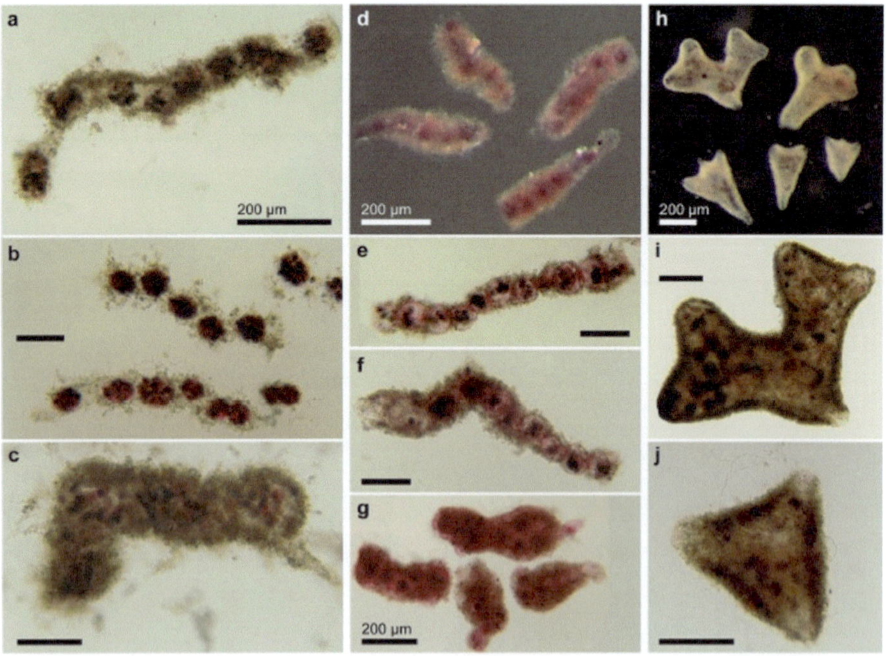

Fig. 5 Examples of uncountable (difficult to count) foraminifera. (Photographs are referred from Goineau and Gooday 2017)

on the information that the effects of benthic disturbance were limited both vertically and horizontally, the vertical distribution of the sediment components of 0–1 cm and the correlation between PZ1 and the distance of up to 300 m were taken as the basis for the statistical analysis.

3 Results

3.1 Sediment Properties

In the image of the one-shot camera installed in the sediment sampler, the tow track of the benthic disturber 17 years ago and the redeposition conditions on the seabed can be confirmed. The vertical profiles of the concentrations of TOC, $CaCO_3$, and TN are shown in Fig. 6. All profiles exhibit an exponential reduction from the surface layer to the deeper layer; however, a statistically significant change could only

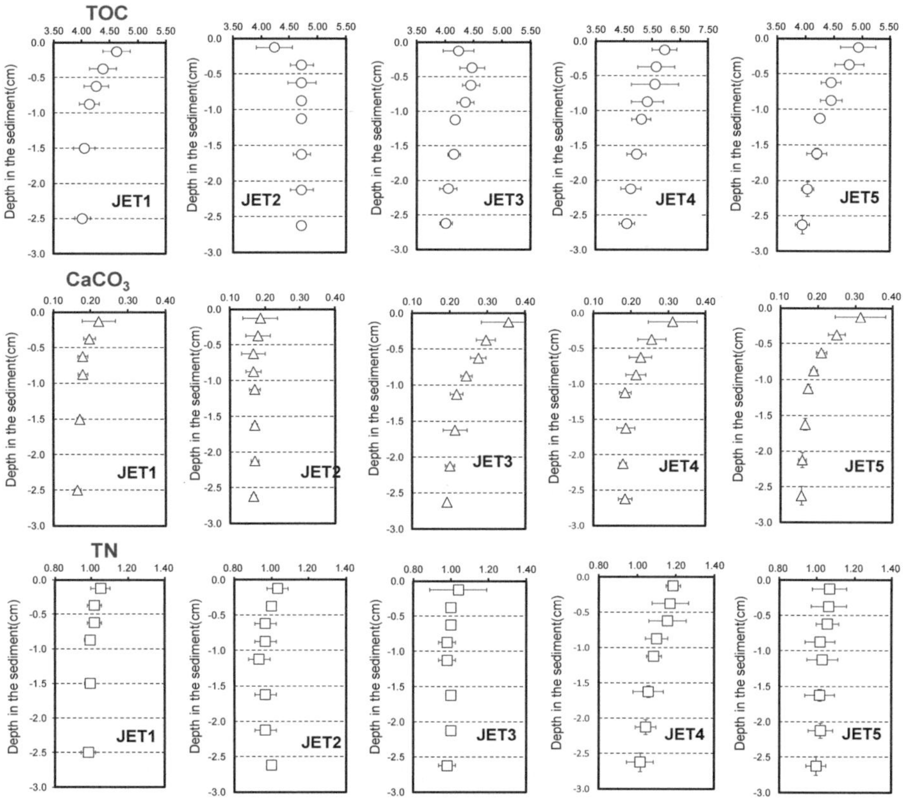

Fig. 6 Vertical profiles of total organic carbon (TOC), calcium carbonate ($CaCO_3$), and total nitrogen (TN) in the sediment. (After Fukushima and Tsune 2018)

Table 2 Probabilities of vertical reductions in chemical component parameters were tested by analysis of variance (ANOVA)

		df	*F* value	*P* value	Results
JET1 (1994)	TOC	11	10.47	2.55E-05	**
	$CaCO_3$	11	7.66	3.16E-04	**
	TN	11	4.17	0.01	*
JET2 (1994)	TOC	9	0.26	0.85	*ns*
	$CaCO_3$	9	0.46	0.71	*ns*
	TN	9	0.05	0.98	*ns*
JET3 (1995)	TOC	9	1.18	0.33	*ns*
	$CaCO_3$	9	12.25	1.12E-05	**
	TN	9	1.23	0.31	*ns*
JET4 (1996)	TOC	7	1.08	0.38	*ns*
	$CaCO_3$	7	7.61	7.15E-04	**
	TN	7	2.64	0.07	*ns*
JET5 (2011–2012)	TOC	8	3.31	0.03	*
	$CaCO_3$	8	17.76	0.00	**
	TN	8	0.46	0.71	*ns*

be seen for TOC and $CaCO_3$ (ANOVA, $p < 0.01$, respectively). Although all are not significant, as with JET1, if compared with the results from the previous JET surveys, JET5 result is closer to the state of JET1 than those of JET2, JET3, and JET4 (Table 2). With the relationship between PZ1 and the distance from the disturbance area, no significantly linearly regressing relationship was seen such as that for JET2 or JET3, even in the layer that was closest to the surface (Fig. 7).

J2: 0–0.25 cm layer	$Y = 0.024X - 1.218$**
J2: 0.25–0.5 cm layer	$Y = 0.0073X - 0.801$*
J3: 0–0.25 cm layer	$Y = 0.002X - 0.9925$*

No significant trend can be found in JET4 and JET5
$^{**}P < 0.01$, $^{*}p < 0.05$

3.2 Meiobenthos

The total abundance of meiobenthos was 92 ± 40inds /10 cm^3 (mean±SD), while those of the nematodes and crustaceans were 82 ± 38inds /10 cm^3 (mean±SD) and 8 ± 5inds /10 cm^3 (mean±SD), respectively. This is slightly less than that of JET1 and is equivalent to twice that of JET2; however, there were no significant differences in either case (Fig. 8). The meiobenthos, nematodes, and crustaceans were primarily distributed in the 0–0.25 cm layer and their abundance reduced with depth

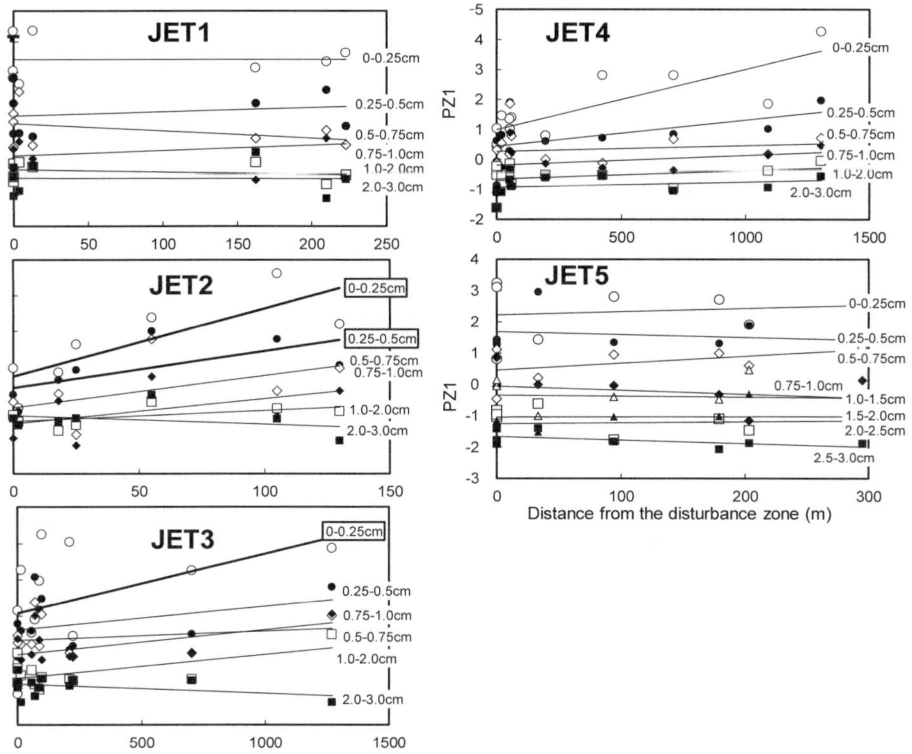

Fig. 7 The relationship between the first principal component (PZ1) and the distance from the disturbance zone. (After Fukushima and Tsune 2018)

from the surface layer (Fig. 9). This is the same as the tendency seen in previous JET surveys. Nematodes were the most prevalent of the meiobenthos and made up 91% of the total population, while crustaceans accounted for 8% and other organisms constituted 1% (Fig. 10). This composition ratio was virtually the same as the results from previous JET surveys. Other than nematode and crustaceans, organisms such as *Gastrotricha*, *Tardigrada*, and *Loricifera* were present. On the other hand, as for species diversity of nematodes, the index value was closer to that of JET4 than that of JET1. As already mentioned, it is known that the species diversity of JET4 became higher than that of JET1. This suggested that the higher species diversity has continued from the time of the impact (Fig. 11).

4 Discussion

JET was an experiment conducted to estimate the effects of resuspension and resettlement of seabed sediment, and the disturber used was a device developed to generate resuspension and resettlement. The working hypotheses relating to the

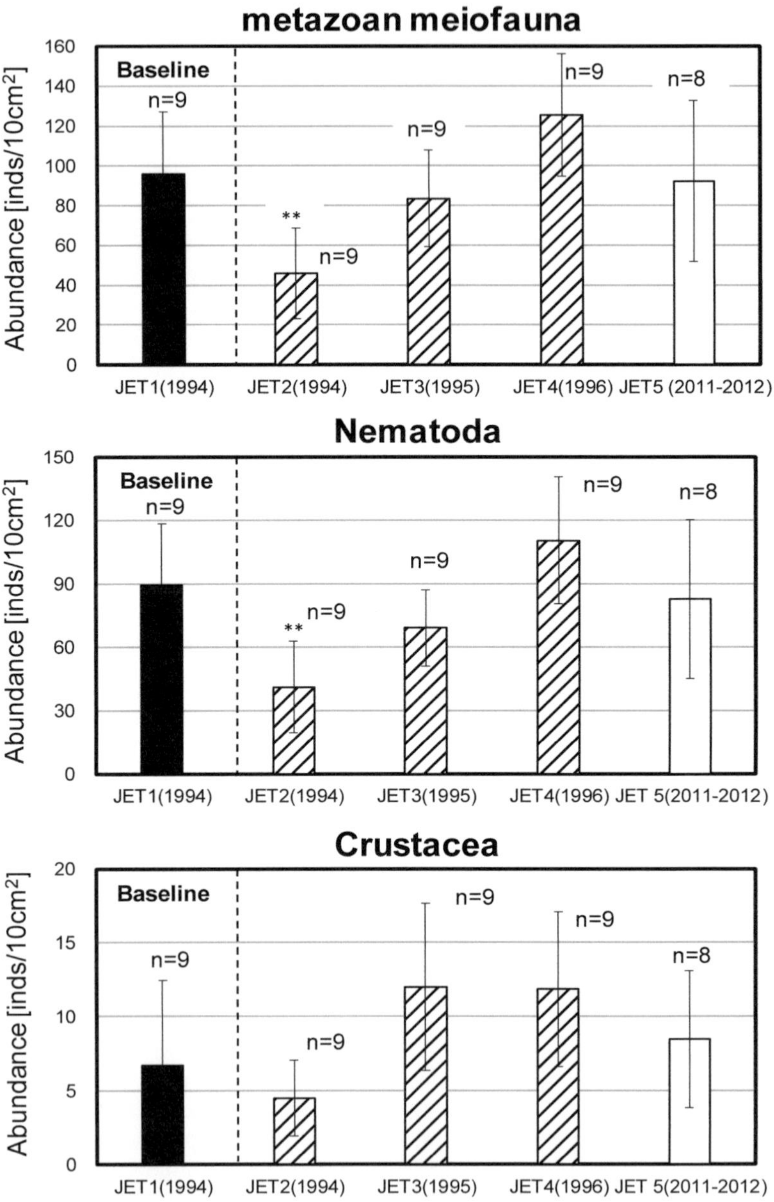

Fig. 8 Abundance of total number of metazoan meiofauna, nematodes, and crustaceans during the previous JET survey and JET5. (After Fukushima and Tsune 2018)

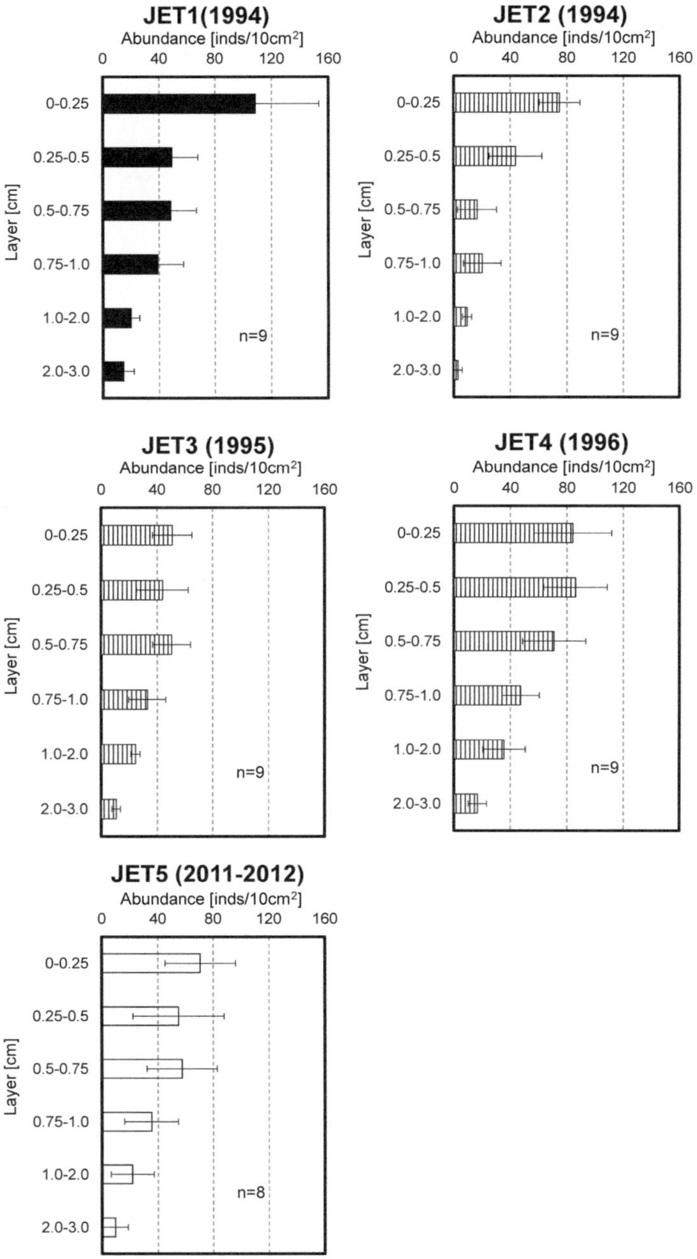

Fig. 9 Vertical distributions of metazoan meiobenthos during the previous JET survey and JET 5. (After Fukushima and Tsune 2018)

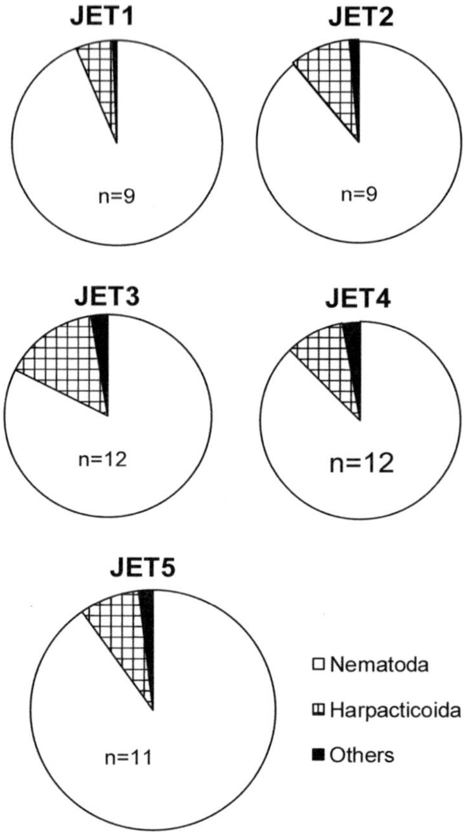

Fig. 10 Faunal compositions of metazoan meiobenthos during previous JET survey and JET5. (After Fukushima and Tsune 2018)

resettlement effects are: (1) primarily due to the resuspension and resettlement of inorganic sediment, the concentration of organic material in the sediment surface area is diluted, deteriorating the food environment for deposit feeders who depend on it; and, (2) with a faster than usual resettlement, suspension feeders could experience clogging of the oral apparatus, which would negatively impact their ability to gather food. As these mechanisms are hypothesized, it is essential to compare the numbers of organisms as well as to study the resettlement of sediment, which changes the availability of food material in the environment.[3]

Under natural conditions, there were significant changes in the vertical profiles of TOC, TN, and $CaCO_3$. However, immediately after benthic disturbance, the structure changed drastically and the significant vertical declines disappeared. Two years later, only the vertical structure of $CaCO_3$ returned to normal, and, 17–18 years later, the vertical profiles became even closer to the original state, with significant declines also visible in the TOC profile. Moreover, the horizontal distribution of

[3] Since the size of meiobenthos was small, the influence of clogging was excluded in this paper.

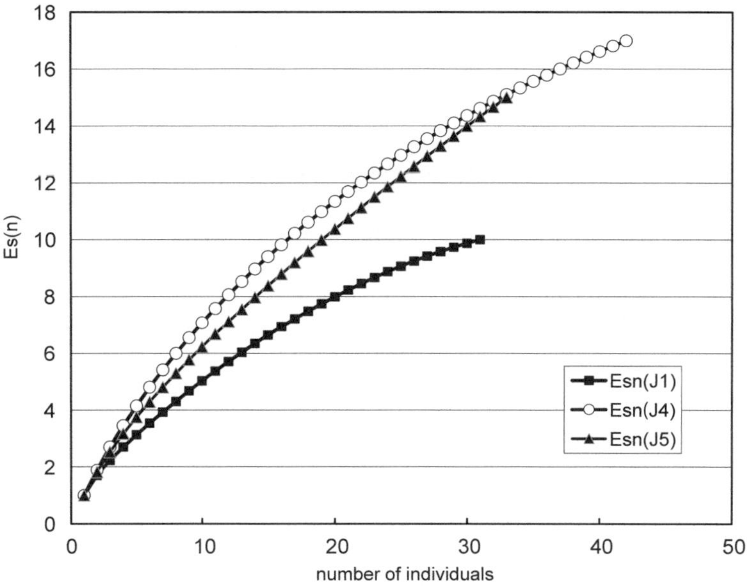

$$Es(n) = \sum_{i=1}^{S}\left(1 - \frac{{}_{N-Ni}C_n}{{}_{N}C_n}\right)$$

S: total number of species
Ni: number of individuals in the ith species in the samples
N: total number of individuals

Fig. 11 Species diversity of nematodes by expected species number (ES(n)). Esn (J1), Esn (J4), Esn (J5): expected species number of nematodes from a population before the disturbance, 2 years after the disturbance, and 17–18 years after the disturbance, respectively

PZ1 immediately after benthic disturbance changed significantly with distance from the disturbance area. Subsequently, 1 year after the disturbance, an environmental gradient was seen limited to 0.25 cm, and 2 years after the disturbance, there was no longer a significant correlation. There was no correlation between the distance and PZ1 17–18 years later. If these vertical and horizontal changes are both taken into consideration, the benthic disturbance implemented in JET2 is thought to have temporarily created a disturbed seabed environment. However, this recovered gradually over 2 years, and there was no longer a specific environment after 17 years.

Similarly, although the abundance of meiobenthos reduced significantly immediately after benthic disturbance, there was no significant difference a year later. In a similar experiment conducted by NOAA, the abundance of nematodes increased immediately afterward and was reported to have reduced a year later (Trueblood et al. 1997). Similarly, with IOM'BIE, the meiobenthos, nematodes, and crustaceans increased 2 years later (Radziejewska 2014). Furthermore, with INDEX, the total number of meiobenthos continues to be below the original state even after 2 years (Sharma et al. 2000). This indicates that although the same equipment for disturbance was used, there is variability among experiments.

In JET5, the abundance of meiobenthos 17 years later was greater than that immediately after the disturbance and a year later. However, this was still lower than the original level and the level 2 years later. In other words, even if it is lower than the original level, the fact that it is neither the minimum nor the maximum value since the disturbance experiment suggests that the meiobenthos abundance 17 years later is at a level within the variability range subsequent to the benthic disturbance.

Previous JET surveys had suggested that artificial resuspension and resettlement of sediment impacted the abundance of meiobenthos[4] (Fukushima and Tsune 2018). In contrast, when the similar sampling approach was taken 17–18 years later, no results were obtained that enabled the same suggestions to be made. This is in agreement with the results from the aforementioned sediment property analysis, which also suggests that the effect of the disturbance does not continue. However, the diversity of nematodes was lower before the disturbance and higher 2 years after the disturbance. It remained high even after 17–18 years. Although, because the phenomenon seen here is based on limited survey data and different interpretations of the results are possible, the authors propose the following scenario from this study.

> Due to the rapid redeposition of particles, which are mainly composed of inorganic matter, the organic matter on the surface of the sediment that forms the food bait of benthic organisms is diluted, the carrying capacity decreases, and the biomass of benthic communities decreases in the short term. However, after 17 years, both the amount of organic matter and the biomass regained their original levels, and there was no evidence of a sustained effect on biomass. Moreover, under the natural (undisturbed) conditions a niche is fixed by competitive exclusion on the deep sea floor where the environment is stable, and, in general, the species diversity of biological communities is low. However, once an event like artificial disturbance occurs, if the niche is disturbed, the species diversity will temporarily increase due to the invasion of new organisms, such as opportunistic species. Niche fixation as seen before the disturbance takes more time than biomass recovery, which may require a relatively short time of 17 years.

5 Future Tasks for Deep-Sea Impact Experiments

BIE type of experiment is only one of the tools for obtaining information necessary for environmental impact assessment as it mainly evaluates the effects of redeposition and does not cover the impact of other influencing factors related to deep-sea mining. Furthermore, there are limitations of the target environmental parameters and the accuracy of the data. Therefore, it is necessary for researchers to share these constraints and discuss the benefits of deep-sea mineral resource development and

[4] The redeposition area is divided into heavy, medium, light, and no deposition areas; the duration is classified into before, soon after, 1 year after, and 2 years after. Furthermore, separate comparisons are done for a group of organisms depending on organic matter in sediments and others. As a result of comparing it separately, it was possible to make a reasonable estimation.

the magnitude of environmental impacts while reporting the results. Unless the stakeholders share such conditions, it is impossible for the reader to understand the purpose and scale of the experiment so as to avoid any misunderstanding, regardless of the amount of survey data accumulated. Further, the authors would like to make the following observations for BIE type of experiments.

5.1 Sample Number and Representation

Although the data demonstrated that the impact of benthic disturbance had not continued for 17 years, in order to conclude that there was no continued effect of benthic disturbance, further discussion may be needed regarding the number of samples required to arrive at that conclusion. In 2001, ISA held a workshop on sampling strategies, where among others Etter (2001) proposed 216 samplings of 12 years, without considering implementation constraints, whereas the number of samples in our experiment is insufficient compared to his sampling design. In the deep seabed, there is limited knowledge regarding spatial variability in the environment. In addition, in the case of meiobenthos surveys, since the survey area is represented by a sample of about 10 cm^2, surveys are always accompanied by discussions on the validity of the sample size. On the other hand, given the fact that the EIA is part of an economic activity, interpretation should not be abandoned because the survey effort is not sufficient. That is, it is necessary to make efforts to obtain some recommendations.

This study presumed that the place where the seafloor was disturbed was no longer a special place as a result of the passage of 17 years, then analyzed the abundance of meiobenthos, and concluded that no continuation of impact was recognized. As mentioned above, although the problem about appropriate sample number remains, it should be carefully examined in the future from the results of the IOM'BIE survey and DISCOL surveys.

5.2 Baseline and Natural Variation

The word "environmental baseline" is commonly used, but there is no rigorous definition. Rather, due to its ambiguity, the word may be used frequently.

Even natural conditions in the ocean environment are known to exhibit fluctuation at seasonal, annual, decadal scales, and so on. It is understood that the environmental baseline refers to a condition that becomes a "base," which is invariable even when such a change is excluded. However actually, it is necessary to understand changes at seasonal, annual, or decadal levels. Furthermore, the environment we are researching can be thought of as one section of the process converging in a certain direction. An "Intermediate Disturbance Hypothesis,"[5] which explains species diversity of deep-sea organisms, also assumes that the community will change. This means that it will not return to its original condition, whether there is an artificial impact or not. It takes a long time to understand the environmental baseline. Thus, whether the environmental baseline is a concept suitable for EIA for an economic objective is a matter of debate. It is necessary to keep in mind that the submarine disturbance experiment is one that presupposes a comparison between baseline and disturbance without resolving these questions.

5.3 Impact and Evaluation

Deep-sea impact experiments have been designed to provide quantitative comparisons between conditions before and after an artificial impact resembling mining activities. However, to understand the impact, there are various indicators for comparison. For example, among biological communities, there are size categories such as megabenthos, macrobenthos, meiobenthos, and nanobenthos, each of which has different comparison parameters such as biomass, diversity, distribution, or genetic diversity, as well as functions including ecosystem services. However, there is no agreement about what parameters to compare for impact assessment as well as monitoring the recovery of the ecosystem. Even when the four BIE-type disturbance experiments are taken as an example, the number of occurrences of meiobenthos is common, but there are cases where either macrobenthos or megabenthos sampling is not carried out. Similarly, there is a need to bring about uniformity in ISA's environmental guidelines as well as other templates for environmental impact assessment (e.g., Clark et al. 2017).

Although it is ideal to target as many parameters as possible, actual investigation is carried out under various restrictions. Under these circumstances, it should be recognized that benthic impact experiments exist to accumulate many environmental facts until definite standards are presented by the regulating agencies.

5.4 Accuracy of Data

Advancement of analytical techniques has made it possible to acquire more detailed data. Therefore, it has become easy to detect the difference in data between two samples whether it is sediment or benthic fauna. This also applies to data comparisons before and after impact in benthic impact experiments. In other words, with technical advancement, even if it was concluded that there was no difference in the past, differences may be recognized as time goes by. Not limited to advances in

[5]The theory that the biodiversity of deep sea benthic organisms is not too strong, is not too large, and is maintained by the impact repeated with moderate frequency (Connell 1978).

science and technology, in general, there is a time gap between technical discovery, scientific cognition, and stakeholder acceptance. Considering this, it is better not to confuse scientifically detectable changes with changes required by consensus formation of EIA.

5.5 *Scientific Approach and Environmental Management*

Needless to say, scientific knowledge is important, both in environmental impact assessment and in environmental preservation measures. On the other hand, if the "scientific" aspect is given too much weightage, environmental policies cannot be advanced, because the purpose of science is to know the truth about nature, not to control environmental policies. A scientific approach is to assemble the facts and to find the laws of nature that explain these facts. The basis of science is the induction approach, which is to create a "theory" from the known "information" and to infer new conclusion based on this theory. However, the induction method can provide new information, but it is not always right. Even if we guess based on 99 observed facts, the 100th cannot be said to be absolutely right. That is, the results provided by scientists are always subject to modification. Therefore, scientists may hinder the progress of development, policy planning, and legislation. On the contrary, because science makes unexpected new discoveries and corrects the past, we realize our ignorance and are humbled by nature. In any case, in decision-making, science is important but not universal—an appropriate balance is necessary.

6 Conclusion

The BIE-type experiment was one of the few in situ experiments conducted to evaluate the environmental impacts and recovery process associated with manganese nodule development. Therefore the knowledge from the short-term monitoring survey, which has been done after the artificial disturbance, was important and valuable to evaluate the environmental impacts. Moreover, the long-term monitoring survey conducted in this study is also important to bring consensus between the stakeholders of ocean mining. However, there are operational restrictions in the experiment. In addition, considering natural environmental fluctuations, examples of the monitoring survey should be repeated, and then knowledge should be improved accurately. For the future environmental impact evaluation associated with deep-sea mineral resource development, in consideration of the above matters, careful and comprehensive efforts are necessary.

Acknowledgment This paper is based on results obtained from a project commissioned by the Ministry of Economy, Trade and Industry, Japan. And we would like to thank Editage (www.Editage.jp) for English language editing.

References

Barnett, B. G., & Yamauchi, H. (1995). *Deep sea sediment resuspension system used for the Japan deep sea impact experiment* (Proc of the 1st Ocean Mining Symposium) (pp. 175–180). Tsukuba: ISOPE.

Clark, M.R., Rouse, H.L., Lamarche, G., Ellis, J.I., Hickey, C. (2017). Preparation of environmental impact assessments: General guidelines for offshore mining and drilling with particular reference to New Zealand. NIWA Science and Technology Series No. 81, 110. Aukland: NIWA.

Connell, J. H. (1978). Diversity in tropical rain forests and coral reefs. *Science, 199*, 1302–1309.

Etter, R. (2001). General sampling design for baseline studies. In: ISA (Ed.), *Standardization of environmental data and information* (pp. 539). Kingston: ISA.

Fukushima, T. (1995). *Overview "Japan deep-sea impact experiment = JET"* (Proc of the 1st Ocean Mining Symposium) (pp. 47–53). Tsukuba: ISOPE.

Fukushima, T. (2004). Ecological characteristics of Deep-sea benthic organisms in relation to manganese nodules development practices. Thesis, Kyoto University. pp. 155.

Fukushima, T., & Imajima, M. (1997). *A study of a macrobenthos community in a deep sea resedimentation area* (Proceedings of the International Symposium on Environmental Studies for Deep Sea Mining) (pp. 331–336). Tokyo: MMAJ.

Fukushima, T., & Nishijima, M. (2017). Taxonomic problems in environmental impact assessment (EIA) linked to ocean mining and possibility of new technology developments. In R. Sharma (Ed.), *Deep-sea mining* (p. 535). Berlin: Springer.

Fukushima, T., & Tsune, A. (2018). *Environmental condition 17–18 years after the benthic impact experiment* (Proc of the 28th ISOPE Conference). Sapporo: ISOPE.

Fukushima, T., Shirayama, Y., & Kuboki, E. (2000). The characteristics of deep-sea epifaunal megabenthos two years after an artificial rapid deposition event. *Publications of the Seto Marine Biological Laboratory, 39*, 17–27.

Fukushima, T., Harada, K., Shibamoto, Y., Kuboki, E. (2001). Chemical compositions of deep-sea sediment where artificial rapid deposition event occurred. Proc of the 11th Int Offshore and Polar Eng Conf, pp. 541–547. Stavanger: ISOPE.

Fukushima, T., Harada, K., Shibamoto, Y., & Kuboki, E. (2002). *Chemical compositions of sediment plume discharged from the benthic disturber* (Proceedings of the 12th International Society of Offshore and Polar Engineering Conference) (pp. 443–446). Kitakyusyu: ISOPE.

Goineau, A., & Goody, A. (2017). Novel benthic foraminifera are abundant and diverse in an area of the abyssal equatorial Pacific licensed for polymetallic nodule exploration. *Scientific Reports, 7*, 45288. https://doi.org/10.1038/srep45288.

JPI Oceans. (2016). Long-term impacts of deep-sea mining –latest results from marine scientific research. JPI Oceans. p. 2. Brussels: JPI Oceans.

Kotlinski, R. (1995). *InterOceanMetal organization: Achievements and challengers* (Proc of the 1st Ocean Mining Symposium) (pp. 5–8). Tsukuba: ISOPE.

National Oceanic and Atmospheric Administration (NOAA). (1981). *Deep seabed mining final programmatic environmental impact statement 1* (p. 295). Washington, DC: NOAA/OME.

NOAA. (1982). *Deep seabed mining-marine environmental research plan 1981–85* (p. 140). Washington, DC: National Oceanic and Atmospheric Administration.

Ozturgut, E. (1981). *Suspended particulate measurement techniques and accuracy and precision of chemical analyses used in the DOMES project* (Final Report to NOAA) (p. 12). Seattle: Science Applications Incorporated.

Ozturgut, E., Anderson, G. C., Burns, R. E., Lavelle, J. W., & Swift, S. A. (1978). *Deep ocean mining of manganese nodules in the North Pacific: Pre-mining environmental conditions and anticipated mining effects* (NOAA Technical Memorandum ERL MESA-33) (p. 133). Colorado: NOAA.

Ozturgut, E., Trueblood, D. D., & Lawless, J. (1997). *An overview of the United States' benthic impact experiment* (Proc of the Int. Sym. on Environmental Studies for Deep Sea Mining) (pp. 23–32). Tokyo: MMAJ.

Radziejewska, T. (2014). *Meiobenthos in the sub-equatorial Pacific Abyss* (p. 105). Heidelberg: Springer.
Sharma, R., Nath, B. N., Valsangkar, A. B., Parthiban, G., Sivakholundu, K. M., & Walker, G. (2000). Benthic disturbance and impact experiment in the Central Indian Basin. *Marine Georesources and Geotechnology, 18*(3), 209–221.
Smith, C. R., Bennett, B. A., & Brumsickle, S. J. (1988). *Assessment of benthic faunal sensitivity to rapid sediment burial at DOMES site C-1* (Final Report for NOAA Contract #2-5676-05 #1) (p. 14). Seattle: University of Washington.
Thiel, H. (1991). From MESEDA to DISCOL: A new approach to deep-sea mining risk assessment. *Marine Mining, 10*(4), 369–386.
Trueblood, D. D. (1993). *US cruise report for BIE 2 cruise* (NOAA Technical Memorandum OCRS 4). Maryland: NOAA. 51 pp.
Trueblood, D. D., Ozturgut, E., Pilipchuk, M., & Gloumov, I. F. (1997). *The ecological impacts of the joint US-Russian benthic impact experiment* (Proc of the 7th ISOPE Conference) (pp. 237–243). Honolulu: ISOPE.
Vanreusel, A., Hilario, A., Ribeiro, P. A., Menot, L., & Arbizu, P. M. (2016). Threatened by mining, polymetallic nodules are required to preserve abyssal epifaunal. *Scientific Reports, 6*(26808), 6. https://doi.org/10.1038/srep26808.
Wilson, G. D. F. (1990). *RUM3-R/V New Horizon Cruise Quagmire 2: 23 April- 17 May, 1990* (Post Cruise Report. SIO Reference 90-25). San Diego: SIO. 43 pp.

Dr. Tomohiko Fukushima is working for Research and Development (R&D) Center of Submarine Resources, Japan Agency for Marine-Earth Science and Technology (JAMSTEC). He has graduated from the faculty of fisheries, Tokyo University, in 1984 (bachelor degree) and graduate school of fisheries science, Tokyo University, in 1986 (master degree). In 2004, he has obtained D. Sci. Degree from Kyoto University. The title of doctoral thesis is "Ecological Characteristics of Deep-Sea Benthic Organisms in Relation to Manganese Nodules Development Practices." After being awarded the doctoral degree, he continued his research in the Ocean Policy Research Foundation as a Research Fellow, University of Tokyo, and as an Associate Professor and then went to the incumbent.

Dr. Akira Tsune works in Deep Ocean Resources Development, Co., Ltd. (DORD), Japan. Currently, he serves as a manager of environmental surveys or a chief geologist in the manganese nodule project in DORD. He graduated from the Department of Earth Sciences, Faculty of Science, Kanazawa University, in 1999 (bachelor's degree) and Graduate School of Natural Science & Technology, Kanazawa University, in 2001 (master degree). In 2005, he got Doctor of Science degree from the Department of Earth and Planetary Sciences, Graduate School of Sciences, Kyushu University.

Metal Mobility from Hydrothermal Sulfides into Seawater During Deep Seafloor Mining Operations

Shigeshi Fuchida, Jun-ichiro Ishibashi, Tatsuo Nozaki, Yoshitaka Matsushita, Masanobu Kawachi, and Hiroshi Koshikawa

Abstract Seafloor hydrothermal sulfides, which are expected to be a future resource for metals, could be a potential source for metal contamination in the seawater around mining sites. In this chapter, we illustrate the potential for metal leaching of both non-oxidized (non-exposed to atmosphere; before and during exploitation) and oxidized (exposed to atmosphere; after lifting and recovery) hydrothermal sulfides to seawater under different temperatures and redox conditions. One of the crucial findings was that metal dissolution behaviors differed significantly according to the specific areas and/or the initial oxidation states of

S. Fuchida (✉) · H. Koshikawa
Marine Environment Section, Center for Regional Environmental Research, National Institute for Environmental Studies (NIES), Tsukuba, Ibaraki, Japan
e-mail: fuchida.shigeshi@nies.go.jp; koshikaw@nies.go.jp

J.-i. Ishibashi
Department of Earth and Planetary Sciences, Faculty of Sciences, Kyushu University, Nishi-ku, Fukuoka, Japan
e-mail: ishibashi.junichiro.779@m.kyushu-u.ac.jp

T. Nozaki
Research and Development (R&D) Center for Submarine Resources, Japan Agency for Marine-Earth Science and Technology (JAMSTEC), Yokosuka, Kanagawa, Japan

Frontier Research Center for Energy and Resources, The University of Tokyo, Tokyo, Japan

Department of Planetology, Kobe University, Kobe, Hyogo, Japan

Ocean Resources Research Center for Next Generation, Chiba Institute of Technology, Narashino, Chiba, Japan
e-mail: nozaki@jamstec.go.jp

Y. Matsushita
Biodiversity Resource Conservation Office, Center for Environmental Biology and Ecosystem Studies, National Institute for Environmental Studies (NIES), Tsukuba, Ibaraki, Japan
e-mail: Matsushita.Yoshitaka@nims.go.jp

M. Kawachi
Research Network and Facility Services Division, National Institute for Materials Science (NIMS), Tsukuba, Ibaraki, Japan
e-mail: kawachi.masanobu@nies.go.jp

R. Sharma (ed.), *Environmental Issues of Deep-Sea Mining*,
https://doi.org/10.1007/978-3-030-12696-4_8

the sulfide surfaces. Once the non-oxidized sulfide chips were ground to particulates and mixed with seawater, Zn and Pb were preferentially released even though these metals were included as minor components of the sulfides. For Zn, the dissolution rate increased under the oxic and higher temperature (20 °C) conditions when compared to the anoxic and lower temperature (5 °C) conditions, but the absolute rate was relatively moderate. These findings suggest that instantaneous metal release from sulfides into seawater will not occur before or during the crushing and lifting processes of seafloor mining. In contrast to the non-oxidized sulfides, the oxidized sulfides rapidly released large amounts of various metals and metalloids (e.g., Mn, Fe, Zn, Cu, As, Sb, and Pb) into seawater. The different metal dissolution behaviors between the non-oxidized and oxidized hydrothermal sulfides suggest the importance of the implementation of appropriate environmental measures to prevent leakage of the lifted sulfides to the marine surface.

Keywords Deep-sea mining · Metal mobility · Hydrothermal sulfides

1 Introduction

Toxic metal release from hydrothermal sulfides is a potential problem that could be caused by seafloor metal-mining operations. Metal sulfides, which are present in abundance in submarine hydrothermal deposits, have become a major source of metal contamination via release of metal cations by oxidation of the sulfides (Tsang and Parry 2004; Cook et al. 2009; Hageman et al. 2015). These released metal cations are easily introduced into aquatic ecosystems, where they gradually accumulate in the bodies of living organisms (Eggleton and Thomas 2004; Simpson and Spadaro 2016).Thus, the anthropogenic releases of sulfide minerals and metal-contaminated seawater into marine environments by mining operations are likely to have negative effects on marine ecosystems.

Different metal dissolution potentials of sulfides are predicted in each mining process because the reaction rates can vary depending on oxidation states of the sulfide surface (non-oxidized or oxidized) and marine environmental conditions such as benthic and surface environments (Bilenker et al. 2016). Natural hydrothermal sulfides beneath the seafloor are often covered by insoluble oxides and/or sulfates and unconsolidated sediments known as gossan (Feely et al. 1987; Fallon et al. 2017). Under such states, the oxidative dissolution of hydrothermal sulfides is suppressed by the insoluble layer. However, metal release is likely to occur when sulfides are crushed and their fresh interfaces are exposed to seawater by seafloor mining operations (Fig. 1a). After the crushed sulfides are lifted with seawater from the seafloor to a mining support vessel (Fig. 1b), the sulfide surface will be gradually oxidized by atmospheric oxygen and rainwater during onboard storage (Fig. 1c).

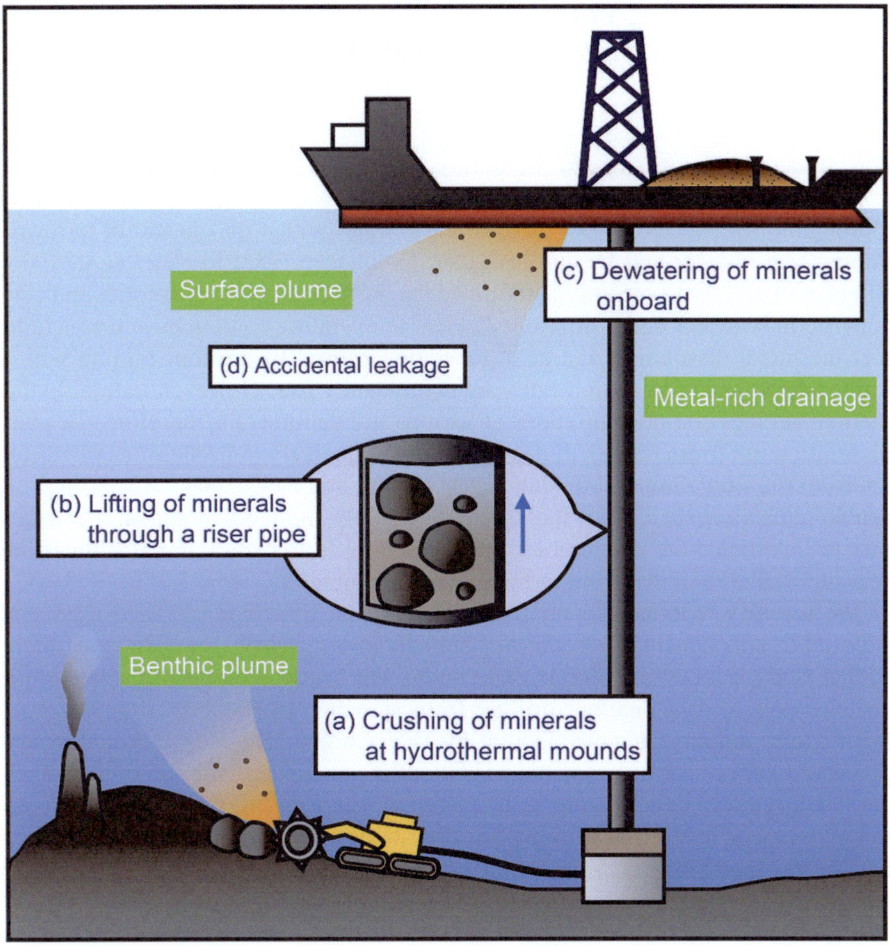

Fig. 1 Seafloor mining model and possible impact on marine environments

If the oxidized sulfides are discharged as tailings either operationally or accidentally from the vessel to the surface seawater, the fine particulates of hydrothermal sulfide in the surface plume are likely to release toxic metals (Fig. 1d). To enable adequate and realistic evaluation of the impact of such mining activities on marine environments, it is important to clarify the potential for metal leaching and dissolution processes of both non-oxidized and oxidized hydrothermal sulfides into seawater under different environmental conditions. In this chapter, we show the results of seawater leaching experiments using various hydrothermal sulfide samples and discuss the possibility of metal release in each mining process.

2 Metal Leaching Potential and Possible Mechanism of Dissolution from Non-oxidized Hydrothermal Sulfides into Seawater

Few studies have investigated dissolution of metals from single sulfide minerals (e.g., pyrite, sphalerite, and galena) into seawater, even though this topic has been thoroughly investigated in low ionic strength water within the context of terrestrial mining (e.g., Steger and Desjardins 1980; Goldhaber 1983; Michael et al. 1986). Bilenker et al. (2016) recently determined the oxidation rate of pyrrhotite and chalcopyrite in seawater under different P_{O2} and temperature conditions and concluded that the metal dissolution and acid generation caused by seafloor mining will be limited because the oxidation rates are significantly low. However, natural hydrothermal sulfides comprise mixtures of various sulfide minerals; therefore, the metal dissolution rate from these sulfides into seawater is likely to be significantly different from the single mineral oxidation rate (Tsang and Parry 2004). Accordingly, to enable realistic evaluation of the impact of seafloor mining operations, it is necessary to investigate the potential for metal leaching from hydrothermal sulfides into seawater rather than that from individual sulfide minerals.

We recently reported the metal dissolution rate from non-oxidized (i.e., non-exposed to atmosphere) hydrothermal sulfides into seawater (Fuchida et al. 2018). In that study, a metal dissolution experiment was conducted onboard using sulfide samples collected by seafloor drilling from an active hydrothermal mound at the Izena Hole, middle Okinawa Trough by D/V Chikyu. The sulfide samples were powdered under inert gas to avoid oxidation by atmospheric air and then immediately subjected to experiments under different temperature (5 °C and 20 °C) and redox (oxic and anoxic) conditions. For that study, oxic and anoxic conditions were considered high (air-saturated, initial at room temperature: 5–6 mg/L) and low (degassed and nitrogen-purged, initial at room temperature: 1 mg/L) dissolved oxygen concentrations, respectively. Metal concentrations in the reacted solution were quantified by inductively coupled plasma-mass spectrometry (ICP-MS). The details of experimental methods are described in Fuchida et al. (2018).

Chemical compositions and mineral assemblages of the four sulfide samples (CKL-1, CKL-2, CKL-3, and CKL-4) used for the experiments are shown in Table 1. These sulfide samples contained various metals (i.e., Mn, Fe, Cu, Zn, Cd,

Table 1 Chemical compositions and mineral assemblages of hydrothermal sulfides

	Concentration in sulfide samples (mmol/kg)						
Sample	Mn	Fe	Cu	Zn	Cd	Pb	Major minerals
CKL-1	3.5	2500	96	4600	11	970	Sphalerite, galena
CKL-2	5.8	5800	56	1700	3.2	370	Pyrite, sphalerite
CKL-3	1.9	7600	100	370	0.43	12	Pyrite
CKL-4	7.9	5700	56	3200	8.6	170	Sphalerite, marcasite, pyrite

Modified after Fuchida et al. (2018)

and Pb), whereas Zn and Pb were preferentially released from the sulfide samples into seawater under all experimental conditions. Large monotonic increases in Zn concentrations were observed for two samples (CKL-3 and CKL-4), and their dissolution rates per unit surface area (mol m^{-2} s^{-1}) were clearly higher under the oxic and 20 °C condition than the anoxic and 5 °C condition (Fig. 2). The final concentrations of Zn in seawater differed according to the sulfide sample; however, they did not correlate with those in the sulfide samples under all experimental conditions (see Fig. 6 in Fuchida et al. 2018). This lack of correlation indicates that simple

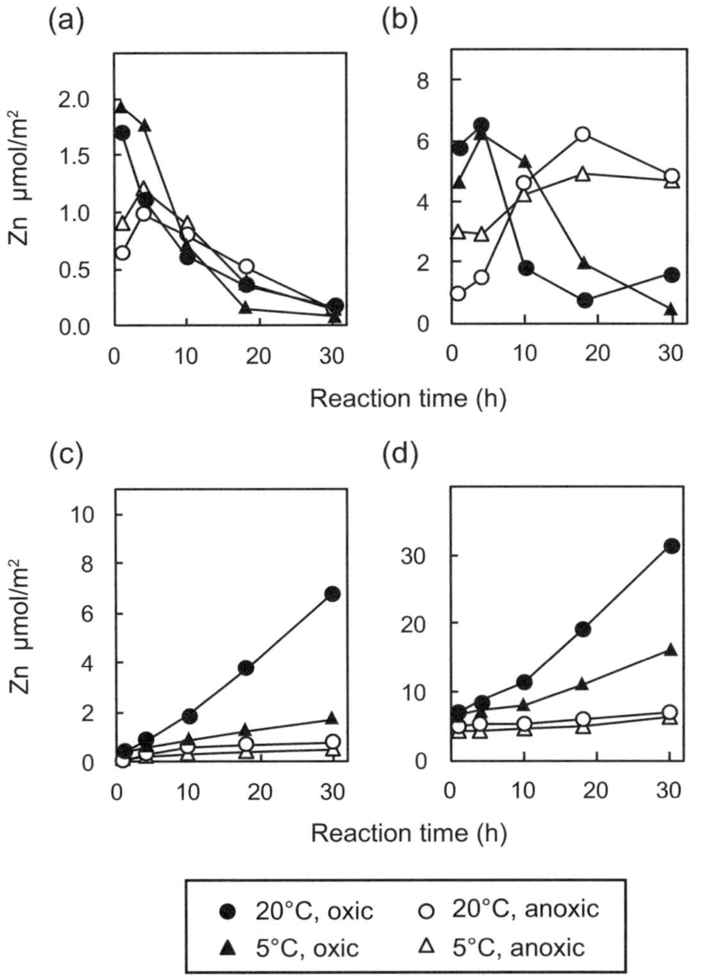

Fig. 2 Time-course changes in concentrations of dissolved Zn for (**a**) CKL-1, (**b**) CKL-2, (**c**) CKL-3, and (**d**) CKL-4 under different redox and temperature conditions (Modified after Fuchida et al. 2018)

oxidation reaction of sphalerite (i.e., $ZnS + 2O_2 \rightarrow Zn^{2+} + SO_4^{2-}$) made a minor contribution to Zn dissolution from hydrothermal sulfides into seawater.

Several studies have reported that galvanic effects generated between different sulfide minerals can greatly accelerate their oxidative dissolution (Kwong et al. 2003; Tsang and Parry 2004; Fallon et al. 2017). Sulfide minerals generally have semiconducting properties, and direct contact between sulfide minerals with different resting potentials produces galvanic effects (Cruz et al. 2005; Liu et al. 2008; Fallon et al. 2017). The minerals with the highest and lowest rest potentials act as a cathode and anode, respectively. Cathodic minerals can be galvanically protected, whereas anodic minerals are easily dissolved through electronic interactions (Chopard et al. 2017). The resting potentials for some sulfide minerals measured in 1.0 M H_2SO_4 solution at room temperature are listed in Table 2 (Kwong et al. 2003 and references therein). Sphalerite and galena have relatively low resting potentials when compared to disulfide minerals (i.e., pyrite and chalcopyrite) and can be anodically dissolved when they come into contact with anodic disulfide minerals.

Scanning electron microscopy (SEM) backscattered images of the particulate fragments of hydrothermal sulfides before the experimentation showed the presence of different mineral couplings (i.e., adjacent placement of iron disulfide and other sulfide minerals) in the high metal release samples (i.e., CKL-3 and CKL-4). In contrast, such couplings were less abundant or coated with silicates in other samples with less Zn and Pb releases (i.e., CKL-1 and CKL-2, see Fig. 5 in Fuchida et al. 2018). These microscopic observations implied that the galvanic couplings of sphalerite and galena with iron disulfides greatly affected the Zn and Pb dissolution from the hydrothermal sulfides into seawater.

The different solubility of each metallic compound in seawater also influences the metal composition of the leachates. The solubility of ferric ion is estimated to be only 0.2–0.3 nM in seawater. This is because ferrous ions (Fe^{2+}) released from sulfide minerals are instantaneously oxidized to ferric ions (Fe^{3+}) and precipitated as FeO(OH) in seawater (Liu and Millero 1999). The finding that the Fe concentration was less than the detection limit in our leachates was likely because of the low solubility of FeO(OH). Lead also has low solubility in seawater (3 μM) and can be removed by formation of $PbSO_4$ and $PbCO_3$ (Angel et al. 2016). In our experiment,

Table 2 Resting potentials (V vs. standard hydrogen electrode, SHE) of individual sulfide minerals measured in 0.1 M H_2SO_4 solution at room temperature (Kwong et al. 2003 and references therein)

Sulfide minerals		Resting potential (V) vs. SHE
Pyrite	FeS_2	0.63
Chalcopyrite	$CuFeS_2$	0.52
Chalcocite	Cu_2S	0.44
Covellite	CuS	0.42
Galena	PbS	0.28
Sphalerite	ZnS	−0.24
Pyrrhotite	$Fe_{1-x}S$	−0.28

the Pb concentrations in the seawater reacted with CKL-2 and CKL-4 increased initially but then greatly decreased from 1 to 10 h (Fuchida et al. 2018). This decrease can be explained by the formation of insoluble precipitates. Conversely, the Zn concentration is predicted to be high, even in seawater, because most of the secondary Zn minerals are highly soluble (Kwong et al. 2003). Thus, the removal of Fe and Pb by precipitate formation (i.e., secondary mineral formation) could play a large role in the preferential increase of Zn in seawater after the reaction with hydrothermal sulfides.

The metal concentrations in seawater were higher under the oxic and 20 °C conditions than the anoxic and 5 °C conditions in our experiment. This was likely because both the galvanic dissolution of sulfide minerals was promoted and the metal solubility increased at higher temperature and P_{O2} condition. Based on these results, metal dissolution from hydrothermal sulfides can be greatly accelerated when the minerals are transported to areas with warm oxic conditions, such as surface environments, although little metal dissolution from crushed sulfides is expected to occur under cold anoxic conditions, such as deep seafloor environments.

3 Evaluation of Leachable Metals and Metalloids from Hydrothermal Sulfides After Exposed to Atmosphere into Seawater

Metal leaching potential from "oxidized" hydrothermal sulfides into seawater has been reported in several studies, even though our study is the only one known to have investigated metal release from non-oxidized hydrothermal sulfides to date (Fuchida et al. 2018). The "oxidized" was taken to indicate that the samples were exposed to air and/or oxic water for a long time; the original assemblage of constituent minerals could be partly changed to be secondary/altered phase. The results of a leaching experiment using the oxidized sulfide samples showed different metal release patterns from those of non-oxidized sulfide samples and indicated that various metals and metalloids can be rapidly released in large quantities from oxidized sulfide samples into seawater. Here, we show the results of leaching experiments conducted using various oxidized sulfides and the characteristics of metal compositions in the leachates.

Two studies (i.e., Simpson et al. 2007; Parry 2008) have investigated the metal leaching potentials of seafloor hydrothermal sulfides collected from active/inactive chimney structures in the East Manus Basin hydrothermal field (Papua New Guinea) as part of the Solwara 1 project. In these studies, blocks (large chips of 25 mm diameter) of sulfide samples containing large amounts of metals (Mn, Fe, Cu, Zn, and Pb) and As (Table 3a) were reacted with natural seawater at a liquid-to-solid ratio of 10:1 for a maximum of 24 h. Representative results of the leaching experiments are shown in Table 3b. Large amounts of Zn and Mn followed by Cu, Pb, and As were released from those oxidized samples; however, the metal and metalloid compositions in the leachates were significantly different from those in the sulfide samples.

Table 3 (**a**) Chemical compositions of hydrothermal sulfides and (**b**) results of leaching experiments in the Solwara 1 project

(a) Concentrations of metals and As in sulfide samples (mmol/kg)							
Sample		Mn	Fe	Cu	Zn	As	Pb
Sediment rock*	(Simpson et al. 2007)	7.0	1933.9	420.2	428.3	31.6	7.1
Active chimney	(Simpson et al. 2007)	2.3	324.1	1636.6	1173.1	112.4	46.2
Inactive chimney	(Simpson et al., 2007)	11.9	891.8	21.4	647.0	52.5	24.0
Fe-rich sample	(Parry, 2008)	1.7	5658.5	807.3	56.1	26.2	2.4

(b) Concentrations of released metals and As in leachate (μmol/L)								
Sample		Mn	Fe	Cu	Zn	As	Pb	(Experimental conditions)
Sediment rock*	(Simpson et al. 2007)	2.7–10.7	<0.1–5.6	1.3–6.6	<0.1–6.2	<0.1–0.7	<0.1–0.4	(20 °C, 0.2–24 h)
Active chimney	(Simpson et al. 2007)	4.9	n.d.*	<0.1	78.0	10.6	0.6	(22 °C, 0.2 h)
Inactive chimney	(Simpson et al. 2007)	168.0	n.d.	0.9	428.3	0.6	1.6	(22 °C, 0.2 h)
Fe-rich sample	(Parry 2008)	2.9	0.0	7.6	145.6	0.1	1.2	(24 °C, 3 h)

Sediment rock*, referred to as "weathered chimney" in Simpson et al. (2007)
*n.d.**, no data

Our recent study reported the metal leachability of four different hydrothermal sulfides collected from the Okinawa Trough hydrothermal fields (Japan) (sample no. 1–4 in Table 4) (Fuchida et al. 2017). We used fine particulate matters (<1/16 mm) of those samples, which would contribute to plume formation, although previous studies used large block sample chips (<25 mm). Approximately 3 g powdered sample was stirred into 30 mL artificial seawater (Daigo's SP, Nihon Pharmaceutical Co. Ltd., Tokyo, Japan; pH = 8.2; dissolved oxygen = 5.7 mg/L) in an acid-cleaned polypropylene centrifuge tube (50 mL) for a liquid-to-solid ratio of 20:1, and then the tube was reciprocally shaken at 200 rpm per min at 25 °C for 5 min, 6 h, or 18 h. After shaking, the solid phase was separated by centrifugation and collected by filtration through a polyvinylidene difluoride membrane filter (0.45 μm). The leachate was preserved in a polypropylene tube with HNO_3. Metals and metalloids present at detectable levels in the leachates (i.e., Mn, Fe, Cu, Zn, As, Cd, Sb, and Pb) were selected after screening by ICP-atomic emission spectroscopy and quantified by ICP-MS (modified after Fuchida et al. 2017).

The chemical compositions and mineral assemblages of the sulfide samples are shown in Table 4. As the results of reaction, significant amounts of metals and metalloids were released into the seawater (Fig. 3); the metals and metalloid concentrations in the leachates increased significantly within 5 min of shaking and showed little change between 5 min and 6 or 18 h of shaking. A similar rapid initial release of metals from oxidized sulfide samples within 5 min was also reported by Parry (2008) (Table 3b). The chemical compositions of the leachates differed depending on the mineral assemblages and chemical compositions of the sulfide samples. Specifically, Fe–Zn–Pb-rich, Ba-rich, and Fe-rich mineral samples released abundant amounts of Zn + Pb, As + Sb, and Zn + Cu into seawater, respectively. There were no correlations

Table 4 Chemical compositions of sulfides and their leachates and major constituent mineral assemblages collected from various seafloor hydrothermal sites

		Concentration in leachate mmol/L (in hydrothermal sulfide sample mmol/kg)								pH of leachate	Major minerals	Location (year)
		Mn	Fe	Cu	Zn	As	Cd	Sb	Pb			
1	Fe–Zn–Pb-rich	**0.0198**	**5.62**	**b.d.**	**6.69**	**b.d.**	**0.03**	**b.d.**	**0.231**	4.6	Sphalerite, galena, anglesite	Izena Hole, Okinawa Trough (2011)
	(HPD1313G04)	(3.78)	(4650)	(37.1)	(2260)	(1.91)	(5.30)	(0.715)	(458)			
2	Ba-rich	**0.0129**	**0.100**	**b.d.**	**1.03**	**0.459**	**b.d.**	**0.101**	**b.d.**	6.6	Barite, realgar	Izena Hole, Okinawa Trough (2011)
	(HPD1313G05)	(0.275)	(39.1)	(0.675)	(29.3)	(179)	(0.0290)	(26.9)	(0.944)			
3	Fe-rich	**0.164**	**b.d.**	**1.19**	**62.1**	**b.d.**	**0.19**	**b.d.**	**b.d.**	4.9	Pyrite	Izena Hole, Okinawa Trough (2011)
	(HPD1311G06)	(5.95)	(5370)	(84.0)	(1240)	(70.8)	(3.94)	(1.77)	(1.85)			
4	Zn-Pb-rich zero-age	**0.0837**	**b.d.**	**b.d.**	**3.67**	**b.d.**	**0.01**	**b.d.**	**0.0887**	6.8	Sphalerite, galena, anglesite, anhydrite	Iheya Knoll, Okinawa Trough (2012)
	(HPD1355R04)	(47.3)	(1590)	(491)	(4280)	(16.2)	(6.75)	(3.40)	(514)			
5	Ba-rich	**0.0213**	**b.d.**	**b.d.**	**0.00152**	**0.150**	**b.d.**	**0.0376**	**b.d.**	8.2	Barite	Daisan Kume Knoll, Okinawa Trough (2015)
	(KK673R01-1a)	(0.313)	(2.45)	(0.910)	(4.61)	(29.7)	(0.0246)	(60.8)	(1.11)			
6	Zn–Ba-rich	**b.d.**	**b.d.**	**b.d.**	**b.d.**	**0.535**	**b.d.**	**0.00344**	**b.d.**	7.7	Barite, sphalerite	Daisan Kume Knoll, Okinawa Trough (2015)
	(KK673R03)	(0.0824)	(4.24)	(0.205)	(2480)	(254)	(6.04)	(51.0)	(1070)			
7	Ba-rich	**0.197**	**1.64**	**b.d.**	**0.807**	**1.27**	**0.00136**	**0.178**	**0.00726**	3.8	Barite, realgar, stibnite, pyrite	Iheya Knoll, Okinawa Trough (2015)
	(KK1857R01)	(11.9)	(1490)	(4.89)	(58.0)	(280)	(0.0584)	(386)	(13.1)			
8	Dolomite	**0.0126**	**b.d.**	**0.0239**	**0.419**	**b.d.**	**0.00149**	**b.d.**	**0.209**	5.9	Dolomite, quartz	Minami-Ensei Knoll, Okinawa Trough (2011)
	(HPD1328G02)	(2.09)	(614)	(92.8)	(661)	(0.987)	(1.28)	(0.691)	(256)			
9	Zn–Ba-rich	**0.0136**	**b.d.**	**b.d.**	**3.21**	**b.d.**	**0.00912**	**0.00230**	**0.108**	6.4	Barite, sphalerite	Hatoma Knoll, Okinawa Trough (2014)
	(HPD1621R02)	(4.56)	(62.2)	(542)	(2050)	(66.6)	(2.53)	(21.5)	(250)			

(continued)

Table 4 (continued)

		Concentration in leachate mmol/L (in hydrothermal sulfide sample mmol/kg)								pH of leachate	Major minerals	Location (year)
		Mn	Fe	Cu	Zn	As	Cd	Sb	Pb			
10	Ba-rich	**0.00610**	**b.d.**	**b.d.**	**0.00912**	**0.807**	**b.d.**	**0.00148**	**b.d.**	7.3	Barite, quartz, magnesium catena-silicate	Hatoma Knoll, Okinawa Trough (2015)
	(HPD1621R03)	(0.167)	(1.72)	(5.97)	(6.45)	(27.5)	(0.0181)	(1.40)	(0.848)			
11	Zn–Ba-rich	**0.0253**	**b.d.**	**b.d.**	**4.69**	**b.d.**	**0.0446**	**b.d.**	**0.00198**	6.8	Barite, sphalerite, pyrite	Bayonnaise Knoll, Izu-Ogasawara Arc (2014)
	(HPD1647R05)	(1.32)	(1190)	(669)	(4450)	(10.9)	(18.3)	(3.75)	(4.14)			
12	Zn–Pb-rich	**0.205**	**b.d.**	**b.d.**	**8.57**	**b.d.**	**0.00535**	**b.d.**	**0.129**	6.4	Sphalerite, anglesite, barite	Iheya Knoll, Okinawa Trough (2015)
	(HPD1857R03)	(22.4)	(1650)	(144)	(3260)	(3.20)	(3.17)	(2.30)	(939)			
13	Gypsum	**0.0215**	**b.d.**	**b.d.**	**0.0588**	**b.d.**	**0.00181**	**b.d.**	**0.00533**	7.9	Gypsum	Iheya Knoll, Okinawa Trough (2015)
	(HPD1859R05)	(1.86)	(45.7)	(13.7)	(80.8)	(0.824)	(0.126)	(0.143)	(4.25)			
14	Zn–Pb-rich ore	**0.0151**	**b.d.**	**b.d.**	**0.0120**	**0.450**	**b.d.**	**0.00754**	**b.d.**	7.4	Quartz, sphalerite, galena	Daisan Kume Knoll, Okinawa Trough (2015)
	(KK673R01-b)	(0.249)	(4.33)	(0.206)	(1980)	(347)	(3.59)	(26.3)	(1070)			

Powdered samples were digested with $HCl/HClO_4/HF/HNO_3$ ($n = 2$), and metals and metalloids were quantified by inductively coupled plasma-mass spectrometry (ICP-MS; 8800 ICP-QQQ, Agilent Technologies, Inc., Santa Clara, CA, USA). Dissolved metals and metalloids in the leachate samples were analyzed by ICP-MS; 1–4, the results from Fuchida et al. (2017), values are means of the triplicate analyses for which the relative standard deviation was within ±5%; 5–14, the results of additional leaching experiments and the values are the means of duplicate analyses. The mineral assemblages of the samples were determined by X-ray diffraction (XRD; MiniFlex600, Rigaku, Tokyo, Japan) with Ni-filtered monochromatic Cu Kα radiation at 2θ angles between 5 and 80°

*b.d.**, below detection limit

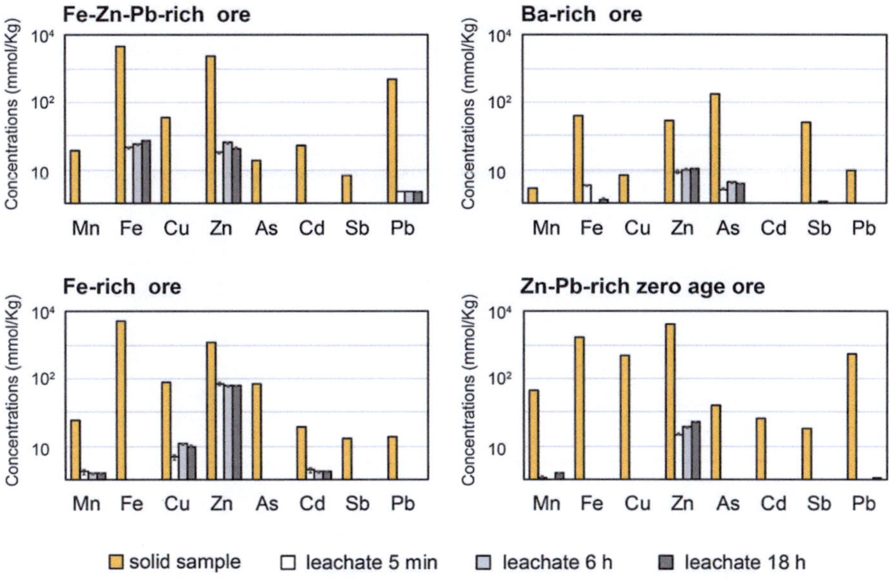

Fig. 3 Metal and metalloid concentrations in both leachate (n = 3± SD) and sulfide samples (n = 2) (Fuchida et al. 2017)

between the chemical compositions of leachates and those of sulfide samples, similar to the results reported by Simpson et al. (2007) and Parry (2008).

For further discussion of leachable metal and metalloid compositions in different hydrothermal sulfides, we conducted leaching experiments using ten sulfide (chimney) samples with the same method of our previous study (liquid-to-solid ratio of 20:1 and reacted for 6 h) (Fuchida et al. 2017). These mineral samples were collected from different hydrothermal sites of the Okinawa Trough (sample nos. 5–10 and 12–14) and Izu-Ogasawara Arc (sample no. 11). The chemical compositions of leachates and sulfide samples and mineral assemblages are summarized in Table 4, and the metals and metalloids are indicated in four colors (red, blue, yellow, and white) in Fig. 4 based on their concentrations in both the leachate and sulfide samples.

Among all released metals and metalloids, Zn was the most easily leached from oxidized hydrothermal sulfides, regardless of chemical compositions and mineral assemblages of sulfide samples (represented by red in Fig. 4). These results agree with those of our (Fuchida et al. 2017) and other previous studies (Simpson et al. 2007; Parry 2008). Several samples released high amounts of Mn and Cd, despite their being present in low levels in the sulfide samples (represented by yellow in Fig. 4). For example, Mn and Cd could be co-released with Zn into seawater because they are generally contained in Zn-bearing sulfide minerals (i.e., sphalerite) as impurities (Cook et al. 2009). Moreover, Pb was released in high levels into seawater when high amounts of Pb-bearing minerals such as galena and anglesite were present in the sulfide samples, e.g., Fe–Zn–Pb-rich and Zn–Pb-rich samples. For several samples from which high levels of Zn and Pb

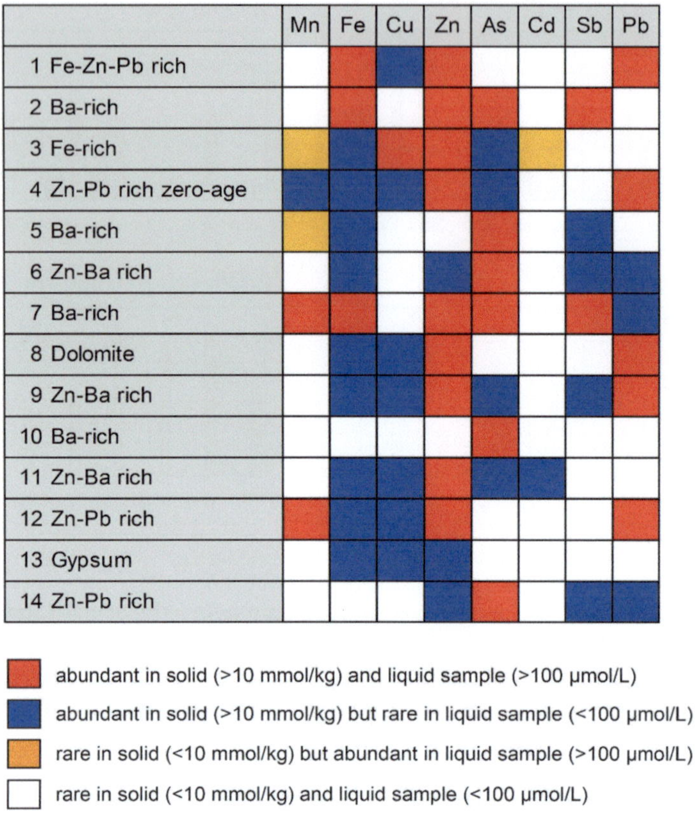

Fig. 4 Classification of leachable metals and metalloids from oxidized hydrothermal sulfide (chimney) samples

were released, the pH of their leachates decreased to four to six after the reaction (Table 4). Zn- and Pb-sulfide minerals are known to be acid soluble (Heidel et al. 2013), indicating that dissolution of these sulfide minerals would be promoted as pH decreases (i.e., acid generation).

Arsenic was released from Ba-rich samples (Ba-rich and Zn–Ba-rich) because Ba-rich sample is often associated with As-bearing sulfide minerals such as realgar (As_4S_4) and orpiment (As_2S_3). Antimony was also released with As from Ba-rich samples, whereas Zn–Ba-rich samples did not release Sb.

In contrast to Zn and Pb, Fe and Cu were released at lower levels, even though they were present in high abundance in the sulfide samples (represented by blue in Fig. 4). Fe and Cu in the solid samples were mainly present as disulfide minerals, i.e., pyrite and chalcopyrite, based on XRD analysis. In addition, Fe was often contained in the sphalerite as an impurity, and as a result it can be released with Zn and other impurities (i.e., Mn and Cd) into seawater. However, Fe was absent in most of the leachates, while Zn, Mn, and Cd were present, indicating that Fe can be rapidly removed from seawater by the formation of insoluble oxyhydroxides.

The results of our experiments and previous studies revealed that oxidized hydrothermal sulfides have high metal leaching potentials and that the compositions of leachable metals and metalloids differ significantly from those of solid samples. The chemical compositions and mineral assemblages of hydrothermal sulfides are highly dependent on the geological settings and fluid chemistry (e.g., Von Damm 1995); thus, the chemical compositions of leachates are expected to differ among different types of the hydrothermal sulfide samples. However, our results using 14 sulfide samples indicate that Zn, Pb and As are generally relatively more leachable than Fe and Cu, even if they are present at high levels in the sulfide samples. Therefore, this information would be useful for estimation of reference values for the maximum amounts of leachable metals and metalloids that might be leached from hydrothermal sulfides.

4 Metal Release from Hydrothermal Sulfides During Mining Operations

Our and other leaching experimental results demonstrated that metal dissolution from both non-oxidized and oxidized hydrothermal sulfides into seawater can occur. One of the crucial findings of these studies is that the metal dissolution behavior in seawater differed significantly during the initial oxidation of sulfide surfaces (i.e., oxidized or non-oxidized). For example, metal dissolution from non-oxidized samples was relatively gradual (Fig. 2), while that from oxidized samples was rapid, occurring within a few minutes (Fig. 3). Furthermore, the main components of metals released from non-oxidized samples were only Zn and Pb, whereas the oxidized samples released various metals and metalloids. Physicochemical parameters of seawater such as redox conditions and temperature also control the metal dissolution rate; thus, different metal contents of sulfide-contact seawater are predicted in between the surface and seafloor environment and between mining processes.

Based on the leaching experimental results of non-oxidized sulfide, severe metal releases from the minerals would not likely occur near the seafloor due to its low temperature, at least in the short term, even if a benthic plume with a large quantity of fine sulfide particulates was formed by a seafloor mining operation (Fig. 1a). Once the crushed sulfides are lifted to the vessel through the riser pipe, it is predicted that the metal release would be accelerated because of changes in the environmental condition to oxic and higher temperature (Fig. 1b). However, the reaction rate would be rather slow relative to that of fully oxidized sulfides. In cases in which sulfides lifting from the seafloor to the vessel occur within several tens of minutes, the metal concentration in the seawater with the crushed sulfides would not increase greatly.

However, once the sulfides are exposed to air and rainwater after lifting, their surface will be gradually oxidized and transformed into more soluble compounds. If these well-oxidized sulfides are discharged and/or accidentally spilled from the vessel into the oxic and warm surface environment, a rapid dissolution reaction of the minerals would occur. Moreover, if the discharged minerals were composed of very fine particulates for lifting up from seafloor, their suspension and metal dissolution reaction would persist in the plume for a long time.

When the collected hydrothermal sulfides are carried to shore by shuttle barge and temporarily stored on port for processing (Collins et al. 2013), the exposure of the sulfide to air-oxygen is continued for a long period of time. If the well-oxidized sulfides are accidentally discharged to soil and groundwater on port and other places, the metal contamination will extend to shore and land areas. Therefore, it should pay particular attention to the treatment of the oxidized sulfides during the mining activity.

5 Summary

We described the results of leaching experiments using both non-oxidized and oxidized hydrothermal sulfides. Metal dissolution behaviors differed depending on initial oxidation states of the sulfide surface, mineral configurations (galvanic couplings), specific surface areas, and other physical parameters (e.g., temperature and redox potential), indicating that metal leachability and their major released elements will differ depending on marine environmental conditions, which can vary drastically according to the sequential progression of the mining process. The properties of metal dissolution from hydrothermal sulfides into seawater are summarized as follows:

1. Zn and Pb were the main components in the seawater that reacted with non-oxidized sulfides that were underwater, whereas large amounts of various metals and metalloids were released from oxidized sulfides exposed to atmosphere.
2. Metals dissolution from non-oxidized sulfides was greatly accelerated at high temperature and oxic conditions close to the surface relative to the low temperature and anoxic conditions close to the seafloor.
3. The dissolution of metals from non-oxidized sulfides was gradual, while that from oxidized sulfides was rapid, occurring within several minutes.
4. The presence of iron disulfide minerals (e.g., pyrite and marcasite) would induce galvanic metal dissolution from hydrothermal sulfides into seawater.

On the basis of these findings, it is estimated that little metal-rich seawater is produced when crushing and lifting hydrothermal sulfides during mining because the sulfides have not been oxidized and there is little opportunity or the minerals to be exposed to air-oxygen for a long period of time. However, once the minerals have been oxidized through contact between the lifted hydrothermal sulfides on the vessel and air, oxic seawater, and rainwater, various metals and metalloids can be released rapidly if the oxidized sulfides are spilled into the ocean. Thus, surface plumes of spilled material could cause more serious problems than benthic plumes produced by the crushing process of the seafloor mine. As reported in previous studies (Satoh et al. 2005; Caroppo et al. 2006; Fuchida et al. 2017), metals and metalloids released from hydrothermal sulfides could be toxic to marine organisms living in the surface environment, such as marine phytoplankton species, which are a significant primary producer in the ocean. Therefore, we suggest that disposal of tailings and mining wastes below the oxygen minimum zone may be better to minimize the impacts of metal leaching from hydrothermal sulfides on the marine surface environments. Also, adequate monitoring of water quality, drainage treatment, and impact assessment of accidental leakage of the tailings, mining wastes, and drainages to the surface environment are essential.

Acknowledgments We are grateful to the crew members of the NT11-15, NT11-20, NT12-06, KY14-02, NT14-06, NT15-13, and KR15-16 cruises. This study was financially supported by the Cross-ministerial Strategic Innovation Promotion Program (SIP), Next-Generation Technology for Ocean Resource Exploration (funding agency: JAMSTEC). We thank Jeremy Kamen, MSc, from Edanz Group (www.edanzediting.com/ac) for editing a draft of this manuscript.

References

Angel, B. M., Apte, S., Batley, C. E., & Raven, M. D. (2016). Lead solubility in seawater: An experimental study. *Environmental Chemistry, 13*, 489–495.

Bilenker, L. D., Romano, G. Y., & McKibben, M. A. (2016). Kinetics of sulfide mineral oxidation in seawater: Implications for acid generation during in situ mining of seafloor hydrothermal vent deposits. *Applied Geochemistry, 75*, 20–31.

Caroppo, C., Stabili, L., Aresta, M., Corinaldesi, C., & Danovaro, S. (2006). Impact of heavy metals and PCBs on marine picoplankton. *Environmental Toxicology, 21*, 541–551.

Chopard, A., Plante, B., Benzaazoua, M., Bouzahzah, H., & Marion, P. (2017). Geochemical investigation of the galvanic effects during oxidation of pyrite and base-metals sulfides. *Chemosphere, 166*, 281–291.

Collins, P. C., Croot, P., Carlsson, J., Colaço, A., Grehan, A., Hyeong, K., Kennedy, R., Mohn, C., Smith, S., Yamamoto, H., & Rowden, A. (2013). A primer for the environmental impact assessment of mining at seafloor massive sulfide deposits. *Marine Policy, 42*, 198–209.

Cook, N. J., Ciobanu, C. L., Pring, A., Skinner, W., Shimizu, M., Danyushevsky, L., Saini-Eidukat, B., & Melcher, F. (2009). Trace and minor elements in sphalerite: A LA-ICPMS study. *Geochimica et Cosmochimica Acta, 73*, 4761–4791.

Cruz, R., Luna-Sánchez, R. M., Lapidus, G. T., González, I., & Monroy, M. (2005). An experimental strategy to determine galvanic interactions affecting the reactivity of sulfide mineral concentrates. *Hydrometallurgy, 78*, 198–208.

Eggleton, J., & Thomas, K. V. (2004). A review of factors affecting the release and bioavailability of contaminants during sediment disturbance events. *Environment International, 30*, 973–980.

Fallon, E. K., Petersen, S., Brooker, R. A., & Scott, T. B. (2017). Oxidative dissolution of hydrothermal mixed-sulphide ore: An assessment of current knowledge in relation to seafloor massive sulphide mining. *Ore Geology Reviews, 86*, 309–337.

Feely, R. A., Lewison, M., Massoth, G. J., Robertbaldo, G., Lavelle, J. W., Byrne, R. H., Vondamm, K. L., & Curl, H. C. (1987). Composition and dissolution of black smoker particulates from active vents on the Juan-De-Fuca Ridge. *J Geophys Res-Solid, 92*, 11347–11363.

Fuchida, S., Yokoyama, A., Fukuchi, R., Ishibashi, J.-i., Kawagucci, S., Kawachi, M., & Koshikawa, H. (2017). Leaching of metals and metalloids from hydrothermal ore particulates and their effects on marine phytoplankton. *ACS Omega, 2*, 3175–3182.

Fuchida, S., Ishibashi, J., Shimada, K., Nozaki, T., Kumagai, H., Kawachi, M., Matsushita, Y., & Koshikawa, H. (2018). Onboard experiment investigating metal leaching of fresh hydrothermal sulfide cores into seawater. *Geochemical Transactions*. https://doi.org/10.1186/s12932-018-0060-9., in press.

Goldhaber, M. B. (1983). Experimental study of metastable sulfur oxyanion formation during pyrite oxidation at pH 6-9 and 30 degrees C. *American Journal of Science, 283*, 193–217.
Hageman, P. L., Seal, R. R., Diehl, S. F., Piatak, N. M., & Lowers, H. A. (2015). Evaluation of selected static methods used to estimate element mobility, acid-generating and acid-neutralizing potentials associated with geologically diverse mining wastes. *Applied Geochemistry, 57*, 125–139.
Heidel, C., Tichomirowa, M., & Junghans, M. (2013). Oxygen and sulfur isotope investigations of the oxidation of sulfide mixtures containing pyrite, galena, and sphalerite. *Chemical Geology, 342*, 29–43.
Kwong, Y. T. J., Swerhone, G. W., & Lawrence, J. R. (2003). Galvanic sulphide oxidation as a metal-leaching mechanism and its environmental implications. *Geochemistry: Exploration, Environment, Analysis, 3*, 337–343.
Liu, X., & Millero, F. J. (1999). The solubility of iron in sodium chloride solutions. *Geochimica et Cosmochimica Acta, 63*, 3487–3497.
Liu, Q., Li, H., & Zhou, L. (2008). Galvanic interactions between metal sulfide minerals in a flowing system: Implications for mines environmental restoration. *Applied Geochemistry, 23*, 2316–2323.
Michael, A., McKibben, M. A., & Barnes, H. L. (1986). Oxidation of pyrite in low temperature acidic solutions: Rate laws and surface textures. *Geochimica et Cosmochimica Acta, 50*, 1509–1520.
Parry, D. L. (2008). Solwara 1 project elutriate report phase 1 and 2.
Satoh, A., Vudikaria, L. Q., Kurano, N., & Miyachi, S. (2005). Evaluation of the sensitivity of marine microalgal strains to the heavy metals, Cu, As, Sb, Pb and Cd. *Environment International, 31*, 713–722.
Simpson, S. L., & Spadaro, D. A. (2016). Bioavailability and chronic toxicity of metal sulfide minerals to benthic marine invertebrates: Implications for deep sea exploration, mining and tailings disposal. *Environmental Science & Technology, 50*, 4061–4070.
Simpson, S. L., Angel, B., Hamilton, I., Spadaro, D. A., & Binet, M. (2007). Water and sediment characterization and toxicity assessment for the Solwara 1 Project, CSIRO Land and Water Science Report. Coffey Natural Systems Pty Ltd.
Steger, H. F., & Desjardins, L. E. (1980). Oxidation of sulfide minerals. v. galena, sphalerite and chalcocite. *Canadian Mineralogict, 18*, 365–372.
Tsang, J. J., & Parry, D. L. (2004). Metal mobilization from complex sulfide ore concentrate: Effect of light and pH. *Australian Journal of Chemistry, 57*, 971–978.
Von Damm, K. L. (1995). Controls on the chemistry and temporal variability of seafloor hydrothermal fluids. *Geophysical Monograph, 91*, 222–247.

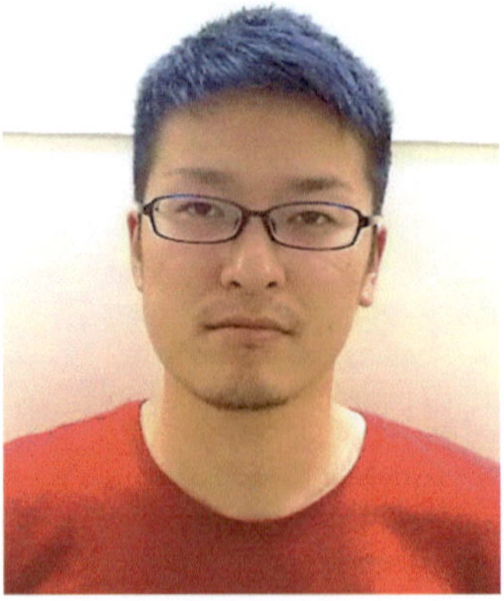

Shigeshi Fuchida, born in 1986, studied marine geochemistry at the Osaka City University. Since his doctorate in 2015, he has been engaged in environmental impact assessment of deep-sea mining at the National Institute for Environmental Studies (NIES). His main scientific fields are organic geochemical process in hydrothermal site, interaction of mineral-water interface, and speciation and biogeochemical cycling of metals and metalloids in seafloor.

Hiroshi Koshikawa is working for the Center for Regional Environmental Research of the National Institute for Environmental Studies (NIES), Japan, as a head of Marine Environmental Section. His research background is in phytoplankton ecology, and he has focused on anthropogenic alterations of marine environments and their impacts on the algal production and food webs such as eutrophication, oil pollution, and climate change issues. Since 2014, he has worked on the development of the EIA protocol especially for the marine surface environment at the seafloor mining site where there are risks of heavy metal contamination. He was born in 1968 and obtained the Doctor of Engineering from the Science University of Tokyo in 1997.

Mining in Hydrothermal Vent Fields: Predicting and Minimizing Impacts on Ecosystems with the Use of a Mathematical Modeling Framework

Kenta Suzuki and Katsuhiko Yoshida

Abstract Accelerated demand for exploration of minerals and development of mining technologies over the past decade could lead to commercial mining of the deep seafloor in the near future. The campaign for conservation of biological diversity claims that there will be impacts of seabed mining to the deep-sea community and suggests a precautionary approach. In this chapter, we summarize the basic characteristics of communities in hydrothermal vent fields and describe the potential impact of resource mining as well as some previous observations on the effect of natural disturbances. We then introduce a model-based approach to determine the resilience of vent communities, thereby predicting if the communities will be vulnerable or robust to disturbances. Resilience of ecological systems is assessed by measuring the time required to recover to the original state prior to being disturbed. A mathematical model capable of predicting resilience would represent an important contribution to the management of these unique ecosystems. However, compared to most terrestrial and shallow water ecosystems, information regarding hydrothermal vent ecosystems, which are typically found at depths of over 1000 m, is limited. We thus focused on connectivity of vent communities through larval dispersal as a key factor for resilience. We will show how our framework can be used as a practical tool to characterize, understand, or predict resilience. The framework presented here can help assess ecological impacts and develop mitigation strategies associated with deep-sea resource mining. We also discuss what will need to be developed further to better achieve these objectives.

Keywords Hydrothermal vent fields · Community responses · Ecological impacts · Mathematical model

K. Suzuki (✉) · K. Yoshida
Center for Environmental Biology and Ecosystem Studies, National Institute for Environmental Studies, Tsukuba, Ibaraki, Japan
e-mail: suzuki.kenta@nies.go.jp

R. Sharma (ed.), *Environmental Issues of Deep-Sea Mining*,
https://doi.org/10.1007/978-3-030-12696-4_9

1 Introduction

1.1 Importance of Protecting Deep-Sea Chemosynthetic Communities

Accelerated exploration of minerals and development of mining technologies over the past decade could realize commercial mining of the deep seabed in the near future (Urabe et al. 2015; Coffey Natural Systems 2008a). One of the targets for commercial mining is the seafloor massive sulfide (SMS) deposits formed around hydrothermal vents, which is a highly attractive source of copper, zinc, lead, gold, and silver ores (Hoagland et al. 2010; Herzig 1999; Binns and Scott 1993; Halbach et al. 1989). Hydrothermal vents host chemosynthetic communities as well as the metal-rich ores. The chemosynthetic communities consist of endemic invertebrate species specifically adapted to the vent environment via microbial chemoautotrophic primary production (van Dover 2010). These species have provided new scientific insights into the mechanisms by which organisms adapt to the extreme environment (Jannasch and Wirsen 1979). Furthermore, as reviewed by Le et al. (2016), ecological function and services of these communities range from providing habitat and refuge for other species including non-endemic species (Levin et al. 2016; Govenar 2010), playing a key role in global carbon, sulfur, and heavy metal cycling (Jeanthon 2000; D'Arcy et al. 2013) and offering new biomolecules that could contribute to industrial development (Terpe 2013; Mahon et al. 2015). Hence, both quantitative evaluations of potential impacts on ecological systems and consideration of effective mitigation strategies are necessary to avoid or minimize the disruptive effect of anthropologic activities in the deep-sea environment (van Dover 2010; Collins et al. 2013; van Dover et al. 2014; Boschen et al. 2016).

1.2 Potential Impact of Mining Activities on Deep-Sea Chemosynthetic Communities

Mining of seafloor massive sulfide deposits potentially changes the physicochemical environment of a vent community through the loss of sulfide habitat, degradation of sulfide habitat quality, modification of fluid flux regimes, and exposure of surrounding seafloor habitats (including non-sulfide habitats) to sedimentation and heavy metal deposition (International Seabed Authority 2007; Van Dover 2014). This will directly affect the ecological community by removing and reclaiming organisms, reducing the amount of habitable substrate, and changing resource supply. Physicochemical models and organism distribution data have been integrated to estimate the potential area of sedimentation (Coffey Natural Systems 2008b). However, after the instantaneous effects of a disturbance, the ecological community will reach a new equilibrium state within the disturbed environment (Ives and Carpenter 2007). Hence, potential impacts of artificial disturbances, including how they may cause extinction and modify community structure at different spatial scales (local, regional, and global) and decrease diversity at different biological levels (genetic, specific, and phylogenetic), will be understood by considering both direct impacts of mining activities and subsequent ecological responses. Environmental impact assessments (EIAs) that lack this point of view might severely underestimate the potential risks of anthropological activities.

1.3 Difficulty in Assessing the Effect of Disturbances

One of the challenges in conducting EIAs for seafloor mining activities is the lack of historical data (Boschen et al. 2013). It is difficult to estimate for a specific vent community how a natural or anthropologic disturbance will change its equilibrium. As in other ecological systems, incorporating ecological modeling is expected to address this challenge (Collins et al. 2013; Wathern 2013). An ecological model is an abstract mathematical representation of an ecological system that is designed based on literature and observational data. Model systems are then applied to analyze the dynamics and response of real ecological systems. For example, a unified modeling framework, Ecopath with Ecosim (Christensen and Walters 2004), has been used to assess past and future impacts of fishing and environmental disturbances on exploited aquatic ecosystems (Christensen and Walters 2005; Villasante et al. 2016). However, most of these existing modeling frameworks require detailed information on the target community (e.g., basal growth rate and predation rate). Developing an ecological model for a vent community and specifically modeling the response to disturbances are challenges because of inaccessibility and lack of historical data (Boschen et al. 2013).

1.4 Resilience of Vent Ecosystems

Instead of developing a full model that can reproduce detailed population dynamics of organisms with in an ecological system, a simpler mathematical model could be used to obtain predictions on key characteristics of the system, such as stability of the system to disturbances. Recoverability, or resilience, refers to persistence of ecosystems in the face of natural or anthropogenic disturbances (Holling 1973). Resilience of vent communities indicates whether they are vulnerable or robust to natural disturbances, such as volcanic activity, and anthropogenic disturbances, such as mineral resource mining. It can be quantified as the recovery time to the original state before disturbance (Holling 1996). For chemosynthetic communities in hydrothermal vent fields (HVFs), observations on recovery from disruption caused by volcanic activities suggest that most of the diversity and biomass recovered within 5 years after the disturbance (Tunnicliffe et al. 1997: Shank et al. 1998; Marcus et al. 2009; Gollner et al. 2015, 2017). For example, total mega- and macrofaunal species richness at the vents in Juan de Fuca Ridge reached 75% of the pre-disturbance values 3 years after the 1998 eruption (Marcus et al. 2009) and 90% 2 years after the 1993 eruption (Tunnicliffe et al. 1997), representing about 30–60% of the species from the larger regional species pool. At the East Pacific Rise (EPR), total mega- and macrofaunal species richness reached 69% of pre-disturbance values 4.6 years after the 1991 eruption (Shank et al. 1998). After the EPR 2006 eruption, recovery reached 55% for macrofaunal and 48% for meiofaunal species after 4 years (Gollner et al. 2015), with 39% of the macro and 42% of meiofaunal species recovered. These observations provide an expectation that both biomass and diversity of communities in HVFs can almost recover within 5 years. However, recoverability would be different depending on the regional connectivity between HVFs (Gollner et al. 2017).

1.5 Dispersal: A Key Factor for Resilience of Communities in HVFs

Vent ecosystems are typically dominated by benthic invertebrate taxa (e.g., *Bathymodiolus* mussels, *Shinkaia* squat lobsters, and *Alviniconcha* gastropods in Okinawa Trough; Watanabe et al. 2010; Desbruyères et al. 2006; Podowski et al. 2010) that host symbiotic, chemoautotrophic microorganisms. Deep-sea vents are ephemeral habitat islands from the moment of their discovery (Macdonald et al. 1980; Van Dover 2014). The frequency of disruptive natural disturbances to vent communities can range from several decades to several hundred years. Faunal adaptations for colonizing new vent fields are thus important aspects of the sustainability of these communities, especially since neighboring vent fields are often separated by 10 s or 100 s of kilometers. Vent-restricted taxa are characterized by rapid growth rates, early maturation, large reproductive output, and well-developed dispersal capabilities (Grassle 1986). These characteristics are shared by opportunistic marine invertebrate species that persist despite frequent local extinctions and are divergent from those of deep-sea species in low-disturbance regimes (Grassle and Sanders 1973; Van Dover 2014). Local vent communities form regional meta-communities nested within several biogeographic provinces linked by pelagic larval dispersal.

Recent studies have assessed vent field connectivity by looking at genetic differentiation among vent populations (Hurtado et al. 2004; Johnson et al. 2008, 2014; Thaler et al. 2014; Yahagi et al. 2015; Mitarai et al. 2016). Genetic data imply that back-arc basin populations are forming well-mixed genetic pools (Watanabe et al. 2005; Nakamura et al. 2014). In contrast, populations in distant regions (~3000 km apart) are genetically distinct, although occasional migrations may have occurred over the course of several hundred thousand generations (Thaler et al. 2011). For example, population genetics of *Neoverruca* barnacles show no significant genetic differentiation among populations within the Okinawa Trough

(Watanabe et al. 2005). Furthermore, haplotype analyses of the barnacles imply that *Neoverruca* populations inhabiting the Izu-Bonin region originated in the Okinawa Trough, although these two populations are genetically distinct (Watanabe et al. 2005).

To understand the pattern of larval dispersal and gene flow of vent species by ocean circulation, Mitarai et al. (2016) quantified potential larval dispersal among 131 HVFs in the western Pacific basin. The estimates were based on a physical model of deep-ocean circulation that was validated through a deep-ocean profiling float experiment and considered temperature dependency of larval development that can control duration of pelagic larval stages. They demonstrated that vent fields within back-arc basins could be well connected when there was a lack of directionality, whereas basin-to-basin dispersal is expected to occur infrequently, once in tens to hundreds of thousands of years, when clear dispersal barriers and directionality associated with ocean currents are present.

2 A Model-Based Data Analysis

A mathematical model capable of predicting resilience would be an important contribution to the management of these unique ecosystems. One of the ways to assess resilience of ecological systems is through development of a detailed model that can describe population dynamics of the focal system. This model can be applied to measure the time required to recover to the original state just prior to disturbance. However, as we discussed above, compared to most terrestrial and shallow water ecosystems, data regarding hydrothermal vent ecosystems are limited because these ecosystems are typically found at depths of over 1000 m. Thus, resilience is estimated by connectivity of vent communities through larval dispersal, a key factor in determining the ability of a vent community to recover after disturbance. Recently, Suzuki et al. (2018) introduced a model-based approach to determine the resilience of vent communities by integrating a meta-population model and the estimated larval dispersal between HVFs (Mitarai et al. 2016). By simulating disturbances to vent fields, Suzuki et al. (2018) mapped recoverability (recovery time) of communities in 131 hydrothermal vent fields in the western Pacific Ocean. We introduce the approach and present some extended framework used for the model-based data analysis. HVFs may include both active and inactive vent systems. In this study, we focused on the recoverability of communities in active vent systems. Our results can be applied to chemosynthetic communities in inactive vent systems, if they are included in or located close to the HVFs included in the analysis. However, because of the limited resource supply, recoverability of inactive vent systems could be lower than active vent systems (Gollner et al. 2017).

2.1 Potential Larval Dispersal Between Hydrothermal Vent Fields

The results of Mitarai et al. (2016) were based on a unified model that accounts for the temperature dependence of larval development in marine animals (O'Connor et al. 2007) as well as a physical model describing deep-ocean circulation. Dispersal distance will depend both on the speed of ocean currents and expected duration of larvae at a given depth. For example, larvae that use shallower depths can disperse further because of fast ocean currents, although duration of the pelagic larval stage would be shortened when water temperature is higher (O'Connor et al. 2007). In the following analysis, we assumed a dispersal depth of 1000 mas in Suzuki et al. (2018). It is supported by published data on the hydrographic structure of water columns obtained from observation and simulation models (England 1992; Gupta and England 2007; Nakamura et al. 2013). The biogeographic studies of larval dispersal at hydrothermal vents have a consensus on the transport mechanism by deep-sea advection and effluent layers (Marsh et al. 2001; Tyler and Young 2003), where physicochemical parameters, physiographic features of a region, and seafloor topography are recognized as possible barriers to dispersal (Won et al. 2003). The water mass below 1000 m depth generally has stable physicochemical parameters, such as temperature and salinity, and may offer suitable conditions for larval survival in advection above the vent area and lateral transport in the effluent layer of the deep sea. For example, there is a discontinuity in water temperature and salinity at 500–700 m in the Okinawa Trough and at 1000 m in the western and southern Pacific Ocean (Talley 2007). Moreover, two chemosynthetic communities found at a difference of more than 1000 m depth had significant differences in their community structure (Watanabe et al. 2010), suggesting the effect of environmental barriers (but see Arellano et al. (2014) and Yahagi et al. (2017) for recent findings).

2.2 Differential Equation Model

We have used a differential equation for our analysis, which is described as:

$$\frac{\mathrm{d}x_i}{\mathrm{d}t} = f_i\left(x_i\right). \tag{1}$$

Generally, a differential equation describes the time evolution of the variable x_i (here, it represents population size of organisms in HVF *i)* based on some processes implemented as a mathematical representation in function f_i. Specifically, we used the following function:

$$f_i(x_i) = \left(1 - \frac{x_i}{K_i}\right)\left(rA_{ii}x_i + r\Sigma_{j\neq i}A_{ji}K_j\right). \quad (2)$$

Here, K_i is the carrying capacity (equilibrium population size of vent field i), r is the reproduction rate defined as the number of larvae that one individual produces per year, and A_{ij} is the dispersal rate that a larva produced at vent field j will migrate into i per year, where A_{ii} corresponds to self-recruitment. The dispersal networks were implemented as a dispersal matrix A whose elements are A_{ii} and A_{ij}. In Eqs. (1) and (2),

$$rA_{ii}x_i,$$

represents the recruitment of larvae from the HVFi and,

$$r\Sigma_{j\neq i}A_{ji}K_j,$$

represents the recruitment of individuals from other vent fields within the same region, assuming that other vent fields are in equilibrium ($x_j = K_j$). A negative dependency of recruitment on population size x_i is introduced as:

$$\left(1 - \frac{x_i}{K_i}\right),$$

following the standard formulation of ecological systems limited by a carrying capacity. Here, we assumed that K_i is determined by the total amount of resource supply in vent field i. Our model does not include the duration of pelagic larval phase as a parameter because it is included in the calculation of A (O'Connor et al. 2007; Mitarai et al. 2016). We assumed that x_i is the population size of a species or group of species that share the same niche in vent field i. The species or group of species is assumed to distribute across all vent fields in the same region (connected by dispersal) with sufficient abundance. This assumption is realistic because vent communities frequently have a dominant taxon that constitutes most of the biomass at a regional scale, although the dominance-diversity relationships may depend on the environmental conditions such as fluid flux intensity and sediment types (Portail et al. 2015). Among different regions however, it would be reasonable to interpret x_i as the abundance of different species or group of species that accounts for a similar proportion of biomass and having the same growth and dispersal characteristics. We do not directly consider the effect of disturbances on biological diversity because our model accounts for only one species. However, we expect that recoverability of these representative species would be a proxy for the recoverability of other infrequent community members; thus we regarded it as the recoverability of the community as a whole. For example, this is supported by observations of recovery after eruptions (Tunnicliffe et al. 1997; Shank et al. 1998; Marcus et al. 2009; Gollner

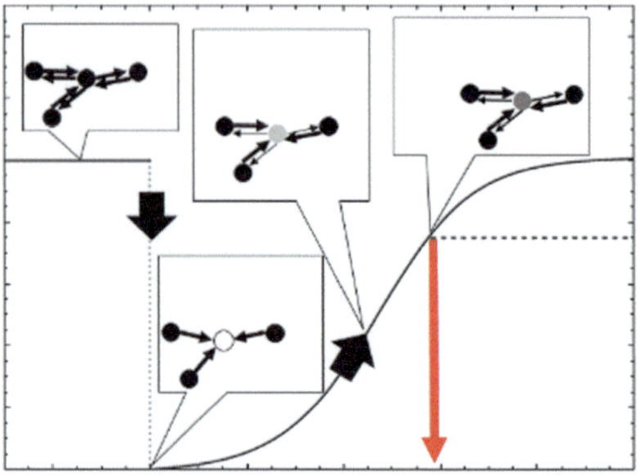

Fig. 1 Schematic distribution of a numerical experiment to estimate recovery time

et al. 2015, 2017) which showed a concurrence of recovery in total organism density and species richness.

Using Eqs. (1) and (2), we conducted a numerical experiment to calculate the recovery time of HVFs (Fig. 1). We defined τ_i as the mean recovery time of a vent field i for the ensemble of various spatial distributions of K_i, where recovery time is the time required for x_i to recover 75% of its original abundance (equal to K_i by definition) after it was temporarily reduced to zero. To calculate τ_i, we assumed that K_i is assigned from a log-normal distribution with $\mu = 6.9$ ($e^{\mu} = 1000$) and $\sigma = 1$ and all vent fields except for i are in equilibrium. However, the following results are independent of K_i if K_i values are identical across vent fields (Suzuki et al. 2018). Furthermore, they can be referred to as the mean recovery time in the ensemble of potential spatial heterogeneity in K_i as long as K_i of all vent fields follows the same probability distribution (Suzuki et al. 2018). We assumed that K_i values are constant throughout time. Our results may underestimate recovery time if population size is highly constrained by the amount of suitable habitat that increases slowly along with the reestablishment of vent fields after disturbances. For example, the loss and recovery of mussel beds may restrict the abundance of associated small invertebrate species (Turnipseed et al. 2003). However, we focused on the resilience of HVFs depending only on the larval supply as this has the strongest influence on recoverability.

2.3 *Resilience of HVFs Around Japan*

Recovery Time of HVFs

A dispersal matrix of HVFs around Japan is shown in Table 1. The dispersal matrix includes seven HVFs in Okinawa region (Iheya Ridge, Irabu Knoll, Izena Cauldron, Natsushima 84-1 Knoll, North Knoll Iheya Ridge, Dai-Yon Yonaguni Knoll, and Hatoma Knoll) and four HVFs in Izu-Bonin region (Mokuyo Seamount, Myojin Knoll, Suiyo Seamount, and Sumisu Rift). There are no dispersals connecting

Table 1 Dispersal matrix of HVFs around Japan. Units are *larva/adults/year*

	Okinawa; Iheya Ridge	Okinawa; Irabu Knoll	Okinawa; Izena Cauldron	Okinawa; Natsushima 84-1 Knoll	Okinawa; North Knoll Iheya Ridge	Okinawa; Dai-Yon Yonaguni Knoll	Okinawa; Hatoma Knoll	Izu-Bonin; Mokuyo Seamount	Izu-Bonin; Myojin Knoll	Izu-Bonin; Suiyo Seamount	Izu-Bonin; Sumisu Rift
Okinawa; Iheya Ridge	0.002281	0.00234	0.005349	0.002376	0.001407	0.000567	0.000994	0	0	0	0
Okinawa; Irabu Knoll	0	0.006708	0	0	0	0.006281	0.008571	0	0	0	0
Okinawa; Izena Cauldron	0.003016	0.001548	0.008017	0.003775	0.001366	0.000275	0.000756	0	0	0	0
Okinawa; Natsushima 84-1 Knoll	0.003174	0.001994	0.007631	0.003973	0.001512	0.000345	0.000507	0	0	0	0
Okinawa; North Knoll Iheya Ridge	0.003676	0.001555	0.004608	0.002347	0.001805	0.000479	0.000568	0	0	0	0
Okinawa; Dai-Yon Yonaguni Knoll	0	0.004336	0	0	0.00021	0.004978	0.003714	0	0	0	0
Okinawa; Hatoma Knoll	0	0.0061	0	0	0.000131	0.006814	0.0086	0	0	0	0
Izu-Bonin; Mokuyo Seamount	0	0	0	0	0	0	0	0.000794	0	0.00066	0

(continued)

Table 1 (continued)

	Okinawa; Iheya Ridge	Okinawa; Irabu Knoll	Okinawa; Izena Cauldron	Okinawa; Natsushima 84-1 Knoll	Okinawa; North Knoll Iheya Ridge	Okinawa; Dai-Yon Yonaguni Knoll	Okinawa; Hatoma Knoll	Izu-Bonin; Mokuyo Seamount	Izu-Bonin; Myojin Knoll	Izu-Bonin; Suiyo Seamount	Izu-Bonin; Sumisu Rift
Izu-Bonin; Myojin Knoll	0	0	0	0	0	0	0	0.000083	0.000443	0.000107	0.000327
Izu-Bonin; Suiyo Seamount	0	0	0	0	0	0	0	0.000822	0	0.000671	0.000001
Izu-Bonin; Sumisu Rift	0	0	0	0	0	0	0	0.000288	0.000192	0.000315	0.000172

Okinawa and Izu-Bonin region when a dispersal depth of 1000 m is assumed. Thus, the two regions were separated in the analysis. It is clear from the matrix that not all the HVFs are mutually connected within the region. For example, in Okinawa, dispersal from Irabu Knoll to North Knoll Iheya Ridge and all dispersals from Irabu Knoll, Dai-Yon Yonaguni Knoll, and Hatoma Knoll to Iheya Ridge, Izena Cauldron, and Natsushima 84-1 Knoll are unidirectional. In Izu-Bonin region, dispersal from Mokuyo Seamount to Myojin Knoll and Sumisu Rift and Myojin Knoll to Suiyo Seamount is unidirectional. These directionalities in larval transport are associated with directionality in ocean currents in these regions.

By using this matrix as the dispersal matrix in Eqs. (1) and (2), we estimated the recovery time (τ_i) of HVFs in Okinawa and Izu-Bonin region as in Fig. 2, Tables 2 and 3, respectively. τ_i in Okinawa was between 3.6 (Natsushima 84-1 Knoll) to 6.1 years (Dai-Yon Yonaguni Knoll) and 67.15 (Suiyo Seamount) to 91.9 years (Myojin Knoll) in Izu-Bonin. Thus, τ_i in Izu-Bonin is larger than Okinawa by more than one order of magnitude. Both self-recruitment and recruitment between HVFs are lower in Izu-Bonin than Okinawa. However, larval supply to HVFs in Izu-Bonin

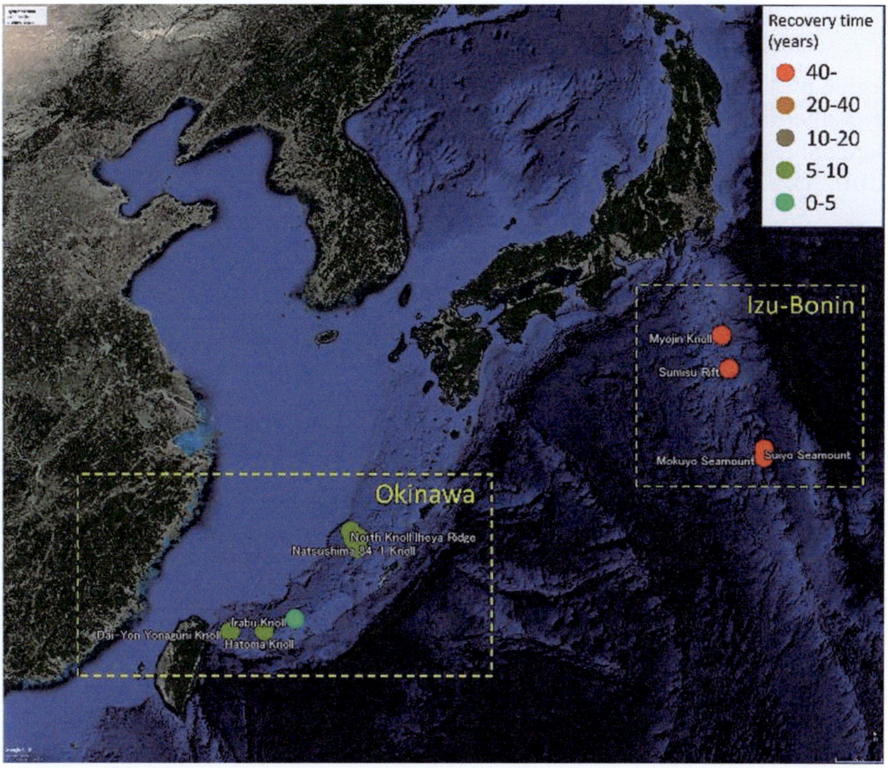

Fig. 2 Hydrothermal vent fields around Japan (The map was generated from digital information available at Google Earth Pro v7.3.0.3832 (https://www.google.com/intl/en/earth/; Map data: Google Earth, Image Landsat/Copernicus, Data SIO, NOAA, U.S. Navy, NGA, GEBCO))

Table 2 Recovery time of HVFs in Okinawa shown with other basic properties. Following Suzuki et al. (2018), we set r = 17.4 to calculate recovery time. Here, "In-degree" is the number of incoming links, "Self-recruitment" is A_{ii}, and "Recruitment from other HVFs" is the sum of $A_{ij}(j \neq i)$

HVFs	τ_i (years)	95% CI of τ_i	In-degree	Self-recruitment (larva/adults/year)	Recruitment from other HVFs (larva/adults/year)
Iheya Ridge	4	0.5 28.9	6	0.002281	0.013033
Irabu Knoll	3.9	0.4 18.9	2	0.006708	0.014852
Izena Cauldron	4.1	0.7 18.7	6	0.008017	0.010736
Natsushima 84-1 Knoll	3.5	0.4 19.1	6	0.003973	0.015163
North Knoll Iheya Ridge	4.3	0.5 32.4	6	0.001805	0.013233
Dai-Yon Yonaguni Knoll	6.2	0.8 30.1	3	0.004978	0.00826
Hatoma Knoll	4.25	0.5 18.6	3	0.0086	0.013045

region may be overestimated because we did not consider larval flow from seeps to hydrothermal vents. The frequency of species shared between vents and seeps is high around Japan where vents and seeps occur in close proximity (Watanabe et al. 2010; Kiel 2016).

Simultaneous Disturbance to HVFs

Ongoing development of deep-sea resource mining technologies raises the possibility that natural and anthropogenic factors will simultaneously disturb multiple HVFs in a region. We evaluated recovery time for disturbance on multiple HVFs as τ_C, where $C = \{X, Y, \ldots\}$ is a possible combination of disturbed HVFs in a region. We tested all possible combinations. The procedure to calculate recovery time is the same as previous analyses except population abundance in HVFs included in C were simultaneously reduced to zero. τ_C is the largest recovery time of HVFs included in C, i.e., $\tau_c = \max(\{\tau_i\}_{X \in C})$. For simplicity, we did not consider variation of K in this analysis and assumed $K = 1000$ for all HVFs.

In Fig. 3, τ_Cs is shown as a function of the number of simultaneously disturbed HVFs. The mean recovery time steadily increases with the number of disturbed HVFs. We found that some combinations substantially delay the recovery as compared to the others. For example, when more than three HVFs were disturbed in Okinawa, τ_C was larger than 20 years if Hatoma Knoll, Dai-Yon Yonaguni Knoll, and Irabu Knoll were included in the disturbance, whereas recovery times of other combinations were less than 10 years. In Fig. 3a, this observation is illustrated by the appearance of the first point showing a recovery time greater than 20 years when three HVFs are simultaneously disturbed (Fig. 3a). The dispersal matrix for this region (Table 1) shows that if all three HVFs are disturbed, recovery will depend on dispersal from Iheya Ridge to Hatoma Knoll and Dai-Yon Yonaguni Knoll, and dispersal via these links is more than ten times smaller than the mean dispersal between vents in this region (Table 1).

The combination of HVFs having longer recovery time is systematically found using the *disconnectivity graph* (Fig. 4). The *disconnectivity graph* is obtained by removing links between HVFs based on a threshold level of dispersal rate. After removing these links, dispersal network is divided into the subsets of HVFs that we regard as the *maximal subsets of connected components*. A subset of HVFs is a *maximal subset of connected components* if there are mutual paths between any pairs of HVFs in the subset and inclusion of additional HVFs results in a violation to this criterion. Any dispersal network will be expressed as the union of such subsets. For example, if A→B, B→C, and C→A, then the network is expressed as {{A,B,C}}, and if C→A is removed, then it becomes {{A},{B},{C}}. By changing the threshold

Table 3 Recovery time of HVFs in Izu-Bonin shown with other basic properties. Following Suzuki et al. (2018), we set $r = 17.4$ to calculate recovery time. Here, "In-degree" is the number of incoming links, "Self-recruitment" is A_{ii}, and "Recruitment from other HVFs" is the sum of $A_{ij}(j \neq i)$

HVFs	τ_i (years)	95% CI of τ_i	In-degree	Self-recruitment (larva/adults/year)	Recruitment from other HVFs (larva/adults/year)
Mokuyo Seamount	81.1	6.2 278.4	1	0.000794	0.00066
Myojin Knoll	91.45	11.5 353.8	3	0.000443	0.000517
Suiyo Seamount	71	6.4 317.1	2	0.000671	0.000823
Sumisu Rift	66.7	6.7 468.	3	0.000172	0.000795

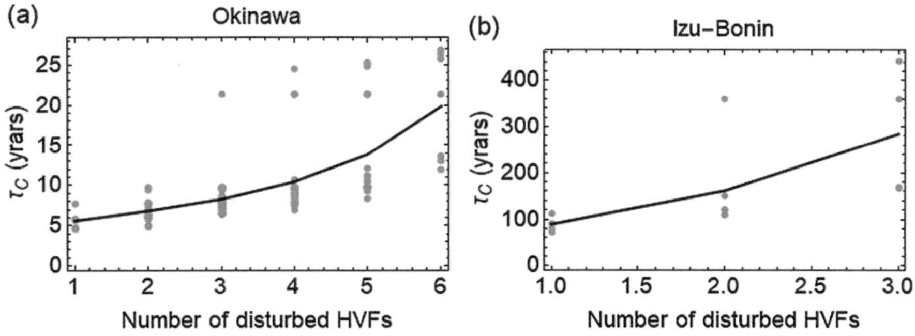

Fig. 3 Result of simultaneous disturbances to multiple vent fields for Okinawa (**a**) and Izu-Bonin (**b**). To calculate τ_C, we set $r = 17.4$. Points in figure show τ_C of a combination of HVFs, and lines indicate the mean

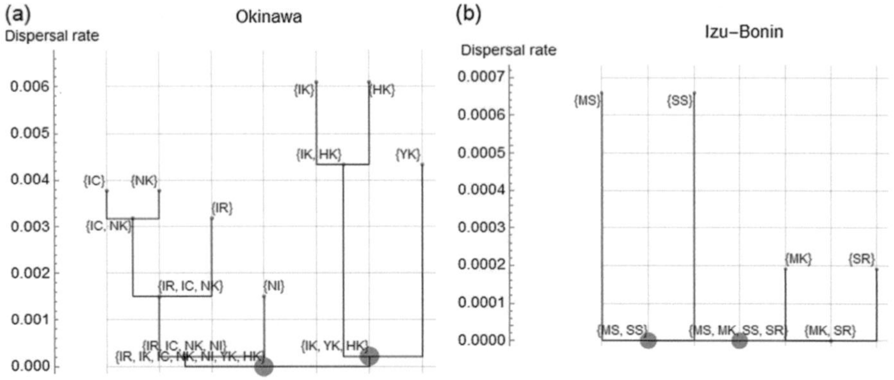

Fig. 4 Disconnectivity graph for Okinawa (**a**) and Izu-Bonin (**b**). Size of points indicates normalized recovery time of simultaneous disturbance to the corresponding combination of HVFs. Here, IR, Iheya Ridge; IK, Irabu Knoll; IC, Izena Cauldron; NK, North Knoll Iheya Ridge; NK, Natsushima 84-1 Knoll; YK, Dai-Yon Yonaguni Knoll; HK, Hatoma Knoll; MS, Mokuyo Seamount; MK, Myojin Knoll; SS, Suiyo Seamount; and SR, Sumisu Rift

level from zero to the maximal rate included in the dispersal matrix, *maximal subsets of connected components* are calculated at each step. The disconnectivity graph shows hierarchical relationships of the connected components between the two extremes above. In Okinawa, the HVFs are separated into two subgraphs: (1) Hatoma Knoll, Dai-Yon Yonaguni Knoll, and Irabu Knoll and (2) the others. At a small threshold level (0.00021 larva/adults/year), simultaneous disturbance to the former three HVFs resulted in a long recovery time. This indicates that larval supply from other HVFs to this subset of HVFs is limited. Similarly, in Izu-Bonin region, the disconnectivity graph shows that there is limited larval supply from Myojin Knoll and Sumisu Rift to Mokuyo Seamount and Suiyo Seamount. This approach would be especially helpful when many HVFs are included in the analysis.

Contribution of a HVF to the Resilience of Other HVFs

As we mentioned above, HVFs that are less resilient to disturbance will be an important target for protection. However, HVFs should also be protected if they have significant contribution to the resilience of other HVFs. Here, we show how the contribution of HVF i on the resilience of other HVFs (I_i) is quantified by a numerical experiment, before calculating recovery time of HVFs as we explained above, wherein we selected one of the HVFs and removed all the links from the HVF to other HVFs. We then calculated I_i as the mean increase of recovery time in HVFs in the region except for i. For simplicity, we did not consider variation of K in this analysis and assumed $K = 1000$ for all HVFs.

Figure 5 indicates the importance of HVFs in terms of their resilience (τ_i; x-axis) and contribution of HVFi to the resilience of other HVFs (I_i; y-axis). Although the results should be cautiously interpreted, HVFs having longer recovery time and higher I_i may be important targets for protection. One of the criteria is considering the sum of τ_i and I_i which is indicated by the vertical distance to a line $y(x) = -x + \frac{1}{n}\Sigma_n(\tau_i + I_i)/2$ (black lines in Fig. 5). We found that in Okinawa region, Dai-Yon Yonaguni Knoll would be the most important target for protection followed by Izena Cauldron. In Izu-Bonin region, Suiyo Seamount and Mokuyo Seamount were indicated to be important than Sumisu Rift and Myojin Knoll.

Optimal Scheduling

There were combinations of HVFs that significantly delay recovery, for example, Irabu Knoll, Dai-Yon Yonaguni Knoll, and Hatoma Knoll (here we call Irabu, Yonaguni, and Hatoma for simplicity) in Okinawa region. In the previous result, we assumed that these HVFs are simultaneously disturbed. However, if the causes of disturbances are anthropogenic factors, such as resource mining, it should be possible to schedule the sequence of mining and interval of disturbances. The recovery time would be minimized if the sequence and interval of disturbances is optimally scheduled. Here, using the disturbance to Irabu, Yonaguni, and Hatoma as an example, we show a procedure for optimal scheduling.

Let us define a schedule $\boldsymbol{T}$ as

$$T = \{T_{\text{Irabu}}, T_{\text{Yonaguni}}, T_{\text{Hatoma}}\},$$

where $Tx(X \in \{\text{Irabu, Yonaguni, Hatoma}\})$ specifies the timing at which X is disturbed. We assumed that at least one of the Tx is 0 which is the starting point of the disturbances. If Tx > 0, X is disturbed after Tx year from the first disturbance. We

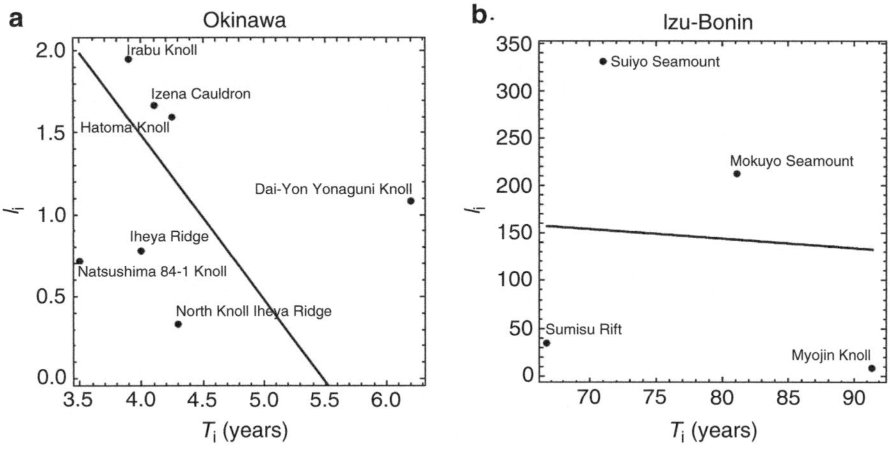

Fig. 5 Plot of τ_i and I_i for HVFs in Okinawa (**a**) and (**b**). Black line indicates $y(x) = -x + \frac{1}{n}\Sigma_n(\tau_i + I_i)/2$

defined that the recovery time is the time interval between the first disturbance and 75% recovery of the HVF that is disturbed last.

We used simulated annealing (Box 1; Kirkpatrick et al. 1983) to find the optimal schedule. Simulated annealing is a probabilistic technique for approximating the global optimum of a given function. Specifically, it is a metaheuristic approach to approximate global optimization in a large search space. In the algorithm, $p(t)$ is the time-dependent acceptance rate of a new schedule when it does not improve the fitting result. This acceptance of worse parameters helps the algorithm find global minima by searching globally at the early stage while locally at the later stage. A 3000 step SA ($t_{max} = 3000$) was sufficient to find the best $\boldsymbol{T}$.

Box 1 Simulated Annealing

1. Set iteration step $t = 0$, set the initial $\boldsymbol{T}$ by assigning 0 for one randomly selected element among the three elements. The list of elements is assigned a uniform distribution (0,20).
2. Calculate the recovery time τ_C with $\boldsymbol{T}$.
3. Increment t by 1.
4. Replace one of the elements in $\boldsymbol{T}$, Tx, by λTx where λ is a random value drawn from a uniform distribution (0.8,1.2), and set the alternative schedule as $\boldsymbol{T}^*$.
5. Calculate the recovery time τ_C^* with $\boldsymbol{T}^*$.
6. If $\tau_C^* < \tau_C$, or $\tau_C^* \geq \tau_C$ and **rand**$< p(t)$, set $\boldsymbol{T} = \boldsymbol{T}^*$ and $\tau_C = \tau_C^*$. Here, **rand** is a random value drawn from a uniform distribution (0,1). $p(t)$ is the acceptance rate of a new parameter vector when it does not improve the fitting result and is $p(t) = \exp\left(-\left(\tau_c^* - \tau_c\right) / \omega\alpha^t\right)$. Here, we set $\alpha = 0.998$ and $\omega = 1000$.
7. Back to 4 if $t < t_{max}$, else terminate the algorithm.

When the schedule is given as {1.4, 0, 4.5}, the recovery time (τ_C, C={Irabu, Hatoma, Yonaguni}) was reduced to 11.3 years, which is about one half the recovery rate when these vents are simultaneously disturbed (21.6 years, Fig. 6). Thus, if the set of targets for resource mining is given (e.g., based on the amount of mineral resources required), the scheduling algorithm could find an optimal schedule to minimize the recovery time.

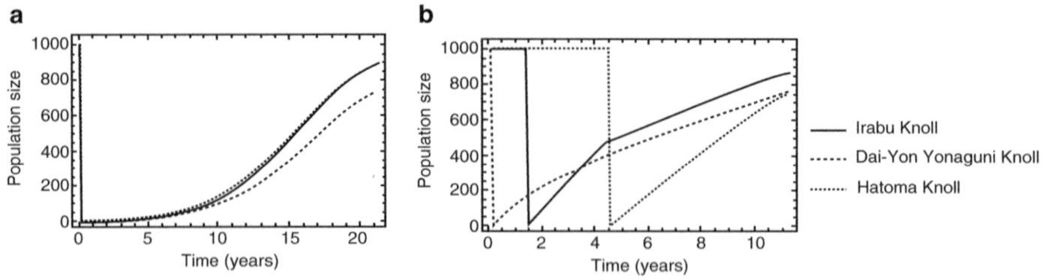

Fig. 6 Recovery of population size in the three HVFs with simultaneous disturbances (**a**) and scheduled disturbances (**b**)

Individual-Based Model

While differential Eqs. (1 and 2) include basic mechanisms that can explain recovery, it overlooks the fact that the processes are actually based on the dispersal of individuals. When describing dispersal processes using a differential equation, propagules are expressed as a small real number (e.g., 10^{-2} individuals) that can disperse to connected HVFs with an arbitrary short time interval. This may not be appropriate to describe dispersal if recovery is limited by a small dispersal rate. For example, arrival of 10^{-2} individuals/year to a HVF in a differential equation is interpreted as arrival of 1 individual per 100 years in an individual-based model. Estimated recoverability of a HVF may be different if using a differential equation relative to an individual-based model. To test the effect of individuality, we transformed the differential Eqs. (1 and 2) to an equivalent stochastic formulation using the Gillespie algorithm (Gillespie 1977). In this approach, the equation is decomposed into three processes weighted by w_1, w_2, and w_3 as follows: first, reproduction of organisms (self-recruitment) weighted by $w_1 = rA_{ii}x_i$ which adds one individual to the population; second, recruitment of individuals from other HVFs weighted by $w_2 = (1 - x_i/K_i)(+r\Sigma_{j \neq i}A_{ji}K_j)$ which also adds one individual to the population; and finally, mortality weighted by $w_3 = rA_{ii}x_i^2 / K_i$ which removes one individual from the population. At each time step, one of these processes is selected based on their weights, and time t is incremented by $\Delta t = \log(\epsilon^{-1})/\Sigma_i w_i$, where ϵ is a random value drawn from a uniform distribution (0,1).

Comparison of the distribution of recovery time calculated by the differential equation and the individual-based model (Fig. 7) shows that there are no significant

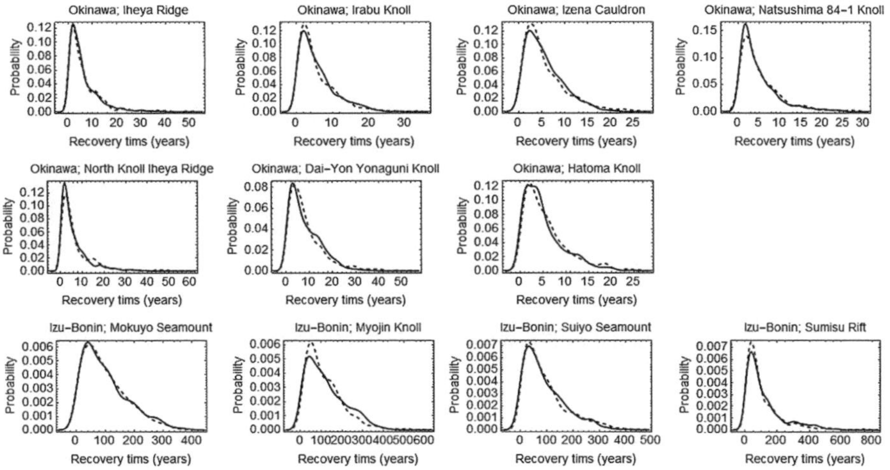

Fig. 7 Smooth kernel distribution of recovery time calculated for the ensemble of 1000 K values. Recovery time calculated from a stochastic model (solid) and a deterministic model (dashed) had no significant difference

differences between their distribution. This suggests that at least in Okinawa and Izu-Bonin region, the differential Eqs. (1 and 2) well approximate the individual-based population dynamics. Because most of the HVFs are mutually connected (Table 1), self-recruitment would be the determinant of the recovery time in these regions. The effect of individuality would be more significant if the connection between HVFs is sparse and the supply of individuals rather than self-recruitment is a determinant of recovery.

3 Future Directions

In future studies, researchers should attempt to reveal dispersal networks in a broader region of the world ocean, with a wider range of taxonomic and habitat types, which will extend the coverage of this analysis. This will allow us to map the resilience of different species and enable comparative studies on the relationship between resilience and species traits, and, when the model is modified to multispecies system, it will further help us to understand community dynamics, such as succession. However, presented results were obtained by omitting various biological details. Information about biological traits (e.g., larval development, ontogenetic vertical migration, and settling behaviors) is essential to accurately predict larval dispersal distance. Species-specific resilience should be predicted by implementing quantitative biological information. Furthermore, we did not consider the activity and configuration of individual vents in each HVF. It is important to integrate information regarding the detailed spatial structure and the predictions obtained by fine-scale physical modeling, such as the amount of sedimentation caused by resource mining, to accurately predict the relationship between disturbances and recovery time at more detailed spatial scales. It also should be mentioned that there is still a relative lack of empirical data from disturbance-recovery studies that can support our results. All previous studies have been carried out on either the Juan De Fuca Ridge (Tunnicliffe et al. 1997; Marcus et al. 2009) or the East Pacific Rise (Shank et al. 1998; Gollner et al. 2015). Because both are in a fast-spreading ridge where biological communities are frequently disturbed (e.g., ~15 years; Tolstoy et al. 2006), it would be controversial whether these results represent recovery of HVFs in slow-spreading ridges or arc-back-arc basins where disturbance to communities is relatively infrequent (Gollner et al. 2017). However, our results still suggest substantial differences in recovery time among HVFs, which can span two orders of magnitude, highlighting the importance of understanding connectivity among HVFs to assess their recoverability.

4 Conclusion

Limited accessibility to HVFs and rare opportunities to observe natural disturbances hinders the ability of researchers to assess the recoverability of chemosynthetic communities in the environment. With some numerical analyses, we here showed how our approach provides quantitative evaluations of potential impacts of disturbances on ecological systems by indicating regions or groups of HVFs as well as individual HVFs that are less resilient to disturbances. We also presented a methodology to optimize a schedule for resource development that can minimize recovery time of ecological communities. These results suggest potential applications of our approach to avoid or minimize the disruptive effect of anthropogenic activities in the deep-sea environment. However, more extensive efforts would be needed to use it for applied purposes. We hope that the benefit of integrating observational data and mathematical models is widely acknowledged and stimulates further development of methodology, observational technologies, and application of these studies.

Acknowledgments We would like to thank Satoshi Mitarai for providing published data sets. We also would like to thank Hiromi Watanabe, Hiroyuki Yamamoto, Masanobu Kawachi, Hiroshi Koshikawa, Hironori Higashi, and Hiroyuki Yokomizo for their support and Robert G. Jenkins for his guidance on vent ecosystems. This work was supported by Council for Science, Technology and Innovation (CSTI), Cross-ministerial Strategic Innovation Promotion Program(SIP), "Next-generation technology for ocean resources exploration."

References

Arellano, S. M., Van Gaest, A. L., Johnson, S. B., et al. (2014). Larvae from deep-sea methane seeps disperse in surface waters. *Proceedings of the Royal Society London B, 281*, 20133276.

Binns, R. A., & Scott, S. D. (1993). Actively forming polymetallic sulfide deposits associated with felsic volcanic rocks in the eastern Manus back-arc basin, Papua New Guinea. *Economic Geology, 88*, 2226–2236.

Boschen, R. E., Rowden, A. A., Clark, M. R., et al. (2013). Mining of deep-sea seafloor massive sulfides: A review of the deposits, their benthic communities, impacts from mining, regulatory frameworks and management strategies. *Ocean and Coastal Management, 84*, 54–67.

Boschen, R. E., Rowden, A. A., Clark, M. R., et al. (2016). Seafloor massive sulfide deposits support unique megafaunal assemblages: Implications for seabed mining and conservation. *Marine Environmental Research, 115*, 78–88.

Christensen, V., & Walters, C. J. (2004). Ecopath with Ecosim: Methods, capabilities and limitations. *Ecological Modelling, 172*, 109–139.

Christensen, V., & Walters, C. J. (2005). Using ecosystem modeling for fisheries management: Where are we. ICES CM/M: 19.

Coffey Natural Systems. (2008a). *Environmental impact statement* (Solwara 1 project. Nautilus Minerals Niugini Limited, Executive Summary). Brisbane: Coffey Natural Systems.

Coffey Natural Systems. (2008b). *Environmental impact statement* (Solwara 1 project. Nautilus Minerals Niugini Limited, Main Report). Brisbane: Coffey Natural Systems.

Collins, P. C., Croot, P., Carlsson, J., et al. (2013). A primer for the environmental impact assessment of mining at seafloor massive sulfide deposits. *Marine Policy, 42*, 198–209.

D'Arcy, R., Amend, J. P., & Osburn, M. R. (2013). Microbial diversity and potential for arsenic and iron biogeochemical cycling at an arsenic rich, shallow-sea hydrothermal vent (Tutum Bay, Papua New Guinea). *Chemical Geology, 348*, 37–47.

Desbruyères, D., Hashimoto, J., & Fabri, M.-C. (2006). Composition and biogeography of hydrothermal vent communities in western Pacific back-arc basins. In D. M. Christie, C. R. Fisher, S.-M. Lee, et al. (Eds.), *Back-arc spreading systems: Geological, biological, chemical, and physical interactions* (pp. 215–234). Washington, DC: American Geophysical Union.

England, M. H. (1992). On the formation of Antarctic intermediate and bottom water in ocean general circulation models. *Journal of Physical Oceanography, 22*, 918–926.

Gillespie, D. T. (1977). Exact stochastic simulation of coupled chemical reactions. *The Journal of Physical Chemistry, 81*, 2340–2361.

Gollner, S., Govenar, B., Arbizu, P. M., et al. (2015). Differences in recovery between deep-sea hydrothermal vent and vent-proximate communities after a volcanic eruption. *Deep Sea Research, Part I, 106*, 167–182.

Gollner, S., Kaiser, S., Menzel, L., et al. (2017). Resilience of benthic deep-sea fauna to mining activities. *Marine Environmental Research, 129*, 76–101.

Govenar, B. (2010). Shaping vent and seep communities: Habitat provision and modification by foundation species. In S. Kiel (Ed.), *The vent and seep biota* (pp. 403–432). Dordrecht: Springer.

Grassle, J. F. (1986). The ecology of deep-sea hydrothermal vent communities. *Advances in Marine Biology, 23*, 301e362.

Grassle, J. F., & Sanders, H. (1973). Life histories and the role of disturbance. *Deep Sea Research, 20*, 643e659.

Gupta, A. S., & England, M. H. (2007). Evaluation of interior circulation in a high-resolution global ocean model. Part II: Southern hemisphere intermediate, mode, and thermocline waters. *Journal of Physical Oceanography, 37*, 2612–2636.

Halbach, P., Nakamura, K. I., Wahsner, M., et al. (1989). Probable modern analogue of Kuroko-type massive sulphide deposits in the Okinawa Trough back-arc basin. *Nature, 338*, 496–499.

Herzig, P. M. (1999). Economic potential of sea–floor massive sulphide deposits: Ancient and modern. *Philosophical Transactions of the Royal Society London A, 357*, 861–875.

Hoagland, P., Beaulieu, S., & Tivey, M. A. (2010). Deep-sea mining of seafloor massive sulfides. *Marine Policy, 34*, 728–732.

Holling, C. S. (1973). Resilience and stability of ecological systems. *Annual Review of Ecology and Systematics, 4*, 1–23.

Holling, C. S. (1996). Engineering resilience versus ecological resilience. In National Academy of Engineering (Ed.), *Engineering within ecological constraints* (pp. 31–44). Washington, DC: National Academies Press.

Hurtado, L. A., Lutz, R. A., & Vrijenhoek, R. C. (2004). Distinct patterns of genetic differentiation among annelids of eastern Pacific hydrothermal vents. *Molecular Ecology, 13*, 2603–2615.

International Seabed Authority. (2007). Polymetallic sulphides and cobalt-rich ferromanganese crusts deposits: Establishment of environmental baselines and an associated monitoring programme during exploration, https://www.isa.org.jm/files/documents/EN/Workshops/2004/Proceedings-ae.pdf.

Ives, A. R., & Carpenter, S. R. (2007). Stability and diversity of ecosystems. *Science, 317*, 58–62.

Jeanthon, C. (2000). Molecular ecology of hydrothermal vent microbial communities. *Antonie Leeuwenhoek, 77*, 117–133.

Johnson, S. B., Warén, A., & Vrijenhoek, R. C. (2008). DNA barcoding of *Lepetodrilus* limpets reveals cryptic species. *Journal of Shellfish Research, 27*, 43–51.

Johnson, S. B., Warén, A., & Tunnicliffe, V. (2014). Molecular taxonomy and naming of five cryptic species of *Alviniconcha* snails (Gastropoda: Abyssochrysoidea) from hydrothermal vents. *Systematics and Biodiversity, 13*, 278–295.

Jannasch, H. W., & Wirsen, C. O. (1979). Chemosynthetic primary production at East Pacific sea floor spreading centers. *Bioscience, 29*, 592–598.

Kiel, S. (2016). A biogeographic network reveals evolutionary links between deep-sea hydrothermal vent and methane seep faunas. *Proceedings of the Royal Society B, 283*, 20162337.
Kirkpatrick, S., Gelatt, C. D., & Vecchi, M. P. (1983). Optimization by simulated annealing. *Science, 220*, 671–680.
Le, J. T., Levin, L. A., & Carson, R. T. (2016). Incorporating ecosystem services into environmental management of deep-seabed mining. *Deep Sea Research, Part II, 137*, 486–503.
Levin, L. A., Baco, A. R., Bowden, D. A., et al. (2016). Hydrothermal vents and methane seeps: Rethinking the sphere of influence. *Frontiers in Marine Science, 3*, 72.
Macdonald, K., Becker, K., Spiess, F. N., et al. (1980). Hydrothermal heat flux of the 'black smoker' vents on the East Pacific Rise. *Earth and Planetary Science Letters, 48*, 1e7.
Mahon, B. P., Bhatt, A., Vullo, D., et al. (2015). Exploration of anionic inhibition of the α-carbonic anhydrase from *Thiomicrospira crunogena* XCL-2 gammaproteobacterium: A potential biocatalytic agent for industrial CO2 removal. *Chemical Engineering Science, 138*, 575–580.
Marcus, J., Tunnicliffe, V., & Butterfield, D. A. (2009). Post-eruption succession of macrofaunal communities at diffuse flow hydrothermal vents on axial volcano, Juan de Fuca Ridge, Northeast Pacific. *Deep Sea Research, Part II, 56*, 1586–1598.
Marsh, A. G., Mullineaux, L. S., Young, C. M., et al. (2001). Larval dispersal potential of the tubeworm *Riftia pachyptila* at deep-sea hydrothermal vents. *Nature, 411*, 77–80.
Mitarai, S., Watanabe, H., Nakajima, Y., et al. (2016). Quantifying dispersal from hydrothermal vent fields in the western Pacific Ocean. *Proceedings of the National Academy of Sciences, 113*, 2976–2981.
Nakamura, H., Nishina, A., Liu, Z., et al. (2013). Intermediate and deep water formation in the Okinawa Trough. *Journal of Geophysical Research, Oceans, 118*, 6881–6893.
Nakamura, M., Watanabe, H., Sasaki, T., et al. (2014). Life history traits of *Lepetodrilusnux* in the Okinawa Trough, based upon gametogenesis, shell size, and genetic variability. *Marine Ecology Progress Series, 505*, 119–130.
O'Connor, M. I., Bruno, J. F., Gaines, S. D., et al. (2007). Temperature control of larval dispersal and the implications for marine ecology, evolution, and conservation. *Proceedings of the National Academy of Sciences, 104*, 1266–1271.
Podowski, E. L., Ma, S., Luther, G. W. I. I. I., et al. (2010). Biotic and abiotic factors affecting distributions of megafauna in diffuse flow on andesite and basalt along the Eastern Lau Spreading Center, Tonga. *Marine Ecology Progress Series, 418*, 25–45.
Portail, M., Olu, K., & Escobar-Briones, E. (2015). Comparative study of vent and seep macrofaunal communities in the Guaymas Basin. *Biogeosciences, 12*, 5455–5479.
Shank, T. M., Fornari, D. J., Von Damm, K. L., et al. (1998). Temporal and spatial patterns of biological community development at nascent deep-sea hydrothermal vents (9 50′ N, East Pacific Rise). *Deep Sea Research, Part II, 45*, 465–515.
Suzuki, K., Yoshida, K., Watanabe, H., et al. (2018). Mapping the resilience of chemosynthetic communities in hydrothermal vent fields. *Scientific Reports, 8*, 9364.
Talley, L. D. (2007). *Hydrographic atlas of the world ocean circulation experiment (WOCE): Volume 2: Pacific Ocean*. Southampton: WOCE International Project Office. 2007.
Terpe, K. (2013). Overview of thermostable DNA polymerases for classical PCR applications: From molecular and biochemical fundamentals to commercial systems. *Applied Microbiology and Biotechnology, 97*, 10243–10254.
Thaler, A. D., Zelnio, K., Saleu, W., et al. (2011). The spatial scale of genetic subdivision in populations of *Ifremeria nautilei*, a hydrothermal-vent gastropod from the Southwest Pacific. *BMC Evolutionary Biology, 11*, 372.
Thaler, A. D., Plouviez, S., Saleu, W., et al. (2014). Comparative population structure of two deep-sea hydrothermal- vent-associated decapods (*Chorocaris* sp. 2 and *Munidopsis lauensis*) from southwestern Pacific back-arc basins. *PLoS One, 9*, e101345.
Tolstoy, M., Cowen, J. P., Baker, E. T., et al. (2006). A sea-floor spreading event captured by seismometers. *Science, 314*, 1920–1922.
Tunnicliffe, V., Embley, R. W., Holden, J. F., et al. (1997). Biological colonization of new hydrothermal vents following an eruption on Juan de Fuca Ridge. *Deep Sea Research, Part I, 44*, 1627–1644.

Turnipseed, M., Knick, K. E., & Lipcius, R. N. (2003). Diversity in mussel beds at deep-sea hydrothermal vents and cold seeps. *Ecology Letters, 6*, 518–523.
Tyler, P. A., & Young, C. M. (2003). Dispersal at hydrothermal vents: A summary of recent progress. *Hydrobiologia, 503*, 9–19.
Urabe, T., Ura, T., Tsujimoto, T. et al. (2015). Next-generation technology for ocean resources exploration (Zipangu-in-the-Ocean) project in Japan. In *Proceedings of OCEANS 2015-Genova, IEEE*, (pp. 1–5), https://ieeexplore.ieee.org/document/7271762.
Van Dover, C. L. (2010). Mining seafloor massive sulphides and biodiversity: What is at risk? *ICES Journal of Marine Science, 68*, 341–347.
Van Dover, C. L. (2014). Impacts of anthropogenic disturbances at deep-sea hydrothermal vent ecosystems: A review. *Marine Environmental Research, 102*, 59–72.
Van Dover, C. L., Aronson, J., Pendleton, L., et al. (2014). Ecological restoration in the deep sea: Desiderata. *Marine Policy, 44*, 98–106.
Villasante, S., Arreguín-Sánchez, F., Heymans, J. J., et al. (2016). Modelling marine ecosystems using the Ecopath with Ecosim food web approach: New insights to address complex dynamics after 30 years of developments. *Ecological Modelling, 331*, 1–4.
Watanabe, H., Tsuchida, S., Fujikura, K., et al. (2005). Population history associated with hydrothermal vent activity inferred from genetic structure of neoverrucid barnacles around Japan. *Marine Ecology Progress Series, 288*, 233–240.
Watanabe, H., Fujikura, K., Kojima, S., et al. (2010). Japan: Vents and seeps in close proximity. In S. Kiel (Ed.), *The vent and seep biota* (pp. 379–402). Dordrecht: Springer.
Wathern, P. (Ed.). (2013). *Environmental impact assessment: Theory and practice*. Oxford: Routledge.
Won, Y., Young, C. R., Lutz, R. A., et al. (2003). Dispersal barriers and isolation among deep-sea mussel populations (Mytilidae: *Bathymodiolus*) from eastern Pacific hydrothermal vents. *Molecular Ecology, 12*, 169–184.
Yahagi, T., Watanabe, H., Ishibashi, J., et al. (2015). Genetic population structure of four hydrothermal vent shrimp species (Alvinocarididae) in the Okinawa Trough, Northwest Pacific. *Marine Ecology Progress Series, 529*, 159–169.
Yahagi, T., Kayama, H., Watanabe, H., et al. (2017). Do larvae from deep-sea hydrothermal vents disperse in surface waters? *Ecology, 98*, 1524–1534.

Kenta Suzuki is an theoretical biologist at the National Institute for Environmental Studies (NIES) specialized in ecological modeling and theoretical ecology. He has published academic papers and consulted for problems/techniques as diverse as mathematical modeling, numerical analysis of mathematical models, and data analysis using both mechanistic and statistical models. He engages in a project "Next-generation technology for ocean resources exploration," supported by Council for Science, Technology and Innovation (CSTI) cross-ministerial Strategic Innovation Promotion Program (SIP), the Cabinet Office of Japan. He received a Ph.D. from the University of Tokyo in 2010.

Katsuhiko Yoshida is a theoretical ecologist at the National Institute for Environmental Studies (Japan). He is mainly researching on the response of ecosystems to environmental disturbances by means of computer simulations. He is also participating in several conservation research projects.

Ecotoxicological Bioassay Using Marine Algae for Deep-Sea Mining

Takahiro Yamagishi, Shuhei Ota, Haruyo Yamaguchi, Hiroshi Koshikawa, Norihisa Tatarazako, Hiroshi Yamamoto, and Masanobu Kawachi

Abstract A new bioassay method using delayed fluorescence (DF) intensity in marine cyanobacterium has been developed. This method offers several advantages for marine environmental risk assessment in deep-sea mining areas: DF-based bioassay uses smaller amounts of a test substance or wastewater and takes less time and space than the standard bioassay method. We selected the marine cyanobacterium *Cyanobium* sp. (NIES-981) as our test algal species and demonstrated that use of this species was valid in standard growth inhibition testing based on OECD guideline criteria. Standard inhibition tests and shorter testing using DF were performed on NIES-981 by using five chemicals (3,5-DCP, simazine, diflufenican, $K_2Cr_2O_7$, and $CuSO_4$), and their EC_{50} and low-toxic-effect values (EC_{10}, EC_5, and NOEC) were determined from dose-response curves. On the basis of comparisons of the two dose-response curves and the EC_{50} values, we concluded that DF intensity was useful as an endpoint for rapid estimation of EC_{50} in NIES-981. In addition, a delayed fluorescence-based bioassay using *Cyanobium* sp. NIES-981 was used to evaluate the toxicity of core samples obtained from drill holes at the Izena Hole, Middle Okinawa Trough, East China Sea. The results revealed that unexpected leakage of recovered minerals and mining wastewater from the mining plant could result in heavy metal contamination of the surface water. Moreover, on the basis of the results

T. Yamagishi (✉) · H. Yamamoto
Center for Health and Environmental Risk Research, National Institute for Environmental Studies, Tsukuba, Ibaraki, Japan
e-mail: yamagishi.takahiro@nies.go.jp

S. Ota · H. Yamaguchi · M. Kawachi
Center for Environmental Biology and Ecosystem Studies, National Institute for Environmental Studies, Tsukuba, Ibaraki, Japan

H. Koshikawa
Center for Regional Environmental Research, National Institute for Environmental Studies, Tsukuba, Ibaraki, Japan

N. Tatarazako
Center for Health and Environmental Risk Research, National Institute for Environmental Studies, Tsukuba, Ibaraki, Japan

Faculty of Agriculture, Ehime University, Matsuyama, Ehime, Japan

R. Sharma (ed.), *Environmental Issues of Deep-Sea Mining*,
https://doi.org/10.1007/978-3-030-12696-4_10

of microplate-based assay using various marine algae, we suggest using eukaryotic marine algae such as *Emiliania huxleyi* NIES-1310, *Micromonas pusilla* NIES-1411, and *Bathycoccus prasinos* NIES-2670 in addition to *Cyanobium* sp. NIES-981 for management of seawater quality at deep-sea mining sites because sensitivity to lead in eukaryotic marine algae are more sensitive than cyanobacteria.

Keywords Bioassay · Marine algae · Deep-sea mining

1 Introduction

Commercial use of metals such as copper, nickel, and cobalt has increased markedly in recent years, and this has resulted in increased interest in deep-sea mining of mineral resources. Recent deep-sea surveys have identified rich and massive seafloor sulfide deposits associated with hydrothermal vents; these deposits may be economically viable, given sufficient development of the deep-sea mining industry (Fujita 2001). Several studies have examined the impacts of human activities on the deep-sea environment (Ahnert and Borowski 2000; Thiel 2003; Smith et al. 2008); however, little is known about the direct or long-term effects of human activities on the upper ocean zones, although heavy metals such as copper, lead, and zinc can easily leach from sulfide minerals around hydrothermal fields and may pollute the upper ocean layers (Simpson et al. 2007). Fuchida et al. (2017) have also suggested that there was a risk of unexpected leakage of recovered minerals and mining wastewater from mining plants, potentially resulting in heavy metal contamination of surface waters. In addition, the International Seabed Authority states that environmental impact assessments should address not only areas directly affected by mining but also the wider region affected by discharged plumes and materials released during mineral transport to the surface (ISA/LTC 2013). Considering the above, methods for assessing the effects of metals released by deep-sea mining operations on the ocean environment are urgently required.

Bioassays are commonly used to monitor water quality and assess the ecological risks posed by chemicals. Chemical analyses are also important for identifying the chemicals present in a particular environment, but they do not provide data directly related to the bioavailability of these chemicals. Ecotoxicological testing protocols using a variety of aquatic organisms have been published by organizations such as the International Organization for Standardization (ISO), the Organisation for Economic Co-operation and Development (OECD), and the US Environmental Protection Agency (US EPA); but time and space onboard vessels are limited in deep-sea mining operations, making these bioassays often unsuitable when animals such as fish are used.

In contrast, algal bioassay requires only simple handling and needs less time and space than bioassays using animals. A general method for testing the toxicity of chemicals to algae is described in OECD test guideline No. 201 (TG201; Freshwater Algal Growth Inhibition Test) (OECD 2011), in which growth inhibition 72 h after exposure is used to estimate the effective concentration (i.e., EC_{50}) of a chemical.

However, this method requires the counting of cells every 24 h for 72 h of exposure or longer, using an electronic particle counter or under a microscope. It is thus still too intricate and time-consuming for onboard work.

2 Test Algal Species

Currently, large phylogenetic groups, including some classes of algae, are used for bioassays of toxicity. For example, two species of green algae (*Pseudokirchneriella subcapitata* and *Desmodesmus subspicatus*), one species of diatom (*Navicula pelliculosa*), and two species of cyanobacteria (*Anabaena flos-aquae* and *Synechococcus leopoliensis*) are test species recommended in TG201 (OECD 2011). However, there are no tests using marine algal species in this guideline. The marine diatoms *Phaeodactylum tricornutum* and *Skeletonema costatum* have been recommended as test species in ISO 10253 (ISO 2006) and USEPA 850.5400 (USEPA 1996) and by the American Society for Testing and Materials (ASTM E1218-97a; ASTM 1997). These are the most common marine species used in bioassays of toxicity. Although bioassays using other marine microalgal species have also been reported, as yet there are no standard protocols for marine algae other than *P. tricornutum* and *S. costatum*. Because different taxa respond differently to chemical toxicants, it is important to develop standard protocols for a wide range of marine algal species, aside from diatoms, for marine environmental risk assessment. In addition, we need to develop methods for using new marine algal species to monitor deep-sea mining, because it is difficult to handle the culture of *P. tricornutum* and *S. costatum* onboard ships. The use of cryopreserved species may simplify the handling of algal cultures.

Recently, we developed bioassay method using a marine alga, the cyanobacterium *Cyanobium* sp. (NIES-981), obtained from microbial collection at the National Institute for Environmental Studies (Fig. 1) with the aim of management of seawater quality at deep-sea mining sites, where time and space are extremely limited

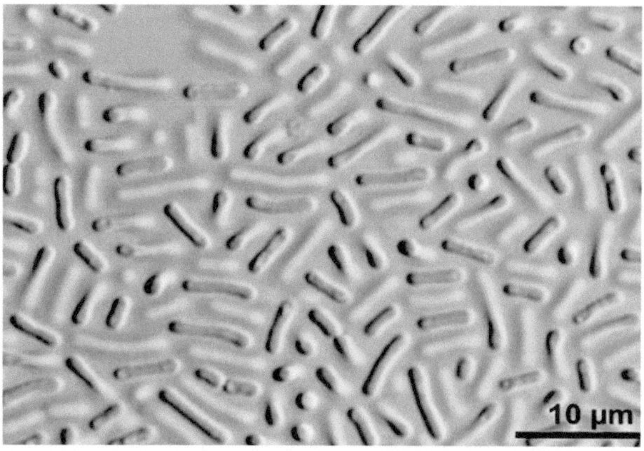

Fig. 1 Light microscopic image of *Cyanobium* sp. (NIES-981)

(Yamagishi et al. 2016). NIES-981 is very useful as a test algal species and is appropriate for use in deep-sea mining areas for the following reasons: (I) NIES-981 exhibits rapid and stable growth; (II) cryopreservation is available; (III) NIES-981 is closely related to *Synechococcus* and *Prochlorococcus* species, which are among the major primary producers both offshore and close to the coast; and (VI) the complete genome of NIES-981, which can provide a basis for developing an ecotoxicological bioassay using this strain, has been sequenced (Yamaguchi et al. 2016). Annotation revealed that the genome codes 3268 proteins with 46 tRNA genes and 3 sets of rRNA genes.

Rapid and stable growth is one of the most important requirements of a test algal species. NIES-981 grows exponentially by a factor of approximately 40 within 4 days under test conditions, namely, at 23 ± 1 °C under continuous illumination with white fluorescent light (60–80 $\mu E \cdot m^{-2} s^{-1}$) with orbital agitation (100 rpm) in ASW-SN (Table 1), a medium exclusively for NIES-981 (Fig. 2); this corresponds to a natural logarithmic growth rate of 1.16 day^{-1} (Fig. 2). The mean coefficient of variation (CV) for section-by-section growth rates (0–24, 24–48, and 48–72 h, for 72-h tests) in culture was 14.6% (Yamagishi et al. 2016). The CV of the average specific growth rates during the whole test period in replicate cultures was 1.6%. These values fulfill the validity requirements for test algal species as defined by the OECD's TG201, indicating that NIES-981 exhibits fast and stable growth.

The sensitivity of NIES-981 to typical toxicants (3,5-DCP, $Cu^{2+}/CuSO_4 \cdot 5H_2O$, Cr(VI)/$K_2Cr_2O_7$) and two herbicides, namely, CAT (simazine) and DFF (diflufenican), with the standard growth inhibition test (72 h) is sufficiently high for it to be used as a test algal species. Its sensitivities to the toxicants 3,5-DCP and CAT are as high as those of *P. subcapitata* (Table 2).On the other hand, its sensitivities to Cu^{2+} and Cr(VI) are markedly lower than those of *P. subcapitata* (Table 1), but marine algae are generally more tolerant to heavy metals than are freshwater algae: the EC_{50}ofCu^{2+}isapproximately 700–4000 times greater in marine algae (Vignati et al. 2010; Ebenezer and Ki 2013). Nevertheless, although there are limited data on the toxicity of heavy metals to marine algae, the EC_{50} values of Cu^{2+}in NIES-981 are markedly lower than the values previously reported for Cu^{2+} in marine algae, with an EC_{50} approximately 25 times lower than that of *Prorocentrum minimum*, 73 times lower than that of *Tetraselmis suecica*, and 13 times lower than that of *Heterocapsa triquetra* (Millán de Kuhn et al. 2006; Ebenezer and Ki 2013). These results indicate that NIES-981 is a useful species for algal ecotoxicity in marine environment.

3 Rapid Bioassay Method Using Delayed Fluorescence

A much faster alternative for testing algal growth is the noninvasive measurement of in vivo fluorescence, which facilitates the monitoring of activity changes by affecting photosystem II (PS II) (Thompson 1997). For example, pulse amplitude modulation of prompt fluorescence (PF) is a well-established method for detecting the

Table 1 Reagents for ASW-SN

ASW-SN		Nutrients stock solution		Trace metal stock solution		Tris stock solution	
NaCl	25.0 g	$NaNO_3$	75 g	$Na_2EDTA \cdot 2(H_2O)$	580 mg	Tris	100 g
$MgCl_2 \cdot 6(H_2O)$	2.0 g	$K_2HPO_4 \cdot 3H_2O$	3.0 g	$FeCl_3 \cdot 6(H_2O)$	422 mg	Deionised water	1000 ml
KCl	0.5 g	Deionised water	1000 ml	$ZnSO_4 \cdot 7(H_2O)$	2.93 mg		
$CaCl_2 \cdot 2(H_2O)$	0.5 g			$CoCl_2 \cdot 6(H_2O)$	1.33 mg		
$MgSO_4 \cdot 7(H_2O)$	3.5 g			$MnCl_2 \cdot 4(H_2O)$	24.0 mg		
Nutrients stock solution	10 ml			Na_2SeO_3	2.30 mg		
Trace metal stock solution	100 µl			$Na_2MoO_4 \cdot 2(H_2O)$	0.839 mg		
Tris stock solution	10 ml			$NiCl_2 \cdot 6(H_2O)$	0.37 mg		
Deionised water	1000 ml			Deionised water	100 ml		
pH	8.2					pH	8.2

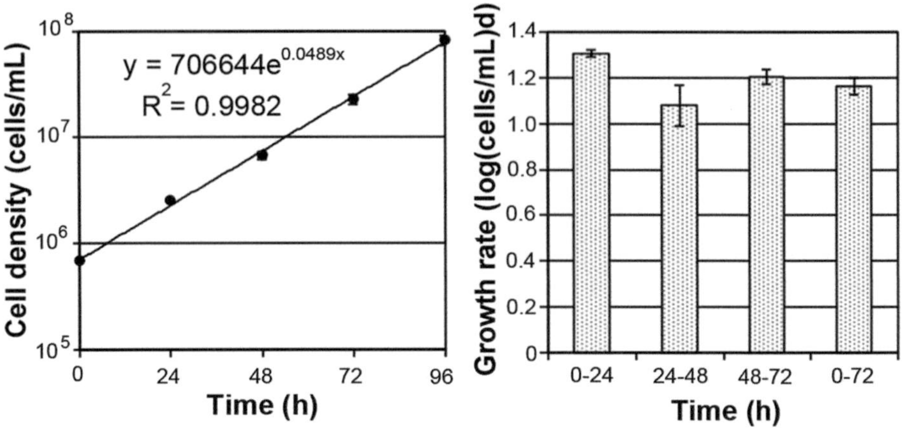

Fig. 2 Growth rates of *Cyanobium* sp. (NIES-981) for 96 h after culture under control conditions (left) and section-by-section-specific growth rates (0–24, 24–48, and 48–72 h) (right). Data are means of three independent experiments. Error bars show standard deviation

Table 2 Comparisons of EC_{50} and low-toxic-effect values (EC_{10}, EC_5, and NOEC) between *Cyanobium* sp. and *Pseudokirchneriella subcapitata*

	Cyanobium sp.				*P. subcapitata*		
	72-h EC_{50} (95% confidence limit)	72-h EC_{10} (95% confidence limit)	72-h EC_5 (95% confidence limit)	NOEC	72-h EC_{50}	72-h EC_{10}	NOEC
3,5-DCP (mg/L)	1.71	0.940	0.793	0.625	1.8–2.3[a, b]	0.91[b]	0.75[a]
	(1.60–1.82)	(0.934–0.947)	(0.534–0.896)				
CAT (μg/L)	105	26.8	18.2	12.5	100–297[c, d, e]	–	10[e]
	(94.7–118)	(24.2–28.4)	(15.3–20.3)				
DFF (μg/L)	4.10	1.26	0.902	0.750	0.27–1.23[f, g]	–	–
	(3.70–4.62)	(1.23–1.28)	(0.733–1.25)				
Cu^{2+} (mg/L)	0.550	0.139	0.0947	0.113	0.0075–0.047[h, i]	0.012[i]	0.0018–0.01[h, i]
	(0.475–0.632)	(0.0793–0.203)	(0.045–0.135)				
Cr(VI) (mg/L)	4.61	2.19	1.77	1.76	0.30–0.488[a, b]	0.092[b]	0.064[a]
	(4.24–4.96)	(2.06–2.23)	(1.27–2.13)				

EC effective concentration, *NOEC* no observed effect concentration, *3,5-DCP* 3,5-dichlorophenol, *CAT* simazine, *DFF* diflufenican

[a]Comber et al. (1995)
[b]Mayer et al. (1998)
[c]Okamura et al. (2000)
[d]Pérez et al. (2011)
[e]Sbrilli et al. (2005)
[f]Katsumata et al. (2009)
[g]Weyman et al. (2012)
[h]Franklin et al. (2002)
[i]Radix et al. (2000)

effects of chemicals on algae (Conrad et al. 1993; Schreiber et al. 2002; Escher et al. 2008; Vallotton et al. 2008). However, although PF is very sensitive to the inhibition of primary photosynthetic reactions, experiments on duckweed indicate that it has limitations when used to estimate growth inhibition due to the effects of copper, cadmium, and zinc (Drinovec et al. 2004). Delayed fluorescence (DF) is the very weak fluorescence signal from photosynthetically active cells that are transferred from light to dark; it occurs with a time delay from milliseconds to minutes and is also known as delayed luminescence or delayed light emission (Strehler and Arnold 1951; Arnold and Davidson 1954). This long emission time is explained by repopulation of the excited states of chlorophyll from stored energy after charge separation – that is, the back-reaction of accumulated charges across the thylakoid membrane in the electron transport chain (Joliot et al. 1971). On the basis of the mechanism of DF emission, it has been suggested that DF is an indicator of the state of electron transfer within the photosynthetic apparatus and is a sensitive intrinsic probe of photosynthetic activity (Jursinic 1986; Schmidt and Senger 1987a, b). Its use has thus been proposed to achieve rapid bioassay for screening for chemical toxicity in algae and plants (Gerhardt and Kretsch 1989; Katsumata et al. 2006, 2017; Berden-Zrimec et al. 2010).

However, it should be noted that there is an important difference between 72-h algal growth inhibition tests and tests with a shorter time period using PF or DF. Although 72-h growth inhibition tests involve multiple cell generations, tests that use shorter periods of time determine only the effects of the chemicals on one generation. Nevertheless, Katsumata et al. (2006, 2009) have demonstrated that the relationship between a test with a shorter time period using DF and a 72-h growth inhibition test was sufficient for rough toxicity testing of at least six test chemicals. They concluded that the DF intensity 24 h after exposure was a possible endpoint for rapid estimation of the EC_{50} values obtained with the conventional 72-h growth inhibition test (Katsumata et al. 2009).

We adapted a DF-based bioassay method for use in the marine cyanobacterium *Cyanobium* sp. NIES-981 (Yamagishi et al. 2016). The results for five typical test chemicals using this method approximated those from the standard growth inhibition test, despite some variations between the standard inhibition test and the DF-based bioassay in the case of DFF and CAT (Figs. 3 and 4) (Yamagishi et al. 2016). These results suggest that the DF-based method is a viable alternative to the standard growth inhibition test (Yamagishi et al. 2016). In addition, because this method uses smaller amounts of test substances or wastewater and takes less time than the standard bioassay method (Fig. 5), it may be useful for testing on onboard deep-sea mining support vessels, where time and space are limited, for assessing environmental risk. Recently, we performed a DF-based onboard bioassay using a cryopreserved culture of the *Cyanobium* sp. NIES-981 to evaluate the toxicity of three sulfide core samples obtained from three drill holes at the Izena Hole, Middle Okinawa Trough, East China Sea, during a drilling program conducted by the D/V *Chikyu* (CK16-05 cruise or Exp. 909: 11 December 2016 to 17 November 2016) (Kumagai et al. 2017). Leachates from two of the cores (C9027B 1X-CC and C9028A 7S-CC) contained high concentrations of zinc and lead, and they markedly

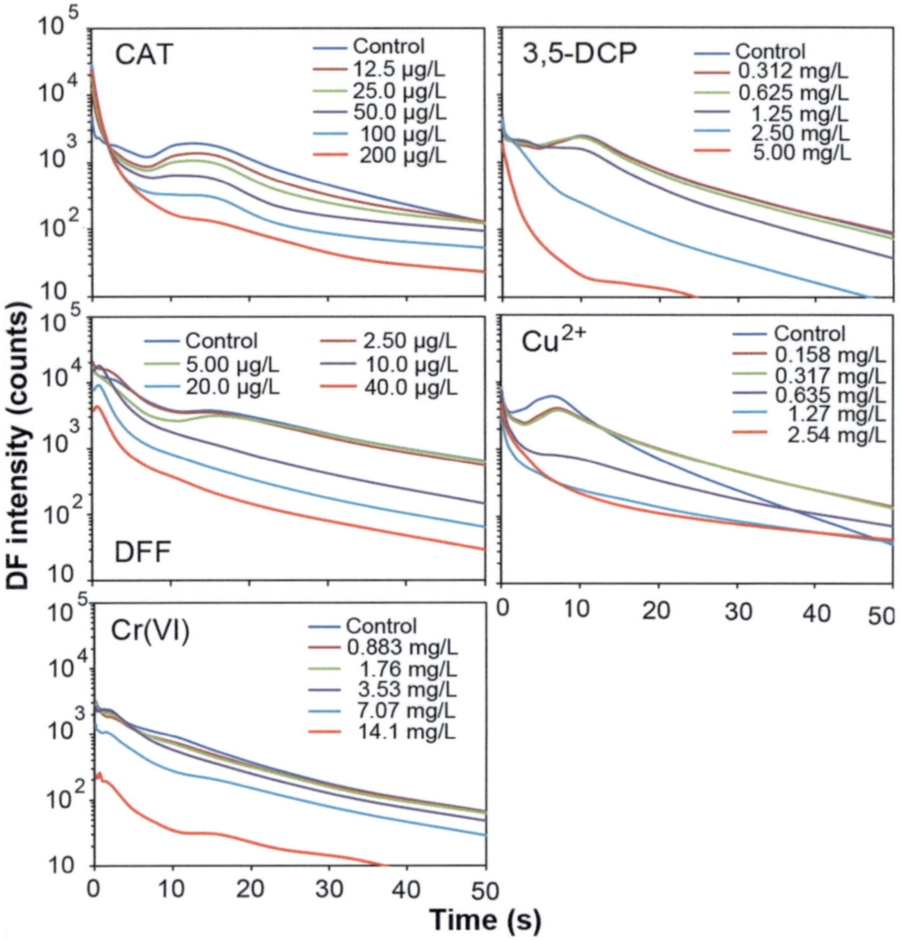

Fig. 3 Delayed fluorescence (DF) intensity curves in *Cyanobium* sp. after 24 h exposure to CAT, 3,5-DCP, DFF, Cu^{2+}, or Cr(VI)

inhibited DF in NIES-981 (Fig. 5 and Table 3). This result suggests the importance of examining the leaching of metals from sulfide-rich rock collected during deep-sea mining and establishing methods to determine the effects of this leaching on the upper ocean zones. During deep-sea mining operations, it is likely that metals will leach from sulfide minerals into the surrounding marine environment. It is also possible that unforeseen events will result in the spillage of sulfide-rich rock into the ocean, where long-term metal leaching could increase the impacts on the environment.

Our findings demonstrate that the DF-based bioassay using NIES-981 is sufficiently simple and easy for onboard bioassays. On the other hand, although the use of cryopreserved culture simplifies the handling of algal cultures, the results show

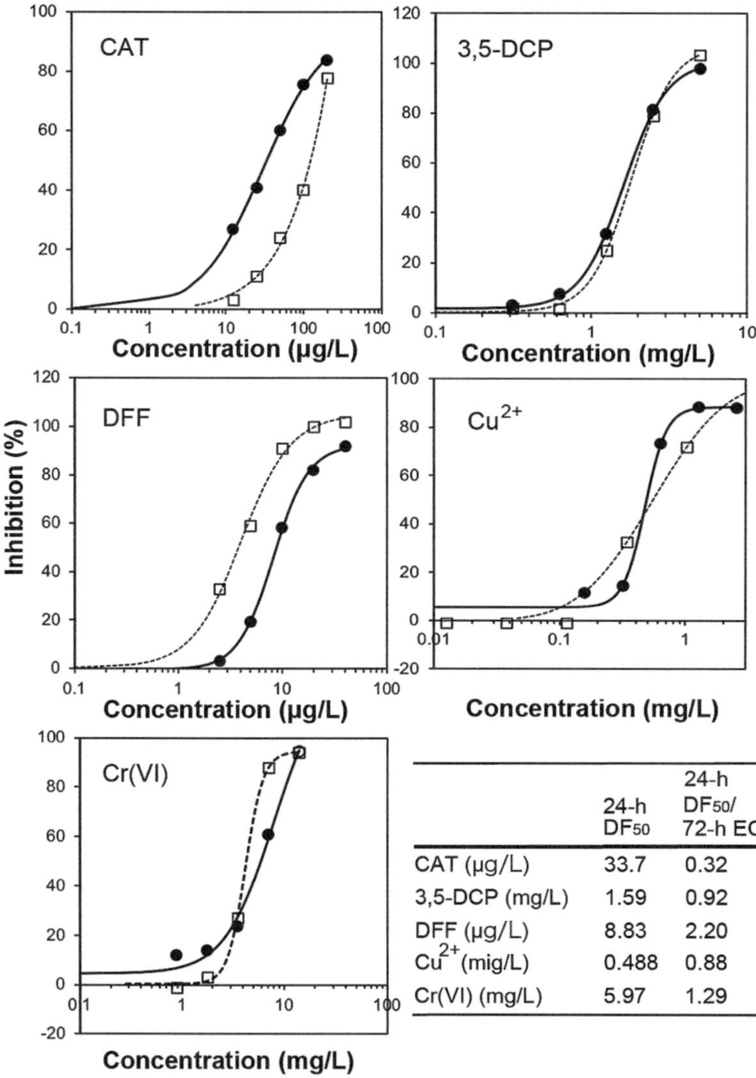

	24-h DF50	24-h DF50/ 72-h EC50
CAT (μg/L)	33.7	0.32
3,5-DCP (mg/L)	1.59	0.92
DFF (μg/L)	8.83	2.20
Cu^{2+}(mig/L)	0.488	0.88
Cr(VI) (mg/L)	5.97	1.29

Fig. 4 Dose–response curves of exposed samples relative to control samples in the standard growth inhibition test (72-h growth inhibition test) (dotted line) and the test using a shorter time period and delayed fluorescence (DF) intensity at 24 h (solid line) [In the case of DFF, the DF intensity at 48 h was used for the endpoint because saturation for the calculation of EC_{50} value could not be reached within 24 h. The inset table compares EC_{50} values from the standard inhibition test (72-h EC_{50}) and the test with a 24-h time period using DF (24-h DF_{50})]

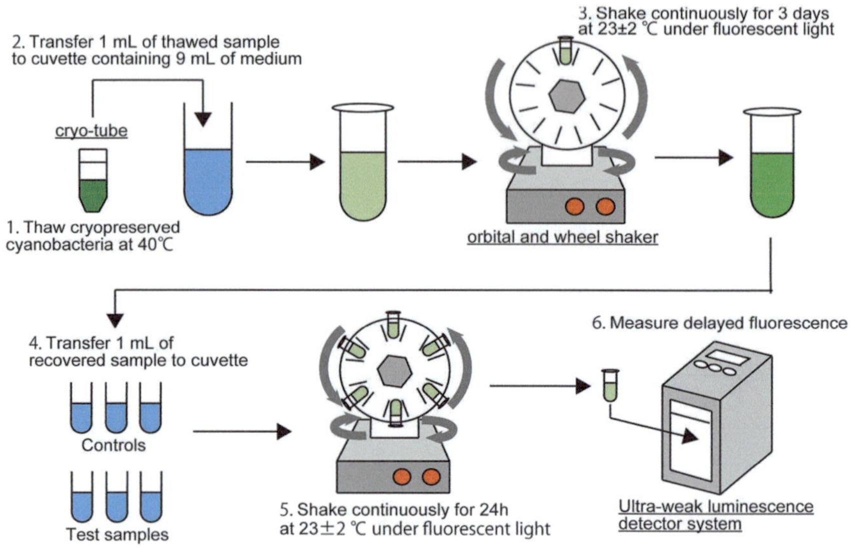

Fig. 5 Schematic of the procedure used in the onboard delayed fluorescence (DF)-based bioassay. 1. Thaw vials of the cryopreserved cyanobacteria with the thawing device (Allutox Starter: Hamamatsu Photonics, K.K., Shizuoka, Japan) at 40 °C until the cell suspension thaws completely. 2. Transfer thawed sample to cuvette (φ 25 × 85 mm) (Hamamatsu Photonics) containing 9.0 mL of fresh culture medium. 3. Shake continuously by an orbital and wheel shaker (Hamamatsu Photonics) for 3 days at 23 ± 1 °C under white fluorescent light (10–20 μE $m^{-2}S^{-1}$) to promote recovery from freezing damage. 4. Transfer recovered algal sample to cuvette (φ 25 × 85 mm) (Hamamatsu Photonics) containing 9.0 mL of control or test medium. 5. Shake continuously with an orbital and wheel shaker (Hamamatsu Photonics) for 24 h at 23 ± 2 °C under white fluorescent light (60–80 μE $m^{-2}S^{-1}$). 6. Measure delayed fluorescence light of test samples with ultra-weak luminescence detector system (Type 7100 or 7600, Hamamatsu Photonics)

Table 3 Effective concentration (EC_χ) values of the core leaches in cryopreserved and non-cryopreserved *Cyanobium* sp. NIES-981

	Cryopreserved		Non-cryopreserved	
Leaches	EC_{50}	EC_{10}	EC_{50}	EC_{10}
C9027B 1X-CC	3.52	1.10	0.587	0.0891
	(3.48–3.56)	(1.08–1.13)	(0.585–0.589)	(0.0885–0.0898)
C9028A 7S-CC	2.73	0.649	0.143	0.00180
	(2.70–2.76)	(0.632–0.665)	(0.141–0.145)	(0.00175–0.0019)

The numbers in parentheses indicate 95% confidence limits

that the non-cryopreserved cyanobacteria were more sensitive than the cryopreserved cyanobacteria to the toxicity of the core leaches for the reason of freezing damage during cryopreservation (Fig. 6) (Table 3). Although this issue could be addressed by increasing the time allotted for the culture to recover after thawing, further examination of the cryopreservation conditions to minimize freezing damage is warranted.

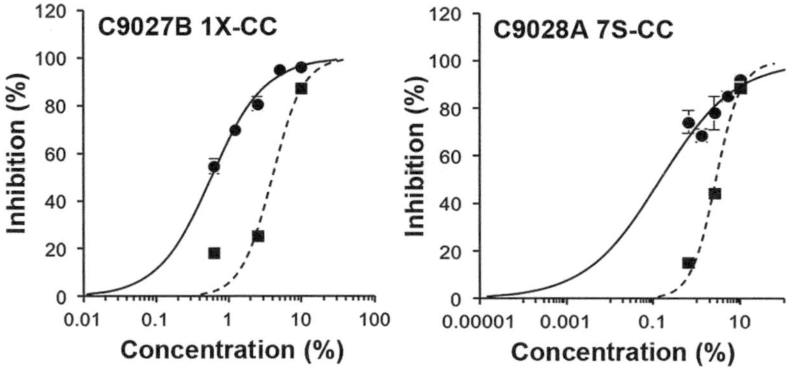

Fig. 6 Dose–response curves of cryopreserved or non-cryopreserved *Cyanobium* sp. exposed to core leaches [Cryopreserved or non-cryopreserved cyanobacteria were exposed to leach from C9027B 1X-CC and C9028A 7S-CC, and delayed fluorescence was measured at 24.Solid (closed circles) and dotted lines (closed squares) indicate curves for the assays using non-cryopreserved cyanobacteria and cryopreserved cyanobacteria, respectively.]

4 Impact of Heavy Metals on Marine Algal Species

It is expected that heavy metal species diffusing in the developmental area vary depending on the type of hydrothermal ore and internal ultrastructure and that the composition of leachates does not reflect the chemical composition of hydrothermal ores (Fuchida et al. 2017). Hence, it is necessary to evaluate the sensitivities to various metal species on NIES-981 for marine environment assessment at deep-sea mining area. The results of DF-based bioassay using various metal species indicate that the sensitivity of NIES-981 to zinc (EC_{50}: 71.1 ppb) is equivalent to that of *P. subcapitata* whereas greatly tolerant to lead (EC_{50}: 9.34 ppm) comparing to that of *P. subcapitata*. The high sensitivity of NIES-981 to zinc was also demonstrated by microplate-based bioassay using various marine algal species, 3 strains of marine cyanobacteria and 11 strains of eukaryotic algae (Table 4, Fig. 7). The growth of 3 strains of marine cyanobacteria (two *Synechococcus* and NIES-981) was suppressed with relative cell density of 0.2-fold or less by exposure to 0.4 ppm of zinc (Fig. 7). Although zinc more than 0.2 ppm strongly suppressed growth of three strains of marine cyanobacteria, NIES-981 was most sensitive strain to zinc among three strains of marine cyanobacteria (Fig. 7). On the contrary, growth in ten strains of eukaryotic marine algae was not affected by exposure to zinc more less 0.4 ppm (Fig. 7).Only *Micromonas pusilla* NIES-2672 was suppressed with relative cell density of 0.5 by exposure to 0.4 ppm of zinc (Fig. 7), suggesting that marine eukaryotic algae tend to be tolerant to zinc.

The DF assay using representative marine algal strains showed that eukaryotic marine algae generally more sensitive to lead than marine cyanobacteria including NIES-981 (Fig. 8). Three strains (NIES-1411, NIES-1310, and NIES-2670) and six

Table 4 List of strains tested in this study

NIES no.	Species	Class
323	*Skeletonemamarinoi-dohrnii*	Bacillariophyceae
613	*Amphidinium klebsii*	Dinophyceae
969	*Synechococcus* sp.	Cyanophyceae
971	*Synechococcus leopoliensis*	Cyanophyceae
981	*Cyanobium* sp.	Cyanophyceae
1000	*Gephyrocapsa oceanica*	Prymnesiophyceae
1001	*Imantonia rotunda*	Prymnesiophyceae
1310	*Emiliania huxleyi*	Prymnesiophyceae
1411	*Micromonas pusilla*	Prasinophyceae
2304	*Florenciella* sp.	Dictyochophyceae
2587	*Microchloropsis gaditana*	Eustigmatophyceae
2590	*Isochrysis galbana*	Prymnesiophyceae
2670	*Bathycoccus prasinos*	Prasinophyceae
2672	*Micromonas pusilla*	Prasinophyceae

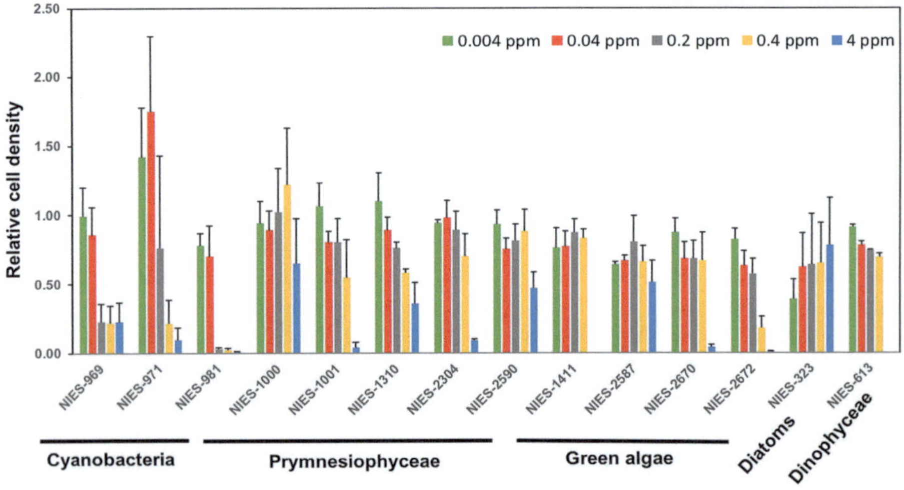

Fig. 7 Growth inhibition assay using various marine algae [mean + s.e.m. relative cell density (vs. non-exposure plot) for 3 marine cyanobacteria and 11 marine eukaryotic algae exposed to 0–4 ppm zinc (n = 3 independent biological replicates)]

strains (NIES-613, NIES-1000, NIES-1001, NIES-1310, NIES-1411, and NIES-2670) of eukaryotic marine algae showed relative DFI approximately of 0.2 and 0.5 by exposure to 1 ppm of lead, respectively (Fig. 8). On the contrary, three strains of cyanobacteria (Fig. 8) were more tolerant to lead than eukaryotic algae with relative DFI approximately 1.1 in NIES-981 and 0.5 in NIES-971 and NIES-969 by exposure to 1 ppm of lead (Fig. 8). Although the mechanism of lead tolerance in cyanobacteria remains unclear, recent study showed that the minimum inhibitory

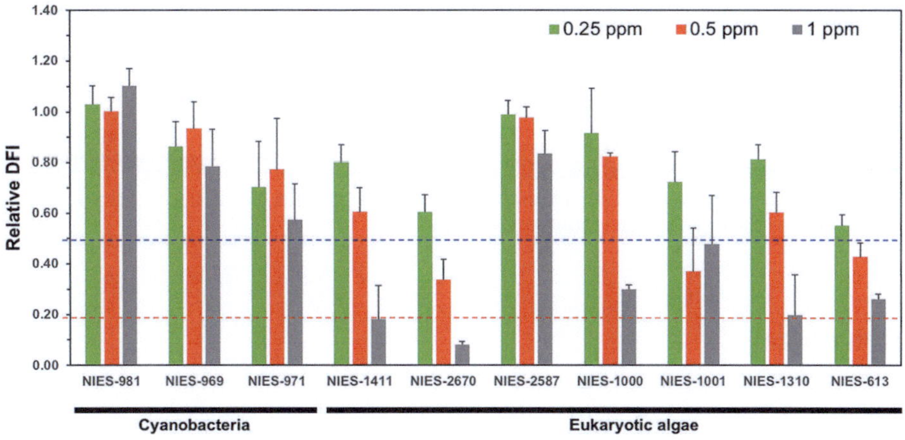

Fig. 8 DF-based assay using selected strains [mean + s.e.m. relative DFI (vs. non-exposure plot) for three marine cyanobacteria and seven marine eukaryotic algae exposed to 0.25–1 ppm lead. Blue and red bars indicate thresholds of DFI $\leqq 0.5$ and DFI $\leqq 0.2$, respectively ($n = 3$ independent biological replicates)]

concentration of zinc is lower than that of lead among different bacterial species (Perelomov et al. 2018), and this is consistent with our results. Taken together, the present study revealed that marine cyanobacteria tended to be more sensitive to zinc than to lead, while eukaryotic algae were more sensitive to lead than to zinc. Simultaneously, these results suggest that lead and zinc may widely impact prokaryotic and eukaryotic microbial communities in deep-sea mining area considering that lead and zinc easily leach from hydrothermal ores (Fuchida et al. 2017).As a corollary to our study, it is necessary to use a wide range of algal species (e.g., NIES-1310, NIES-1411, NIES-2670, as well as NIES-981) for marine environment assessment at deep-sea mining area.

5 Conclusion

In this study, we developed a new DF-based bioassay method using marine cyanobacterium, *Cyanbium* sp. (NIES-981). This method would be very useful for marine environment assessment at deep sea mining area, where time and space are limited, because it takes less time and it is easy to handle. The usefulness of this method was also evaluated with the onboard test performed during the D/V *Chikyu* (CK16-05 cruise). On the other hand, we found that cyanobacteria including NIES-981 are very sensitive to zinc but tolerant to lead. This result indicates that it is necessary to use some other species of alga in addition to NIES-981 for marine environment assessment. Screening test for selecting other useful algal species using microplate showed that some of eukaryotic algal species are more sensitive to lead than cyanobacteria. Therefore, we suggest using not only NIES-981 but also eukaryotic alga (such as *Emiliania huxleyi* NIES-1310, *Micromonas pusilla* NIES-1411, and *Bathycoccus prasinos* NIES-2670) for marine environment assessment at deep sea mining area.

Acknowledgments This work was supported in part by the Japanese Council for Science, Technology, and Innovation Cross-ministerial Strategic Innovation Promotion Program, "Next-generation Technology for Ocean Resources Exploration."

References

Ahnert, A., & Borowski, C. (2000). Environmental risk assessment of anthropogenic activity in the deep sea. *Journal of Aquatic Ecosystem Stress and Recovery, 7*, 299.

Arnold, W., & Davidson, J. B. (1954). The identity of the fluorescent and delayed light emission spectra in chlorella. *The Journal of General Physiology, 37*, 677–684.

ASTM. (1997). *Standard guide for conducting static 96-h toxicity tests with microalgae (ASTM E1218-97A)*. American Society for Testing and Materials; 1997.

Berden-Zrimec, M., Drinovec, L., & Zrimec, A. (2010). Delayed fluorescence. In D. J. Sugget, O. Prasil, & M. Borowitzka (Eds.), *Chlorophyll a fluorescence in aquatic sciences: Methods and applications* (pp. 293–309). Dordrecht: Springer.

Comber, M. H. I., Smyth, D. V., & Thompson, R. S. (1995). Assessment of the toxicity to algae of colored substances. *Bulletin of Environmental Contamination and Toxicology, 55*, 922–928.

Conrad, R., Büchel, C., Wilhelm, C., Arsalane, W., Berkaloff, C., et al. (1993). Changes in yield of in-vivo fluorescence of chlorophyll a as a tool for selective herbicide monitoring. *Journal of Applied Phycology, 5*, 505–516.

Drinovec, L., Drobne, D., Jerman, I., & Zrimec, A. (2004). Delayed fluorescence of *Lemna minor*: A biomarker of the effects of copper, cadmium, and zinc. *Bulletin of Environmental Contamination and Toxicology, 72*, 896–902.

Ebenezer, V., & Ki, J. (2013). Quantification of the sub-toxicity of metals and endocrine-disrupting chemicals to the marine green microalga *Tetraselmissuecica. Fisheries and Aquatic Sciences, 16*, 187–194.

Escher, B. I., Bramaz, N., Quayle, P., Rutishauser, S., & Vermeirssen, E. L. M. (2008). Monitoring of the ecotoxicological hazard potential by polar organic micropolutants in sewage treatment plants and surface waters using a mode-of-action based test battery. *Journal of Environmental Monitoring, 10*, 622–631.

Franklin, N. M., Stauber, J. L., Apte, S. C., & Lim, R. P. (2002). Effect of initial cell density of the bioavailability and toxicity of copper in microalgal bioassays. *Environmental Toxicology and Chemistry, 21*, 742–751.

Fuchida, S., Yokoyama, A., Fukuchi, R., Ishibashi, J., Kawagucci, S., Kawachi, M., & Koshikawa, H. (2017). Leaching of metals and metalloids from hydrothermal ore particulates and their effects on marine phytoplankton. *ACS Omega, 2*, 3175–3182.

Fujita, R. (2001). Protecting deep sea ecosystems from a new era of mining. *Environmental Defence, 5*, 7–15.

Gerhardt, V., & Kretsch, G. (1989). Development of a delayed fluorescence based biotest for detection of water pollutants. Dortmunder Eiträgezur Wasserforschung, Grundlagen und Anwendungs-bereiche der Chlorophyll fluoreszenz; Veröffentlichungen des Institutsfür Wassersforchung GmbH Dortmund und der Hydrobiologischen Abteilung der Dortmunder Stadtwerke AG 37, 87–95.

ISA/LTC (2013). Recommendations for the guidance of contractors for the assessment of the possible environmental impacts arising from exploration for marine minerals in the area. In I.S. Authority (Ed.), ISBA/19/LTC/8.

ISO. (2006). *Water quality–marine algal growth inhibition test with Skeletonemacostatum and Phaeodactylumtricornutum*. Geneva: International Organization for Standardization.

Joliot, P., Joliot, A., Bouges, B., & Barbieri, G. (1971). Studies of system-II photocenters by comparative measurements of luminescence, fluorescence, and oxygen emission. *Photochemistry and Photobiology, 14*, 287–305.

Jursinic, P. A. (1986). Delayed fluorescence: Current concepts and status. In J. A. Govindjee & D. C. Fork (Eds.), *Light emission by plants and bacteria* (pp. 291–328). New York: Academic.

Katsumata, M., Koike, T., Nishikawa, M., Kazumura, K., & Tsuchiya, H. (2006). Rapid ecotoxicological bioassay using delayed fluorescence in the green alga *Pseudokirchneriellasubcapitata*. *Water Research, 40*, 3393–3400.

Katsumata, M., Koike, T., Kazumura, K., Takeuchi, A., & Sugaya, Y. (2009). Utility of delayed fluorescence as endpoint for rapid estimation of effect concentration on the green alga *Pseudokirchneriellasubcapitata*. *Bulletin of Environmental Contamination and Toxicology, 83*, 484–487.

Katsumata, M., Ikushima, Y., Bennet, K., Sato, Y., Takeuhi, A., Tatarazako, N., & Hakamata, T. (2017). Validation of rapid algal bioassay using delayed fluorescence in an interlaboratory ring study. *Science of the Total Environment, 605–606*, 842–851.

Kumagai, H., Nozaki, T., & Ishibashi, J., (2017). The Cross Ministerial Strategic Innovation Promotion Program (SIP) expedition 909 hydrothermal sediments in Okinawa Trough III, Cruise report.

Mayer, P., Frickmann, J., Christensen, E. R., & Nyholm, N. (1998). Influence of growth conditions on the results obtained in algal toxicity tests. *Environmental Toxicology and Chemistry, 17*, 1091–1098.

Millán de Kuhn, R., Streb, C., Breiter, R., Richter, P., Neeβe, T., & Häber, D. P. (2006). Screening for unicellular algae as possible bioassay organisms for monitoring marine water samples. *Water Research, 40*, 2695–2703.

OECD. (2011). *Alga, growth inhibition test (201) OECD guideline for testing of chemicals*. Paris: Organization for Economic Co-operation and Development. 2011.

Okamura, H., Aoyama, I., Liu, D., Maguire, R. J., Pacepavicius, G. J., & Lau, Y. L. (2000). Fate and ecotoxicity of the new antifouling compound irgarol 1051 in the aquatic environment. 34:3523–3530

Perelomov, L. V., Sarkar, B., Sizova, O. I., Chilachava, K. B., Shvikin, A. Y., Perelomova, I. V., & Atroshchenko, Y. M. (2018). Zinc and lead detoxifying abilities of humic substances relevant to environmental bacterial species. *Ecotoxicology and Environmental Safety, 151*, 178–183.

Pérez, J., Domingues, I., Soares, A. M. V. M., & Loureiro, S. (2011). Growth rate of *Pseudokirchneriellasubcapitata* exposed to herbicides found in surface waters in the alqueva reservoir (Portugal): A bottom-up approach using binary mixtures. *Ecotoxicology, 20*, 1167–1175.

Radix, P., Léonard, M., Papantoniou, C., Roman, G., Saouter, E., Gallotti-Schmitt, S., Thébaud, H., & Vasseur, P. (2000). Comparison of four chronic toxicity tests using algae, bacteria, and invertebrates assessed with sixteen chemicals. *Ecotoxicology and Environmental Safety, 47*, 186–194.

Sbrilli, G., Bimbi, B., Cioni, F., Pagliai, L., Luchi, F., & Lanciotti, E. (2005). Surface and ground waters characterization in Tuscany (Italy) by using algal bioassay and pesticide determinations: Comparative evaluation of the results and hazard assessment of the pesticides impact on primary productivity. *Chemosphere, 58*, 571–578.

Schmidt, W., & Senger, H. (1987a). Long-term delayed luminescence in *Scenedesmus obliquus* I. Spectral and kinetics properties. *Biochemistry and Biophysics, 890*, 15–22.

Schmidt, W., & Senger, H. (1987b). Long-term delayed luminescence in *Scenedesmus obliquus* II. Influence of exogenous factors. *Biochemistry and Biophysics, 981*, 22–27.

Schreiber, U., Muller, J. F., Haugg, A., & Gademann, R. (2002). New type of dual channel PAM chlorophyll fluorometer for highly sensitive water toxicity biotests. *Photosynthesis Research, 74*, 317–330.

Simpson, S. L., Angel, B., Hamilton, I., Spadaro, D. A., & Binet, M. (2007). Water and sediment characterization and toxicity assessment for the Solwara 1 Project. In *Environmental Impact Statement*; Natural Minerals Inc.: Australia; Appendices 4–7, Vol. B.

Smith, C. R., Levin, L. A., Koslow, A., Tyler, P. A., & Glover, A. G. (2008). The near future of the deep-sea floor ecosystems. In N. Polunin (Ed.), *Aquatic ecosystem* (pp. 334–351). Cambridge, UK: Cambridge University Press.

Strehler, B. L., & Arnold, W. (1951). Light production by green plants. *The Journal of General Physiology, 34*, 809–820.

Thiel, H. (2003). Anthropogenic impacts on the deep sea. In P. A. Tyler (Ed.), *Ecosystems of the world* (Ecosystems of the Deep Ocean) (Vol. 28, pp. 427–472). Amsterdam: Elsevier.

Thompson, J. A. (1997). Cellular fluorescence capacity as an endpoint in algal toxicity testing. *Chemosphere, 35*, 2027–2037.

USEPA. (1996). Ecological effects test guidelines OPPTS 850.5400 algal toxicity, Tires I and II. United States Environmental Protection Agency; 1996.

Vallotton, N., Iida, R., Eggen, L., Escher, B. I., Krayenbuhl, J., et al. (2008). Effect of pulse herbicidal exposure on *Scenedesmusvacuoleatus*: A comparison of two photosystem II inhibitors. *Environmental Toxicology and Chemistry, 27*, 1399–1407.

Vignati, D. A. L., Dominik, J., Beye, M. L., Pettine, M., & Ferrari, B. J. D. (2010). Chromium (VI) is more toxic than chromium(III) to freshwater algae: A paradigm to revise. *Ecotoxicology and Environmental Safety, 73*, 743–749.

Weyman, G. S., Rufli, H., Eltje, L., Salinas, E. R., & Hamitou, M. (2012). Aquatic toxicity tests with substances that are poorly soluble in water and consequences for environmental risk assessment. *Environmental Toxicology and Chemistry, 31*, 1662–1669.

Yamagishi, T., Katsumata, M., Yamaguchi, H., Shimura, Y., Kawachi, M., Koshikawa, H., Horie, Y., & Tatarazako, N. (2016). Rapid ecotoxicological bioassay using delayed fluorescence in the marine cyanobacterium *Cyanobium* sp. (NIES-981). *Ecotoxicology, 25*, 1751–1758.

Yamaguchi, H., Shimura, Y., Suzuki, S., Yamagishi, T., Tatarazako, N., & Kawachi, M. (2016). Complete genome sequence of *Cyanobium* sp. NIES-981, a potentially useful marine strain for ecotoxicological bioassay. *Genome Announcements, 4*, e00736–e00716.

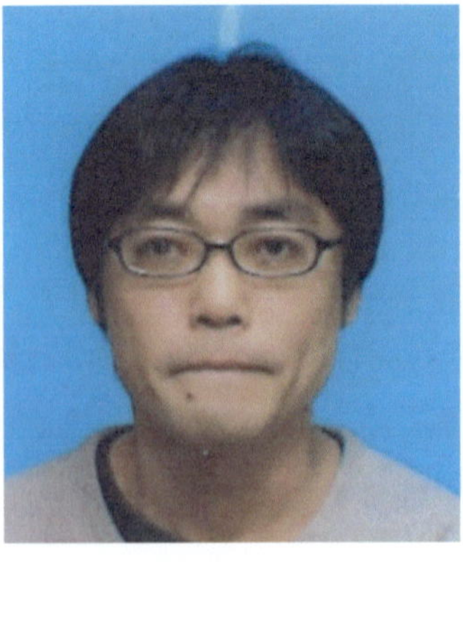

Takahiro Yamagishi is a senior researcher at the National Institute for Environmental Studies, Tsukuba, Japan, since 2015. Earlier he served as Assistant Professor at the Organization of Advanced Science and Technology in Kobe University, Kobe, Japan, from 2008 to 2015 and Postdoctoral Researcher at the Research Center for Inland Seas in Kobe University, Kobe, Japan, from 2004 to 2008. Takahiro holds a PhD degree in Life Sciences from Tohoku University, Sendai, Japan. His research interests include ecotoxicology and cell biology in algae and plants.

Shuhei Ota is a Research Associate at the Center for Environmental Biology and Ecosystem Studies, National Institute for Environmental Studies, Tsukuba, Japan, since 2017. Earlier he has served as an Assistant Professor and Project Researcher in University of Tokyo from 2011 to 2017, as well as a Postdoctoral Fellow in the University of Oslo, University of Paris, and University of Tsukuba from 2007 to 2011. He has a PhD degree in Life Sciences from Kanazawa University, Japan. His research interest includes taxonomy and cell biology in plants and algae.

Part III
Environmental Data Standardization and Application

New Techniques for Standardization of Environmental Impact Assessment

Yasuo Furushima, Takehisa Yamakita, Tetsuya Miwa, Dhugal Lindsay, Tomohiko Fukushima, and Yoshihisa Shirayama

Abstract The exploitation of deep-sea mineral resources has not yet begun. In order to realize it in the future, various issues need to be addressed. One such issue is establishing an appropriate environmental impact assessment (EIA) technique and a supporting environmental research method. At present, knowledge on the deep-sea environment is quite poor, and existing environmental research methods rely on various methods used in ocean science. However, the goals of oceanography do not always align with those for EIA. Furthermore, the purposes of EIA and scientific research are different. Considering economic performance and technical convenience, scientific research methods are not necessarily suitable for EIA. In this context, this paper introduces three new technologies of turbulence measurement, niche modeling, and genetic connectivity survey method and suggests approaches for their standardization for the purpose of EIA.

Keywords Environmental impact assessment (EIA) · Turbulence · Habitat mapping · Larval dispersal techniques

1 Introduction

The development of deep-sea mineral resources is a topic of potential interest in the future. Regarding areas beyond national jurisdiction (ABNJ), state parties or entities applying for exploration areas to the International Seabed Authority (ISA) are increasing (Fukushima and Nishijima 2017), whereas testing of various technologies for ocean mining is being conducted within areas of national jurisdiction. For example, Japan Oil, Gas and Metals National Corporation (JOGMEC) has successfully conducted a continuous rising test of seafloor massive sulfide (SMS) at a depth of 1600 m in a site within the Japanese exclusive economic zone (EEZ)

Y. Furushima · T. Yamakita · T. Miwa · D. Lindsay
T. Fukushima (✉) · Y. Shirayama
Japan Agency for Marine-Earth Science and Technology (JAMSTEC), Yokosuka, Kanagawa, Japan
e-mail: fukushimat@jamstec.go.jp

R. Sharma (ed.), *Environmental Issues of Deep-Sea Mining*,
https://doi.org/10.1007/978-3-030-12696-4_11

(JOGMEC 2017). In Papua New Guinea (PNG), Nautilus Minerals Inc. has started to take action toward SMS development at Solwara 1 site (Nautilus 2008). In addition, approaches are being taken for phosphate development in Namibia (190–250 m water depth) and Chatham Rise (350–450 m water depth) (Namphos 2012; CRP 2014; Miyata and Fukushima 2018). As described above, developments of various minerals have started in different areas, under varying topographic settings and water depths.

Along with international movements, social demands for environmental impact assessment (EIA) concerning the development of ocean mineral resources are becoming more strict. The importance of EIA for ocean mineral resources development was declared in the G7 summit at Elmau (Germany) in 2015. In addition, nongovernment organizations (NGOs) associated with environmental issues, such as the International Union for Conservation of Nature and Natural Resources (IUCN), are voicing their concern (IUCN 2018). These indicate the increasing societal pressure on EIA. In other words, with increased diversity of the target ore, development area, and water depth, more strict and detailed regulations are being formulated. Therefore, the ISA reconsidered the contents of "Recommendations for the Guidance of Contractors for the Assessment of the Possible Environmental Impacts Arising from Exploration for Marine Minerals in the Area" (hereafter the Environmental Guidelines) (ISA 2013) and "Environmental Management Needs for Exploration and Exploitation of Deep Sea Minerals" (EMP) (ISA 2011). Moreover, the ISA is currently developing regulations for the exploitation phase, and in the process, several steps are being taken particularly addressing environmental issues.

To improve EIA, we must focus on technology development as well as changes in regulations and guidelines. Previous EIA technologies relied on the information derived through ocean science studies and methodologies formulated according to marine technologies. However, the purpose of EIA differs from that of regular oceanographic research, and so scientific research methods may not be suitable for EIA. It is clear that requirements of seabed mineral resource development, which is expected to begin in the near future, cannot be met by simply waiting for scientific progress.

EIA is a process for consensus building and is supported by correct information, efficient methodology, and unbiased evaluation. Therefore, these conditions are required in EIA technologies. Correct information can be obtained by appropriately interpreting data derived from research, observation, and analysis with high precision, wide range, and long time scale. For example, considering natural fluctuations of the environment, more accurate data can be expected through more frequent and continuous observations. Considering the development of seabed mineral resources as an economic activity, it is also important to develop an efficient methodology. This implies that EIA cannot expect for more effort that exceeds the benefits derived from development. Finally, objective information is required to achieve unbiased evaluation. Therefore, in order to guarantee reproducibility, a data cross-check system should be secured. According to the background mentioned above, Japan is working on the development of EIA methods with a view of future mineral resource development through a national project called "Zipangu in the Ocean" (JAMSTEC 2017a).

Within this project, some technologies have reached phases of practical use. For example, Edokko Mark 1 is a deep-sea observatory system, and it has already been applied to environmental studies in areas of cobalt-rich ferromanganese crust and seafloor massive sulfide (Fig. 1) (Fukuba et al. 2018). By installing this system on the seabed with a free falling method, conditions of the seabed can be observed for half a year or 1 year. Therefore, using this system, the activity characteristics of benthic organisms or demersal fish on daily, monthly, or seasonal cycles can be studied according to the preset photographing interval (Fig.2). In addition, short-

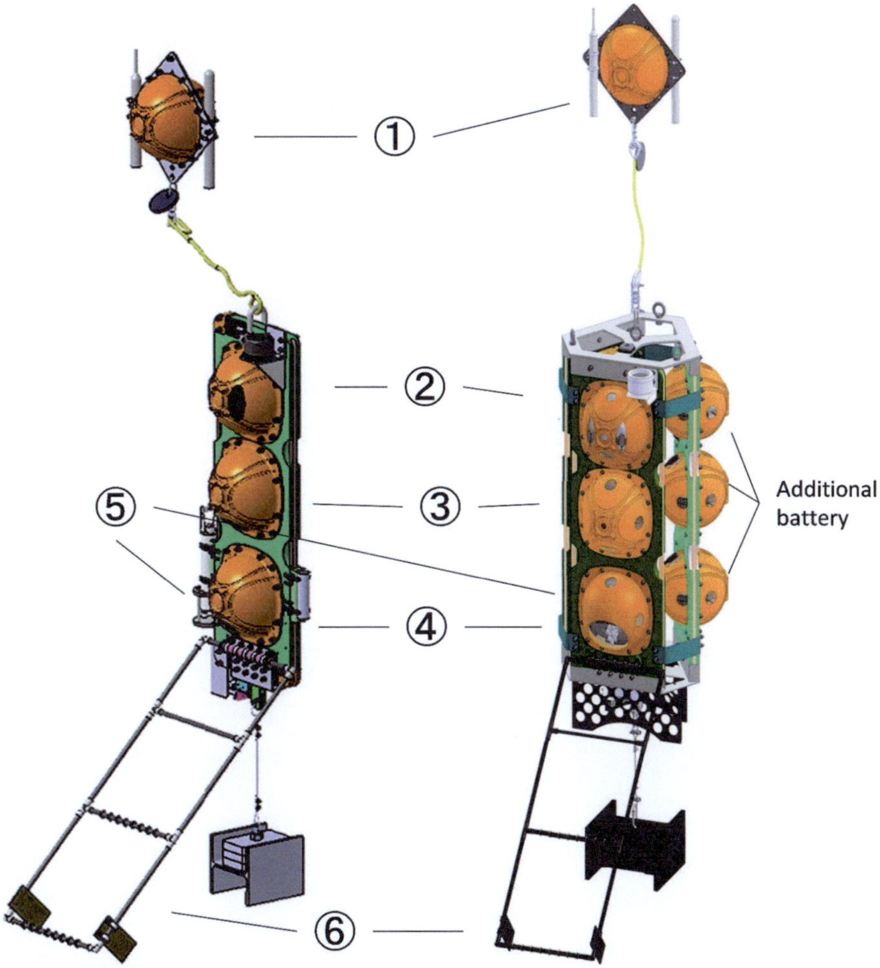

Fig. 1 External view of Edokko Mark 1. The left side is a lightweight model. The right side is a long-term observation model. From the top, it consists of offshore communication balls used for collection (1), transponder balls used for detachment (2), illumination balls used for LED lighting (3), and photographing balls by time-lapse camera (4). CTD measurement is installed (5). Shooting will observe the bottom of the sea from the altitude of 1 m (6)

term biological observation has already been performed using this system with baited trap, which can lure demersal scavengers (Sugishima et al. 2018). Moreover, because the scale is set at the position visible from the viewfinder, it is effective as a quantitative observation tool (Fig. 3) (Matsui et al. 2018).

In developing this equipment, the design was created on the premise that resource development is an economic activity, considering cost performance and convenience, and to satisfy the requirements of ISA's environmental guidelines (Fig. 4) (Fukushima 2017). In addition, this system is registered in the repository of IOC-UNESCO's technical database for international standardization (JAMSTEC 2017b).

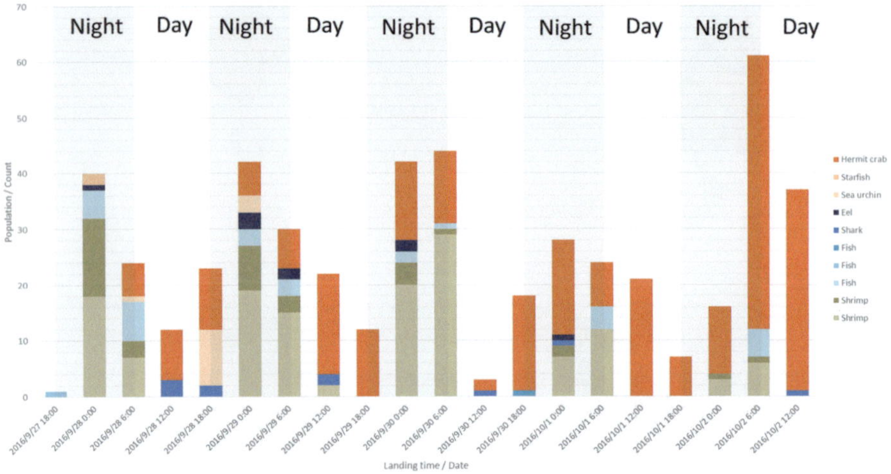

Fig. 2 Changes in species and number of organisms observed by Edokko Mark 1 at Omuro Hole, off Izu (depth 260 m). The time-lapse shot was taken every 30 min for 1 min. Each column represents the number of organisms integrated in 6 h

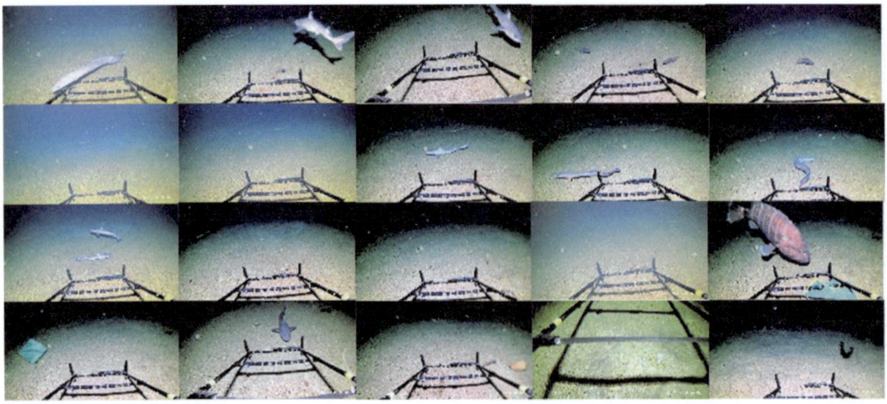

Fig. 3 Screenshots observed at the seabed. From the upper left to the lower right, a photo of each of the 1 h has been placed

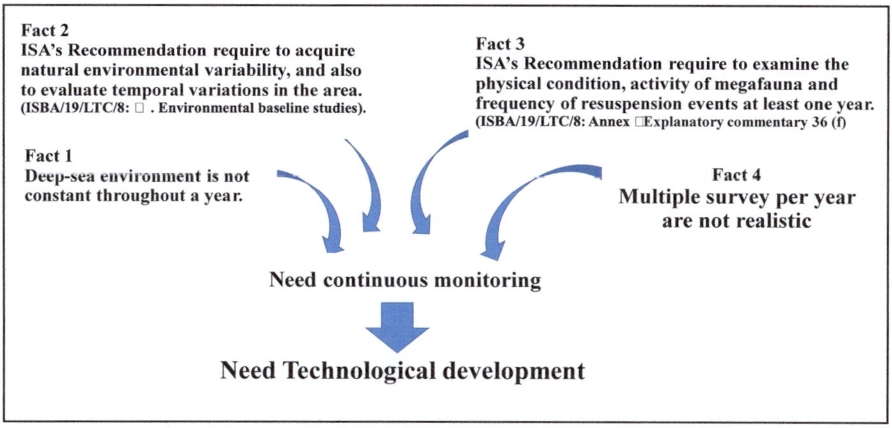

Fig. 4 Rationales of technical developments of seabed continuous monitoring systems

The development of the abovementioned technologies is almost complete. In the next section, we introduce potential technologies that currently face some hurdles for practical application but can be expected to contribute significantly to future EIA.

2 Technique for Turbulence Measurements

2.1 Necessity of Deep-Sea Turbulence Measurements

Flows of fluids, such as seawater and air, that fluctuate irregularly over both time and space are called turbulence flows. Such flows are widespread in nature (Tennekes and Lumley 1972), for example, smoke from a chimney and wind passing through the gap between two buildings. In the ocean, the Kuroshio Current and hot water spouting from hydrothermal vents are also turbulent flows. Turbulent flows cause vertical mixing under the influence of bottom topography (ridges, seamount, etc.) surrounded by complex tidal flows near the deep-sea floor, contributing to the lifting, transportation, and diffusion of suspended particles. However, it is difficult to grasp the actual conditions from flow velocity observations on the seabed using an acoustic Doppler current profiler (ADCP) or other current meters. Therefore, verification through a numerical model is considered to be effective, but a general analysis method is yet to be established for the closure problem. As for the equation describing turbulent flow, there are more unknown numbers than the number of equations. Therefore, we cannot solve the equation. This is called a "closure problem." We can close the equation by adding modeled values and estimates based on intuition and experience (Tennekes and Lumley 1972). Various attempts have been made to improve the vertical mixing calculation method (scheme) (Furuichi et al.

2012; Higashi et al. 2017). However, most models have mainly been developed to improve the surface mixed layer, accommodate deep-sea turbulence, and improve modeling accuracy. Therefore, it is indispensable to properly measure the distribution of spatiotemporal turbulence intensity in the deep sea and the dynamics of the bottom mixed layer.

In the development of submarine mineral resources, the main factor impacting the surrounding environment is the disturbance of the seabed (Fig. 5) as the deep-sea ecosystem is directly damaged by the drilling and movement of the mining machine. Furthermore, the emission of turbid water (plumes) from the mining machine and the transportation, diffusion, and redeposition of suspended particles have been suggested to adversely affect the habitats of organisms living on the deep seabed and deep-sea-specific ecosystems (Furushima et al. 2016, 2018). The ISA environmental guidelines describe investigation methods that assume redeposition. This issue has been garnering attention for some time, and verification tests have been repeatedly performed since the 1990s. However, the actual measurement of redeposition is difficult. Sediment traps are only installed at 2 m directly above the seabed at most, and thus the behavior within 2 m from the seabed cannot be deci-

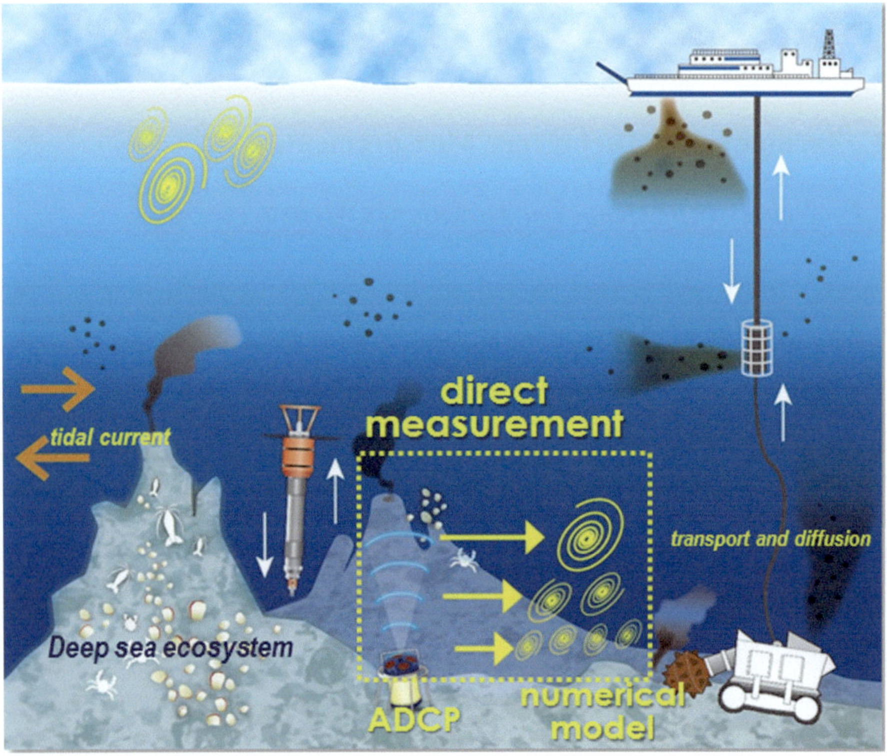

Fig. 5 Schematic view of fluid environment observations for the environmental impact assessment of seafloor submarine mining

phered. Furthermore, it is unlikely that the particles will remain in place once they arrive at the ocean floor, and it is not possible to estimate their behavior using data from existing current direction/velocity meters. Therefore, it is also difficult to compare measured values and predicted values based on numerical models.

Diffusion of plumes and lifting/redeposition of suspended particles are governed by the flow velocity of the bottom mixed layer and turbulence (vertical mixing) intensity. However, it is impossible to distinguish the actual turbulence intensity from measurements of vertical water temperature/salinity content by conventional CTD (conductivity-temperature-depth) and from current measurement by ADCP (acoustic Doppler current profiler). Therefore, direct measurement of deep-sea turbulence using a turbulence meter is indispensable in order to discern the turbulence intensity and depth/range at which vertical mixing occurs. Direct measurement of deep-sea turbulence has been made possible in recent years by new technologies. Therefore, this turbulence measurement technique will contribute to a study to characterize a mechanism of the deep-sea turbulence.

In light of the above, direct measurement of deep-sea turbulence for the evaluation and prediction of plume dispersion dynamics and the lifting and redeposition of suspended particles is ongoing, and research is being conducted aiming at improving the accuracy of numerical models.

2.2 *Example Measurement of Deep-Sea Turbulence*

The VMP-X, an expendable vertical microstructure profiler capable of measuring turbulence flow between the surface of the ocean and immediately above the deep-seabed, was recently developed by Rockland Scientific International Inc. (RSI), Canada (Xiaodong et al. 2017) (Fig. 6). The VMP-X consists of a data logger (Fig. 6a) and a sensor (Micro-XM, manufactured by JFE Advantech Co., Ltd., Japan, Fig. 6b). When dropped from a vessel, the VMP-X free-falls at a speed of 0.6–0.7 m s-1 and measures the vertical profile of the flow velocity shear (by two sensors) and water temperature at 1250 Hz until it reaches the seabed (Fig. 7). This device can measure up to a maximum water depth of 6000 m. Upon landing on the ocean floor, the sensor automatically detaches, and the data logger returns to the surface (Fig. 7). The detached sensor (Fig. 6b) is somewhat expensive, constraining multiple observations, which is a problem. After conducting spectral calculations, the obtained flow velocity shear data are converted into turbulence energy dissipation rate ε (W kg-1), and the vertical fluctuation is taken as the turbulence intensity.

Figure 8 shows the vertical distribution of the turbulence intensity and water temperature from the ocean surface layer to directly above the seabed obtained off Itoh in Sagami Bay (Fig. 9a). A very strong turbulence intensity of $\varepsilon = 10^{-6}$ W kg-1 was observed in the vicinity of the layer 20 m above the seabed (Furushima et al. 2017). This value is comparable to the turbulence intensity in the ocean surface mixed layer, suggesting that the same level of turbulence is generated near the seabed. Direct measurement of the turbulence energy dissipation rate has been carried out in coastal areas,

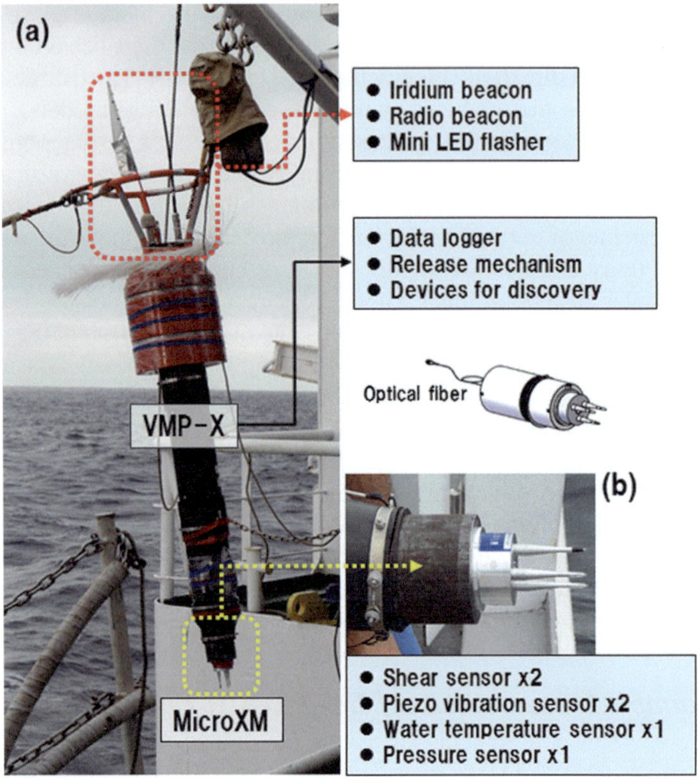

Fig. 6 Expendable vertical microstructure profiler system. (**a**) VMP-X (**b**) MicroXM

such as the Ariake Sea (Saita et al. 2008), Hiroshima Bay (Hashimoto and Takasugi 1998), Seto Inland Sea (Nagao et al. 2004), Ago Bay (Nagao et al. 2005), and Tokyo Bay (Higashi and Maki 2012). However, examples of studies conducted in the vicinity of the deep seabed are rare, although the results obtained within are very interesting.

Figure 10 shows the turbulence energy dissipation rate (ε) and the vertical water temperature distribution obtained in the hydrothermal venting area (Iheya small ridge) of the Okinawa Trough (Fig. 9b). Similar to Sagami Bay, in the Okinawa Trough, a distribution of turbulence intensity (ε) of about 10^{-7} W kg-1 was observed in a layer 30 m above the seabed and in deep layers of 940 and 1060 m (Fig. 10a, b). Furthermore, near the seabed, the water temperature increased slightly (Fig. 10a). This may suggest the influence of vertical mixing due to submarine hydrothermal fluids from nearby hydrothermal vents. However, vertical mixing near the seabed varies greatly with the deep-sea tidal current. The isolated turbulence intensity data obtained here shows incredibly interesting results, but the temporal representation may be extremely small. Therefore, in order to ensure the usefulness of measured deep-sea turbulence intensity values, it is necessary to conduct multiple observations in accordance with the tidal cycle. At the same time, if physical environment data from flow near the seabed or CTD/turbidity data, etc. can be obtained, the turbulence intensity can be potentially estimated.

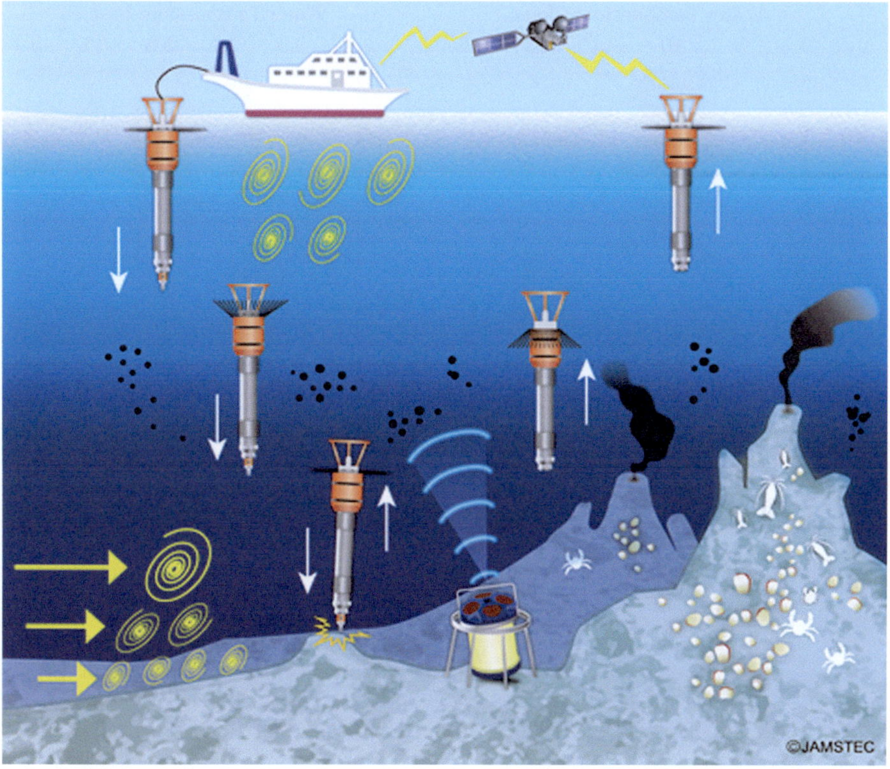

Fig. 7 Schematic view of VMP-X observation

Direct observation of turbulence in the deep sea has been actively carried out in recent years, but as a new technology of EIA for the development of submarine mineral resources, this research is the first Japanese attempt to introduce deep-sea turbulence flow measurements by VMP-X. The results of the turbulence flow measurements shown here represent only an overview of the turbulence intensity in each marine area. Currently, multiple turbulence observations are being conducted, incorporating spatiotemporal and tidal fluctuations and accumulating and analyzing field turbulence data to clarify the relationship between changes in tidal current and turbulence intensity and their relationship with the mixed seabed layer.

2.3 Adaptation to a Deep-Sea Turbulence EIA Prediction Model

In EIA methods for the development of submarine mineral resources, numerical simulations based on physical models are as essential as field measurements for evaluation and prediction. A numerical model is required to make preliminary

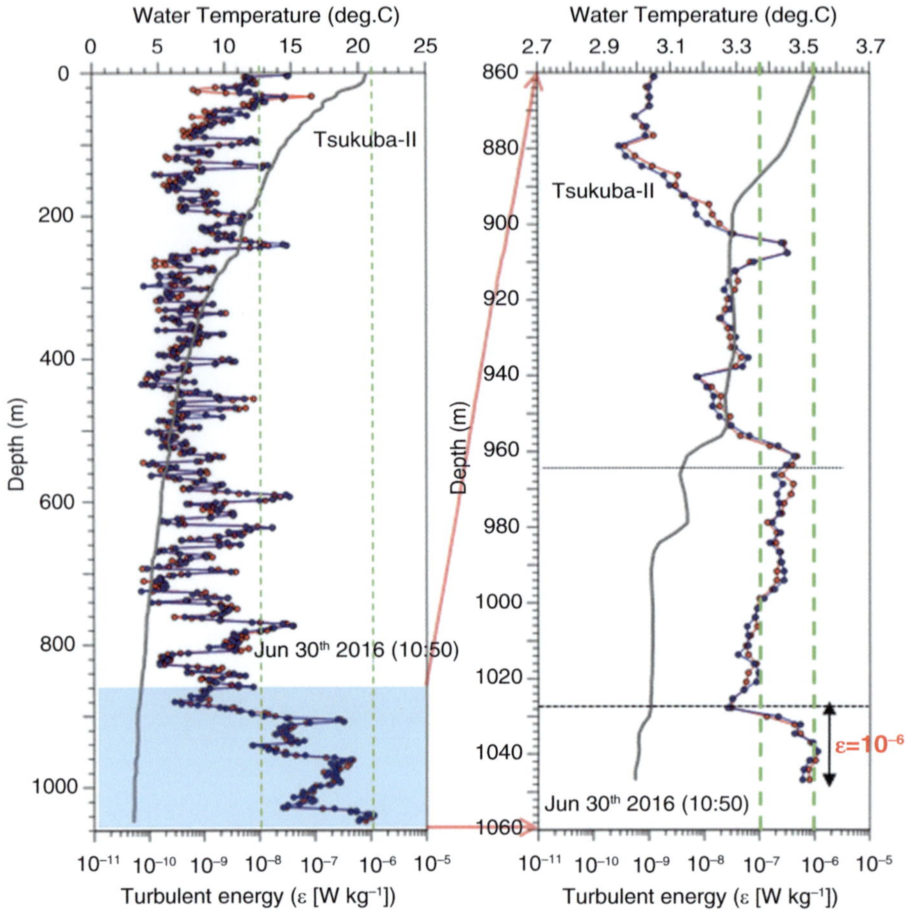

Fig. 8 Vertical profiles of turbulence intensity (ε) and water temperature off Itoh in Sagami Bay. The blue and red lines show ε obtained from the two shear sensors. The black line in the figures shows the vertical fluctuation in the water temperature. The right figure shows the spread and the fluctuation in the energy dissipation rate (ε) near the seabed. The value of the turbulence intensity ($\varepsilon = 10^{-6}$ W kg-1) around the bottom was equal to the surface mixed layer

predictions regarding the effect of environmental conservation activities for which field verification tests are difficult to perform and to predict environmental recovery after development. Development studies on dynamic prediction models for suspended particles have been carried out since the 1980s (Jankowski et al. 1996; Nakata et al. 1997; Doi et al. 1999). These models are three-dimensional region model combining a flow model that predicts the flow of seawater and a dynamic model that considers the transport, diffusion, sedimentation, deposition, and resuspension of suspended particles. However, turbulence modeling near the seabed which dominates a dynamics of suspension particles did not have sufficient

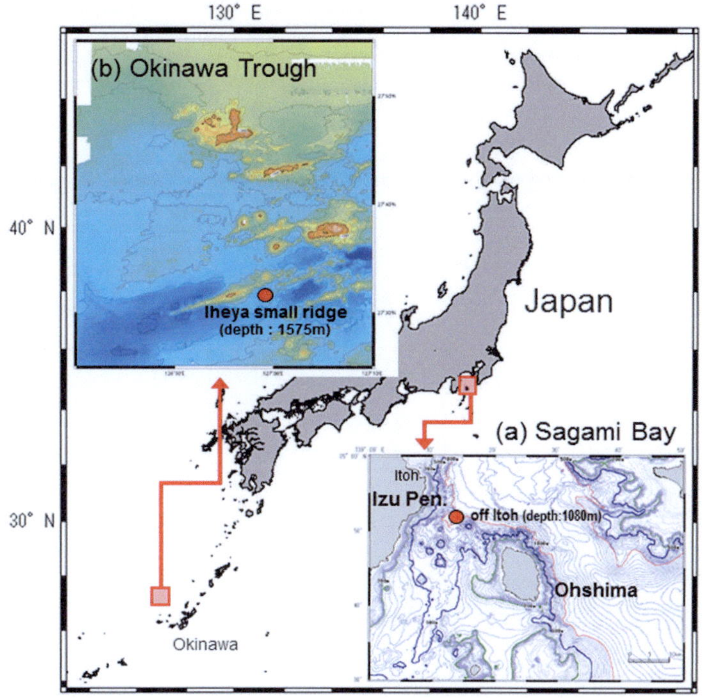

Fig. 9 Location map of the observation points. (**a**) Sagami bay (**b**) Okinawa Trough

accuracy. Furuichi et al. (2016) constructed a numerical simulation model consisting of a large-eddy simulation (LES) model (Furuichi et al. 2012; Furuichi and Hibiya 2015; Furuichi and Higashi 2014) that predicts the turbulent mixing process in the deep-seabed boundary layer and a particle tracking model that predicts the transportation/diffusion of suspended particles using obtained flow velocity field data.

Figure 11 shows a distribution map (instantaneous diagram) of the water temperature field using the LES model. Along with the tidal fluctuation, the bottom mixed layer developed from 30 to 60 m from the seabed. The result of this model analysis is generally consistent with the field flow measurement by ADCP in the Okinawa Trough, in which the field flow could only be between 30 and 60 m above the seabed (Furushima and Yamamoto 2015, 2018; Furuichi et al. 2016). Therefore, it was inferred that in the Okinawa Trough, the bottom mixed layer developed due to the half-day flow (tidal current) cycle in the vicinity of the deep seabed, forming an environment in which current measurements could be taken by ADCP. In addition, Furuichi et al. (2016) placed particles at heights of 15 m, 30 m, and 45 m above the seabed to resemble suspended particles and carried out a particle tracking experiment. The majority of the particles were found to be finally distributed in the bottom mixed layer rather than being diffused into the upper layer region.

Thus, extremely interesting results have been obtained from numerical simulations of suspended particle dynamics. However, the collection of more deep-sea

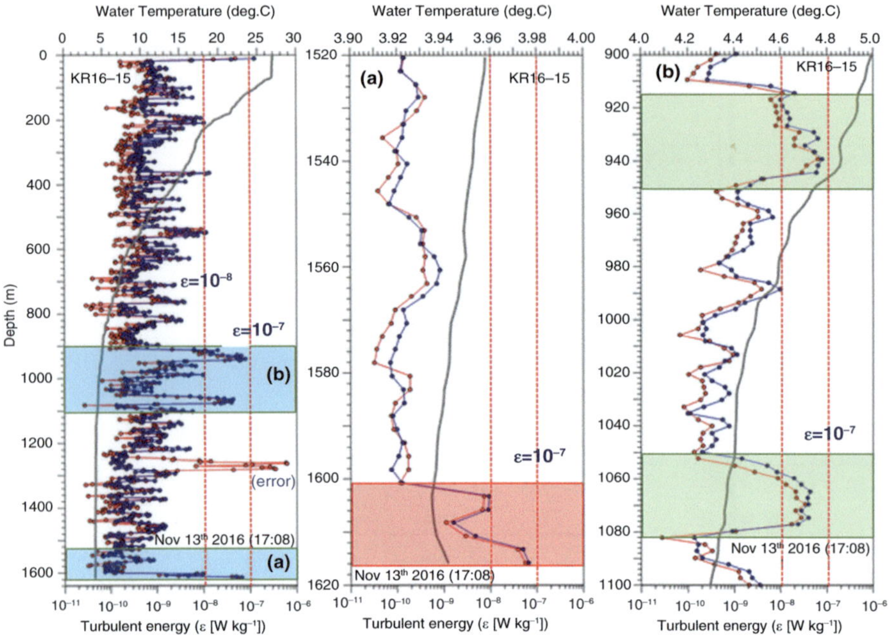

Fig. 10 Vertical profiles of turbulence intensity (ε) and water temperature at the Iheya small ridge in the Okinawa Trough. The blue and red lines show the ε values obtained from the two shear sensors. The black line in the figures shows the vertical fluctuation in the water temperature. Figure 6a, b show the spread and the fluctuation in the energy dissipation rate (ε) near the bottom and at 1000 m depth. The turbulence flow was strong from the bottom to a height of around 20 m from the bottom, and water temperature rose slightly

flow and turbulence observation data remains indispensable for improving the accuracy of the prediction model for suspended particulate dynamics at the bottom of the ocean. Furthermore, such data can be used to verify the validity of the numerical simulation results. The current LES model (Furuichi et al. 2012, 2016; Furuichi and Higashi 2014; Furuichi and Hibiya 2015) assumes a flat submarine topography, but turbulence near the seabed develops under the influence of complicated ocean floor topography. Therefore, model improvements, incorporating submarine topography, are urgently required. Furthermore, as previously noted and suggested in Fig. 10a, hot hydrothermal fluids from seafloor vents also cause turbulence. Therefore, it will probably be necessary to consider the influence of hot water emissions.

The aforementioned observations of turbulence, related to EIA methods for the development of submarine mineral resources, can be used as effective data if the dynamics leading to the transportation, diffusion, and redeposition of suspended particles in the vicinity of the seabed are known. However, ultimately, combination with a numerical model is indispensable. Depending on the amount of data required by the numerical model, problems, such as the need for a direct investigation of deep-sea turbulence using a turbulence meter, still remain.

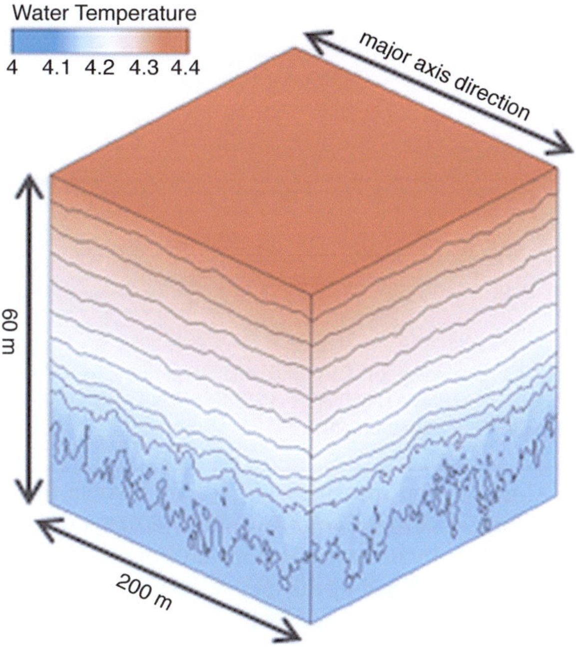

Fig. 11 Illustration of the water temperature distribution provided by the LES (large-eddy simulation) model

3 Technique for Habitat Mapping

3.1 Necessity of Habitat Mapping for EIA

In this section, we describe techniques to produce a suitable habitat map of organisms. Several methods for estimating species distribution, biomass, and/or biodiversity have become very popular in recent years (Franklin 2010). Estimated species distribution or other indicators can also be used as data to extract important marine areas in broad-scale analysis. The extraction of ecologically and biologically significant areas (EBSAs) (Clarke et al. 2006; Yamakita et al. 2017) or strategic environment assessment (SEA) (Glasson et al. 1995) are well known examples of such applications. In addition, if the density of the data is sufficient, this type of analyses also can be applied to narrower (local) spatial ranges, such as the scope of a single natural resource development plan. The produced habitat map in the local area can be used to measure the area affected by development after implementing the decision of any project. Using the model for producing the habitat map, the habitat suitability of potential restoration area can be evaluated, or future distribution after the alteration of the environment can be predicted. To explain such applications through an example, we will show a case study in a seamount in three to six. We also show an outlook of advanced spatial data acquisition using automated survey, creation of maps, and recognition of organisms the last of this section (Fig. 12).

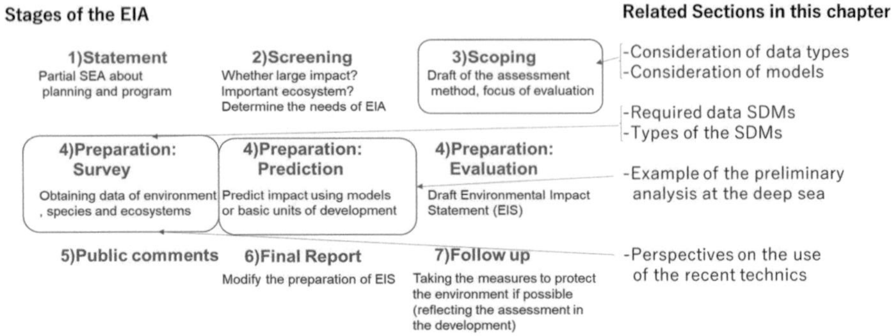

Fig. 12 Overview of EIA and related paragraphs of this subsection. (see the broacher of ENVIRONMENTAL IMPACT ASSESSMENT IN JAPAN for more detail https://www.env.go.jp/en/policy/assess/pamph.pdf)

Conservation measures in the EIA involve three stages, avoidance, reduction, and compensation. Use of quantitative maps of species habitat is crucial in every step of EIA. In the case of Japan, developers legally execute EIA within their practical scope as defined by the law. Although some guidelines and existing cases can be referred to, EIA is limited by the budget and time of the developers. Therefore, low-cost and efficient methods will promote the implementation of EIA to derive better survey and conservation measures. This is one of the reasons why producing habitat maps is advantageous in such cases.

3.2 Overview of Data Types Required for Producing Habitat Maps

In order to produce good habitat maps, a complete count survey of all species (such as national census) is ultimately desired. Although most marine researchers might find such an idea ridiculous or unrealistic, some countries have achieved extensive bird census or green census (i.e., census of nature and environment). Such nearly complete census is not always feasible because of limitations, such as budget, time, and availability of taxonomic specialists. It is especially difficult in the case of EIA funded by developers. In such cases, appropriate survey designs that will reduce the necessary number of sampling and interpolation using model predictions will help in generating appropriate habitat maps for use in EIA. The necessary number and distribution of the data will vary depending on the distribution of environment variables, mobility of the target species, relationship between the environmental factors and species, accuracy of data, area of analysis, and resolution of data (Guisan et al. 2007). In addition, in order to find significant changes in the biomass, considering formal field experiment planning is ideally recommended (Underwood 1997).

Surveys carried out with a limited budget will have data gaps or sparse density of data. Even in the case of the overall census with sufficient funds, it is still important to know bias of data and spatial accuracy of the result. If the model used for future prediction, the model should ideally be tested for sampling bias and accuracy of the results using existing data. However, available cases of spatial and temporal data with similar density are very rare. Finally, it is important to fully consider the disadvantages or shortcomings of the model applied. Even with the use of high-level statistical models, the model cannot provide more information beyond the data supplied, and assumptions of the model may limit the interpretation. Recognition of certain assumptions and evaluating the accuracy considering including sample size are important factors to be kept in mind.

3.3 Types of Models Used to Map the Habitat of Species

Several types of indicators are used in generating habitat maps. Estimated biomass, number of individual organisms, species richness, and surrogate of any biological indicators, such as calculation using topography, are common examples. This section describes the estimation of species distribution (prediction of the presence of species), which has often been used in recent evaluations of biodiversity or distribution of indicator species including the distribution of endangered species. Various techniques can be used for the purpose of predicting species distribution (i.e., presence or absence), and these techniques can be broadly classified into three types as follows (Gallien et al. 2010): (1) "analytical method," which is based on the fitting of the mathematical model (Suzuki et al. 2018); (2) "simulation," which describes known biological processes (Yamakita and Nakaoka 2009; Rochette et al. 2012; Nishikawa et al. 2015); and (3) "statistical models," which correlate the pattern of distribution and explanatory factors.

As statistical methods, various approaches have been used in recent years. Among them, "species distribution models (SDMs)" are commonly used for estimating the spatial distribution of organisms using habitat suitability. Statistical methods have prospective points compared to other methods based on simulation or simple equation techniques. Because this method handles the statistical relationships of data, even species without ecological details can be estimated in a realistic range of the distribution. The model also provides importance of the factors simultaneously. Therefore, with the ability to include species for which much information is not available, statistical methods are suitable for estimating the potential habitat of organisms.

3.4 Required Data for SDMs

To perform statistical estimation, data of explanatory variables in the target waters are required as spatial data. Furthermore, biological data (such as presence/absence or biomass) of the focused species are required as response (dependent) variables. This section summarizes such types of data. The technical aspect for obtaining such spatial data through field survey will be explained separately later. Nevertheless, the emphasis is on the utmost importance of appropriate density and location of sampling point in producing accurate results of the habitat mapping, following the philosophical expression "nothing comes from nothing."

In deep waters, it is not easy to obtain spatial data of explanatory variables (i.e., environment data). Therefore, interpolation using sparse point data, prediction using other environment data, such as topography, and estimations using simulation models, are popularly employed to fill this spatial data gap. Accuracy assessment and clarifying the data gaps of these environment data are also recommended for the production of effective habitat maps, but they are neglected in many cases.

The selection of explanatory variables depends on the purpose of the analysis, but the latter factors are popularly used in the estimation of the marine benthic species and organisms at the intermediate layer of water. Water depth and seafloor topographical features as well as its derivatives are most popular, such as slope, orientation of the slope, benthic positioning index, and surface roughness (rugosity). With remote sensing data of surface water temperature, the surface layer of productivity, mixture of current, and slope of sea surface height (as a representative of current) are also used. If models or databases are available for a region, waves, tide, current (average seasonal variations, the maximum value, the threshold value greater than or equal), and sediment types are also popular variables in large-scale analyses.

As biological data of the response variable, the presence or absence of species is extracted from videos, images, specimens, or records of on-site observation. If quantitative values such as biomass, or number of species need to be estimated, response variables should be continuous for counting the number or area per survey area. Records of databases of biodiversity in recent years, such as OBIS or GBIF, are also increasing. Such databases are useful in some cases, but they do not usually cover deep-sea or offshore waters. Consequently, new surveys should be planned extensively. In selecting the spatial location for new sampling, the purpose of the survey plays a major role. Some popular methods include random sampling, gridded sampling (equal intervals), line transect, and stratification by environmental conditions. Therefore, it is advisable to consider field experiment and spatial analysis to decide the appropriate sampling method.

Practically, in the deep seafloor, most investigations are likely to be based on line transects along with depth gradient according to the operation of instruments. It is ideal to combine multiple sampling. Avoiding spatial autocorrelation and large missing values is advisable statistical modeling. Such problems can be avoided by not only selecting appropriate sampling plans but also resampling data or selecting a complex model.

3.5 Types of Models in SDMs

Many prediction models have been made available in recent years (Franklin 2010). In general, common methods are fitting of simple rules, statistical regression model, models using machine learning, comparison, and fusion of several models. Simpler models provide easier understanding of mechanisms inside the model. For higher-accuracy requirements, recent increase in machine power has enabled the application of machine learning or more general models.

Methods using the convex hull and BIOCLIM are examples of the fitting of simple rules. Regarding regression models, linear and nonlinear models are available. The generalized linear model (GLM), which includes multiple regression, is classically used. Particularly in the analysis of species presence/absence, binomial distribution, which can treat discrete values over 0, is usually used. When treating continuous values, such as size or continuous percentage coverage, normal distribution, gamma distribution, and lognormal distribution can be used. In addition, if there are numerous zero values, the zero inflated model is often used. As shown here the distribution, the type of GLM should be selected depending on the nature and characteristics of the data.

When the relationship between the environmental factors and species distribution is not linear, a simple way is to include some variables converted nonlinearly and applied within the linear model. Fitting of the model to the data is implemented as the generalized additive model (GAM). Some models also consider spatial autocorrelation such as conditional autoregressive model (CAR) and simultaneous spatial autoregressive model (SAR). Plausible models can also be achieved by machine learning. For example, neural network (NN), genetic algorithm (GA), support vector machine (SVM), and random forest (RF) are popularly used for modeling species distribution.

Among them, the maximum entropy modeling (Maxent) has the advantage of not requiring explicit absence data. It also features a GUI tool for easy operation. Some studies have demonstrated its robustness with a small number of samples (Anderson and Gonzalez 2011; Shcheglovitova and Anderson 2013). Thus, we also used Maxent as an example case (later part of this chapter).

To compare several models some modeling platforms are available, such as openModeller, the sdm package or the dismo package in R. openModeller was implemented in a web-based SDM tool in a mapping platform of the OBIS Japan node named BISMaL mapper (Fig. 13). Although the BISMaL mapper can treat only certain predefined environmental variables, it is suitable for comparing the results of different models.

Higher accuracy can be achieved by not simply using a single model but by using several models. One of the simplest ways to select the best model is by comparing different models. Another way is to average different models depending on their accuracy (model averaging). The idea of this method is based on the fact that we cannot determine statistically significant weaknesses of the model that showed the closest accuracy to the best model. This type of method will increase the calculation load and increase machine power requirement. However, the accuracy will also increase in general. In addition there is additional caution in model averaging. Each of the original result shows different types of model and variable, making it difficult to directly compare the model structure of different results. But it might be not serious problem as far as used in similar model. Some machine learning models, such

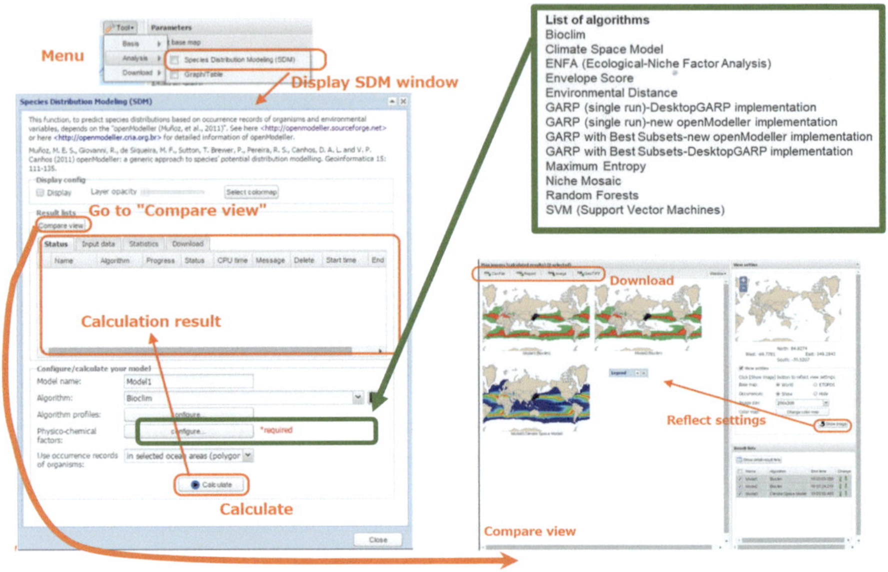

Fig. 13 The possible models shown in a window of BISMaL mapper. (From http://www.godac.jamstec.go.jp/mapper/help.jsp; Yamamoto et al. 2012; Tanaka et al. 2014)

as RF or bagging, already use a similar method in comparing model parameters within a process that repeating same analysis method in the model.

Considering characteristics of the models and machine effort, we can apply SDM to a dataset. Then we can visualize the potential distribution of species as habitat for use in EIA.

3.6 *Example of Preliminary Analysis in the Deep Sea*

This section presents an example of a preliminary analysis of species distribution in an area covered by cobalt-rich crust around the Takuyo-Daigo Seamount, which is located 1800 km southeast of the Honshu main island of the Japanese archipelago (Cruise No. NT 13-13 Dive number HPD 1544). Differences in the preference of their habitats were investigated considering two sessile species and one mobile species. Because a large number of individuals were observed, only the following species (taxons precisely) were targeted: conger (Synaphobranchidae), thin-shaped hexactinellid sponge (Lyssacinosida), and soft coral (Corallimorpharia), which are mapped in the left side of Fig. 14.

Environment factors limiting the distribution of these organisms are suitable climate conditions, appropriate food sources, less predation or completion, and suitable conditions of reproduction or attachment of the larvae. Although it is not possible to fully cover variables directly corresponding to these ultimate causes, we extracted

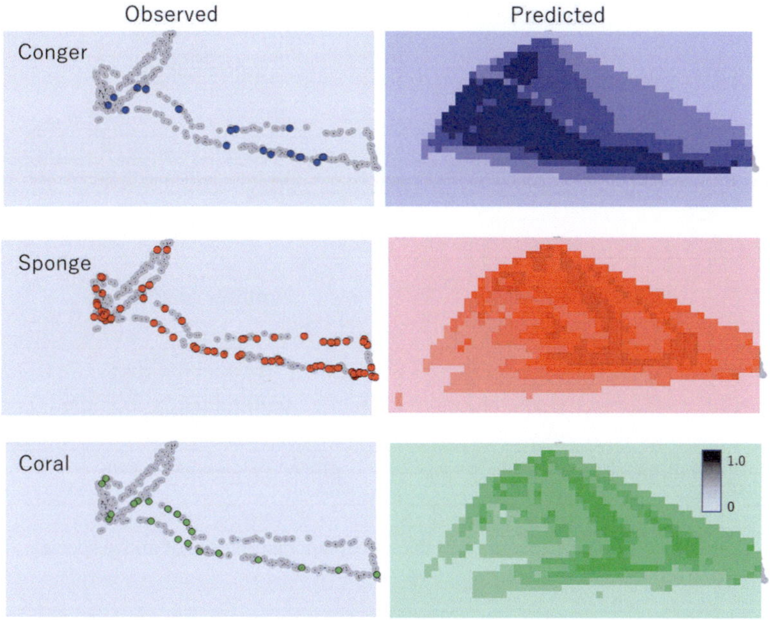

Fig. 14 Distribution of conger (Synaphobranchidae), thin-shaped hexactinellid sponge (Lyssacinosida), and soft coral (Corallimorpharia) (left) and predicted probability of presence (right).This data was created in corroboration with Katsunori Fujikura, Yukiko Nagai, and Masashi Tsuchiya

the following environmental factors: percentage of sand, water depth, water temperature, and number of discovered species (Fig. 15). Other than the number of species, all variables showed similar trends with water depth, and thus, these values were not correctly separated. All values were extracted from image and CTD data of the survey using ROV and linearly interpolated using package akima in R. Thus, the prediction of high-resolution spatial heterogeneity was limited in our area of focus.

The potential distribution of each species was estimated using Maxent. Firstly, the accuracy of the produced model was evaluated using an area under receiver operation curve (AUC). A receiver operation curve (ROC) represents the amount of increase in the model explanation power depending on the number of samples when the samples are eliminated randomly from 100% to 0%. If the model does not have information (i.e., just random), the ROC remains as a diagonal line (equal ratio of eliminated percentage of samples will be explained by the model). In that case AUC shows 0.5 fraction of the area in the plot (under a black line in Fig. 16) and considered as undistinguished from a random value. In Fig. 16, the area under the red curve means the AUC of this model, and area between the black line and red curve means the increase of the information because of the model. Conventionally AUC over 0.7 was considered as a meaningful model and over 0.9 considered as high degree of model accuracy. In this case, all modes showed over 0.7 AUC. When the number of species was removed, the AUC decreased to below 0.7 for sponge and coral.

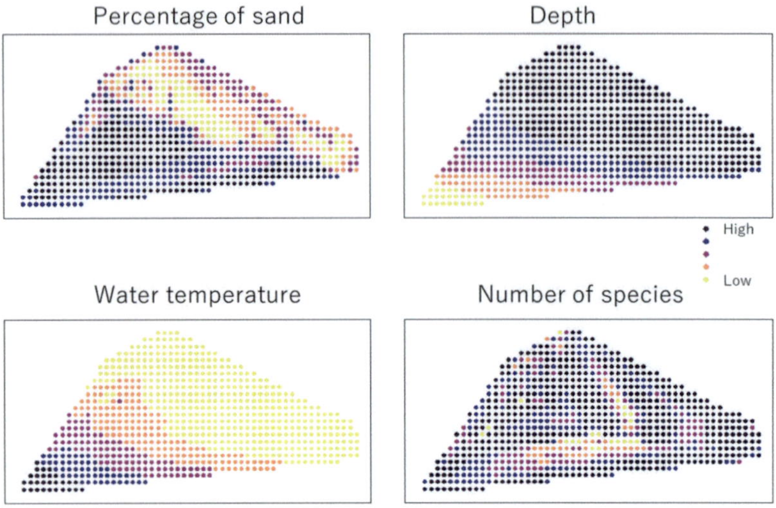

Fig. 15 Distribution of environmental variable to explain distribution of above species

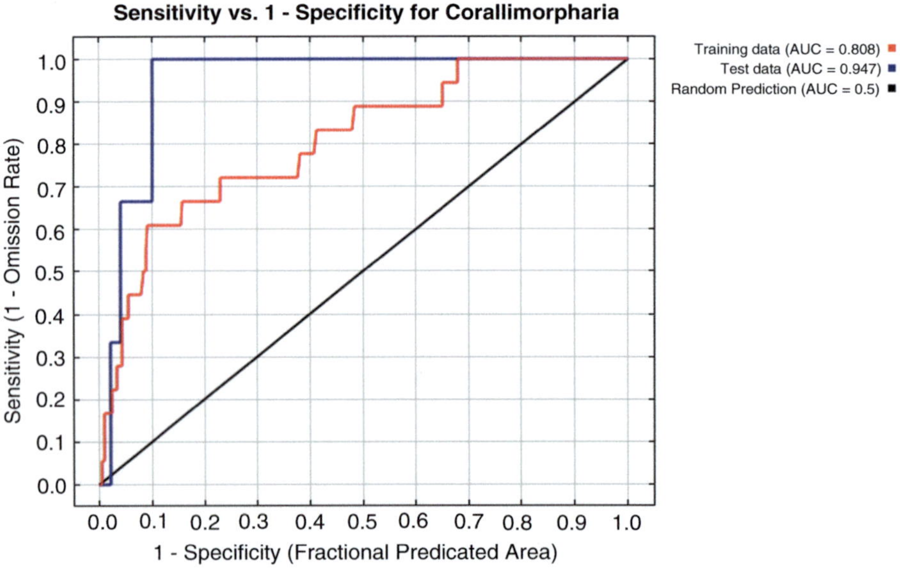

Fig. 16 An example of the calculation of AUC. The result of the prediction of soft coral distribution was used as example

The contribution of factors in each model is shown in percentages according to the results of Maxent. For the conger prediction, water temperature and water depth contributed 66% and 18%, respectively. Because the separation of correlated factors is not guaranteed, the potential distribution of conger increased with water temperature, which decreased with depth. For hexactinellid sponge, species richness and temperature contributed 52% and 42%, respectively. For soft coral, species richness and percentage of sand contributed 50% and 30%, respectively.

Response curves of the estimated relationships between environment factors and probability of the presence of species are presented in Fig. 17, with soft coral as an example. According to this curve, the potential distribution of hexactinellid sponge and soft coral increased with higher number of species. Because the shape of the response curve is also almost linear, a trend in the environment factor may be identified with the limited numbers of data which shows narrow variation of the values in limited locations. Although the model was extremely simple or not of very high accuracy, it could benefit from machine learning, and the map of species distribution could be easily obtained.

Comparing the estimated potential distribution maps (Fig. 14 right), the distribution could be basically allocated according to the water depth and temperature gradient. In particular, higher potential of conger could be explained by colder and shallower areas. Higher potential of hexactinellid sponges and soft corals was attributable to the number of species.

There are still several things to be considered before practical application. For example, the extent of this example is very limited, and the range of environment variables showed relatively small variations. Therefore, as the next steps, data acquisition from wider areas or more accurate measurement of directly related environmental variables at higher resolutions will help to produce more effective maps. Even for areas with sufficient available data, it is not easy to separate environmental variables that are correlated. Comparing different locations, testing through field experiments is another option. Estimations using AUC will be overestimated when a large number of obviously non-appropriate areas exist for the habitat. In addition, biases in data will also affect the results under certain conditions. These should be addressed by seeking the help of experts. If such issues are addressed, the feasibility of easy on demand analysis will have revolutionary implications on both environmental assessments and scientific researches.

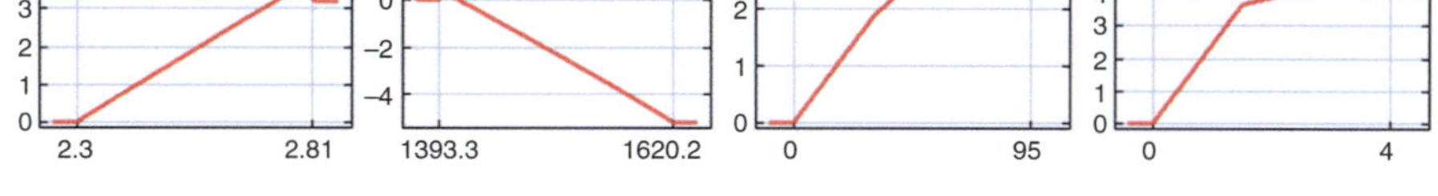

Fig. 17 Response curve of the probability of the species distribution depending on the values of the environmental variables, soft coral as example

3.7 Perspectives on the Use of the Recent Techniques

The most significant gap in mapping the deep-sea habitat is the limitation of spatial data in both the biological and environmental aspects. To overcome this problem, two new image analysis techniques developed recently were selected. They are related to the recent increase of automatic surveys using drones. One is the SfM-MVS (Structure from Motion and Multi View Stereo), which is used to create three-dimensional maps from drone images, and the other is the automatic recognition of images using machine learning, which includes deep learning.

Perspectives on Data Acquisition Feasibility for Mapping Sea Bottom

Automatic image acquisition over terrestrial areas using multi-copter platforms (UAVs, drones) is becoming popular even by normal consumers. In case of marine surveying, automatic underwater vehicle (AUV) or automatic sea surface vehicle (ASV) has also been used but only for business or professional use. As the autopilot mode is also popular in the operation of large marine ships, automatic surveying has already been undergoing technological evolution for application in marine areas. It has high potential for wide use if significant improvements in usability, cost, and customizability (open sourcing of information) are achieved. Such increase of trade in new markets attributable to lowered costs and usability has already occurred in terrestrial aerial surveying. For example, automatic helicopter surveys have existed and been used in some specific business segment before drones became popular. However, the market was totally changed after the emergence of trade edition drones, which were sold under 10% of the original price of the business edition.

Recently, drone companies especially in China are actively developing remotely operated underwater vehicles (ROVs), which have drone-like usability and prices, under the name of aquatic drones (Ito 2017). There are also waterproof versions of terrestrial multi-copter-shaped drones or drones being used as platforms to throw down sensors and cameras in water (Akamatsu et al. 2017). Although there are still several advantages of professional ROVs, the potential of aquatic drones emerging as more effective tools, especially in shallow waters, is very high in the near future. We also should consider that the formats, protocols, user interface, and data center (cloud) of widespread consumer tools will become standard in many cases. Sharing such information with each other or standardization before the intervention of new forces will be more effective for industry-wide success rather than competition among companies. Considering these situations, the use of automatic observation is an attractive option for acquiring images or sampling over wide areas in the near future or even now if sufficient funds can be invested. Nevertheless, processing of numerous images is an important technical issue in this era of big data.

Along with the automatic image acquisition using UAVs, automatic map reconstruction from numerous small-extent images is another major progress in terrestrial surveying. A technology supporting this approach is SfM-MVS (Structure from

Motion and Multi-View Stereo; Fig. 18). In this technique, the same feature points are firstly identified from several images. Using such points, three-dimensional locations of the camera are predicted, distortion is corrected, three-dimensional surfaces are produced, and textures from images are pasted over the surface of the 3D model. This type of the modeling has been used to create CAD data. Recent software intended for use in remote sensing and the increase in machine power has enabled the processing of numerous images.

For example, in the deep sea, it is not easy to take a picture of a whole sunken drift wood. Light will not reach that far, and turbidity and marine snow/small species will cause noises on the image. Therefore, we used SfM-MVS to reconstruct the wood, and we attempted to evaluate the local distribution of organisms around the sunken drift wood, which was observed in 2013 in a survey after the Great East Japan Earthquake that occurred on March 11, 2011. During the survey, we repeatedly passed over the wood and took over 40 images. Using these images, a three-dimensional model of the wood was developed, as shown in Fig. 19. Using this image and after rectifying two dimensionally merged images, we could extract species around this wood. In the results, over 90% (12) feather stars were found to be attached to the wood. Most snails were observed within 10 cm. More mobile organisms, such as fish and sea stars, were observed 20 cm and 22 cm from the wood on average, respectively. Using this information, the amount of organisms accumulated around a single wood debris and their rate of decrease with distance could be estimated.

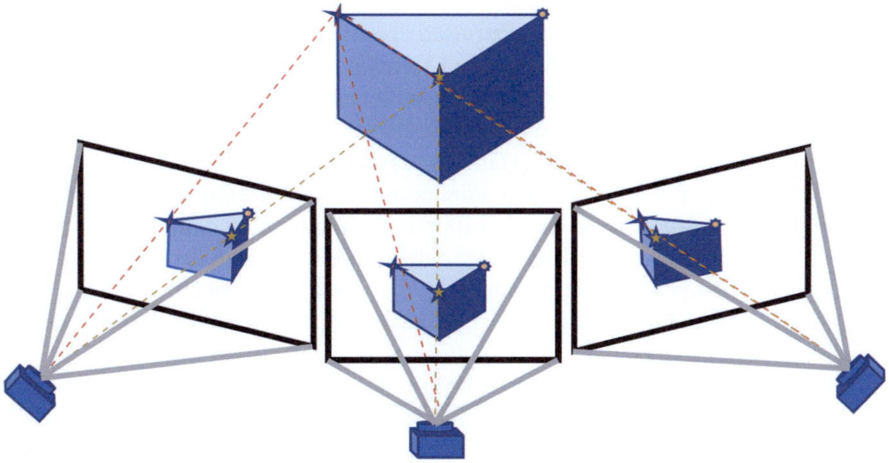

Fig. 18 Fundamental mechanism of the SfM-MVS which rebuilding triangle pole using three images. (CC-BY T. Yamakita from e-Species Biology; Yamakita 2018)

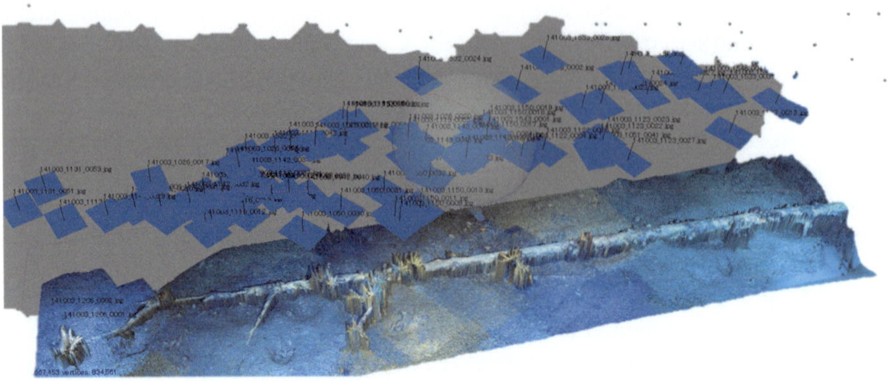

Fig. 19 A single sunken tree debris reconstructed from numerous images using SfM-MVS. (CC-BY T. Yamakita from e-Species Biology; Yamakita 2018)

Perspectives on Feasibility of Automatic Extraction of Organisms

To extract organisms from the deep-sea images, human interpretation is still implemented and considered to be the most accurate method. However, recent developments of automatic identification can be useful even for marine organisms. Automatic extraction techniques for extracting organisms have already been developed in the field of plankton survey. FlowCAM and ZooScan, which image materials in water, are among the most popular tools. These equipments feature very simple detection method using thresholds from the background to extract the organisms. To conduct more accurate and practical extraction or identification of organisms, a model is required for further analysis. For example, difference in colors and similarity of shapes are popular examples of classical analysis in image classification. In case of the remote sensing of land use from aerial photographs, difference in colors are used in pixel-based supervised classification, and shapes are also considered in the object-based classification (Yamakita et al. 2011, under review).

Unlike such simple extraction in the classical machine learning models for image recognition, feature extraction and matching with extracted features in supervised images are used as the basic ideas. The simplest way is to perform template matching, which attempts to match same feature points of two different images containing same parts by considering geometric differences. When more complex features are produced, the model will consider feature distribution within a certain size of grids (window) not just points. Using features in the window not just the recognition of types of organisms but detection of organisms from large images is possible with the sliding window method. Several feature patterns, such as Haar-like, Bags, and cascade classifiers, are used for this purpose and especially applied to face detection. We applied this method for brittle star detection in the next example. With this method, several types of images can be detected by another cascade recognition. This approach is known as bag of features (Bag of Visual Words, BoVW), which was applied in the next example together with deep learning.

Methods based on "AI" in recent years are fully automatic image recognition models that do not require the specification of any of the features to differentiate each category. Because this model is characterized by the multiple layers of neural networks, known as convolutional neural network (CNN), the model can be called a deep learning method.

We investigated the level of accuracy at which single species can be extracted by applying the automatic recognition techniques. We tested deep-sea bottom images taken after the Great East Japan Earthquake. Here, we used classical image recognition techniques using the cascade classifier model with Haar-like features and detected organisms using OpenCV, a free open-source program, to treat image data. We first produced a dataset of supervised data on organisms and inorganic materials that are frequently misclassified (Yamakita et al. 2018). Using these supervised data, we build the classification model and applied it to images of the bottom area dominated by Ophiuroidea (brittle stars). The results showed an accuracy of approximately 70% for extracting these brittle stars (Fig. 20).

Furthermore, multiple object recognition was applied using both BoVW (Bag Of Visual Words) and recent deep learning techniques (GoogleNet) to the recognition

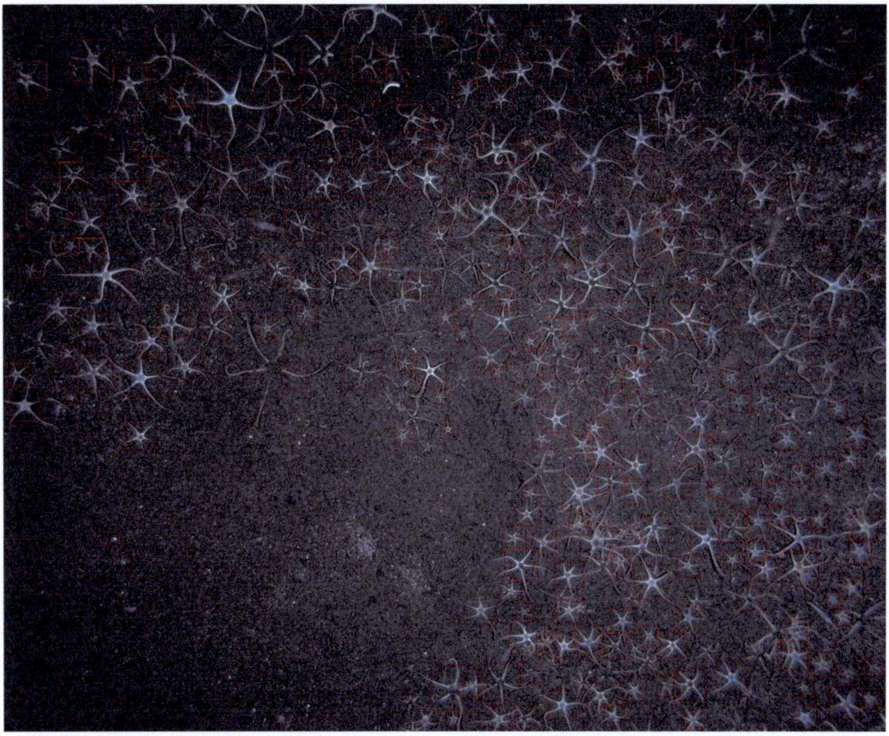

Fig. 20 Detected brittle stars (red rectangles) using the model using data published in (Yamakita et al. 2018). The model was created in collaboration with Idea Inc. Hiroyuki Yokooka and Tadayuki Kurokawa

of sediment types. We prepared 300 labeled images of sand, rocks, and bivalves (*Calyptogena* sp.) for each category. We also prepared ten randomly selected test images for each category and applied two classification models of BoW to separate rocks and bivalves at first. Then, we applied three classification models using both classical and recent techniques. To produce the features in the classical model, KAZE, which has a tendency to pick edges in images, was used. In the case of two classes, the classical model using KAZE correctly classified 70% of the bivalve images but not rock images. In the case of three classes, the accuracy of the classical model decreased to 50% for bivalves. In addition, in the case of sand images, features extracted using KAZE did not have enough recognizable features for further classification because of the relatively flat surface. The deep learning method could correctly classify at least over 80% of the images (Fig. 21).

As mentioned earlier, automation and simplification of field surveys have been developed recently. Availability of distribution data of species or more quantitative data will be dramatically improved in the near future. The use of both advanced survey techniques and advanced models for species mapping is essential, like wheels on a bus, to increase our knowledge. In deep-sea surveying, more applications and discussions on using such techniques are expected to contribute to efficient impact assessment.

4 New Technology/Approach for Investigating Planktonic Larvae

Most EIA surveys focus solely on the benthic fauna, even though it is known that mining impacts can affect the structure and dynamics of water column communities as well (Clark et al. 2017). A large proportion of the benthic fauna (e.g., bivalves, snails, and crustaceans) experience a period spent as planktonic larvae before they settle to the seafloor. Deep-sea habitats that may be targeted for mineral extraction, such as hydrothermal vent fields, are in fact a network of different communities, occurring in patches connected by larval dispersal. Such systems are referred to as metacommunities (Mullineaux et al. 2018). Meroplankton are plankton that have a benthic phase in their life history, some examples of which are the planktonic larvae of benthic animals and the adult medusa stages of jellyfish. Plankton that lack a benthic stage are referred to as holoplankton. Investigations of water column communities often focus on solely one or the other, even though each may eat, be eaten by, or otherwise interact with one or the other. To elucidate the processes involved in (re)colonization of the benthic fauna at proposed mining sites, we suggest that a combination of surveys for both holoplankton and meroplankton, as well as benthic fauna, are needed. This should lead to elucidation of the processes involved in larval dispersal, survival, and (re)colonization, which is very important when trying to assess the resilience of the community.

		BoW using KAZE (2 classes) Rock/Bivalve	BoW using KAZE (3 classes) Rock/Bivalve/sand	Deep Learning (3 classes) Rock/Bivalve/sand
Rock		Fail	Success	Fail
		Fail	Fail	Success
		No features	No features	Success
Bivalve		Success	Success	Success
		Success	Fail	Success
		Fail	Fail	Success
Sand		Not tested	No features	Success
		Not tested	No features	Success
		Not tested	No features	Fail

Fig. 21 Result of classification of sediment types using classical classification (BoVW, Bag of Visual Words) and deep learning. (Modified from Blue Earth 155; JAMSTEC 2018)

Several attempts have been made to gather data on the distributions and compositions of the planktonic community in the water column above hydrothermal vents fields. The traditional method for faunal investigation of the water column is by towing nets (summarized by Wiebe and Benfield 2003). In one example, the planktonic communities above deep-sea hydrothermal vent fields on the Juan de

Fuca Ridge were investigated using a MOCNESS (multiple opening/closing net and environmental sensing system) with a 0.25 m^2 net opening and 64 μm mesh (Mullineaux et al. 1995). This study found that larvae from vent-associated gastropods, such as the genus *Lepetodrilus* and the family Peltospiridae, were entrained by the buoyant hydrothermal plume located about 200 m above the seafloor (Mullineaux et al. 1995). MOCNESS operations have also revealed that the distribution of the larvae of deep-sea mollusks belonging to the genera *Bathymodiolus* and *Bathynerita* occur in surface waters (Arellano et al. 2014). MOCNESS appears to be a good choice for characterizing the vertical distribution of meroplankton in open-ocean environments. In more enclosed environments, such as within submarine calderas, which may sometimes have a diameter of only a few kilometers, multiple vertical tows of a net capable of sampling discrete depth layers, such as a MultiNet or a VMPS (vertical multiple plankton sampler), may be a better choice. Once the depth range at which larvae occur, and presumably disperse, has been narrowed down, larval supply and settlement from/to the deep seafloor can be assessed using large-volume pumps and sediment traps (Beaulieu et al. 2009). The pump system allows the collection of live larvae, which can then be used for further laboratory experiments (Beaulieu et al. 2013). Plankton samplers can also be deployed on AUVs (SyPRID sampler; Billings et al. 2017). The methods listed above all collect "bulk" samples at arbitrary depth ranges or time periods, even though plankton are known to be distributed in layers or patches along environmental gradients. During "bulk" sampling, environmental parameters are usually recorded only once every second, and the sampling scales for plankton and their associated environmental parameters are therefore not consistent and cannot be correlated. The continuous plankton sampler (CPR) or zooplankton sampler (ZPS) can both provide fine-scale or time-series samples. However, the number of samples that can be collected is still limited, and the sampling scales of environmental parameters still do not match those for the plankton.

This problem can be solved through image-based investigations as high-resolution images can be taken every few seconds or even multiple times per second, depending on the resolution. These images can be sufficient for observing detailed morphological characters and identifying species (Lindsay et al. 2017). The autonomous visual plankton recorder (A-VPR) obtains color images of plankton at a resolution of 1024 × 1024 pixels and a rate of 12–15 Hz. The A-VPR is a better choice for correlating environmental data with the fine-scale distribution of plankton. Unfortunately, the volume of water imaged by this system is quite small (ca. 518 mL in total, 255 mL in focus), and the pressure rating of this system is only 1000 m (Lindsay et al. 2013). There are some other commercial image-based methods for imaging plankton, such as the underwater video profiler (UVP), which images approximately 1 liter per grayscale image at a rate of 10 Hz and can be used up to a depth of 6000 m, and the Deep focus Plankton Imager (DPI), which images approximately 484 mL per in-focus grayscale image (2456 × 2058 pixels) at a rate of 1 Hz and can be deployed to depths of 2000 m. High-resolution video cameras (e.g., 8 K) can acquire 7680 × 4320 pixel color images at a rate of 60 Hz, allowing the acquisition of both large image volumes and high resolutions. A combination of various camera and imaging systems deployed on a single platform, such as towed camera system, may be optimal for plankton surveys, even in the benthopelagic layer and/or within submarine calderas.

There are several indirect methods of estimating the range of planktonic larval dispersal or connectivity within a metapopulation. Population genetic analyses help characterize the local community reliably, because genetic data from a population can allow assessment of their diversity and provide information on their history, such as whether there have been demographic contractions, bottlenecks, or migration events. DNA barcoding of a portion of the mitochondrial coxI gene has been applied successfully in a range of marine animals, including deep-sea species (Vrijenhoek 2010). Such analyses applied to the deep-sea chemosynthetic fauna have illuminated the existence of dispersal barriers or genetic gaps between local populations aligned discretely along a seafloor-spreading center, such as at a microplate junction, at a fault across an ocean spreading center, and in trenches in the Eastern Pacific (Vrijenhoek 2010). However, features determining connectivity appear to vary among ocean basins. At the slow-spreading mid-Atlantic Ridge, local populations are connected, even between populations located north and south of a transform fault (Teixeira et al. 2012). In the Indian Ocean, local populations are genetically connected along the Central Indian Ridge, but they are separated between the Central and the Southwestern Indian Ridges (Beedessee et al. 2013; Chen et al. 2015).

During the process of DNA barcoding of individuals of the same species to evaluate population connectivity, it has become apparent that there are many cryptic and synonymous species at various deep-sea hydrothermal vent fields. The environments at hydrothermal vent fields are highly heterogeneous, and the animals exhibit different morphologies according to their developmental stages and/or the environments they inhabit (e.g., the shrimp *Rimicaris*, tubeworms; Buckeridge et al. 2013). On the other hand, sometimes no morphological differences were able to be observed between animals that had clearly different DNA sequences, even with very careful observations. These animals may have evolved physiological rather than morphological differences leading to speciation, and/or there may be other invisible differences at the metaorganism level, such as differences in symbiotic bacteria (e.g., Suzuki et al. 2006). Genetic analyses can reveal these "invisible" population differences that can occur in the deep sea.

Genetic analyses using so-called next-generation sequencers (NGSs) have become standard procedure, and they can also be used for population genetics analyses. Population genetics analyses using genome-wide, single-nucleotide substitutions (SNPs), including restriction site-associated DNA sequencing (RAD-seq; reviewed by Davey et al. 2011 or 2b-RAD-seq; Wang et al. 2012) and multiplex ISSR genotyping by sequencing (MIG-seq; Suyama and Matsuki 2015), have become more mainstream than analyses using microsatellites and/or Sanger sequencing datasets. In most cases, analyses of the datasets generated by NGSs enable the detection of genetic differences between individuals and can provide more detailed information than Sanger datasets, as illustrated by recent works on

Bathymodiolus mussels (Kyuno et al. 2009; Miyazaki et al. 2010; Xu et al. 2017). Metabarcoding techniques, such as the MiFish pipeline developed for fish studies (Miya et al. 2015), could also serve as useful methods for characterizing faunal composition in the vast deep sea because DNA is probably degraded more slowly in the low-temperature environments of the deep sea than in warmer surface waters. Unfortunately, the number of reference datasets and sequences with taxonomic identification performed by trained taxonomists is still limited. Sequences from meroplankton that could plausibly be the larvae of deep-sea hydrothermal vent animals were identified during a metabarcoding analysis of plankton at station ALOHA in Hawaii (Sommer et al. 2017), although their sequence identity was lower than 97% and no morphological data were collected to corroborate the identifications. Population genetic analyses using NGSs are a powerful tool to estimate the direction of gene flow and the scale of connectivity between local populations, although they should ideally be carried out in combination with direct morphological observations of the plankton. We note that NGS analyses could also reveal physiological characteristics if transcriptomic analyses are included (Sun et al. 2017), and this could help elucidate "invisible" faunal changes, such as the occurrence patterns of morphologically cryptic species with different physiologies.

Culture experiments on larvae to characterize their behaviour and physiology can aid in the interpretation of population genetic analyses of their benthic forms. Suitable environments for planktonic larval development and the duration of planktonic larval stages vary according to the metabolic rates of larvae, reflecting species-specific traits that determine their optimal distributional range during dispersal (Marsh et al. 2001; Pradillon et al. 2001; Watanabe et al. 2006; Arellano and Young 2009; Yahagi et al. 2017). Although vertical migrations by the planktonic larvae of deep-sea animals can be important, very little quantitative data are available (Beaulieu et al. 2013; Yahagi et al. 2017). Methods to quantify the migration and settlement of the larvae of deep-sea hydrothermal vent animals in situ have also been developed. Traditional settlement experiments, with the deployment of settlement plates, are still useful in deep-sea studies for identifying areas or environments that may be recolonized and the composition of the recruits. Unfortunately, it can be difficult acquiring the ship time necessary for retrieval of the settlement plates at ideal intervals. Larval supply after a natural environmental disturbance (undersea volcanic eruption) at a hydrothermal vent field in the East Pacific Rise has been investigated, with a striking change in the species composition of larvae and colonists after the eruption (Mullineaux et al. 2010). The study found that larval supply can change markedly after removal of local source populations, enabling recolonization via immigrants from distant sites with different species composition. A settlement experiment at a deep-sea hydrothermal vent field in the Okinawa Trough, where scientific drilling had been carried out, revealed that newly established communities exhibited lower species richness than the originally occurring community; nevertheless, the populations had high genetic connectivity (Nakamura et al. 2018).

Modeling is another useful tool for investigating species distributions. Information on suitable environments and planktonic larval durations can be input into a physical oceanographic model for estimating the dispersal range of the planktonic larvae of deep-sea hydrothermal vent animals (Mullineaux et al. 2012; Mitarai et al. 2016). Further modeling for resilience (e.g., Suzuki et al. 2018) can also be carried out.

5 Conclusion

Standardized EIA technologies for ocean mining need to be established for which the major contributions are expected to be made through science and technology. However, the goal of science is to understand the principles of the natural world. Therefore, in the decision-making process, such as for making policies, laws, or regulations, scientific findings may provide convenient information, but they can also lead to inconvenient truths. Scientists maintain a spirit of criticism, and thus, they do not hesitate to modify past theories and laws. Therefore, even if a finding appears absolute and such a finding has major significance for decision-making, they may be retracted following the development of alternative theories or hypothesis by other scientists. This may hinder the progress of ocean development, policy-making, and legislation, although the scientific practices are reasonable and valid. Therefore, technological development specialized for EIA is required. In order to develop such technologies, it is necessary to not only obtain accurate results but also achieve implementation of the technology in an economic manner while meeting the requirements of legal compliance, comprehensibility, usability, etc. Taking these factors into consideration, the "Zipangu in the Ocean" project is implementing technological developments for EIA.

Acknowledgments This work was supported by Council for Science, Technology and Innovation (CSTI), Cross-ministerial Strategic Innovation Promotion Program (SIP), "Next-generation technology for ocean resources exploration" (Lead agency: JAMSTEC). We would like to thank the Ministry of Education, Culture, Sports, Science, and Technology (MEXT), Tohoku Ecosystem-Associated Marine Sciences (TEAMS), Grant-in-Aid for Scientific Research on Innovative Areas TAIGA (20109003), the Environment Research and Technology Development Fund (S-15 Predicting and Assessing Natural Capital and Ecosystem Services; S-9) of the Ministry of the Environment, and the Canon Foundation, for their partial support of this research. Additionally, we thank the members of each project for helping in data collection and supporting our survey. The authors would like to thank Hiromi Kayama Watanabe (JAMSTEC) who helped make this paper and also Fumiaki Sodeyama, Katsunori Fujikura, Yukiko Nagai, Masashi Tsuchiya (JAMSTEC), Sachiko Ono (Marine Works Japan, Ltd.), Hiroyuki Yokooka, and Tadayuki Kurokawa (IDEA Consultants, Inc.) for supporting the creation of data for habitat mapping.

References

Akamatsu, Y., Watanabe, Y., Goto, M., Inui, R., Katano, I., Nagano, M., Takahara, T., & Minamoto, T. (2017). Water sampling for environmental DNA surveys by using an unmanned aerial vehicle. *Limnology and Oceanography: Methods, 15*(11.) Wiley Online Library), 939–944.

Anderson, R. P., & Gonzalez, I., Jr. (2011). Species-specific tuning increases robustness to sampling bias in models of species distributions: An implementation with Maxent. *Ecological Modelling, 222*(15.) Elsevier), 2796–2811.

Arellano, S. M., & Young, C. M. (2009). Spawning, development, and the duration of larval life in a deep-sea cold-seep mussel. *The Biological Bulletin, 216*(2), 149–162. https://doi.org/10.2307/25470737.

Arellano, S. M., Van Gaest, A. L., Johnson, S. B., Vrijenhoek, R. C., & Young, C. M. (2014). Larvae from deep-sea methane seeps disperse in surface waters. *Proceedings of the Royal Society B: Biological Sciences, 281*(1786), 20133276. https://doi.org/10.1098/rspb.2013.3276.

Beaulieu, S. E., Mullineaux, L. S., Adams, D. K., & Mills, S. W. (2009). Comparison of a sediment trap and plankton pump for time series sampling of larvae near deep-sea hydrothermal vents. *Limnology and Oceanography: Methods, 7*, 235–248.

Beaulieu, S. E., Baker, E. T., German, C. R., & Maffei, A. (2013). An authoritative global database for active submarine hydrothermal vent fields. *Geochemistry, Geophysics, Geosystems, 14*(11), 4892–4905. https://doi.org/10.1002/2013GC004998.

Beedessee, G., Watanabe, H., Ogura, T., Nemoto, S., Yahagi, T., Nakagawa, S., Nakamura, K., Takai, K., Koonjul, M., & Marie, D. E. P. (2013). High connectivity of animal populations in deep-sea hydrothermal vent fields in the Central Indian Ridge relevant to its geological setting. *PLoS One, 8*(12), e81570. https://doi.org/10.1371/journal.pone.0081570.

Billings, A., Kaiser, C., Young, C. M., Hiebert, L. S., Cole, E., Wagner, J. K. S., & Van Dover, C. L. (2017). SyPRID sampler: A large-volume, high-resolution, autonomous, deep-ocean precision plankton sampling system. *Deep-Sea Research II, 137*, 297–306.

Buckeridge, J. S., Linse, K., & Jackson, J. A. (2013). Vulcanolepas scotiaensis sp. nov., a new deep-sea scalpelliform barnacle (Eolepadidae: Neolepadinae) from hydrothermal vents in the Scotia Sea, Antarctica. *Zootaxa, 3745*, 551–568.

Chen, C., Copley, J. T., Linse, K., & Rogers, A. D. (2015). Low connectivity between 'scaly-foot gastropod' (Mollusca: Peltospiridae) populations at hydrothermal vents on the Southwest Indian Ridge and the Central Indian Ridge. *Organisms Diversity and Evolution, 15*, 663–670.

Clark, M. R., Rouse, H., Lamarche, G., Ellis, J., & Hickey, C. (2017). Preparation of environmental impact assessments: General guidelines for offshore mining and drilling with particular reference to New Zealand. *NIWA Science and Technology Series* (pp. 105).

Clarke, C. L., Jamieson, G. S., & Road, H. B. (2006). Identification of ecologically and biologically significant areas in the Pacific North Coast Integrated Management Area: Phase II. *Final Report Pacific Biological Station Canadian Technical Report of Fisheries and Aquatic Sciences* (pp. 2686).

CRP. (2014). *Marine consent application and environmental impact assessment* (pp. 452). Wellington: Chatham Rock Phosphate Limited.

Davey, J. W., Hohenlohe, P. A., Etter, P. D., Boone, J. Q., Catchen, J. M., & Blaxter, M. L. (2011). Genome-wide genetic marker discovery and genotyping using next-generation sequencing. *Nature Reviews Genetics, 12*, 499–510.

Doi, T., Nakata, K., Kubota, M., & Aoki, S. (1999). Environmental study on the deep-sea mining of manganese nodules in the northeastern tropical Pacific –modeling the sediment-laden negative buoyant flow. *Proceedings of the Third Ocean Mining Symposium* (pp. 163–168).

Franklin, J. (2010). *Mapping species distributions: Spatial inference and prediction (ecology, biodiversity and conservation)*. Cambridge, UK: Cambridge University Press. 338 pp.

Fukubat, et al. (2018). Lander observatory with non-contact power supply and communication interfaces for long-term ecosystem monitoring in deep-sea. *Proceedings of the Twenty-eighth*

(2018) International Ocean and Polar Engineering Conference Sapporo, Japan, June 10–15, 2018 (pp. 124–127).

Fukushima, T. (2017). Technical feasibility- brief orientation-. Dialogue on practical utility of EIA technology – proponent, potential users, administrator and independent organization. *ISA Technical Study NO 18* (pp. 42–43). International Seabed Authority.

Fukushima, T., & Nishijima, M. (2017). Taxonomic problems in environmental impact assessment (EIA) linked to ocean mining and possibility of new technology developments. In R. Sharma (Ed.), *Deep-sea mining* (p. 535). Berlin: Springer.

Furuichi, N., & Hibiya, T. (2015). Assessment of the upper-ocean mixed layer parameterizations using a large eddy simulation model. *Journal of Geophysical Research: Oceans, 120*, 2350–2369.

Furuichi, N., & Higashi, H. (2014). A numerical study on suspended particulate matter dynamics in the coastal ocean bottom boundary layer using LES. *Japan Society of Civil Engineers, Ser. B2, (Coastal Engineering), 70*(2), I_1096–I_1100. (in Japanese).

Furuichi, N., Hibiya, T., & Niwa, Y. (2012). Assessment of turbulence closure models for resonant inertial response in the oceanic mixed layer using a large eddy simulation model. *Journal of Oceanography, 68*, 285–294.

Furushima, Y., & Yamamoto, H. (2015). Periodic behavior of deep sea current in the Hatoma Knoll hydrothermal system. In J.-I. Ishibashi et al. (Eds.), *Subseafloor biosphere linked to hydrothermal systems: TAIGA concept* (pp. 625–637). Tokyo: Springer Japan.

Furushima, Y., Furuichi, N., Higashi, H., Koshikawa, H., & Yamamoto, H. (2016). Periodic behaviors of deep-ocean flow and turbulent mixing in hydrothermal field on the Okinawa Trough, Japan. *Proceedings of the OCEANS 2016 MTS/IEEE MONTEREY Conference & Exhibition.* ISBN CD-ROM:978-1-5090-1538-2.

Furushima, Y., Higashi, H., Fukuhara, T., Furuichi, N., & Yamamoto, H. (2017). Direct measurement of near-bottom turbulence in deep ocean for predicting behavior of suspended particles in Sagami Bay and the Okinawa Trough, Japan. *Proceedings of the OCEANS 2017 MTS/IEEE Anchorage Conference & Exhibition.* ISBN CD-ROM: 978-0-692-93559-0.

Furushima, Y., Higashi, H., Fukuhara, T., Matsuda, T., Yamamoto, H., Furuichi, N., & Fukushima, T. (2018). Periodic behavior of abyssal flow in Okinawa trough hydrothermal fields. In *Proceedings of the 28th International Ocean and Polar Engineering Conference* (Vol. 1, pp. 119–123). Sapporo: ISOPE. isbn:978-1-880653-87-6.

Furuuchi, N., Higashi, H., & Furushima, Y. (2016). Towards modeling dispersion of suspended particulate matter caused by deep-ocean mining activity –a large-eddy simulation study on tidal bottom boundary layer. *Aquabiology, 38*(2), 145–150. (in Japanese).

Gallien, L., Münkemüller, T., Albert, C. H., Boulangeat, I., & Thuiller, W. (2010). Predicting potential distributions of invasive species: Where to go from here? *Diversity and Distributions, 16*(3), 331–342. https://doi.org/10.1111/j.1472-4642.2010.00652.x.

Glasson, J., Therivel, R., & Chadwik, A. (1995). Introduction to environmental impact assessment. *Landscape and Urban Planning, 32*(3.) Routledge), 197–198. https://doi.org/10.1016/0169-2046(95)90007-1.

Guisan, A., Zimmermann, N. E., Elith, J., Graham, C. H., Phillips, S., & Peterson, A. T. (2007). What matters for predicting the occurrences of trees: Techniques, data, or species' characteristics? *Ecological Monographs, 77*(4.) Wiley Online Library), 615–630.

Hashimoto, E., & Takasugi, Y. (1998). Measurement of the vertical mixing strength in an inner bay by floating-MSP. *Journal of Coastal engineering, 45*, 966–970. (in Japanese).

Higashi, H., & Maki, H. (2012). Field observations on oxygen-poor water masses and vertical mixing strength in Tokyo Bay. *Japan Society of Civil Engineers, Ser. B2, (Coastal Engineering), 68*(2), I_966–I_970. (in Japanese).

Higashi, H., Furushima, Y., Furuichi, N., & Fukuhara, T. (2017). Performance of vertical mixing schemes based on in-situ measurement of deep sea turbulence in Sagami Bay. *Japan Society of Civil Engineers, Ser. B2, (Coastal Engineering), 73*(2), I_79–I_84. (in Japanese).

International Seabed Authority. (2011). Environmental Management Plan for the ClarionClippertonZone. https://www.isa.org.jm/sites/default/files/files/documents/isba-17ltc-7_0.pdf. Accessed 13 Sept 2018.

International Seabed Authority. (2013). Recommendations for the guidance of contractors for the assessment of the possible environmental impacts arising from exploration for marine minerals in the area. https://www.isa.org.jm/sites/default/files/files/documents/isba-19ltc-8_0.pdf. Accessed 13 Sept 2018.

International Union for Conservation of Nature. (2018). Issues brief. http://cmsdata.iucn.org/resources/issues-briefs/deep-sea-mining. Accessed 18 Spt. 1018.

Ito, A. (2017). Current status of China Drone business: Underwater drone venture from China appears one after another is "DJI underwater" appear? *Drone Journal*. https://www.watch.impress.co.jp/headline/docs/extra/drone/1092800.html Accessed 24 Nov 2017.

JAMSTEC. (2017a). Acquisition of long-term monitoring images near the deep seafloor by Edokko Mark I, OceanBestPractices, UNESCO/IOC Project office for IODE Oostende. https://www.oceanbestpractices.net/handle/11329/326. Accessed 25 Sept 2018.

JAMSTEC. (2017b). Next-generation technology for ocean resources exploration (Zipangu in the Ocean). https://www.jamstec.go.jp/sip/en/index.html. Accessed 18 Sept 2018.

JAMSTEC. (2018). Automatic classification of the ocean floor by deep learning and draw species distribution maps. *Blue Earth, 155*, 16–19. (in Japanese).

Jankowski, J. A., Malcherek, A., & Zielke, W. (1996). Numerical modeling of suspended sediment due to deep-sea mining. *Journal of Geophysical Research, 101*(C2), 3545–3560.

JOGMEC. (2017). Exploration and technological development for deep seabed polymetallic sulphides. http://www.jogmec.go.jp/english/stockpiling/metal_10_000002.html Accessed 18 Sept 2018.

Kyuno, A., Shintaku, M., Fujita, Y., Matsumoto, H., Utsumi, M., Watanabe, H., Fujiwara, Y., & Miyazaki, J.-I. (2009). Dispersal and differentiation of deep-sea mussels of the genus bathymodiolus (Mytilidae, Bathymodiolinae). *Journal of Marine Biology, 2009*, 625672. https://doi.org/10.1155/2009/625672.

Lindsay, D. J., Yoshida, H., Ishibashi, S., Umetsu, M., Yamaguchi, A., Yamamoto, H., Nishikawa, J., Reimer, J.D., Watanabe, H., Fujikura, K., & Maruyama, T. (2013). The uROV PICASSO, the Visual Plankton Recorder, and other attempts to image plankton. *Proceedings of the International Symposium on Underwater Technology,* 2013: doi: https://doi.org/10.1109/UT.2013.6519854.

Lindsay, D. J., Nishikawa, J., Sunahara, K., Fujiwara, Y., & Yamaguchi, A. (2017). First record of the doliolid genus *Paradoliopsis* in the Pacific Ocean. *Plankton and Benthos Research, 12*(1), 66–70.

Marsh, A. G., Mullineaux, L. S., Young, C. M., & Manahan, D. T. (2001). Larval dispersal potential of the tubeworm Riftiapachyptila at deep-sea hydrothermal vents. *Nature, 411*(6833), 77–80. https://doi.org/10.1038/35075063.

Matsui, T., Sugishima, H., Okamoto, N., & Igarashi Y. (2018). Evaluation of turbidity and resedimentation through seafloor disturbance experiments for assessment of environmental impacts associated with exploitation of seafloor massive sulfides mining. Proc. of the Twenty-eighth (2018) *International Ocean and Polar Engineering Conference* Sapporo, Japan, June 10–15, 2018, (pp. 124–127).

Mitarai, S., Watanabe, H., Nakajima, Y., Shchepetkinc, A. F., & McWilliams, J. C. (2016). Quantifying dispersal from hydrothermal vent fields in the western Pacific Ocean. *Proceedings of the National Academy of Science, 113*(11), 2976–2981.

Miya, M., Sato, Y., Fukunaga, T., Sado, T., Poulsen, J. Y., Sato, K., Minamoto, T., Yamamoto, S., Yamanaka, H., Araki, H., Kondoh, M., & Iwasaki, W. (2015). MiFish, a set of universal PCR primers for metabarcoding environmental DNA from fishes: Detection of more than 230 subtropical marine species. *Royal Society Open Science, 2*, 150088. https://doi.org/10.1098/rsos.150088.

Miyata, Y., & Fukushima, T. (2018) Comparison of components and description for benthic organisms between existing environmental impact statements and administrative frameworks. *Proceedings of the Twenty-eighth (2018) International Ocean and Polar Engineering Conference Sapporo*, Japan, June 10–15, 2018, (pp. 128–135).

Miyazaki, J.-I., Martins, L. D. O., Fujita, Y., Matsumoto, H., & Fujiwara, Y. (2010). Evolutionary process of deep-sea <italic>Bathymodiolus</italic> mussels. *PLoS One, 5*(4), e10363. https://doi.org/10.1371/journal.pone.0010363.

Mullineaux, L. S., Wiebe, P. H., & Baker, E. T. (1995). Larvae of benthic invertebrates in hydrothermal vent plumes over Juan de Fuca Ridge. *Marine Biology, 122*, 585–596.

Mullineaux, L. S., Adams, D. K., Mills, S. W., & Beaulieu, S. E. (2010). Larvae from afar colonize deep-sea hydrothermal vents after a catastrophic eruption. *Proceedings of the National Academy of Sciences of the United States of America, 107*(17), 7829–7834. https://doi.org/10.1073/pnas.0913187107.

Mullineaux, L. S., Le Bris, N., Mills, S. W., Henri, P., Bayer, S. R., Secrist, R. G., & Siu, N. (2012). Detecting the influence of initial pioneers on succession at deep-sea vents. *PLoS One, 7*(12), e50015.

Mullineaux, L. S., Metaxas, A., Beaulieu, S. E., Bright, M., Gollner, S., Grupe, B. M., Herrera, S., Kellner, J., Levin, L. A., Mitarai, S., Neubert, M. G., Thurnber, A. M., Tunnicliffe, V., Watanabe, H. K., & Won, Y.-J. (2018). Exploring the ecology of deep-sea hydrothermal vents in a metacommunity framework. *Frontiers in Marine Science, 5*, 49. https://doi.org/10.3389/fmars.2018.00049.

Nagao, M., Hashimoto, E., & Takasugi, Y. (2004). Measurement of the vertical mixing strength in the Seto Inland Sea. *Journal of Coastal engineering, 51*, 946–950. (in Japanese).

Nagao, M., Hashimoto, E., Takasugi, Y., Chiba, S., & Yamagata, Y. (2005). Measurement of the vertical mixing strength in the Ago Bay. *Journal of Coastal engineering, 52*, 341–345. (in Japanese).

Nakamura, M., Nakajima, Y., Watanabe, H. K., Sasaki, T., Yamamoto, H., & Mitarai, S. (2018). Spatial variability in recruitment of benthos near drilling sites in the Iheya North hydrothermal field in the Okinawa Trough. *Deep-Sea Research I, 135*, 63–75. https://doi.org/10.1016/j.dsr.2018.03.009.

Nakata, K., Kubota, M., Aoki, S., & Taguchi, K. (1997). Dispersion of resuspended sediment by ocean mining activity -modelling study. *Proceedings of International Symposium on Environmental Studies for Deep-Sea Mining* (pp. 169–186).

Namphos. (2012). Environmental impact assessment for the proposed dredging of phosphate enriched sediments from marine license area no. 170. (pp. 99). Cape Town: Namibian Marine Phosphate (Pty), Ltd.

Nautilus Minerals Niugini Limited (Nautilus). (2008). Environmental impact statement, Solwara 1 Project. Nautilus Minerals Niugini Limited, Main Report, Nautilus Minerals Niugini Limited.

Nishikawa, H., Toyoda, T., Masuda, S., Ishikawa, Y., Sasaki, Y., Igarashi, H., Sakai, M., Seito, M., & Awaji, T. (2015). Wind-induced stock variation of the neon flying squid (Ommastrephesbartramii) winter–spring cohort in the subtropical North Pacific Ocean. *Fisheries Oceanography, 24*(3).Wiley Online Library), 229–241.

Pradillon, F., Shillito, B., Young, C. M., & Gaill, F. (2001). Deep-sea ecology: Developmental arrest in vent worm embryos. *Nature, 413*(6857), 698–699.

Rochette, S., Huret, M., Rivot, E., & Pape, O. L. (2012). Coupling hydrodynamic and individual-based models to simulate long-term larval supply to coastal nursery areas. *Fisheries Oceanography, 21*(4.) Wiley Online Library), 229–242.

Saita, T., Yano, S., Tai, A., Shiki, S., Shigeta, S., & Komatsu, T. (2008). Field observation of vertical mixing strength in Ariake Sea during winter. *Journal of Coastal engineering, 55*, 421–425. (in Japanese).

Shcheglovitova, M., & Anderson, R. P. (2013). Estimating optimal complexity for ecological niche models: A jackknife approach for species with small sample sizes. *Ecological Modelling, 269*. Elsevier, 9–17.

Sommer, S. A., Van Woudenberg, L., Lenz, P. H., Cepeda, G., & Goetze, E. (2017). Vertical gradients in species richness and community composition across the twilight zone in the North Pacific Subtropical Gyre. *Molecular Ecology, 26*, 6136–6156.

Sugishima, H., Kato, S., Matsui, T., Naito, K., Maeda, N., & Miwa, T. (2018). Utilization of a free-fall deep-sea camera lamder (Edokko Mark1) in environmental baseline survey of the deep sea mining exploration area. *Proceedings of the 27th Ocean Engineering Symposium*, Tokyo, August 7–8, 2018 (in Japanese).

Sun, J., Zhang, Y., Xu, T., Zhang, Y., Mu, H., Zhang, Y., Lan, Y., Fields, C. J., Hui, J. H. L., Zhang, W., Li, R., Nong, W., Cheung, F. K. M., Qiu, J.-W., & Qian, P.-Y. (2017). Adaptation to deep-sea chemosynthetic environments as revealed by mussel genomes. *Nature Ecology and Evolution, 1*, 0121. https://doi.org/10.1038/s41559-017-0121.

Suyama, Y., & Matsuki, Y. (2015). MIG-seq: An effective PCRbased method for genome-wide single-nucleotide polymorphism genotyping using the next -generation sequencing platform. *Scientific Reports, 5*, 16963. https://doi.org/10.1038/srep16963.

Suzuki, Y., Kojima, S., Sasaki, T., Suzuki, M., Utsumi, T., Watanabe, H., Urakawa, H., Tsuchida, S., Nunoura, T., Hirayama, H., Takai, K., Nealson, K. H., & Horikoshi, K. (2006). Host-symbiont relationships in hydrothermal vent gastropods of the genus Alviniconcha from the Southwest Pacific. *Applied and Environmental Microbiology, 72*(2), 1388–1393. https://doi.org/10.1128/AEM.72.2.1388-1393.2006.

Suzuki, K., Yoshida, K., Watanabe, H., & Yamamoto, H. (2018). Mapping the resilience of chemosynthetic communities in hydrothermal vent fields. *Scientific Reports, 8*, 9364. https://doi.org/10.1038/s41598-018-27596-7.

Tanaka, K., Arai, T., Kishira, T., Yamauchi, T., Saito, H., Hanafusa, Y., Sonoda, A., Furushima, Y., Yamamoto, H., Fujikura, K., & Maruyama, T. (2014). Publication and application of marine biological information. *JAMSTEC Report of Research and Development, 18*, 81–88.

Teixeira, S., Serrao, E. A., & Arnaud-Haond, S. (2012). Panmixia in a fragmented and unstable environment: The hydrothermal shrimp Rimicarisexoculata disperses extensively along the Mid-Atlantic Ridge. *PLoS One, 7*(6), e38521.

Tennekes, H., & Lumley, J. L. (1972). *A first course in turbulence* (pp. 1–26). Cambridge, MA: The MIT Press.

Underwood, A. J. (1997). *Experiments in ecology: Their logical design and interpretation using analysis of variance*. Cambridge, MA: Cambridge University Press.

Vrijenhoek, R. C. (2010). Genetic diversity and connectivity of deep-sea hydrothermal vent metapopulations. *Molecular Ecology, 19*(20), 4391–4411. https://doi.org/10.1111/j.1365-294X.2010.04789.x.

Wang, S., Meyer, E., McKay, J. K., & Matz, M. V. (2012). 2b-RAD: A simple and flexible method for genome-wide genotyping. *Nature Methods, 9*(8), 808–810. http://www.nature.com/nmeth/journal/v9/n8/abs/nmeth.2023.html#supplementary-information.

Watanabe, H., Kado, R., Kaida, M., Tsuchida, S., & Kojima, S. (2006). Dispersal of vent-barnacle (genus Neoverruca) in the Western Pacific. *Chaiers de Biologie Marine, 47*, 353–357.

Wiebe, P. H., & Benfield, M. C. (2003). From the Hensen net toward four-dimensional biological oceanography. *Progress in Oceanography, 56*, 7–136.

Xiaodong, S., Qi, Y., Chen, G., Liang, C., Rolf, G. L., Brett, P., & Li, H. (2017). An expendable microstructure profiler for deep ocean measurements. *Journal of Atmospheric and Oceanic Technology, 34*, 153–165.

Xu, T., Sun, J., Lv, J., Watanabe, H. K., Li, T., Zou, W., Rouse, G. W., Wang, S., Qian, P.-Y., Bao, Z., & Qiu, J.-W. (2017). Genome-wide discovery of single nucleotide polymorphisms (SNPs) and single nucleotide variants (SNVs) in deep-sea mussels: Potential use in population genomics and cross-species application. *Deep-Sea Research II, 137*, 318–326.

Yahagi, T., Watanabe, H. K., Kojima, S., & Kano, Y. (2017). Do larvae from deep-sea hydrothermal vents disperse in surface waters? *Ecology, 98*, 1524–1534.

Yamakita, T. (2018). Change of the ocean after the Great East Japan earthquake -utilization of Geographic Information System and GIScience-. *E-Research in Species Biology* 2(1). The Society for the Study of Species Biology: in press.

Yamakita, T., & Nakaoka, M. (2009). Scale dependency in seagrass dynamics: How does the neighboring effect vary with grain of observation? *Population Ecology, 51*(1.) Springer Japan), 33–40. https://doi.org/10.1007/s10144-008-0119-z.

Yamakita, T., Watanabe, K., & Nakaoka, M. (2011). Asynchronous local dynamics contributes to stability of a seagrass bed in Tokyo Bay. *Ecography, 34*(3), 519–528. https://doi.org/10.1111/j.1600-0587.2010.06490.x.

Yamakita, T., Sudo, K., Jintsu-Uchifune, Y., Yamamoto, H., & Shirayama, Y. (2017). Identification of important marine areas using ecologically or biologically significant areas (EBSAs) criteria in the East to Southeast Asia region and comparison with existing registered areas for the purpose of conservation. *Marine Policy, 81*, 273–284. https://doi.org/10.1016/j.marpol.2017.03.040.

Yamakita, T., Yokooka, H., Fujiwara, Y., Kawato, M., Tsuchida, S., Ishibashi, S., Kurokawa, T., & Fujikura, K. (2018). Image dataset of ophiuroid and other deep sea benthic organisms in 2015 extracted from the survey off Sanriku, Japan, by the research following the Great East Japan Earthquake 2011. *Ecological Research, 33*(2), 285–285. https://doi.org/10.1007/s11284-018-1571-7.

Yamakita, T., Sodeyama, F., Wanpetch, N., Watanabe, K., & Nakaoka, M. (under review). Application of deep learning techniques for determining the spatial extent and classification of seagrass beds, Trang, Thailand submitted to Botanica Marina.

Yamamoto, H., Tanaka, K., Fujikura, K., & Maruyama, T. (2012). BISMaL: Biological Information System for Marine Life and role for biodiversity research. In *The Biodiversity Observation Network in the Asia-Pacific Region* (pp. 247–256). New York: Springer.

Dr. Yasuo Furushima is a senior engineer at the Research and Development (R&D) Center of the Submarine Resources Environmental Impact Assessment Research Group, Japan Agency for Marine-Earth Science and Technology (JAMSTEC). He graduated from the School of Marine Science and Technology, Tokai University, in 1988 (bachelor's degree) and Graduate School of Oceanography, Tokai University, in 1990 (master's degree). In 1996, he got his PhD in agriculture in the Graduate School of Agricultural and Life Sciences, University of Tokyo (UT). He was employed as a research scientist in JAMSTEC in 1997. His specialty is fisheries oceanology and physical oceanography. He is studying dynamics of bottom mixed layer using an expendable vertical microstructure profiler.

Dr. Takehisa Yamakita is a spatial ecologist at the JAMSTEC as scientist. He also gives a lecture on marine data processing as associate professor of Hiroshima University. He acquired his PhD in spatial dynamics of seagrass in Chiba University with a President's Award for excellent grades. As postdoctoral researcher at the University of Tokyo, Fishery Research Agency, and Forestry & Forest Products Research Institute, he started to predict marine biodiversity over the East and Southeast Asia to support conservation policy. Recently he worked as a sub-leader of the marine group in a project aimed at predicting the future status of ecosystem services (S15; PANCES) and also has several adjunct roles such as lead author of IPBES Regional assessment, one of the grand marshal members of the marine group in Asia Pacific BON (AP-BON), and unit subleader of earthquake and TSUNAMI assessment in Tohoku Ecosystem-Associated Marine Sciences (TEAMS).

Dr. Tetsuya MIWA is the group leader in the Instrumentation and Robotics Group, Department of Engineering for Geo-Marine Sciences, Marine Technology and Engineering Center, JAMSTEC. He is a technology researcher with a focus on marine engineering and also serves concurrently as a researcher of the Ecosystem Observation and Evaluation Methodology Research Unit, Project Team for Development of New-generation Research Protocol for Submarine Resources, Next-generation Technology for Ocean Resources Exploration, and Cross-ministerial Strategic Innovation Promotion Program (since 2014). His recent interest is the development of a small lander system "EDOKKO MarkI." He was born in 1964 and obtained his PhD in Engineering from the Tokyo Institute of Technology in 1991.

Dr. Dhugal Lindsay (Australia), a marine biologist, is a senior research scientist of the Japan Agency for Marine-Earth Science and Technology (JAMSTEC) and holds adjunct professorships at the Yokohama Municipal University and Kitasato University. Dr. Lindsay's research focuses on mid-water ecology, particularly concentrating on gelatinous organisms, and on the development of new survey tools for investigating the mid-water biota. He is currently involved with the Deep Ocean Stewardship Initiative (DOSI) by serving as the head of the New Technologies Working Group and with the Scientific Committee on Ocean Research (SCOR) by serving on the Working Group "Translation of Optical Measurements into Particle Content, Aggregation & Transfer (TOMCAT)." Dr. Lindsay served as a steering committee member (CMarZ) and as a National Regional Implementation Committee member (Japan) for the Census of Marine Life and has also served on the National Academies of Science Committee "Future Needs in Deep Submergence Science." In addition to his role as editor-in-chief for the scientific journal *Plankton and Benthos Research*, he also serves on the Editorial Boards of the *Marine Technology Society Journal* (USA) and *Scientia Marina*. Dr. Lindsay has participated in over 50 research cruises, both as chief scientist and as a member of multidisciplinary teams, aboard various Japanese research vessels and cruises on Australian, US, and Spanish vessels from the Arctic to Antarctic, and has dived 21 times in crewed submersibles. His publication record includes over 125 scientific papers.

Dr. Tomohiko Fukushima works for the Research and Development (R&D) Center of Submarine Resources, Japan Agency for Marine-Earth Science and Technology (JAMSTEC). He has graduated from the Faculty of Fisheries, Tokyo University, in 1984 (bachelor's degree) and Graduate School of Fisheries Science, Tokyo University, in 1986 (master's degree). In 2004, he obtained his D.Sc. from Kyoto University. The title of his doctoral thesis is "Ecological Characteristics of Deep-Sea Benthic Organisms in Relation to Manganese Nodules Development Practices." After receiving his doctoral degree, he continued his research in the Ocean Policy Research Foundation as a research fellow and at the University of Tokyo as associate professor and then went to the incumbent.

Dr. Yoshihisa Shirayama born in 1955 in Tokyo, Japan, obtained his D.Sc. from the Graduate School of Science, University of Tokyo (UT), in 1982. He then served as assistant and then associate professor at the Ocean Research Institute, UT. In 1997, he became a professor of Seto Marine Biological Laboratory, Faculty of Science, Kyoto University. In 2003, the laboratory moved to Field Science Education and Research Center. He served as director of the center from 2007. After that, he has worked as the executive director of Research at Japan Agency for Marine-Earth Science and Technology (JAMSTEC) for 7 years starting in April 2011 and has served as both the associate executive director and the director of the Global Oceanographic Data Center (GODAC) of JAMSTEC from April 2018. His major research field is marine biology, especially taxonomy and ecology of deep-sea meiobenthos. He also is working on the marine biodiversity and the impact of ocean acidification upon it. He was awarded "Okada Prize" from Oceanographic Society of Japan in 1988, Minister of Environment Japan Recognition in 2011, and "Oceanic State Promotion Contributors Award", one of the Prime Minister's Award in Japan, in August 2018. He also was awarded Cosmos International Prize as a member of Scientific Steering Committee of Census of Marine Life in 2011.

Environmental Factors for Design and Operation of Deep-Sea Mining System: Based on Case Studies

Rahul Sharma

Abstract This chapter brings together environmental factors that are expected to influence the design and operation of deep-sea mining system for polymetallic nodules. The study analyzes possible effects of various environmental factors from nodule-bearing areas in the Central Indian Ocean Basin (CIOB) and compares it with available data from the Pacific Ocean. Similar applications can be envisaged for different mineral deposits in other ocean basins as well.

Keywords Environmental factors · Design considerations · Deep-sea mining system

1 Introduction

The term "environment" in relation to deep-sea mining is generally applied for the purpose of assessing and monitoring the environmental impact of various mining activities, for which International Seabed Authority has established environmental guidelines for different deep-sea minerals (ISA 2013, which are being amended in 2018) and is in the process of formulating environmental regulations (ISA 2017). However, environment and mining share a reverse relationship as well; that is how various environmental factors could impact on (or influence) the design and operation of different components of the deep-sea mining system such as the miner on the seafloor, lifting mechanism in the water column, as well as mining platform on the surface (Table 1). This study evaluates different environmental parameters from some of the potential mining areas in the Indian Ocean and the Pacific Ocean (Fig. 1). It is expected that evaluation of these environmental factors could be extended to other ocean basins for different mineral deposits as well.

R. Sharma (✉)
CSIR-National Institute of Oceanography, Dona Paula, Goa, India
e-mail: rsharma@nio.org

R. Sharma (ed.), *Environmental Issues of Deep-Sea Mining*,
https://doi.org/10.1007/978-3-030-12696-4_12

Table 1 Possible influence of environmental factors on different components of deep-sea mining system

Components	Environmental factors	Possible influence
(a) *Surface components* Mining platform including handling equipment, power generation, processing plant and transport vessels	*Atmospheric and sea surface* – sea surface temperature, wind, rainfall, cyclone, waves	Operating conditions for personnel and equipment and stability of platform as well as transport vessels
(b) *Subsurface components*[a] Ore lifting mechanism Intermediate buffer(s)	*Water column* – currents, temperature, salinity	Behavior of lifting pipes, umbilical and pumping mechanisms as well as leakage/spillage and discharge
(c) *Seafloor components* Nodule miner (hydraulic/mechanical/ submersible-type pickup devices)	*Seafloor* Mineral props. – size, coverage, abundance Substrates – sediment thickness, physical properties, rock outcrops Topography, micro-topography, slopes	Efficiency of miner, weight/ buoyancy, design of pickup device due to variable sizes and inhomogenous distribution of nodules, occurrence of sediments, rock outcrops, seafloor undulations and slopes

[a]In case of submersible type of miner, subsurface components will be eliminated

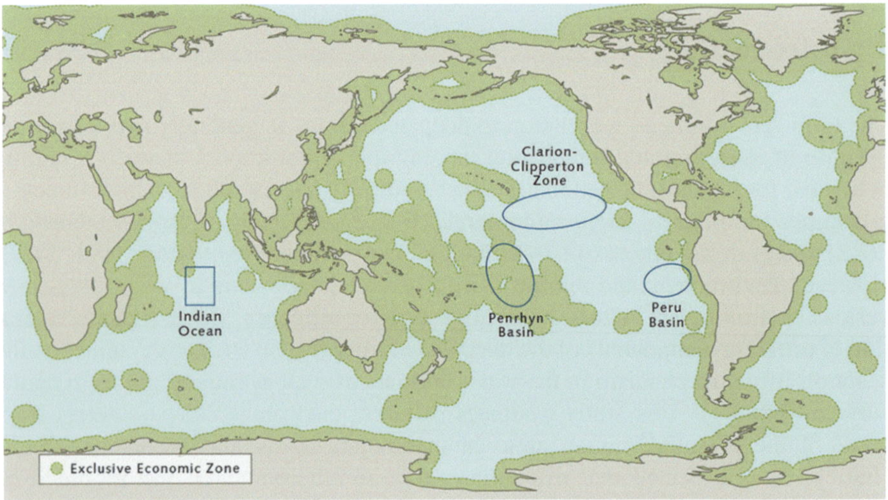

Fig. 1 Areas of potential deep-sea mining in the Central Indian Ocean Basin and Pacific Ocean. (Source: Mukhopadhyay et al. 2018)

2 Atmospheric and Sea Surface Parameters

Various factors that could influence the mining operations on or above the water column include sea surface temperature, wind, rainfall, cyclones, and waves. This section looks at historical data for almost 10-year period for sea surface

Table 2 Variation in temperature for different months (in degrees Celsius)

Year/month	Jan	Feb	Mar	Apr	May	Jun	Jul	Aug	Sep	Oct	Nov	Dec
1988	28.35	29.29	29.43	29.5	28.10	26.4	25.41	25.93	21.71	26.24	26.35	26.95
1989	28.59	28.82	28.75	28.09	27.02	25.33	21.10	23.88	24.27	25.20	26.08	27.33
1990	28.66	28.20	28.93	28.12	27.48	25.73	25.23	24.49	24.76	25.39	26.40	27.26
1991	28.14	28.71	28.83	28.63	28.36	27.45	25.74	24.42	24.33	24.75	26.38	27.13
1992	28.34	28.61	28.76	28.74	27.63	25.94	25.08	24.78	24.44	25.00	25.55	26.52
1993	27.20	27.82	28.61	28.67	27.10	26.01	25.26	25.23	25.08	25.23	26.21	27.09
1994	27.28	27.91	28.31	27.86	27.26	25.15	25.00	24.70	24.91	25.52	26.36	27.74
1995	23.61	28.45	28.76	28.30	27.56	26.63	25.31	24.71	24.67	25.01	26.07	27.16
1996	28.00	28.46	28.65	28.28	27.64	26.72	25.28	24.56	24.74	25.06	26.42	27.33
1997	28.17	28.03	28.34	28.54	27.38	26.26	25.12	24.81	25.30	26.33	27.43	28.26
1998	29.07	28.96	29.48	29.16	28.33	26.90	25.53	24.63	24.44	24.74	25.18	25.92
Average	27.76	28.47	28.80	28.53	27.62	26.22	24.91	24.74	24.42	25.31	26.22	27.15
Minimum	23.61	27.82	28.31	27.86	27.02	25.15	21.10	23.88	21.71	26.33	25.18	25.92
Maximum	29.07	28.96	29.48	29.16	28.36	27.45	25.74	25.93	25.30	24.74	27.43	28.26
Standard deviation	2.850	0.459	0.587	0.650	0.671	1.150	2.474	1.029	1.871	0.805	1.126	1.170

temperature, wind speed, and rainfall (1988–1998), cyclones (1997–2007), and waves (1953–1961) from global sources for the Central Indian Ocean Basin (Grass et al. 2000; www.metoffice.gov.uk; Hogben and Lumb 1967).

2.1 Sea Surface Temperature

Sea surface temperature evaluated for the period (Table 2 and Fig. 2) shows that average temperature between December and May is between 27 and 29° C that corresponds to Austral summer, whereas the period from June to November has an average temperature between 24 and 26° C corresponding with Austral winter. In terms of maximum and minimum temperatures, also the temperatures are generally within the range of +/−2° C (Grass et al. 2000). Hence, the temperature range in the area is very small throughout the year with no extreme temperatures that could have any serious effects on the deep-sea mining operations.

2.2 Wind Speed

Wind speed data collected during the period (Table 3 and Fig. 3) shows that mean wind speed is relatively higher (6.8–8.4 m/s) from June to November with August and September recording the highest wind speeds (maximum of 9.39 m/s and 9.53 m/s, respectively), whereas these are lower (4.8–6.5 m/s) from December to May with February and March having the lowest wind speeds (minimum of 3.38 and 3.88 m/s) during different years. Although, in general, wind speeds are not very strong throughout the year, winds are responsible for generation of waves and can have serious impact on operations during the occurrence of cyclones and hurricanes (see Sect. 2.4).

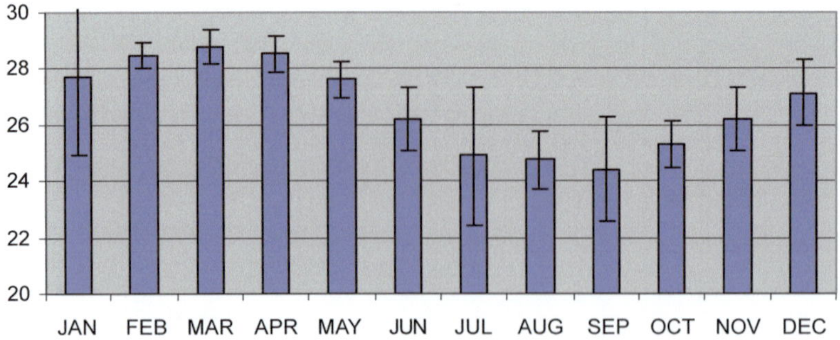

Fig. 2 Variation in average temperature for different months (in degree Celsius)

Table 3 Variation in mean wind speed for different months (in m/s)

Year/month	Jan	Feb	Mar	Apr	May	Jun	Jul	Aug	Sep	Oct	Nov	Dec
1988	5.362	3.3895	3.881	3.93	5.897	6.219	7.755	7.223	7.156	7.306	6.625	7.306
1989	4.764	5.756	4.486	5.787	7.029	7.427	8.220	8.350	7.995	7.206	6.075	6.23
1990	5.480	4.257	5.072	5.398	6.201	7.270	6.654	7.866	8.275	7.164	6.690	5.869
1991	6.172	4.996	5.661	6.120	5.780	7.011	8.722	8.311	9.530	8.698	7.390	6.683
1992	4.755	6.202	4.599	5.073	6.803	7.540	8.245	8.719	8.371	8.447	7.423	6.600
1993	6.610	4.852	4.314	5.713	7.127	7.812	7.929	7.959	8.05	8.200	6.990	6.751
1994	6.027	4.900	5.328	6.212	6.381	7.681	8.983	9.39	8.004	7.830	7.610	6.277
1995	5.620	5.364	5.420	6.324	6.358	7.205	8.500	9.029	8.811	7.450	6.844	6.130
1996	5.690	5.915	5.536	6.753	6.185	7.740	7.933	8.580	8.035	7.847	6.740	5.039
1997	5.149	5.381	4.593	5.867	6.826	8.001	8.811	8.63	7.529	7.199	6.381	5.420
1998	5.567	5.199	4.166	5.101	6.990	7.922	8.227	8.68	9.05	8.708	6.502	7.483
Average	5.563	5.110	4.823	5.661	6.507	7.438	8.179	8.430	8.255	7.823	6.842	6.344
Minimum	4.755	3.389	3.881	3.93	5.897	6.219	6.654	7.223	7.156	7.164	6.075	5.039
Maximum	6.610	5.915	5.661	6.753	7.127	8.001	8.983	9.39	9.530	8.708	7.610	7.483
Standard deviation	0.930	1.290	0.890	1.423	0.615	0.910	1.182	1.085	1.188	0.774	0.767	1.222

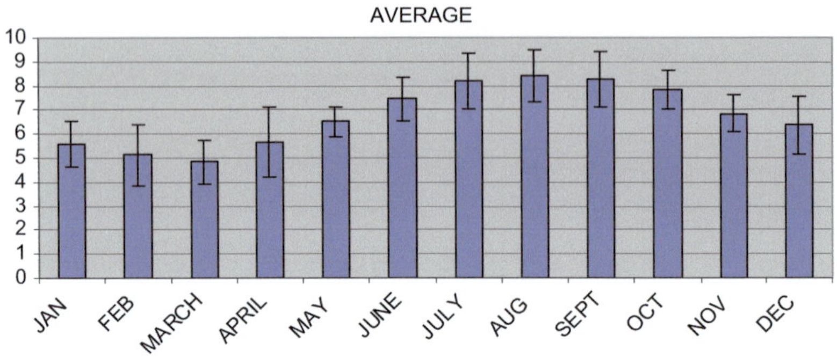

Fig. 3 Average wind speed for different months (in m/s)

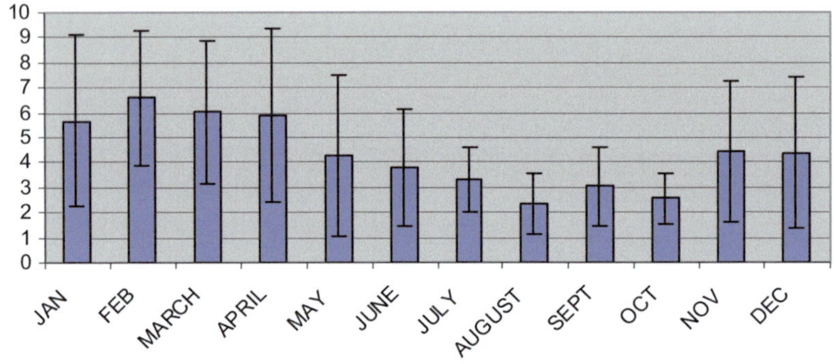

Fig. 4 Variation in average rainfall for various months (in mm/day)

2.3 *Rainfall*

Analysis of mean rainfall data for different months during the period (Fig. 4 and Table 4) shows that although the area does not appear to receive very heavy rains throughout the year, the months from January to April received relatively higher average rainfall (5.679–6.577 mm/day), whereas July to October have recorded lower average rainfall (2.327–3.311 mm/day), and the intervening months (May to June and November to December) had intermediate rainfall. However, there could be episodic heavy rainfall coinciding with the events of cyclones as described in the next section.

Table 4 Variation in rainfall for various months (in mm/day)

Year/month	Jan	Feb	Mar	Apr	May	Jun	Jul	Aaug	Sep	Oct	Nov	Dec
1988	7.925	6.31	7.335	4.391	6.004	2.6875	2.62	2.0225	4.027	3.898	7.37	2.1285
1989	4.839	6.248	6.815	9.07	6.039	3.840	3.391	3.37	3.722	3.511	3.505	3.991
1990	4.32	3.435	8.135	5.732	3.901	5.148	2.421	4.003	4.033	2.6025	5.168	4.991
1991	6.885	6.6	2.472	6.335	4.880	5.136	4.273	1.611	4.572	1.709	6.71	7.57
1992	5.52	7.985	4.776	2.089	2.764	1.483	3.067	2.569	1.955	1.644	2.382	6.565
1993	6.975	5.71	6.79	6.995	3.350	3.384	3.541	1.855	2.5	1.371	3.179	4.605
1994	7.33	8.77	4.113	4.595	2.134	2.093	2.544	1.851	1.459	2.093	7.09	1.851
1995	4.599	6.85	7.8	5.249	3.346	1.784	2.04	2.284	1.818	2.467	5.44	3.076
1996	6.355	7.935	7.38	6.96	3.674	5.875	4.566	1.959	3.615	3.553	2.388	5.755
1997	6.475	7.989	6.32	5.112	2.364	6.12	4.656	2.10	2.54	3.112	3.654	6.145
1998	1.249	4.522	4.465	8.09	8.48	4.086	3.309	1.977	3.177	1.92	1.762	1.544
Average	5.679	6.577	6.036	5.874	4.266	3.785	3.311	2.327	3.038	2.541	4.422	4.383
Minimum	1.249	3.435	2.472	2.089	2.134	1.483	2.04	1.611	1.459	1.92	7.37	1.544
Maximum	7.925	8.77	8.135	9.07	8.48	6.12	4.656	4.003	4.572	3.898	1.762	7.57
Standard deviation	3.397	2.681	2.862	3.494	3.229	2.318	1.308	1.227	1.556	1.0115	2.805	3.014

2.4 Cyclones

Cyclones are one of the major meteorological factors that could influence the operation of a mining platform. In this section, occurrence of cyclones has been analyzed in terms of frequency and speed for a 10-year period (1997–2007) for the region. Cyclones are classified into four types on the basis of speed (www.metoffice.gov.uk), as follows:

i. Tropical depressions (<34 knots)
ii. Tropical storms (34–63 knots)
iii. Hurricane/ typhoon (64–127 knots)
iv. Severe hurricane/super typhoon (>127 knots)

Data for different types of cyclones in the Indian Ocean (Fig. 5 and Table 5) shows that the months which are more prone to cyclones (more than ten occurrences in a month of any type of cyclone) are January to March, whereas the months with lesser no. of cyclones (i.e., less than ten occurrences in a month) are April to December, during which no cyclones were recorded in August and September.

Although, most destructive category of cyclone, i.e., super hurricane, did not occur during these years, hurricanes are more prominent during the months January to March, which have been also registered as the months having maximum number of cyclones. In the category of storms, although January and February witnessed maximum number of storms, April, June, July, October, November, and December also witnessed number of storms in fewer numbers. Depressions are more in the months of March, April, October, and November, but they do not pose a serious threat in comparison to hurricanes and super hurricanes, which may cause severe damage to the mining platform. During August and September, there were no cyclones of any intensity in the area.

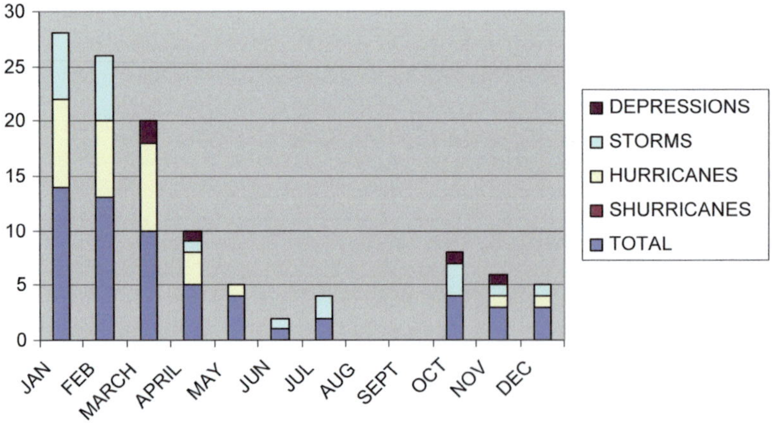

Fig. 5 Occurrence of cyclones for different months between 1997 and 2007. (Source: www.metoffice.gov.uk)

Table 5 Frequency of occurrence and speed of cyclones during the years

	Jan		Feb		Mar		Apr		May		Jun		Jul		Aug		Sep		Oct		Nov		Dec	
Month/year	N	Km	N	Km	N	Km	N	Km	N	Km	N	Km	N	Km	N	Km	N	Km	N	Km	N	Km	N	Km
1997	4	55, 45, 65, 115	3	90, 60, 125	1	75	0	0	3	nd	0	0	1	45	0	0	0	0	0	0	0	0	0	0
1999	1	55	2	40, 40	3	25, 90, 30	1	115	0	0	0	0	1	45	0	0	0	0	0	0	0	0	0	0
2001	3	100, 80, 105	0	0	1	75	2	30, 60	0	0	1	40	0	0	0	0	0	0	2	30, 55	1	35	1	nd
2003	2	75, 55	3	105, 80, 45	2	115, 140	1	65	1	65	0	0	0	0	0	0	0	0	1	45	1	105	1	65
2005	3	45, 100, 35	1	60	1	65	nd	nd	0	0	0	0	0	0	0	0	0	0	1	35	1	45	1	35
2007	1	65	4	115, 55, 125, 70	2	105, 115	1	110	0	0														
Total number of cyclones	14		13		10		5		4		1		2		0		0		4		3		3	
Tropical depressions	0		0		2		1				0		0		0		0		1		1		0	
Tropical cyclones	6		6		0		1				1		2		0		0		3		1		1	
Hurricanes/typhoons	8		7		8		3		1		0		0		0		0		0		1		1	
Super hurricanes/supertyphoons	0		0		0		0				0		0		0		0		0		0		0	

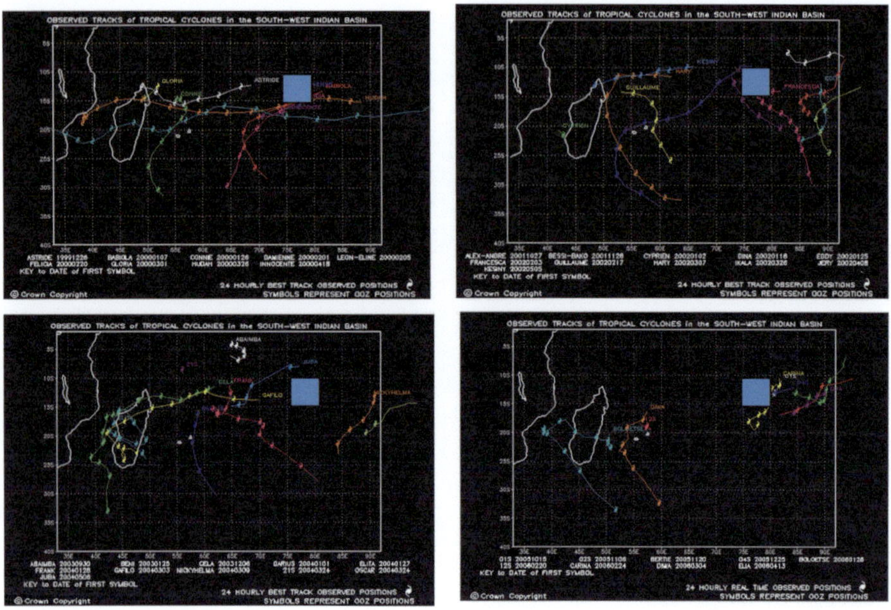

Fig. 6 Cyclone tracks during the study period: 1999–2000 (top left), 2001–2002 (top right), 2003–2004 (bottom left), and 2005–2006 (bottom right). Central Indian Ocean basin is marked with blue box

The months from April to December appear to be the most favorable months for carrying out the seabed mining operations, since these months are only subjected to none or low-speed storms and depressions, without registering any major hurricanes. Hence, about a 90–100-day period between January and early April appears to be unfavorable in terms of total number of cyclones and hurricanes in the area. Cyclones play a major role in perturbing the stability of any offshore platform. Also, it is rather difficult to conduct operations on platforms because cyclones, especially hurricanes, can have adverse effects on working as well as structural parts. Moreover, analysis of cyclone tracks (Fig. 6) shows that most of the cyclones were concentrated in the western parts of the Indian Ocean, whereas very few cyclones actually crossed the Central Indian Ocean Basin, making it favorable area for deep-sea mining operations.

2.5 Waves

Wave data was analyzed from ocean wave atlas (Hogben and Lumb 1967) for block no. 35 (0–10° S, 60–90° E) in the Central Indian Ocean Basin. The data is classified into three seasons – Northeast monsoon (December to February), Southwest monsoon (May to September), and transition period (March to April, October to

November). The maximum number (1000–1500 waves) recorded throughout the year irrespective of seasons was 1 m high, followed by <1 m and 1–2.5 m (200–1000 waves) and very few (<100 waves) of 3–8 m height (Fig. 7 and Table 6). Similarly, maximum number (700–2000 waves) had a direction between 80° and 160° during the period from May to September followed by direction between 230°

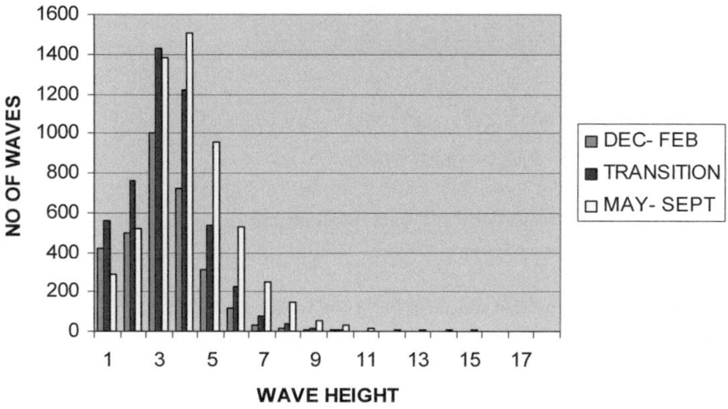

Fig. 7 Number of waves of different heights for various seasons (in meters)

Table 6 Number of waves of different heights for various seasons

	December to February (Northeast monsoon)	March to April, October to November (transition months)	May to September (Southwest monsoon)
Wave height (m)	Total no. of waves	Total no. of waves	Total no. of waves
0.25	420	556	284
0.50	494	765	520
1.00	1004	1432	1386
1.50	719	1223	1507
2.00	314	536	959
2.50	116	224	530
3.00	34	75	248
3.50	19	37	149
4.00	5	14	54
4.50	7	8	31
5.00	1	3	12
5.50	1	1	6
6.00	0	0	5
6.50	1	0	7
7.00	1	0	4
7.50	0	1	3
>8.00	0	0	3

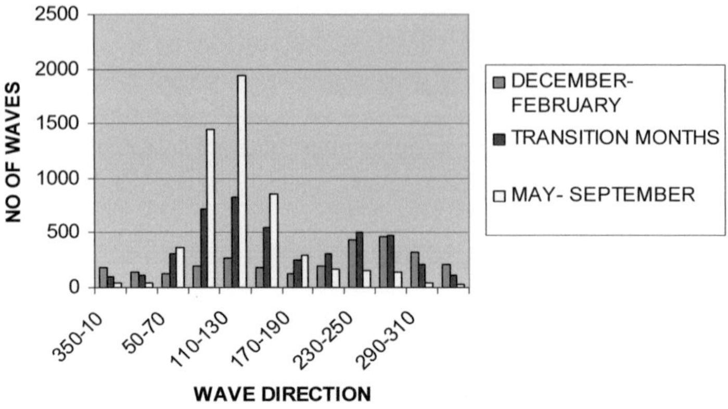

Fig. 8 Number of waves in different directions (in degrees)

Table 7 Number of waves in different directions for various seasons

	December to February (Northeast monsoon)	March to April, October to November (transition months)	May to September (Southwest monsoon)
Wave direction (degrees)	Total no. of waves	Total no. of waves	Total no. of waves
350–10	182	94	42
20–40	135	111	38
50–70	122	310	361
80–100	200	717	1441
110–130	266	831	1932
140–160	177	542	855
170–190	131	255	290
200–220	194	313	169
230–250	429	511	150
260–280	462	474	140
290–310	325	208	48
320–340	206	108	30

and 280° (400–500 waves) during the period from December to February, whereas very few (<300 waves) were recorded with directions ranging from 0 to 360 throughout the year (Fig. 8 and Table 7).

In terms of wave period (the average of time interval between passages of successive crests or troughs of waves), the majority of waves were of the time period of 5–9 s. Very few waves have been reported for the time period >14 s. However, some waves have been recorded for the time interval of 10–13 s. In case of the total number of waves recorded, the duration from May to September has witnessed the maximum number of waves, followed by transition months (Table 8 and Fig. 9).

Table 8 Number of waves for different periods

	December to February (Northeast monsoon)	March to April, October to November (transition months)	May to September (Southwest monsoon)
Wave period (secs)	Total no. of waves	Total no. of waves	Total no. of waves
Calm waves	307	401	212
5 or less	1382	2083	1730
6 or 7	825	1254	1727
8 or 9	367	630	1192
10 or 11	146	319	532
12 or 13	42	108	215
14 or 15	18	38	42
16 or 17	4	6	22
18 or 19	1	2	2
20 or 21	3	5	2
Over 21	41	29	32

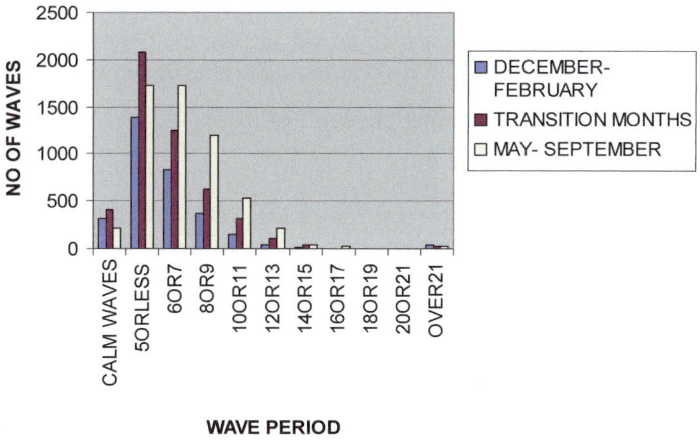

Fig. 9 Number of waves for different waves for different periods for various months (in seconds)

From the above observations, we can interpret that wave heights are generally less than 2.5 m throughout the year with few waves having more heights (3–8 m). Also the wave direction varied between 80–160° and 230–280° during different seasons which will have to be considered during designing of the mining platform. Winds are responsible for initiating the waves, that is, the higher the wind velocity, the higher the waves, and the time interval between the two successive crests (or troughs) of waves will be high.

Table 9 Summary and implications of meteorological conditions for mining platform

Parameters	Characteristics	Implications
1. Surface – temperature	25–29° C	Design of mining platform, loading/unloading of ore, operations of mining
Rainfall	2–6 mm/day	
Wind speed	5–8 m/s (18–29 km/h)	
2. Waves – height	Generally <2.5 m (max. = 8 m)	Equipment, transportation to shore and back
Direction	Variable (0–360°)	
Period	Generally <10/s	
3. Cyclones – none	0 – August to September	As above
Depressions	<34 km/h – Jan to Dec	
Storms	34–63 km/h. – Jan to Dec	
Hurricanes	64–127 km/h – Jan to Mar	
Severe hurricanes	>127 km/h – Nil	As above

Source: 1. Grass et al. (2000), 2. Hogben and Lumb (1967), 3. www.metoffice.gov.uk (n.d.)

2.6 *Summary of Atmospheric and Sea Surface Factors*

Atmospheric and sea surface data such as sea surface temperature, wind, rainfall, cyclones, and waves evaluated for 10-year period (Table 9) shows the following:

- Average sea-surface temperature range = 25–29°.
- Average wind speed = 5–8 m/s.
- Average rainfall range = 2–6 mm/day.
- Temperature and rainfall follow a similar pattern (high during December to Feb and low during May to September), whereas wind speed is opposite.
- Temperature and rainfall vary over a small range throughout the year.
- Variable (seasonal) wind speed, wave ht., direction, and period.
- Depressions are common throughout the year.
- More hurricanes during January to March, but no super hurricanes.

These environmental factors (Table 8) could be effectively used in designing and planning operations of deep-sea mining platform and transportation.

3 Water Column Characteristics

Long-term deployment of (~270 days) current meters at three locations and different depths ranging from ~500 to 5000 m during 1995–1996 in the Central Indian Ocean Basin (Murty et al. 2001) revealed that mean current velocities at shallower depths (450–670 m) vary between 3.12 cm/s (during April to September) as against 5.23 cm/s (during September to January) at the northern location (10° S, 75° E), whereas these were considerably low between 2.21 and 2.29 cm/s (during both deployments) at the southwestern location (15° S, 72° E) and still lower between 52 and 0.65 cm/s (during both deployments) at the southeastern location (15° S, 77° E). At deeper depths, the mean current velocities recorded were very low (<1 cm/s) throughout the period for the entire water column (1200–5100 m) at all the three locations with the exception of one deployment (September to January) at the northern location (10° S, 75° E) where it ranged between 1 and 2 cm/s (Table 10). Conductivity, temperature, and depth records in the area (Fig. 10) have also shown that water temperature ranges between 26° C (at the surface) and 1.0° C (at 5000 m depth), whereas salinity is constant at 34.7 ppm (±0.2 ppm) throughout the water column during different seasons (Babu et al. 2001). As the entire deployment mechanism along with cables for transmission of signals as well as power and as also lifting of the ore from seafloor to mining platform will be through the water column, these characteristics will be essential in the design and behavior of the mechanism as well as in dispersion of sediment plume generated by the mining operation.

Table 10 Mean currents in Central Indian Ocean Basin

Station period of measurements	Depth (m)	*V* (cm/s)[a]	*R* (°)[a]
MS-1 Sep 95 to Jan 96 (spring to summer)	500	5.23	329
	1200	2.93	269
	3500	1.28	271
	4900	1.72	240
MS-1A Apr 96 to Sep 96 (fall to winter)	500	3.12	106
	1200	0.46	149
	3500	0.21	93
	4900	0.15	132
	5100	0.08	170
MS-2 Sep 95 to Jan 96 (spring to summer)	600	2.21	76
	1300	1.23	117
	3600	0.07	132
	4400	0.11	277
MS-2A Apr 96 to Sep 96 (fall to winter)	670	2.29	184
	3670	0.02	84
	4270	0.07	100
	4470	0.56	107
MS-3 Sep 95 to Jan 96 (spring to summer)	1200	0.52	144
	4900	0.18	244
	5100	0.51	254
MS-3A Apr 96 to Sep 96 (fall to winter)	450	0.65	352
	1150	1.23	51
	3450	0.14	124

[a]*V* and *R* are the phase–mean current velocity and mean direction (Source: Murty et al. 2001)

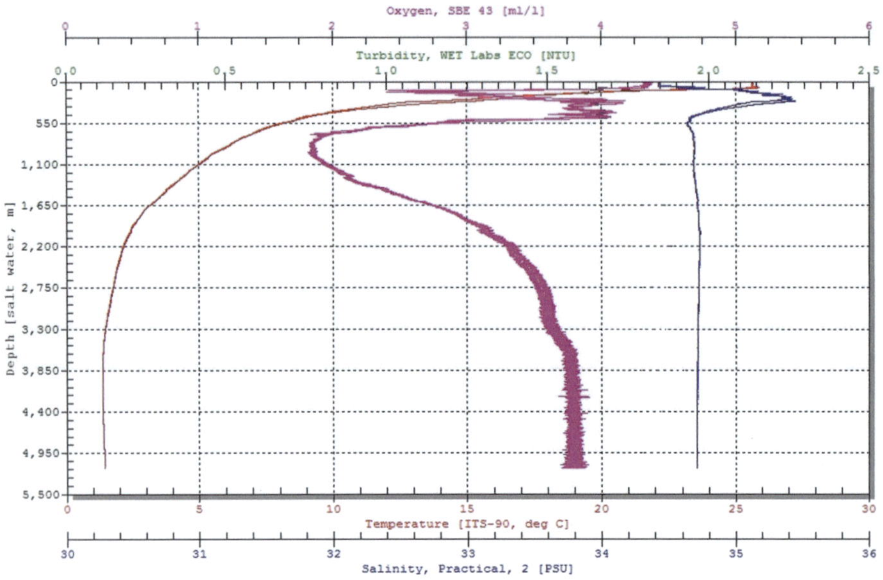

Fig. 10 Typical profile of water column characteristics in the Central Indian Ocean Basin. (Source: Cruise Report of RV-Sindhu Sadhana Cruise 13, 2015, NIO/PMN-EIA project)

4 Seafloor Features

4.1 Nodule Distribution Characteristics

Among the seafloor features, nodule characteristics are the key factor that would influence the design and operation of the entire mining system from the nodule miner on the seafloor to the lifting mechanism in the water column, as well as handling, storage, pre-processing, effluent discharge, and transportation on the surface. Hence, data on nodule characteristics such as size, coverage, and abundance are compiled from different studies in the Central Indian Ocean Basin (Table 11) as well as different basins of the Pacific Ocean (Table 12) and described in this section.

Distribution of Nodule Sizes

Among size classes ranging from <2 to >8 cm (Fig. 11), the dominant size class observed in the CIOB was 2–4 cm (67%), followed by 4–6 cm (20%) and <2 cm (9%) from analysis of ~9000 nodules (Sharma 1998). However, another study showed dominant size class to be 4–6 cm (53%), followed by 2–4 cm (23%) and 6–8 cm (16%) from analysis of 171 nodules (Sarkar et al. 2008). Similarly in the

Table 11 Frequency (%) distribution of nodule characteristics in Central Indian Ocean

Parameter/source (no. of samples)	Classification (nos. in % samples)									
Size (cm)	*<2*	*2–4*	*4–6*	*6–8*	*>8*					
Sharma (1998) (9000)[a]	9[b]	67	20	3	1					
Sarkar et al. 2008 (171)[a]	1	23	53	16	7					
Wt%/size	*<2*	*2–4*	*4–6*	*6–8*	*>8 cm*					
Sharma (1998) (9000)[a]	9[b]	40	35	11	5					
Coverage (%)	*0*	*<10*	*10–20*	*20–30*	*30–40*	*40–50*	*50–60*	*60–70*	*70–80*	***>80***
Sharma (1998) (988)[c]	67	18	6.5	3	2.8	1.8	0.6	0.5	0.1	0.1
Sharma et al. (2010) (20516)[c]	65	13	4.1	2.9	4.7	4.0	2.8	2.5	0.7	0.0
Abundance (kg/m^2)[d]	*<5*	*5–10*	*10–15*	*15–20*	*20–25*					
Sharma (1998) (725)[e]	56	25	13	5.5	0.5					
Kodagali and Sudhakar (1993) (479)[e]	67	23	8	2	–					
Sarkar et al. (2008) (47)[e]	38	19	28	15	–					
Jauhari et al. (2001) (23)[e]	43	48	9	–	–					

[a]No. of nodules
[b]Including broken fragments
[c]No. of photos
[d]Average abundance at each location
[e]No. of operations

Pacific Ocean, majority of nodules (49–58%) belong to 2–4 cm size class followed by <2 cm (40%) in two out of three basins (Usui 1986, 1994), whereas the third basin shows 4–6 cm as the next dominant class (31% – Usui and Nakao 1984). This indicates that average (or dominant) sizes could vary within a single basin and between different basins that could also depend on sample size as well as local conditions. In terms of weight percent also the dominant sizes were 2–4 and 4–6 cm (40 and 35%) in CIOB. Similar study in part of the Pacific Ocean also showed that ~3 cm size of nodules accounted for 50% of the total mass of nodules sampled (Handa and Yamazaki 1986) signifying the dominant contribution of 2–4 cm size class in the overall weight among nodules that could form a critical factor while designing the nodule miner from the seafloor as well as the crushing and lifting mechanism through water column.

Distribution of Nodule Coverage on the Seafloor

Analysis of nodule coverage (percent area covered by nodules) from seafloor photographs (Fig. 12) reveals that a majority (65–67%) of the locations/photographs do not show the presence of exposed nodules, whereas most photographs having

Table 12 Frequency (%) distribution of nodule characteristics in Pacific Ocean

Parameter/location (no. of samples) (source)	Classification (nos. in % samples)									
Size (cm)	*<2*	*2–4*	*4–6*	*6–8*	*>8*					
Central Pacific Basin (3512)[a] (Usui (1986))	40	49	9	1.5	0.5					
Magellan trough (61)[a] (Usui and Nakao1984)	11	49	31	7	2					
Penrhyn basin (84)[a] (Usui 1994)	40	58	1	1	–					
Coverage (%)	*0*	*<10*	*10–20*	*20–30*	*30–40*	*40–50*	*50–60*	*60–70*	*70–80*	*>80*
Central Pacific Basin (125)[b] (Usui 1986)	76	10	8	3	2	1				
Central Pacific Basin (340)[b] (Yamazaki and Sharma 1998)	nd	53	3	4	5	4	5	26	0.6	–
Magellan trough (57)[b] (Usui and Nakao1984)	14	4	26	21	9	14	9	4	–	–
Penrhyn Basin (68)[b] (Usui 1994)	0	4	15	10	15	10	3	13	19	10
Abundance(kg/m^2)	*<5*	*5–10*	*10–15*	*15–20*	*20–25*	*>25*				
Central Pacific Basin(90)[c] (Usui 1986)	67	14	12	7	–	–				
Magellan trough (61)[c] (Usui and Nakao1984)	36	16	12	8	16	11				
Penrhyn basin (84)[c] (Usui 1994)	32	13	12	17	5	21				

Key: [a]No. of nodules, [b]No. of photos, [c]No. of operations, nd = no data

Fig. 11 Typical nodules of different sizes

exposed nodules (13–18%) have <10% coverage in CIOB (Sharma 1998; Sharma et al. 2010). Similar studies on distribution of nodule coverages across Central Pacific Ocean based on seafloor photographs also showed that most locations (53–86%) have nil to few exposed nodules (0–10% coverage), followed by higher coverages (Usui 1986; Yamazaki and Sharma 1998). However, other basins of the Pacific Ocean have also shown higher coverages (up to 50%) from seafloor photos (Usui and Nakao 1984; Usui 1994) implying influence of local conditions on distribution of nodules.

Here, it is important to recognize that this data is on basin scale, whereas it is most likely that areas identified within the basin as first-generation mine sites would have many (if not all) locations/photographs with higher nodule coverages as their demarcation would be based on higher nodule abundances. Also as discussed later in the chapter, even if few nodules are exposed on the seafloor, it is quite likely that many of these are buried under thin sediment-water interface and are collected in the grab samples that are used for abundance estimates (Fewkes et al. 1979; Felix 1980; Sharma 1989; Sharma et al. 2013), and so inclusion of buried nodules is critical for resource estimation and demarcation of potential mining sites.

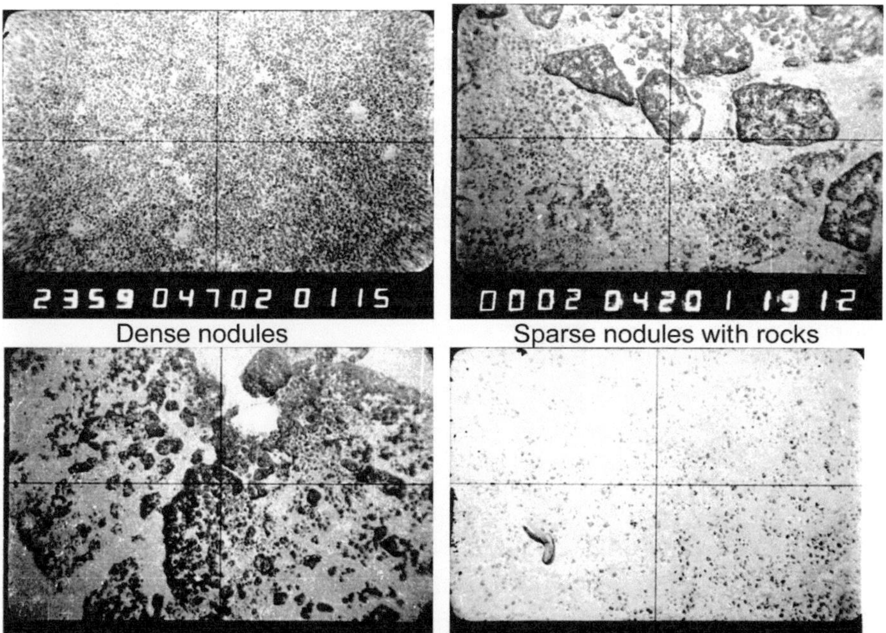

Fig. 12 Distribution of nodule coverage on the seafloor

Distribution of Nodule Abundance on the Seafloor

Evaluation of nodule abundance (weight per unit area) from grab or core samples (Fig. 13) from different studies in the CIOB shows that a majority of the sampled locations (38–67%) have <5 kg/m^2 nodules, followed by locations with 5–10 kg/m^2 of nodules. Similarly, basin scale evaluation of nodule abundance in Pacific Ocean also showed majority of the locations (32–67%) with <5 kg/m^2 of nodules followed by 5–10 kg/m^2 (13–16%) and 10–15 kg/m^2 (12%) and remaining locations with >15 kg/m^2 abundance (Usui and Nakao 1984; Usui 1986, 1994). However, these studies being on a regional scale give a wide variation in nodule abundance, whereas areas identified as first-generation mine sites would normally have higher average nodule abundance of at least 8–10 kg/m^2 or more for the resource to be economically viable. It should be noted that the cutoff abundance for a nodule deposit to be economically viable is 5 kg/m^2 (UNOET 1987).

4.2 Sediment Properties

Effect of Sediment Cover on Nodule Burial

Variable nodule coverage and abundance depends on certain intrinsic as well as extrinsic factors contributing to their population and distribution. Whereas the intrinsic factors may be the formation process and the required environmental settings such as proximity to source of elements, degree of oxidation, nature and age of substrates, availability of nucleating material, sedimentation rates, and bottom currents (Cronan 1980); the extrinsic factors are the influence of associated substrates such as thickness of sediments under which the nodules are either partly or fully buried, rock outcrops that are strewn within a nodule field, as well as macro- and micro-topography of the seafloor.

Fig. 13 Partially buried nodules on top of box core

An analysis of ~20,000 seafloor photographs from CIOB showed a majority (65%) of them having nil nodule coverage, some (17%) of them with low nodule coverage (<20%), and the remaining (18%) having moderate (20–50%) and high (50–80%) coverage (Sharma et al. 2010). However, physical sampling using grabs in the same area has shown average nodule abundance ranging from 3.84 to 8.23 kg/m^2 confirming the presence of nodules in the entire area implying that nodules at many locations in CIOB could be buried under the top 20–25 cm of sediment cover to which grabs can penetrate (Sharma et al. 2013). A ground truth confirmation of nodule occurrence from close to 1000 locations in the CIOB (Fig. 14) showed that whereas nodules were recorded in both photographs and grabs at 44% locations, no nodule were recorded in either of them at 21% locations indicating no nodules, and at 29% of the remaining 35% locations, nodules were collected in grabs only (Sharma 2017) indicating not only the prevalence of buried nodules but also patchy distribution within the nodule fields.

Studies in the Pacific Ocean have also reported that nodules are buried under sediment-water interface boundary (SWIB) layer that are obscured from camera view and cannot be accounted for in the photographs but are collected in the grabs (Fewkes et al. 1979) and that the extent of burial depends on the size and shape of nodules as also thickness of the SWIB layer (Cronan and Tooms 1967; Felix 1980). Although most nodules are known to occur in the top 1–2 m (Stoffers et al. 1982) of sediment, a few have also been collected at deeper depths of >5 m in the Pacific Ocean (Usui 1986; Cronan 2000) and Indian Ocean (Pattan and Parthiban 2007).

This calls for designing the nodule collector such that it is capable of not only picking up nodules from the surface but also collecting the buried nodules from at least a few tens of centimeters below so as to be efficient. However, "digging"

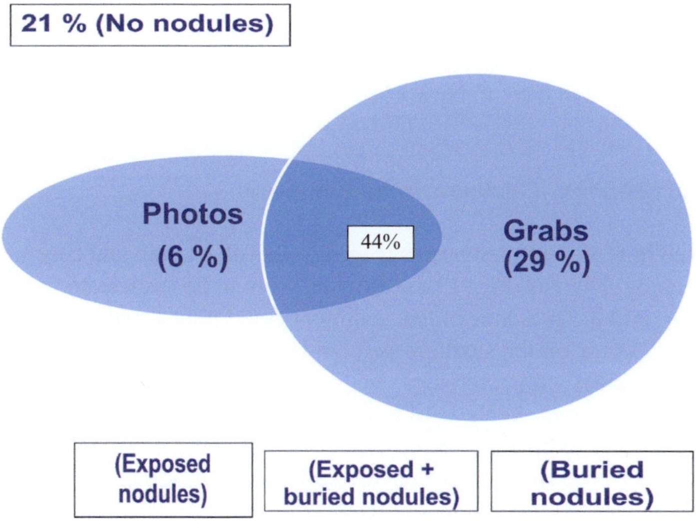

Fig. 14 Venn diagram of nodule occurrence in grabs and photographs (Sharma 2017)

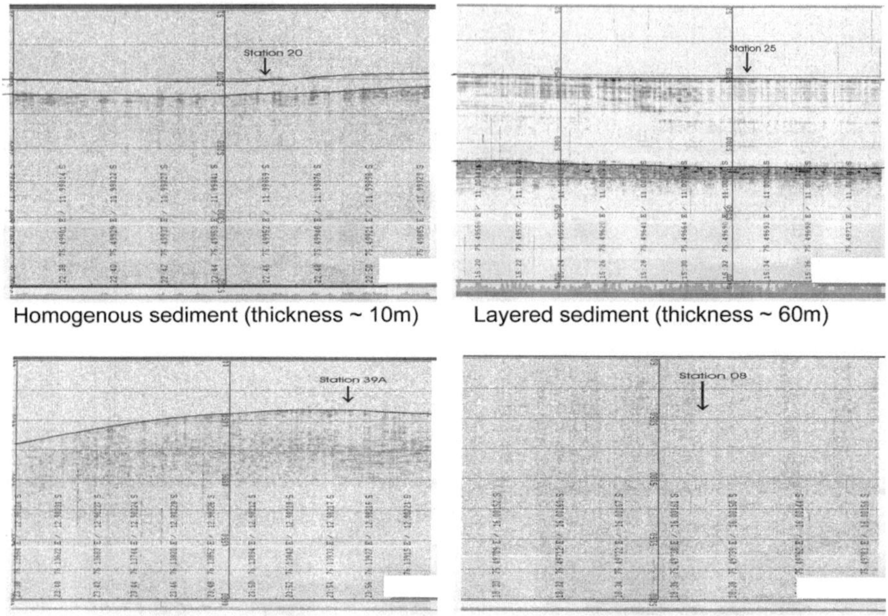

Homogenous sediment (thickness ~ 10m) Layered sediment (thickness ~ 60m)

Compact homogenous sediment Non reflecting sediment

Fig. 15 Records of sub-bottom profiling

deeper into the sediment could pose a difficulty in locomotion due to higher friction and also be risky in terms of destroying the benthic organisms and creating a sediment plume that could remain in suspension for a long time and either migrate to the adjoining areas or settle in the same area but alter the physicochemical characteristics of the seafloor. Hence, optimization of sinkage of the nodule miner versus collection efficiency will have to be worked out for designing the system.

Analysis of Sediment Thickness in Nodule Areas

As observed in seafloor photographs (Fig. 12) as well as grab and core samples collected from nodule areas (Fig. 13), sediments seem to be associated with nodules at all locations, and the presence of this sediment could play a major role in the operation of the collector on the seafloor as it could sink depending on its weight/buoyancy and physical characteristics of the sediment including thickness and geotechnical properties. Analysis of records from sounding data such as sub-bottom profiler shows the sediment as acoustically transparent layer (ATL) having either homogenous or layered structure on the seafloor with variable thickness ranging from a few meters to several tens of meters (Fig. 15).

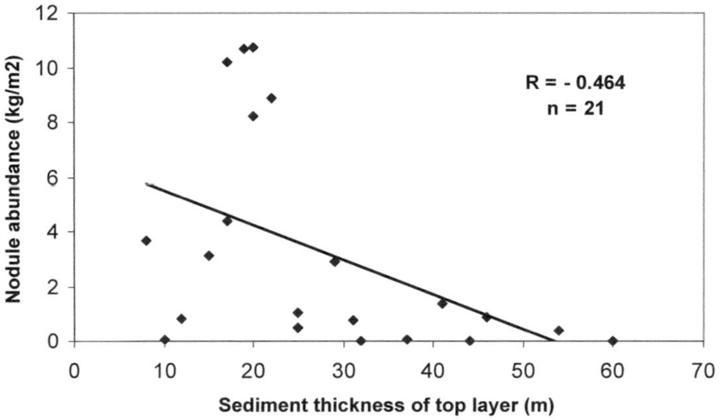

Fig. 16 Nodule abundance vs sediment thickness

Correlation of ATL with different topographic settings suggests that at many of the locations (50%), its thickness is small (<10 m) on the abyssal hills; whereas in the valleys, the majority of locations (>60%) have higher thickness (30–60 m), and slopes have variable thickness indicating that thickness of ATL is controlled by topographic features (Sharma et al. 2013). The study further shows that although nodule abundance does not have significant correlation with thickness of ATL (Fig. 16), the sediment layer is responsible for partial to complete burial depending upon either size or abundance causing problems in estimating nodule populations from seafloor photographs as also observed in the Pacific Ocean (Fewkes et al. 1979; Felix 1980).

Physical Properties of Sediments and Nodules

Physical properties of sediments could be one of the key factors influencing the mobility and collection efficiency of the nodule miner on the seafloor. Most of the sediments associated with nodules in CIOB are clayey silts or silty-clays having dominant proportions of clays (>50%) of minute particle size (<4 μm), lesser proportion of silts (<40%), and small proportion (<10%) of sand-sized particles. These generally contain four common clay minerals, viz., montmorillonite, illite, kaolinite, and chlorite with quartz and feldspars as accessory minerals wherein montmorillonite has a high water-holding capacity resulting in the sediments having high water content (300–400% on wet basis on the surface) that decreases with depth (Khadge 1998). These sediments also indicate medium to high plastic nature (average liquid limit = 207%, average plastic limit = 109%, and average plasticity index = 99%) with almost constant porosity (88–90%), density (~1.2 g/cm^3), and specific gravity (2.2–2.3) with variable shear strength (2.2–13 kPa) increasing with depth down to 4 m (Khadge 1998).

Similar study of nodule-associated sediments sampled from cores in the Pacific Ocean also reports clay-sized particles (<2 μm), high water content (200–300%), and shear strength increasing with depth (2–8 kPa) down to 50 cm (Yamazaki 2017). It is critical to note that strength reduction of sediments due to physical disturbance could be an important factor for design of propulsion mechanism of the seafloor miner (Richards and Chaney 1981). Several studies have attempted to evaluate the dynamic characteristics of deep-sea sediments that could be extrapolated for the design of the deep-sea miner (Yamazaki et al. 1995).

Besides sediments, geotechnical properties of nodules such as bulk density, moisture content, and porosity as well as hardness, compressive strength, and tensile strength vary with nodule type and are key factors in collection, crushing, and lifting of nodules during mining (Yamazaki 2017).

4.3 Occurrence of Rock Outcrops

Several seafloor photographs have shown the presence of rock outcrops that could also be coated with ferromanganese oxides, within nodule fields (Fig. 12). Studies have shown that their coverages range from 1% to 100% on the seafloor depending upon their size and exposure, that could extend over several meters on the seafloor, with a majority of them (63% in the Pacific Ocean and 48% in Indian Ocean) covering 20% area on the seafloor (Yamazaki and Sharma 1998; Sharma et al. 2010). However, their coverage increases with undulating topography especially along

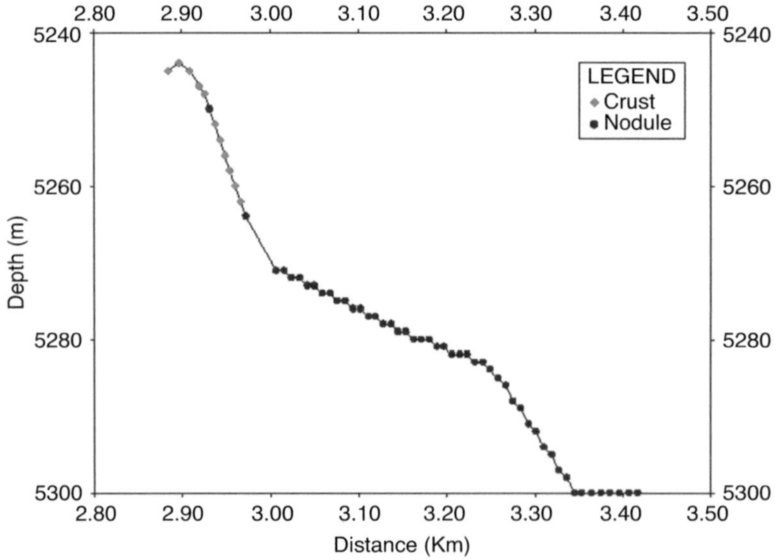

Fig. 17 Distance vs depth profiles of nodules and crusts (Sharma et al. 2010)

steep slopes and tops of seamounts and abyssal hills where there is less accummulation of sediment particles due to gradient and higher currents. These outcrops would act as "obstructions" within a nodule field which the nodule collector will have to avoid either by being driven around or flown over (Sharma 2017).

4.4 Influence of Seafloor Topography

Studies have shown that concentration of nodules, accummulation of sediments, as well as exposure of rock outcrops on the seafloor are controlled by the macro- and micro-topographic undulations (Fig. 17) with variable slopes (<3 to >15°) wherein crusts/rock outcrops are dominantly associated with steep slopes of seamounts and abyssal hills; large nodule deposits are concentrated along gentle slopes and undulating topography where they grow over millions of years with small sediment accumulation, and thick sediments are associated with flat seafloor and valleys that either obscure the nodules from collection or observation or inhibit them from growing (Yamazaki and Sharma 1998; Sharma et al. 2010). Hence in order to perform efficiently and also for the safety of nodule miner, maneuverability along variable topography and balancing it with nodule collection would be a key to the performance of the system (Sharma 2017).

Table 13 Influence of seafloor features on mining design and operation

Seafloor features	Characteristics	Implications
Nodule abundance	1–10 kg/m^2	Collection efficiency of miner due to inhomogenous distribution of nodules
Coverage	1–90%	
Size	<1->8 cm, avg. 2–4 cm (2–4 cm-67%)	
Weight	2.8–236 g, avg. wt.= 24 g (2–4 cm-40%, 4–6 cm-35%)	Design of pickup device due to variable size
		Decision on crushing capability
Sediment cover	Nodule burial (0–100%)	Penetration of collector device
Size	Clay <4 μ (60%)	Picking capability of collector
Shear strength	2 (2 cm)–10 kPa (40 cm)	Sinking/buoyancy of miner
Water content	700 (2 cm)–300% (40 cm)	
Rocks/crusts	Coverage (1–100%) coexisting with nodules	"Obstacles" during mining, damage to collection device, avoid if possible
Topography	10–500 m relief	Maneuverability of miner
Micro-topography	1–10 m undulations	
Slopes	1–35°	

4.5 Summary of Seafloor Features

Several seafloor conditions (Table 13) would play a key role in the design and performance of the nodule miner. The ability of the miner to negotiate and maneuver through heterogeneous distribution of nodule abundance, coverage, size, burial, as well as their association with sediments and rock exposures, controlled by macro- and micro-topographic features, would determine the versatility in design as well as efficiency in mining optimum quantities of nodules at a steady rate without damage or malfunction.

5 Conclusions

Several experimental studies have been conducted to assess the design and behavior of riser (lifting) mechanism and nodule miner (Chung 1997; Hong 1997; Xia et al. 1997; Li and Zhang 1997; Hong et al. 1999; Deepak et al. 2001, Yoon et al. 2003), and a few studies have been conducted on the design parameters for the mining platform or handling and transfer of ore (Amann 1982; Herrouin et al. 1991).

This study has evaluated long-term atmospheric and sea surface conditions in Central Indian Ocean Basin that suggests rather uniform and favorable conditions throughout the year in terms of average sea surface temperature (25–29° C), average rainfall (2–6 mm/day), average wind speed (5–8 m/s), and local depressions (<34 km/h) and storms (34–63 km/h) throughout the year with hurricanes (64–127 km/h) mainly occurring between January and March, indicating adverse weather conditions for about 90 days during the year. On the other hand, wave heights are generally less than 3 m throughout the year which occasionally could go as high as 8 m with heavy winds and rainfall during the occurrence of hurricanes. Such meteorological conditions are key to ascertain the availability of working days at sea as literature suggests between 250 and 300 days of fair weather per year would be ideal for commercial mining (UNOET 1987).

Evaluation of long-term (~270 days) mean currents shows seasonal variations at shallow depths (500 m) with lower velocity (3.12 cm/s) during April to September and higher velocities (5.23 cm/s) during September to January which is also reflected at higher depths (1200–5000 m). Such seasonal variations are not seen at the other two mooring locations 550 km south of the previous mooring where mean currents are generally low (0.6–2 cm/s) at shallow depths (450–670 m) and further less (<1 cm/s) at higher depths (1200–5100 m) throughout the year. However, variable current directions at different depths throughout the year could pose a serious problem for the stability of the riser system as also causing drag on the nodule miner.

Operations on the seafloor will be affected by variable nodule sizes (<1 to >8 cm) with an average nodule size (2–4 cm), and the mining system will have to be designed accordingly for efficient recovery and pumping to the surface, suggesting the use of a crusher on the nodule miner in maintaining the consistency in size of the nodules in the slurry to be lifted to the surface. Due to heterogeneous abundance of nodules (<5 to >25 Kg/m^2), the use of buffer in storing them at an intermediate level before pumping them in fixed quantities to the mining platform for energy conservation could be considered. Occurrence of higher concentration of nodules along undulating topography as compared to the low nodule concentrations in the valleys and plains, coupled with topographic variations and sediment thickness, would require that the collector system should be capable of detecting the zones of higher nodule concentrations while sweeping the seafloor.

The nodule miner will encounter substrates such as sediments and rocks associated with nodules on the seafloor that would pose problems in nodule recovery, and thus the miner will have to be designed to penetrate within the sediments to collect the buried nodules as well in order to be more efficient. The binding strength of nodules to the sediment could be a critical factor in the design of the nodule collector for which the use of water jets may be required to dislodge the nodules from the seabed. Pumping of large quantities of sediments along with nodules will not only increase the energy consumption but also cause a major environmental problem if disposal of debris is at or close to the surface. Hence, the sediments may have to be discharged in deeper areas, at least below the photic zone or deeper, to reduce the impact on marine life in the water column. Alternatively, there could be a mechanism to wash them out near the seafloor to enable pumping of nodules only (Sharma 2017).

Rock outcrops often occur in the nodule fields and could act as obstructions to the nodule miner. Such areas will have to be mapped in advance in order to identify locations where the mining system may not be able to operate or is likely to get damaged due to the occurrence of hard substrates, steep slopes that it cannot negotiate. Hence, the nodule miner should be capable of sensing such zones, as well as being "driven" around or "flown" over these outcrops in order to avoid any damage to the mining equipment (Sharma 2017). Morphometric analysis in CIOB has shown that a majority (92%) of the area has 0–3° slope, the remaining areas have higher slopes (up to 15°) (Kodagali 1989), whereas slope angle studies in a nodule field in the Pacific Ocean recorded <3° slope angles as being nodule dominant, 3–7° slope angles as sediment dominant, 7–15° slope angles as a transition zone, and >15° slope angles as rock-/crust-dominant zones (Yamazaki and Sharma 2000).

Detailed geotechnical properties such as bulk density, porosity, compressive strength, and tensile strength of nodules, would be critical in designing the pickup from the seafloor and abrasion of minerals in the lifting mechanisms, whereas static properties of associated sediments such as particle size, shear strength, sensitivity, water content, and dynamic properties such as displacement, area, mass, and acceleration would be critical in assessing the motion and impact of the miner and also calculating the drag, separating force, and plume generation on the seafloor (Yamazaki 2017).

It is envisaged that various environmental factors described in this study could be applied for designing and operation of different components of deep-sea mining systems for different mineral deposits taking into account the local environmental conditions and the mineral type.

Acknowledgments This study was carried out under the polymetallic nodule program of Ministry of Earth Sciences, Government of India. The data on atmospheric and sea surface parameters was compiled by Mr. Gagan Deka during his dissertation with the author. Figure 13 is obtained with courtesy from project "EIA studies for polymetallic nodules in CIOB," National Institute of Oceanography, Goa, India.

References

Amann, H. (1982). Technological trends in ocean mining. *Philosophical Transactions of the Royal Society, 307*, 377–403.

Chung, J. S. (1997). An articulated pipe system with thrust control for deep-ocean mining. *Proceedings of 2nd ISOPE – Ocean Mining Symposium, Korea (24–26 November, 1997).* International Society of Offshore and Polar Engineers (pp. 82–90).

Cronan, D. S. (1980). *Underwater minerals* (p. 361). London: Academic.

Cronan, D. S. (Ed.). (2000). *Marine mineral deposits handbook*. Boca Raton: CRC Press.

Cronan, D. S., & Tooms, J. S. (1967). Subsurface concentration of manganese nodules in Pacific sediments. *Deep-Sea Research, 14*, 112–119.

Deepak, C. R., Shajahan, M. A., Atmanand, M. A., Annamalai, K., Jeyamani, R., Ravindran, M., Schulte, E., Handschuh, R., Panthel, J., Grebe, H., & Schwarz, W. (2001). Developmental tests on underwater mining system using flexible riser concept. *Proceedings of 4th ISOPE – Ocean Mining Symposium, Szczecin, Poland, (23–27 September, 2001)* (pp. 88–93). International Society of Offshore and Polar Engineers.

Felix, D. (1980). Some problems in making nodule abundance estimates from seafloor photographs. *Marine Mining, 2*, 293–302.

Fewkes, R. H., McFarland, W. D., Reinhart, W. R., & Sorem, R. K. (1979). *Development of a reliable method for evaluation of deep sea manganese nodule deposits* (pp. 91). Bureau of Mines Open File Report 64-80, US Department of the Interior Bureau of Mines.

Grass, H., Jost, V., Ramesh Kumar, M. R., Schulz, J., Bauer, P., & Schluessel, P. (2000). The Hamburg Ocean Atmospheric Parameters and Fluxes from Satellite Data (HOAPS): A climatological ATLAS of satellite derived air- sea interaction parameters over the oceans. Max Planck Report. No. 312, 130pp.

Handa, K., & Yamazaki, T. (1986). Physical and engineering properties of manganese nodules and deep-sea sediments in GH81-4 area. In S. Nakao (Ed.), *Marine geology, geophysics and manganese nodules around deep-sea hills in the Central Pacific Basin* (pp. 84–97). Tsukuba: Geological Survey of Japan.

Herrouin, G., Lenoble, J. P., Charles, C., Mauviel, F., Bernard, J., & Taine, B. (1991). French study indicates profit potential for industrial manganese nodule venture. *Transactions of Society for Mining, Metallurgy, and Exploration, 288*, 1893–1899.

Hogben, N., & Lumb, F. E. (1967). *Ocean wave statistics, national physical laboratory* (p. 263). London: D.R.Hillman and Sons Ltd., Frome.

Hong, S. (1997). 3D dynamic analyses of lifting pipe system in deep-sea mining. *Proceedings of 2nd ISOPE – Ocean Mining Symposium, Korea (24–26 November, 1997)* (pp. 75–81). International Society of Offshore and Polar Engineers.

Hong, S., Choi, J. -S., Kim, J. -H., & Yang, C. -K.. (1999). Experimental study on hydraulic performance of hybrid pickup device of manganese nodule collector. *Proceedings of 3rd ISOPE – Ocean Mining Symposium, Goa (8–10 November, 1999)* (pp. 69–77). International Society of Offshore and Polar Engineers.

ISA. (2013). *Recommendations for the guidance of contractors for the assessment of the possible environmental impacts rising from exploration for marine minerals in the area*. Jamaica: International Seabed Authority. ISBA/19/LTC/8, pp. 32.

ISA. (2017). Discussion Paper to advance stakeholder discussion in connection with the development and drafting of Regulations on Exploitation for Mineral Resources in the Area

(Environmental Matters). International Seabed Authority. Retrieved on May 15th, 2017. (https://www.isa.org.jm/).

Jauhari, P., Kodagali, V. N., & Sankar, S. J. (2001). Optimum sampling interval for evaluating ferromanganese nodule resources in the Central Indian Ocean. *Geo-Marine Letters, 21*, 176–182.

Khadge, N. H. (1998). Physical properties of sediment core from Central Indian Basin. *The Journal of Indian Geophysical Union, 2*, 1–6.

Kodagali, V. N. (1989). Morphometric studies on a part of Central Indian Ocean. *Journal of the Geological Society of India, 33*, 547–555.

Kodagali, V. N., & Sudhakar, M. (1993). Manganese nodule distribution in different topographic domain of the Central Indian Basin. *Marine Georeources and Geotechnology, 11*, 293–209.

Li, L., & Zhang, J. (1997). China's manganese nodule miner. *Proceedings of 2nd ISOPE – Ocean Mining Symposium, Korea (24–26 November, 1997)* (pp. 95–99). International Society of Offshore and Polar Engineers.

Mukhopadhyay, R., Ghosh, A. K., & Iyer, S. D. (2018). *Indian Ocean nodule field: Geology and resource potential*. Amsterdam: Elsevier. 413 pp.

Murty, V. S. N., Savin, M., Ramesh Babu, V., & Suryanarana, A. (2001). Seasonal variability in the vertical current structure and kinetic energy in the Central Indian Ocean Basin. *Deep Sea Research, Part II, 48*, 3309–3326.

Pattan, & Parthiban. (2007). Do manganese nodules grow or dissolve after burial? Results from the Central Indian Ocean Basin. *Journal of Asian Earth Sciences, 30*, 696–705.

Ramesh Babu, V., Suryanarana, A., & Murty, V. S. N. (2001). Thermohaline circulation of the Central Indian Basin during austral summer and winter periods of 1997. *Deep-Sea Research, II*(48), 3327–3342.

Richards, A. F., & Chaney, R. C. (1981). Present and future geotechnical research needs in deep ocean mining, *Marine Mining, 2*(4), 315–337.

Sarkar, C., Iyer, S. D., & Hazra, S. (2008). Inter-relationship between nuclei and gross characteristics of manganese nodules, Central Indian Ocean basin. *Marine Georesources and Geotechnology, 26*, 259–289.

Sharma, R. (1989). Effect of sediment-layer boundary layer on exposure and abundance of nodules: A study from seabed photographs. *Journal of the Geological Society of India, 34*, 310–317.

Sharma, R. (1998). Nodule distribution characteristics and associated seafloor features: Factors for exploitation of deep-sea Fe-Mn deposits. *Society for Mining, Metallurgy and Exploration INC., 304*, 16–22.

Sharma, R. (2017). Assessment of distribution characteristics of polymetallic nodules and their implications on deep-sea mining. In R. Sharma (Ed.), *Deep-sea mining – Resource potential, technical and environmental considerations* (pp. 229–258). Switzerland: Springer International publishing AG.

Sharma, R., Sankar, S. J., Samanta, S., Sardar, A. A., & Gracious, D. (2010). Image analysis of seafloor photographs for estimation of deep-sea minerals. *Geo-Marine Letters, 30*, 617–626.

Sharma, R., Khadge, N. H., & Jai Sankar, S. (2013). Assessing the distribution and abundance of seabed minerals from seafloor photographic data in the Central Indian Ocean Basin. *International Journal of Remote Sensing, 34*, 1691–1706.

Stoffers, P., Sioulas, A., Glasby, G. P., & Thijssen, T. (1982). Geochemical and sedimentological studies of box core from western sector of the Peru Basin. *Marine Geology, 48*, 225–240.

UNOET. (1987). *Delineation of mine sites and potential in different sea areas*. London: UN Ocean Economics and Technology Branch and Graham & Trotman Limited.

Usui. (1986). Local variability of manganese nodule deposits around the small hills in the GH81-4 area. In S. Nakao (Ed.), *Marine geology, geophysics and manganese nodules around deep-sea hills in Central Pacific basin* (pp. 98–159). Report of Geological Survey of Japan no. 21.

Usui, A. (1994). Manganese nodule facies in the western part of the Penrhyn Basin, south Pacific (GH83-3 Area). In A. Usui (Ed.), *Marine geology, geophysics and manganese nodules in the Penrhyn Basin, South Pacific* (pp. 87–164). Report of Geological Survey of Japan no. 23.

Usui, A., & Nakao, S. (1984). Local variability of manganese nodule deposits in GH80-5 area. In S. Nakao, T. Moritani (Eds.), *Marine geology, geophysics and manganese nodules in the northern vicinity of Magellan Trough* (pp. 106–121). Report of Geological Survey of Japan no. 20.

www.metoffice.gov.uk. (n.d.). Referred in 2007 for cyclone data.

Xia, J., Xie, L., Zou, W., Tang, D., Huang, J., & Wang, S. (1997). Studies on reasonable hydraulic lifting parameters of manganese nodules. *Proceedings of 2nd ISOPE – Ocean Mining Symposium, Korea (24–26 November, 1997)* (pp. 112–116). International Society of Offshore and Polar Engineers.

Yamazaki, T. (2017). Fundamental geotechnical considerations for design of deep-sea mining systems. In R. Sharma (Ed.), *Deep-sea mining – resource potential, technical and environmental considerations* (pp. 259–304). Switzerland: Springer International publishing AG.

Yamazaki, T., & Sharma, R. (1998). Distribution characteristics of co-rich manganese deposits on a seamount in the Central Pacific Ocean. *Marine Georesources and Geotechnology, 16*, 283–305.

Yamazaki, T., & Sharma, R. (2000). Morphological features of co-rich manganese deposits and their relation to seabed slopes. *Marine Georesources and Geotechnology, 18*, 43–76.

Yamazaki, T., Tsurusaki, K., & Inagaki, T. (1995). Determination of dynamic geotechnical properties of very fine and weak clayey soils by using a collision test. *Journal of the Japanese Geotechnical Society, 47*(12), 23–27. (in Japanese).

Yoon, C-H., Park, Y-C., Lee, D-K., Kwon, K-S., & Sung, W-M. (2003). Behaviour of deep-sea mining pipe and its effect on internal flow. *Proceedings of 5th ISOPE – Ocean Mining Symposium, Tsukuba, Japan (15–17 September, 2003)* (pp. 76–82). International Society of Offshore and Polar Engineers.

Dr. Rahul Sharma (rsharma@nio.org, rsharmagoa@gmail.com) retired as Chief Scientist from the CSIR-National Institute of Oceanography in Goa, India, with a career spanning 35 years in the field of exploration and exploitation of marine minerals. He has led a multidisciplinary group on "Environmental studies for marine mining." He has a master's degree in geology and a doctorate in marine science. His professional interests include application of exploration and environmental data to deep-sea mining. He has edited 3 special issues of journals, published 37 scientific papers, authored 22 articles and 41 technical reports, and presented more than 50 papers at national and international conferences. He has also edited a book titled *Deep-Sea Mining* published by Springer International Publishers in 2017 that has chapters contributed by experts from around the world.

His international assignments include visiting scientist to Japan, visiting professor to Saudi Arabia, member of the UNIDO mission "to assess the status of Deep-sea mining technologies" in Europe, the USA, and Japan, and invited speaker and consultant for the International Seabed Authority, Jamaica. He has contributed to the "World Ocean Assessment report I' of the United Nations and has also been invited to contribute a chapter on "Potential impacts of deep-sea mining on marine ecosystem" for the Oxford Encyclopedia for environmental science. In addition to his research career, he has been involved with several activities relating to science communication and outreach, as well as training programs for international participants, professionals, and students.

Part IV
Environmental Management

Environmental Policy for Deep Seabed Mining

Michael W. Lodge, Kathleen Segerson, and Dale Squires

Abstract This chapter examines the issue of developing deep-sea mining (DSM) while managing the impact upon deep seabed environmental assets. It reviews pertinent background information relating to DSM and the environment; develops a suite of potential policy instruments, including both direct and incentive-based regulation; develops additional incentive-based policy instruments, largely drawn from terrestrial conservation, for potential consideration; briefly touches upon technology and innovation to address mitigation of adverse environmental impacts from DSM; and then provides a concluding discussion.

Keywords Deep seabed mining · Environmental assets · Instruments · Liability · Financing

The views expressed in this paper are those of the authors and do not necessarily reflect the position of the International Seabed Authority or any of its member States.

M. W. Lodge (✉)
International Seabed Authority, Kingston, Jamaica
e-mail: mlodge@isa.org.jm

K. Segerson
Department of Economics, University of Connecticut, Storrs, CT, USA
e-mail: kathleen.segerson@uconn.edu

D. Squires
Department of Economics, University of California, San Diego, CA, USA
e-mail: dsquires@ucsd.edu

R. Sharma (ed.), *Environmental Issues of Deep-Sea Mining*,
https://doi.org/10.1007/978-3-030-12696-4_13

1 Introduction

Deep seabed mining (DSM) necessarily entails some degree of adverse environmental impact upon the deep sea's biodiversity, ecosystem, and ecosystem services.[1,2] Both mineral and environmental resources are "natural assets" that potentially contribute to human well-being. The International Seabed Authority (ISA), as it regulates DSM in the seabed beyond national jurisdiction, faces balancing DSM's negative environmental impact upon human welfare with the positive impact upon human welfare from mineral royalties and other payments received by the ISA to be disbursed for the benefit of mankind in accordance with the principles expressed in the United Nations Convention on the Law of the Sea. Trade-offs are likely involved, and in principle, an optimum balance can be achieved between mineral extraction and environmental degradation to attain the maximum current and future human welfare and when considering the equitable distribution of net benefits from DSM both intra-temporally (e.g., across potential beneficiaries) and intertemporally (i.e., across current vs. future generations). Achieving these goals requires developing the appropriate regulatory mechanism and regulations for governing extraction and its impact upon the environment.

In developing mechanisms to protect environmental resources and address their potential degradation, the ISA needs to achieve two objectives: (1) reduce the likelihood and magnitude of environmental damage in a least-cost application of the mitigation hierarchy and (2) provide funds for response, possible restoration, and compensation.[3,4]

A sound legal requirement supports these goals. Article 145 of the United Nations Law of the Sea Conference (LOSC) recognizes that exploitation of mineral resources in the deep seabed can disrupt the marine environment. The article requires the ISA to establish rules, regulations, and procedures to protect and conserve natural resources in the area of its jurisdiction, the Area Beyond National Jurisdiction

[1] Ecosystem services are defined to be the "benefits people obtain from ecosystems." See, for example, Dasgupta (2001), MEA (2005), and Madureira et al. (2018).

[2] For an overview of the DSM environmental issues, see Madureira et al. (2018).

[3] The mitigation hierarchy is comprised of four steps: avoidance of environmental damage, minimizing environmental damage after avoidance, restoration/rehabilitation of environmental damage, and addressing the residual after the first three steps have been undertaken, with no net loss of biodiversity and ecosystem services and potentially even a net gain in biodiversity and ecosystem services. See, for example, ten Kate and Crowe (2014). Below we raise a discussion of whether or not no net loss is applicable to an ecosystem that operates on almost geological time scales or at least very long time periods and consistency, whether the no net loss objective applies at all and whether an alternative objective is more appropriate (Griffiths et al. 2018; Kotchen 2009). Besides altering the level of biodiversity loss that is acceptable, the applicability of no net loss also impacts the size of the residual after the first three steps of the mitigation hierarchy have been undertaken and the applicability and role of biodiversity offsets.

[4] Cost-effective or least-cost conservation of the mitigation hierarchy's four steps can lead to greater overall conservation given limited funds for conservation (Squires and Garcia 2014).

(Area). This requirement thus recognizes the potential for DSM to reduce or impact the provision of ecosystem services.

This chapter examines the issue of developing DSM while managing the impact upon deep seabed environmental assets. The balance of this chapter is organized as follows. The next section reviews background information relating to DSM and the environment pertinent to the rest of the chapter. The third section develops a suite of potential policy instruments, including both direct and incentive-based regulations, both of which have been in the potential policy mix. The fourth section develops additional incentive-based policy instruments, largely drawn from terrestrial conservation, for potential consideration. The fifth section briefly touches upon technology and innovation to address mitigation of adverse environmental impacts from DSM. The final section concludes.

2 Background Information and Issues

2.1 Nonrenewable Resources

For all practical purposes, both deep-sea mineral resources and the deep sea's environmental resources are nonrenewable or at best only renewable over very long time periods.[5] In some cases, replenishment or recovery is technically possible, but not over relevant time scales (Madureira et al. 2018). For example, growth rates of polymetallic nodules, the mineral most likely to be exploited in the foreseeable future, are typically only several mm per million years (Hein et al. 2013), implying that the recovery of the landscape in mining areas of abyssal plains will occur only over a geological time scale. Similarly, much of the deep-sea environment is fragile and either nonrenewable or only renewable over very long time periods. Thus, the recovery of the landscape in most mining areas should not be expected at the scale of human life.

This nonrenewability of both mineral resources and much of the environment implies that mineral extraction by current generations comes at the expense of both mineral extraction and benefits from the environment by future generations. Future generations can exercise their ownership rights and claims to the benefits from deep-sea mineral resources in one of two ways. The first occurs when the current generation leaves an intergenerationally equitable share of the resource in situ for extraction by future generations. However, future generations can also share in the benefits from current DSM if some portion of the royalties from current exploitation are saved and invested to provide increased consumption for future generations. Increasing future consumption requires reducing current consumption of DSM royalties earned now, which in turn increases savings and investment in economic

[5]The impact of the nonrenewability of deep-sea mineral resources is somewhat mitigated by the large size of the Area and expected increase in new reserves as exploration and technological progress proceed. Nonetheless, these resources should be viewed as effectively nonrenewable resources.

growth, and thereby future consumption. Some portion of DSM royalties can be saved to build up a stock of non-resource assets. The return on these assets could sustain the spending annuity after DSM has ended, thereby insuring that future generations enjoy some share of the benefits from DSM and promoting sustainable development.[6] In a sense, the question is: Which form of capital should the benefits from minerals and the ecosystem be held, natural capital held in situ or financial and physical capital to generate the economic growth that allows current and future generations to improve their standard of living?

A similar issue and set of considerations arise with the intergenerational distribution of benefits from a sound ecosystem and the environmental costs due to DSM. These environmental resources can provide a flow of services continually over time, but the provision of environmental services to future generations depends on protection of the deep-sea environment by the current generation. In addition, although proceeds from the investment of royalties can help to ensure that future generations reap a share of the benefits derived from deep-sea mining by current generations, they will not replace the lost future environmental services that future generations would otherwise enjoy. Of course, future generations would also forego those services if resources are left in situ now but mined by future generations.

Thus, when considering both current and future generations, the ultimate question is how to balance consumption fueled by mining royalties (now and/or in the future) and the enjoyment of largely nonmarket benefits from a sound ecosystem, both within each time period and across current vs. future generations. Because the nonrenewable nature of the deep-sea minerals and environment means that some decisions are irreversible, the extent to which DSM is allowed to occur now will largely determine the answer to that question, not only for the current generation but for future generations as well.[7]

---

[6] Sustainable development is development that "meets the needs of the present without compromising the ability of future generations to meet their own needs," taking into account the capacity of the natural environment to sustain, indefinitely, the quantity and quality of ecosystem services, biological diversity, and ecological integrity. Sustainable development with nonrenewable mineral resources requires transforming the nonrenewable natural capital into sustainable financial, human, and physical capital that can lead to sustainable consumption and well-being for future generations (World Commission on Environment and Development 1987). Sustainability, stemming from depleting resource revenues, is not as central for the ISA as it is for many individual mining-dependent countries with short reserve horizons, given the size of the Area and expected increase in new reserves.

[7] This trade-off is further complicated by the potential for improved information over time about the demand for mineral resources, the environmental impacts of mining, and the availability of new technologies. Having better information can improve decision-making but may come at the cost of foregone net benefits that stretch across up to many generations (an opportunity cost). The trick is to achieve the optimum mix of waiting and not waiting. Waiting too long penalizes current and future consumption, but "hasty" decision-making without adequate scientific information can also turn out to be suboptimal.

2.2 Equitable Sharing of Benefits

In addition to the balance regarding the benefits from mineral resources and environmental services, DSM raises complicated questions related to how those benefits should be shared, both within and across generations. Equitable sharing can reflect either perceived rights or an implicit or explicit desire to redistribute income or wealth, presumably from wealthier groups (e.g., States) or generations to poorer ones.[8]

If the goal is redistribution, then, within a given time period, shares should be distributed based on some indicator of a state's priority in the redistribution goal and would, typically, embody some form of progressivity that favors lower-income states in the distribution scheme. Progressivity can be defined in various ways. For example, it can mean (i) that the share of rents received by a low-income state is higher than the share received by a high-income state or (ii) that the total amount received as a percentage of income is higher for low-income states than for high-income states. The first definition is more favorable to low-income states, but either implies a redistribution of income or wealth relative to what would be required by a proportional distribution scheme based solely on ownership rights.

Equitable allocations across generations can be based on some implied principle of justice embodied in an inter-temporal social welfare function that assigns weights to the well-being of different generations, where those weights reflect an implicit "utility-based" social discount rate.[9] However, even with equal utility-based weights across all generations, society may choose to weight consumption differently for different generations based on differences in income or wealth (or other socioeconomic criteria that may be deemed important). If future generations are likely to be wealthier (due to technological advances and economic growth), then a progressive approach to intergenerational allocation would assign greater weight to consumption by current generations because they are less well off than future generations.[10] Conversely, if future generations are likely to be poorer than current generations (e.g., due to resource degradation), then a progressive approach would put more weight on consumption by future generations by, for example, placing greater weight on investment that leads to higher future consumption when evaluating

---

[8] A redistributive goal could be based on a variety of ethical premises, including Rawlsian redistributive justice or a utilitarian foundation with decreasing marginal utility of income. See, for example, Rawls (2009) and Sen and Williams (1982). Note that these principles can be applied in redistributing income either inter-temporally or across generations.

[9] See Ramsey (1928) and Koopmans (1960) for the utilitarian approach and Rawls (1972) for justice among generations; Arrow et al. (2012) and Nordhaus (2008) provide related discussion on climate.

[10] A similar result applies to intra-temporal allocations, where equal utility weights, but differences in income levels and the associated marginal utility of income can provide a rationale for putting more weight on consumption by low income groups, thereby justifying redistribution of income from high- to low-income households or states (Squire and van der Tak (1975).

policy options. Unfortunately, the LOSC provides little, if any, guidance on the appropriate allocation of DSM benefits across generations.

2.3 Market Prices for Minerals and Environmental Damage Costs

Market prices for minerals from all mining, including DSM, do not include the accompanying costs to humanity from adverse environmental impacts.[11] When these "external costs" are excluded from DSM mineral market prices, mining firms and subsequent users of the minerals and consumers of final products made from these minerals do not take into account these adverse environmental impacts. Because the market price and the costs of DSM are both lower than optimal to society when these "external costs" of environmental damage are excluded, the scale of DSM production, adverse environmental impacts, and final consumption all potentially exceed the socially optimal amounts. Similarly, the mix of minerals mined and consumed, and the consequent adverse environmental impacts may not be the socially optimal combinations. This failure of minerals markets to allocate resources both now and in future generations to the socially optimal levels of DSM and accompanying adverse environmental impacts due to incomplete or missing markets means that the ISA has a clear role to play in environmental management.

2.4 Uncertainty

The ISA faces considerable uncertainty across multiple dimensions, including the nature and dimensions of the environmental impacts of DSM. These impacts are imperfectly understood: regulators will encounter uncertainties as to the magnitude of damages; difficulties in monitoring, measuring, and quantifying; little evidence of effective habitat restoration strategies; and sparse knowledge of spatial gradients, nonlinear damage-to-the-environment functions and thresholds. These information deficits are compounded by the heterogeneity of impacts across the various ISA potential mining sites and types of minerals mined and by the rapidly changing technology of DSM and environmental mitigation.

When the ISA considers incentive-based policy instruments, or combinations of instruments, the ISA should therefore make allowance for the incomplete character of the data that would inform its estimates of the costs of environmental compliance, enforcement, and remediation. Moreover, due to this uncertainty and newness of DSM, application of the precautionary principle and its combination with increased marine scientific research becomes important (Madureira et al. 2018;

[11] Although this chapter does not cover the payment regime or the economics of the mining process itself, see Roth (2018).

Lallier and Maes 2016). The ISA also must identify the fundamental principles that would guide its choices of incentive-based policy instruments. Those principles should inform discussions of the allocation of costs and risks, incentives to achieve protection goals, and responsibilities for monitoring and enforcement expenses.

2.5 Economic Issues Requiring Consideration

One important issue that requires consideration when crafting DSM environmental policy and regulations is alternative measures to ensure that the benefits from DSM incorporate humanity's concomitant net benefits from the deep sea's environmental resources both now and in future generations. Specifically, this consideration firstly entails achieving the optimum exploitation of mineral resources for current and future generations that account for society's direct net benefits from DSM while also accounting for the additional benefits to society from the environment and the costs from adverse environmental impacts that market prices do not account for plus the direct costs that accompany environmental mitigation and management.

Second, and as a consequence, the ISA must develop environmental regulations to govern DSM and address the accompanying adverse environmental impacts, both on the ocean floor and in the water column, in a practicable and cost-effective manner. These environmental regulations can broadly come through some combination of direct regulation and incentive- or market-based approaches. This chapter considers these broad approaches, their interactions, their main strengths and weaknesses, and allocation of costs and risk.[12]

Third, when DSM leads to environmental damage, key questions include who will be responsible and under what terms? Since the start of discussions on the legal regime for DSM, there have been suggestions that contractors exploring and mining the deep sea should assume responsibility for environmental damage and that the international regulatory authority (in this case, the ISA) considers setting funds aside to "provide for additional insurance against the risks of pollution" (United Nations 1971 (n 15) at p. 8).[13] The ISA, however, has yet to establish mechanisms

[12] This chapter focuses upon policy approaches that the ISA could implement, but another approach seeks to harness market pressures exerted by consumers who are willing to pay for products made in environmentally friendly ways. This approach addresses insufficient information in markets, where the amount and quality of information typically decline from producers through firms in the supply chain to consumers of final products. Different forms of certification and eco-labeling provide examples. Nonetheless, because minerals from the deep sea are typically used as inputs in the production of other goods and services, eco-labeling requires tracing the minerals' origins throughout the entire supply chain. The effectiveness of eco-labeling depends upon the extent to which final consumers of the products produced by these minerals are aware of and willing to pay a price premium when the products are produced with DSM that are minimally harmful to the environment as much as possible and when the price premium is actually transmitted to producers to incentivize their behavior and decision-making to reduce environmental damage from their activities. See Kotchen (2013) and Segerson (2013).

[13] In its *Advisory Opinion on the Responsibilities and Obligations of States Sponsoring Persons*

to specify the nature and extent of contractor responsibility for environmental damage that might occur, despite compliance with any applicable environmental regulations.[14] Similarly, how should funds that might be needed for redress or compensation be financed? Should this financing be based upon the "polluter pays principle" by contractors and sponsoring states or the "beneficiary pays principle" by the ISA and society as a whole or some sharing of costs among these groups, both at a given point in time and across generations?[15] If costs are shared, how should the appropriate costs shares be determined?

Fourth, developing the environmental regulations grapples with considerable risk and uncertainty over future DSM royalties and costs, and future DSM technology – both the technology to exploit the minerals and to reduce the remaining adverse environmental impacts – where some impacts are irreversible, and others are not. Waiting to acquire additional information has advantages in reducing this risk and uncertainty, but only at the cost of foregone DSM royalties and the net benefits they confer to humanity, including both current consumption and future consumption from enhanced economic growth fueled by mining royalties.[16] The impact of this waiting includes the impact of discounting that lowers the present value to humanity of expected net benefits received at a later date and the foregone expected net benefits from earlier economic growth that is otherwise postponed. This chapter, most importantly of all, raises issues to consider and potential trade-offs between DSM and its benefits with adverse environmental impacts.

and Entities with Respect to Activities in the Area (Advisory Opinion), Case No. 17, 1 February 2011, the Seabed Disputes Chamber of the International Tribunal for the Law of the Sea suggested that consideration could be given to a fund to provide compensation for damage to the marine environment that could not otherwise be compensated by contractors or by sponsoring states, either to fill the "liability gap" or where the contractor or sponsoring state are impecunious.

[14] The allocation of costs and the allocation of risks are potentially two different issues. As an example, an environmental fee could impose fully expected costs on contractors but little risk, because the fee's amount that would have to be paid would be very predictable. In contract, strict liability imposes full actual costs, but also significant risk, because the amount for which the contractor would ultimately be held responsible is unknown beforehand. See Craik (2018) for additional discussion.

[15] With the polluter pays principle, the party inflicting biodiversity loss pays for it. This party has willingness to pay (WTP), and the affected party or society has willingness to accept (WTA) compensation. Maximum WTP for net gain and minimum WTA for no net loss in biodiversity or ecosystem services bound the size of economically rational compensation in monetary values for any voluntary payments. The beneficiary or user pays principle is just the opposite in that it holds that agents who have benefited from the processes that cause loss in biodiversity and ecosystem services should pay the costs of addressing its harms. Just the opposite of the polluter pays principle, the WTP holds for the "impacted" party and WTA for the "impacting" party.

[16] The economic value of waiting until additional information is available with an irreversible investment is called quasi-option value (Arrow and Fisher 1984). For an economic analysis and discussion of quasi-option value with mining, see Costello and Kolstaad (2015).

2.6 Regulatory Issues When Selecting Environmental Policy Instruments

Key issues when choosing policy instruments include (1) uncertainty about the magnitude of the damages; (2) uncertainty and difficulty in monitoring and measuring impacts; (3) insufficient information on the means to reduce impacts; (4) inability to restore impacted habitats; (5) spatial gradients, nonlinear damage-to-the-environment functions, and thresholds; (6) heterogeneity of impacts across mining sites; (7) whether the contractor response to mitigate environmental damage is due to the policy instrument or would have been done regardless ("additionality"); (8) policy instruments with regulatory outcomes conditional upon actual performance ("conditionality"); (9) costs of implementation (monitoring, control, surveillance); (10) compliance; and (11) enforcement.

2.7 Basic Economic Principles to Consider When Choosing Environmental Policy Instruments

Basic economic principles to consider when choosing environmental policy instruments include:

1. Who should bear the cost of ensuring adequate protection? Should the "polluter pays principle" or the "beneficiary pays principle" apply? The latter implies sharing of costs among contractors, sponsoring states, and the ISA.
2. Who bears the environmental risks, contractors, sponsoring states, or society collectively (ISA)? Should society collectively bear some of those risks in return for a share of the monetary returns? Different policy instruments differ in how they allocate risks.
3. Consideration of least-cost incentives to protect the environment. Policy instruments based upon market incentives create stronger incentives and afford contractors the flexibility to address environmental damage in their own way as long as environmental targets are satisfied. Least-cost conservation mitigates environmental damages at lowest cost for a given environmental target, which potentially allows saved funds to be used for other activities, including environmental mitigation that leads to higher welfare for society.
4. Who should bear the costs of monitoring and enforcement? Are these costs borne by the regulated parties (contractors and/or sponsoring states) as charges separate from payment for mining, or does society collectively, through the ISA, assume the burden of these costs through, for example, royalty payments?
5. Policy instruments can be based upon performance or outcomes of the environmental regulation or on the process of the mining activity, i.e., on the process of production, transportation, and refining. Process-centered policy instruments affect the choice and state of technology and the choice and use of inputs in the

mining activity. Performance-based approaches generally provide contractors greater flexibility in meeting environmental protection goals than those based upon process. Performance-based approaches tend to create stronger and more direct economic incentives because they directly address the desired policy outcome. Process-centered incentives are more indirect because only some of the inputs and practices are regulated, and the relationship between the regulated inputs and the expected outputs can be indirect and more uncertain. Process-centered incentives are consequently weaker because they are more indirect. Nonetheless, in some instances, performance-centered policy instruments may be more difficult and more costly to verify than process-centered approaches, especially in DSM where production occurs in deep waters far out at sea.

3 Suite of Possible Policy Instruments for Protecting Marine Environment

Alternative policy approaches can be classified as direct regulation and incentive-based. The approaches differ in terms of (1) types, strength, and directness of incentives they create (for care per unit of output and scale of activity or extent of operation), (2) cost to the mining entity and implied property rights (implementation of the "polluter pays principle" or the "beneficiary pays principle"), (3) allocation of risk (financial and environmental), (4) generation of funds for restoration and compensation, (5) utilization of contractor information unknown to the regulator, and (6) administrative costs (including monitoring and enforcing). Which of the policy approaches is best and whether or not approaches should be combined remains to be addressed. We review each of the policy approaches in turn.

3.1 Direct Regulation

Under direct regulation, mining contractors are required to comply with certain requirements regarding how they conduct their mining activities (a process-related approach) or environmental outcomes (a performance-related approach) in order to protect the marine environment. These requirements can be (1) technology standards, which are mandatory design and equipment requirements and include operating standards, (2) performance standards, which are a standard on outcomes requiring achievement of a target, but not specifying how the target is to be met, or (3) process standards, which are a standard on the mining process.[17]

[17] Madureira et al. (2018) discuss many potential technology standards, adaptive use of technology standards as knowledge is gained, and potential process standards, notably protected areas and accompanying scientific research to evaluate the impacts of DSM.

Direct regulation has a number of advantages. These include the known impact on mining contractor behavior if the contractor is compliant, low levels of risk for mining contractors when the regulatory requirements are well defined and established, and relatively low administrative costs if compliance can be easily monitored and enforced.

Direct regulation as the sole regulatory approach has a number of disadvantages that reflect information, cost, and incentive compatibility issues faced by regulators and mining entities. First, direct regulation does not use all of the information that can potentially engage all mining entity channels to mitigate environmental damage that is otherwise available and is constantly evolving as mining entities learn by doing. Direct regulation only uses the information on environmental mitigation held by the regulatory body. However, mining entities hold information, sometimes quite subtle, that the regulator does not typically know and hence does not use. This information grows as miners gain experience and learn. Mitigation of environmental damage can entail multiple, ongoing adjustments in mining, and accompanying environmental mitigation that is taken individually may have varying and even small impacts but collectively can have a significant impact. In short, direct regulation, which is top-down, does not utilize all of the available information that would otherwise be potentially used – and even deliberately created through adaptive management and experimentation – in a bottom-up approach.

Second, direct regulation does not impose responsibility upon contractors for any degradation that might occur despite compliance with those regulations. It implies only partial implementation of the "polluter pays principle," since mining entities pay the cost of complying with regulatory requirements but not for residual damages occurring despite compliance. Thus, direct regulation shares conservation-related costs between contracts (for avoidance, minimization, and restoration) and those other stakeholders who also suffer from the residual loss of biodiversity and environmental services and related risks. Third, direct regulation does not generate any funds for compensation of either anticipated or unanticipated environmental damages. Society instead bears the full cost and the full risk of any resulting significant adverse environmental impact despite compliance.

Fourth, when costs of environmental mitigation vary across contractors, uniform or "one-size-fits-all" regulation is not cost-effective, because it does not create economic incentives to meet environmental performance targets in a least-cost way. The cost-minimizing mitigation approaches can vary by contractor and even across mining sites.

Fifth, because direct regulation does not directly price residual environmental damage, it fails to create a mining cost that incorporates the "external cost" of the remaining environmental damage associated with each unit of minerals mined, although the mined minerals average production costs reflect the variable (operating) costs due to direct regulation. This rise in unit costs of production creates a "crude" incentive to reduce environmental damage through reducing the scale of production. However, because the remaining environmental damage is not given a price and hence a cost, the mining unit costs do not incorporate the full social costs of the mining activity (which include the residual environmental costs), and the mining level can be higher than is socially efficient.[18] Moreover, by failing to engage all environmental damage reduction channels across and within steps of the mitigation hierarchy and across all contractors and mining sites, mining areas will not be selected and mined in a socially optimal way. Finally, direct regulation fails to incentivize contractors to exceed their regulatory requirements set through either performance or technology standards. For all these reasons, reliance solely upon direct regulation to conserve the marine environment is likely to fall short on several of the criteria that the ISA might use in evaluating alternative policy approaches.

3.2 Incentive-Based Approaches to Regulation

Incentive-based approaches create strong incentives for contractors to reduce potential environmental damage in a cost-effective way, i.e., through methods that best suit the specific contractor and site. Incentive-based approaches potentially incentivize least-cost mining activity across all channels of environmental damage mitigation and contractors and across and within all steps of the biodiversity mitigation hierarchy. They can also induce contractors to technologically innovate to alter production techniques in ways to lower the environmental damage per unit of ore produced. The larger the market for the mineral, the stronger are the incentives to technologically innovate (Acemoglu 2002).

Incentive-based approaches to regulation incentivize contractors to minimize a targeted limit of environmental damage in their own way by engaging all mitigation channels across and within steps of the mitigation hierarchy and across mining sites. Incentive-based approaches place a price upon the residual environmental damage – the "external cost" faced by contractors – corresponding to the standard set by the regulator. By pricing residual environmental damage, the unit cost of mined ores rises, and contractors reduce the scale of production to the socially optimal level. To the extent that residual environmental damage varies by mineral mined and the location of mines, the mix of minerals and even grades of ores extracted could also change when residual environmental damage receives a price and hence cost. The external costs corresponding to the environmental damage target are accounted for by markets and borne by society through a sharing between contractors, firms in the supply chain, and consumers.

[18] The socially efficient level of an activity is the level that maximizes the net social benefits of that activity, which equals the benefits derived from the activity minus all social costs (including environmental costs and opportunity costs) that result from the activity. When environmental costs are not fully considered by producers, the activity level that emerges in a competitive market exceeds that level. In addition, the mix of minerals mined, areas mined, and methods of mining can be socially inefficient. All channels to mitigate environmental damage within and across steps of the mitigation hierarchy and across contractors will also be socially suboptimal. Finally, activities across steps of the mitigation hierarchy will not be socially optimal, with excessive mitigation in one step (e.g., avoidance) and insufficient mitigation in another step (e.g., minimization). See Squires and Garcia (In press).

The extent by which a mineral price rises under an incentive-based policy compared to without the incentive-based environmental regulation, and the corresponding impact upon the scale and mix of production, depends upon a number of factors. An important determinant is the extent to which producers and purchasers can shift the cost increase onto others due to conditions such as the supply and grade of minerals, the time suppliers have to respond, other substitute minerals a purchaser can purchase, the number of suppliers or purchasers of minerals, or the extent to which the fee increases the cost of each unit of production.[19]

A number of potentially suitable incentive-based policy approaches are available. We discuss in order environmental taxes or fees, including "double dividend" ones; environmental liability; insurance; environmental (assurance) bonds; environmental trust funds; cap-and-trade habitat impact quotas; conservation easements; biodiversity offsets; and payments for ecosystem services. Some policy instruments are more likely than others, but discussing a full suite of potential incentive-based environmental policy instruments supports a wide-ranging and informed basis upon which the ISA can decide.

3.3 Environmental Taxes or Fees

Environmental taxes or fees, a performance-based approach, directly price and hence create a cost for the environmental damage – the external cost that is otherwise not incorporated into the ore's market price. This type of tax, in contrast to many other types, increases society's economic welfare by changing miners' decision-making and behavior to produce at the socially optimal scale and mix of production. It implements the "polluter pays principle" when each contractor's payment covers the full expected environmental cost of its mining. Payment of environmental taxes or fees presumably occurs on a regular basis throughout the mining.

Determining the size of the tax or fee can be very difficult, since measuring the environmental damage at each site for site-varying fees is itself difficult. In contrast to other environmental contexts, such as air and water pollution where the performance is typically defined in terms of emissions or discharges of pollutants, little is known about how DSM's environmental performance would be measured on an ongoing basis and hence upon what to base a performance-based fee. An emissions tax, for example, is typically levied on some measure of the emissions, but there is no accepted comparable unit of environmental damage from DSM to serve as the basis to calculate a performance-based environmental fee.

The extent to which taxes or fees create strong incentives depends upon how they are structured (as well as the size of the fee). Fixed fees, for example, generate revenue, but they are not tied to environmental performance and thereby do not impact

[19] In the language of technical economics, these factors are the degree of competition in the market and the relative price elasticities of supply and demand. (Price elasticities give the proportional change in quantity supplied or demanded to a 1% change in price.)

contractor's behavior and decision making to alter the scale and mix of production consistent with the extent of the environmental damage. Fixed fees may not even alter the scale of production over short time periods. Conversely, fees that vary with performance incentivize contractors to reduce the scale and mix of environmental damage by altering production techniques and/or locations.

The revenue from an environmental tax or fee can serve a second purpose that mitigates the environmental damage or in other ways increases society's welfare. The funds could help finance innovations to reduce environmental damage, since innovators do not typically enjoy the full benefits of such innovations (unless there are patents and licenses) and hence do not face incentives to innovate to the full socially optimal level. The revenue could be placed into an environmental fund and used in different ways to reduce environmental damage or enhance environmental quality somewhere else. Such funds should be "ring-fenced," i.e., kept separate from a resource fund that is financed through royalty or other payments from extracting the minerals, because the purposes and impacts of the funds are different. A resource fund finances payments to resource owners for the right to extract and sell the minerals, whereas an environmental fund finances compensation or mitigation for environmental damage resulting from mining.

3.4 Environmental Liability

Environmental liability is a performance-based approach to protect the environment, under which contractors and/or sponsoring states can be held liable for environmental damages exceeding some baseline[20]. Environmental liability is typically triggered in some period by some event or condition.

A key issue in imposing environmental liability with DSM is defining the threshold or event that triggers the liability. As with environmental damage under a tax or fee, defining, measuring, and monitoring the triggers pose challenges and must be something that can be implemented with relative ease and low cost. The current ISA exploration regulations refer to "serious harm" to the marine environment.[21] Nonetheless, except for unexpected accidents, exceeding a threshold for "significant damage" is likely to occur gradually, and the point at which a contractor would become liable could be difficult to determine with sufficient precision and certainty to withstand a legal challenge, even with a "zero tolerance" threshold, under which environmental damage of any type would be enough to trigger potential responsibility. However, because mining necessarily impacts the seabed and water column,

[20] For extensive legal discussions, see Craik (2018).

[21] Regulations on Prospecting and Exploration for Polymetallic Nodules in the Area (ISBA/6/A/18, 13 July 2000, as amended by ISBA/19/A/9 and ISBA/19/A/12, 25 July 2013, and ISBA/20/A/9 24 July 2014)

some means of defining "damage" and determining when it has occurred are necessary to employ a liability approach.[22]

The two general forms of liability attribution that can be applied are strict liability and negligence-based liability. Under strict liability, a producer is responsible for damage regardless of the amount of care taken to avoid damage. In contrast, under a negligence rule, a producer is not held responsible for damage unless the producer is negligent in conducting its operations. Under negligence liability, full compliance with existing regulations might be viewed as de facto evidence of non-negligence and thereby absolve the contractor of responsibility for any residual environmental damage.

Current LOSC rules, as implemented in the ISA exploration regulations, impose a fault-based liability due diligence standard (which can also be viewed as a legal as opposed to environmental threshold question) (Craik 2018). Nonetheless, it remains open for the ISA or sponsoring states to revisit this legal standard in relation to the development of new rules. Other civil liability regimes in international law have favored no-fault approaches to liability, typically combined with channeled liability and mandatory insurance.

The liability framework under the LOSC identifies that liability for wrongful acts resulting in damages shall be for the actual amount of damage, but the LOSC does not define which types of damages are compensable and how that damage is to be quantified (Craik 2018). Damages to persons and property are well understood in international and domestic legal settings, but environmental harm is less settled. Unresolved questions include whether there should be limits on damages for remediation and whether "pure" environmental losses are commensurable.

A number of unresolved issues remain (Craik 2018). The scope of compensable damages requires identification and should reflect the particular features of the Area's marine environment as well as the status of the Area and its resources as the common heritage of mankind. Key issues in determining the scope of compensable damages include the following: (1) whether, in order to require compensation, damages must exceed a threshold, such as "serious" or "significant" harm; (2) whether pure environmental losses will be recoverable and, if so, whether they can be quantified; (3) which parties have standing to claim for DSM environmental damages, notably damages to the Area and its resources; (4) the adequacy of existing dispute settlement mechanisms and fora for claims; and (5) how to ensure that compensable funds are available.

The incentive, cost, and risk implications of imposing liability for DSM environmental damage depend upon the form of liability employed. Incentives arising from strict reliability are weakened if contractors could become judgment-proof (such as through bankruptcy) or otherwise avoid payments for which they are legally responsible. Incentives are also weakened the longer the length of time before liability might be imposed, which is the expected case with DSM due to its long production period. Strict liability creates a potentially high risk for contractors, because their potential exposure and ultimate costs are unknown when DSM begins.

[22] Craik (2018) and Madureira et al. (2018) provide additional discussion related to damages.

Strict liability and environmental fees both apply the "polluter pays principle," but they differ along several important dimensions. Strict liability holds contractors liable for actual environmental costs, while an environmental fee is based upon expected environmental costs. A contractor's environmental costs under strict liability are necessarily site-specific, while levying environmental fees by location might be difficult – although fees could be progressively graded according to significance or sensitivity of locations. The two approaches also differ in the amount of risk they impose on contractors, with strict liability imposing substantially more risk.

Under a negligence rule, non-negligent contractors are not liable for any residual environmental damage. This implied property right, and the associated allocation of costs between society and contractors contrasts with strict liability and implies only partial implementation of the "polluter pays principle." As a consequence, the scale and cost of production and price of minerals for non-negligent contracts do not incorporate the full social cost of mining (including the external cost of environmental damage), which is similar to the outcome under direct regulation. Thus, the outcomes under a negligence rule and direct regulation would be very similar if the relevant due standard of care used to define negligence is based solely on compliance with existing regulations.[23] Moreover, as long as contractors know the determination of the due standard of care at the time they undertake mining activities and they choose to comply with that standard, they will bear little risk from potential environmental losses. Those risks, instead, will be borne by society in the form of uncompensated residual environmental damage. Contractors thus bear both lower expected costs and lower risk under a negligence rule than under strict liability. A negligence rule also creates a weaker incentive for technological innovation designed to reduce environmental risks, although incentives to reduce the compliance costs with the due standard of care still exist.

3.5 *Insurance*

An additional incentive-based policy instrument is the requirement that contractors carry private environmental liability insurance for any residual DSM environmental damage. This requirement combines elements of an environmental fee with some elements of a strict liability rule. As with an environmental fee, contractors would make payments, through insurance premiums, based on expected environmental damage. Contractors thereby carry the full, expected social costs of DSM consistent with the "polluter pays principle," and insurance payouts provide a source of compensation revenue (even if the contractor is judgment-proof). The payment amounts can be expected to be set prior to when production begins, so that contractors know

[23] This outcome presumes that the penalties for violation of existing regulations are comparable to the magnitude of the potential liability contractors face if they are negligent and that both face the same likelihood of enforcement.

the payments and contractors thereby face little risk. The insurer instead bears the risk.

The incentives created by insurance alter contractor behavior and decision-making to reduce DSM's environmental impacts but depend upon how the premiums are set. The strength of the incentive depends upon the extent to which the insurance is performance-based.

The incentives created by insurance also face a classic problem of differing amounts of information held by the contractor and the insurer – what is called asymmetric information. This asymmetric information problem in turn gives rise to two problems, moral hazard and adverse selection. Adverse selection, which arises prior to when the insurance contract is written and issued, arises because private insurance markets typically function best when risks can be accurately estimated and spread across a large population of insured parties. However, because the insurer typically has less ability to evaluate the environmental risks at individual mining sites accurately, insurance premiums based upon average risks will make that insurance unattractive to contractors with lower-risk sites. When insurance coverage is voluntary, these lower-risk contractors may choose to not take the coverage. In contrast, when insurance is mandatory, lower-risk contractors in effect subsidize higher-risk contractors when both pay premiums based upon average risk. This adverse selection problem is compounded in a nascent industry such as most DSM in which the technologies of both extraction and environmental damage mitigation are still under development, and there is little experience with either DSM or environmental damage mitigation and indeed of the very nature of the damage and its measurement. This considerable uncertainty compounds the problem of accurately setting the premium.

More hazard, which arises after the insurance contract is written, issued, and paid, arises when the insurance contract does not fully and accurately cover all possible contingencies pertaining to environmental damage. The contractor then faces a weak incentive to be careful in its operations because the insurance covers the environmental damage. For this reason, insurance policies typically include a deductible clause, by which the insured party must first pay an agreed fixed amount prior to receiving insurance payments. The deductible effectively creates a shared liability between the insured party – here the contractor – and the insurer. The deductible limits the insured party's liability. This deductible would then incentivize the contractors to reduce environmental damage, but nonetheless does not fully address the moral hazard problem resulting from the insurance.

Insurance payments, just like the case of liability, typically require an event or situation to trigger. An observable trigger or threshold needs to be clearly defined and measurable to allow insurers to write the corresponding policies and for the demand for coverage and payouts by the insurance company to begin. Issues with defining the trigger that were discussed above apply to insurance as well, which is compounded by the need to enter an enforceable contract at a very early stage of production.

3.6 Environmental Bonds

Environmental bonds, also known as assurance bonds, are effectively a form of self-insurance. These bonds typically require the resource user to place funds up front prior to mining that is equivalent to the potential environmental damage or, more commonly for land-based mining, the remediation cost.[24] Funds not needed to cover environmental damage are then refunded. Environmental bonds are thus inherently performance-based and run into the problems of defining, observing, and measuring the environmental damage. They require the mining contractor to bear the full social cost of production, including all environmental costs, because they require the contractor to set aside funds equivalent to the maximum potential damage from mining or to the cost of remediation. The "polluter pays principle" thus applies.

Some bond details require consideration. These include the form of the guarantee, such as cash versus financial security, and whether the bond is posted entirely up front or by installments over time.

Environmental bonds are similar to strict liability, but without the judgment-proof problem that arises under strict liability. The incentives are similar to those under strict liability, but they are stronger when the potential for contractors to become judgment-proof is considered. Environmental bonds ensure that funds are available for restoration and/or compensation even if the contractor is subsequently declared bankrupt or insolvent. Like strict liability, an environmental bond requirement imposes substantial risk upon contractors, although the bond limits the extent of this risk.

Environmental bonds also differ from strict liability along several other dimensions. First, the bond's amount must be specified up front, whereas the payment under strict liability is determined at the time that damage occurs and liability is imposed. Determining the appropriate up-front amount, given the current understanding of DSM's environmental impacts, poses a challenge and creates considerable risk and uncertainty – which can increase the de facto cost of the bond. The determination will likely be made on a case-by-case, site-specific basis, with considerable learning over time as experience is gained. Second, an environmental bond could be used to cover a variety of costs, including costs from abandonment prior to closure and post-closure activities, along with post-closure monitoring, emergency responses, and liability for economic damage. This contrasts with liability payments that are typically focused upon providing compensation for environmental damage. The bond's specific uses require clear specification up front. Third, the posting of a DSM environmental bond constitutes a contract between the contractor and presumably the ISA and thereby suffers from common contract-related issues, such as asymmetric information between the contractor (who typically holds

[24] The bonds can take a variety of forms: cash deposit, parent company guarantee, state guarantee, financial institution letter of credit, and associated investment grade of any issuer. The terms of its release and what can be deducted against the deposit can also vary.

better information) and the ISA, monitoring and measuring issues, and moral hazard once the contract enters into force.

4 Additional Incentive-Based Policy Instruments for Potential Consideration

In this section, we sketch out several incentive-based policy instruments that could conceivably be implemented at some point, although they are currently receiving less attention than the policy instruments discussed in the previous section. These policy instruments are increasingly applied in terrestrial conservation and to a lesser extent other applications of marine conservation for living resources and habitat.

4.1 Transferable Habitat Impact Quotas

Transferable habitat impact quotas, which operate through a cap-and-trade system, are a form of rights-based management that can be applied to benthic habitat (Holland and Schnier 2006; Wallace et al. 2015).They incentivize the avoidance and minimization steps of the mitigation hierarchy in a potentially least-cost manner. Contractors can potentially be incentivized to balance mining sites that are the most profitable with the potential environmental damage that varies by site. The world's first habitat impact program was recently implemented in a British Columbia, Canada, trawl fishery (Wallace et al. 2015). This incentive-based policy instrument is a performance-oriented approach that implements the "polluter pays principle" if they are auctioned off, but it does not hold if they are distributed freely.

Transferable habitat impact quotas in a cap-and-trade system are distributed to contractors with the total allowable habitat impact quota (TAHIQ) set to maintain a target habitat over some defined area. Each year a total number of habitat impact units are allocated to contractors, and the unused portion can be transferred to other contractors. The right is typically a percentage or share of the TAHIQ, and then when the TAHIQ is periodically set or reset, the allocated share is multiplied by the TAHIQ to give an amount. A contractor with multiple mining sites can trade among the contractor's own sites. A given total stock of habitat impact units might be comprised of a combination of totally and partially regenerated areas. The quantity of habitat impact units in the quota that are allocated each year would depend upon the habitat regeneration time, the standing stock of habitat, and the target level of the habitat stock. A proxy serves to measure marginal habitat damage. Due to the difficulty of directly monitoring actual marginal habitat damage, a proxy could be based upon area mined.

In contrast to rights-based management systems within national continental shelves, the property right is common property administered by the ISA. The right that is allocated to contractors is a use right of some duration. Aligning the impact right's duration with the contractor's mining site duration should strengthen conservation incentives.

A number of challenges would arise in the application of this approach. These include spatial gradients, nonlinear damage functions, and thresholds. Impacts at a given site are likely to vary spatially and could potentially increase nonlinearly with the scale of production, possibly crossing a threshold at some point. A uniform proxy may not accurately measure spatially differentiated habitat. Hence, the standard might be differentiated to the extent possible, rather than uniform, to account for these spatial differences in impacts.

4.2 Payments for Ecosystem Services

Payments for ecosystem services are voluntary transactions paid by the buyer to the seller to provide a desired conservation outcome. They reflect the "beneficiary pays principle," since the party that benefits from mitigation of environmental damage compensates the contractor that creates the damage. This payment, which is conditional upon a well-defined and monitored conservation outcome ("conditionality") that would not otherwise be provided ("additionality"), creates a market price that more closely aligns the incentives of contractors and society and thereby induces changes in contractor behavior and decision-making to mitigate environmental damage. Their sustainability can be jeopardized by lack of sustainable financing and changes in government policies. They are difficult to value, since biodiversity markets are missing or incomplete, but they should exceed the opportunity costs of foregone profits of the contractors.

Environmental nongovernmental organizations or states could conceivably make payments for ecosystem services to contractors to mitigate environmental damage through avoiding mining of particularly vulnerable marine ecosystems or to adopt mining techniques that may be less profitable but also less destructive of the marine environment. In addition, they can be performance or process-oriented. For example, a payment to avoid mining a specific area is performance-oriented, while a payment to mine or mitigate with a specific technique or required equipment is process-oriented. Uncertainty and monitoring are also case-by-case, depending upon the circumstances of each situation.

Payments for ecosystem services are subject to issues of asymmetric information between the buyer and the ecosystem service provider. The provider could offer mitigation services that would have been offered anyway or offer a site that provides lower quality or fewer ecosystem services than purported, creating a lack of "additionality." The provider could also evade protection or provision of services if monitoring is incomplete or inaccurate or if the contract is incomplete (lack of "conditionality").

4.3 Conservation Easements and Concessions

A conservation easement is a voluntary, legally binding agreement that limits certain types of uses or prevents development from taking place on a piece of property or area over some agreed upon time period to protect the property's biodiversity and ecological services (Parker and Thurman 2013). The owner of the property retains use of the property that can be for, but not necessarily limited to, commercial purposes, but these purposes are attenuated according to the conservation easement. Conservation easements can be flexible policy instruments.

A conservation easement selectively targets only those rights necessary to protect specific conservation purposes and is individually tailored to meet the area leaseholder's requirements. The remainder of the lease remains intact, allowing the easement property to continue to generate private benefits.

Initial conservation easements protected terrestrial landscapes from development and other human impacts, but they are evolving to become tailored to specific conservation objectives. DSM easements, for example, might include a management plan that clearly specifies a conservation objective, such as protecting the most ecologically vulnerable part of a deep seabed mining tract, and allow more flexibility in the way the goal is achieved. A conservation easement could be applied to the extent of a mine's operation or space or time, a mine's location, or the type of technology employed for mining and/or loading and transportation of ore. Such applications may be more costly, since they account for the environmental impact and may yield less ore or lower-quality ore (i.e., production costs rise since these costs now account for external costs). Conservation easements are a form of compensation for these higher costs.

Conservation concessions appear to be ideally suited to address the issue of what economists call adverse selection and what the conservation literature calls lack of additionality.[25] The mining companies might propose locations and sizes of protected areas in which no mining occurs. Self-selection of protected areas, however, can lead to contractors choosing protected areas that are economically unimportant rather than those that are environmentally important or choosing protected areas that would not have been mined in any case. A conservation concession would then compensate and incentivize contractors to place protected areas in locations that would have otherwise been adversely impacted. The concept might be especially well suited to apply at a regional level in the context of REMPs (Regional Environmental Management Plans). Then, society, through the ISA, decides to put certain vulnerable areas off limits to mining in the form of a protected area (e.g.,

[25] Adverse selection occurs in some instances when one party in a negotiation has relevant information the other party lacks, i.e., it is a form of asymmetric information that occurs before "the contract," i.e., the agreement. In this case, the contractor has more and higher-quality information about where the sites for mining and environmental protection are to be located that the other parties lack. By then placing an environmentally protected area in a location that would not have been mined and would have been implicitly protected regardless, there is no additional conservation beyond what would have occurred without the protection (i.e., lack of additionality).

active hydrothermal vents), thereby forgoing revenue that would otherwise benefit mankind, rather than locating the protected areas in places where there are no known resources. Of course, determining these areas and potential actions presents a challenge. Even contractors' voluntary application of the precautionary principle through selection of mining sites and/or methods and/or timing of mining could be incentivized through conservation concessions.

4.4 *Biodiversity Offsets*

Offsets are measurable conservation outcomes resulting from actions designed to (at least partially) compensate for significant adverse residual impact on biodiversity arising from project development or other activities such as mining after appropriate actions to mitigate environmental damage using the first three steps of the mitigation hierarchy (avoidance, minimization, restoration) have been taken. Offsets have been defined as conservation actions intended to compensate for the residual, unavoidable harm to biodiversity caused by development projects (i.e., what remains after everything possible has been done to avoid inflicting that harm), so as to ensure no net loss (or even net gains) of biodiversity (ten Kate and Pilgrim 2014; Kate and Crowe 2014). Ten Kate and Pilgrim (2014), Squires et al. (2018), and Squires and Garcia (2018) have argued that this orthodox view of offsets might not be economically effective and proposed a least-cost approach to their use, integrating them in the mitigation plan from the beginning and not only as a measure of last resort (Squires et al. 2018; Squires and Garcia In press).

Offsets are thus designed to compensate, with either no net loss or even a net gain, the residual impacts after the first three steps are followed.[26] Compensation is "in-kind" or "like-for-like" within a biogeographical region. Biodiversity offsets have the implied rights and costs of the "beneficiary pays principle."

Van Dover et al. (2017) argue that, when offsets cannot be located in the area in which the affected biodiversity is found and where the affected biodiversity is important for geographically restricted functions such as connectivity, "in-kind" offsets are an inappropriate mitigation strategy. Determining "like-for-like" or ecological equivalence and accounting for ecological risk pose complex difficulties. "Out-of-kind" offsets, such as restoring coral reefs in exchange for loss of deep-sea biodiversity, implicitly assume that loss of largely unknown deep-sea biodiversity is acceptable. They further argue that compensating biodiversity loss in the Area with biodiversity gains in national waters transfers net benefits away from the international

[26] In the economics literature, offsets are voluntary contributions to a public good (here biodiversity and ecosystem services) that are motivated, in part, by a desire to compensate for (i.e., offset) the detrimental impacts of other activities such as mining (Vicary 2000; Kotchen 2009). The objective or baseline for offsets is not necessarily no net loss of biodiversity, as posed in the conservation biology literature, but rather the level of offsets that yields an optimal level of voluntary provision of the public good. Such a level balances the benefits of offsetting degradation (i.e., the benefits of the resulting increase in environmental services) and the associated costs.

community and is therefore inconsistent with the principle that the resources of the Area must be utilized for the benefit of mankind as a whole. They conclude that residual biodiversity loss from mining cannot be offset with no net loss because of the very slow natural recovery rates in affected ecosystems.

An open question that we pose, and one for which we do not have a conclusive answer, is whether or not the whole concept of no net loss, and indeed the mitigation hierarchy itself, is applicable in what is, except on geological time scales or over very long time periods, an exhaustible or nonrenewable resource – rather than a renewable resource – because of the very slow natural recovery rates in deep-sea ecosystems (Madureira et al. 2018).[27] The mitigation hierarchy with its concept of no net loss was developed for terrestrial ecosystems and especially for the terrestrial habitat and ecosystems, such as forests and grasslands, that are inherently renewable and often on human generational time scales. The mitigation hierarchy also implicitly accepts that damage to environmental resources will occur through economic activity, which in our case is that mining will occur. Applying the mitigation hierarchy with no net loss as a baseline to deep seabed mining and ecosystems would imply near-complete, if not complete, avoidance of mining. While the mitigation hierarchy itself likely remains pertinent, a more relevant baseline to no net loss with "nonrenewable" resources may be something closer to a minimized biodiversity loss according to the guidance established by the Convention, which prioritizes the use of resources for the "benefit of mankind."

4.5 Conservation Credits and Biobanking

Conservation banking refers to the process of establishing species or habitat credits via a banking agreement and the trading (using, buying, selling, loaning, renting, leasing, etc.) of these credits (Fox and Nino-Murcia 2005). Conservation credits traditionally address the fourth step in the mitigation hierarchy, the residual and compensatory offsets. Conservation banks establish – in advance of any losses – conservation credits that may be used to offset them (ten Kate et al. 2014). Conservation banks require up-front investment in the credit site.

Conservation banks are usually designed to supply – over time and for several or many different projects – offsets of residual losses or adverse impacts that occurred elsewhere on species, habitats, biodiversity, and ecosystem services in general. Banks cannot sell credits until they have successfully purchased or restored habitat for the species in question. The legality and applicability of conservation banking is uncertain in a LOSC framework.

A conservation or biodiversity bank can be viewed as a parcel of privately or publicly held property that is conserved and managed in perpetuity under a

[27] See Griffiths et al. (2018) for a broader discussion of the applicability of no net loss to the mitigation hierarchy and Kotchen (2009) for additional discussion of the objective of offsets within an economics context.

conservation easement for the benefit of rare species (Fox and Nino-Murcia 2005). The government grants credits to the party holding the conservation easement for the land's species and habitat value. A conservation bank owner may use or sell the credits within a pre-designated service area to address required mitigation. In some programs, credits may be earned through restoration of biodiversity and more generally ecosystem services, i.e., Step 3 of the mitigation hierarchy. Mining entities that need to compensate for adverse environmental impacts upon species, as subject to the Law of the Sea, or habitat may purchase credits in the secondary market from the conservation bank owners. Conservation banking is possible for species within the same population and ecosystem.

Conservation credits are units of gain that can be traded in the secondary market – the offset market (ten Kate et al. 2014). Credits must meet all the requirements for gain as specified in the offset policy of the jurisdiction. Credits must thus be measured using the same metrics as established in the no net loss or net gain policy.

Incentives are created through the flexibility, creation of an asset with market value and price, and opportunity to realize a profit through supporting populations of species or habitat.

The conservation bank provides an alternative for mining entities or firms in the supply chain who are seeking a rapid, legitimate, and cost-effective mitigation option. Conservation banking allows a mining entity or firm in the supply chain to transform a former legal liability (the species or habitat) into a financial asset (the credit).

Conservation banking has elements of conservation easements, by catch credit systems, offsets, and payments for ecosystem services. Although conservation banking could easily be classified as a form of biodiversity offsets, we distinguish the two because of the explicit use of credits. Nonetheless, many of the questions that arise with offsets, such as the currency or numeráire, "like-for-like," arise. When the conservation banking system finances annual management of habitats for environmental services, conservation banking could be considered a payment for ecosystem services (USFWS 2009).

The benefits and risks of conservation banks and aggregated offsets can be summarized as (ten Kate et al. 2014) follows: (1) a number of offsets can be consolidated into a large contiguous site that can have higher habitat and security values; (2) the conservation effort can be concentrated into one project that can facilitate more specialist input to offset design and management; (3) a conservation bank can have landscape-/seascape-scale benefits by providing connectivity and preempting future fragmentation; (4) there can be cost savings from economies of scale and reduced transaction costs; and (5) conservation banks can provide those whose activities require offsets with immediate access to credits and thereby reduce the information and transaction costs required to find the required offset. The increased risks associated with conservation banking include the following: (1) the risk from a natural disaster or other failures of the banks is magnified with a number of offsets at the same location and (2) pressure on the offsetting policy manager to relax offsetting requirements and conditions in order to increase financial viability.

5 Combining Policy Approaches

Individual policy instruments, whether from direct or incentive-based regulation, may be insufficient, because there are multiple types of behavior and decision-making that affects environmental performance, multiple spillover effects, and multiple policy objectives. Instead, combining two or more policy approaches may be necessary.[28] When combining instruments, however, the ISA should consider whether the different approaches being combined form substitutes or complements. Policy instruments that are substitutes can create redundancies without any improvement in protection of the marine environment, which also raises costs and can even be counterproductive. Combining approaches that are complementary can lead to a better overall outcome than use of a single approach in isolation.

Consider, for example, the use of a combination of regulation, environmental fees (put into an environmental fund), and negligence-based liability. Because of the difficulty of designing performance-based, site-specific fees in the context of DSM (beyond differentiation based on broad categories, such as within or outside of environmentally significant or sensitive areas), the environmental fees would provide relatively little incentive effect. However, they could provide a source of revenue for remediation, restoration, or compensation and ensure that contractors face the full expected costs of their mining activities (thereby implementing the "polluter pays principle"). Environmental regulations could then be designed to ensure that mining activities are undertaken in an environmentally responsible way, with negligence-based liability serving to provide further incentives for diligence. For this purpose, it would have to be clear that the standard of care that would be applied would not simply be based on compliance with existing regulations but would incorporate appropriate behavior given the specific circumstances and information available at the time for that mining location. Such an approach would provide some certainty for contractors regarding the extent of their associated costs and also provide the resources (through the environmental fund) for compensatory and other related actions. Thus, the combination of approaches can better address the dual goals of (i) creating incentives to reduce the likelihood and/or magnitude of environmental damage from DSM and (ii) providing resources for compensating for damage that does occur.

6 Technology and Innovation

Technological change may potentially be central to the development and direction by which to avoid and minimize many of the DSM environmental impacts. Technological change can lower the proportional impact of DSM upon the environment and the overall scale of this impact. Because of the nascent state of

[28] In general, typically a single policy instrument is not sufficient for providing efficient incentives to address multiple-faceted problems, unless the problems are closely linked (Bennear and Stavins 2007).

DSM technology and potentially long-lasting environmental impacts, the cumulative impact of technological change is likely to be large.

Technology and its development are typically undersupplied (Jaffe et al. 2003). Suppliers of new technology incur all the costs of research and development but also create benefits for others. Once the new technology is developed and implemented, other contractors can utilize all or part of this technology without having to pay for it – what are called "free riders." This reduces the innovating contractor's incentive to invest in the research and development relative to what would be optimal from a broader social perspective. This inability of contractors to enjoy the full benefits of new technology to avoid and minimize environmental damage from DSM means that investment in the development of technology to reduce environmental damage will be less than the socially optimal level. In short, private contractors face incentives to under-provide investment in new technology to reduce the adverse environmental impact of DSM.

This private contractor under-provision in turn requires a public technology policy through the ISA. Legal, payment, licensing, technology, environmental, and other policies directly and cumulatively create dynamic economic incentives that affect the research, development, supply, direction, rate, diffusion, and adoption of new technology with long-lasting, cumulative impacts (Jaffee et al. 2005). Policies relating to patents and licenses for new technology and to initial contractors, as opposed to those entering later, are particularly important. Even then, however, a demonstration effect can still potentially confer benefits to other contractors who do not pay these fees but still learn some facets of the new technology.

A second public option is the public, either through the state-funded contractors (such as the national institutions sponsored by Germany, Korea, China, and Japan) or the ISA or some mix, directly funding research and development that lowers the environmental impact of DSM or even undertaking some of the research itself, perhaps funded through fines, fees, taxes, etc. from some of the environmental policy instruments or from a direct technology development levy. In fact, a majority of current research and development, including basic research and development, is currently provided by contractors that are states or state enterprises, with a view to later commercialization.[29] Only a minority of ISA contractors at the present time are genuine private sector contractors funded by private capital. The question then follows whether such state and/or state enterprise-funded research is provided to the socially optimum amount, since its aim is ultimately commercial, in competition with other private and public contractors, and essentially private rather than public in intent. A related question is whether the ISA and/or states and state enterprises should fund more basic applied scientific research directly related to lowering the environmental impact of DSM and from which all contractors can benefit. Such research may be under-provided from society's standpoint because it may not have an immediate and

[29] In this regard, even though the supplying parties are public, the situation is closer to that of a classic public good in which the supply is the result of the collective supply of individual private providers, which in turn is subject to the classic public good issues of under-provision compared to the socially optimal level, benefits that created by each provider that they do not fully receive (external benefits), and free riding.

direct payoff to mining contractors but is nonetheless beneficial to their research and development of new technology that reduces DSM's adverse environmental impact. Many national governments routinely fund or conduct themselves such comparatively basic research, leaving private contractors to build upon these results.

Such basic research from which all contractors, whether private or public, and indeed all of humanity can benefit might be undersupplied due to the provider's inability to enjoy sufficient benefits, plus coordination, cooperation, and free-riding issues among transnational parties (Hirschleifer 1983; Barrett 2007). Technology (and basic scientific research and development) that reduces the environmental impact per tonne of ore mined and that does not appreciably and directly bear upon commercial competitiveness may be a best-shot technology.[30] A best-shot technology only needs to be supplied once and then becomes freely available to all. Because the providing party can capture enough of the local benefits to provide a level of supply that is close to the global optimum, the independent actions of parties, whether private or public, will be adequate. The incentive to supply is strong enough that the provider develops and supplies the technology to close to the socially optimal level (or even the socially optimal level) without regard to potential free riding by other parties. In contrast, when local benefits to the supplying party lead to a level of technology development and supply that leaves global demand unsatisfied, international cooperation or coordination will be required. Examples of best-shot technologies include vaccines that need to be developed and supplied only once or environmentally friendly fishing gear. Other types of technology supply, such as additive or aggregate, may be applicable in other situations, such as development of commercial mining technology.

Once the best-shot technology is developed and supplied, different parties may subject it to their own tests, but coordinated actions among these parties are sufficient to implement the technology rather than the more demanding multilateral coordination that would be undertaken through the ISA and be more strictly subject to international law and protocols and diplomatic norms.[31] In this regard, states can pursue

[30] There are several types of technology that can provide public goods (Hirschleifer 1983; Barrett 2007). Let Q denote the total level of public good supplied, and let q_i denote the quantity of public good supplied by party i, where $i = 1, 2, \ldots, M$, and there are M total parties. Then for (1) a best-shot technology, $Q = \max(q_1, q_2, \ldots, q_M)$; (2) a weakest link technology, there are issues of weak supply incentives and free riding, and supply is determined by the weakest link or smallest provider of the public good, $Q = \min(q_1, q_2, \ldots, q_M)$; (3) an aggregate technology $Q = f(q_1, q_2, \ldots, q_M)$, where $f(\bullet)$ denotes some general function that aggregates the individual supplies q_i and supply depends upon the combined efforts of all parties and issues of weak supply incentives and free riding apply and each provider's supply is an imperfect substitute for every other supplier's contribution; (4) an additive technology, a form of aggregate technology, $Q = q_1 + q_2 + \bullet \bullet \bullet + q_M$, so that one provider's supply is a perfect substitute for another's supply of the public good and faces the issues of supply incentives and free riding; (5) a weighted average technology, a form of aggregate technology, $Q = \alpha_1 q_1 + \alpha_2 q_2 + \bullet \bullet \bullet + \alpha_M q_M$, where the α_i denote weights and $\sum_{i-1}^{M} \alpha_i = 1$, and some providers' supply contributes more to the public good than others; and (6) a weaker link technology, $Q = \alpha_1 q_1 + \alpha_2 q_2 + \bullet \bullet \bullet + \alpha_M q_M$, where one of the α_i denotes the weaker link so that greater efforts by one party i can offset the weaker effort by another party j, $i \neq j$, and there are issues of weak incentives to supply the public good and free riding.

[31] For further discussion of cooperation versus coordination in the international environmental area, see Barrett (2016).

collective action through what comes more easily, complementary and voluntary coordination agreements (informal or formal such as memorandum of understanding), rather than what is more difficult, such as a negotiated multilateral cooperation that is formally binding upon states. Because states and private parties coordinate (and cooperate) more effectively with the aid of institutions, the existing institutional framework can facilitate such collective action through lowering transactions and information costs and clarifying the nature of property rights (Libecap 2014).

Several other issues complicate the development of new technology to reduce the adverse environmental impact from DSM. The costs or value of new technology to one user may depend upon how many others have adopted the technology, so that users are better off when others use the same technology, creating network effects (Arthur 2009).Thus the number of mining firms using new technologies affects the direction and rate of technological change. The size and types of markets for minerals and inputs into the DSM production process, which affect mineral and input prices, also affect the direction and rate of technological change (Acemoglu 2002). New technologies are more likely to be developed when they are more profitable due to a larger market, as well as when the produced minerals command higher prices or when their cost of production rises due to a rising cost of the adverse environmental impacts (the so-called price effect) (Acemoglu 2002; Acemoglu et al. 2012). For example, a relative price increase of one production factor – here the environment – will induce innovations that scale down the need for this factor (Hicks 1932).

Future DSM and environmental policy interventions further complicate the process of technological change. An open question is which environmental policy instrument or instruments, whether direct regulation or incentive-based, provide the strongest incentives for DSM contractors to innovate to develop more effective and lower-cost conservation technologies and operations (Goulder and Parry 2008). As a general rule, mining contractors should save on payments, tax payments, permits, etc. along with conservation and abatement cost when using incentive-based policy instruments, whereas under direct regulation they save only on conservation or abatement cost.

Choice of environmental policies and technology policies may work best in tandem. Increasing the spending on research and development for technology that reduces environmental damage rather than reducing current mining activities affects intergenerational equity and environmental damage. These questions and issues are just some of the many that development of DSM faces.

7 Concluding Remarks

The ISA faces the challenge of regulating a new industry that uses new technology and even technology that has yet to be invented while also ensuring the protection of the environmental resources that could be adversely affected by DSM. This will involve consideration of a number of issues that arise, including the nonrenewability of the mineral and environmental assets in the deep sea, the value of waiting for additional information, the opportunity cost of foregone welfare to current and future generations that is weighted toward lower income and wealth people, trade-offs with foregone welfare (opportunity costs) between policy options, time discounting so that benefits and costs received in the future receive less weight than those received sooner (giving a present value and the discount rate determines each time period's weight), whether or not the polluter or beneficiary pays principle holds, and that there are limits to current and future generations' willingness to pay or accept compensation with any environmental asset (especially when extinction is not entailed). All of these considerations are relevant for economically rational decision-making and policy design.

Environmental policies must be put in place prior to commencement of commercial operations, both to ensure adequate environmental protection and to give contractors as much certainty as practicable through supplying them with needed information about the potential environment-related costs they will encounter. Alternative policy approaches are available, which include environmental regulations, taxes/fees, liability, insurance, and bonds, all of which have been under discussion. These approaches differ in terms of the incentives, risks, and costs that they entail for contractors and society as a whole. Additional policy instruments that can be thrown into the mix of discussion, largely drawn from terrestrial conservation but some from the marine realm, include transferable habitat impact quotas, conservation easements and concessions, payments for ecosystem services, biodiversity offsets, and conservation credits and biobanking.

Because there are multiple types of behavior affecting environmental performance, multiple spillover effects that need to be considered, and multiple policy objectives, it is possible that combining two or more policy approaches might be preferred. The ISA will need to consider which approach or combination of approaches will best meet the dual goals of promoting DSM while also protecting the marine environment.

Innovation and technological development can play a key role in determining not only the economic viability of the industry but also the production processes that are adopted and the resulting environmental impacts of that production. The ISA should recognize that the governance structures that it puts in place now will affect the incentives for development and adoption of more efficient and environmentally friendly production technologies and processes. Thus, the ISA will need to balance the industry's need for some certainty and predictability regarding the "rules of the game" with the need to both promote and adapt to changes in technology (and other market conditions) over time. The question of the development and sharing of new technology with developing countries also arises.

References

Acemoglu, D. (2002). Directed technical change. *The Review of Economic Studies, 69*, 781–809.

Acemoglu, D., Aghion, P., Bursztyn, L., & Hemous, D. (2012). Directed technical change and the environment. *American Economic Review, 102*(1), 131–166.

Arrow, K., Dasgupta, P., Goulder, L. H., Mumford, K. J., & Olesop, K. (2012). Sustainability and the measurement of wealth. *Environment and Development Economics, 17*(3), 317–353.

Arrow, K. J., & Fisher, A. C. (1984). Environmental preservation, uncertainty, and irreversibility. *The Quarterly Journal of Economics, 89*, 312–319.

Arthur, W. B. (2009). *The nature of technology: What it is and how it evolves*. New York: Free Press. 235 pp.

Barrett, S. (2007). *Why cooperate? The incentive to supply public goods*. Oxford: Oxford University Press. 258 pp.

Barrett, S. (2016). Coordination vs. voluntarism and enforcement in sustaining international environmental cooperation. *Proceedings of the National Academy of Sciences, 113*(51), 14515–14522.

Bennear, L. S., & Stavins, R. N. (2007). Second-best theory and the use of multiple policy instruments. *Environmental and Resource Economics, 37*(1), 111–129.

Costello, C., & Kolstaad, C. D. (2015) *Mining and quasi-option value*. NBER Working Paper 21325, US National Bureau of Economic Research. Available at (Accessed 28 Aug 2018). http://www.nber.org/papers/w21325.

Craik, N. (2018). *Deep seabed mining liability: Potential legal pathways*. Briefing Paper 01/2018. Available at (Accessed 28 Aug 2018). https://www.isa.org.jm/sites/default/files/files/documents/bp012018-ilrp-rev.pdf

Dasgupta, P. (2001). *Human well-being and the natural environment*. Oxford: Oxford University Press.

Fox, J., & Nino-Murcia, A. (2005). Status of species conservation banking in the United States. *Conservation Biology, 19*, 996–1007. https://doi.org/10.1111/j.1523-1739.2005.00231.x.

Goulder, L., & Parry, I. (2008). Instrument choice in environmental policy. *Review of Environmental Economics and Policy, 2*, 152–174.

Griffiths, V. F., JW Bull, J. B., & Milner-Gulland, E. J. (2018. in press). No net loss for people and diversity. *Conservation Biology*. https://doi.org/10.1111/cobi.13184.

Hein, J., Mizell, K., Koschinsky, A., & Conrad, T. (2013). Deep-ocean mineral deposits as a source of critical metals for high- and green-technology applications: Comparison with land-based resources. Ore Geology Reviews, 51, 1–14. https://doi.org/10.1016/j.oregeorev.2012.12.001. Available at (accessed August 28 2018) https://pubs.er.usgs.gov/publication/70043020.

Hicks, J. (1932). *The theory of wages* (2nd ed.). London: Macmillan.

Hirschleifer, J. (1983). From weakest link to best-shot: The voluntary provision of public goods. *Public Choice, 41*(3), 371–386.

Holland, D., & Schnier, K. (2006). Individual habitat quotas for fisheries. *Journal of Environmental Economics and Management, 51*, 72–92.

Jaffe, A., Newell, R., & Stavins, R. N. (2003). Technological change and the environment. Chapter 11. In K.-G. Mäler & J. R. Vincent (Eds.), *Handbook of environmental economics* (Vol. I). North Holland: Elsevier Science B.V.

Jaffee, Newell, R., & Stavins, R. N. (2005). A tale of two market failures: Technology and environmental policy. *Ecological Economics, 54*, 164–174.

Kate, K., & Crowe, M. L. A. (2014). *Biodiversity offsets: Policy options for governments. An input paper for the IUCN technical study group on biodiversity offsets*. Gland: IUCN. 91pp. Available at (accessed August 28 2018): https://portals.iucn.org/library/sites/library/files/documents/2014-028.pdf.

Koopmans, T. C. (1960). Stationary ordinal utility and impatience. *Econometrica, 28*(2), 287–309.

Kotchen, M. (2009). Voluntary provision of public goods for Bads: A theory of environmental offsets. *The Economic Journal, 119*(Issue 537), 883–899.

Kotchen, M. (2013). Voluntary- and information-based approaches to environmental management: A public economics perspective. *Review of Environmental Management: A Public Economics Perspective, 7*, 276–295.

Lallier, L. E., & Maes, F. (2016). Environmental impact assessment procedure for deep seabed Mining in the Area: Independent expert review and public participation. *Marine Policy, 70*, 212.

Libecap, G. (2014). Addressing global environmental externalities: Transaction cost considerations. *Journal of Economic Literature, 52*(2), 424–479.

Madureira, P., Brekke, H., Cherkashov, G (2018). Deep sea mining in the area: Scales of destruction, rehabilitation and protection of the marine environment.

MEA (Millennium Ecosystem Assessment). (2005). *Ecosystems and human well-being: General synthesis*. Available at (Accessed 28 Aug 2018): https://www.millenniumassessment.org/documents/document.356.aspx.pdf

ten Kate, K., Pilgrim, J., Brooks, T., Gibbons, P., Hughes, J., Mackey, B., Manuel, J., McKenney, B., Mehra, S., Quétier, F., & Watson, J. (2014). *Biodiversity offsets technical paper*. Gland: IUCN. 89 pp.

Parker, D. P., & Thurman, W. N. (2013). Conservation easements: Tools for conserving and enhancing ecosystem services. In J. F. Shogren (Ed.), Encyclopedia of energy, natural resources, and environmental economics (Vol. 3, pp. 133–143). Amsterdam: Elsevier.

Ten Kate, K., & Pilgrim, J. (2014). https://www.iucn.org/ru/content/biodiversity-offsets-technical-study-paper

Nordhaus, W. (2008). *A question of balance: Weighing the options on global warming policies*. New Haven: Yale University Press.

Ramsey, F. (1928). A mathematical theory of saving. *Economic Journal, 38*(4), 543–559.

Rawls, J. (1972). *A theory of justice*. Cambridge, MA: The Belknap Press of Harvard University Press.

Rawls, J. (2009). *A theory of justice, revised edition*. Cambridge, MA: Harvard University Press.

Roth, R. (2018) Understanding the economics of seabed mining for polymetallic nodules, presentation to the International Seabed Authority Council Meeting, Kingston, Jamaica.

Segerson, K. (2013). When is reliance on voluntary approaches in agriculture likely to be effective? *Applied Economic Perspectives and Policy, 35*(4), 565–592.

Sen, A., & Williams, B. (Eds.). (1982). *Utilitarianism and beyond*. Cambridge, UK: Cambridge University Press.

Squire, L., & van der Tak, H. G. (1975). *Economic analysis of projects*. Baltimore: Johns Hopkins University Press.

Squires, D., & Garcia, S. M. (2014). Ecosystem-level impacts of fisheries bycatch on marine megafauna: Biodiversity conservation through mitigation, policy, economic instruments, and technical change. Report of an IUCN-CEM-FEG Scientific Workshop, Gland, Switzerland, 7–10 Oct 2013: 81 p. http://ebcd.org/wp-content/uploads/2014/11/353-Squires_et_Garcia-2014-_Mitigating_ecosystem_level_impacts_of_bycatch-FEG_Meeting_report.pdf or https://euagenda.eu/upload/publications/untitled-865-ea.pdf

Squires, D., & Garcia, S. M. (2018). Economic efficiency and the biodiversity mitigation hierarchy with a focus on marine and fishery issues. *Conservation Biology, 32*(5), 989–997.

Squires, D., Restrepo, V., & Garcia, S. M. (2018). Fisheries bycatch reduction within the least-cost biodiversity mitigation hierarchy: Conservatory offsets with an application to sea turtles. *Marine Policy, 93*, 55–61.

Squires, D., & Garcia, S. M. (In press). Economic efficiency and the biodiversity mitigation hierarchy with a focus on marine and fishery issues. *Conservation Biology*. https://doi.org/10.1111/cobi.13155

United Nations. (1971). Possible methods and criteria for the sharing by the international community of proceeds and other benefits derived from the exploitation of the resources of the area beyond the limits of national jurisdiction, United Nations General Assembly, Committee on the peaceful uses of the sea-bed and the ocean floor beyond the limits of national jurisdiction, A/AC.138/38, mimeo, 1971.

USFWS (United States Fish and Wildlife Service). (2009). *Conservation banking: Incentives for stewardship*. Washington, DC: WSFWS. Available at (Accessed 28 Aug 2018): https://www.fws.gov/southwest/es/Documents/R2ES/conservation_banking.pdf.

Van Dover, C. L., Ardron, J. A., Escobar, E., Gianni, M., Gjerde, K. M., Jaeckel, A., Jones, D. O. B., Levin, L. A., Niner, H. J., Pendleton, L., Smith, C. R., Thiele, T., Turner, P. J., Watling, L., & Weaver, P. P. E. (2017). Biodiversity loss from deep sea mining. *Nature Geoscience, 10*(7), 464–465.

Vicary, S. (2000). Donations to a public good in a large economy. *European Economic Review, 44*, 609–618.

Wallace, S., Turris, B., Driscoll, J., Bodtker, K., Mose, B., & Munro, G. (2015). Canada's Pacific groundfish trawl habitat agreement: A global first in an ecosystem approach to bottom trawl impacts. *Marine Policy, 60*, 240–248.

World Commission on Environment and Development, Our Common Future ("Brundtland Report"). (1987). Oxford: Oxford University Press.

Michael W. Lodge is a British national. He received his LLB from the University of East Anglia and has an MSc in marine policy from the London School of Economics and Political Science. He is a barrister of Gray's Inn, London. Prior to his election as Secretary-General of the International Seabed Authority in July 2016, he had served as Deputy to the Secretary-General and Legal Counsel. Other professional experiences include serving as Legal Counsel to the ISA (1996–2003); Counsellor to the Round Table on Sustainable Development, OECD (2004–2007); and Legal Counsel to the South Pacific Forum Fisheries Agency (1991–1995). He has also held appointments as a Visiting Fellow of Somerville College, Oxford (2012–2013); an Associate Fellow of Chatham House, London (2007); and a member of the World Economic Forum's Global Agenda Council on Oceans (2011–2016).

With 28 years of experience as a public international lawyer, Michael Lodge has a strong background in the field of law of the sea as well as 10 years' judicial experience in the UK and South Pacific. He spent many years living and working in the South Pacific and was one of the lead negotiators for the South Pacific Island States of the 1995 UN Fish Stocks Agreement. He has also worked as a consultant on fisheries and environmental and international law in Europe, Asia, Eastern Europe, the South Pacific, and Africa.

With extensive knowledge of the United Nations and other international organizations, Michael Lodge has facilitated high-level multilateral and bilateral negotiations at international and regional level. His significant achievements include his pivotal role in the ISA from its inception in 1996 and in helping to create and implement the first international regulatory regime for seabed mining. He also contributed to the future security of global fish stocks by leading the process to create the Western and Central Pacific Fisheries Commission from concept to its establishment as the largest regional fisheries management organization in the world, also serving as the interim executive director of the Commission. He was instrumental in advising the Pew Charitable Trusts on their support for the Global Ocean Commission and also acted as an adviser to the Commission on international law of the sea and ocean policy.

Mr. Lodge has published and lectured extensively on the international law of the sea, with over 25 published books and articles on law of the sea, ocean policy, and related issues.

Dr. Kathleen Segerson is a Board of Trustees Distinguished Professor and UConn Alumni Association Distinguished Professor of Economics at the University of Connecticut. She previously held the Philip E. Austin Endowed Chair and served as department head in the Department of Economics. She received a BA in mathematics from Dartmouth College (Phi Betta Kappa) and a PhD in agricultural and resource economics from Cornell University. She is a Fellow of the Association of Environmental and Resource Economists (AERE) and of the Agricultural and Applied Economics Association (AAEA). Dr. Segerson's research focuses on the incentive effects of alternative environmental policy instruments, with particular emphasis on the application of legal rules and principles to environmental problems. Specific research areas include the impact of legal liability for environmental damages in a variety of contexts, including groundwater contamination, hazardous waste management, and workplace accidents; land use regulation and the takings clause; voluntary approaches to environmental protection; the impacts of climate change on US agriculture; and incentives to control nonpoint pollution from agriculture.

Dr. Segerson is currently a member of the Board on Agriculture and Natural Resources of the National Research Council of the National Academy of Sciences and served from 2005 to 2011 as a member of the Chartered Board of the US Environmental Protection Agency's Science Advisory Board (SAB). She is also past-president of AERE and has served as Vice-President and a member of the AERE Board of Directors. She is currently a co-editor of the Journal of the Association of Environmental and Resource Economists and has previously served as a co-editor and an associate editor of the American Journal of Agricultural Economics and an associate editor of the Journal of Environmental Economics and Management.

Dale Squires is an Adjunct Professor of Economics at the University of California San Diego, where he teaches three classes, and Senior Scientist with US NOAA Fisheries. He is the co-author or co-editor of 10 books, about 110 refereed journal articles, about 20 refereed book chapters, and about 35 book chapters. He is on the Scientific Advisory Committees of the International Seafood Sustainability Foundation and International Pole and Line Foundation and works with the Food and Agriculture Organization of the United Nations, the International Seabed Authority, International Whaling Organization, Regional Fisheries Management Organizations, foundations, industries, national governments and universities around the world, and nongovernmental organizations, primarily on issues of global fishing and biodiversity conservation.

Ecosystem Approach for the Management of Deep-Sea Mining Activities

Roland Cormier

Abstract Within a regulatory context, an environmental impact assessment is more than an assessment of impacts of a given project's footprint. It involves a review of the impacts that could occur in addition to the potential impacts from catastrophic events and the need for emergency responses. In most cases, there are a number of regulatory requirements that fall within a broad range of jurisdictions and competent authorities. Environmental impact assessments are done primarily to identify ecosystem and socio-economic issues that become the basis for management measures to prevent and mitigate the impacts as conditions for project approvals.

This chapter discusses the need to reconcile the scale of ecosystem vulnerabilities with the impacts generated by deep-sea mining activities as the basis for an ecosystem approach. It introduces the use of international risk management standards and techniques already in use in the industry to bridge the assessment of these vulnerabilities with the regulatory requirements of this industry. It discusses the need for an assessment of effectiveness of the controls, measures and procedures in preventing and mitigating ecosystem impacts.

Keywords Ecosystem approach · Environmental management · Deep-sea mining

1 Introduction

Ecosystem-based or ecosystem approach to management advocates the need to integrate the management of human activities to maintain the functions of the ecosystem and to sustain development (McLeod et al. 2005). Ecosystem-based policies, protocols and practices need to be implemented to manage human activities (Christensen et al. 1996) where adaptive management should be implemented through monitoring and research (Holling 1978). Integration of macroeconomic and sector-specific policies combined with management actions that control the sources and effects of environmental change is also advocated to achieve environmental

R. Cormier (✉)
Helmholtz-Zentrum Geesthacht, Institute for Coastal Research, Geesthacht, Germany
e-mail: Roland.Cormier@hzg.de

R. Sharma (ed.), *Environmental Issues of Deep-Sea Mining*,
https://doi.org/10.1007/978-3-030-12696-4_14

objectives (Elliott et al. 2017). Considerable science has gone into the causal relationship between industry activities and their effects on ecosystem components in environmental impact assessments (MacKinnon et al. 2018). Although governance, management and integration are recurring themes of the ecosystem approach (Sardà et al. 2014; Long et al. 2015; Link and Browman 2017), it is the reduction of the root causes of environmental effects through the implementation of measures in industry operations that are considered as the solution to sustainable development (Cormier et al. 2017).

Governance and management are sometimes confused as the same thing (Green 2015). Governance processes establish the reasons 'why' a given strategy is being undertaken to reach long-term goals (Ackoff 1990), whereas management control processes identify 'what' are the short-term objectives needed to reach those goals (Anthony and Dearden 1980). However, it is operational control processes that determine 'how' to achieve objectives effectively and efficiently through the implementation of measures, procedures and tasks (Giraud et al. 2011; Hupe and Hill 2016). Thus, an ecosystem approach to governance establishes the scope and the context for the environmental objectives developed by management, whereas it is an ecosystem approach to the development and implementation of operational controls that ultimately achieves the environmental objectives (Gavaris 2009; Cormier et al. 2017). Operational controls are typically implemented through regulations, standards or codes of practice that stipulate the conditions found in licences and permits (Gouldson et al. 2009). Based on Article 145 of the United Nations Convention on the Law of the Sea (UNCLOS 1982), an ecosystem approach to the management of deep-sea mining activities implies that measures be implemented to ensure effective protection of the marine environment from their harmful effects (Sharma 2017; Danovaro et al. 2017; Boetius and Haeckel 2018). More importantly, such an approach would also help identify the key provisions for environmental management plans and for the development of rules, regulations and procedures for deep-sea mining in the *Area* based on environmental impact assessments (Durden et al. 2018).

This chapter discusses the differences between ecosystem approaches to management and operational control processes. It uses the UNCLOS and selected parts of the International Seabed Authority (ISA) regulations, policies and environmental management plans (ISA 2010, 2011, 2012, 2013a, b) to demonstrate the importance of the scope and context of such policies in an ecosystem approach. It further discusses the application of risk management processes based on ISO 31000 (ISO 2018) as well as the Bow-tie analysis of IEC/ISO 31010 (IEC/ISO 2009) to show how an operational ecosystem approach is nested within a policy hierarchy of management and governance processes.

2 Uncertainty and Managing Risk

Uncertainties contribute to the risks of not achieving an objective and ultimately reaching a goal. Although we may rightfully perceive that our ability to address environmental concerns is stymied by the uncertainties of the scientific knowledge that underpins environmental impact assessments (DFO 2014), we often leave the management and operational uncertainties to implementation after the assessment is completed (Hupe and Hill 2016) with the understanding that monitoring will tell us if things are working (Behn 2003). However, management control processes also contribute uncertainties. For example, management uncertainties could be the result of coordination impediments between governance, management and operations as well as communication and misinterpretations of legislation and policies by the parties involved (Milliken 1987; Rowe 1994). Operational control processes further contribute to uncertainties that may lie in the design and effectiveness of the controls or the capacity, ability and technology to implement the controls as well as the lack of quality assurance needed to monitor and take the corrective actions (Behn 2003; Ferdous et al. 2012; Cormier et al. 2015). In risk management, management and operational control processes are used to address these uncertainties to reduce the risks of not achieving objectives (Cormier et al. 2018).

In ISO 31000, risk is defined as the effect of uncertainty on objectives. The purpose of the risk management process of the standard is to reduce the uncertainties by taking a course of action that either eliminates the sources of risk, reduce the likelihood of risk or reduce the consequences of risk. The standard requires that the risk assessment process be conducted within the scope and context of the policies to ensure that the risks are framed within the span of control provided by those policies (Cormier et al. 2018). Such risk assessment must identify the root causes and the consequences of risk and analyse their effects on objectives to evaluate the management options that could reduce those risks to a level as low as reasonably practicable. As for implementation, risks are then treated through the selection and implementation of operational controls. Without the scope and context of the policy, risks can be whatever is being perceived or assessed as risk by someone (Holsman et al. 2017). In addition, management cannot manage an unlimited number risks given the compounding effects of uncertainties involved and the underlying limits to human and technological resources to manage them. Given the need to provide a systematic and transparent process for all parties involved in deep-sea mining (Ardron et al. 2018), the standard also formalizes such a process (Ciocoiu and Dobrea 2007). The standard has been adapted to environmental risk management (SAL 2012) and bridged to ecosystem approaches to risk management (Cormier et al. 2013).

For deep-sea mining, UNCLOS was the output of the governance process led by United Nations General Assembly that established the scope and context for the activities to be managed in the *Area*. As goals outlined in the preamble of UNCLOS, activities should be managed for the 'peaceful uses of the seas and oceans, the equitable and efficient utilization of their resources, the conservation of their living resources, and the study, protection and preservation of the marine environment'. Under the authority of the ISA, Article 1 establishes the scope of the risk management process as it relates to the deleterious effects of substances and energy introduced by the exploration, exploitation and dumping activities in the *Area* and the potential for causing 'harm to living resources and marine life, hazards to human health, hindrance to marine activities, including fishing and other legitimate uses of the sea, impairment of quality for use of sea water and reduction of amenities'. The scope of the risks represents a much broader set of environmental concerns than those that would be considered for an ecosystem approach. Each would need to be assessed, evaluated and managed individually as part of a comprehensive risk management strategy. The ecosystem risk management context is further detailed in Part XI Section 2 of UNCLOS. Article 145 stipulates that the ISA shall adopt rules, regulations and procedures to prevent, reduce and control pollution and other hazards from causing harmful effects to the marine environment as well as protect and conserve the natural resources of the *Area* and prevent the damage to the flora and fauna. Annex III, Article 17 established the expected outcomes of the mining standards and practices for mineral extraction and shipboard processes. An ecosystem risk management approach to mineral extraction and shipboard processing activities need to identify the risks of these activities on the components and functions of the marine ecosystem, analyse the standards and practices that could reduce those risks and ultimately implement them based on an evaluation of their effectiveness and feasibility.

ISO 31000 does not replace the various assessments currently being used in environmental management. In fact, these assessments inform different steps of the risk management process. For example, a strategic environmental assessment can help establish the scope and context as these are typically used for policy and strategic considerations (Bina 2007). Environmental assessments can help identify the environmental risks to ecosystems, health and safety, or socio-economic considerations as these are conducted within the scope and context of environmental legislation and policies (Cormier and Suter II 2008). Environmental impact assessments are, for the most part, an analysis of the likelihood and consequences of project proposals (Durden et al. 2018; MacKinnon et al. 2018). Coupled with cumulative effects assessments (Stelzenmüller et al. 2018) and controls assessments of IEC/ISO 31010, the effectiveness of various management options can then be analysed and evaluated to select the courses of action considered adequate to reduce the risks. Cost and benefits analysis (Shaffer 2010) and regulatory impact assessments (Radaelli 2008) inform the implementation process of the operational controls used to treat the risks. Regulatory reviews and approvals are also part of the risk treatment function of ISO 31000 (Black 2005).

3 Carrying into Effect the Ecosystem Objectives

As mentioned earlier, it is the implementation of operational controls that carries into effect ecosystem objectives. Here, the Bow-tie analysis of IEC/ISO 31010 is used to structure and analyse the operational controls used to manage mineral extraction and shipboard processing activities to reduce harmful effects and to protect and conserve the marine environment from serious harm. The Bow-tie analysis is a controls assessment technique of the IEC/ISO 31010. As a qualitative analytical tool, the Bow-tie diagram combines a fault tree analysis (left side) and an event tree analysis (right side). The Bow-tie analysis has its origins in the early 1980s in the petrochemical industries as a tool to identify the operational tasks, procedures and accountabilities that are needed to prevent and mitigate risk within a management system (Lewis and Hurst 2005; Cockshott 2005). The technique is currently used in wide variety of industries (Saud et al. 2014; van Thienen-Visser et al. 2014; Abimbola et al. 2014). Recently, the technique has been adapted to environmental management policies and planning (Creed et al. 2016; Elliott et al. 2017; Kishchuk et al. 2018; Cormier et al. 2018) and stakeholder engagement (Gerkensmeier and Ratter 2018).

The central knot of a Bow-tie sets the risk management context linking the risk source to an event of concern (Fig. 1). On the left side of the Bow-tie, the causes of the event are identified in relation to the risk source. The expected outcome of the prevention controls implemented between the causes and the event is analysed to determine their effectiveness at reducing the likelihood of the event. On the right side of the Bow-tie, the consequences of the event are identified. The expected outcomes of the mitigation controls implemented between the event and the consequence are analysed to determine their effectiveness in reducing the magnitude of the consequence, whereas the expected outcomes of the recovery controls are analysed to determine their effectiveness to recover from the consequences that cannot be mitigated. These are the operational controls used to reduce the risks. Escalation factors that can undermine the effectiveness of any control are managed by adding escalation controls to reduce the effect of the factor on the effectiveness of the control. These are management controls that are used to ensure that the operational controls are being implemented effectively. It should be noted that all risks can only be eliminated by eliminating the risk source. Figure 1 also links the elements of the Bow-tie to the three steps of the ISO 31000 risk assessment process.

The ecosystem considerations from the Clarion-Clipperton Zone Environmental Management Plan (ISA 2011), the regulations for prospecting different seabed minerals (ISA 2010,2012, 2013a) and environmental guidelines for contractors (2013b) are used to demonstrate the use of the Bow-tie analysis of operational and management controls within an ecosystem approach (Fig. 2). The articles of UNCLOS and the ISA are referenced for each element of the Bow-tie where possible to show the role of the legislation, regulations and policies in setting the scope and the context.

Prevention, reduction and control of pollution and other hazards are the expected outcome that needs to be met by the prevention controls implemented for mineral extraction and shipboard processing operations. Although under the authority of the ISA, the prevention controls need to be developed and implemented by the contractor to reduce the likelihood of harmful effects from his operations. Licencing the area of operations for contractors and the designation of reserve areas are the mitigation controls implemented by the ISA for the whole of the deep-sea mining sector to reduce the magnitude of the consequences to the natural resources in the *Area*. The designation of areas of particular environmental interest (AEPI) is one of the mitigation controls implemented by the ISA to reduce the magnitude of the damage to the flora and fauna of the marine environment.

In Fig. 3, the significant volumes of deep-sea water used to extract the minerals could entrain associated deep-sea biota and generate sediment plumes near the

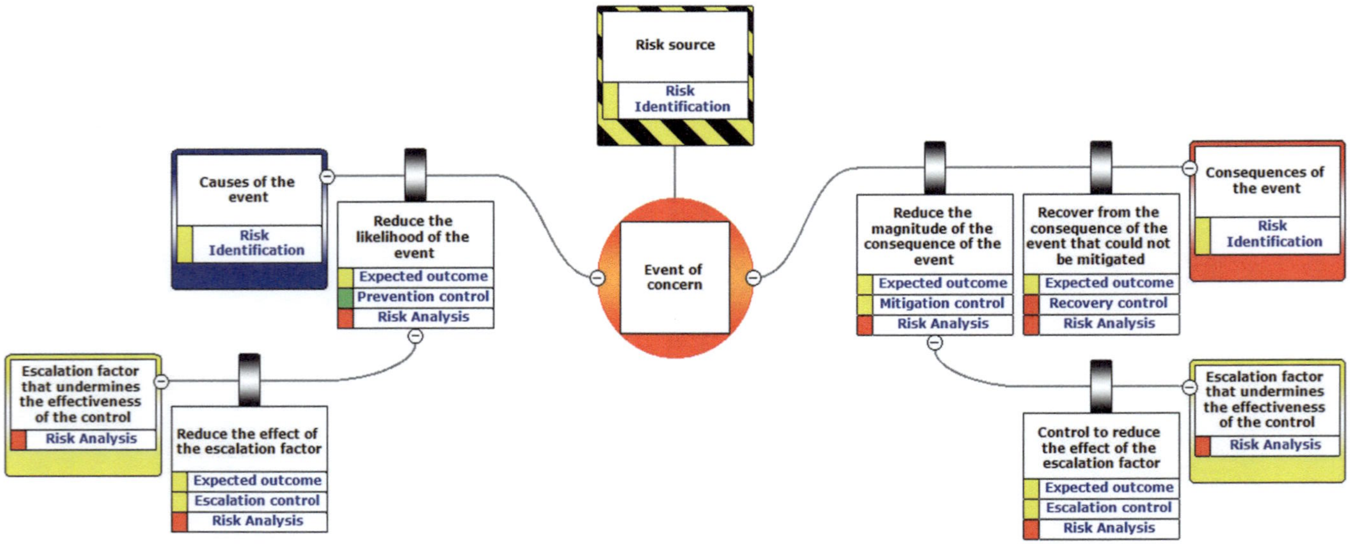

Fig. 1 Bow-tie analysis (BowTieXP version 9.0.10.1 adaptation of EIC/ISO 31010)

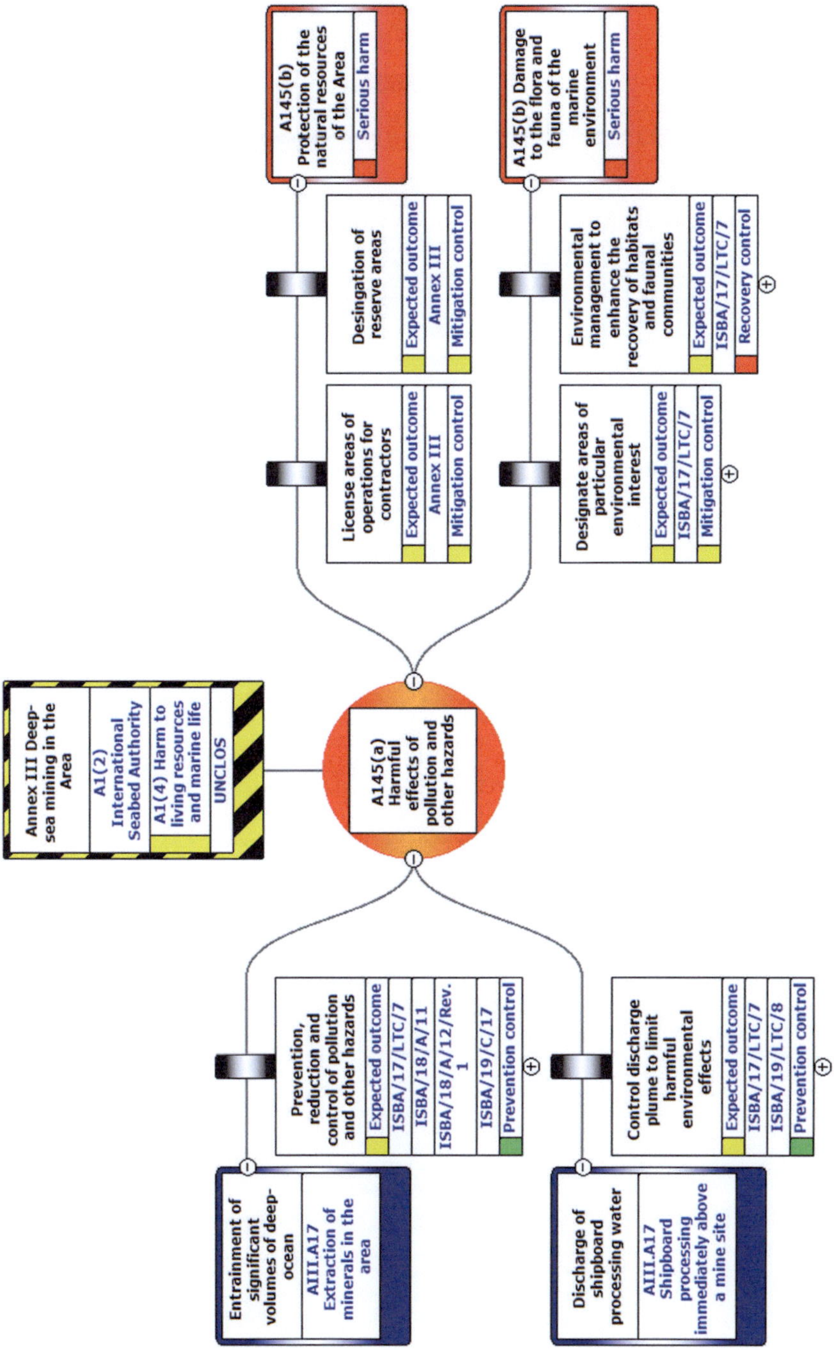

Fig. 2 Bow-tie analysis of an ecosystem approach to operational and management deep-sea mining activities

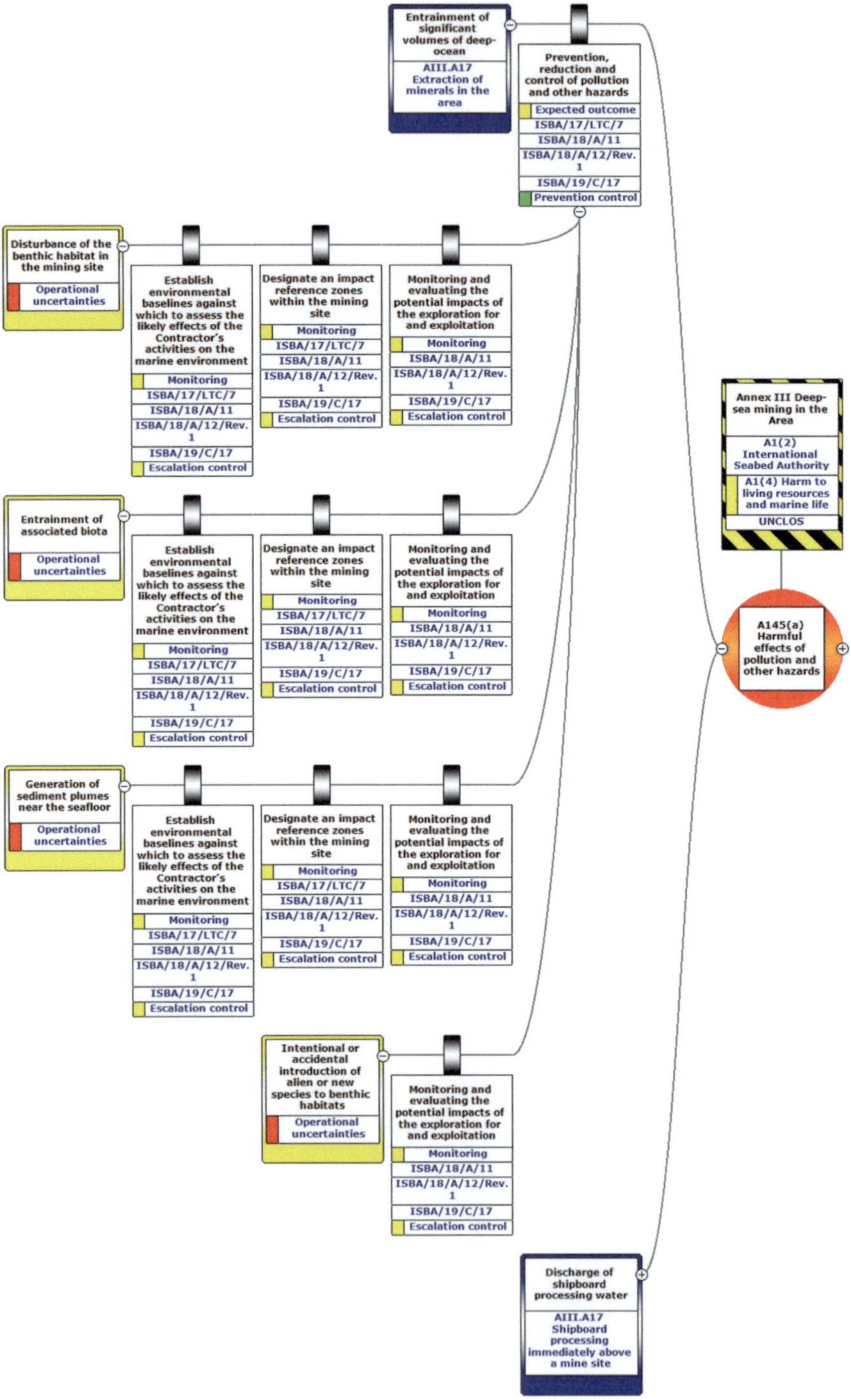

Fig. 3 Escalation factors related to extraction of mineral in the Area

seafloor and cause disturbances to the benthic habitat. As escalation factors, these introduce operational uncertainties that could undermine the effectiveness of the prevention controls. Additional escalation controls would then be required to establish baselines and designate impact reference zones to monitor and evaluate the impacts of the mineral extraction operations. Given the concerns for the introduction of alien or new species (Miller et al. 2018), monitoring would also include the introduction of alien or new species from the surface to the deep-sea environment.

In Fig. 4, shipboard processing could discharge water containing fine-grained sediments and nodule fragments that could generate sediment plumes in the surface water. As escalation factors, these could also introduce operational uncertainties that could undermine the effectiveness of the prevention controls. As with mineral extraction, additional escalation controls would then be required to designate impact reference zones to monitor and evaluate the impacts of the shipboard operations. Given the concerns for the introduction of alien or new species (Miller et al. 2018), monitoring would include the introduction of alien or new species from the deep-sea environment to the surface waters.

However, there are management and scientific uncertainties that could also undermine the effectiveness of the mitigation controls that would be used to reduce the magnitude of the damage to the flora and fauna of the marine environment. In Fig. 5, the escalation factors that may undermine the effectiveness of areas of particular interest could include the management uncertainties of the area needed to maintain minimum population sizes. As an escalation control, the establishment of a 200 × 200 km zone would be enhanced by another 100 km buffer zone for adjacent activities. Given the scientific uncertainties in the understanding of the harmful effects of mining activities to enhance the recovery of habitats and faunal communities, assembling and disseminating data from the operational activities would be used to assess the impacts. However, scientific research would also be needed to enhance the knowledge and understanding of the ecosystem structure and functions for all mitigation and recovery controls. Escalation factors were not identified for the protection of the natural resources given the designation of areas for mining activities and reserve areas.

Although this Bow-tie is not intended to be a comprehensive analysis of all the management and operational processes outlined in UNCLOS, the ISA regulations and the Environmental Management Plan, it illustrates that the ecosystem approach to management and operational control processes address different spatial scales in terms of human activities and ecosystem consideration. The prevention controls of the left side of the Bow-tie deals with the ecosystem impacts within the mining site and the ecosystem effects that could occur in the vicinity of the site. Monitoring in the left side of the Bow-tie are the management controls used to ascertain the effectiveness of the prevention, reduction and control measures needed to address harmful effects within the site. This is in contrast to the mitigation and recovery controls of the right side of the Bow-tie that deals with the ecosystem scale effects within the management area in relation to the collective mining activities for that sector. Data assembled and disseminated by contractors combined with the knowledge generated by scientific research are the management controls used to ascertain if the

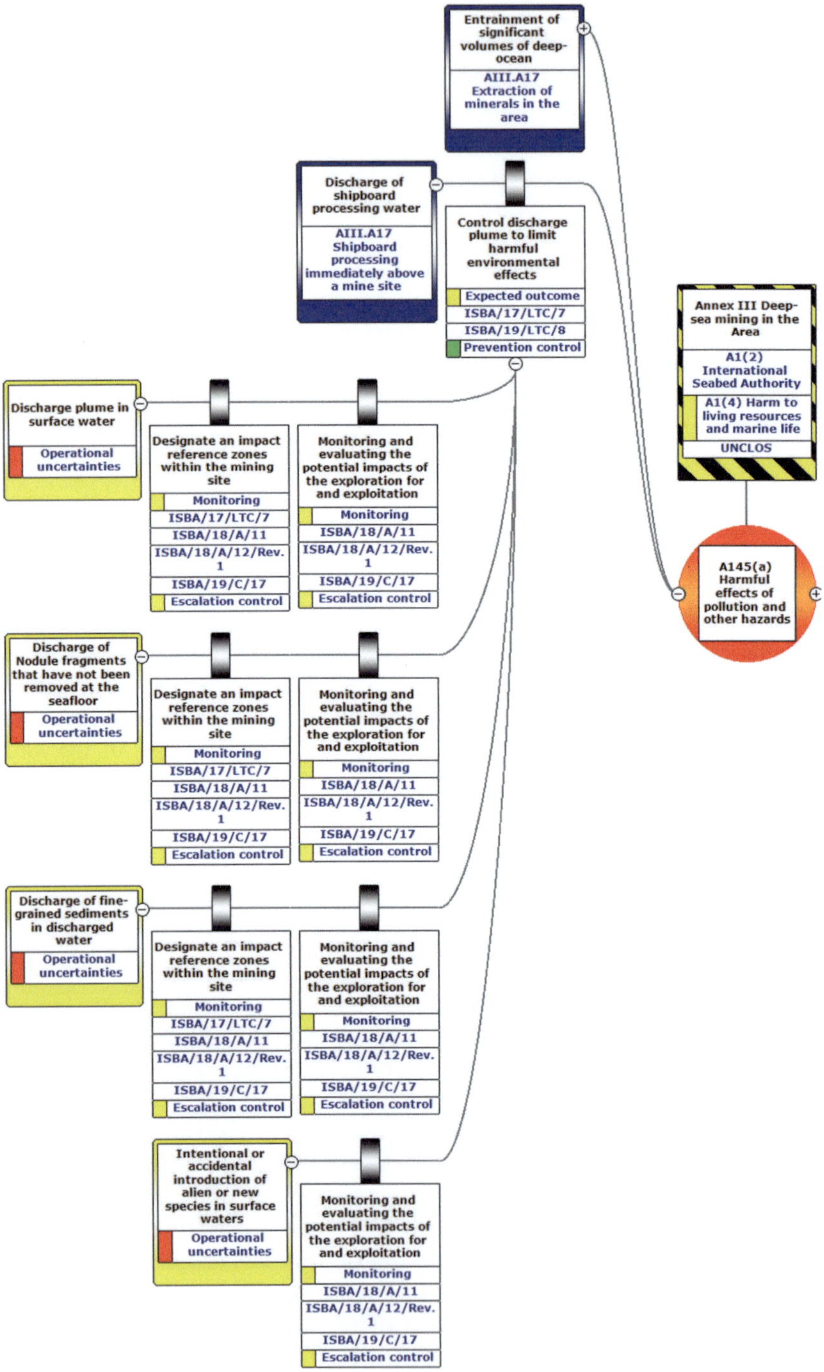

Fig. 4 Escalation factors related to shipboard processing immediately above a mine site

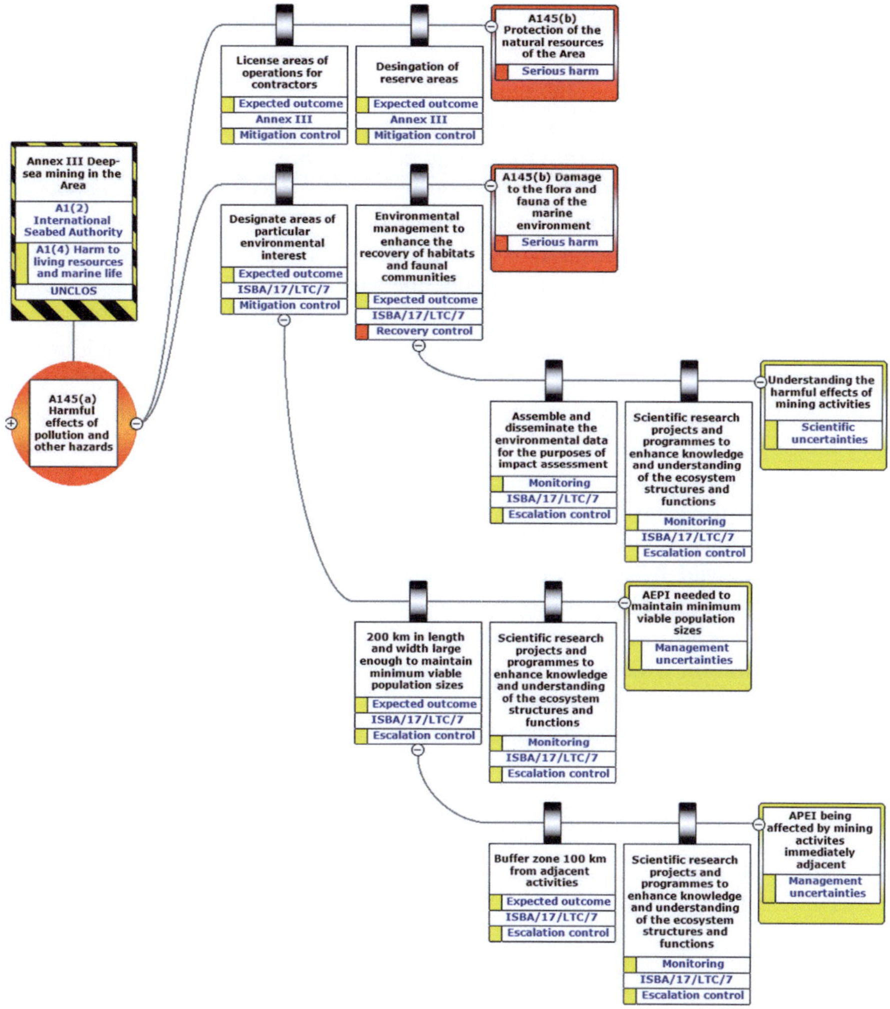

Fig. 5 Escalation factors for the mitigation and recovery of damage to flora and fauna of the marine environment

preservation and conservation strategies are providing the necessary precautionary approach for the development of that sector. In a nutshell, prevention controls address the pressures on ecosystem components that are changed within the mining site, whereas the mitigation and recovery controls address the cumulative effects of the deep-sea mining sector within the management area. This Bow-tie reflects the scope of UNCLOS and the management context of deep-sea mining in the management area. The effects that natural processes and climate change could have on ecosystem components and functions are not accounted for in this analysis. These would be reflected as additional escalation factors.

4 Pressures, Components, Functions and Harm

Pressures are the mechanism by which deep-sea mining activities can cause physical change, chemical interferences or biological disturbances on ecosystem functions (Elliott et al. 2017). In a policy setting, pressures are the root causes of harmful effects that can lead to serious harm (Levin et al. 2016). Pressures change ecosystem components and ultimately may have an effect on the ecosystem functions that support the life cycles of species. DFO (2015) outlines three possible ecosystem states. As pressures increase on the components, the function moves from a maintaining state to a changing state and, ultimately, to a reduced state (Fig. 6).

From an ecosystem risk management context, prevention controls should reduce the pressures to avoid reaching the transition point A where the function starts changing. Prior to that point, the functions are maintained even though the pressures are changing the component because it resists or compensates in the face of perturbation. Upon exceeding transition point A, mitigation controls are then used to avoid reaching the transition point B where the function is reduced or lost. During the changing state, recovery of the function is still possible if the pressures are reduced or removed. At the reduced state, the function is no longer supported by the component. The component is degraded to a point that recovery is no longer possible even if the pressures are removed. Recovery could include artificial rehabilitation and restoration where they have been proven to work. In the Bow-tie context, mitigation and recovery from the consequences of an event cannot return to the conditions of their initial state (de Dianous and Fiévez 2006). Therefore, once the component has been degraded, it would not return to its original natural state.

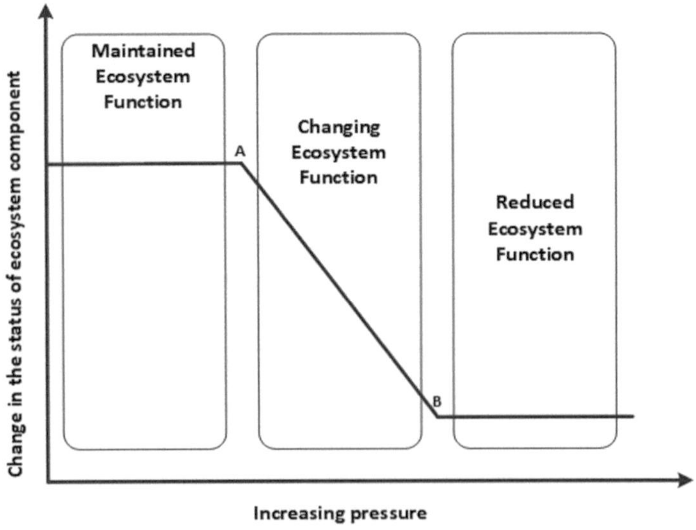

Fig. 6 Conceptual relationship between pressures, ecosystem components and functions

In deep-sea mining, the prevention, reduction and control of pollution and other hazards should be able to avoid harmful effects by maintaining the ecosystem functions provided by the ecosystem components. Harmful effects could imply that the functions are changing due to the pressures generated by mining activities. Serious harm could be considered for situations where the function is reduced or lost. The three states (e.g. maintaining, changing and reduced) could help characterize the level of risk based on the ecological responses to the pressures of deep-sea mining. Quantitative methods could then be used to assess the effectiveness of the prevention controls in relation to the three states (Sharma 2015, 2017; Levin et al. 2016; Miller et al. 2018). Based on this Bow-tie example, the effectiveness of the prevention controls would be assessed from within and in the vicinity of the mining site, whereas effectiveness of the mitigation and recovery controls would be assessed from within the boundaries of the management area. The cumulative effects of deep-sea mining would have to be assessed in combination with the collective pressures of other human activities taking into account the influences generated by natural processes and the effects of climate change (Elliott 2011; Jones 2016; Levin et al. 2016). Several factors would also have to be considered in such assessments including:

- Spatial and temporal overlap relating to both the ecological component and the pressure
- Zone of influence of the pressure
- Characteristics of the pressure
- Sensitivity of the ecological component
- Ecological significance of the component
- Recoverability and resilience of the ecosystem

The significance of the ecosystem component and the functions that it provides is one of the key factors in the determination of harmful effects and serious harm. Criteria for ecologically and biologically significant areas of the Convention on Biological Diversity (CBD 1993) and Vulnerable Marine Ecosystems (FAO 2009) form the basis for marine conservation. In fact these criteria are considered for the high seas and for areas beyond national jurisdictions to identify areas that are most vulnerable to human activities – at risk so to speak (Clark et al. 2014; Ardron et al. 2014; Watling and Auster 2017). However, these criteria do not provide the functions that are supported by ecosystem components that would be of concern in an impact assessment. The Canadian version of the ecologically and biologically significant areas (EBSA) (DFO 2004) and ecologically significant species (ESS) (DFO 2006) provides an outline of functions that could guide such assessment. Significance in the Canadian criteria implies that a perturbation of an EBSA or ESS would result in greater ecological consequences than an equivalent perturbation outside such areas. These criteria could be used to link the functions to the components that would be impacted by deep-sea mining activities that could result in ecosystem scale effects in the management area. Table 1 summarizes the functions that could be supported by deep-sea ecosystem components such as the function of structural habitat features provided by manganese nodules to benthic species (Levin et al. 2016).

Table 1 Ecosystem functions based on ecologically and biologically significant areas and species (DFO 2004, 2006)

Ecologically and biologically significant areas	Ecologically significant species
Spawning and breeding	Key trophic species
Nursery and rearing	Structure providing species
Feeding	Properties at the community level
Migration	
Seasonal refugia	
Physical oceanographic features	
Structural habitat features	

5 Monitoring and Review

Rick management involves a broader set of activities than environmental or ecosystem monitoring (Danovaro et al. 2017) such as:

- Effectiveness and reliability of the implemented measures, procedures and controls in reducing the pressures as expected when they were designed (Tobias and Trindade 2012)
- Conformity assessments of the measures, procedures and controls to determine if these are being implemented as specified (Pendrill 2014)
- Compliance surveillance of those that have to implement the regulatory conditions to determine if they are complying to the requirements (Shimshack and Ward 2005)
- Audits of environmental management plans to ensure that they meet their certification requirements (Tung et al. 2014)
- Performance evaluations to ascertain if their management processes and practices are achieving their objectives (Helm and Sprinz 2000)
- Reviews of strategic plans and their goals to ascertain if these are being reached adequately (Jabnoun et al. 2003)

Each of the activities above answers very different governance and management questions spanning a full range of issues (Bouckaert and van de Walle 2003; Baehler 2003). Given that the intent of operational controls is to reduce the root causes of risks (Burgess et al. 2016), monitoring and review activities build upon the information gathered from effectiveness studies, conformity assessments, compliance surveillance and audits to evaluate the performance and ultimately review the strategies and course of actions chosen.

Using an ecosystem approach, operational controls should reduce the likelihood of harmful effects of pollution and other hazards and to avoid the damage to the flora and fauna of the marine environment. Management processes have to provide assurance that deleterious effects will not likely result in harm to living resources and marine life, create hazards for human health and cause hindrance to marine activities, including fishing and other legitimate uses of the sea, impairment of quality for use of sea water and reduction of amenities. The governance process has to ascertain that the activities in the Area are managed to benefit mankind and reach their goals of peaceful uses of the seas and oceans, equitable and efficient utilization of their resources, conservation of their living resources and protection and preservation of the marine environment while maintaining the common heritage to mankind.

Therefore, goals set in UNCLOS by the UN General Assembly will not be reached if the environmental objectives established by ISA cannot be achieved because the measures cannot prevent, reduce or control pollution and other hazards in the marine environment. In risk management, the lack of effectiveness in managing the root causes of risk can result in a chain of events that can ultimately affect goals (Cormier et al. 2018). Figure 7 is a Bow-tie analysis of the chain of events along a pathway of risk of deep-sea mining activities. It implies that the consequences of harmful effects from pollution and other hazards can lead to deleterious effects of pollution. The consequences of such deleterious effects can eventually affect benefits to mankind and to the aspirational goals of UNCLOS in the preamble.

An operational ecosystem approach to mineral extraction and shipboard processing activities is ensured by the implementation of the measures to prevent, reduce and control the harmful effects of pollution, whereas the licencing and designation of reserve areas and APIE represent the approach for protecting natural resources and reducing the damage to the flora and fauna in the management area. The management ecosystem approach is reflected in the environmental objectives of UNCLOS (Article 1) for avoiding harm to living resources and marine life from deleterious effects. However, these objectives cannot be achieved without measures that provide effective protection of the marine environment (Article 145). In a chained Bow-tie, prevention controls cannot be assigned for cases where the consequences of another Bow-tie become the cause of a subsequent Bow-tie (red line arrows). Only mitigation and recovery controls can be implemented to reduce the magnitude of the consequences as in the case of deleterious effects in Fig. 7. The governance ecosystem approach is reflected in the preamble of UNCLOS as a goal of conservation of living resources and protection and preservation of the marine environment. Similar to the management ecosystem approach, only mitigation and recovery would be possible in cases where the consequences of deleterious effects would undermine the benefit to mankind by activities in the *Area*.

Elements of monitoring and review activities (i.e. arrows in Fig. 7) are found in UNCLOS, the ISA regulations and the Environmental Management Plans. For example, testing of mining operations could help answer the questions regarding the effectiveness and conformity of the prevention controls. Monitoring of biological communities impacted by mining activities could help answer questions regarding environmental effects coupled with research to assess the magnitudes of those effects. Annual reports on the activities in the Area on compliance submitted to the ISA would contribute to compliance surveillance. Ultimately, periodic reviews undertaken by the Review Conference in collaboration with commissions and the ISA would enable performance evaluations and strategic reviews needed to improve the legislation, regulations, policies and procedures and, thus, adhere to the principle of adaptive management.

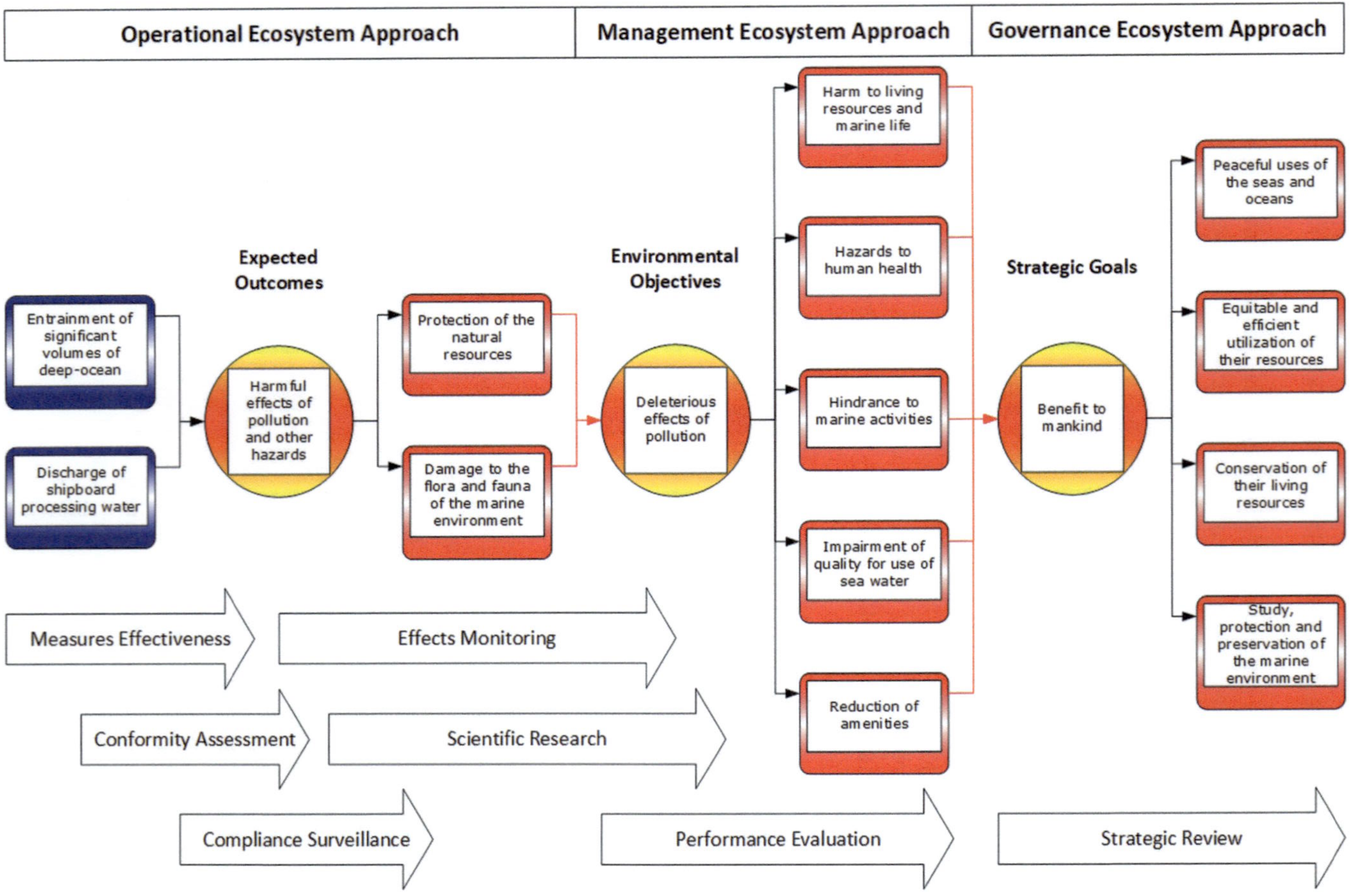

Fig. 7 Bow-tie chain of events

This section does not attempt to cover all aspects of monitoring and review. Figure 7 illustrates the role of monitoring, surveillance and research activities that would be needed to determine the effectiveness of the operational controls. Only monitoring the effects and conducting studies on the magnitude of these effects would not necessarily determine effectiveness, conformity or compliance issues to ensure that corrective actions and improvements address the right problem. Although the monitoring, surveillance and research results are key inputs into performance evaluations, such evaluations may also consider administrative procedures, organizational structures, consultation procedures and management decision processes. Monitoring, surveillance and research could have excellent results while there could be improvement needed in these management functions. Recommendations for changes and improvement can be provided in strategic reviews which are based on the operational and management monitoring and review processes.

As implied by SMART objectives (Rice et al. 2005), the objectives should be specific to clearly specify the state to be achieved and be interpreted unambiguously by all stakeholders. More importantly, they should relate to measurable properties of ecosystems and human societies, so that indicators and reference points can be developed to measure progress towards the objective. These two characteristics also apply to expected outcomes to gauge the effectiveness of operational controls. It implies that operational monitoring, surveillance and research activities would use quantitative approaches as they deal with the root causes of risk. However, aspirational goals as for benefit to mankind and common heritage to mankind are not as specific and therefore not measurable (Fig. 7) (Cormier and Elliott 2017). Although qualitative criteria in terms of policy could be developed for these, an evaluation or review is a process of consensus that is still based on the quantitative data and information from the operational monitoring, surveillance and research activities. It also includes, however, a broader socio-economic and policy analysis. In the UNCLOS, such a process is reflected in Article 154 Periodic review and Article 155 the Review Conference. Paraphrased here from Article 155, the Review Conference would ensure the 'maintenance of the principle of the common heritage to mankind' and the 'equitable exploitation of the resources of the Area' as well as 'exclusion of claims or exercise of sovereignty over any part of the Area' including the 'prevention of monopolization of activities'. As shown in Fig. 7, the review would also include 'whether they have benefited mankind as a whole' and 'protection of the marine environment and protection of human life'.

6 Concluding Remarks

As shown in this chapter, a risk management process starts with the strategic goals as reflected in the preamble of UNCLOS, whereas the risk management context sets the scope of the risks as outlined in Article 1 of UNCLOS. In an operational ecosystem approach, however, the expected outcomes of the prevention controls have to reduce the pressures on ecosystem components and mitigate the effects on the functions of these components. This requires an analysis of Article 145 and Annex III of UNCLOS as well as the ISA regulations and EMP. In risk management, it is the policy context that drives the risk assessment process instead of the risk assessment driving the policy. Scientific research can inform policy to set the scope and establish the context. Ecosystem risk assessments, environmental impact assessments or cumulative effects assessments, to name a few, provide the necessary scientific advice as to the ecosystem risks and the need for operational controls as well as monitoring. Technical advice is needed for the development of measures, procedures and controls given that these have to be implemented in everyday operations.

The ecosystem approach is primarily an approach to deal with environmental concerns as a result of human activities as they relate to marine habitat and the species that are part of the biological diversity of the marine environment. As discussed in this chapter, an ecosystem approach to governance, management and operations does not address the same concerns and questions, even the type of science and technical advice needed is not the same. An ecosystem approach from a governance perspective is through policymaking, whereas management is through protection and conservation objectives under the direction provided by the governance. However, it is the operational ecosystem approach that provides assurance for achieving protection and conservation objectives.

References

Abimbola, M., Khan, F., & Khakzad, N. (2014). Dynamic safety risk analysis of offshore drilling. *Journal of Loss Prevention in the Process Industries, 30*, 74–85. https://doi.org/10.1016/j.jlp.2014.05.002.

Ackoff, R. L. (1990). Strategy. *Systemic Practice, 3*, 521–524.

Anthony, R. N., & Dearden, J. (1980). In D. Richard (Ed.), *Management control systems* (4th ed.). Homewood: Irwin. isbn:0-256-02325-5. 724 pp.

Ardron, J. A., Clark, M. R., Penney, A. J., et al. (2014). A systematic approach towards the identification and protection of vulnerable marine ecosystems. *Marine Policy, 49*, 146–154. https://doi.org/10.1016/j.marpol.2013.11.017.

Ardron, J. A., Ruhl, H. A., & Jones, D. O. B. (2018). Incorporating transparency into the governance of deep-seabed mining in the area beyond national jurisdiction. *Marine Policy, 89*, 58–66. https://doi.org/10.1016/j.marpol.2017.11.021.

Baehler, K. (2003). "Managing for outcomes": Accountability and trust. *Australian Journal of Public Administration, 62*, 23–34. https://doi.org/10.1111/j..2003.00346.x.

Behn, R. D. (2003). Why measure performance? Different purposes require different measures. *Public Administration Review, 63*, 586–606. https://doi.org/10.1111/1540-6210.00322.

Bina, O. (2007). A critical review of the dominant lines of argumentation on the need for strategic environmental assessment. *Environmental Impact Assessment Review, 27*, 585–606. https://doi.org/10.1016/j.eiar.2007.05.003.

Black, J. (2005). The emergence of risk based regulation and the new public risk management in the UK. *Public Law, 32*, 1–42.

Bouckaert, G., & van de Walle, S. (2003). Comparing measures of citizen trust and user satisfaction as indicators of "Good Governance": Difficulties in linking trust and satisfac-

tion indicators. *International Review of Administrative Sciences, 69*, 329–343. https://doi.org/10.1177/0020852303693003.
Burgess, M. G., Clemence, M., McDermott, G. R., Costello, C., & Gaines, S. D. (2016). Five rules for pragmatic blue growth. *Marine Policy, 87*, 331–339. https://doi.org/10.1016/j.marpol.2016.12.005.
CBD. (1993). Convention on biological diversity of 29 December 1993. Accessed 2018-06-07 https://www.cbd.int/doc/legal/cbd-en.pdf.
Christensen, N. L., Bartuska, A. M., Brown, J. H., et al. (1996). The report of the Ecological Society of America Committee on the Scientific Basis for Ecosystem Management. *Ecological Applications, 6*, 665–691. https://doi.org/10.2307/2269460.
Ciocoiu, C. N. & Dobrea, R. C. (2007). The role of standardization in improving the effectiveness of integrated risk management. *Advanced Risk Management* 1–19.
Clark, M. R., Rowden, A. A., Schlacher, T. A., et al. (2014). Identifying Ecologically or Biologically Significant Areas (EBSA): A systematic method and its application to seamounts in the South Pacific Ocean. *Ocean and Coastal Management, 91*, 65–79. https://doi.org/10.1016/j.ocecoaman.2014.01.016.
Cockshott, J. (2005). Probability bow-ties – a transparent risk management tool. *Process Safety and Environment Protection, 83*, 307–316.
Cormier, R., & Elliott, M. (2017). SMART marine goals, targets and management – is SDG 14 operational or aspirational, is 'Life Below Water' sinking or swimming? *Marine Pollution Bulletin, 123*, 28–33. https://doi.org/10.1016/j.marpolbul.2017.07.060.
Cormier, S. M., & Suter, G. W., II. (2008). A framework for fully integrating environmental assessment. *Environmental Management, 42*, 543–556. https://doi.org/10.1007/s00267-008-9138-y.
Cormier, R., Kannen, A., Elliott, M., Hall, P., & Davies, I. M. (2013). Marine and coastal ecosystem-based risk management handbook. ICES Coop Res Rep 317. 60 pp.
Cormier, R., Kannen, A., Elliott, M., & Hall, P. (2015). Marine spatial planning quality management system. ICES Coop Res Rep 327. 106 pp.
Cormier, R., Kelble, C. R., Anderson, M. R., et al. (2017). Moving from ecosystem-based policy objectives to operational implementation of ecosystem-based management measures. *ICES Journal of Marine Science, 74*, 406–413. https://doi.org/10.1093/icesjms/fsw181.
Cormier, R., Elliott, M., & Kannen, A. (2018). IEC/ISO 31010 Bow-tie analysis of marine legislation: A case study of the marine strategy framework directive. *ICES Coop Res Rep, 324*, 63. https://doi.org/10.17895/ices.pub.4504.
Creed, I. F., Cormier, R., Laurent, K. L., et al. (2016). Formal integration of science and management systems needed to achieve thriving and prosperous great lakes. *Bioscience, 66*, 408–418. https://doi.org/10.1093/biosci/biw030.
Danovaro, R., Aguzzi, J., Fanelli, E., et al. (2017). An ecosystem-based deep-ocean strategy. *Science (80-), 355*, 452–454. https://doi.org/10.1126/science.aah7178.
deDianous, V., & Fiévez, C. (2006). ARAMIS project: A more explicit demonstration of risk control through the use of bow-tie diagrams and the evaluation of safety barrier performance. *Journal of Hazardous Materials, 130*, 220–233. https://doi.org/10.1016/j.jhazmat.2005.07.010.
DFO. (2004). Identification of ecologically and biologically significant areas. DFO Can Sci Advis Sec Ecosystem Status Rep 2004/006.
DFO. (2006). Identification of ecologically significant species and community properties. DFO Can Sci Advis Sec Sci Advis Rep 2006/041.
DFO. (2014). Science advice for managing risk and uncertainty in operational decisions of the fisheries protection program. DFO Can Sci Advis Sec Sci Advis Rep. 2014/015.
DFO. (2015). A science-based approach to assessing the impact of human activities on ecosystem components and function. DFO Can Sci Advis Sec Sci Advis Rep 2015/020.
Durden, J. M., Lallier, L. E., Murphy, K., et al. (2018). Environmental impact assessment process for deep-sea mining in 'the Area'. *Marine Policy, 87*, 194–202. https://doi.org/10.1016/j.marpol.2017.10.013.

Elliott, M. (2011). Marine science and management means tackling exogenic unmanaged pressures and endogenic managed pressures-a numbered guide. *Marine Pollution Bulletin, 62*, 651–655. https://doi.org/10.1016/j.marpolbul.2010.11.033.
Elliott, M., Burdon, D., Atkins, J. P., et al. (2017). "And DPSIR begat DAPSI(W)R(M)!" – a unifying framework for marine environmental management. *Marine Pollution Bulletin, 118*, 27–40. https://doi.org/10.1016/j.marpolbul.2017.03.049.
FAO. (2009). International guidelines for the management of deep-sea fisheries in the high seas, 2009. Accessed 2018-02-15 http://www.fao.org/docrep/011/i0816t/i0816t00.htm.
Ferdous, R., Khan, F., Sadiq, R., Amyotte, P., & Veitch, B. (2012). Handling and updating uncertain information in bow-tie analysis. *Journal of Loss Prevention in the Process Industries, 25*, 8–19. https://doi.org/10.1016/j.jlp.2011.06.018.
Gavaris, S. (2009). Fisheries management planning and support for strategic and tactical decisions in an ecosystem approach context. *Fisheries Research, 100*, 6–14. https://doi.org/10.1016/j.fishres.2008.12.001.
Gerkensmeier, B., & Ratter, B. M. W. (2018). Multi-risk, multi-scale and multi-stakeholder – the contribution of a bow-tie analysis for risk management in the trilateral Wadden Sea Region. J *Coast Conserv 22*, 145–146. https://doi.org/10.1007/s11852-016-0454-8.
Giraud, F., Zarlowski, P., Saulpic, O., & Lorain, M. -A. (2011) Fundamentals of management control, 1st ed (pp. 382). Pearson Education France.
Gouldson, A., Morton, A., & Pollard, S. J. T. (2009). Better environmental regulation – contributions from risk-based decision-making. *Science of the Total Environment, 407*, 5283–5288. https://doi.org/10.1016/j.scitotenv.2009.06.013.
Green, P. E. J. (2015). *Enterprise risk management: A common framework for the entire organization*. Elsevier. 241 pp.
Helm, C., & Sprinz, D. (2000). Measuring the effectiveness of international environmental regimes. *Journal of Conflict Resolution, 44*, 630–652.
Holling, C. S. (1978). *Adaptive environmental assessment and management*. London: Wiley. 377 pp.
Holsman, K., Samhouri, J., Cook, G. S., et al. (2017). An ecosystem-based approach to marine risk assessment. *Ecosystem Health and Sustainability, 3*, e01256. https://doi.org/10.1002/ehs2.1256.
Hupe, P. L., & Hill, M. J. (2016). 'And the rest is implementation.' Comparing approaches to what happens in policy processes beyond great expectations. *Public Policy and Administration, 31*, 103–121. https://doi.org/10.1177/0952076715598828.
IEC/ISO. (2009). Risk *management – risk assessment techniques*. Edition 1.0. International Electrotechnical Commission. IEC/ISO 31010 2009-11.
ISA. (2010). Decision of the Assembly of the International Seabed Authority relating to the regulations on prospecting and exploration for polymetallic sulphides in the Area. ISBA/16/A/12/Rev.1, 15 November, 2010. Accessed 2018-06-07 https://www.isa.org.jm/documents/isba16a12-rev-1.
ISA. (2011). Environmental Management Plan for the Clarion-Clipperton Zone, ISBA/17/ LTC/7, 13 July, 2011. Accessed 2018-06-07 https://www.isa.org.jm/sites/default/files/files/documents/isba-17ltc-7_0.pdf.
ISA. (2012). Decision of the Assembly of the International Seabed Authority relating to the regulations on prospecting and exploration for cobalt-rich ferromanganese crusts in the area. ISBA/18/A/11, 27 July, 2012. Accessed 2018-06-07 https://www.isa.org.jm/documents/isba18a11.
ISA. (2013a). Decision of the Council of the International Seabed Authority relating to amendments to the regulations on prospecting and exploration for polymetallic nodules in the area and related matters. ISBA/19/C/17, 22 July, 2013. Accessed 2018-06-07 https://www.isa.org.jm/documents/isba19c17.
ISA. (2013b). Recommendations for the guidance of contractors for the assessment of the possible environmental impacts arising from exploration for marine minerals in the Area. ISBA/19/

LTC/8, 1 March 2013. Accessed 2018-06-07 https://www.isa.org.jm/sites/default/files/files/documents/isba-19ltc-8_0.pdf.
ISO. (2018). Risk management – guidelines 2nd edition. International Organization for Standardization. ISO 31000:2018(E).
Jabnoun, N., Khalifah, A., & Yusuf, A. (2003). Environmental uncertainty, strategic orientation, and quality management: A contingency model – ASQ. *Quality Management Journal, 10*, 17–31.
Jones, F. C. (2016). Cumulative effects assessment: Theoretical underpinnings and big problems. *Environmental Reviews, 24*, 187–204. https://doi.org/10.1139/er-2015-0073.
Kishchuk, B. E., Creed, I. F., Laurent, K. L., et al. (2018). Assessing the ecological sustainability of a forest management system using the ISO Bowtie Risk Management Assessment Tool. *The Forestry Chronicle, 94*, 25–34. https://doi.org/10.5558/tfc2018-005.
Levin, L. A., Mengerink, K., Gjerde, K. M., et al. (2016). Defining "serious harm" to the marine environment in the context of deep-seabed mining. *Marine Policy, 74*, 245–259. https://doi.org/10.1016/j.marpol.2016.09.032.
Lewis, S., & Hurst, S. (2005). Lessons learned from real world application of the bow-tie method. In *7th Professional Development Conference and Exibition* (p. 9).
Link, J. S., & Browman, H. I. (2017). Operationalizing and implementing ecosystem-based management. *ICES J Mar Sci, 74*, fsw247. https://doi.org/10.1093/icesjms/fsw247.
Long, R. D., Charles, A., & Stephenson, R. L. (2015). Key principles of marine ecosystem-based management. *Marine Policy, 57*, 53–60. https://doi.org/10.1016/j.marpol.2015.01.013.
McLeod, K. L., Lubchenco, J., Palumbi, S. R., & Rosenberg, A. A. (2005). Scientific consensus statement on marine ecosystem-based management. Signed by 221 academic scientists and policy experts with relevant expertise and published by the Communication Partnership for Science and the Sea. 21 pp.
Miller, K. A., Thompson, K. F., Johnston, P., & Santillo, D. (2018). An overview of seabed mining including the current state of development, environmental impacts, and knowledge gaps. *Frontiers in Marine Science* 4:.doi: https://doi.org/10.3389/fmars.2017.00418.
Milliken, F. J. (1987). Three types of perceived uncertainty about the environment: State, effect, and response uncertainty. *Academy of Management Review, 12*, 133–143. https://doi.org/10.5465/amr.1987.4306502.
MacKinnon, A. J., Duinker, P. N., & Walker, T. R. (2018). The application of science in environmental impact assessment. Routledge. New York, NY. 141 pp.
Pendrill, L. R. (2014). Using measurement uncertainty in decision-making and conformity assessment. *Metrologia, 51*, S206–S218. https://doi.org/10.1088/0026-1394/51/4/S206.
Radaelli, C. M. (2008). Evidence-based policy and political control: What does regulatory impact assessment tell us? Soc. Res. (New York). 11–16.
Rice, J., Trujillo, V., Jennings, S., Hylland, K., Hagstrom, O., Astudillo, A., & Jensen, J. N. (2005). Guidance on the application of the ecosystem approach to management of human activities in the European Marine Environment. *ICES Cooperative Research Report, 273*, 22.
Rowe, W. D. (1994). Understanding uncertainty. *Risk Analysis, 14*, 743–750.
SAL. (2012). Managing environment-related risk. Standards Australia Limited/Standards New Zealand. HB 203:2012 Handbook. ISBN 978-1-74342-059-1. 97pp.
Sardà, R., O'Higgins, T., Cormier, R., Diedrich, A., & Tintoré, J. (2014). A proposed ecosystem-based management system for marine waters: Linking the theory of environmental policy to the practice of environmental management. *Ecological Society, 19*, 51. https://doi.org/10.5751/ES-07055-190451.
Saud YE, Israni K (Chris), Goddard J (2014) Bow-tie diagrams in downstream hazard identification and risk assessment. Process Safety Progress 33:26–35. https://doi.org/10.1002/prs.11576.
Shaffer, M. (2010). *Multiple account benefit-cost analysis*. Toronto: University of Toronto Press. 152 pp.
Sharma, R. (2015). Environmental issues of deep-sea mining. *Procedia Earth Planet Science, 11*, 204–211. https://doi.org/10.1016/j.proeps.2015.06.026.

Sharma, R. (2017). *Deep-sea mining*: Resource Potential, Technical and Environmental Considerations. Springer International Publishing, Gewerbestrasse 11, 6330 Cham, Switzerland. 535 pp. https://doi.org/10.1007/978-3-319-52557-0.

Shimshack, J. P., & Ward, M. B. (2005). Regulator reputation, enforcement, and environmental compliance. *Journal of Environmental Economics and Management, 50*, 519–540. https://doi.org/10.1016/j.jeem.2005.02.002.

Stelzenmüller, V., Coll, M., Mazaris, A. D., et al. (2018). A risk-based approach to cumulative effect assessments for marine management. *Science of the Total Environment, 612*, 1132–1140. https://doi.org/10.1016/j.scitotenv.2017.08.289.

Tobias, P. A., & Trindade, D. (2012). *Applied reliability*. 3rd edition. CRC Press. 564 pp.

Tung, A., Baird, K., & Schoch, H. (2014). The relationship between organisational factors and the effectiveness of environmental management. *Journal of Environmental Management, 144C*, 186–196. https://doi.org/10.1016/j.jenvman.2014.05.025.

UNCLOS. (1982). United Nations Convention on the Law of the Sea of 10 December 1982. Accessed 2018-06-07 http://www.un.org/depts/los/convention_agreements/texts/unclos/unclos_e.pdf.

van Thienen-Visser, K., Hendriks, D., Marsman, A., et al. (2014). Bow-tie risk assessment combining causes and effects applied to gas oil storage in an abandoned salt cavern. *Engineering Geology, 168*, 149–166. https://doi.org/10.1016/j.enggeo.2013.11.002.

Watling, L., & Auster, P. J. (2017). Seamounts on the high seas should be managed as vulnerable marine ecosystems. *Frontiers in Marine Science* 4. https://doi.org/10.3389/fmars.2017.00014.

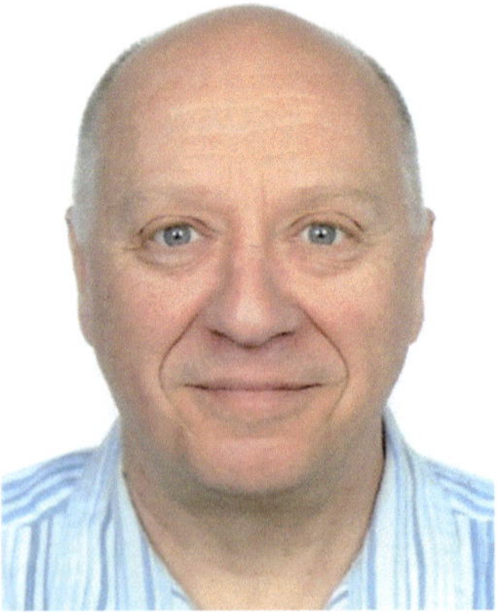

Roland Cormier is a guest scientist at Helmholtz-Zentrum Geesthacht and an associate of the Institute of Estuarine and Coastal Studies at the University of Hull. He is also a member of the International Council for the Exploration of Sea (ICES) working group on marine planning and coastal zone management and the Group of Experts on Risk Management in Regulatory Systems of the United Nations Economic Commission for Europe. His current research is on risk-based approaches to legislation and policies in marine management and conservation. He is currently active as a consultant in environmental risk management in Europe, Canada, and the United States as well as a lecturer in universities in Canada, United States and Europe.

Improving Environmental Management Practices in Deep-Sea Mining

D. S. M. Billett, D. O. B. Jones, and P. P. E. Weaver

Abstract As the business of deep-sea mining develops, greater attention is being paid to the ways in which the impacts from mining on the environment might be minimised and controlled. The management of environmental impacts is highly complex encompassing a wide variety of topics including physical oceanography, sediment characteristics, particle sinking velocities, particle aggregation, sediment geochemistry, toxic discharges, chemical contamination, biological studies from microbes to mammals and from the sea surface to the subseabed, studies of biodiversity, genetic connectivity, ecosystem functioning, the value of ecosystem services, hydrodynamic plume modelling and noise and light hazards. Owing to the history of how contractors in deep-sea mining have developed, there are a wide variety of approaches to environmental management. This chapter seeks to assist contractors in achieving a more consistent approach by making some key recommendations and highlighting some outstanding issues. It covers why environmental baseline studies are necessary and how various levels of environmental assessment will help contractors achieve a suitable standard when submitting Environmental Impact Assessments. In particular it highlights the use of the 'mitigation hierarchy' as a suitable way to consider environmental management issues. Apart from detailing ways in which environmental impacts might be minimised and avoided altogether, at the local and regional scales, the chapter calls for greater consideration of ways to assist ecosystems to recover more quickly through enhancing natural ecosystem processes. The restoration of marine ecosystems, as occurs after a mine has been closed on land, should be included in environmental planning but requires experimental research to be undertaken during the exploration phase and before an

D. S. M. Billett (✉)
Deep Seas Environmental Solutions Ltd., Southampton, UK
e-mail: david.billett@deepseasolutions.co.uk

D. O. B. Jones
National Oceanography Centre, University of Southampton, Southampton, UK

P. P. E. Weaver
Seascape Consultants Ltd, Romsey, Hampshire, UK

R. Sharma (ed.), *Environmental Issues of Deep-Sea Mining*,
https://doi.org/10.1007/978-3-030-12696-4_15

Environmental Impact Statement is submitted to gain an environmental and social licence to exploit deep-sea minerals.

Keywords Environmental management · Mitigation hierarchy · Deep-sea mining

1 Introduction

As deep-sea mining moves from exploration to the exploitation of mineral resources, it is timely to review progress in environmental management practices in ocean mining and how they might be improved. While the presence of minerals in the deep ocean has been known for some 150 years, for polymetallic nodules, and in more recent times for cobalt-rich ferromanganese crusts and polymetallic sulphides, the business of deep-sea mining is still in its earliest stages (Weaver and Billett 2019 this volume). Consequently, there are large differences in approaches to collecting environmental data. This is likely to lead to a variety of environmental management plans being submitted by mining contractors depending on (1) their knowledge, (2) the staff engaged, (3) an organisation's management structure and skills, (4) the organisation's operational priorities and (5) the resource being addressed. The aim of this chapter is to suggest ways in which environmental management might be made more consistent between deep ocean mining contractors.

Currently there is no active commercial-scale mining in the deep ocean. However, Nautilus Minerals produced an Environmental Impact Statement (EIS) for mining polymetallic sulphides within the waters of Papua New Guinea more than 10 years ago (Coffey Natural Systems 2008) although no active mining has taken place since then. In addition, test mining for polymetallic sulphides has occurred within the exclusive economic zone of Japan (Japan Times 2017), but without open reporting of environmental data and only some aspects of the EIA being publically available (Narita et al. 2015). In international waters several benthic impact experiments for polymetallic nodule mining have taken place (Jones et al. 2017), and two contractors have submitted Environmental Impact Assessments (EIAs) for testing a pre-prototype mining system for consideration by the International Seabed Authority (ISA) (GSR 2018; BGR 2018). However, the vast majority of contractors to the ISA are only in the initial stages of producing environmental baseline studies (Weaver and Billett 2019 this book). This paper focuses on mining in international waters for the three principal sources of minerals (nodules, sulphides and crusts).

In international waters, each exploration contract with the International Seabed Authority requires the contractor to undertake environmental baseline studies (International Seabed Authority 2013a, Regulation 32) as detailed by ISA Legal and Technical Commission's (LTC) 'Recommendations for the guidance of contractors for the assessment of the possible environmental impacts arising from exploration for marine minerals in the Area' (International Seabed Authority 2013b), hereafter referred to as the 'LTC Recommendations'. The 'LTC Recommendations' are framed to advise contractors on the main elements required in environmental baseline studies while at the same time allowing contractors flexibility in developing these studies depending on the setting of a particular contract area and the resource being explored.

Each contractor, therefore, has developed their own approach to gathering environmental data. In some cases contractors at the start of operations had limited knowledge of the deep ocean environment, and the teams engaged produced work of variable quality. The ISA Legal and Technical Commission commented on the different stages of development of baseline studies by the contractors and concluded that while work undertaken by some contractors was becoming sufficient to proceed to developing an Environmental Impact Statement, other contractors had generated few useful data (International Seabed Authority 2015b). Some contractors have combined their work on resource assessment with the gathering environmental baseline data throughout the contract period, including the mobilisation of sophisticated technologies such as remotely operated vehicles (ROVs); in other cases work has focussed almost exclusively on evaluating mineral resources.

The objective of this chapter is to suggest how a standardised approach to environmental management might be achieved across all contractors. The aims are to review (1) why environmental baseline studies are needed, (2) the need to generate environmental baselines to a common high standard in order to assess environmental impacts and their effects on the environment over a variety of scales, (3) how the data generated might assist in cost-effective measures to avoid and minimise environmental impacts and (4) how modern technologies might be used to gather large quantities of high-quality data economically. In addition, the chapter comments on the importance of mining contractors contributing standardised data for large-scale Regional Environmental Management Plans (REMPs) and the benefits of REMPs to contractors in setting their individual mining impacts within a wider spatial and temporal context. The chapter highlights areas of environmental management where greater attention may have to be given in the future. Consequently, there is a particular focus on the restoration of marine ecosystems, set in the context of the 'mitigation hierarchy' to avoid and minimise impacts and to restore marine ecosystems when a mine site is closed, where possible.

2 Environmental Assessment

Environmental management aims to maximise human benefits of resource utilisation while minimising environmental degradation owing to human activities (Jones et al. 2018a). This is done by describing and monitoring expected or actual environmental changes during project duration, and beyond, to enable good decisions to be made or actions to be put in place to reduce impacts. There are a wide range of tools for environmental management (Jones et al. 2019), and most legislative regimes, including those being developed for deep-sea mining, mandate the use of several approaches (Durden et al. 2017). For DSM specifically, an Environmental Impact Assessment (EIA) process is required that is documented in a report (Environmental Impact Statement or EIS). The EIA enables the ISA to decide if a project can go ahead and allows discussion on the key risks to the environment and on how the impacts of the project should be reduced (Clark et al. 2017; Durden et al. 2018). In addition to the EIA, an Environmental Management and Monitoring Plan (EMMP) is decided upon before the project starts and is implemented throughout the mining activity. Finally, a plan is put in place for the decommissioning and closure of a mine (Closure Plan). A contractor is required to submit these documents not only to fulfil their legal obligations but also to facilitate stakeholder engagement in order to obtain an environmental and social licence to operate. These assessments are reliant on good quality data and representative information on the mining activities, development site and its environment, including baseline data.

3 What Are Environmental Baseline Studies for?

In 2014 and 2015, the ISA organised two stakeholder consultations to guide the development of the draft regulations for the exploitation of marine minerals (International Seabed Authority 2015a). These consultations requested feedback on a range on draft environmental provisions, including the need to establish a baseline of environmental conditions. Responses by some contractors to the consultation indicated that there was some confusion as to the value and necessity of collecting comprehensive environmental baseline data.

Contracts with the International Seabed Authority for exploration in the 'Area' require a contractor to collect oceanographic and environmental baseline data and to establish baselines to assess the effects on the environment of its programme of work for exploration (International Seabed Authority 2013a). In addition to this contractual obligation, environmental data have considerable value to contractors in (1) demonstrating due diligence in identifying and quantifying all impacts and their effects on the environment and so reducing the overall risk of the project by better understanding potential environmental liabilities; (2) enhancing the corporate image of individual contractors, and indeed the deep-sea mining sector as a whole; (3) producing cost-effective and convincing cases for monitoring strategies during exploitation, as required for the Environmental Management and Monitoring Plan and the mine site Closure Plan; (4) building engineering solutions at an early stage in the design of mining equipment to minimise impacts; (5) as a result of thought-through mitigation and monitoring plans, reducing environmental monitoring costs over the lifetime of a mining project; and (6) contributing data to the ISA's Regional Environmental Management Plans. If restoration of the original environment is to be achieved, baseline surveys will determine the broad biodiversity and ecosystem functioning of an area, as well as the actions required to repair the physical environment and allow the characteristic fauna and ecosystem processes to be re-established following mining (Aronson et al. 2016). In addition, having a clear business focus on the eventual uses of environmental baseline data focuses effort onto the most meaningful data thereby preventing wasted effort and costs in collecting and analysing data of less relevance during the exploration phase.

There are a number of challenges for contractors in how environmental baseline studies are planned and conducted. At the lowest level, environmental baseline studies may be carried out simply to satisfy annual reporting requirements. This may fulfil contractual obligations in the early stages of a contract, but the studies are likely to be inadequate for an Environmental Impact Assessment (EIA) during any subsequent application for test mining and commercial exploitation of minerals. As can be seen in the EIAs which have been published online (Coffey Natural Systems 2008; Narita et al. 2015; GSR 2018; BGR 2018), they are complex and highly technical documents with data and analysis covering a wide variety of topics, including physical oceanography, sediment characteristics, particle sinking velocities, particle aggregation, sediment geochemistry, toxic metal discharges, chemical contamination, biological studies from microbes to mammals and from the sea surface to the subseabed, studies of biodiversity measures, genetic connectivity, ecosystem functioning (e.g. rates of nutrient remineralisation), the valuation of ecosystem services, hydrodynamic plume modelling, noise and light hazards, etc. This presents a great challenge for some contractors whose original organisation structure, operational priorities and management expertise were, and may still be, focussed almost exclusively on resource evaluation and with little experience of (1) the variety of data and the standards required for statistically robust environmental sampling and analysis and (2) the eventual uses of the environmental data. In particular, the amount of time and funding required for environmental sampling in the field and subsequent analysis in the laboratory may be far greater than managers have experienced previously and are equivalent to, or greater than, the amount of time and effort spent on resource evaluation. As a consequence, environmental baseline studies may not be organised to deliver the right quality and quantity of data for developing a business case for exploitation. Large parts of environmental surveys may have to be repeated, increasing costs and delaying operations.

Collecting baseline data on a wide variety of variables allows a contractor to demonstrate that they understand the different impacts their activities might have on the environment at the sea surface, in mid-water and on the seabed. These are essential for EIA and represent key elements of an EIS. Based on the baseline information and knowledge of the scale, intensity and duration of pressures and the potential effects they may have on the environment in space and time, contractors are able to make accurate predictions of environmental impacts and offer engineering and operational solutions to avoid and minimise those impacts. Knowledge gained through baseline studies also allows contractors to select the most appropriate, cost-effective and accurate methods for monitoring the environment during mining and as part of a long-term Closure Plan. The monitoring programme may use only a subset of the original baseline data collected, but the baseline data will be used to justify the best and most economical methods for monitoring the environment in a consistent fashion. The time required for monitoring as part of the Closure Plan will depend to some extent on the adequacy of the engineering and operational mitigation solutions applied before and during mining, including, for instance, restricting the spread of sediment plumes and reducing sediment compaction. Considerable cost savings may be made by addressing environmental concerns in the design of the mining equipment.

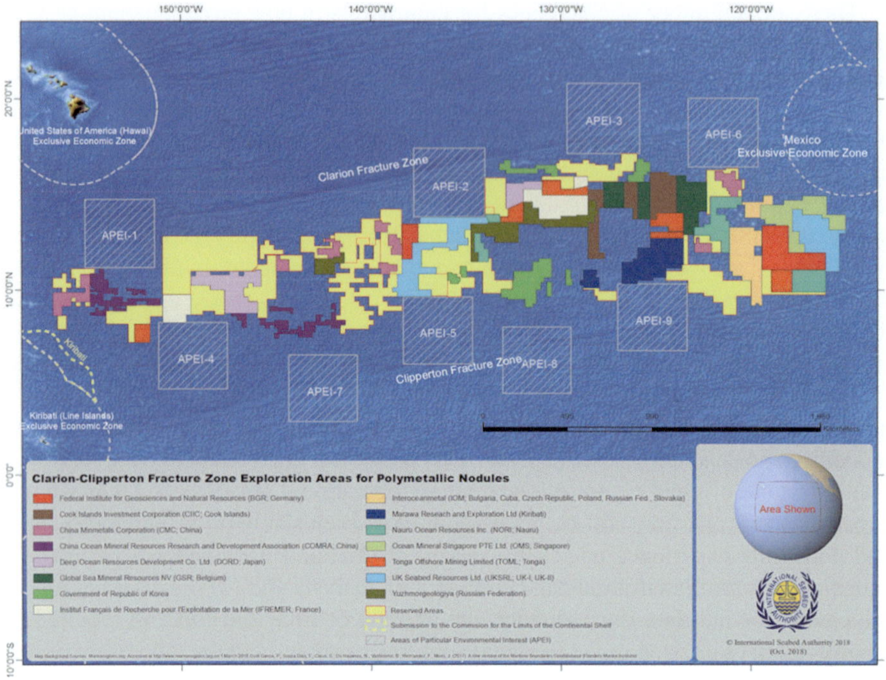

Fig. 1 Chart showing ISA contract areas, reserved areas and Areas of Particular Environmental Interest (APEIs) in the Clarion-Clipperton Fracture Zone in the equatorial Eastern Pacific Ocean. (www.isa.org.jm)

Owing to the distances between individual mining sites, and the wide variety of environmental settings encountered, each contract area will require its own environmental baseline study. It will not be possible to use published data from other mine sites or for a contractor with multiple mine sites to copy one set of data from one site and apply it to another. With reference to the map of contract areas issued by the International Seabed Authority for the polymetallic nodule province in the Clarion-Clipperton Fracture Zone (Fig. 1), the majority of contractors have multiple and separate areas where they intend to mine, which will require multiple baseline surveys (one for each area) even though a contractor may have just one contract with the ISA. Within a large contract area, multiple surveys may be required in space and time depending on habitat heterogeneity. Similarly for polymetallic sulphide and cobalt crust contracts, separate baseline studies will be required for potential mine sites along a ridge segment or for separate seamounts, respectively.

A key objective of an environmental baseline study is to enable a contractor to designate impact reference zones (IRZs) and preservation reference zones (PRZs) (International Seabed Authority 2017a): areas that will be used to assess the magnitude of impacts and recovery rate of ecosystems following mining impacts (Jones et al. 2018b). The exact nature of IRZs and PRZs is a topic of ongoing discussion (Jones et al. 2018b; International Seabed Authority 2018c).

They provide a method by which the impact of mining on ecosystems and the environment can be monitored. To be most effective, the PRZs should be located in the vicinity of the mine site being monitored and include examples of all the different habitats that have been impacted by mining in the area. Baseline data will be essential to determine and justify the size, number and location(s) of PRZs and the size of the buffer zones around them. These data need to be collected over a number of years prior to mining to characterise the stability and natural variability of the PRZs before mining commences (e.g. International Seabed Authority 2018c).

In addition, if transboundary effects have to be taken note of (e.g. UNCLOS Article 195), baseline data are required to determine how large a buffer zone should be around the entire perimeter of a contract area, or subareas, to ensure that plumes do not affect operations and set aside areas in adjacent contract areas, reserved areas or areas which might yet be approved by the ISA for future contracts. Figure 2 shows a hypothetical arrangement of a number of separate contract areas, each with a 30 km square PRZ (10 × 10 km core area and 10 km buffer zone around the core) and a 10 km buffer zone around the perimeter of the contract subareas (where these occur at different locations across the Clarion-Clipperton Fracture Zone (CCFZ). The size and number of PRZs required in each contract area have yet to be determined. The 10 km buffer zone is based on modelling (Aleynik et al. 2017) and may need to be greater unless plumes are constrained through mitigation measures. Figure 2 shows that the smaller a mine size area is, a greater proportion of the area

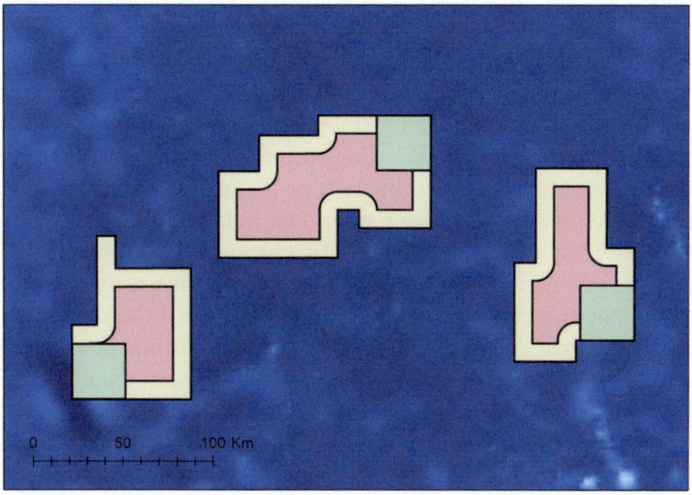

Fig. 2 A hypothetical representation (based loosely on ISA contract areas in part of the CCFZ) showing how the area available for mining is greatly reduced by planning for a 10 km square preservation reference zone and 10 km buffer zone around the PRZ (combined together and shown in green) and the potential requirement for a 10 km buffer zone (shown in yellow) around the perimeter of, and within the boundaries of, a contract area. The remaining part of the contract area available for mining is shown in pink. The precise location of a PRZ should be based on environmental criteria. Scale bar 100 km

will be required for environmental and other planning measures. It also shows that the more irregular the shape a contract area is, the smaller potentially the area will be for mining. In some cases it may be possible for adjacent contractors to work together to reduce the proportion of the area required for buffer zones. In addition, the perimeter buffer zone may be incorporated into the PRZ, although the positioning of the PRZ should be based solely on environmental criteria.

The way in which a contractor plans environmental management and monitoring during exploitation will depend upon the differences in the environment between their area and the wider region. The greater the differences, the more likely it is that mining activities may lead to serious environmental harm and greater environmental controls will be required. Clearly, these decisions require both baseline data in a contractor's area and the wider region.

This highlights the importance of all contractors contributing their environmental baseline data to the ISA's Regional Environmental Management Plans (REMPs). Combining environmental baseline studies allows a better understanding of cumulative impacts and how mining impacts at a local scale relate to the wider distributions of species across a region. At present, owing to the limited levels of knowledge of marine ecosystems on a regional scale, a high level of precaution in operations will be needed, and this may be reflected in the environmental planning at local and regional scales. As knowledge increases, especially by combining the data from different contractors, it may be possible to reduce the level of precaution and hence the costs associated with local environmental planning and monitoring. Contractors are likely to make considerable cost savings by contributing all of their environmental baseline data to the ISA in a timely fashion before an EIS is presented and mining commences, especially where several contractor areas are located close to each other, such as in the CCFZ.

The importance of understanding the uniqueness of contractor areas is particularly important in the cases of polymetallic sulphides on mid-ocean ridges (Van Dover et al. 2011; Dunn et al. 2018). Contractors are able to select 100 blocks for mineral exploration spread over a distance of about 1000 km along a mid-ocean ridge segment (the greatest dimension permitted in the polymetallic sulphide regulations) (Fig. 3). Only a reduced number of blocks (c. 25) are retained for more detailed study after a period of 10 years. Owing to the time restrictions imposed by the relinquishment process, contractors are drawn to focus their operations primarily on resource assessment. Only limited work on environmental baseline studies may be undertaken until the blocks to be retained have been identified. This means that environmental baseline data will be generated at only a small part of the original contract area. As a result, the ability to relate environmental data at individual mine sites to the wider ocean ridge environment will be limited making it difficult to determine the true distributions of species and habitats. Species distributions may appear localised when in fact they are widespread leading to a greater precautionary approach than would be necessary if information, notably on biological communities and their connectivity, was available from the wider area. Contractors on mid-ocean ridges should consider if some types of environmental data should be collected

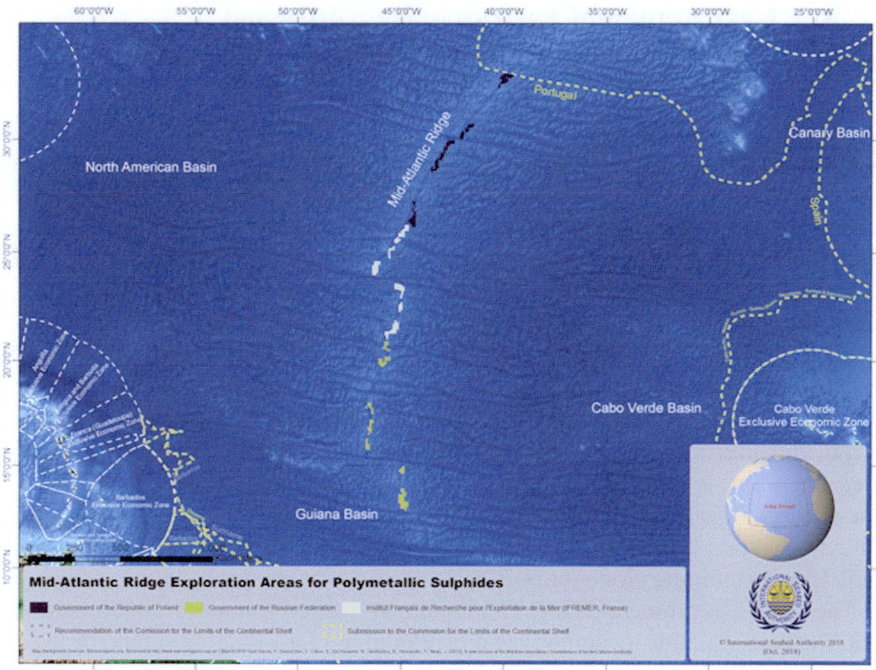

Fig. 3 Chart with ISA contract areas on the Mid-Atlantic Ridge in the North Atlantic showing the long linear distances over which 100 individual mining blocks are distributed within each contract area. (www.isa.org.jm)

at all sites during exploration and how this might assist them in their subsequent applications for exploitation contracts. On a larger scale, linking environmental information from a number of contract areas along a ridge axis might provide a more complete picture of the distribution of organisms and habitats at the regional scale, but large variations in the depth of a ridge within a basin might limit this approach.

All the issues described above point to the importance of collecting good environmental baseline data for convincing Environmental Impact Statements (EISs), Environmental Management and Monitoring Plans (EMMPs) and Closure Plans. Unless the environmental baseline studies have been set up at the outset in a coherent way, it is unlikely that the data generated will have the statistical power to be able to be used in subsequent planning and monitoring. The degree and types of environmental information required for an EIS will depend on the resource being addressed Certain environmental factors are likely to be more important than others, depending on the setting and how the data will be used.

4 Key Environmental Factors to Be Addressed in an Environmental Baseline Study

The ISA Legal and Technical Commission (LTC) makes recommendations to the ISA Council on issues relating to the protection of the marine environment (UNCLOS Article 165(e)). This includes recommending standards and protocols for environmental baseline studies for all three mineral resources (nodules, sulphides and crusts) (International Seabed Authority 2013b) (the 'LTC Recommendations'). The 'LTC Recommendations' define the biological, chemical, geological and physical measurements to be made and the procedures to be followed. Every plan of work for the exploration must not only provide environmental baseline data but also include a plan for monitoring to ensure that no serious harm is caused to the marine environment during exploration; this includes monitoring during and after tests of prototype mining systems. An important aim of a baseline study therefore is to determine the natural conditions of the contract area prior to test mining, including documenting natural changes in faunal succession over time. Studying natural change allows a better understanding of the recovery of ecosystems after test mining (Jones et al. 2017).

The 'LTC Recommendations' (International Seabed Authority 2013b) stipulate that 'the best available technology and methodology for sampling should be used in establishing baseline data for environmental impact assessments'. Best available technologies (BAT) and best environmental practices (BEP), which are not specified but implied, mean that data of sufficient quality are required for effective environmental evaluation and the development of a representative Environmental Impact Statement. This includes very high-resolution GIS mapping and the generation of statistically significant results for environmental variables. In the latter case, for biological samples, statistical power may be influenced by (1) the number and size of samples, (2) the abundances of individual species and (3) the heterogeneity of the seabed at a variety of scales. All depend on the use of appropriate technologies and methods, which, of course, change over time as new technologies and work practices develop.

BAT and BEP have progressed significantly since the first contractors to the ISA started collecting environmental baseline data in 2001. In some cases sampling equipment used 15 years ago may not be acceptable for an EIA today. ISA workshop reports (e.g. International Seabed Authority 2015c) detail the latest types of instruments and equipment, sample collection procedures, treatment and preservation techniques, quality control, data processing methods, statistical analyses and reporting. Where appropriate specific procedures have been recommended, such as issues relating to the size category of the benthic fauna being studied (microbial communities, meiofauna, macrofauna, megafauna, necrophage scavengers, fish). The level of advice can be highly detailed, including methods of sieving sediment samples at sea using filtered, chilled seawater or the way in which morphological and genetic taxonomic identifications should be made (Glover et al. 2016).

In some cases the level of detail recommended might seem to a contractor as being rather onerous; data collected 30 or 40 years ago using old methods and procedures might appear to be sufficient. The challenge is that, as time progresses, scientific research exposes the inadequacies of many old systems and procedures and drives change to better ways of working. While older data might be useful in providing a broad picture of the environment and how it might have changed over time, the old ways of working will not be suitable for the monitoring of impacts and their effects in the future, especially when each mine site and post-closure monitoring might last for tens of years. Some new technologies, such as autonomous underwater vehicles (AUVs), while appearing to be novel now, offer ways in which monitoring costs will be reduced significantly in the near future (BP 2016; Meyer 2016; Send et al. 2013; Wynn et al. 2014). BP plans that all subsea inspections in offshore oil and gas operations will be by Marine Autonomous Systems by 2025.

Below, we consider each of the main categories of environmental data required for a baseline study, offering a few specific thoughts to augment details provided in the 'LTC Recommendations' Annex (International Seabed Authority 2013b) and in Weaver and Billett (2019 this book) on why the various elements in the 'LTC Recommendations' are required operationally.

4.1 Physical Oceanography

The 'LTC Recommendations' detail the standards and approaches to be used in physical oceanography, especially in regions where there are large lateral gradients (e.g. areas in proximity to geomorphological features) (Thurnherr 2011). The data are required to assess how plumes may spread (Weaver and Billett, this volume).

Natural sedimentation rates in the CCZ are very low, less than 10 mm per thousand years (Mewes et al. 2014). In some areas mining plumes will deposit sediment layers orders of magnitude greater than this background level. They will smother organisms, clog the filter-feeding in corals and gelatinous zooplankton and cause alterations in sediment characteristics for seabed deposit feeders. Depending on the mining process, effects of plumes may be experienced up to 100 km from the mine site (Wedding et al. 2013). Sites considerably shallower and deeper than the mine site may also be affected depending on the environmental setting. The use of models to predict the fate of suspended matter and chemicals in plumes will be crucial in designing monitoring strategies (Aleynik et al. 2017). The complex nature of plumes, which may be sinuous, layered and patchy, will require monitoring during mining with good spatial or temporal coverage to address the predictions from plume modelling (MIDAS 2016a, b).

Physical oceanographic models that are accepted by the international ocean modelling community should be used for plume dispersal studies near the seabed. Standard models are not specified in the 'LTC Recommendations' owing to the rapid changes occurring in model design, but it would be expected that state-of-the-art models would be used in an EIS (MIDAS, 2016a, b). Satellite data analyses are recommended for understanding synoptic-scale large-scale features at the sea surface (such as eddies) (Aleynik et al. 2017) and to identify large-scale natural events (such as El Nino cycles) (e.g. Racault et al. 2017). Physical oceanographic data are also important in relation to assessing larval dispersion pathways and genetic connectivity between populations. Studies of interannual variability in the direction and strength of currents near the seafloor used to drive numerical physical oceanographic models have produced high variability in the model outputs that could lead to highly significant effects on larval dispersal ranges and connectivity between populations (Amorim et al. 2017).

4.2 Geological Data

For geological data Geographic Information System (GIS) maps are required with high-resolution bathymetry produced at a scale appropriate to the resource and habitat variability. High-resolution bathymetry, such as that obtained by AUV, can enable assessment of fine-scale variations (<5 m horizontally, <1 m vertically) in the seabed, which may be correlated with polymetallic nodule coverage (Peukert et al. 2018). The interaction between the geological setting and bottom currents also seems to be important in determining the spatial variation of nodule coverage and the seabed fauna (Simon-Lledó et al. 2019). The fine detail now possible from high-resolution mapping using an AUV platform surveying close to the seabed has important consequences for mining operations, not least in detecting seabed features tens of metres in size which might be hazardous for mining equipment (Figs. 23 and 24 in GSR 2018). The scale of environmental heterogeneity now evident in new high-resolution bathymetric maps is also important in understanding biological distributions within the sediment and on nodules (Vanreusel et al. 2016).

4.3 Chemical Oceanography

For chemical oceanography the 'LTC Recommendations' require data to be collected on metals and other hazardous substances which might be released at the seabed or in the discharge plume following initial processing of minerals at the sea surface. In the case of polymetallic sulphides, 'fresh' sulphide mineral surfaces will be exposed to seawater during mining resulting in the oxidation of the sulphides, especially if the sulphides are stockpiled on the seabed (Gjerde et al. 2016). The crushing of the ore into finer particles at the seabed to facilitate the transportation of the ore up a pipe to the surface vessel will increase greatly the surface area over which oxidation may occur. Release of metals can start very quickly on contact with oxygenated seawater releasing dissolved metal ions into the environment, decreasing pH and lowering oxygen (Knight and Roberts 2016). In experiments chalcopyrites from sulphides were found to remain reactive for at least 3 days after exposure (Knight et al. 2018).

Some metals that will be released, such as cadmium, antimony and arsenic, can be harmful in low concentrations. They are scavenged, especially onto the surfaces of fine particles (e.g. Simpson 1978). The processes used in dewatering the ore at the sea surface will release very fine particles into the environment as part of the discharge plume (Coffey Natural Systems 2008). Dilution of the discharge plume will be rapid, but the very fine particles may remain in the water column for a long period. Owing to bioaccumulation, metals might build up in the tissues of seabed and pelagic fauna over time. There is currently little information on this potential impact from mining. It is uncertain how these mining impacts compare with natural hydrothermal vent plumes and their potential toxicity to marine fauna.

4.4 Biological Oceanography

The collection and analysis of biological data are probably the greatest challenges to contractors. There are no published international standards to follow. To address this gap, the ISA organised three standardisation workshops (International Seabed Authority 2014, 2015c, d). The workshops recommended new protocols for the reporting of biological data including the use of multiple preservation methods.

As sampling progresses the importance of nodules as a substratum for attached fauna is becoming more evident (Gooday et al. 2015, 2017; Kamenskaya et al. 2015; Vanreusel et al. 2016). For example, of the 170 morphotypes of megafauna identified in the eastern Clarion-Clipperton, as many as half of them occurred only on polymetallic nodules (Amon et al. 2016). Some taxa, such as alcyonacean and antipatharian corals, occur almost exclusively in areas of the seabed with nodules owing to the importance of the nodules as a substratum (Vanreusel et al. 2016; Simon-Lledó et al. 2019). Moreover, benthic impact experiments have shown that attached sessile fauna recover only very slowly following simulated mining activities (Bluhm 2001; Thiel 2001; Jones et al. 2017). Environmental baseline data on nodule-attached fauna are scarce. Greater information will be required for management and mitigation actions in an EIS.

Nauru Ocean Resources Ltd in response to the 2015 ISA stakeholder consultation suggested that 'activities in the Area should be contextualised with reference to what is generally considered an "acceptable impact" by society on land'. It was stated that on land environmental work related generally to 'higher' forms of life, such as mammals, reptiles, amphibians and birds, and that impacts on 'lower forms of life, such as microfauna', were not studied in detail. It is not clear how widely such a view may be held in the deep-sea mining industry, but it shows a lack of understanding of how different and unique deep-sea ecosystems are, especially in contrast to those on land.

In the marine environment, as one moves from coasts to the abyss, major changes occur in how ecosystems are structured as depth increases. This is because the vast majority of species in the deep ocean, pelagic and benthic, rely for their nutrition on organic matter created initially through photosynthesis in sunlit surface waters.

The few exceptions to this rule occur in chemosynthetic environments, such as hydrothermal vents and cold seeps, where new bacterial production using reduced compounds derived from below the seabed form the basis of specialised food chains. However, for the majority of species that are reliant on organic matter created by photosynthesis, organic matter is rapidly consumed and reworked by a wide variety of organisms as the detrital particles fall through the water column. The result is that only about 1% of the organic matter formed originally at the sea surface reaches depths of 4000–5000 m. The lack of food in the deep sea causes large changes in the size and composition of the deep-sea fauna (Billett 1991; Carney 2005). While microbial and meiofaunal biomass changes little with increasing depth, there are exponential decreases in the biomass and abundance of the larger fauna (Lampitt et al. 1986; Wei et al. 2010). The smaller fauna become much more important in ecosystem functioning within deep-sea ecosystems, and so particular regard should be paid to them in environmental baseline studies. It is possible, however, that larger fauna are still important in structuring deep-sea communities. Indeed, monitoring megafauna (the larger invertebrates) using AUVs, as a possible indicator of ecosystem health, may provide a good cost-effective long-term monitoring tool.

There is a major gap in standards and protocols for sampling and studying the pelagic fauna. The types of pelagic organisms present will change with increasing depth from the epipelagic zone (0–200 m), through the mesopelagic (200–1000 m), bathypelagic (c. 1000–3000 m) and abyssopelagic (>3000 m) realms (Angel 2003). In addition, specialist organisms live in the benthic boundary layer (BBL) up to about 100 m above the seabed surface, known as the benthopelagic zone (Billett et al. 1985; Gebruk et al. 1997) (Fig. 4). Within the water column in certain areas of the world's oceans, significant changes in species may occur also in rela-

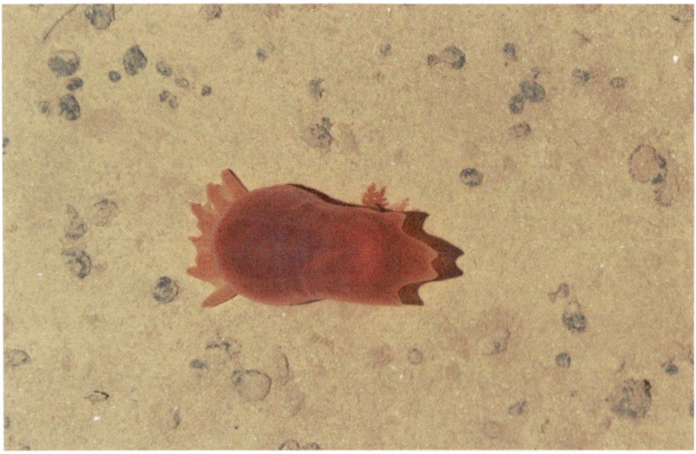

Fig. 4 The benthopelagic holothurian *Peniagone leander* feeding on the seabed among polymetallic nodules in the Clarion-Clipperton Fracture Zone. Daniel Jones, National Oceanography Centre, Southampton

tion to an oxygen minimum zone (OMZ), which typically lies between about 100 and 1000 m water depth in some areas (Karstensen et al. 2008). Many organisms in the mesopelagic zone undertake large diurnal migrations from depths as great as 1500 m up into food-rich surface waters at night (Roe et al. 1984; Herring 2002) linking deep waters with the epipelagic zone. Within the deepwater column, a wide variety of gelatinous zooplankton occur which depend for their nutrition on capturing fine particles in large mucous nets (Robison et al. 2005). Operational plumes at the seabed have the potential to affect benthopelagic organisms, while the discharge plume following dewatering of the ore by the surface vessel could affect a wide variety of zooplankton, especially gelatinous zooplankton, depending on the depth of the discharge. The pelagic realm, however, is vast, and it is likely that pelagic organisms have wide geographic ranges at least at the ocean basin scale (Sutton et al. 2017).

The particle loads of the operational and discharge plumes are not the only environmental considerations for the pelagic realm. Large volumes of cold, nutrient-rich, deep-sea water will be brought to the sea surface during mining operations. In addition, the potential introduction of high levels of nutrients into oligotrophic (low nutrient) waters above the thermocline may lead to changes in biodiversity, community structure, ecosystem function and food chains unless there is careful management of water flows and discharges. However, combined physical, biogeochemical and biological modelling of Ocean Thermal Energy Conversion (OTEC) plants, which will bring much larger volumes of cold deep water to the sea surface, show only very small effects on the productivity and composition of surface phytoplankton (Grandelli et al. 2012). While at the surface, the deep water will warm and surface water may be mixed in; controls will be needed to ensure that discharges, following dewatering, into the cold deep waters do not impact ecosystems owing to the sensitivity of deep-sea species to even small increases in temperature (Somero 1983).

4.5 Geochemical Effects on Biological Systems

Toxicity testing in shallow water aquatic organisms is usually considered one metal ion at a time, in a fixed oxidation state, under standard laboratory conditions of temperature and hydrostatic pressure, over short time intervals using well-known 'indicator species'. These conditions poorly represent the environmental conditions that will prevail for deep-sea mining (MIDAS 2016a). It will not be possible to extrapolate toxicity information and practices from shallow water areas to deep-sea situations. Rather, individual mineral deposits at each deep-sea mine site should be assessed for their toxicity using deep-sea key taxa under appropriate temperature and pressure conditions (Hauton et al. 2017). This requires technologies to sample and retain deep-sea organisms at in situ temperature and pressure and the use of pressurised aquaria for laboratory experiments (Shillito et al. 2014; Auguste et al. 2016) (Fig. 5).

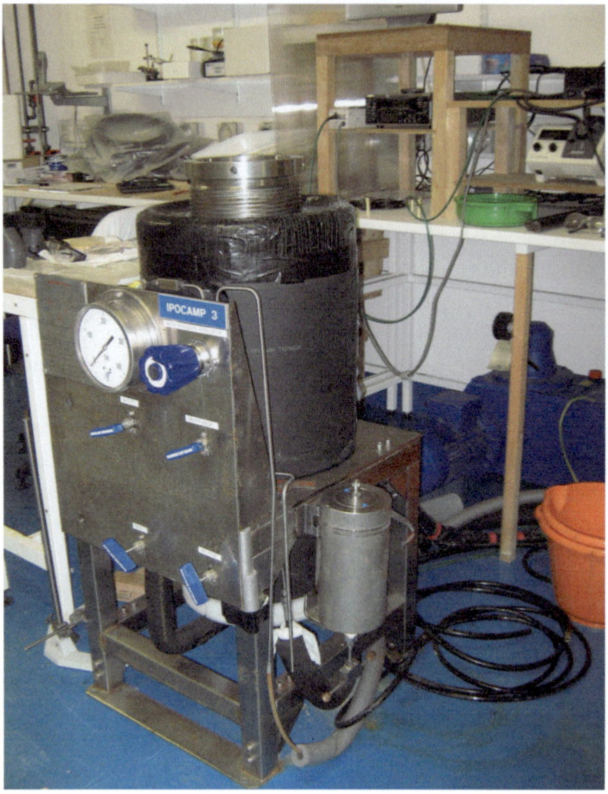

Fig. 5 Pressurised aquaria, such as the IPOCAMP system, can retain deep-sea animals at in situ temperature and pressure within a flow-through water circulation chamber and be used for ecotoxicology experiments. Christopher Hauton, University of Southampton

4.6 *Using the Baseline Data in Environmental Management*

The issues described above and those detailed in Weaver and Billett (2019 this book) demonstrate the wide variety and complexity of subjects that will need to be addressed in an EIS and why good environmental baseline data are required. Apart from specialised technical nature of the various subjects identified, contractors will need to set this within a business context, prioritising the most important factors to be addressed, and how they will be mitigated within the context of the precautionary approach and in a cost-effective manner. To deliver a convincing EIS requires detailed planning of environmental baseline studies so that they deliver the right information and in a way that can be used for subsequent monitoring. Below we consider how contractors might approach good environmental management procedures through (1) value-based business development models, (2) templates for EISs and (3) the 'mitigation hierarchy' now being applied by many industries.

5 Value-Based Business Development

The International Marine Mineral Society (IMMS) brings together leading players in marine mining from industry, government departments and scientific institutions. As part of its work, the IMMS has published a 15-page voluntary code to guide enterprises in their environmental responsibilities (International Marine Minerals Society 2011). The ISA encourages contractors operating in the 'area' to apply the IMMS Code for Environmental Management of Marine Mining (International Seabed Authority 2011a). Companies adopting the IMMS Code voluntarily commit themselves to (1) apply fit-for-purpose procedures (cf. best environmental practice), (2) observe the precautionary approach, (3) consult stakeholders, (4) maintain an environmental quality review programme and (5) ensure transparency in their environmental activities and reporting. In many ways the IMMS Code foresees the development of new value-driven hybrid business models integrating, and giving equal consideration to, environmental, economic and social matters in planning, decision-making and management (Boyd et al. 2009; International Seabed Authority 2013c).

Major environmental accidents by land-based mining companies are reported annually in the press worldwide. While there are many examples of good environmental management practices on land, lapses in environmental control tarnish the whole mining industry, including that in the deep ocean. In the deep sea, where best practices have yet to be developed and knowledge about how mining will affect ecosystems is partly lacking, there should be an emphasis on strict environmental controls.

The International Council for Mining and Metals (ICMM) (https://www.icmm.com/) have committed to 'mining with principles', notably the integration of sustainable development in corporate strategy and decision-making processes. The ICMM and the IMMS Environmental Code anticipate advantages in companies showing management commitment to communication and implementing transparent environmental management systems.

To address shortcomings in how environmental data are collected and how decisions are made by contractors, the IMMS Environmental Management Code sets out clear management responsibilities and commitments to integrate environmentally responsible and sustainable management practices into all operations, including (1) exploration and baseline studies, (2) the design and construction of mining equipment, (3) mining operations, (4) mineral processing, (5) waste disposal and (6) mine site rehabilitation and decommissioning. The value of integrating environmental management decisions from the start is seen as vital. However, in practice companies at present tend to plan in sequence, firstly for resource assessment, secondly for mining equipment design and then environmental baseline studies. This produces particular challenges for collecting some environmental data over a large area and over an extended period of time (e.g. to assess natural change in deep-sea communities in and around a mine site). It will take a significant amount of sampling and analysis time to address issues such as (1) the spatial and temporal variability of currents in and around mine sites, (2) the spatial and temporal variability of biological communities and (3) the genetic connectivity of species across the region.

To overcome these shortcomings, the IMMS Environmental Code concluded that within each company or organisation, a senior executive environmental manager should be appointed to monitor the legal requirements on environmental matters for all marine mining activities, including monitoring internal environmental performance targets and communicating these to employees and regulators. In addition, working together, environmental managers in different companies, could set, improve and update industry-led and industry-wide environmental policies and standards.

6 Environmental Impact Statements

The collection of quality data in environmental baseline studies is the key to the submission of an Environmental Impact Statement (EIS) to gain an environmental and social licence to operate. The EIS will be a public document, and it is likely the data supporting the assertions made in the document will be scrutinised very closely. Mitigation solutions to avoid and minimise impacts will be made. In addition, where significant impacts persist despite these actions, there may be restoration procedures that should be followed to assist the faster recovery of the ecosystems, as, for instance, occurs in mining on land (see Mitigation Hierarchy below).

To assist contractors the ISA has introduced a draft template for the structure of an EIS as part of its 'Draft Regulations on Exploitation of Mineral Resources in the Area' (International Seabed Authority 2018a, b). The template builds on a number of recent initiatives including (1) a New Zealand National Institute of Water and Atmospheric Research (NIWA) project to develop guidelines for the preparation of Environmental Impact Assessments for offshore mining and hydrocarbon drilling (Clark et al. 2017) and (2) several ISA-sponsored workshops (International Seabed Authority 2012, 2017b, c).

The NIWA guidelines provide detailed descriptions of each element of an EIA/EIS. It notes the need for rigorous environmental baseline studies in order to justify decisions on hard data. Where continuing levels of uncertainty exist, it details how they might be addressed through 1) a precautionary approach, 2) statistical and probability analyses and/or 3) predictive models. Where models are applied, the main features of the model should be described including how the model and the input data used to initiate and drive the model have been validated and tested against other models (Clark et al. 2017).

The evaluation of an EIS will include an assessment of whether the right technologies and methods have been used in gathering environmental baseline data. It will be important, therefore, that environmental baseline studies use the latest technologies and methods as they are developed. This poses questions whether some environmental data collected during the exploration phase might still have value to the contractor if sampling and processing methodologies change over time. While the value of data collected tens of years ago may be of questionable quality, and may be of little value in future comparisons with monitoring plans during mining, all data will have value in developing knowledge within a contract area. Some data may still be of value for monitoring by using particular statistical analyses.

While the development of technologies and methods may require some environmental variables to be collected again, or at least the collection of new data to validate the accuracy of information gathered using old methods, the new technologies and methods are likely to be much more cost-effective in the long term. For instance, multi-robotic surveys, including robotic seabed crawlers, are now being conducted (Smith et al. 1997, 2017; Schmidt Ocean Institute 2018). Environmental DNA (e-DNA) may become another fast, accurate and cost-effective way of monitoring impacts at the seabed and in the water column (Goodwin et al. 2017; Stat et al. 2017). The rate at which new technologies and methods are developing, especially genetic analyses, indicates that specialised knowledge of best environmental practice and best available technologies with a business focus is required within contractor teams, especially during the period when environmental baseline data are being gathered.

An Environmental Impact Statement should evaluate the exploitation of minerals in relation to other ecosystem services of value to humankind. Minerals are not the only ocean resource. To realise all the benefits the oceans supply as part of the Common Heritage of Mankind (CHM), planning for the exploitation of deep-sea minerals should take into account the trade-offs in exploiting one resource (1) over another and (2) in relation to the 'life-support' ecosystem services. Ecosystem services are the direct and indirect benefits that people derive from ecosystem processes and functions (Armstrong et al. 2012; Thurber et al. 2014; Le et al. 2017; TEEB 2010, 2012). Valuing ecosystem services, where possible by placing monetary values on them, allows for better decision-making (Hattam et al. 2015). In some cases potential monetary benefits, such as from biotechnological products derived from deep-sea fauna for medicines and other commercial applications (Yao et al. 2010; Snelgrove 2016; Jaspers et al. 2016; Harden-Davies 2017; Barzkar et al. 2018; Xu et al. 2018), are relatively easy to estimate. However, ecosystem services that relate to 'life support' for our everyday lives, such as the regeneration of nutrients that drive new productivity and oxygen production at the sea surface, will present a greater, but not impossible, challenge. 'Life-critical' ecosystem services should be valued in relation to the area of seabed impacted by deep-sea mining; the cumulative impacts of polymetallic nodule mining over large areas, for instance, may be significant in the value of ecosystem services within a region.

An evaluation of the benefits and values of ecosystem services is important because all deep-sea ecosystems have slow recovery rates in terms of biodiversity, ecosystem structure and ecosystem functioning (e.g. Jones et al. 2017; Gollner et al. 2017; Stratmann et al. 2018a, b); ecosystem services once lost will be slow to return. The slow rate of recovery is related to longer generation times in many deep-sea species and is an adaptation to the low food inputs to deep-sea ecosystems (described above). Recognising that deep-sea ecosystems will recover only slowly means that extra care will be needed when applying methods to avoid and minimise impacts. Novel restoration measures to assist the recovery of ecosystems at mine sites may also be required as part of mitigation measures detailed below.

7 Mitigation Hierarchy

The 'mitigation hierarchy' provides a structured approach to considering how mining impacts might be controlled and form an important part of an Environmental Impact Statement. In the 'mitigation hierarchy', a number of elements are considered but in a strict order (Avoid → Minimise → Restore → Offset). The initial stage examines whether impacts can be avoided altogether and at what cost. Then, methodological and engineering solutions to minimise impacts, such as limiting the spread of plumes, should be considered. Where impacts are unavoidable and cannot be minimised to an adequate level, consideration should be given to methods by which ecosystems might be restored, or remediated, once mining has ceased. The rates at which ecosystems will return to pre-impact health and the expenditure required for restoration may depend particularly on the extent to which impacts were minimised. Finally, as in many land-based assessments, and increasingly in coastal marine environments, where impacts lead to permanent changes and where restoration is not possible, contractors may offset and/or compensate for biodiversity and ecosystem functioning loss through the creation of new, similar habitats at a nearby location. This method is usually applied where permanent changes have occurred in the environment, such as the construction of a port, and the replacement of a salt marsh habitat which would otherwise be lost. Even in coastal ecosystems, newly created habitats may take tens of years to approach the dynamics and biodiversity of the original natural habitat that has been lost. This aspect of the resilience of ecosystems, and of the time taken for them to return to states resembling or comparable to habitat lost, will be a particular challenge in the deep ocean. It will also be unlikely that the restored ecosystems will match exactly the deep-sea ecosystems that have been destroyed (Niner et al. 2018).

While the mitigation hierarchy approach is only now being considered in the management of deep-sea ecosystems (Niner et al. 2018; Cuvelier et al. 2018), the Cross-Sector Biodiversity Initiative (CSBI), including the International Council of Mining and Metals (ICMM), and a number of other organisations, has provided practical guidance already on the implementation of the mitigation hierarchy for extractive industries (Cross Sector Biodiversity Initiative 2015). While the examples given in the CSBI guidance document relate almost exclusively to land-based mining, the principles espoused for one part of the mining business are clearly transferrable to all mining sectors.

7.1 Avoid

It has been proposed that the need for deep-sea mining might be avoided altogether by increasing rates of recycling of metals and stimulating investments in the development of alternative technologies (Teske et al. 2016). However, whether this would meet future demand for metals, such as for renewable power generation technologies and electric vehicles, and, whether alternative technologies can be developed in time, is open to question. These issues are likely to be weighed by potential investors in deep-sea mining. While recycling and alternative technologies might reduce market demand, other factors may have a greater influence on how and when mineral resources in the deep ocean are exploited. The first is that the many and varied member states of the International Seabed Authority, and especially developing nations, will benefit from the generation of royalties by deep-sea mining. Secondly, island states, notably in the SW Pacific, now have rights to exploit mineral resources within their exclusive economic zones (EEZs) as national assets. These are strong drivers for the exploitation of minerals in the deep ocean. It is likely, therefore, that once the market prices of minerals are favourable, deep-sea mining will occur.

Aside from the complete avoidance of mining, many approaches are possible to 1) avoid mining in specific areas, such as those of conservation importance 2) avoid specific mining practices, which are particularly harmful, or 3) or avoid mining at specific times, such as during breeding seasons (Cross Sector Biodiversity Initiative 2015). These avoidance measures may be combined in a way that is beneficial for environmental management, but at the same time have little impact on the costs and planning of mining operations, for example, mining one area of the block during one season and then moving to another area for the rest of the year. Avoidance should be considered as early as possible in planning because it gets harder and more expensive to implement avoidance measures as the project develops. Avoidance can operate at many scales, from the setting aside of large regional areas where no mining is allowed to avoiding mining in small areas where particular species may occur in particularly high abundances within a mine contract area. Avoidance will be an important strategy for dealing with uncertainty and provides a tool for passive adaptive management, because areas set aside to avoid impacts might be reconsidered for exploitation as a project progresses and more information becomes available. Spatial management measures will be important in reducing the overall impacts from mining on habitats, such as the loss of nodules and their importance as a hard substrate to support a wide variety of attached species.

An important part of the approach for avoidance is setting aside representative areas at a regional scale to ensure that the effects of industrial activities on regional biodiversity are minimised. The International Seabed Authority (2011a) introduced an environmental management plan for the nodule province in the Clarion-Clipperton Fracture Zone and is now in the process of formulating further Regional Environmental Management Plans (REMPs) for other regions (International Seabed Authority 2018d). Among the aims of the ISA REMPs are (1) to maintain regional biodiversity, ecosystem structure and ecosystem function across a region and (2) to preserve representative and unique marine ecosystems. While localised impacts are unavoidable, the use of spatial management should minimise ecosystem effects at the regional scale, including species extinctions (Niner et al. 2018). REMPs should be (1) of an appropriate size to maintain viable populations, (2) not affected by mining operations and (3) contain broadly the same species pool as areas that will be affected by mining. The challenge is that the present level of knowledge to build REMPs is still very poor owing to the cost, remoteness, immense spatial scales and the lack of expertise in deep-sea ecology in several countries (Niner et al. 2018). In addition, the slower rate over which processes occur in the deep ocean means that most current funding schemes for research are wholly inadequate for studying long-term changes. This is a critical element and difference; the time required for research and experimentation on deep-sea ecosystems is not served by most grant awarding schemes which are aimed primarily at land-based and shelf sea ecosystems. While the development of REMPs will minimise regional impacts, their introduction initially will require the very broad application of the precautionary approach owing to the many current uncertainties about ecosystem structure and functioning in deep-sea ecosystems.

The development of REMPs will be vital for many contractors if avoiding and minimising impacts at the regional scale are to be addressed through spatial management measures. REMPs are an important part of facilitating the environmental management of deep-sea mining. Their success will depend primarily on the timely submission of comprehensive and standardised environmental data generated by all contractors.

In addition to regional-scale management, avoidance of mining impacts to specific areas within a mining block is likely to comprise part of the EMMP, particularly if pre-existing specific sites of scientific interest or conservation value have been identified. Even without these, it may be prudent to protect a range of representative areas from mining to reduce the chances of causing serious harm to the environment and the loss of ecosystem services. Although set up specifically for monitoring, it may be advantageous to maintain PRZs in the long term as conservation areas in addition to their use as control sites for monitoring (Jones et al. 2018b).

7.2 *Minimise*

The nature of mining means that few impacts can be avoided altogether across the mining area. Consideration will then move to how specific impacts from deep-sea mining can be minimised, for instance, whether engineering solutions can limit the creation and spread of plumes during mining operations on the seabed. Exploring how impacts can be minimised is a vital part of EIA. An EIS should document how approaches for minimising impacts have been considered and costed, including balancing the cost of introducing various engineering solutions against the benefits derived from the ecosystems affected.

At present, there is little evidence that the avoidance or minimisation of environmental impacts has been considered in equipment design. Clearly technological solutions might lead initially to greater costs, but over the lifetime of a mining project, engineering designs should reduce costs considerably for contractors, considering the magnitude and longevity of environmental monitoring that may be required and in gaining social licence for DSM activities. The proper valuation of ecosystem services in driving engineering innovations to minimise impacts is critical. Unless contractors include ecosystem services, and the costs associated with their loss or impairment, as part of their decision-taking, and as part of an ethical approach to ensuring the health of the oceans for the Common Heritage of Mankind, it is unlikely that resource and engineering managers will be stimulated to devise technical solutions to reduce environmental harm.

Weaver and Billett (2019 this book) detail the main impacts that will be caused by mining in the deep sea: (1) habitat destruction and/or modification, (2) the dispersal of sediment-laden plumes generated by mining systems, (3) impacts related to plumes released in the water column after dewatering of the ores on the ship and transport barges and (4) impacts related to other factors such as sound and light. For seafloor massive sulphide mining there is also the potential to alter the natural conduits and flow rates of the hydrothermal fluids. The relative importance of these main impacts may vary depending on the resource being considered and the ecosystems affected. For issues relating to plumes, sound and light, engineering solutions will be important, possibly in combination with spatial management measures.

Niner et al. (2018) suggest that the use of shrouds on collecting and cutting systems and the development of methods to reduce the creation of fine particulate materials might reduce the impacts of plumes both in terms of duration and spatial extent. The management of plumes is probably the single most important impact that could be solved through design innovations of mining vehicles and equipment. Methods for enhancing the flocculation of particles within plumes using non-toxic materials may be possible (Cuvelier et al. 2018). For the mining of certain polymetallic sulphides, where there is a risk that oxidation will lead to the release of toxic metals that are attracted to fine particulate material, careful management of how the sulphide is brought to the surface and then moved to transportation barges may be required. For instance, the MIDAS Project concluded that certain sulphide minerals should not be stockpiled on the seabed before being piped to the surface vessel (MIDAS 2016a, b). The assessment of an Environmental Impact Statement will require evidence that the major impacts, and alternatives for the minimisation of those impacts, have been considered and that the conclusions made are justified by experimental evidence.

For each mine site, the engineering solutions introduced to minimise impacts will influence the time, and therefore the cost, required to monitor sites following mining. In addition, the time and costs of long-term monitoring may also be reduced by undertaking restoration activities following mine closure, as occurs in many cases on land (Cross Sector Biodiversity Initiative 2015).

7.3 Restore

The practice of restoring ecosystems has advanced so rapidly in recent years that 'International Standards for the Practice of Ecological Restoration – Including Principles and Concepts' have been published by the Society for Ecological Restoration (McDonald et al. 2016). While the Standards have been formulated principally from experiences in restoring habitats on land, they are broadly applicable to terrestrial, freshwater, coastal and marine environments. In the mining context, restoring ecosystems following the closure of a mine is often standard practice. In contrast, the restoration of marine ecosystems in a fashion analogous to the restoration of land-based mine sites is still in its infancy (Van Dover et al. 2014). There are considerable challenges in the restoration of marine ecosystems as a whole and especially in the deep sea.

Ecological restoration is the process of assisting the recovery of an ecosystem that has been degraded, damaged or destroyed and seeks to 'assist recovery' of a natural or seminatural ecosystem (McDonald et al. 2016). Restoration therefore has a focus on helping nature to recover naturally but perhaps at a faster rate. While the aim might be of re-establishing an ecosystem to the state of biodiversity and functioning that existed prior to its impairment, ecosystems are always in a state of flux, and so restoration aims to reinstate natural ecological progression (Aronson et al. 2016). There may be a family of restorative activities operating at different levels, from removing impacts and allowing natural recovery to active methods, to the transplantation of fauna or the use of artificial reefs to kick-start and speed up recovery.

In coastal marine ecosystems, there have been some notable failures in restoration measures, which have led to the impression in some minds that restoration in the sea is of little utility. However, in a review of good and poor practices in shallow water and coastal ecosystem restoration, Mitsch (2014) concluded that some restoration practices have been very successful in situations where ecologists and coastal engineers worked together to find simple engineering approaches to kick-start natural recovery processes. The greatest successes occurred where engineering actions assisted 'Mother Nature and Father Time' to rehabilitate at a faster pace than might otherwise have occurred. The challenge for the restoration of deep-sea ecosystems is that 'Father Time' is particularly slow (Cuvelier et al. 2018) and it is unclear (1) how rates of recovery in seamount, ridge and abyssal sediment ecosystems might be accelerated and (2) the directions in which recovery might progress (Niner et al. 2018).

These unknowns are particular challenges to contractors and environmental scientists alike. Does this mean that considering restoration measures should be ignored, or should mining companies, policymakers, regulators, sponsoring states and the scientific community stimulate experimental research to address them? The slow recovery rates of deep-sea ecosystems, and, in the case of nodule mining, the large areas over which mining will occur, oblige all these groups to consider how some elements of restoration could be included in Environmental Management and Monitoring Plans and Closure Plans. The scale of some mining impacts and the time required for recovery of ecosystems will be of particular societal concern. The research required to test and verify marine ecosystem restoration methods will be specific to each mineral resource (cobalt crusts, sulphides and polymetallic nodules) and the ecosystems that will be affected (seamounts, ridges and abyssal plains, respectively).

The Society of Ecological Restoration (SER) has developed a set of nine Attributes of Restored Ecosystems to determine what is meant by 'recovery' (Society for Ecological Restoration International Science & Policy Working Group 2004). These include (1) having a characteristic assemblage in the reference ecosystem; (2) stimulating primarily, or exclusively, indigenous species; and having (3) all the functional groups necessary for the development and stability of an ecosystem; (4) a physical environment capable of sustaining reproducing populations; (5) normal ecosystem functioning; (6) abiotic and biotic interactions, flows and exchanges with the wider landscape; (7) impacts eliminated; (8) resilience to normal periodic stress; and (9) self-sustaining capability. These were developed with terrestrial systems in mind, but they are generally applicable to the deep sea.

Cobalt Crust Mining and Restoring Ecosystems on Seamounts

Cobalt crust mining is likely to occur initially across 'flat-topped' guyots where mining systems will be easier to operate. In the western Pacific, where there are four exploration contracts with the ISA, the seamount summits of interest to mining contractors lie at depths between c. 1100 and 1600 m. Considerable work needs to be carried out to characterise the biodiversity of the seamounts and the degree of connectivity in populations across different spatial scales. Previous research on seamounts in the region has indicated that the dominant large invertebrate taxa will include foraminiferans, sponges, corals, squids, echinoderms (sea stars, sea cucumbers, feather stars), crabs and sea squirts (Fukushima 2007; International Seabed Authority 2011b; Schlacher et al. 2013) (Fig. 6). Of these, foraminiferans have been found to be conspicuously abundant and diverse (Mullineaux 1987). Seamounts may support commercially important fish stocks depending on the depth. If whole seamount summits are exploited for minerals, then significant impacts will occur on the faunal communities, especially if a significant number of seamounts within a region are all mined. The restoration of faunal communities on each seamount may be considered to be important environmentally and ethically.

The environmental management of cobalt crust mining is likely to be approached through a combination of 1) spatial management, 2) engineering solutions to reduce impacts and 3) methods for the re-introduction of fauna to the mine site when it is closed. The four contractors to the International Seabed Authority for the development of cobalt crust mining on seamounts in the western Pacific are working with the ISA to generate a Regional Environmental Management Plan (REMP). The creation of set aside areas, perhaps of whole seamounts, in each contractor area, is likely to be required for the protection and preservation of biodiversity of all taxa and all depths. However, it is not the only measure, and experiments on how

Fig. 6 A large sponge on the flanks of a seamount on the Mid-Atlantic Ridge at a depth of about 800 m. National Oceanography Centre, Southampton

communities might be reintroduced to mine sites in a way that accelerates natural recolonisation and recovery should be conducted in parallel.

Studies of genetic connectivity and how they might guide environmental management decisions in the deep sea are increasing rapidly. They are leading to a greater understanding of larval dispersal and the ways in which recruitment to the benthos occurs in a variety of environmental settings (e.g. Baco et al. 2006, 2016; Hilario et al. 2015; Zeng et al. 2017). The combination of physical oceanographic modelling and genetic studies is also leading to new insights in how variability in the physical environment might lead to interannual, and perhaps longer-term decadal, changes in larval dispersal characteristics (e.g. MIDAS 2016a, in this case related to hydrothermal vent mussels). Recolonisation on some seamounts may occur only through episodic recruitment events. Recruitment to areas impacted by mining will depend initially on physical oceanographic processes in association with differences in reproductive characteristics of species (such as egg size and dispersal capability). There is evidence in genetic studies of significant self-recruitment within an area once a small population has become established (Zeng et al. 2017). Restoration measures, therefore, might include the transplantion of mature individuals of key species or, preferably, established whole communities. Natural processes would then drive self-recruitment at sites in need of rehabilitation.

Restoration of coral communities in coastal and shelf seas has been successful in reinstating some corals following bleaching events. Methods include underwater nurseries, mobile laboratories for the fertilisation, larval settlement and initial outgrowing of coral recruits and transplantations (e.g. van Koningsveld et al. 2017). Applying these methods to the deep sea might be possible using pressurised aquaria for maintenance of deep-sea fauna (e.g. Shillito et al. 2014) (Fig. 5) and the use of remotely operated vehicles for the in situ attachment of corals. The costs of these activities may be high, owing to ship and technology hire (Van Dover et al. 2014), but costs could be driven down by using standby time in operational ROV activities, such as occurs in the offshore oil and gas industry (Gates et al. 2017). In addition, combining the skills of ecologists and engineers, as in shallow water (van Koningsveld et al. 2017), could lead to novel approaches for deep-sea restoration. For instance, cost-effective solutions are being tested using a variety of frame structures to move and reintroduce coral ecosystems into the deep sea (http://www.merces-project.eu/?q=content/wp4-field-work-activities).

Potentially, a new approach of transplanting whole artificial reefs on frames might prove successful if engineering solutions are developed on (1) ways to transport mature artificial reef structures/frames from healthy communities in set aside areas to the site in need of rehabilitation and (2) methods to retain the diverse communities which have colonised the artificial reef while they are being moved. Reef structures with sufficient rugosity and complexity (Cuvelier et al. 2018) could be emplaced initially in mature set aside conservation areas where colonisation by a wide variety of fauna is more likely to occur owing to proximity to mature individuals of different species. The reef structures would be cheap allowing a large number to be used. When a mine site is closed, the artificial reefs could be moved to the mine site, forming a distributed network, enhancing local recruitment. Such a method would accelerate recolonisation of whole, diverse communities.

Diverse communities on artificial reefs may take time to become established, and so innovative solutions would need to be tested and proven ahead of the submission of an Environmental Impact Statement. This indicates that contractors should address potential restoration measures early during exploration activities because the success of the various restoration techniques in the deep sea might not be apparent for 5–10 years following deployment.

Polymetallic Sulphide Mining and Restoring Ecosystems on Mid-ocean Ridges

Areas with seafloor massive sulphides (SMS) have a variety of habitats including hydrothermally active vents and chimneys, relict chimneys in (currently) inactive locations, lava flows and bedrock (Van Dover 2011; Boschen et al. 2013, 2016) (Fig. 7). Detailed mapping at one SMS site has revealed considerable heterogeneity in the distributions of various megafaunal assemblages (Boschen et al. 2016) with the indication that some specific assemblages may be associated with inactive vent sites. The dominant species around inactive vent chimneys do have wider distributions, and further work is required to determine whether the assemblage is related to the inactive vents or to differences in microbiology, food supply and current flows which might influence local recruitment. Many of the taxa that are not specific to active vents appear to have wider distributions in the region. These different assemblages may require to be managed in different ways.

Fig. 7 An example of a seafloor massive sulphide deposit: the Jabberwocky edifice, a mineral-rich hydrothermal vent chimney about 5 m high and at depth of 2757 m in an area of the Southwest Indian Ocean Ridge licenced by the ISA. Jon Copley, University of Southampton; https://www.youtube.com/watch?v=y6iK19xaYJg

Mining at or adjacent to active hydrothermal vents, if it occurs, will need to be managed carefully. Mining will have consequences for biological communities that are highly localised and separated by large distances along a ridge axis. Methods for avoiding and minimising impacts may be sufficient, including ensuring that existing conduits and flow rates of the hydrothermal vent fluids are maintained. However, in some cases, restoration of hydrothermal vent communities may be required. Methods may include the placing of artificial chimney structures over new seafloor vents to simulate and enhance the precipitation and formation of sulphide chimneys and their various microhabitats (Coffey Natural Systems 2008; Section 9.6.2; Cuvelier et al. 2018). New chimney structures at different stages of temporal development might need to be created. Restoration actions might also include creating new vent fluid outlets by drilling holes to ensure fluid flow and allow for vent fauna to recolonise artificial chimney structures (Cuvelier et al. 2018).

Van Dover et al. (2014) developed a scenario for the restoration of hydrothermal vent communities using artificial chimneys at a proposed massive sulphide mine site to support fauna associated with actively venting (e.g. holobiont provannid snails) and inactive sulphide deposits (e.g. bamboo corals). A cost model that included the use of remotely operated vehicles (ROVs) and other autonomous vehicles and the creation of artificial chimneys was produced. At least 80% of the costs were related to shiptime and the use of the ROV technologies. The costs of restoration were estimated to be two to three orders of magnitude larger than restoration

Fig. 8 The octocoral *Metallogorgia* sp., at a depth of 2600 m on the Mid-Atlantic Ridge, is an example of non-hydrothermal vent organisms that could be impacted by seafloor massive sulphide mining. Daniel Jones, National Oceanography Centre, Southampton

actions in shallow water. As indicated above for cold-water corals, hydrothermal vent restoration could be reduced considerably by planning the activity in and around other seabed mining operations using ROV systems or by using smaller and less costly vessels.

The restoration of fauna normally found on rocky surfaces, such as octocorals (Fig. 8), and in sediments, such as enteropneusts (Jones et al. 2013; Fig. 9), in areas with seafloor massive sulphides should be considered separately. Rocky slope and sediment fauna in and around massive sulphides may have wide distributions along a ridge system, although they are likely to be restricted to certain depths and may be abundant only within a depth range of a few hundreds of metres (e.g. Billett 1991; Carney 2005). In addition, sediment fauna and possibly rocky slope fauna, on the Mid-Atlantic Ridge (MAR), have been shown to have close affinities to the fauna found at similar depths and latitudes on continental slopes to the west or the east of the MAR in the Atlantic Ocean, including fish (e.g. Alt et al. 2013; Priede et al. 2013). It is possible that many species on mid-ocean ridges will have wide distributions on an ocean basin scale.

It is possible that the small footprint of a sulphide mine site, and the likely wide-ranging distributions of the rocky slope fauna, means that investment in restoration actions may not enhance natural unassisted recovery rates significantly. More information is needed from baseline data and scientific studies on species distributions and connectivity. If particular communities are found to be unique and restricted in their distributions (e.g. Boschen et al. 2016), then the artificial reef and transplantation techniques described for cobalt crust mining above may be effective. Large clumps of ecosystem engineer species, and associated attached or sessile fauna,

Fig. 9 An acorn worm (Hemichordata, Enteropneusta) at a depth of about 2500 m on the Mid-Atlantic Ridge. Daniel Jones, National Oceanography Centre, Southampton

could be transplanted away from active mining operations and potentially returned to the mined area as part of a Closure Plan (Coffey Natural Systems 2008, Section 9.6.2; Cuvelier et al. 2018). The use of artificial reefs/frames is likely to be cost-effective solution but will require experimentation and research to demonstrate their utility during the exploration phase for seafloor massive sulphides.

Polymetallic Nodule Mining and Restoring Ecosystems on Abyssal Plains

The restoration of benthic ecosystems following mining for polymetallic nodules may target (1) fauna attached to the nodules that may be dependent on the nodules as a substratum and (2) the sediments, which will be disturbed during the collection of the nodules. Impacts will also occur on pelagic ecosystems, especially in the benthic boundary layer (within c. 100 m of the seabed). It is unlikely that restoration actions could be devised for pelagic ecosystems.

The removal of a large part of the hard substratum (nodules) on abyssal plains in polymetallic nodule provinces, especially the removal of areas with dense nodule cover, is likely to have a significant environmental impact on obligate nodule-attached fauna (Amon et al. 2016; Vanreusel et al. 2016). The nodules act as substratum for a wide variety of taxa, including sea anemones, a wide variety of soft and stony corals, sponges, sea lilies, komokiaceans and xenophyophores (giant protozoans) (Fig. 10). As nodules grow only exceptionally slowly, their replacement will occur only over millennia; essentially they will be lost on ecological timescales (Amon et al. 2016). Not all nodules will be mined, but the areas with high nodule abundance are likely to be targeted for mining, and it is possible that areas with high nodule abundance are important for the maintenance of

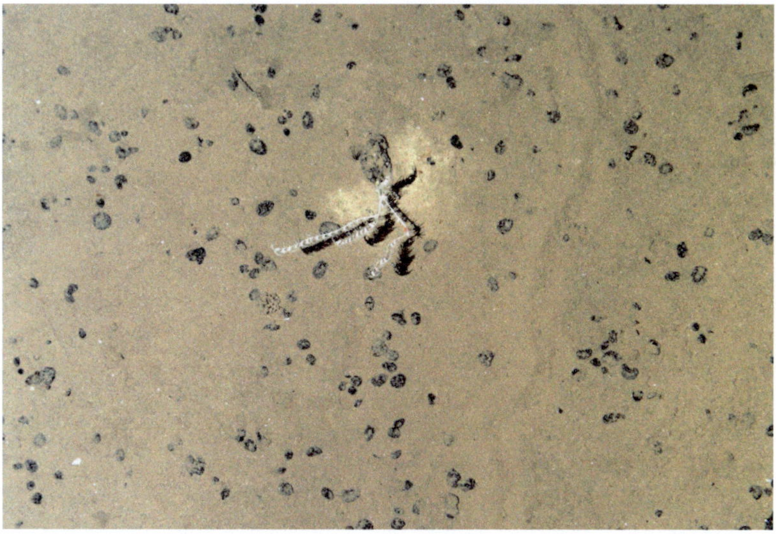

Fig. 10 The octocoral *Bathygorgia* sp. and associated fauna, attached to a large polymetallic nodule at a depth of about 4000 m in the Clarion-Clipperton Fracture Zone, eastern equatorial Pacific. Daniel Jones, National Oceanography Centre, Southampton

natural populations of a wide variety of taxa (Vanreusel et al. 2016). In addition, nodules will also become covered by sediment deposited from plumes created during mining operations. Owing to the large areas over which nodule mining will occur (c. 200 km^2 area per mining machine per year; Smith et al. 2008), there may be a compelling need to replace hard substrata on the seafloor in order to ensure that species do not go extinct. While areas with nodules may be retained through current spatial management plans, for instance, within Areas of Particular Environmental Interest (APEI) (International Seabed Authority 2011a; Wedding et al. 2013; Lodge et al. 2014), this may not be sufficient to maintain healthy self-replicating populations, particularly if the APEIs show environmental differences to the mined areas (Vanreusel et al. 2016).

A potential restoration strategy is for contractors to replace a significant proportion of the nodules mined with artificial nodules of a similar size, shape, rugosity, porosity and (gross) chemistry to those that have been lost (Cuvelier et al. 2018). This poses questions as to the cost-effective materials that might be used and that would not be considered as pollutants of the abyssal environment. One proposal would be to utilise some of the waste material extracted from the nodules following processing on land. While, for instance, the manganese in the nodules has commercial worth, the amount produced from nodule processing will outweigh the commercial needs thus potentially leaving large volumes of manganese that could be used to make artificial nodules with similar composition to those which have been mined. Considerable research is needed on the relationship between nodule density and the population maintenance of obligate-attached fauna, and the materials and methods by which artificial nodules might be introduced to the abyssal environment. The reduction in sponge stalks and coral stands, an important habitat for some associated species (Purser et al. 2016), may also need to be replaced (Cuvelier et al. 2018) although novel deployment methods would be required. Artificial sponge stalks have been shown to be successful in attracting fauna (Beaulieu 2011). Owing to the large areas impacted, the use of artificial reef structures/frames, conditioned in advance in set aside areas, as described above for seamount fauna restoration following cobalt crust mining, may be required also to assist in connectivity across large expanses of seabed.

The restoration of sediment communities is a particular challenge owing to the very large areas of seabed which will be affected and the slow recovery rates of abyssal ecosystems. It is likely that the main actions for the protection and preservation of the marine environment will be through spatial management measures and the minimisation of impacts by engineering solutions. However, there may be some restoration actions that could be undertaken to assist natural recovery rates. These relate to ways in which the physical structure of the surface sediment layer (top 5 cm of the seabed) might be reinstated to allow natural processes and recolonisation to proceed faster, bearing in mind that the principal contributors to ecosystem functioning will be the microbial and meiofaunal communities. These communities may also be the components of sediment communities that are more resilient to mining operations.

Where previous experiments have removed and compacted the seabed using a sledge 2.5 m wide, there has been very little recolonisation of the sediment by meiofauna (Miljutin et al. 2011). It is likely this is related to changes in the physical nature of the sediment (grain size, organic content and texture) (Gollner et al. 2017). The supply rate of particulate organic carbon (organic material, food) and its interaction with the seabed structure may also be important. Stratmann et al. (2018a) suggested that in an experimental study of the response of deep-sea megafauna to the large-scale disturbance of the seabed, megafaunal taxa were slow to recolonise the disturbance site for a period of at least 7 years following the impact, probably owing to the poorer food conditions and changes in the sediment particle size and composition. In the mining context, it is likely that sediments compacted by mining vehicles will be covered by a layer of unconsolidated sediment falling out from the plume caused by operations on the seafloor (Weaver and Billett 2019 this book). However, if the underlying sediment has been compacted, this may still have an effect on the sediment communities lying above. The design of mining systems might therefore consider reducing compaction to a minimum.

In some places the unconsolidated surface layer remaining after mining may be resuspended by near seabed currents, potentially caused by eddy systems emanating from storms at the sea surface (Aleynik et al. 2017). Sediments on the move are also less likely to be recolonised by the diverse species assemblages normally found in abyssal sediments around polymetallic nodules before mining, even though the microbial and meiofaunal communities may survive the mining process, at least initially. Restoration of the sediment fabric might therefore be considered important to resist resedimentation and to provide a suitable habitat for faster recolonisation and rehabilitation of sediment communities. Engineering solutions at the time of mining may be able to assist in this process. Alternatively, stimulation of the microbial community and the production of exopolymers to consolidate the sediment surface, a natural process known in intertidal mudflats (Yallop et al. 2000), could be stimulated by applying a fine rain of organic particulate material (potentially several times) following mining. While this may lead to changes in the composition of the microbial community initially, it may assist the natural long-term sediment community to re-establish itself. Abyssal sediment communities are not immutable over time and respond to natural pulses in organic inputs (e.g. Billett 2001, 2010; Ruhl 2007, 2008; Smith et al. 2009). Over time, sediment communities will stabilise, although whether the original assemblages will be replicated is unknown. A state of biodiversity and ecosystem functioning similar to that which existed prior to mining may, however, be set on course and at a faster rate than would be achieved by making no interventions. In order to determine if any unintended consequences might occur, it is vital that research is undertaken during the exploration phase, possibly using ongoing benthic impact experiments to test potential restoration measures.

7.4 *Offset/Compensate*

Offsetting and compensation for biodiversity and habitat loss through man's impacts is the last consideration of the mitigation hierarchy, once all approaches in avoiding minimising and then restoring ecosystems have been exhausted. In the deep ocean, compensation/offsetting should aim to provide a net gain for biodiversity (Niner et al. 2018). This can be achieved in theory in several ways: 1) 'like for like' averted loss of biodiversity through the protection of a similar habitat to that mined; 2) 'like for like' habitat restoration to create/restore new, additional and equivalent biodiversity of a similar type in a different location to that lost; 3) 'out of kind' habitat restoration to create new biodiversity of a different type, such as in shallow water; and 4) 'additional actions' that do not seek equivalence or provide biodiversity gains but compensate in another way, such as capacity building (Niner et al. 2018). Niner et al. (2018) argue that all of these strategies have major barriers to application in the deep sea. For example, in the deep sea, 'like for like' offsetting may only be possible in relation to seamounts and cobalt crust mining, where the restoration of seamount ecosystems impacted by bottom trawling in the past might be considered. However, as Niner et al. (2018) point out, actions taken by one international agency within the activities regulated by another may not be successful. A solution for cross-sector environmental management might be arranged through the Memoranda of Understanding the ISA has with other UN bodies. Another potential problem is that generally there is little information on the communities that occurred on seamounts prior to being impacted by bottom trawling. Methods using artificial reefs and frames deployed on healthy seamount ecosystems or set aside areas at the same depth and within the same region as those being mined, as described for cobalt crusts above, may be suitable in this case. This indicates the importance of set aside areas being created before the exploitation of minerals and for the potential restoration measures to be investigated during the exploration phase. The many barriers to compensation in the mitigation hierarchy suggest that, while continued research is important, it may not be a desirable strategy for most mining operations.

8 Conclusions

Other potential restoration measures suggested by the MIDAS Project (MIDAS 2016a, b) have been reviewed by Cuvelier et al. (2018) and may offer additional solutions to improving the rate at which deep-sea ecosystems might recover following mining. However, all the methods require rigorous testing and experimentation in order to avoid unintended consequences and to test their practicality and cost-effectiveness. It is surprising, therefore, that very little attention has been given to identifying potential restoration measures in the deep sea. Owing to the time required to measure significant changes between restoration actions and natural processes, experiments are required sooner than later during the exploration phase in order to be included in Environmental Impact Statements. Good, cost-effective and simple restoration actions have the potential to lead to considerable cost savings for contractors over the life time of a mining project and will be important for the maintenance of important ecosystem services in the deep sea. Realising the need for restoration and its costs may also play an important role in cost-benefit analyses of engineering solutions to minimise impacts while mining and so reduce the needs for subsequent restoration (Durden et al. 2017). Deciding on environmental trade-offs in deep-sea mining, especially in relation to biological factors, will be complex, including morphological taxonomy, genetics, connectivity, ecotoxicology, bioaccumulation, ecosystem functioning and ecosystem services. It will be important for contractors to have environmental business-focussed decision takers within their organisations at a senior level who understand these complexities and to make best use of ships, technologies, methods, people and time.

Key recommendations for improving environmental management practices are:

- Contractors should appoint a senior environmental manager at the start of the exploration phase.
- Mining equipment should be designed at the prototype stage to avoid and minimise environmental damage and reduce long-term monitoring costs.
- Contractors should use the latest best available technologies and best environmental practices.
- Contractors should plan environmental management using the 'mitigation hierarchy'.
- Contractors should work together to produce industry standards and protocols.
- All contractor environmental data should be archived and submitted to the ISA for the development of Regional Environmental Management Plans (REMPs).
- Contractors should sponsor some expeditions focussed primarily on environmental management issues.

- Contractors should publish their annual reports and data on environmental baseline surveys and monitoring studies online.
- Experimentation on, and planning for, ecosystem restoration should be undertaken during exploration activities.

Acknowledgments The research leading to these results has received funding from the European Community's Horizon 2020 Research and Innovation Programme for Blue Growth under grant agreement No. 688975 (Blue Nodules) for PPEW and grant agreement No. 689518 (Marine Ecosystem Restoration in Changing European Seas, MERCES) for DSMB and DOBJ. This work also received funding from the European Union Seventh Framework Programme (FP7/2007–2013) under the MIDAS project, grant agreement 603418. DOBJ was funded for this work by the Climate Linked Atlantic Section Science (CLASS) programme and was supported by UK Natural Environment Research Council National Capability funding, grant number NE/R015953/1.

References

Aleynik, D., Inall, M., Dale, A., & Vink, A. (2017). Impact of remotely generated eddies on plume dispersion at abyssal mining sites in the Pacific. *Scientific Reports, 7*, 16959.

Alt, C. H. S., Rogacheva, A., Boorman, B., Hughes, J. A., Billett, D. S. M., Gooday, A. J., & Jones, D. O. B. (2013). Trawled megafaunal invertebrate assemblages from bathyal depth of the Mid -Atlantic Ridge (48°N–54°N). *Deep Sea Research, Part II, 98*, 326–340.

Amon, D. J., Ziegler, A. F., Dahlgren, T. G., Glover, A. G., Goineau, A., Gooday, A. J., Wiklund, H., & Smith, C. R. (2016). Insights into the abundance and diversity of abyssal megafauna in a polymetallic-nodule region in the eastern Clarion-Clipperton Zone. *Scientific Reports, 6*, 30492. https://doi.org/10.1038/srep30492.

Amorim, P., Perán, A. D., Pham, C. K., Juliano, M., Cardigos, F., Tempera, F., & Morato, T. (2017). Overview of the ocean climatology and its variability in the Azores Region of the North Atlantic including environmental characteristics at the seabed. *Frontiers in Marine Science, 4*, 56. https://doi.org/10.3389/fmars.2017.00056.

Angel, M. V. (2003). The pelagic environment of the open ocean. In P. A. Tyler (Ed.), *Ecosystems of the world. Volume 28 Ecosystems of the deep oceans* (pp. 39–79). Amsterdam: Elsevier.

Armstrong, C. W., Foley, N. S., Tinch, R., & van den Hove, S. (2012). Services from the deep: Steps towards valuation of deep sea goods and services. *Ecosystem Services, 2*, 2–13.

Aronson, J., Clewell, A., & Moreno-Mateos, D. (2016). Ecological restoration and ecological engineering: Complementary or indivisible? *Ecological Engineering, 91*, 392–395.

Auguste, M., Mestre, N. C., Rocha, T. L., Cardoso, C., Cueff-Gauchard, V., Le Bloa, S., Cambon-Bonavita, M.-A., Shillito, B., Zbinden, M., Ravaux, J., & Bebianno, M. J. (2016). Development of an ecotoxicological protocol for the deep-sea fauna using the hydrothermal vent shrimp *Rimicaris exoculata*. *Aquatic Toxicology, 175*, 277–285.

Baco, A. R., Clark, A. M., & Shank, T. M. (2006). Six microsatellite loci from the deep- sea coral *Coralliumlauuense* (Octocorallia: Coralliidae) from the islands and seamounts of the Hawaiian archipelago. *Molecular Ecology Notes, 6*, 147–149.

Baco, A. R., Etter, R. J., Ribeiro, P. A., von der Heyden, S., Beerli, P., & Kinlan, B. P. (2016). A synthesis of genetic connectivity in deep-sea fauna and implications for marine reserve design. *Molecular Ecology, 25*, 3276–3298.

Barzkar, N., Homael, A., Hemmati, R., & Patel, S. (2018). Thermostable marine microbial proteases for industrial applications: Scopes and risks. *Extremophiles, 22*, 335–246.

Beaulieu, S. E. (2011). Colonisation of habitat islands in the deep sea: Recruitment to glass sponge stalks. *Deep Sea Research, Part I, 48*, 1121–1137.

BGR. (2018). Environmental Impact Assessment for the testing of a pre-prototype manganese nodule collector vehicle in the Eastern German license area (Clarion-Clipperton Zone) in the framework of the European JPI-O Mining Impact 2 research project. https://www.isa.org.jm/files/documents/EN/EIA/BGR/EIA_BGR.pdf

Billett, D. S. M. (1991). Deep-sea holothurians. *Oceanography and Marine Biology: An Annual Review, 29*, 259–317.

Billett, D. S. M., Hansen, B., Huggett, Q. J. (1985). Pelagic Holothurioidea (Echinodermata) of the northeast Atlantic. In: B. F. Keegan, B. D. S. O'Connor (Eds.), *Echinodermata: Proceedings of the 5th International Echinoderms Conference* (pp. 399–411). Galway.

Billett, D. S. M., Bett, B. J., Rice, A. L., Thurston, M. H., Galéron, J., Sibuet, M. & Wolff, G. A. (2001). Long-term change in the megabenthos of the Porcupine Abyssal Plain (NE Atlantic). *Progress in Oceanography* (pp. 399–411).

Billett, D. S. M., Bett, B. J., Reid, W. K. D., Boorman, B & Priede, M. (2010). Long-term change in the abyssal NE Atlantic: The 'Amperima Event' revisited. *Deep-Sea Research II 57* https://doi.org/10.1016/j.dsr2.2009.02.001.

Bluhm, H. (2001). Re-establishment of an abyssal megabenthic community after experimental physical disturbance of the seafloor. *Deep Sea Research, Part II, 48*, 3841–3868.

Boschen, R. E., Rowden, A. A., Clark, M. R., & Gardner, P. A. (2013). Mining of deep-sea seafloor massive sulphides: A review of the deposits, their benthic communities, impacts from mining, regulatory frameworks and management strategies. *Ocean and Coastal Management, 84*, 54–67.

Boschen, R. E., Rowden, A. A., Clark, M. R., Pallentin, A., & Gardner, P. A. (2016). Seafloor massive sulfide deposits support unique megafaunal assemblages: Implications for seabed mining and conservation. *Marine Environmental Research, 115*, 78–88.

Boyd, B., Henning, N., Reyna, E., Wang, D., Welch, M., & Hoffman, A. J. (2009). *Hybrid organisations. New business models for environmental leadership* (p. 148). London: Routledge.

BP. (2016). How a new robot fleet is monitoring the underwater world. BP Magazine May 2016. https://www.bp.com/en/global/corporate/bp-magazine/innovations/ocean-monitoring-with-robot-technology.html.

Carney, R. S. (2005). Zonation of deep biota on continental margins. *Oceanography and Marine Biology: An Annual Review, 43*, 211–278.

Clark, M.R., Rouse, H.L., Lamarche, G., Ellis, J.I., Hickey, C. (2017). *Preparation of environmental impact assessments: General guidelines for offshore mining and drilling with particular reference to New Zealand.* NIWA Science and Technology Series No 81. 110 pp.

Coffey Natural Systems (2008). Environmental impact statement: Solwara 1 Project. Nautilus Minerals Niugini Limited, http://www.cares.nautilusminerals.com/Downloads.aspx

Cross Sector Biodiversity Initiative (2015). A cross-sector guide for implementing the mitigation hierarchy. https://www.icmm.com/website/publications/pdfs/biodiversity/cross-sector-guide-mitigation-hierarchy.

Cuvelier, D., Gollner, S., Jones, D.O.B., Kaiser, S., Martinez Arbizu, P., Menzel, L., Mestre, N.C., Morato, T., Pham, C.K., Pradillon, F., Purser, A., Raschka, U., Sarrazin, J., Simon-Lledo, E., Stewart, I.M., Stuckas, H., Sweetman, A.K., Colaço, A. (2018). Potential mitigation and restoration actions in ecosystems impacted by seabed mining. *Frontiers in Marine Science* 5 (467).

Dunn, D. C., Van Dover, C. L., Etter, R. J., Smith, C. R., Levin, L. A., Morato, T., Colaço, A., Dale, A. C., Gebruk, A. V., Gjerde, K. M., Halpin, P. N., Howell, K. L., Johnson, D., Perez, J. A. A., Ribeiro, M. C., Stuckas, H., Weaver, P. P. E., & SEMPIA Workshop Participants. (2018). A strategy for the conservation of biodiversity on mid-ocean ridges from deep-sea mining. *Science Advances, 4*(7), eaar4313. https://doi.org/10.1126/sciadv.aar4313.

Durden, J. M., Murphy, K., Jaeckel, A., Van Dover, C. L., Christiansen, S., Gjerde, K., Ortega, A., & Jones, D. O. B. (2017). A procedural framework for robust environmental management of deep-sea mining projects using a conceptual model. *Marine Policy, 84*, 193–201.

Durden, J. M., Lallier, L. E., Murphy, K., Jaeckel, A., Gjerde, K., & Jones, D. O. B. (2018). Environmental impact assessment process for deep-sea mining in 'the Area'. *Marine Policy, 87*, 194–202. https://doi.org/10.1016/j.marpol.2017.10.013.

Fukushima, T. (2007). Amounts of megabenthic organisms in areas of manganese nodules, cobalt-rich crusts and polymetallic sulphides occurrences. In: (Eds. ISA) Polymetallic sulphides and cobalt-rich ferromanganese crust deposits: establishment of environmental baselines and an associated monitoring programme during exploration, Proceedings of the International Seabed Authority's workshop, Kingston, Jamaica, 6–10 September 2004, 356–368.

Gates, A. R., Benfield, M. C., Booth, D. J., Fowler, A. M., Skropeta, D., & Jones, D. O. B. (2017). Deep-sea observations at hydrocarbon drilling locations: Contributions from the SERPENT Project after 120 field visits. *Deep Sea Research, Part II, 137*, 463–479. https://doi.org/10.1016/j.dsr2.2016.07.011.

Gebruk, A. V., Tyler, P. A., & Billett, D. S. M. (1997). Pelagic juveniles of the deep-sea elasipodid holothurians: New records and review. *Ophelia, 46*(2), 153–164.

Gjerde, K. M., Weaver, P. P. E., Billett, D. S. M., Paterson, G. L. J., Colaço, A., Dale, A., Greinert, J., Hauton, C., Jansen, F., Jones, D. O. B., Martinez Arbizu, P., Murphy, K., Sweetman, A. (2016). *Implications of the MIDAS results for policy makers: recommendations for future regulations*. (p. 43). http://www.eu-midas.net/sites/default/files/downloads/MIDAS_recommendations_for_policy_lowres.pdf.

Glover, A. G., Dahlgren, T. G., Wiklund, H., Mohrbeck, I., & Smith, C. R. (2016). An end-to-end DNA taxonomy methodology for benthic biodiversity survey in the Clarion-Clipperton Zone, Central Pacific abyss. *Journal of Marine Science and Engineering, 4*(1), 2. 34 pp. https://doi.org/10.3390/jmse4010002.

Gollner, S., Kaiser, S., Menzel, L., Jones, D. O. B., Brown, A., Mestre, N. C., van Oevelen, D., Menot, L., Colaço, A., Canals, M., Cuvelier, D., Durden, J. M., Gebruk, A., Egho, G. A., Haeckel, M., Marcon, Y., Mevenkamp, L., Morato, T., Pham, C. K., Purser, A., Sanchez-Vidal, A., Vanreusel, A., Vink, A., & Martinez Arbizu, P. (2017). Resilience of benthic deep-sea fauna to mining activities. *Marine Environmental Research, 129*, 76–101.

Gooday, A. J., Goineau, A., & Voltski, I. (2015). Abyssal foraminifera attached to polymetallic nodules from the eastern Clarion Clipperton Fracture Zone: A preliminary description and comparison with North Atlantic dropstone assemblages. *Marine Biodiversity*. https://doi.org/10.1007/s12526-014-0301-9.

Gooday, A. J., Holzmann, M., Caulle, C., Goineau, A., Kamenskaya, O., Weber, A. A.-T., & Pawlowski, J. (2017). Giant protists (xenophyophores, Foraminifera) are exceptionally diverse in parts of the abyssal eastern Pacific licensed for polymetallic nodule exploration. *Biological Conservation, 207*, 106–116.

Goodwin, K. D., Thompson, L. R., Duarte, B., Kahlke, T., Thompson, A. R., Marques, J. C., & Caçador, I. (2017). DNA Sequencing as a tool to monitor marine ecological status. *Frontiers in Marine Science, 4*, 107. https://doi.org/10.3389/fmars.2017.00107.

Grandelli, P., Rocheleau, G., Hamrick, J., Church, M., Powell, D. (2012). *Modeling the physical and biochemical influence of ocean thermal energy conversion plant discharges into their adjacent waters*. (p. 139) United States Department of Energy. Office of Scientific and Technical Information.. doi:https://doi.org/10.2172/1055480.

GSR. (2018). Environmental Impact Statement. Small-scale testing of nodule collector components on the seafloor of the Clarion-Clipperton Fracture Zone and its environmental impact. https://www.isa.org.jm/files/documents/EN/EIA/GSR/GSR-EIS.pdf.

Harden-Davies, H. (2017). Deep-sea genetic resources: New frontiers for science and stewardship in areas beyond national jurisdiction. *Deep Sea Research, Part II, 137*, 504–513.

Hattam, C., Atkins, J. P., Beaumont, N., Börger, T., Böhnke-Henrichs, A., Burdon, D., de Groot, R., Hoefnagel, E., Nunes, P. A. L. D., Piwowarczyk, J., Sastre, S., & Austen, M. C. (2015). Marine ecosystem services: Linking indicators to their classification. *Ecological Indicators, 49*, 61–75.

Hauton, C., Brown, A., Thatje, S., Mestre, N. C., Bebianno, M. J., Martins, I., Bettencourt, R., Canals, M., Sanchez-Vidal, A., Shillito, B., Ravaux, J., Zbinden, M., Duperron, S., Mevenkamp, L., Vanreusel, A., Gambi, C., Dell'Anno, A., Danovaro, R., Gunn, V., & Weaver, P. P. E. (2017).

Identifying toxic impacts of metals potentially released during deep-sea mining – A synthesis of the challenges to quantifying risk. *Frontiers in Marine Science, 4*, 368.
Herring, P. J. (2002). *The biology of the deep ocean* (p. 314). Oxford: Oxford University Press.
Hilario, A., Metaxas, A., Gaudron, S. M., Howell, K. L., Mercier, A., Mestre, N. C., Ross, R. E., Thurnherr, A. M., & Young, C. (2015). Estimating dispersal distance in the deep sea: Challenges and applications to marine reserves. *Frontiers in Marine Science, 2*, 1–14. https://doi.org/10.3389/fmars.2015.00006.
International Marine Mineral Society. (2011). Code for environmental management of marine mining. (p. 15). http://www.immsoc.org/IMMS_code.htm.
International Seabed Authority. (2011a). Environmental management plan for the Clarion-Clipperton Zone. ISBA/17/LTC/7. https://www.isa.org.jm/sites/default/files/files/documents/isba-17ltc-7_0.pdf.
International Seabed Authority. (2011b). Fauna of cobalt-rich ferromanganese crust seamounts. ISA Technical Study No.8.83pp. eBook. ISBN: 978-976-95268-7-7.
International Seabed Authority. (2012). Environmental management needs for exploration and exploitation of deep sea minerals: Report of a workshop held by the International Seabed Authority in collaboration with the Government of Fiji and the SOPAC Division of the Secretariat of the Pacific Community in Nadi, Fiji, from 29 November to 2 December, 2011. ISA Technical Study No 10. 47pp. https://www.isa.org.jm/sites/default/files/files/documents/tstudy10.pdf.
International Seabed Authority. (2013a). Decision of the Council of the International Seabed Authority relating to amendments to the Regulations on Prospecting and Exploration for Polymetallic Nodules in the Area and related matters. Annex. Regulations on Prospecting and Exploration for Polymetallic Nodules in the Area. ISBA/19/C/17. https://www.isa.org.jm/sites/default/files/files/documents/isba-19c-17_0.pdf.
International Seabed Authority. (2013b). Recommendations for the guidance of contractors for the assessment of the possible environmental impacts arising from exploration for marine minerals in the Area. ISBA/LTC/19/8. https://www.isa.org.jm/sites/default/files/files/documents/isba-19ltc-8_0.pdf.
International Seabed Authority. (2013c). Towards the development of a regulatory framework for polymetallic nodule exploitation in The Area. ISA Technical Study No. 11. 83pp. https://www.isa.org.jm/sites/default/files/files/documents/tstudy11.pdf.
International Seabed Authority. (2014). Taxonomic standardisation workshop series. 1. Deep-sea megafauna of the Clarion Clipperton Fracture Zone. ISA Briefing Paper 02/2014. https://www.isa.org.jm/sites/default/files/files/documents/orem-bp2-2014.pdf.
International Seabed Authority. (2015a). Developing a regulatory framework for mineral exploitation in the Area. https://www.isa.org.jm/files/documents/EN/Survey/Report-2015.pdf.
International Seabed Authority. (2015b). Report of the chair of the legal and technical commission on the work of the commission during the twenty-first session of the International Seabed Authority. Annex II. General comments of the Commission on the annual reports of the contractors. ISBA/21/C/16. https://www.isa.org.jm/sites/default/files/files/documents/isba-21c-16_5.pdf.
International Seabed Authority. (2015c). Deep-sea macrofauna of the Clarion-Clipperton Zone (CCZ): Taxonomic standardization workshop, Uljin, South Korea, 23–30 November 2014. ISA Technical Study 13. 41 pp. https://www.isa.org.jm/sites/default/files/files/documents/techstudy13_web_27july.pdf.
International Seabed Authority. (2015d). Workshop on taxonomic methods and standardization of meiofauna in the Clarion Clipperton Zone (December 2015). https://www.isa.org.jm/workshop/workshop-taxonomic-methods-and-standardization-meiofauna-clarion-clipperton-zone-dec-2015.
International Seabed Authority. (2017a). Background note on the legal basis for impact reference zones and preservation reference zones. https://www.isa.org.jm/files/documents/EN/Workshops/2017/IRZ/LegalBasis.pdf.

International Seabed Authority. (2017b). *Environmental assessment and management for exploitation of minerals in the area* (ISA Technical Study No 16). Kingston: International Seabed Authority.

International Seabed Authority. (2017c). Towards an ISA environmental management strategy for the area: Report of an international workshop convened by the German Environment Agency (UBA), the German Federal Institute for Geosciences and Natural Resources (BGR) and the Secretariat of the International Seabed Authority (ISA) in Berlin, Germany, 20–24 March 2017. ISA Technical Study 17. International Seabed Authority, Kingston, Jamaica.

International Seabed Authority. (2018a). Draft regulations on exploitation of mineral resources in the area. ISBA/24/LTC/WP.1. https://www.isa.org.jm/sites/default/files/files/documents/isba24-ltcwp1-en.pdf.

International Seabed Authority. (2018b). Draft environmental impact statement template. ISBA/24/LTC/WP.1/Add.1. https://www.isa.org.jm/sites/default/files/files/documents/isba24-ltcwp1-add1-en.pdf.

International Seabed Authority. (2018c). Design of IRZs and PRZs in deep-sea mining contract areas. Briefing Paper 02/2018. https://www.isa.org.jm/sites/default/files/files/documents/bp02-2018irz-final-18jul.pdf.

International Seabed Authority. (2018d). Preliminary strategy for the development of regional environmental management plans in the Area. ISBA/24/C/3. https://www.isa.org.jm/sites/default/files/files/documents/isba24-c3-e.pdf.

Japan Times. (2017). Japan successfully undertakes large-scale deep-sea mineral extraction. https://www.japantimes.co.jp/news/2017/09/26/national/japan-successfully-undertakes-large-scale-deep-sea-mineral-extraction/#.W3Gr-_ZFxhE.

Jaspers, M., De Pascale, D., Andersen, J. H., Reyes, F., Crawford, A. D., & Ianora, A. (2016). The marine biodiscovery pipeline and ocean medicines of tomorrow. *Journal of the Marine Biological Association of the United Kingdom, 96*, 151–158.

Jones, D. O. B., Alt, C. H. S., Priede, I. G., Reid, W. D. K., Wigham, B. D., Billett, D. S. M., Gebruk, A. V., Rogacheva, A., & Gooday, A. J. (2013). Deep-sea surface-dwelling enteropneusts from the Mid-Atlantic Ridge: Their ecology, distribution and mode of life. *Deep-Sea Research Part II, 98*(B), 374–387.

Jones, D. O. B., Kaiser, S., Sweetman, A. K., Smith, C. R., Menot, L., Vink, A., Trueblood, D., Greinert, J., Billett, D. S. M., Martinez Arbizu, P., Radziejewska, T., Singh, R., Ingole, B., Stratmann, T., Simon-Lledo, E., Durden, J. M., & Clark, M. R. (2017). Biological responses to disturbance from simulated deep-sea polymetallic nodule mining. *PLoS One, 12*(2), e0171750. https://doi.org/10.1371/journal.pone.0171750.

Jones, D. O. B., Amon, D. J., & Chapman, A. S. A. (2018a). Mining deep-ocean mineral deposits: What are the ecological risks? *Elements, 14*(5), 325–330.

Jones, D. O. B., Ardron, J. A., Colaço, A., & Durden, J. M. (2018b). Environmental considerations for impact and preservation reference zones for deep-sea mining. *Marine Policy* https://doi.org/10.1016/j.marpol.2018.10.025.

Jones, D.O.B., Durden, J.M., Murphy, K., Gjerde, K.M., Gebicka, A., Colaço, A., Morato, T., Cuvelier, D.,Billett, D.S.M., (2019). Existing environmental management approaches relevant to deep-sea mining. *Marine Policy* https://doi.org/10.1016/j.marpol.2019.01.006.

Kamenskaya, O. E., Gooday, A. J., Tendal, O. S., & Melnik, V. F. (2015). Xenophyophores (Protista, Foraminifera) from the Clarion-Clipperton Zone with description of three new species. *Marine Biodiversity, 45*, 581–593.

Karstensen, J., Stramma, L., & Visbeck, M. (2008). Oxygen minimum zones in the eastern tropical Atlantic and Pacific oceans. *Progress in Oceanography, 77*, 331–350.

Knight, R. D., & Roberts, S. (2016). Initial results of batch reactor experiments to determine sulphide oxidation rates and trace metal release under seafloor conditions. *Applied Earth Science, 125*, 88–89.

Knight, R. D., Roberts, S., & Cooper, M. J. (2018). Investigating monomineralic and polymineralic reactions during the oxidation of sulphide minerals in seawater: Implications for mining seafloor massive sulphide deposits. *Applied Geochemistry, 90*, 63–74.

Lampitt, R. S., Billett, D. S. M., & Rice, A. L. (1986). The biomass of the invertebrate megabenthos from 500 to 4100m in the North East Atlantic. *Marine Biology, 93*, 69–81.

Le, J. T., Levin, L. A., & Carson, R. T. (2017). Incorporating ecosystem services into environmental management of deep-sea mining. *Deep Sea Research, Part II, 137*, 486–503.

Lodge, M., Johnson, D., Le Gurun, G., Wengler, M., Weaver, P. P. E., & Gunn, V. K. (2014). Seabed mining: International Seabed Authority environmental management plan for the Clarion–Clipperton Zone. A partnership approach. *Marine Policy, 49*, 66–72.

McDonald, T., Gann, G. D., Jonson, J., & Dixon, K. W. (2016). *International standards for the practice of ecological restoration – including principles and key concepts* (p. 48). Washington, DC: Society for Ecological Restoration. https://www.ser.org/page/StandardsDownload.

Mewes, K., Mogollón, J. M., Picard, A., Rühlemann, C., Kuhn, T., Nöthen, K., & Kasten, S. (2014). Impact of depositional and biogeochemical processes on small scale variations in nodule abundance in the Clarion-Clipperton Fracture Zone. *Deep-Sea Research Part I, 91*, 125–141.

Meyer, D. (2016). Glider technology for ocean observations: A review. *Ocean Science Discussions*. https://doi.org/10.5194/os-2016-40.

MIDAS. (2016a). Managing impacts of deep-sea resource exploitation: Research highlights. http://www.eu-midas.net/sites/default/files/downloads/MIDAS_research_highlights_low_res.pdf.

MIDAS. (2016b). Implications of MIDAS results for policy makers: Recommendations for future regulations. http://www.eu-midas.net/sites/default/files/downloads/MIDAS_recommendations_for_policy_lowres.pdf.

Miljutin, D. M., Miljutina, M. A., Martinez Arbizu, P., & Galéron, J. (2011). Deep-sea nematode assemblage has not recovered 26 years after experimental mining of polymetallic nodules (Clarion Clipperton Fracture Zone. Tropical Eastern Pacific). *Deep Sea Research, Part I, 58*, 885–897.

Mitsch, W. J. (2014). When will ecologists learn engineering and engineers learn ecology? *Ecological Engineering, 65*, 9–14.

Mullineaux, L. S. (1987). Organisms living on manganese nodules and crusts: Distribution and abundance at three North Pacific sites. *Deep Sea Research, Part II, 34*, 165–184.

Narita, T., Oshika, J., Okamoto, N., Toyohara, T., & Miwa, T. (2015). Summary of environmental impact assessment for mining seafloor massive sulfides in Japan. *Journal of Shipping and Ocean Engineering, 5*, 103–114.

Niner, H. J., Ardron, J. A., Escobar, E. G., Gianni, M., Jaeckel, A., Jones, D. O. B., Levin, L. A., Smith, C. R., Thiele, T., Turner, P. J., Van Dover, C. L., Watling, L., & Gjerde, K. M. (2018). Deep-sea mining with no net loss of biodiversity – An impossible aim. *Frontiers in Marine Science, 5*, 53. https://doi.org/10.3389/fmars.2018.00053.

Peukert, A., Schoening, T., Alevizos, E., Köser, K., Kwasnitschka, T., & Greinert, J. (2018). Understanding Mn-nodule distribution and related deep-sea mining impacts using AUV-based hydroacoustic sensing and optical observations. *Biogeosciences, 15*, 2525–2549.

Priede, I. G., Bergstad, O. A., Miller, P. I., Vecchione, M., Gebruk, A., Falkenhaug, T., Billett, D. S. M., Craig, J., Dale, A. C., Shields, M. A., Tilstone, G. H., Sutton, T. T., Gooday, A. J., Inall, M. E., Jones, D. O. B., Martinez-Vicente, V., Menezes, G. M., Niedzielski, T., Sigurðsson, Þ., Rothe, N., Rogacheva, A., Alt, C. H. S., Brand, T., Abell, R., Brierley, A. S., Cousins, N. J., Crockard, D., Hoelzel, A. R., Høines, Å., Letessier, T. B., Read, J. F., Shimmield, T., Cox, M. J., Galbraith, J. K., Gordon, J. D. M., Horton, T., Neat, F., & Lorance, P. (2013). Does presence of a Mid Ocean Ridge enhance biomass and biodiversity? *PLoS One, 8*(5), e61550. https://doi.org/10.1371/journal.pone.0061550.

Purser, A., Marcon, Y., Hoving, H.-J. T., Veccione, M., Piatkowski, U., Eason, D., Bluhm, H., & Boetius, A. (2016). Association of deep-sea incirrate octopods with manganese crusts and nodule fields in the Pacific Ocean. *Current Biology, 26*, R1247–R1271. https://doi.org/10.1016/j.cub.2016.10.052.

Racault, M.-F., Sathyendranath, S., Brewin, R. J. W., Raitsos, D. E., Jackson, T., & Platt, T. (2017). Impact of El Nino variability on oceanic phytoplankton. *Frontiers in Marine Science, 4*, 133. https://doi.org/10.3389/fmars.2017.00133.

Robison, B. H., Reisenbichler, K. R., & Sherlock, R. E. (2005). Giant larvacean houses: Rapid carbon transport to the deep sea floor. *Science, 308*(5728), 1609–1611.

Roe, H. S. J., Angel, M. V., Badcock, J., Domanski, P., James, P. T., Pugh, P. R., & Thurston, M. H. (1984). The diel migrations, distributions within a mesopelagic community in the North East Atlantic. 1. Introduction, sampling procedures. *Progress in Oceanography, 13*, 189–199.

Ruhl, H. A. (2007). Abundance and size distribution dynamics of abyssal epibenthic megafauna in the northeast Pacific. *Ecology, 88*(5), 1250–1262. https://doi.org/10.1890/06-0890.

Ruhl, H. A. (2008). Community change in the variable resource habitat of the abyssal northeast Pacific. *Ecology, 89*(4), 991–1000. https://doi.org/10.1890/06-2025.1.

Schlacher, T. A., Baco, A. R., Rowden, A. A., O'Hara, T. D., Clark, M. R., Kelley, C., & Dower, J. F. (2013). Seamount benthos in a cobalt-rich crust region of the central Pacific: Conservation challenges for future seabed mining. *Diversity and Distributions*. https://doi.org/10.1111/ddi.12142.

Schmidt Ocean Institute. (2018). Schmidt Ocean Institute 2018 Expeditions. https://schmidtocean.org/cruises/schmidt-ocean-institute-2018-expeditions/.

Send, U., Regier, L., & Jones, B. (2013). Use of underwater gliders for acoustic data retrieval from subsurface oceanographic instrumentation and bidirectional communication in the deep ocean. *Journal of Atmospheric and Oceanic Technology, 30*, 984–998.

Shillito, B., Gaill, F., & Ravaux, J. (2014). The IPOCAMP pressure incubator for deep-sea fauna. *Journal of Marine Science and Technology, 22*, 97–102.

Simon-Lledó, E., Bett, B. J., Huvenne, V. A. I., Schoening, T., Benoist, N. M. A., Jeffreys, R. M., Durden, J. M., & Jones, D. O. B. (2019). Megafaunal variation in the abyssal landscape of the Clarion Clipperton Zone. *Progress in Oceanography 170*, 119–133.

Simpson, W. R. (1978). A critical review of cadmium in the marine environment. Institute of Oceanographic Sciences Internal Document 44, available online University of Southampton Institutional Repository, National Oceanography Centre, Southampton, UK. https://eprints.soton.ac.uk/id/eprint/392237.

Smith, K. L., Jr., Glatts, R. C., Baldwin, R. J., Ulman, A. H., Horn, R. C., Reimers, C. E., & Beaulieu, S. E. (1997). An autonomous, bottom-transecting vehicle for making long time-series measurements of sediment community oxygen consumption to abyssal depths. *Limnology and Oceanography, 42*(1), 601–1,612. https://doi.org/10.4319/lo.1997.42.7.1601.

Smith, K. L., Jr., Sherman, A. D., McGill, P. R., Henthorn, R. G., Ferreira, J., & Huffard, C. L. (2017). Evolution of monitoring an abyssal time- series station in the northeast Pacific over 28 years. *Oceanography, 30*(4), 72–81. https://doi.org/10.5670/oceanog.2017.425.

Smith, C. R., Levin, L. A., Koslow, A., Tyler, P. A., & Glover, A. G. (2008). The near future of the deep seafloor ecosystems. In N. Polunin (Ed.), *Aquatic ecosystems: Trends and global prospects* (pp. 334–349). Cambridge, UK: Cambridge University Press.

Smith, K. L., Ruhl, H., Bett, B. J., Billett, D. S. M., Lampitt, R. S., & Kaufmann, R. S. (2009). Climate, carbon cycling and deep-ocean ecosystems. *Proceedings of the National Academy of Sciences, 106*(46), 19211–19218.

Snelgrove, P. V. R. (2016). An ocean of discovery: Biodiversity beyond the census of marine life. *Planta Medica, 82*, 790–799. https://doi.org/10.1055/s-0042-103934.

Society for Ecological Restoration International Science & Policy Working Group. (2004). *The SER International Primer on Ecological Restoration*. www.ser.org. Tucson: Society for Ecological Restoration International.

Somero, G. N., Siebenaller, J. F., & Hochachka, P. W. (1983). Biochemical and physiological adaptations of deep-sea animals. In G. T. Rowe (Ed.), *The sea, Vol. 8. Deep-sea biology* (pp. 261–330). New York: Wiley-Interscience.

Stat, M., Huggett, M. J., Bernasconi, R., DiBattista, J. D., Berry, T. E., Newman, S. J., Harvey, E. S., & Bunce, M. (2017). Ecosystem biomonitoring with eDNA: Metabarcoding across the tree of life in a tropical marine environment. *Scientific Reports, 7*, 12240.

Stratmann, T., Lins, L., Purser, A., Marcon, Y., Rodrigues, C. F., Ravara, A., Cunha, M. R., Simon-Lledó, E., Jones, D. O. B., Sweetman, A. K., Köser, K., & van Oevelen, D. (2018a). Abyssal plain fauna carbon flows remain depressed 26 years after a simulated deep-sea mining disturbance. *Biogeosciences, 15*, 4131–4145. https://doi.org/10.5194/bg-15-4131-2018.

Stratmann, T., Voorsmit, I., Gebruk, A., Brown, A., Purser, A., Marcon, Y., Sweetman, A. K., Jones, D. O. B., & van Ouvelen, D. (2018b). Recovery of Holothuroidea population density, community composition and respiration activity after a deep-sea disturbance experiment. *Limnology and Oceanography, 63*, 2140–2153.

Sutton, T. T., Clark, M. R., Dunn, D. C., Halpin, P. N., Rogers, A. D., Guinotte, J., Bograd, S. J., Angel, M. V., Perez, J. A. A., Wishner, K., Haedrich, R. L., Lindsay, D. J., Drazen, J. C., Vereshchaka, A., Piatkowski, U., Morato, T., Błachowiak-Samołyk, K., Robison, B. H., Gjerde, K. M., Pierrot-Bults, A., Bernal, P., Reygondeau, G., & Heinov, M. (2017). A global biogeographic classification of the mesopelagic zone. *Deep Sea Research, Part I, 126*, 85–102.

TEEB. (2010). In P. Kumar (Ed.), *The Economics of Ecosystems and Biodiversity (TEEB). Ecological and economic foundations*. London: Earthscan.

TEEB. (2012). In J. Bishop & W. Evison (Eds.), *The economics of ecosystems and biodiversity in business and enterprise*. London: Earthscan.

Teske, S., Florin, N., Dominish, E., & Giurco, D. (2016) Renewable energy and deep sea mining: Supply, demand and scenarios. Report prepared by Institute for Sustainable Futures (ISF), University of Technology, Sydney, Australia, for J.M.Kaplan Fund, Oceans 5 and Synchronicity Earth, July 2016.

Thiel, H. (2001). The large-scale environmental impact experiment DISCOL – reflection and foresight. *Deep-Sea Research II, 48*, 3869–3882.

Thurber, A. R., Sweetman, A. K., Narayanaswamy, B. E., Jones, D. O. B., Ingels, J., & Hansman, R. L. (2014). Ecosystem function and services provided by the deep sea. *Biogeosciences, 11*, 3941–3963. https://doi.org/10.5194/bg-11-3941-2014.

Thurnherr, A. (2011). Annex II connectivity and dispersal at vents on mid-ocean ridges. In C. L. Van Dover et al. (Eds.), *Environmental management of deep-sea chemosynthetic ecosystems: Justification of and considerations for a spatially-based approach* (ISA Technical Study 9) (p. 90). Kingston: International Seabed Authority.

Van Dover, C. L. (2011). Mining seafloor massive sulphides and biodiversity: What is at risk? *ICES Journal of Marine Science, 68*, 341–348.

Van Dover, C. L., Smith, C. R., Ardron, J., Arnaud-Haond, S., Beaudoin, Y., Bezaury, J., Boland, G., Billett, D., Carr, M., Cherkashov, G., Cook, A., DeLeo, F., Dunn, D., Fisher, C. R., Godet, L., Gjerde, K., Halpin, P., Levin, L., Lodge, M., Menot, L., Miller, K., Milton, D., Naudts, L., Nugent, C., Pendleton, L., Plouviez, S., Rowden, A., Santos, R. S., Shank, T., Smith, S., Tao, C., Tawake, A., Thurnherr, A., & Treude, T. (2011). *Environmental management of deep-sea chemosynthetic ecosystems: Justification of and considerations for a spatially-based approach* (ISA Technical Study 9) (p. 90). Kingston: International Seabed Authority. http://www.isa.org.jm/files/documents/EN/Pubs/TS9/files/index.html.

Van Dover, C. L., Aronson, J., Pendleton, L., Smith, S., Arnaud-Haond, S., Moreno-Mateos, D., Barbier, E., Billett, D. S. M., Bowers, K., Danovaro, R., Edwards, A., Kellert, S., Morato, T., Pollard, E., Rogers, A., & Warner, R. (2014). Ecological restoration in the deep sea: Desiderata. *Marine Policy, 44*, 98–106.

Van Koningsveld, M., ter Hofstede, R., Elzinga, J., Smolders, T., Schutter, M., & Osinga, R. (2017). ReefGuard: A scientific approach to active reef rehabilitation. *Terra et Aqua, 147*, 5–16. https://www.iadc-dredging.com.

Vanreusel, A., Hilario, A., Ribeiro, P. A., Menot, L., & Martinez Arbizu, P. (2016). Threatened by mining, polymetallic nodules are required to preserve abyssal epifauna. *Scientific Reports, 6*, 26808. https://doi.org/10.1038/srep26808.

Weaver, P. P. E., & Billett, D. S. M. (2019). Potential impacts of deep-sea mining on ecosystems. In R. Sharma (Ed.), *Environmental issues of deep-sea mining – impacts, consequences and policy perspectives*. Switzerland: Springer International publishing AG.

Wedding, L. M., Friedlander, A. M., Kittinger, J. N., Watling, L., Gaines, S. D., Bennett, M., Hardy, S. M., & Smith, C. R. (2013). From principles to practice: A spatial approach to systematic conservation planning in the deep sea. *Proceedings of the Royal Society B, 280* 20131684. https://doi.org/10.1098/rspb.2013.1684,

Wei, C-L., Rowe, G.T., Escobar-Briones, E., Boetius, A., Soltwedel, T., Caley, M.J., Soliman, Y., Huettmann, F., Qu, F., Yu, Z., Pitcher, C.R., Haedrich, R.L., Wicksten, M.K., Rex, M.A., Baguley, J.G., Sharma, J., Danovaro, R., MacDonald, I.R., Nunnally, C.C., Deming, J.W., Montagna, P., Lévesque, M., Weslawski, J.M., Wlodarska-Kowalczuk, M., Ingole, B.S., Bett, B.J., Billett D.S.M., Yool, A., Bluhm, B.A., Iken, K., Narayanaswamy, B.E. (2010). Global patterns and predictions of seafloor biomass using Random Forests.PLoSONE5 (12) e15323. https://doi.org/10.1371/journal.pone.0015323

Wynn, R. B., Huvenne, V. A. I., Le Bas, T. P., Murton, B. J., Connelly, D. P., Bett, B. J., Ruhl, H. A., Morris, K. J., Peakall, J. B., Parsons, D. R., Sumner, E. J., Darby, S. E., Dorrell, R. M., & Hunt, J. E. (2014). Autonomous Underwater Vehicles (AUVs): Their past, present and future contributions to the advancement of marine geoscience. *Marine Geology, 352*, 451–468.

Xu, D., Han, L., Li, C., Cao, Q., Zhu, D., Barrett, N. H., Harmody, D., Chen, J., Zhu, H., McCarthy, P. J., Sun, X., & Wang, G. (2018). Bioprospecting deep-sea Actinobacteria for novel anti-infective natural products. *Frontiers in Microbiology, 9*, 787. https://doi.org/10.3389/fmicb.2018.00787.

Yallop, M. L., Paterson, D. M., & Wellsbury, P. (2000). Interrelationships between rates of microbial production, exopolymer production, microbial biomass and sediment stability in biofilms of intertidal sediments. *Microbial Ecology, 39*, 116–127.

Yao, H., Dao, M., Imholt, T., Huang, J., Wheeler, K., Bonilla, A., Suresh, S., & Ortiza, C. (2010). Protection mechanisms of the iron-plated armor of a deep-sea hydrothermal vent gastropod. *PNAS, 107*(3), 987–992.

Zeng, C., Rowden, A. A., Clark, M. R., & Gardner, J. P. A. (2017). Population genetic structure and connectivity of deep-sea stony corals (Order Scleractinia) in the New Zealand region: Implications for the conservation and management of vulnerable marine ecosystems. *Evolutionary Applications, 10*, 1040–1054. https://doi.org/10.1111/eva.12509.

Dr. David Billett is a deep-sea biologist with over 40 years' experience of the deep-sea environment, including mid-ocean ridges, abyssal sediments, seamounts, coral mounds, submarine canyons and continental slopes. He has a particular interest in (1) the potential effects of climate change on deep-sea ecosystems; (2) distinguishing between natural and man-made impacts; (3) the environmental management of offshore deepwater fisheries, oil and gas production and mining; and (4) the spatial and temporal management of man's activities in the deep ocean. David was a member of the ISA Legal and Technical Commission (LTC) between 2007 and 2017 specialising in environmental matters. David is now a Emeritus Research Fellow at the National Oceanography Centre, Southampton, UK, and is coordinating putting industry at the heart of the marine ecosystem restoration agenda as part of a large multinational research project 'Marine Ecosystem Restoration in Changing European Seas' (MERCES) funded by the European Commission (http://www.merces-project.eu/).

Dr. Daniel Jones is a deep-sea biologist working at the National Oceanography Centre in Southampton, UK, where he leads the DeepSeas Group. He specialises in understanding temporal and spatial patterns in deepwater ecosystems, including those impacted by anthropogenic disturbances such as mining. He applies his experience in working in the deep sea to improving environmental management. This includes regular work with the mining and hydrocarbon industries. He has been on research expeditions to the abyssal Pacific, Mid-Atlantic Ridge and many other deepwater areas.

Dr. Phil Weaver is a marine scientist with research background in seabed environments including geological and ecological as well as human aspects. He spent the early part of his career looking at sedimentary processes in the NE Atlantic before broadening his expertise to look at human impacts especially on deep-sea ecosystems. He spent most of his career at the National Oceanography Centre, Southampton and its predecessor the Institute of Oceanographic Sciences, becoming Deputy Director in 2006. In 2010 he left to set up his own company – Seascape Consultants that specialises in environmental marine work. Phil has coordinated a series of EC-funded projects that investigated deep-sea ecosystems from geological and biological perspectives. He coordinated the MIDAS project (Managing Impacts of Deep-sea Resource Exploitation) funded by the European Commission from 2013 to 2016. This project examined the environmental impacts of deep-sea mining. Phil has carried out a number of assignments for the International Seabed Authority.

The Development of Environmental Impact Assessments for Deep-Sea Mining

Malcolm R. Clark

Abstract The chapter provides an account of general Environmental Impact Assessment (EIA) processes and applications, and their role in the developing exploitation of deep-sea mineral resources. It includes aspects such as definitions, the position of EIA as part of a larger process, the structure and content of an EIS and the role of risk assessment in the EIA and considers some of the key elements specific to deep-sea mining that need to be addressed as potential mining progresses from exploration to future exploitation. Elements identified by the ISA, and examples from national assessments, are also reviewed briefly to determine what will be required in future.

Keywords EIA format · EIA process · EIS components · Deep-sea mining

1 Introduction

Environmental Impact Assessment (EIA) is an integral component of the planning, development and management of many human activities. It is a process that is well established in many national jurisdictions and for many industrial developments (Petts 1999; Husky Oil 2001 Glasson et al. 2012). However, the nature and extent of the process and form of environmental management, including EIAs, is not yet fully developed for deep-sea mining (Durden et al. 2017; Ellis et al. 2017).

The requirement for an EIA is embodied in the national legislation of many countries and regions (Government of the USA 1970; Government of Canada 2012; New Zealand Government 2012) and also in the evolving guidelines for the management of seabed mining in "the Area" under the jurisdiction of the International Seabed Authority (ISA). The UN Convention for Law of the Sea (UNCLOS) has general provisions for States to protect and preserve the marine environment from

M. R. Clark (✉)
National Institute of Water & Atmospheric Research, Wellington, New Zealand
e-mail: malcolm.clark@niwa.co.nz

R. Sharma (ed.), *Environmental Issues of Deep-Sea Mining*,
https://doi.org/10.1007/978-3-030-12696-4_16

activities (Articles 145, 192), and more specifically Article 206 states that when planned activities could cause significant and harmful changes, the potential effects of activities should be assessed and reported. Contractors, sponsoring States and other interested States or entities are also required during exploration to cooperate with the ISA in the establishment and implementation of programmes for monitoring and evaluating the impacts of deep seabed mining on the marine environment (ISA 2013a). The need to assess environmental impacts requires the development and implementation of a robust EIA process.

This chapter is structured to describe key elements of EIA, starting with its definition, then moving on to describe the main components of the process and the scope, structure and content of the EIA report. The chapter concludes with addressing some of the main issues that need to be considered for deep-sea mining, utilizing experience from several national EIAs that can be applied in current efforts by, for example, the ISA, to develop and implement best practice in environmental assessment and management.

2 What Is an EIA?

There are many definitions of EIA, but a commonly used one is that of the International Association for Impact Assessment (IAIA) (Senécal et al. 1999) which defines an EIA as "the process of identifying, predicting, evaluating and mitigating the biophysical, social, and other relevant effects of development proposals prior to major decisions being taken and commitments made." Following on from this, the IAIA describes four key objectives of an EIA:

(i) To ensure that environmental considerations are explicitly addressed and incorporated into the development decision-making process
(ii) To anticipate and avoid, minimize or offset the adverse significant biophysical, social and other relevant effects of development proposals
(iii) To protect the productivity and capacity of natural systems and the ecological processes which maintain their functions
(iv) To promote development that generates less destruction and optimizes resource use and management opportunities

EIAs are required for a wide range of activities within a legislative and policy context. While beyond the scope of this chapter, each of these will have reference to the goals of environmental management and the role of EIA within that particular framework, whether national (Wood 2003); regional, e.g. Pacific Islands (Secretariat of the Pacific Community 2012); or international, e.g. Law of the Sea Convention, Convention on Biological Diversity (Jaekel 2017).

3 EIA as a Process

EIA is not a single report but part of a wider process that involves the carrying out of studies to define the existing environment before exploitation occurs and the assessment of impacts and an evaluation of effective mitigation measures.

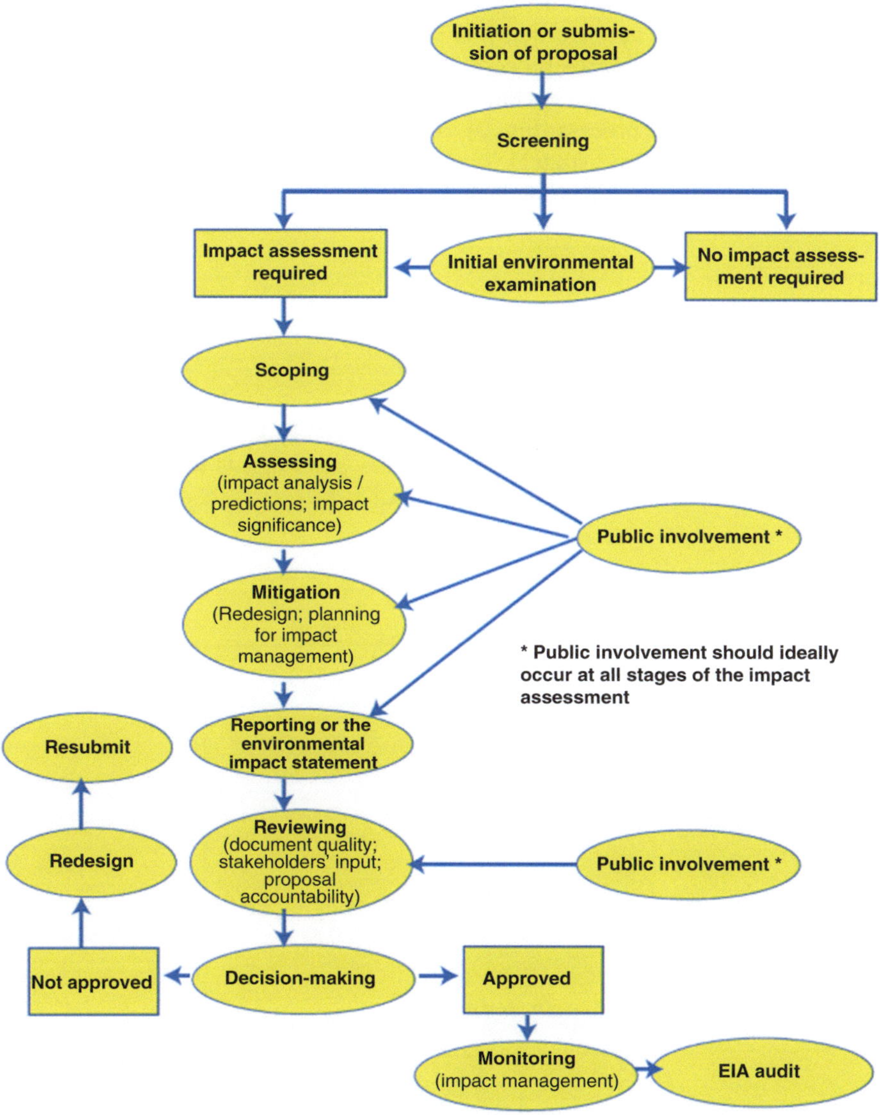

Fig. 1 Example of an EIA process flowchart. (Reproduced with permission from RAMSAR 2010)

There are numerous flowcharts and diagrams outlining variations of this sequence (Collins et al. 2013; ISA 2017; Bradley and Swaddling 2018; Durden et al. 2018). There is nothing exceptional to deep-sea mining in the process as such, but many specific issues come under each step, tailored for the resource and the legislation/policy of the regulatory body. However, a key issue for a developing resource such as deep-sea minerals is the ongoing interaction of policy, review and engagement of interested parties (Fig. 1).

The EIA process is part of a larger picture, and these components are not independent or constant. There are tiers and layers of systems and tools that influence, and are influenced by, the EIA process. Policy initiatives, strategic or regional environmental strategies and assessments and other individual projects all need to be integrated through the process (Fig. 2). An individual project EIA might be nested under multiple higher level strategic or regional plans and can also involve multiple scales of time and space (Glasson et al. 2012).

3.1 *EIA Process Components*

The main components of the process include:

Screening: to determine whether a project requires an EIA, and if so at what level.

Scoping: to identify at an early stage the issues and impacts that are likely to be important and to be emphasized in the EIA.

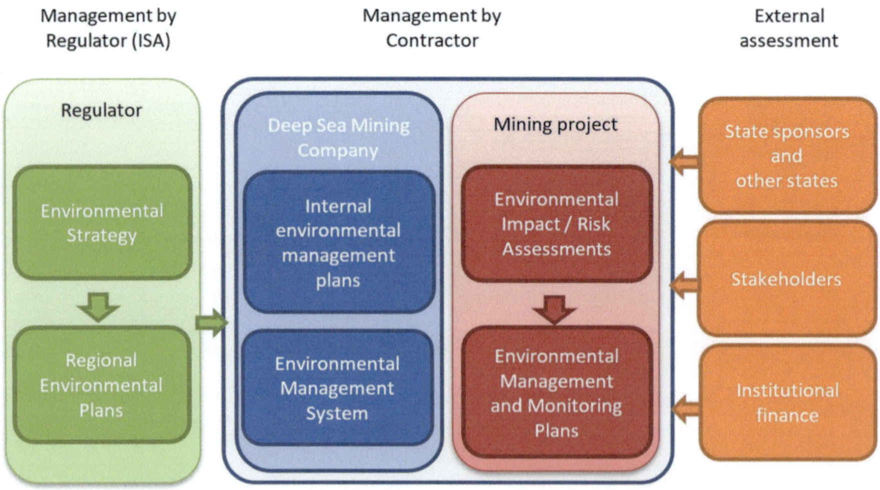

Fig. 2 A schematic of various drivers and plans associated with deep-sea mining, highlighting the interplay between different processes including EIA. (From Jones et al. 2019)

Examination of alternatives: to establish the most feasible and environmentally sound approaches for achieving the project objectives. This includes description of the project methods.

Environmental baseline: to describe the current status of the environment in the absence of the project.

Impact analysis: to identify and predict the likely environmental (and social plus other related) effects.

Mitigation and impact management: to evaluate measures necessary to avoid, minimize or offset predicted adverse effects.

Evaluation of significance: to determine the relative importance and acceptability of residual impacts.

Stakeholder consultation and participation: to incorporate a wide range of views of interested and affected parties and public, to ensure the EIA is comprehensive, complete and has a balance between social and scientific evidence.

Preparation of EIA/EIS reports: to document clearly the impacts of the project, proposed measures for mitigation, significance of effects and concerns raised by consultation (see Box 1 for EIA/EIS).

Review: to determine the content of the EIA/EIS report provides a satisfactory assessment of the project and can contribute to the decision-making process.

Decision-making: to be considered by the relevant authorities and approve/reject and establish terms and conditions for implementation of the project.

Follow-up and audit: to monitor the project and ensure conditions are met, impacts are monitored, and the effectiveness of mitigation and management measures can be assessed.

Box 1 Terminology

The terms EIA (Environmental Impact Assessment) and EIS (Environmental Impact Statement) often cause confusion. Some countries do not use EIS at all and simply have EIA report/s, whereas for others the EIS is the key output of the EIA. At times they are used interchangeably. Confused?

Based on the terms used in NEPA and by the ISA in developing its exploitation regulations, the following explanation may help.

- EIA is the process of undertaking the assessment of environmental impacts, determining the significance of a project's environmental outcomes and looking at alternatives to reduce impact.
- EIS is the formal documentation (i.e. the final EIA study report) which brings together all the parts of the EIA which describe the environmental impacts, and the measures that can be taken to reduce them where possible, and residual impacts that cannot be avoided.

Typically, these are grouped into five "clusters":

1. Screening
2. Scoping
3. EIA (alternatives, baseline, impact analysis, mitigation and management, evaluation of significance, preparation of EIA/EIS report)
4. Decision-making (which includes EIA review)
5. Follow-up and Audit (often part of an Environmental and Monitoring Management Plan)

Preconditions for, and all steps to be taken during, an EIA process prior to permitting mining tests or operations need to be defined at the outset, including the overall scope of the EIA, roles, timelines, scoping procedures, public participation and review, as well as setting performance criteria for environmental reporting and assessment. Funding and institutional mechanisms and procedures need to be clarified in order to ensure an independent EIA.

Importantly, the various stages are a continuum, and they may overlap between regulations. This is an issue for the ISA where there are separate Exploration and Exploitation regulations, yet screening and scoping stages should occur at the start of exploration, with separate EIAs at the stage of test mining/component testing, and then a full EIA is related to the application for an exploitation license.

A review by Sadler (1996), although almost 20 years old, still serves as a good example of some of the good, the bad and the ugly features found in EIAs that should be checked during any proposed EIA process and structure (see Box 2).

Box 2 EIA Process: The Good and the Bad (Adapted from Sadler 1996)
Best-case performance:

- Facilitates informed decision-making by providing clear, well-structured and quantitative analysis of the effects and consequences of proposed actions
- Assists the selection of the best practicable and environmentally friendly techniques and approaches
- Screens out environmentally unsound options and enables a review of feasible options
- Encompasses all relevant issues and factors, including cumulative effects, social issues, etc.
- Directs (but not dictates) evaluation processes and development of terms and conditions on the project
- Uses best available techniques and methods to determine significance of effects
- Includes adaptation and feedback mechanisms to inform future developments

Worst-case performance:

- Is inconsistently applied to, or leaves out, components of the project activities.
- Operates as a "stand-alone" process that is poorly linked to the project cycle as a whole.
- Has weak, or no, follow-up processes for monitoring and assessment.
- Does not consider cumulative effects or social and other risk factors.
- Consultation and engagement (including public) is weak or non-existent.
- EIA reports are voluminous and poorly organized, with inadeqcuate summaries.
- Information is not directed at helping decision-making.
- Understates environmental impacts and mitigation measures are insufficient.

4 The Structure of EIAs

4.1 Scope

The "ideal" EIA process will become incorporated into integrated ecosystem approaches across a range of sustainable development themes which include triple-bottom line, integrated and sustainability assessments (Hacking and Guthrie 2008). However, this starts to become a very complicated matrix of components that is good as a concept and theory but in practice is very difficult for a single project-level EIA to address, let alone in the deep-sea environment with a developing industry. Nevertheless, in recent years there has been a trend to widen the scope of EIAs beyond simple ecological impacts and encompass a wider range of decision-making tools operating at different levels (Jay et al. 2007; Hacking and Guthrie 2008; Morgan 2012). Although this is often more a case of "stretching" an EIA to include other types of assessments than real integration of assessments (Hacking and Guthrie 2008), it is appropriate because an EIA has a role to identify environmental limits and constraints on the project as a whole, not just its impacts on the environment. The trend to broader assessments beyond just biophysical elements is seen, for example, in New Zealand where the expectation is that an "Impact Assessment" will include a "whole of environment" approach (Clark et al. 2017) with a balanced consideration of both biophysical and socio-economic impacts (including cultural). This approach can lead to a triangular "integrated assessment" (biophysical, societal, economic assessments), which includes using EIA to identify environmental limits and constraints on the project rather than just identifying the projects' impacts on the environment (Glasson et al. 2012; Clark et al. 2017). However, the scope of the EIA content needs to be balanced, and there may still be links to more extensive socio-economic and cultural assessments as separate reports.

A number of existing EIA templates for deep-sea minerals mining (Swaddling 2016; Clark et al. 2017; ISA 2017) currently include "social impact assessment" as well as consideration of cultural and economic factors, although these are less developed and extensive than environmental impacts.

In a similar way, linkages between the EIA and Environmental Monitoring and Management Plan (EMMP) vary between countries. Some require this as part of the overall EIA (such as New Zealand which combines EIA and EMMP under an Environmental Assessment), whereas others have it separate. It may be required close to the time of the EIA (e.g. ISA draft regulations) or later (e.g. Papua New Guinea-Nautilus). However, the main issue is that the EIA and EMMP are part of the overall project management system and both together are required in order for managers to evaluate the overall project.

The geographical scope of an EIA needs to comply with the regulations as well as environmental goals of the project. Within national waters, this is generally straightforward as many countries have a single unified management agency. In some cases, however, there may be a need for an EIA to cover several types of legislation when impacts could cross territorial sea-EEZ boundaries, or affect different industries (e.g. mining and fishing) (Clark et al. 2017; Ellis et al. 2017), or multiple EEZs such as when impacts from a mining operation in one jurisdiction could spread into another, and require cooperation between countries (Secretariat of the Pacific Community 2012). EIAs associated with seabed mining in the Area under the jurisdiction of the ISA should not be limited to just the Area, as transport of products or processing will involve other countries. In this case there may need to be several EIAs with a differing emphasis to satisfy the requirements of the country or regulatory body. However, each EIA should cover the full project, so the decision-making processes are linked (ISA 2017).

4.2 EIS Format

EIAs can come in all shapes and sizes, as different contractors, consultants and institutes have their own way of doing things. However, a degree of higher level structural standardization can make the task of contractors and the reviewing regulatory body much easier, because the former know what they need to provide, and the latter know what to expect.

The ISA developed a provisional EIS template for deep-sea mining activities (ISA 2012), partly based on the structure of an Environmental Impact Statement prepared for Seafloor Massive Sulphide (SMS) mining off Papua New Guinea (Coffey Natural Systems/Nautilus Minerals Ltd 2008). The ISA template has since been modified by Swaddling (2016), and it was also used in developing guidelines for EIAs that could bridge the international template and the requirements of the EEZ Act in New Zealand (Clark et al. 2014, 2017). The 2012 ISA version was further modified on the basis of these other initiatives in 2016 (ISA 2017) which formed the basis for the template included in the draft ISA regulations for exploitation (ISA 2018). The key objective of the template is to act as a guide for contractors to achieve consistency in the type of EIA information that would need to be presented. This is also an aid to ensuring that exploration activities can collect the necessary information to carry out the EIA. The high-level heading structure (see Box 3) has many subheadings to help ensure the report covers the key elements that need to be assessed. However, it is important to stress that the EIA process is intended to result in a report that is focused on the significant effects, and these need to be emphasized in the final report over activities and impacts that are of less concern.

Box 3 First Level Headings from the ISA's EIS Template

Table of Contents

Executive summary

1. Introduction
2. Policy, legal and administrative context
3. Description of the proposed development
4. Description of the existing physico-chemical environment
5. Description of the existing biological environment
6. Description of the existing socio-economic environment
7. Assessment of impacts on the physico-chemical environment and proposed mitigation
8. Assessment of impacts on the biological environment and proposed mitigation
9. Assessment of impacts on the socio-economic environment and proposed mitigation
10. Accidental events and natural hazards
11. Environmental management, monitoring and reporting
12. Consultation
13. Product stewardship
14. Glossary and abbreviations
15. Study team
16. References
17. Appendices

The template has deliberately been developed at a high level, so it does not rapidly become complex and complicated with trying to pre-empt the wide range of environmental situations covered by a variety of mineral resource types. In order to check that the content suggested by the headings was appropriate, earlier versions of the template have been reviewed at a workshop involving a range of scientists, managers, industry and environmental groups from Pacific Island countries (Swaddling 2016) and government agencies and industry in New Zealand (Clark et al. 2017). In addition the proposed ISA template structure (ISA 2017) was used by two contractors applying to the ISA for a permit to undertake nodule mining collector trials in the Clarion Clipperton Zone planned for 2019. The EIAs were submitted in 2018, and from both contractor and reviewer comments, the template headings and subheadings were found to be useful in guiding the nature and extent of data and analyses required for the task.

The effectiveness of EIAs has been found to be limited when they have too much focus on overly descriptive baseline work and not enough emphasis on key impacts of the activity (Glasson et al. 2012). Often it is difficult to wade through long-winded accounts of the environment that lack a good interpretation of the relevance to impacts from the proposed activities. In the development of EIA/EIS templates to date, there is a consistent approach that key impacts from offshore mining activities should be structured by "receptor" and depth range to enable an understanding of the source and nature of impacts caused by the various components of the operation at the surface or seafloor and help to focus the EIA on key impacts and potential mitigation measures.

The main principles and criteria of an EIA are often not fulfilled, and it is relatively straightforward for an applicant to keep in mind the basics of environmental assessment and check points off as they review their application. An EIA/EIS should be (after Senécal et al. (1999):

- Purposive: be informative for decision-making
- Rigorous: apply best practicable science
- Practical: result in useful information and outputs
- Relevant: provide useable information
- Cost-effective: achieve EIA objectives within acceptable resource and time limits
- Efficient: process should minimize cost burdens
- Focused: concentrate on significant issues
- Adaptive: adjustable to the specific situation but not compromise the process
- Participative: inform and involve interested and affected parties
- Interdisciplinary: involve multiple techniques and experts across a range of fields
- Credible: a professional process, subject to independent checks/verification
- Integrated: interrelationships of social, economic and biophysical aspects
- Transparent: an open and informative process
- Systematic: consider all relevant information and options

4.3 EIA Guidelines

Given the variability in environmental characteristics between deep-sea mineral resource types and locations, there is a careful balance required between EIA guidelines being highly prescriptive (which may not fit certain situations) and being too general (where adequate standards aren't clear). Countries often produce guidance documents to help companies develop their EIAs (US Department of Energy 2004). Clark et al. (2017) provide more explanation on what should be included under the ISA template headings, provide links to several relevant data and information sources and advice on a number of general issues across EIAs (such as uncertainty, cumulative impacts, adaptive management). It has a focus on New Zealand legislative requirements but is intended also to apply more internationally. However, EIAs are not "one size fits all", and every situation is going to be different in the studies, data and information required to compile a robust assessment. This will be a challenge as the ISA, in particular, plans to develop standards and guidance for contractors as part of the proposed exploitation regulations (ISA 2018). However, such outputs will be a substantial step forward in trying to define and develop best environmental practices.

5 Future Challenges for Deep-Sea Minerals Mining EIA

There is a growing literature on potential impacts of deep-sea mining (Boschen et al. 2013; Gollner et al. 2017; Miller et al. 2018), and together with experience from several seabed mining license application EIAs (Coffey Natural Systems/ Nautilus Minerals Ltd 2008; Chatham Rock Phosphate 2014), some elements that are needed to improve EIA or their uptake in the context of the developing ISA exploitation regulations have been identified (Lallier and Maes 2016; Clark 2017; Durden et al. 2018; Clark et al. 2018). A number of these are in progress within countries or the ISA, especially those of process, and so these are not covered below as any comments could rapidly be outdated. However, issues regarding the technical content of the EIA are longer-term, and some of these are briefly outlined.

5.1 Data Shortcomings

The deep sea is a remote, difficult, dynamic and expensive environment to sample (Clark et al. 2016). Given the complexity of the environment, and a poor understanding of spatial and temporal scales of faunal organization and environmental drivers (Ramirez-Llodra et al. 2010), it is likely that there will always be a constrained understanding of ecosystem structure and function. Even in coastal or terrestrial situations, many EIAs suffer from incomplete data or information that is inadequate to fully assess the impacts of activities on one or more receptors. A long list is possible of issues that EIAs can have in the deep sea, and in practical terms, there may be little one can do about some of them. Scientists and managers have to be prepared to work with a data-limited situation. However, common and avoidable problems relate to:

- Lack of standardization of data or sampling procedures
- Poor integration of all available data
- No assessment of what is an adequate baseline dataset
- Inadequate baseline survey design (often not enough thought)
- Insufficient regional setting for studies done at a smaller-scale site of interest
- Insufficient assessment of potential cumulative impacts
- Limited expression or acknowledgement of uncertainty

5.2 *Adequate Environmental Baseline Information*

Baseline data collection and short-term monitoring studies are important aspects of exploration activities, as they underpin the preparation of an EIA, prior to any application for a full mining permit. It is expected that some information will be available from an area before any exploration occurs, and desktop studies will form the basis of initial screening and scoping of the project. However, available data will invariably be inadequate to describe and characterize the receiving environment of any likely mining site. Hence baseline surveys and targeted scientific studies will be needed to provide information on the pre-mining state of the environment, as well as some monitoring of conditions over time to understand temporal variability of key environmental factors. Such studies will need to cover a wide range of research aspects and be carried out using current "best practice" approaches and methods. The ISA has published two reports that describe and give some advice on the sorts of studies, type of data and nature of sampling required for both baseline measurements and ongoing monitoring. These cover manganese nodule (ISA 1999), seafloor massive sulphide and cobalt-rich crust resources (ISA 2007), with an additional report on sampling standardization (ISA 2002). Many of these recommendations have subsequently been amalgamated (ISA 2013b). Protocols and standards for baseline studies specific to deep-sea mining have also been reviewed as part of the European MIDAS project (Billett et al. 2015) and a further review and recommendations done in the Southwest Pacific (Swaddling et al. 2016). ISA recommendations for assessment of possible environmental impacts (ISA 2013b) are currently being reviewed and updated by the ISA's Legal and Technical Commission.

Every location, resource type and habitat can have different characteristics. Hence baseline studies to support EIAs have to be flexible to ensure they are fit for purpose. However, in collecting data during exploration to support the baseline definition, and subsequent monitoring, there should be a level of consistency so that core deep-sea ecological information demands are met, and these are comparable and can be combined between contractors to form a regional picture. The key aspects include:

- What parameters should be measured from the outset, and how.
- What is measured to acceptable standards (accuracy and precision).
- What are the key ecological indicators that need to be assessed in transitioning from baseline data to measuring/monitoring future changes under the EMMP.
- What level of change might be acceptable in terms of mitigation (against generic ecological limits and thresholds, not management targets).

There are two "categories" of impacts that assessment and monitoring must consider:

- Operational: including mining or drilling location and rate, volumes discharged, hazardous discharge events and quantity of mined material removed
- Effects: including physical (e.g. water column turbidity, sediment deposition on seafloor), chemistry (e.g. analytical suite of contaminants, zone of initial mixing for guideline comparisons, relevant water and sediment quality guidelines,

bioaccumulation assessment) and biological monitoring (e.g. sentinel species, survey approaches)

The second, which is focused on environmental aspects, can build on available environmental assessment guidance (ISA 2013b) and more recent work. The main aspects to be included and parameters to be measured, for both baseline and monitoring survey programmes, are described in guidelines on scientific research developed by the SPC-EU DSM project and NIWA (Swaddling et al. 2016). This report covers survey design, sampling equipment and "best scientific practices" for deep-sea sampling relevant to marine minerals. A summary table from that report on recommended scientific studies, their rationale and methods is reproduced as Appendix 1, which provides a starting point for determining the studies that need to be conducted for assessing and monitoring impacts and environmental changes.

However, while the "why" and "how" are tractable, the "what" is more problematic. Science needs to address questions such as what are the key indicators of system health; can we measure with enough precision to detect change; what is an acceptable level of change; how can we address ecosystem-level responses; etc. These are simple questions to ask but very difficult to answer. Some progress is being made (Levin et al. 2016; Le et al. 2017), but it will be an important part of the EIA and EMMP process to have robust feedback mechanisms and a flexible approach (e.g. adaptive management) as part of the system to ensure that we learn from the initial mining activities to improve future assessment and management and avoid unnecessary risk. Knowledge uncertainty will need to flow through into measures to be adopted in the environmental management process (see below).

The role of test mining under the ISA exploration regulations is also uncertain in the EIA process. However, such testing could be an important element of the transition from exploration to exploitation, as without it there can be only limited understanding of the likely nature and extent of impacts because the spatial scale of most potential mining operations is very large relative to exploration. Such scaled-up trials will not provide all the solutions to address the long-term sustainability of deep-sea ecosystems in the mined region but will start to provide a more realistic and scale-relevant assessment of impacts.

5.3 *Uncertainty*

There are numerous technical/scientific challenges for any deep-sea EIA. The underlying cause of this difficulty is the nature of the environment. The deep sea is difficult to access because of its remoteness and depth. Ecosystems are open, both horizontally and vertically, and community definition and boundaries are difficult to establish. Many ecological processes are slow, and hence natural variability can take a long time to measure. Given these, and other issues, data will invariably be limiting for an EIA in comparison with terrestrial or many coastal marine environments. This makes the issue of expressing uncertainty and confidence particularly important.

As there are many kinds of uncertainties, it can be helpful to define different sources of these uncertainties so that they can be better understood and managed. These can be defined in several groups (Clark et al. 2017):

- Knowledge uncertainty arises where there is incomplete understanding of processes, interactions or system behaviours.
- Unpredictability arises from chaotic (often random) components of complex systems or of human behaviour.
- Structural uncertainty arises from inadequate models, ambiguous system boundaries or over simplification or omission of processes from models.
- Value uncertainty arises from missing or inaccurate data, inappropriate spatial or temporal resolution or poorly known model parameters.
- Uncertain interpretations arise when values or terms are interpreted differently by different user groups.

These types of uncertainties may all be relevant to assessing what effect an impact may have. Many of the techniques used in preparing an EIA, such as models used to make predictions, will have the potential for associated uncertainty, as do monitoring programme measurements and information due to precision of instruments, or if new and less proven technologies are used.

Rouse and Norton (2010) proposed managing scientific uncertainty by making three steps very clear:

1. Identify sources of uncertainty.
2. Reduce uncertainty where possible.
3. Acknowledge and manage the residual (unavoidable) uncertainty.

Scientists conduct the first two steps in the context of their disciplines using a variety of statistical or probabilistic techniques. However, for an EIA as equally important is acknowledging, stating and quantifying residual uncertainty as much as possible, as that will be a key consideration for regulators and managers.

5.4 Cumulative Impacts

Policy and regulatory requirements in many countries include that EIAs identify, analyse and evaluate cumulative effects. However, although they have long been recognised as an important component of EIAs, they are poorly assessed as a rule (Burris and Canter 1997). There are many stressors caused by anthropogenic activities that can affect the marine environment in a number of ways, and there is a large body of literature dealing with this field of research (Glasson et al. 2012; Solan and Whiteley 2016). Results of numerous studies indicate that interactions between stressors can be variable and hard to predict (Crain et al. 2008; Darling and Cote 2008). Nevertheless, cumulative effects should be explored as much as possible given available data and considered early on during the exploration phase so appropriate information can be collected.

The assessment of cumulative impacts needs to consider three key elements:

1. Multiple sources of impact (either different types of mining operation or different sectors)
2. Additive or interactive processes (repetition leading to accumulation of impacts)
3. Different types of cumulative effects (e.g. direct physical, indirect, natural)

There is now increasing guidance on what to describe and evaluate (Smit and Spaling 1995; Crain et al. 2008; Ban et al. 2010). The potential for cumulative impacts needs to be considered at the start of exploration, so the baseline studies can contribute to this aspect in a meaningful way as part of the EIA.

6 The Role of Ecological Risk Assessment (ERA)

An important component of the EIA process is to ensure the EIA focuses on the main sources of impact and does not spend undue time on elements of little risk. There are many approaches and methods to ERA (IEC-ISO 2009), and this ISO standard provides a useful resource to plan for a realistic and practical two-stage risk assessment process for deep-sea mining:

1. At the beginning of the exploration phase, given the often-limited amount of information available, a qualitative (Level 1) assessment should be carried out. This type of assessment commonly uses an expert panel to consider the likelihood of an impact occurring and the consequences if it does (Fletcher 2005; MacDiarmid et al. 2012). The results of this risk assessment should guide data collection during exploration activities. As part of the EIA process, a level 1 risk assessment identifies the main issues as part of the Screening and Scoping stages.
2. A more quantitative assessment will be required before progressing to a mining license application stage. A semi-quantitative level 2 assessment (Hobday et al. 2011) at the stage of preparing the EIA applies a more rigorous evaluation with data collected through the exploration phase on those aspects identified as high risk. This upgraded ERA is an important check to ensure the EIA is focused on the activities and aspects of impact that pose most risk to the environment.

Other options that embed a risk assessment of some form include environmental assessment scoping reports or environmental hazard and impact identifications (ENVID) to identify both accidental events and planned operational procedures related to a mining operation that can impact the environment (Recommended Practice DNVGL-RP-O601, 2016).

7 Concluding Remarks

The importance of environmental assessment is well recognised with any project development, but in particular for the deep sea given the increasing interest in the potential use of seabed minerals for future technologies, and our poor understanding of ecosystem structure and function. The process of EIA is established, and there is a lot that can be used from existing terrestrial and coastal systems and applications. There is, without doubt, more development required, especially by the ISA in the Area, but there is no need to reinvent the wheel and the process, and administrative issues are likely to be tractable in the short-medium term. However, technical and scientific issues are potentially more challenging as they face limited baseline knowledge on the environment and currently a poor understanding of the spatial and temporal scale of potential impacts. Nevertheless, considerable scientific resources are being applied to deep-sea science and evaluating human impacts, which will help fill some of the existing knowledge gaps. Integrated into a robust EIA process, we can look forward to ongoing and productive development of a balanced environmental management system.

Acknowledgements I am grateful to many colleagues working in the area of deep-sea science and management for wide-ranging discussions in recent years on the needs of environmental assessment for deep-sea mining.

Appendix 1: Summary Table of Recommended Scientific Methodologies, Including the Aspects to Be Covered During the Survey Programme, Parameters to Be Measured and Appropriate Methods to Consider Given the Local Environmental Conditions (From Swaddling et al. 2016)

	Aspect	Reason	Main parameters	Sampling
Geology	Topography	Seabed characteristics, classification of habitats for assessment, survey stratification, selection of test and control areas	Bathymetry, morphometry, seafloor type	Shipboard/towed acoustic systems, optical sensors, dredges, box corer, drilling equipment
	Backscatter	Seabed characteristics, classification of habitats for assessment, survey stratification, selection of test and control areas	Acoustic reflectivity	Shipboard/towed acoustic systems; side-scan sonar, hyperspectral imaging
	Sub-seafloor	Petrology, geochemistry and mineralogy for resource characterization	Penetration layers, rock properties, mineral and chemical composition	Seismic, drilling, rock sampling (dredges, coring)
Sediment characteristics	Sediment properties	Sediment plume dynamics, classification of habitats	Substrate type, sediment and pore water measurements: water content, grain size, specific gravity, porosity, depth of toxic layer, carbon content, chemical composition (trace and heavy metals)	Sediment cores (box corer or mulitcorer)
	Bioturbation rates	Natural mixing of sediments	Bioturbation depth, faunal zonation, Pb210 activity	Sediment cores (box corer or mulitcorer)
	Sedimentation rates	Distribution and concentration of natural suspension, settlement rates	Particle flux, suspended particle concentrations, settlement rates	Moorings and sediment traps
Pelagic community	Deepwater pelagic (plankton and nekton)	Impacts of sediment plume and discharges on midwater communities, vertical migrators, and near-bottom hyper-benthos	Species composition, distribution, abundance. Biological characteristics (sensitivity, recoverability parameters)	Opening/closing nets for plankton (remotely operated vehicle (ROV) also possible). Pelagic trawls/commercial records for fish
	Surface fauna	Effects of surface discharges, presence of vessels and equipment	Species composition, distribution, abundance. Biological characteristics (sensitivity, recoverability parameters)	Opening/closing nets, surface plankton nets, remote-sensed data
	Marine mammals/ sea birds	Effects of surface discharges, presence of vessels and equipment	Species composition, distribution, abundance. Biological characteristics (sensitivity, recoverability parameters)	Marine mammal observer protocols

(continued)

(continued)

	Aspect	Reason	Main parameters	Sampling
Seafloor community	Megafauna	Impacts on benthic communities	Species composition, distribution, abundance. Biological characteristics (sensitivity, recoverability parameters)	Photographic surveys from ROV/towed camera; direct sampling from dredge/sled/trawl/ROV
	Macrofauna	Impacts on benthic communities	Species composition, distribution, abundance. Biological characteristics (sensitivity, recoverability parameters)	Mulitcorer or box corer, and epibenthic sled; photographic surveys from ROV/towed camera; direct sampling from dredge/sled/trawl/ROV
	Meiofauna	Impacts on benthic communities	Biodiversity, distribution, abundance	Multicorer or box corer; direct sampling from dredge/sled/trawl/ROV
	Microfauna	Impacts on benthic communities	Biodiversity, distribution, abundance	Sediment cores (box corer or mulitcorer)
	Specific resource fauna	Endemic species or communities, sensitive habitats (including biogenic habitats)	Species composition, distribution, abundance	ROV/towed camera, epibenthic sled; direct sampling by ROV, box corer for nodule environments
	Scavenger/demersal fish	Impacts on benthic communities	Species composition, distribution, abundance	Baited lander, fish trawls, traps, ROV observations
	Ecotoxicity	Impacts of heavy metals/contaminants on benthic communities, accumulation through food chain potential	Tissue samples from representative and abundant fauna	Various direct sampling methods (as above)

Physical oceanography	Currents	Dispersal of impacts, biological connectivity	Current speed, direction, depth variation, tidal dynamics, sea surface temperature (SST), sea surface height (SSH), ocean colour	Conductivity temperature depth profiler (CTD), current meters, acoustic Doppler current profiler (ADCP), remote-sensed data, profiling moorings
	Hydrodynamic modelling	Dispersal of impacts, sediment plume dynamics, biological connectivity	Oceanographic parameters (temperature, salinity, current flow and direction), turbulence, turbidity, bathymetry	Various models applicable, e.g. Regional Ocean Modelling System (ROMS), Hybrid Coordinate Ocean Model (HYCOM), CORMIX (discharges)
Chemical oceanography	Water quality	Effects of discharges, sediment plume	Chemical composition (including heavy metals and toxic contaminants), turbidity, suspended sediment, dissolved oxygen, pH	Water samples (from CTD), surface remote-sensed data, core samples, nephelometer, transmissometer, optical backscatter sensors
	Visual characteristics	Effects of discharges, sediment plume	Optical backscatter, light attenuation, black disc distance	Transmissometer, optical backscatter sensors, remote sensing
	Bottom water chemistry	Effects of sediment/rock disturbance, release of chemicals, effluent discharge	Elutriation for chemical and toxicity testing, pH, trace and heavy metal concentrations, dissolved oxygen	Water samplers (CTD-Niskin bottles), core samples
	Water column chemistry	Effects on chemical characteristics due to sediment plume and discharges	Nutrients (P, N, Si, C), dissolved oxygen, trace and heavy metal concentrations	Water samplers (CTD-Niskin bottles)

References

Ban, N. C., Alidina, H. M., & Ardron, J. A. (2010). Cumulative impact mapping: Advances, relevance and limitations to marine management and conservation, using Canada's Pacific waters as a case study. *Marine Policy, 34*, 876–886.

Billett, D., Jones, D., Murphy, K., Gjerde, K., Calaco, A., Morato, T., Cuvelier, D., Vercruijsse, P., Rolin, J. F., & Ortega, A. (2015). Review of existing protocols and standards applicable to the exploitation of deep-sea mineral resources. Deliverable 8.2 of MIDS project.

Boschen, R. E., Rowden, A. A., Clark, M. R., & Gardner, J. P. A. (2013). Mining of deep-sea seafloor massive sulfides: A review of the deposits, their benthic communities, impacts from mining, regulatory frameworks and management strategies. *Ocean & Coastal Management, 84*, 54–67.

Bradley, M., & Swaddling, A. (2018). Addressing environmental impact assessment challenges in Pacific island countries for effective management of deep-sea minerals activities. *Marine Policy, 95*, 356–362.

Burris, R. K., & Canter, L. W. (1997). Cumulative impacts are not properly addressed in environmental assessments. *Environmental Impact Assessment Review, 17*, 5–18.

Chatham Rock Phosphate. (2014). Marine consent application and environmental impact assessment. https://www.epa.govt.nz/assets/FileAPI/proposal/EEZ000006/Applicants-proposal-documents/EEZ000006-CRPMarine-Consent-Application-EIA.pdf.

Clark, M. R. (2017). Project-level environmental impact assessment. Towards an ISA environmental management strategy for the area: Report of an international workshop convened by the German Environmental Agency (UBA), the German Federal Institute for Geosciences and Natural Resources (BGR), and the Secretariat of the International Seabed Authority (ISA) in Berlin, Germany, 20–24 March 2017. *ISA Technical Study No. 17*: 54–57.

Clark, M. R., Rouse, H. L., Lamarche, G., Ellis, J. I., & Hickey, C. (2014). Preparing environmental impact assessments: Provisional guidelines for offshore mining and drilling in New Zealand. NIWA Client Report WLG2014-67. 86 p.

Clark, M. R., Consalvey, M., & Rowden, A. A. (2016). *Biological sampling in the deep sea*. Oxford: Wiley Blackwell.

Clark, M. R., Rouse, H. L., Lamarche, G., Ellis, J. I., & Hickey, C. W. (2017). Preparation of environmental impact assessments: General guidelines for offshore mining and drilling with particular reference to New Zealand. *NIWA Science and Technology Series No. 81*. 106p.

Clark, M. R., Durden, J. M., & Christiansen, S. (2018). Environmental Impact Assessments for deep-sea mining: can we improve their future effectiveness? *Marine Policy* (in press) https://doi.org/10.1016/j.marpol.2018.11.026.

Coffey Natural Systems/Nautilus Minerals Ltd. (2008). Environmental impact statement. Solwara 1 project. Volume A. Main report.

Collins, P. C., Croot, P., Carlsson, J., Colaco, A., Grehan, A., Hyeong, K., Kennedy, R., Mohn, C., Smith, S., Yamamoto, H., & Rowden, A. A. (2013). A primer for the environmental impact assessment of mining at seafloor massive sulfide deposits. *Marine Policy, 42*, 198–209.

Crain, C. M., Kroeker, K., & Halpern, B. S. (2008). Interactive and cumulative effects of multiple human stressors in marine systems. *Ecology Letters, 11*, 1304–1315.

Darling, E. S., & Cote, I. M. (2008). Quantifying the evidence for ecological synergies. *Ecology Letters, 11*, 1278–1286.

Durden, J. M., Murphy, K., Jaeckel, A., Van Dover, C. L., Christiansen, S., Gjerde, K., Ortega, A., & Jones, D. O. B. (2017). A procedural framework for robust environmental management of deep-sea mining projects using a conceptual model. *Marine Policy, 84*, 193–201.

Durden, J. M., Lallier, L. E., Murphy, K., Jaeckel, A., Gjerde, K., & Jones, D. O. B. (2018). Environmental impact assessment process for deep-sea mining in 'the area'. *Marine Policy, 87*, 194–202.

Ellis, J. I., Clark, M. R., Rouse, H. L., & Lamarche, G. (2017). Environmental management frameworks for offshore mining: Insights from experience in New Zealand's exclusive economic zone. *Marine Policy, 84*, 178–192.

Fletcher, W. J. (2005). The application of qualitative risk assessment methodology to prioritize issues for fisheries management. *ICES Journal of Marine Science, 62*, 1576–1587.

Glasson, J., Therivel, R., & Chadwick, A. (2012). *Introduction to environmental impact assessment*. London: UCL Press Ltd, University College.

Gollner, S., Kaiser, S., Menzel, L., Jones, D. O. B., Brown, A., Mestre, N. C., van Oevelen, D., Menot, L., Colaco, A., Canals, M., Cuvelier, D., Durden, J. M., Gebruk, A., Egho, G. A., Haeckel, M., Marcon, Y., Mevenkamp, L., Morato, T., Pham, C. K., Purser, A., Sanchez-Vidal, A., Vanreusel, A., Vink, A., & Arbizu, P. M. (2017). Resilience of benthic deep-sea fauna to mining activities. *Marine Environmental Research, 129*, 76–101.

Government of Canada. (2012). Canadian Environmental Assessment Act.

Government of the USA. (1970). National Environmental Policy Act of 1969.

Hacking, T., & Guthrie, P. (2008). A framework for clarifying the meaning of triple bottom-line, integrated, and sustainability assessment. *Environmental Impact Assessment Review, 28*, 73–89.

Hobday, A. J., Smith, A. D. M., Stobutzki, I. C., Bulman, C., Daley, R., Dambacher, J. M., Deng, R. A., Dowdney, J., Fuller, M., Furlani, D., Griffiths, S. P., Johnson, D., Kenyon, R., Knuckey, I. A., Ling, S. D., Pitcher, R., Sainsbury, K. J., Sporcic, M., Smith, T., Turnbull, C., Walker, T. I., Wayte, S. E., Webb, H., Williams, A., Wise, B. S., & Zhou, S. (2011). Ecological risk assessment for the effects of fishing. *Fisheries Research, 108*, 372–384.

Husky Oil. (2001). *White Rose Oilfield comprehensive study report*. St. John's: Husky Oil Operations Ltd.

IEC-ISO. (2009). International standard IEC/ISO 31010. Risk management-risk assessment techniques. IEC, Geneva, Switzerland.

ISA. (1999). Deep-seabed polymetallic nodule exploration: Development of environmental guidelines. ISA, Jamaica. https://www.isa.org.jm/node/246.

ISA. (2002). Standardization of environmental data and information-development of guidelines. In *Proceedings of the ISA's workshop, Kingston, Jamaica, 25–29 June 2001. ISA, Jamaica.* https://www.isa.org.jm/documents/standardization-environmental-data-and-informationdevelopment-guidelines.

ISA. (2007). *Polymetallic sulphides and cobalt-rich ferromanganese crust deposits: Establishment of environmental baselines and an associated monitoring programme during exploration*. Kingston: International Seabed Authority, ISA, Jamaica. (https://www.isa.org.jm/documents/polymetallic-sulphides-and-cobalt-richferromanganese-crusts-deposits-establishment).

ISA. (2012). *Regulations on prospecting and exploration for cobalt-rich ferromanganese crusts in the area*. Kingston: International Seabed Authority. ISA Jamaica (ISBA/18/A/11).

ISA. (2013a). Decision of the Council of the International Seabed Authority relating to amendments to the regulations on prospecting and exploration for polymetallic nodules in the area and related matters. ISA Jamaica (ISBA/19/C/17).

ISA. (2013b). Recommendations for the guidance of contractors for the assessment of the possible environmental impacts arising from exploration for marine minerals in the area. ISA, Jamaica (ISBA/19/LTC/8).

ISA. (2017). Environmental assessment and management for exploitation of minerals in the area. *ISA Technical Study No 16*:66. ISA Jamaica. https://www.isa.org.jm/document/environmental-assessment-and-managementexploitation-minerals-area.

ISA. (2018). Draft regulations on exploitation of mineral resources in the area. ISA Jamaica. https://www.isa.org.jm/legal-instruments/ongoing-development-regulations-exploitationmineral-resources-area.

Jaekel, A. L. (2017). The international seabed authority and the precautionary principle: Balancing deep seabed mineral mining and marine environmental protection. *Publications on Ocean Development, 83,* 382p BRILL, Leden, Netherlands.

Jay, S., Jones, C., Slinn, P., & Wood, C. (2007). Environmental impact assessment: Retrospect and prospect. *Environmental Impact Assessment Review, 27*, 287–300.

Jones, D. O. B., Durden, J. Murphy, K., Gjerde, K. M., Gebicka, A., Colaço, A., Morato, T., Cuvelier, D., & Billett, D. S. M. (2019). Existing environmental management approaches relevant to deep-sea mining. *Marine Policy 103*, 172–181.

Lallier, L. E., & Maes, F. (2016). Environmental impact assessment procedure for deep seabed mining in the area: Independent expert review and public participation. *Marine Policy, 70*, 212–219.

Le, J. T., Levin, L. A., & Carson, R. T. (2017). Incorporating ecosystem services into environmental management of deep-seabed mining. *Deep-Sea Research Part Ii-Topical Studies in Oceanography, 137*, 486–503.

Levin, L. A., Mengerink, K., Gjerde, K. M., Rowden, A. A., Van Dover, C. L., Clark, M. R., Ramirez-Llodra, E., Currie, B., Smith, C. R., Sato, K. N., Gallo, N., Sweetman, A. K., Lily, H., Armstrong, C. W., & Brider, J. (2016). Defining "serious harm" to the marine environment in the context of deep-seabed mining. *Marine Policy, 74*, 245–259.

MacDiarmid, A., Beaumont, J., Bostock, H., Bowden, D., Clark, M., Hadfield, M., Heath, P., Lamarche, G., Nodder, S., Orpin, A., Stevens, C., Thompson, D., Torres, L., & Wysoczanski, R. (2012). Expert risk assessment of activities in the New Zealand exclusive economic zone and extended continental shelf. NIWA Client Report 2011/39. 106 p.

Miller, K. A., Thompson, K. F., Johnston, P., & Santillo, D. (2018). An overview of seabed mining including the current state of development, environmental impacts, and knowledge gaps. *Frontiers in Marine Science, 4*, 418.

Morgan, R. K. (2012). Environmental impact assessment: The state of the art. *Impact Assessment and Project Appraisal, 30*, 5–14.

New Zealand Government. (2012). Exclusive economic zone and continental shelf (environmental effects) act 2012.

Petts, J. (1999). *Handbook of environmental impact assessment*. Oxford, UK: Wiley-Blackwell.

Ramirez-Llodra, E., Brandt, A., Danovaro, R., De Mol, B., Escobar, E., German, C. R., Levin, L. A., Arbizu, P. M., Menot, L., Buhl-Mortensen, P., Narayanaswamy, B. E., Smith, C. R., Tittensor, D. P., Tyler, P. A., Vanreusel, A., & Vecchione, M. (2010). Deep, diverse and definitely different: Unique attributes of the world's largest ecosystem. *Biogeosciences, 7*, 2851–2899.

Ramsar. (2010). Impact assessment: Guidelines on biodiversity-inclusive environmental impact assessment and strategic environmental assessment. In *Ramsar handbooks for the wise use of wetlands* (Vol. 16, 4th ed.). Gland: Ramsar Convention Secretariat.

Rouse, H. L., & Norton, N. (2010). Managing scientific uncertainty for resource management planning in New Zealand. *Australasian Journal of Environmental Management, 17*, 66–76.

Sadler, B. (1996). Environmental assessment in a changing world: Evaluating practice to improve performance, final report of the international study of the effectiveness of environmental assessment. Canadian Environmental Assessment Center, Ottawa, Canada.

Secretariat of the Pacific Community. (2012). Pacific-ACP states regional legislative and regulatory framework for deep-sea minerals exploration and exploitation. SPC, Noumea. http://dsm.gsd.spc.int/public/files/2014/RLRF2014.pdf.

Senécal, P., Goldsmith, B., Conover, S.. (1999). Principles of environmental impact assessment best practice. International Association for Impact Assessment. https://www.iaia.org/uploads/pdf/principlesEA_1.pdf.

Smit, B., & Spaling, H. (1995). Methods for cumulative effects assessment. *Environmental Impact Assessment Review, 15*, 81–106.

Solan, M., & Whiteley, N. M. (2016). *Stressors in the marine environment*. Oxford, UK: Oxford University Press.

Swaddling, A. (2016). Pacific-ACP States regional environmental management framework for deep sea minerals exploration and exploitation. Secretariat of the Pacific Community, Noumea, New Caledonia.

Swaddling, A., Clark, M. R., Bourrel, M., Lily, H., Lamarche, G., Hickey, C., Rouse, H., S. Nodder, G. Rickard, Sutton, P., & Wysoczanski, R. (2016). Pacific-ACP States regional scientific research guidelines for deep sea minerals. Secretariat of the Pacific Community, Noumea, New Caledonia.

US Department of Energy. (2004). Recommendations for the preparation of environmental assessments and environmental impact statements. US Department of Energy, Washington, US.

Wood, C. (2003). *Environmental impact assessment: A comparative review* (2nd ed.). Harlow: Prentice Hall.

Dr. Malcolm R. Clark is a Principal Scientist (Fisheries) at the National Institute of Water and Atmospheric Research (NIWA) in Wellington, New Zealand. He has led numerous deep-sea research programmes at NIWA on the biology and ecology of fish and benthic invertebrate communities associated with seamounts and other deep-sea habitats, with an emphasis on evaluating environmental effects of human activities (especially commercial fishing and potential seabed mining) and development of ecological risk and impact assessments to help managers balance the environmental impacts of exploitation with conservation of habitat. He currently leads research investigating the effects of sedimentation plumes from possible deep-sea mining operations. He has published widely, with over 100 journal papers and book chapters, and a similar number of technical reports and articles.

Protection of the Marine Environment: The International and National Regulation of Deep Seabed Mining Activities

Pradeep Singh and Julie Hunter

Abstract This chapter provides an overview of the international and national regulatory framework pertaining to deep seabed mining activities. It begins by discussing the UN Convention on the Law of the Sea, the backdrop for all marine activities – be they national or international – and examines the obligations of states to protect the marine environment from the harmful effects arising from deep seabed mining. Next, the chapter examines the international regime for deep seabed mining (i.e. "activities in the Area"), explaining the "common heritage of mankind" status of the Area (i.e. the international seabed); the functions of the International Seabed Authority (ISA), the international organization established to govern deep seabed mining in the Area; and the concept of state sponsorship of non-state entities (i.e. private actors) for deep seabed mining in the Area. The chapter follows with a discussion of the development of national legislation to regulate deep seabed mining, examining efforts in the Pacific region where many prospective deep-sea mining sites are located. This includes a look at the legislative regimes of several Pacific Island nations, namely, Papua New Guinea, Tonga and the Cook Islands, for whom deep seabed mining may soon become a reality – as well as New Zealand and Japan, countries with comparatively developed rule of law and legislative regimes that have undertaken or considered deep seabed mining in their national waters. Overall, the chapter critically describes and evaluates the current regulatory status in the international and national seabed areas and highlights some salient gaps that require urgent attention in order to ensure marine environmental protection and mitigate impacts on humans.

Keywords Deep-sea mining · Marine environment · National, international regulations

P. Singh (✉)
Faculty of Law, University of Bremen and Center for Marine Environmental Sciences (MARUM), Bremen, Germany
e-mail: pradeep@uni-bremen.de

J. Hunter
Blue Ocean Law, Hagåtña, Guam
e-mail: julie@blueoceanlaw.com

R. Sharma (ed.), *Environmental Issues of Deep-Sea Mining*,
https://doi.org/10.1007/978-3-030-12696-4_17

1 Introduction

Despite resounding scientific evidence that seabed mining could cause significant, adverse harm to the marine environment and resulting impacts on people, commercial interest in harvesting these mineral resources continues to grow. In order to counteract the undesired consequences of this activity, it is necessary to design a robust and precautionary legal framework, both in areas within and beyond national jurisdiction. Although there are different rules, authorities and regimes assigned to these two ocean spaces, ensuring similar levels of environmental protection in both is critical. The seas, its inhabitants and its ecosystems do not recognize the boundary lines and zones as demarcated by nations (Tanaka 2015). Furthermore, the multitude of environmental stressors constantly foisted on the oceans and their impacts are felt all across the ocean space, not just locally (Halpern et al. 2015).

It is becoming increasingly clear that deep seabed mining will impair the natural function of the deep ocean in climate regulation, while also severely impacting the integrity of the seabed and its rich biodiversity (Wedding et al. 2015; Van Dover et al. 2017), leading to potentially dire consequences for coastal communities and humans in general. The conduct of such activity is therefore a matter of common concern to humankind (Hunter et al. 2018), in which another related norm – the "polluter-pays principle", requiring the entity responsible for environmental harm to pay for the damage (Beder 2006) – should be construed concurrently. While the international community is currently taking steps to develop comprehensive regulations pertaining to mineral mining in areas beyond national jurisdiction – including measures to protect the marine environment – the regulation of exploitation of natural resources in areas within national jurisdiction has been left to individual coastal states entirely without any predetermined stipulations (Markus and Singh 2016). This is a matter of concern as the first large-scale commercial deep seabed mining effort in areas within national jurisdiction – specifically the territorial waters of Papua New Guinea – appears likely to commence in 2019 (Miller et al. 2018).

Centred on the protection of the marine environment, this chapter will explore the international and national regulation of deep seabed mining activities. We will begin by introducing and discussing the international obligation to protect the marine environment in both areas beyond national jurisdiction and areas within national jurisdiction. Through this analysis, the zonal practice in the law of the sea, in which separate regimes co-exist and operate within predetermined mandates, will become apparent. Next, we will examine the international deep seabed mining regime for areas beyond national jurisdiction. In this context, we will introduce the International Seabed Authority (ISA), the international organization designated to govern the mineral resources of the international seabed and the regulatory framework that surrounds it. Following that, we will turn our attention to deep seabed mining activities within national jurisdiction. Here, we will inspect the deep seabed mining regulatory approach within the domestic legal setting of several Pacific Island countries (Papua New Guinea, Tonga and the Cook Islands) as well as that of New Zealand and Japan. These countries have been selected based on a high possibility of large-scale commercial mining taking place within their jurisdiction in the near future, as well as past engagement with seabed mining actors. Finally, while the chief purpose of this chapter is to provide a descriptive overview of the two deep seabed mining regimes, we will nevertheless end by identifying some gaps that exist between the two regimes based on the current state of affairs and offer a suggestion to bridge them.

2 The Obligation to Protect the Marine Environment and the Regulation of Deep Seabed Mining Activities

The UN Convention on the Law of the Sea 1982 (UNCLOS) is the starting point for all discourses pertaining to the modern law of the sea. It provides a general legal framework with an overarching aim of harmonizing domestic and global uses of the oceans as well as balancing competing uses of the marine environment, while simultaneously striving to protect and preserve the marine environment. As a legally binding instrument under international law with widespread acceptance, UNCLOS functions to regulate how states (and by extension, entities subject to their sovereignty or control) carry out activities in marine spaces both within and beyond their jurisdiction. In terms of state action and the protection of the marine environment, Harrison (2017) eloquently explains how UNCLOS serves as the foundational basis for, inter alia, jurisdictional mandate, general principles, substantive rules and procedural rules vis-à-vis human endeavours at sea and the protection of the marine environment. Thus, UNCLOS stipulates which states (or international organizations, as the case may be) are seized with jurisdiction to take measures to protect the marine environment; the general principles which expound the responsibility or obligation to do so; descriptive rules of what is expected of them in performing their responsibilities or obligations; and the procedural steps that must be taken in order to fulfil the same.

UNCLOS declares that all "problems of ocean space are closely interrelated and need to be considered as a whole" (UNCLOS Preamble). As seabed mining is slated to join the multitude of activities carried out in the ocean space, it is essential to ensure that it is properly regulated and subjected to good management practices, irrespective of where it is carried out (Verlaan 2018). However, the designation by UNCLOS of separate jurisdictional mandates for different ocean spaces (i.e. maritime areas or zones) creates a situation where a specific activity like deep seabed mining may be subject to wholly disparate and incoherent rules and standards from one zone to another. This poses a significant problem because lenient, compromising measures adopted in one zone will nullify stringent, ironclad measures adopted in the other, not least due to the risk of transboundary harm and aggregate ocean impact from the intended activity. Further, activities carried out in different areas within national jurisdiction also stand to be subject to various national legislations in their respective territories that are not necessarily harmonized with each other.

This section will consider the jurisdictional mandate in relation to deep seabed mining activities, as well as the ensuing obligation to protect the marine environment.

2.1 Jurisdictional Mandate

UNCLOS creates several maritime zones, namely, internal waters, territorial sea, contiguous zone, exclusive economic zone, continental shelf, high seas and the international seabed (i.e. "the Area"), each of which is associated with different forms of prescriptive and enforcement jurisdictions (Churchill and Lowe 1999; Harrison 2011; Tanaka 2015; Rothwell and Stephens 2016; Kaye 2016). A coastal state's sovereignty extends beyond its land territories and internal waters up to a maritime zone defined as the territorial sea (up to 12 nautical miles from its baselines).[1] In the exclusive economic zone or EEZ (declared up to 200 nautical miles from its baselines) and continental shelf (extending up to 350 nautical miles or even more from its baselines)[2], a coastal state possesses sovereign rights to exploit living and nonliving natural resources within those areas. The high seas is specifically referred to as the maritime area beyond the exclusive economic zone,[3] whereas the international seabed (or "the Area") covers the area of the seabed that is not subject to national jurisdiction (i.e. beyond the continental shelf of coastal states). In relation to mineral mining on the seafloor, the zonal approach under UNCLOS gives rise to two separate seabed regimes: areas beyond national jurisdiction (the Area) and areas within national jurisdiction (chiefly, the continental shelf).

In terms of the jurisdictional framework for the international seabed, Article 1(1) of UNCLOS provides the following definition: "the 'Area' means the seabed and ocean floor and subsoil thereof, beyond the limits of national jurisdiction". In other words, the Area begins where national claims to the continental shelf end. Presently, the exact extent of the Area has not been finalized due to the fact that national claims for extended continental shelves have yet to be determined with finality (Franckx 2010). However, there is a general understanding pertaining to the rough estimate of the Area (Chircop 2011). A whole chapter and annex of UNCLOS (namely, Part XI

[1] The contiguous zone is a maritime area contiguous to the territorial sea, in which a coastal state may exercise a number of sovereign acts in matters pertaining to its domestic customs, fiscal, immigration and sanitary laws.

[2] As opposed to the exclusive economic zone, which must be declared, sovereign rights over the continental shelf exist as of right for up to 200 nautical miles since it is seen as an extension of land (Kaye 2016: 11). Rules of delimitation apply if there are overlapping claims from neighboring states, and as such a coastal state's continental shelf may be less than 200 nautical miles. A coastal state may, under certain conditions pursuant to Article 76 of UNCLOS, extend its continental shelf zone for up to 350 nautical miles from its baselines.

[3] It should be noted, however, that certain freedoms enjoyed at the high seas (beyond the rights to exploit natural resources and several others) may also be enjoyed within exclusive economic zones, such as navigation, overflight and the laying of submarine cables (see UNCLOS, Articles 58 and 86).

and Annex III) as well as the 1994 Agreement relating to the Implementation of Part XI of UNCLOS are dedicated to the Area and its resources. Article 134(2) proclaims that Part XI governs the conduct of "activities in the Area", which is defined under Article 1(3) as "all activities of exploration for, and exploitation of, the resources of the Area". The use of the term "resources" in this context is further clarified as "all solid, liquid or gaseous mineral resources in situ in the Area at or beneath the seabed, including polymetallic nodules". Hence, it is clear that living resources do not fall within the ambit of Part XI. Accordingly, the existing legal framework for the Area is synonymous with the international seabed mining regime. Article 137(1) of UNCLOS firmly states that "no State shall claim or exercise sovereignty or sovereign rights over any part of the Area or its resources" and goes on in Article 137(2) to assert that "all rights in the resources of the Area are vested in mankind as a whole, on whose behalf the Authority shall act". As explained in Article 1(2) of UNCLOS, the "Authority" here refers to the International Seabed Authority (ISA). Accordingly, the ISA possesses the mandate to exercise jurisdiction over activities in the Area.

As for the national seabed area (i.e. the continental shelf), coastal states clearly possess the requisite mandate over mineral resources. Article 77(1) of UNCLOS prescribes that the "coastal State exercises over the continental shelf sovereign rights for the purpose of exploring it and exploiting its natural resources" and in Article 77(4) qualifies "natural resources [… to] consist of the mineral and other non-living resources of the seabed […]". This is also the case when a coastal state extends its outer continental shelf as permitted by Article 76; however, the coastal state is required to make payments or contributions in kind to the ISA for all mineral exploitations beyond 200 nautical miles in those instances. Accordingly, the coastal state is clearly seized with jurisdictional mandate over mineral exploitation in its delineated (or delimited, as the case may be) continental shelf.

The following sub-sections will discuss and compare the two separate regimes in the context of the protection of the marine environment. As will be seen, jurisdictional mandate over resources also connotes the obligation to protect the marine environment in the area concerned.

2.2 *Obligation to Protect the Marine Environment*

Starting with the international seabed, Article 145 of UNCLOS is instructive on the general obligation to protect the marine environment from the harmful effects arising from the conduct of activities in the Area. It stipulates the following:

> Article 145: Protection of the marine environment
>
> Necessary measures shall be taken in accordance with this Convention with respect to activities in the Area to ensure effective protection for the marine environment from harmful effects which may arise from such activities. To this end the Authority shall adopt appropriate rules, regulations and procedures for inter alia:

(a) the prevention, reduction and control of pollution and other hazards to the marine environment, including the coastline, and of interference with the ecological balance of the marine environment, particular attention being paid to the need for protection from harmful effects of such activities as drilling, dredging, excavation, disposal of waste, construction and operation or maintenance of installations, pipelines and other devices related to such activities;
(b) the protection and conservation of the natural resources of the Area and the prevention of damage to the flora and fauna of the marine environment.

Article 17(2)(f) of Annex III to UNCLOS goes further and prescribes that "rules, regulations and procedures shall be drawn up in order to secure effective protection of the marine environment from harmful effects directly resulting from activities in the Area or from shipboard processing immediately above a mine site of minerals derived from that mine site, taking into account the extent to which such harmful effects may directly result from drilling, dredging, coring and excavation and from disposal, dumping and discharge into the marine environment of sediment, wastes or other effluents".

Premised on these two key provisions, it is obvious that the ISA has the obligation to take necessary measures, including the adoption of rules and regulations, to protect the marine environment from the harmful effects arising from the conduct of mineral exploitation. This clearly includes coverage of not just the physical seabed but also the ocean surface (particularly from shipboard processing and the discharge of waste) and the water column above the mine site (especially plume and sediment dispersal).

Further to that, Article 209(1) reaffirms that "international rules, regulations and procedures shall be established in accordance with Part XI to prevent, reduce and control pollution of the marine environment from activities in the Area. Such rules, regulations and procedures shall be re-examined from time to time as necessary". It goes on further, in Article 209(2), to require states to "adopt laws and regulations to prevent, reduce and control pollution of the marine environment from activities in the Area undertaken by vessels, installations, structures and other devices flying their flag or of their registry or operating under their authority, as the case may be. The requirements of such laws and regulations shall be no less effective than the international rules, regulations and procedures referred to in paragraph 1".

The 2011 Advisory Opinion on the Responsibilities and Obligations of States Sponsoring Persons and Entities with respect to Activities in the Area by the Seabed Disputes Chamber of the International Tribunal for the Law of the Sea (Advisory Opinion 2011) sheds further light into this topic. The Seabed Disputes Chamber opined that sponsoring states (i.e. states that back private entities to conduct activities in the Area) must meet certain direct and due diligence obligations pertaining to marine environmental protection. This includes adhering to the precautionary principle, ensuring the carrying out of proper environmental impact assessments and the continuous monitoring of mining activities and its environmental impacts during and after its conclusion, facilitating the adoption of best environmental practices in conducting mining activities and assisting the ISA in carrying out its functions.

With respect to seabed mining in areas within national jurisdiction, reference to the exclusive economic zone regime is necessary. In particular, Article 56 of UNCLOS stipulates the following:

Article 56: Rights, jurisdiction and duties of the coastal State in the exclusive economic zone

1. In the exclusive economic zone, the coastal State has:
   (a) sovereign rights for the purpose of exploring and exploiting, conserving and managing the natural resources, whether living or non-living, of the waters superjacent to the seabed and of the seabed and its subsoil, and with regard to other activities for the economic exploitation and exploration of the zone, such as the production of energy from the water, currents and winds;
   (b) jurisdiction as provided for in the relevant provisions of this Convention with regard to:
      (i) the establishment and use of artificial islands, installations and structures;
      (ii) marine scientific research;
      (iii) the protection and preservation of the marine environment;
   (c) other rights and duties provided for in this Convention.
2. In exercising its rights and performing its duties under this Convention in the exclusive economic zone, the coastal State shall have due regard to the rights and duties of other States and shall act in a manner compatible with the provisions of this Convention.
3. The rights set out in this article with respect to the seabed and subsoil shall be exercised in accordance with Part VI.

Thus, Article 56 clearly confers jurisdiction on the coastal state to conserve and protect the marine environment. This is not limited only to the physical seabed (i.e. continental shelf) but also to the surface ocean and water column (i.e. exclusive economic zone). Further to that, Part XII of UNCLOS, in particular Articles 192 and 193, is pertinent. Article 192 lays down the general obligation of states to "protect and preserve the marine environment". Article 193 asserts that states have the "sovereign right to exploit their natural resources pursuant to their environmental policies and in accordance with their duty to protect and preserve the marine environment". This includes, as stipulated under Article 194(3)(c), taking measures to minimize "pollution from installations and devices used in exploration or exploitation of the natural resources of the seabed".

Akin to the position under the international seabed regime, coastal states are also required pursuant to Article 204 of UNCLOS to continuously monitor the environmental harm of activities that they permit and under Article 206 to ensure that environmental impact assessments are carried out prior to the conduct of seabed mining activities. Finally, reference to Article 208, which deals with "pollution from seabed activities subject to national jurisdiction", is essential. Articles 208(1) and (2) provide that coastal states "shall adopt laws and regulations to prevent, reduce and control pollution of the marine environment arising from or in connection with seabed activities subject to their jurisdiction" and "shall take other measures as may be necessary to prevent, reduce and control such pollution", respectively. Article 208(3) goes on to stipulate that "such laws, regulations and measures shall be no less effective than international rules, standards and recommended practices and procedures", while Article 208(4) specifies that states shall "endeavour to harmonize their policies in this connection at the appropriate regional level". Additionally, Article 208(5) instructs that "states, acting especially through competent international organizations or diplomatic conference, shall establish global and regional rules, standards and recommended practices and procedures to prevent, reduce and control pollution of the marine environment referred to in paragraph l. Such rules, standards and recommended practices and procedures shall be re-examined from time to time as necessary".

As will be seen later on in the section dedicated to the national framework of deep seabed mining, the requirements imposed by Articles 208(3), (4) and especially (5) have not been satisfactorily met, thereby resulting in a concern arising from the uncertainty of the conditions in which domestic seabed mining activities will take place.

The following section will discuss the international framework for deep seabed mining in the Area, which is rapidly developing due largely to intensified regulatory efforts and discussions at the ISA.

3 The International Regulation of Deep Seabed Mining Activities

This section attempts to clarify the existing framework pertaining to deep seabed mining in the Area and present a comprehensive overview of the international regulation of deep seabed mining activities. First, it will look at some fundamental features pertaining to the Area, including its "common heritage of mankind" status. Second, it will explore the institutional setting pertaining to activities in the Area and introduce the principal actors involved in it. Third, it will briefly examine the regulatory framework surrounding deep seabed mining activities. Finally, this section will provide an outlook of the activities in the Area in light of other competing uses in the Area.

3.1 The Area and Its Salient Features

Central to the discourse of deep seabed mining in the international seabed is Article 136 of UNCLOS, which declares that the Area and its mineral resources are the common heritage of mankind. The significance of this provision is further reflected through Article 311(6), whereby state parties have agreed that there shall be no derogation from the basic principle of the common heritage of mankind. At its very core, the common heritage of mankind principle affirms the following:

1. There shall be no exercise of sovereignty in the Area (Article 137(1)).
2. Resources of the Area are vested in mankind as a whole and shall be managed solely through the ISA (Article 137(2)).

3. Activities in the area shall be carried out for the benefit of mankind (Article 140(1)).
4. The Area shall be used only for peaceful purposes (Article 141).

Furthermore, as seen earlier from Article 145, the principle also entails the effective protection of the marine environment from harmful effects arising from activities in the Area. Alongside the protection of the marine environment, the common heritage of mankind principle further integrates the concepts of intergenerational equity and sustainable development into the international seabed mining discourse (Jaeckel et al. 2017). As such, the need to preserve the marine environment and to conserve mineral resources for future generations is an integral pillar of the common heritage of mankind. Another implication of the common heritage of mankind principle is the benefit sharing regime. Article 140(2) requires the Authority to provide for the "equitable sharing of financial and other economic benefits derived from activities in the Area through any appropriate mechanism, on a non-discriminatory basis". The equitable dimension in the distribution of benefits arising from such activities is a critical element to the common heritage of mankind principle because the mineral resources of the seabed are non-renewable and will deplete on exploitation (Lodge et al. 2017). Thus, the ISA is essentially tasked to develop the resources of the Area, manage it in a rational and orderly manner and adopt measures to optimize revenues and increase the availability of minerals on the one hand (Article 150 of UNCLOS, Article 13 of Annex III to UNCLOS, 1994 Implementation Agreement) - and to protect the marine environment from the consequential harmful effects, conserve resources for future generations and equitably distribute benefits, on the other. This provides the suitable backdrop for the subsequent analysis on the institutional setting pertaining to activities in the Area.

3.2 Institutional Setting

As can be gleaned from the above, the ISA is the sole body responsible for the governance of the deep seabed mineral resources in the Area (Lodge 2012). Article 139(1) of UNCLOS distinctly provides that only the ISA may permit states or their sponsored entities to conduct activities in the Area. In order to commence exploration or exploitation of mineral resources in the Area, the said interested parties would first need to submit a plan of work for approval and subsequently enter into a contractual relationship with the ISA. In this regard, Article 153(3) is relevant, stating that "activities in the Area shall be carried out in accordance with a formal written plan of work drawn up in accordance with Annex III and approved by the Council after review by the Legal and Technical Commission. In the case of activities in the Area carried out as authorized by the Authority by the entities specified in paragraph 2(b), the plan of work shall, in accordance with Annex III, article 3, be in the form of a contract". At present, a significant majority of exploration contracts involve the ISA and sponsored entities (as opposed to states).

As a starting point, it is useful to mention that private entities may not engage in any form of activity in the Area without a certificate of sponsorship from a sponsoring state. A state, conversely, may engage in activities directly through a contract with the ISA if it so chooses. The rationale behind this is obvious: private contractors are typically not considered subjects of international law (Advisory Opinion 2011).[4] Hence, while a contractual relationship between the ISA and private contractors is necessary for there to be an enforceable recourse under the domestic laws of the sponsoring state, the sponsoring state, a subject of international law, remains responsible for violations (or wrongful acts) under international law (Advisory Opinion 2011). It is critical to note that although there shall be no exercise of sovereignty in the Area, the concept of state sponsorship is an act of sovereignty. A state that chooses to sponsor an entity will incur significant legal responsibilities under international law (Geddis 2017). In this context, a state has the autonomy to decide whether it wishes to sponsor an entity or not. With respect to the prerequisite condition for sponsorship, Article 153(2)(b) of UNCLOS requires that such private entities either possess the nationality of the state in concern or are effectively controlled by the state or their nationals (Advisory Opinion 2011).

It will be recalled that Article 209(2) requires states to enact domestic "laws and regulations to prevent, reduce and control pollution of the marine environment from activities in the Area undertaken by vessels, installations, structures and other devices flying their flag or of their registry or operating under their authority, as the case may be". This includes providing recourse for damages arising from a contractual breach as well as enforcement procedures (Advisory Opinion 2011). Failure to address these matters in an adequate manner may be viewed as a violation of international law. As observed in the 2011 Advisory Opinion, it is incumbent on sponsoring states to enact such national laws in order to give effect to Part XI of UNCLOS. A number of countries (both developed and developing) with an interest in deep seabed mining in the Area, such as Germany, the United Kingdom, China and Singapore, have already done this (Geddis 2017; Jin and Zhang 2017; Egede 2018; Sun 2018).

Another key topic that deserves closer examination is the institutional structure of the ISA. In order to understand the theme of the proceeding sub-section (i.e. the regulatory framework of international deep seabed mining or "DSM"), it is essential to comprehend the operation of the ISA. Several basic points surround this premise. First, Article 156(2) declares that all state parties to UNCLOS are automatically part of the ISA. Second, as provided for in Article 157(1), the ISA, in the form of an international organization, is the vessel for state parties "to organize and control activities in the Area, particularly with a view to administering the resources of the Area". Third, pursuant to Articles 158(1) and (2), there are three principal organs of the ISA, namely, the Assembly, the Council and the Secretariat, and a yet-to-be-established independent organ called the Enterprise.[5] Fourth, Article 159(1)

[4] However, it is acknowledged that there is a growing trend to recognize the role of non-state actors in the realm of international law, especially pertaining to human rights.

[5] Although UNCLOS foresees the establishment of the Enterprise as the independent arm of the

stipulates that all state parties are equal members of the Assembly, which is further determined as the supreme organ of the ISA under Article 160(1). Fifth, in accordance with Article 15 of Section III of the 1994 Implementation Agreement, the Council consists of 36 members of the ISA that are elected by the Assembly. With reference to Article 162(1), the Council, as the executive organ of the ISA, is essentially the decision-making branch of the ISA. Sixth, despite its elevated position, the Council "does not act alone in formulating environmental regulations for the Area"; in this regard, the Legal and Technical Commission, a subsidiary organ to the Council, has "particular responsibility for the protection of the marine environment" (Lodge et al. 2014). Seventh and finally, the Secretariat is responsible for the day-to-day operation of the ISA. While central to the routine administration and functional operation of the ISA, it is useful to recall that the member states – not the Secretariat – are principally and collectively responsible for the management of deep seabed resources. The following sub-section will discuss the regulatory framework for international seabed mining.

3.3 Regulatory Framework

In essence, the regulatory framework for DSM activities in the Area emanates quintessentially from the rule-making feature of the ISA, as discussed above in section 3.2. As mentioned, the ISA is obligated to protect the marine environment from the harmful effects of activities in the Area. To this end, Articles 145 and 209 require the ISA to take necessary measures through the adoption of rules and regulations. Thus, UNCLOS simply lays down the jurisdiction mandate (see Sect. 2.1 above) and provides the general framework for the protection of the marine environment (see Sect. 2.2 above), leaving it to the ISA to develop its own specific set of rules and regulations to precisely govern activities in the Area. This is a unique feature of international law, whereby an organization is assigned full power to create regulations that automatically bind member states with no possibility of opting-out beforehand (Harrison 2011; Jaeckel 2017). However, this can be reconciled with the fact that the Assembly, the supreme organ of the ISA in which all member states are represented, participates in the rule-making function of the ISA at the top of the order (as explained above in Sect. 3.2). This practice enables rule-making to be carried out in a fairly prompt fashion and allows for periodical revision without having to convene a diplomatic conference to amend or modify the parent treaty (Boyle and Chinkin 2007).

The exercise of this rule-making function will eventually result in the collation of a comprehensive dossier known as the "Mining Code", effectively "covering all aspects of mining activities – including prospecting, exploration and exploitation –

ISA, particularly responsible to carry out seabed mining activities on behalf of mankind, the 1994 Implementation Agreement has effectively put this on hold.

and subjecting each of them to specific environmental requirements" (Markus and Singh 2016). To date, three separate sets of regulations covering the exploration of polymetallic nodules, polymetallic sulphides and cobalt-rich crusts, respectively, have been issued alongside numerous guidelines, while a combined set of exploitation regulations encompassing all forms of minerals is currently in the final draft stages. This includes specific regulations for the environment as well as model contractual terms. Given constant developments and the advanced draft stage of the exploitation regulations, an in-depth analysis therein is beyond the scope of this chapter; however, an overview of the current work-in-progress can be found elsewhere (Brown 2018). As the bulk of deep seabed mining activities will involve private entities, the regulations and contractual terms will play a critical function, forming the foundational basis for these actors' obligations and responsibilities, consequently enforceable pursuant to domestic law.

3.4 *The International Regulation of Deep Seabed Mining from the Environmental Protection Perspective: An Outlook*

As a starting point for this overview analysis, it is necessary to acknowledge the general framework for the protection of the marine environment from the harmful effects of activities in the Area under UNCLOS. In this regard, UNCLOS permits the ISA to design a suitable regulatory regime that balances the development of ocean mineral resources on the one hand and the protection of the marine environment on the other. While benign effort is being exerted towards addressing the potential harmful effects of such activities, certain gaps do appear. For instance, apart from the apparent harm to the seabed and its immediate vicinity, other types of potential harm to the marine environment such as shipboard processing of the recovered minerals and the discharge of incidental wastes appear to have been sidelined. Similarly, the onshore processing of minerals obtained from the international seabed, even though beyond the jurisdiction of the ISA, is largely ignored from the current discourse (Markus and Singh 2016). Both these aspects are related concerns as they should be taken into account in determining whether mining the international seabed is feasible and sustainable to begin with. Apart from that, lacklustre knowledge generation attitudes and the absence of a unit dedicated solely to environmental matters (Jaeckel et al. 2017) are an impediment that could easily be resolved. The suitability of using incentives to advance and encourage the adoption of environmentally sound technologies should also be examined (Lodge et al. 2017). Additionally, introducing more intermediate steps between the initial application for approval of a plan of work and the ultimate decision would further support the protection of the marine environment. For instance, the carrying out of a proper pilot mining test and a comprehensive feasibility study indicating positive, net benefit outcomes should be made a prerequisite to approval (Christiansen et al. 2018).

Another critical point to note is that the jurisdiction of the ISA, including its environmental mandate, is restricted to activities in the Area. As such, the ISA does not, strictly speaking, possess the general mandate to protect the international seabed from non-mining activities. In fact, Article 145 of UNCLOS clearly states that "necessary measures shall be taken […] with respect to activities in the Area to ensure effective protection for the marine environment from harmful effects which may arise from such activities". Apart from that, Article 147(1) stipulates that activities in the Area must accommodate other uses of the Area, such as navigation, the conduct of marine scientific research and the laying of submarine cables. In this regard, there is clear overlap between the ISA's jurisdiction over the conduct of activities in the Area and the mandate exercised by other international organizations. For instance, while shipboard processing of minerals immediately above the mining site falls within the jurisdiction of the ISA, the discharge of waste from ships and dumping at sea generally fall under the purview of the International Maritime Organization (IMO). Specifically, the IMO has the mandate to administer the International Convention for the Prevention of Pollution from Ships 1973/78 (MARPOL) and its six annexes, as well as to oversee the implementation of the Convention on the Prevention of Marine Pollution by Dumping of Wastes and Other Matter 1972/1996 (London Convention/London Protocol or LC/LP). Thus, there are still some grey areas with respect to power to regulate and enforce regulations particularly beyond the immediate mining site in the Area, i.e. related conduct in the high seas (Verlaan 2018). Moreover, the transportation and onshore processing of these minerals, a subject matter of significant environmental concern, is beyond the ISA's scope of regulatory control and is, respectively, deferred to international shipping rules, in the case of transportation, and domestic legislation, in the case of onshore processing (Markus and Singh 2016).

Furthermore, growing interest in marine genetic resources of the Area and the imminent possibility of a new instrument to govern living resources may also present a new challenge for the governance of activities in the Area. Likewise, emerging scientific developments pertaining to the function of the deep ocean in climate regulation and the ecosystem services provided by its biogeochemical components should feature more widely in the decision-making processes at the ISA. Increased cooperation in an area in which governance is widely fragmented and diverse proves to be the indispensable ingredient to further the protection of the marine environment with respect to the Area (Singh and Jaeckel 2018). In this regard, it is encouraging to note that the ISA has in fact signed several agreements of cooperation or memoranda of understanding with the International Maritime Organization (IMO), the Intergovernmental Oceanographic Commission (IOC) and the International Cable Protection Committee (ICPC).[6] The extent of the effectiveness of such instruments in practice, however, has yet to be thoroughly studied.

---

[6] See website of the ISA, in particular: https://www.isa.org.jm/files/documents/EN/Regs/IMO.pdf; https://www.isa.org.jm/sites/default/files/documents/EN/Regs/ISA-IOC-MOU.pdf; and https://www.isa.org.jm/sites/default/files/documents/EN/Regs/MOU-ICPC.pdf.

Be that as it may, it is evident that interest in ocean mineral resources is not only confined to the Area. As mentioned earlier, early large-scale commercial mining of the seabed is anticipated to take place in areas within national jurisdiction. This reality necessitates the examination of the national regulation of deep seabed mining activities from the marine environmental protection perspective.

4 National Regulation of Deep Seabed Mining Activities

This section examines national regulatory regimes as they currently exist for deep sea mining in areas of national jurisdiction (typically within the EEZs of coastal states). Although much of the DSM discourse has focused on mining in the Area, many countries also possess significant seabed mineral resources within their own EEZs and have started to move forward with plans to mine their seabed. Pacific Island nations, Japan, New Zealand, Australia, Mexico and Namibia are some of the countries known to be exploring or pursuing various forms of seabed mining within their national waters. Some of these jurisdictions lack discrete seabed mining legislation, relying instead on existing land-based mineral regimes to govern seabed resources. Others have new legislation designed specifically to address DSM.

The following discussion provides a summary overview of the regulatory frameworks of several Pacific Island nations where national DSM is likely, including Tonga, the Cook Islands and Papua New Guinea (where the world's first deep-sea commercial mine is slated to begin production between 2019 and 2020) – as well as a brief look at the regulatory regime of New Zealand, currently considering a phosphate deep-sea mining project in its waters, and Japan, which became the first country to mine its deep seabed in 2017. The overview focuses primarily on states with active deep-sea mineral exploration, excluding shallow seabed or sand mining operations.[7]

4.1 General State Obligations

As discussed above, UNCLOS imposes broad obligations on states to protect the marine environment under their jurisdiction, including establishing laws and regulations "to prevent, reduce and control pollution of the marine environment arising from or in connection with seabed activities subject to their jurisdiction" (Article 208 (1)(2)), in line with international standards (Article 208(3)). Article 214 specifically obligates the enforcement of such domestic legal regimes, established for the purposes of regulating pollution arising from seabed activities.

[7] In general, our definition of DSM entails harvesting mineral deposits in the deep sea at depths ranging from approximately 400–6000 m below sea level (Hunter et al. 2018; Miller et al. 2018).

In addition to the obligations imposed by the UNCLOS, all states are under a wide array of legal obligations with respect to established international environmental law principles, such as the obligation to avoid transboundary harm, the precautionary approach, biodiversity commitments and the need for independent and robust environmental impact assessments (EIAs) and environmental monitoring. To the extent that DSM may cause climate-related impacts (Levin et al. 2016), states are also bound by their commitments to international climate change instruments, including the UNFCCC, the Kyoto Protocol and the Paris Agreement.

Outside the realm of environmental protection, extractive activities which impact human health and other basic human rights (such as the right to work, the right to a livelihood and an adequate standard of living, the right to health, the right to housing and property rights) will trigger further protections under various binding international human rights treaties, widely ratified by most states, including the countries assessed below. Furthermore, indigenous peoples, who are disproportionately impacted by extractive activities – including by impending seabed mining plans, particularly in the Pacific region – should be consulted and their free, prior and informed consent (FPIC) obtained with respect to future development activities threatening to impact them on their traditional territories, regardless of where the actual activity occurs (Szablowski 2011; Anaya 2015). Given that much seabed mining could occur in or near the waters of Pacific Islands as well as other countries with indigenous populations, FPIC should be sought in the seabed mining context (Hunter et al. 2018; Aguon and Hunter 2019).

To a large degree and as evidenced below, most of these relevant principles have not yet been incorporated into domestic DSM regimes (and are also largely absent from current ISA regulations). Basic environmental principles, where included, lack specific requirements or obligations laid on either contractors or states. Coverage of human impact and specific mechanisms for consultation, consent or remedies generally remain absent. In short, domestic DSM regimes tend to share the same characteristics of the international regime, weighted towards facilitating exploitation and a contractual regulatory regime, rather than towards preventing potential impacts or environmental degradation. Below we explore some of the geopolitical realities that have produced this state of affairs in the Pacific Island region.

4.2 The Pacific Context

As the world's largest ocean, the Pacific is home to a high concentration of prospective mineral sites due to geothermal activity around the "Ring of Fire", as well as other geophysical features. In addition to the Clarion-Clipperton Fracture Zone (CCFZ) and other sites in the Area, the seabed areas of New Zealand, Japan and various Pacific Island (PI) nations[8] have garnered the attention of those seeking access to mineral deposits.

[8] In this chapter, PI nations refer generally to the 14 member countries of the SPC-EU DSM Project

Generally speaking, PI nations have limited experience with mining. Where terrestrial mines exist in the region, they have often been associated with disastrous environmental, social and cultural impacts. Although many PI governments are interested in possible revenue accruing from mineral deposits, they are ill-equipped to effectively monitor DSM, enforce regulations and collect taxes and other levies from large multinational companies. Many PI nations lack coast guards or ships to police their own waters from overfishing and other illegal activities and are short-staffed within environmental and ocean agencies where they exist (Blue Ocean Law & Pacific Network on Globalisation 2016). A large number of failed mines and associated environmental disasters, particularly in Papua New Guinea, raise the possibility that even with model legislation, PI nations will have difficulty meeting their obligations for enforcement and avoiding pollution arising from seabed mining under Article 214 of UNCLOS.

Nonetheless, interested parties have proceeded to negotiate directly with PI governments for access to minerals contained within waters of national jurisdiction. The European Union (EU) has been particularly active in this regard, funding the Secretariat of the Pacific Community – European Union Deep Sea Minerals Project (SPC-EU DSM Project),[9] whose objectives include improving the governance and management of PI nations' deep-sea mineral resources in accordance with international law, with particular attention to the protection of the marine environment, and securing equitable financial arrangements for Pacific Island countries and their people. Underlying these stated aims, however, is the EU's desire to access alternative mineral sources. Documents submitted to the European Parliament reveal the EU's dependency on imports of "high-tech" metals such as cobalt, platinum, rare earths and titanium, increasingly essential to the development of technologically sophisticated products. The EU's 2008 "Raw Materials Initiative" seeks to avoid supply crises and diversify access to raw materials beyond somewhat unstable suppliers in Africa, China and South America. Although a 2018 European Parliament resolution calls for a moratorium on commercial DSM exploitation licences until the effects of DSM are better understood,[10]it is non-binding in effect and does not appear to have resulted in an actual policy shift with respect to either the Raw Materials Initiative or European states' national DSM agendas.

In this context, the SPC-EU DSM initiative is better understood as an attempt to establish deep-sea mining frameworks in PI nations in order to gain access to PI minerals.

(excluding only Timor-Leste): Cook Islands, Federated States of Micronesia, Fiji, Kiribati, Marshall Islands, Nauru, Niue, Palau, Papua New Guinea, Samoa, Solomon Islands, Tonga, Tuvalu and Vanuatu.

[9] The EU provides €4.4 million in funding for the project.

[10] European Parliament Resolution of 16 January 2018 on International Ocean Governance: An Agenda for the Future of Our Oceans in the Context of the 2030 SDGs (2017/2055(INI)), P8_TA -PROV(2018)0004, para 42.

4.3 SPC Regional Legislative and Regulatory Framework (RLRF)

The SPC-EU DSM Project has produced several frameworks and guidelines intended to enhance the capacity of PI nations to manage DSM. Among these is the "Pacific-ACP States Regional Legislative and Regulatory Framework for Deep Sea Minerals Exploration and Exploitation" (RLRF), a discussion document designed to assist PI states in their development of national policy and law for DSM. The RLRF has served as a template for national legislation in the region.

The RLRF advises states to incentivize investors by providing a setting that fosters investment (§4.3), encouraging states to provide predictable and stable governance, reasonable taxation and legislation that account for corporate risks and investments (§§10.5–10.7). While emphasizing the benefits of DSM throughout the document, the RLRF characterizes potential adverse effects from DSM-related activities as "extremely minimal" (§20.2) or as having "almost no impact" (§18.6). It claims that an environmental impact assessment (EIA) may or may not be necessary depending on the project size and that different levels of EIAs may also be sought (§18.8), allowing activities that will have a "minor or transitory impact" to proceed without any EIA (§§18.8–18.9). The RLRF also reframes potentially negative impacts as opportunities for research, science and education, while insufficiently addressing any negative impacts of DSM in the initial, prospecting phase (§§4.5–4.8).

Despite numerous references to the precautionary approach, the RLRF's general minimizing of risks seemingly contravenes the goal of the approach. The framework includes no mention of DSM's potential impacts on climate change and related obligations under the UNFCCC or the Kyoto Protocol. The RLRF also mentions indigenous peoples only once (§6.16), despite being designated for use in the largely indigenous Pacific Islands, relying instead on terms like "citizens" and the "public". This obfuscates the special duties owed to indigenous peoples under international law (Anaya 2005). The framework similarly skirts over the idea of FPIC, mentioning informed consent twice, but not directly in relation to indigenous peoples (§§4.7, 16.3).

In short, the RLRF pays lip service to environmental protection while also green-lighting DSM interests in the Pacific. Given this basis, it is unsurprising that individual country legislation regulating DSM in the Pacific is similarly uneven or, in some cases, even more deficient with respect to international environmental law, indigenous rights and other international law standards. The following sections examine three Pacific Island country regulatory frameworks (Papua New Guinea, Tonga and the Cook Islands) in this context.

Papua New Guinea

Papua New Guinea (PNG), the biggest country in the Pacific Island region with a population of around eight million, has been selected for the world's first commercial DSM operation by Nautilus Minerals, a Canadian company and leader in DSM technology. At its proposed mine site, Solwara 1, Nautilus plans to commercially exploit gold and copper deposits associated with deep-sea hydrothermal vents at a depth of around 1600 m in the Bismarck Sea, approximately 30–50 km from coastal and indigenous communities living on the islands of New Ireland and East New Britain, respectively. The project has raised concerns and significant opposition among PNG civil society, including a court case lodged against the government to obtain documents regarding Solwara 1's approval process, and has experienced persistent delays as a result of funding and other setbacks (Roche and Feenan 2013).

To date, PNG has no formal deep-sea mining legislation or framework for the permitting of an offshore mining operation. Rather, DSM in PNG falls primarily under the Mining Act of 1992, as well as the Environment Act of 2000 (Boschen et al. 2013). Nautilus received its initial exploration licences under the Mining Act in 1997 and subsequently again in 2011.

The 1992 Mining Act is a somewhat antiquated law designed primarily to facilitate onshore mining through technical and administrative provisions. It declares all minerals to be owned by the national government (§5), with only one mention of the seabed in the definition of "land" (Mining Act (1992), §2(1d)). The Act contains no mention of the precautionary principle, transboundary harm or FPIC and very little regarding environmental protection generally, with nothing on EIAs and just one mention of the environment in a section on assessing an application for a mining lease (§43). There is no discussion of consent in the context of indigenous peoples or customary resource users. The one provision regarding consultation allows the Mining Minister "to consider the views of those persons whom the Minister believes will be affected by the grant of that special mining lease", including the provincial government, landholders of the land in question, the national government and the mining applicant (§3 "Consultation") – essentially excluding members of local communities, indigenous peoples, civil society organizations (CSOs) and other stakeholders while leaving consultation entirely to the discretion of the Minister. There are no provisions for community revenue-sharing agreements, also known as Impact Benefit Agreements (IBAs), which exist in jurisdictions such as Canada and have proven to alleviate social and environmental ills associated with the extractive industry (Kielland 2015).

The 2000 Environment Act attempts to address some of the gaps of the 1992 Mining Act. Administered by the Department of Environment and Conservation, it requires an environmental impact statement (EIS) prior to permits for mining being granted, with further conditions including installation of monitoring equipment, undertaking an environmental management programme, baseline studies and a rehabilitation programme (Boschen et al. 2013). Under the legislation, companies seeking to obtain a mining lease must complete an Environmental Inception Report and an EIS, to include "physical and social environmental impacts which are likely to result from the carrying out of the activity" (Environment Act (2000), §51(b)). The Director of Environment assesses the EIS, may refer the EIS to other bodies for additional assessment and makes the report available for public review (§§55–57). If the Director accepts the EIS, it passes to the Environmental Council for assessment and recommendation to the Minister (§57). Unfortunately, the Director of Environment under the Act also serves as the Chairperson of the Environmental Council, diminishing the independence of this system of review (§17).

The mining and environmental regimes in PNG, taken together, have been criticized for their inability to stem major environmental disasters and industrial operations, culminating in civil conflict and severe human rights violations in PNG. PNG's Environment (Water Quality Criteria) Regulation 2002 permits the dumping of toxic wastes into PNG's rivers and coastal waters, leading to extensive pollution and water contamination associated with multiple terrestrial mining sites. Corruption, violence and land-grabbing are rampant, making effective governance and enforcement of environmental regulations notoriously difficult (May 2017).

With respect to seabed mining, the regulatory and operational environment in PNG does not bode well for effective enforcement or monitoring. Although a consultation process to discuss amendments and to update the old Mining Act to include offshore mining and grievance mechanisms began in 2013, neither the updated mining policy nor amendments to the Act appear to have been passed or made public. This remains the case despite the fact that full-scale DSM under Nautilus has been imminent for the past couple years (Blue Ocean Law & Pacific Network on Globalisation 2016).

As an operating theatre, PNG's high poverty levels, inequality, civil conflict and insufficient rule of law (UNDP 2014) raise concerns that even with model DSM legislation, a major undertaking in crowded territorial waters close to populated shores would be insufficiently regulated, leading to significant harms with disproportionate impact on indigenous coastal communities.

Tonga

Tonga, a Polynesian country located in the South Pacific, comprises 176 islands scattered over approximately 700,000 km^2 of ocean. With a population of around 107,000, Tonga is the last remaining Polynesian monarchy and still fairly new to the exercise of parliamentary democracy, to which it transitioned in 2010.

Tongan waters contain seafloor massive sulphides (SMS) at depths ranging from 600 to 2000 m below the surface. According to the Ministry of Lands and Natural Resources, around one-fifth of Tongan waters have been licenced for DSM exploration, primarily to Nautilus Minerals, the Korean Institute of Ocean Science and Technology (KIOST) and Bluewater Metals (Blue Ocean Law & Pacific Network on Globalisation 2016).

In August 2014, the previous administration of Tonga passed the Seabed Minerals Act into law. Prior to the law's passage, the government permitted companies to explore without a designated legislative framework (Pulu 2013), issuing contracts under outdated mineral and petroleum mining laws. The 2014 Seabed Minerals Act is the governing regulatory framework that expresses the major aims and guidelines surrounding DSM.[11]

Unlike PNG's Mining Act, Tonga's 2014 Seabed Minerals Act (SMA) is based on the SPC-EU DSM framework. SPC-EU officials and lawyers worked closely with Tongan government officials, providing funding as well as actual draft legislation based on the RLRF, which Tonga then adapted and enacted. In particular, provisions on consultation were shortened, based on the premise that all resources in the Kingdom are vested in the Crown, and therefore consultation or consent from communities was largely unnecessary for extractive projects (Blue Ocean Law & Pacific Network on Globalisation 2016).

While the SMA does emphasize the precautionary principle and the importance of environmental protections, it lacks sufficiently developed protections for indigenous and coastal communities, as well as recognition of the potential harms of DSM and the need for remedy or grievance mechanisms. The legislation also contains provisions which are likely unenforceable due to capacity issues. For instance, the SMA calls for the establishment of a separate Tonga Seabed Minerals Authority, with Minister, CEO and staff to carry out numerous administrative and regulatory functions relative to DSM (Seabed Minerals Act 2014, §§9, 12 "Functions of the Authority"). In practice, Tonga lacks the resources to establish a separate Seabed Minerals Authority. Similarly, the government tends to outsource its EIA work, and lacks sufficient lawyers to enforce regulations and carry out oversight. Tonga has had trouble enforcing the collection of domestic fees and taxes and lacks the capacity to prosecute offenders for non-payment, resulting in underfunding of various government functions, including oversight and monitoring of major development projects (Blue Ocean Law & Pacific Network on Globalisation 2016).

Tonga reportedly possesses only three patrol boats, depending primarily on the New Zealand and Australian air forces for ocean surveillance. One Tongan government observer is sent out on mining vessels in the exploration phase, a practice that is expected to continue during actual mining. According to the Geology Department of the Ministry of Lands and Natural Resources, the companies currently operating in Tonga have been reluctant to provide detailed information to the government regarding the grade of minerals, the specific location of mine sites and ocean floor imaging, despite being legally required to do so. The government is concerned that companies may not share valuable genetic and biodiversity data or centralize data gathering in a way that would be useful to scientists and Pacific communities (Blue Ocean Law & Pacific Network on Globalisation 2016).

In short, while Tonga's DSM regulatory regime appears somewhat inclusive of environmental protections, it falls short with respect to social and consultative provisions. Lack of institutional capacity raises serious concerns of non-enforcement of DSM regulations, leading to irremediable environmental damage in a country already facing dire threats from sea level rise and climate change.

---

[11] A separate, more detailed law implementing the framework and elaborating procedures relating to fees, forms and regulations was reportedly in the works, as of 2016.

Cook Islands

The Cook Islands (CI), an archipelago of small islands with a population of approximately 17,000 people, is exploring the possibility of mining its EEZ for mineral deposits amounting to around ten billion tonnes of polymetallic manganese nodules (MNs), located at depths of 3000–6000 m (Cook Islands Seabed Minerals Authority 2018). The Cook Islands has been actively pursuing the development of a DSM industry, requesting assistance with regard to a comprehensive regulatory framework and sovereign wealth fund from the SPC-EU DSM Project, as well as from the International Monetary Fund and other advisers. CI has an established Seabed Minerals Authority, reporting to the Minister of Finance and comprising a Seabed Minerals Commissioner, a Legal Advisor and a Natural Resources Advisor (funded by the Commonwealth Secretariat). Although a 2016 tender for exploration received no bids, the Cook Islands has since entered into negotiations with multinational companies and foreign governments regarding both exploration in its EEZ and Cook Islands' sponsorship of DSM in the CCFZ. In 2016, CI signed an exclusive agreement with Ocean Minerals, an American company, to prospect for potential new sources of rare earth elements (REE) and scandium in its seabed.

With respect to legislation, the CI Parliament passed the Seabed Minerals Act in 2009, making it one of the world's first national legislations dedicated to regulating seabed mineral activities.[12]

Although CI legislation has been held up as a model framework in the region, it is heavily technical and focuses primarily on facilitating the mining regime. It contains no mentions of the precautionary principle or the avoidance of transboundary harm and very little on consultations with affected communities or the public. A separate instrument, the 2015 Seabed Minerals (Prospecting and Exploration) Regulations, contains a short section requiring DSM companies to apply the precautionary approach; however, it provides no instructions on how to do this in the context of DSM in the Cook Island's EEZ (Seabed Minerals (Prospecting and Exploration) Regulations 2015, §§9, 50). These regulations also have a short section on community consultation, whereby "the Authority *may* consult with the community in relation to any Application for a Title" (§37 emphasis added). This mirrors the permissive language in the original 2009 Seabed Act in which certain environmental provisions are also made optional with respect to the granting of prospecting permits (Seabed Minerals Act 2009, §91 (3)(f)).

In 2017, the Parliament of the Cook Islands passed Marae Moana, a bill creating one of the world's largest ocean sanctuaries. The Marae Moana Act provides zoning for different users, including mining operators, with no seabed mining allowed within 50 nautical miles around all islands of the Cook Islands (Marae Moana Act (2017), §24). The Act has not affected CI's current exploration contracts, however, and does not prohibit seabed mining throughout most of the reserved area, although it does restate the precautionary principle as well as a principle of community

[12] The Seabed Minerals Authority was subsequently established by the CI Government Cabinet in 2012. The Seabed Minerals Act 2009 officially entered into force on March 1, 2013.

participation (§5(c, d)). The Cook Islands appears to now be in the process of amending its seabed mining legislation and policy to be more in line with the principles of the Marae Moana Act and best practice, including possible mention of the precautionary approach and FPIC (Draft Documents 2019); it remains to be seen whether and how these principles will be operationalized and prioritized.

Pacific Islands Summary

The three legislative regimes discussed above, while different, suffer generally from the same fatal tendencies when considering the likelihood of environmental protection in the DSM realm: first, omissions related to mandatory environmental protections, and second, the absence of sufficient provisions on consultation and consent. The lack of an integrated, streamlined approach to regulating DSM – one that can actually be operationalized in the context of small island states with limited capacity – is reflective of legislative frameworks throughout the region. Many states, like PNG, still have old terrestrial mining laws in place from the 1990s, while others, similar to Tonga, possess newer legislation that mirrors the SPC's RLRF but is unlikely to be successfully implemented and enforced given resource constraints (see, e.g. the case of Tuvalu, a country of approximately 11,000 people, with two types of DSM deposits in its waters) (Blue Ocean Law & Pacific Network on Globalisation 2016). Moreover, DSM tends to fall under the purview of multiple government departments, with competing and often conflicting aims, in which better-resourced economic development departments and mineral authorities often win out over comparatively weak environmental or fisheries divisions (a problem faced by many national governments outside the region as well).

In short, even with improved, comprehensive legislation, concerns remain that in practice many of these jurisdictions will be unable to achieve sustainable resource management. Table 1 provides an overview assessment of these three jurisdictions.

4.4 *New Zealand*

DSM is in the exploratory stage in New Zealand (NZ), which contains reserves of SMS deposits and manganese nodules in its EEZ of over four million square kilometres (Lamping 2016). As a jurisdiction with substantially more resources and capacity than other PI nations, as well as established rule of law and comparatively institutionalized protections for its own indigenous peoples, New Zealand is in a better position to effectively legislate DSM and provide some form of regulatory oversight. It has nonetheless faced substantial civil society opposition with respect to proposed seabed mining operations,[13] including Chatham Rock Phosphate Ltd's

[13] Much publicity has centred around Trans-Tasman Resources' application to mine iron sands from the seabed of South Taranaki Bight, located 22–36 km offshore from Patea at depths of

Table 1 Assessment of DSM regulatory regime under norms of international law

	Precautionary principle	Transboundary harm	Polluter-pays principle	FPIC[a]	Other indigenous protections
Papua New Guinea	No mention in Mining Act; one general mention in Environment Act	No mention	Some mention of operator's duty to compensate for environmental harm; unclear how this works in the case of DSM	No mention	Brief mention of customary landowners
Tonga	Mentioned in general, vague terms	No mention	Brief mention of operator's obligation to compensate for and indemnify Tonga from costs relating to environmental harm	Mentioned once, not in relation to indigenous persons	None
Cook Islands	No mention in 2009 Act; mentioned in 2015 Regulations	No mention	Some mention of duty to compensate for environmental harm; regulator has discretion to apply environmental remedies	No mention	None

[a]*FPIC* free, prior and informed consent

attempts to mine a nodular phosphate deposit at 400 m water depth between the east coast of the South Island and the inhabited Chatham Islands (Nielsen et al. 2015). A 20-year seabed mining permit was granted to the company in late 2013, with feasibility studies to be completed shortly thereafter. In 2015, however, the NZ Environmental Protection Authority denied Chatham Rock Phosphate Ltd's application for a marine consent permit, finding that "there would be significant and permanent adverse effects on the existing benthic environment" (Decision on Marine Consent Application Chatham Rock Phosphate Limited (2015), §864) and that the destructive effects of the extractive activity, coupled with the potentially significant impact of the deposition of sediment on the areas adjacent to the mining blocks and on the wider marine environment, could not be mitigated by any set of conditions or adaptive management regime that might reasonably be imposed (Decision Summary, xviii).

---

between 20 and 42 m (thus not "deep-" sea mining, per se). TTR's application was denied by the NZ EPA in 2014, only to be approved in August 2017. The EPA's decision was overturned on by the High Court in Wellington (New Zealand Herald 2018), a decision currently being appealed by TTR and and crcross-appealed by environmental and indigenous groups at the Court of Appeal (Howard 2018).

According to news reports, Chatham Rock Phosphate is planning to resubmit a marine consent application to the EPA, anticipating completing the EPA reapplication process and hearing by early 2020 (Hartley 2018).

Seabed mining in New Zealand currently falls under two pieces of national legislation: the Crown Minerals Act 1991[14] and the Exclusive Economic Zone and Continental Shelf (Environmental Effects) Act (2012) ("the EEZ Act"), which manages the environmental effects of numerous activities, including SMS mining, beyond the 12 nautical mile limit (Boschen et al. 2013). Under the EEZ Act, the Environmental Protection Authority (EPA) is responsible for managing the effects of activities such as seabed mining beyond the territorial sea, specifically defining mining activity to include "areas of the seabed likely to contain mineral deposits" as well as "the taking or extraction of minerals from the sea or seabed, and associated processing of those minerals" (§4(1)). DSM would require a publicly notified marine consent from the EPA, which involves preparing an application and impact assessment, as well as a nationally notified public process in which the public can make submissions, present at a hearing and appeal decisions (§§38–52). In making its decision, a marine consent authority would be required to take into account various criteria, including effects on the environment and human health, biodiversity and species protection, existing interests, economic benefit and other matters (§59). The EEZ Act also includes provisions for a Māori Advisory Committee to provide advice to the EPA and/or to a marine consent authority, "if its advice is sought" (§18).

Although New Zealand's EEZ Act is undoubtedly stronger on environmental protection and inclusion of indigenous input than other legislation in the region, it does not make specific reference to the "precautionary approach", exhorting the Minister to instead "favour caution and environmental protection" if "in relation to the making of a decision under this Act, the information available is uncertain or inadequate" (§34). Although the sentiment is similar to that of the precautionary approach, the wording, as pointed out by both the New Zealand Green and Labour Parties, does not define what "caution" entails and is therefore unclear (Commentary 2011). Both parties also highlight concerns with the bill's provision requiring the Minister to first consider providing for an adaptive management approach if favouring caution means that an activity could be prohibited, given circumstances where adaptive management is inappropriate, as in cases with a risk of significant or irreversible environmental harm. The bill also specifically prevents the marine consent authority from considering "the effects on climate change of discharging greenhouse gases into the air" (§59(5b)), despite the serious risk of seabed mining-associated climate impacts (Levin et al. 2016).

Finally, notwithstanding token language on Māori input (although excluding the indigenous Moriori of the Chatham Islands), the provisions in the EEZ Act do not rise to the level of FPIC and may even fall short as consultative measures, given their optional nature. The Act could be significantly strengthened by including additional

[14] The Crown Minerals Act 1991 is similar to older mining acts from this period and contains only one mention of the seabed in the definition of land (§2(1)).

concrete mechanisms for indigenous participation in decision-making, in conformity with New Zealand's domestic law and international indigenous rights law.[15]

New Zealand's ability to take decisions denying marine consent applications for seabed mining and halting the issuance of new oil and gas offshore exploration permits indicates a regime with stronger rule of law, environmental protections and enforcement ability than many of its Pacific counterparts. That said, New Zealand's DSM legislation needs to clearly delineate the precautionary approach, incorporate climate concerns into marine consent decision-making processes and bolster community and indigenous protections.

4.5 Japan

Japan, one of the world's biggest economies with an ocean area in the top ten, has led efforts to exploit seabed minerals. It has done this in part to reduce its dependency on external imports, being highly reliant on critical metals for domestic manufacturing (particularly in high-tech and consumer electronics). Given its dependency on China for rare earth metals,[16] Japan, as a matter of national urgency, has sought to develop its own resource supply, including in the Area but also in its EEZ. It has made steady progress developing seabed mining technology, in 2012 launching a Rare Earth Research and Technology Centre in Hanoi, Vietnam, as part

[15] See, e.g. more from the Labour and Green Parties: "The Green Party is concerned about the limited way in which the Crown's obligations under the Treaty of Waitangi are implemented. Clause 14 provides for ways in which Māori may participate in decision-making but imposes no broader obligation on the Crown.

The bill should impose a general obligation on the Crown to administer and interpret the Act so as to give effect to the Treaty/Te Tiriti principles as legislation such as the Conservation Act 1987 does. And as iwi such as Ngāi Tahu sought, it should parallel the RMA to require the relationship of iwi and their culture and traditions with the marine environment, including taonga species, to be recognised and provided for in achieving the purpose of the Act"; "Labour is disappointed that iwi were not consulted in the drafting of this legislation. We share the concern of submitters that consultation with, and involvement of, iwi throughout the processes outlined in the bill are only optional and at the discretion of the EPA, especially given the lack of consultation so far. Labour members also support the submission of the Hokotehi Moriori Trust that the bill be amended to include reference to "Māori and Moriori" and "tikanga Māori and tikane Moriori" throughout the bill" (Commentary 2011). Further critiques from commentators include that the marine consent "applications are considered in the absence of a national planning framework for managing the oceans beyond our territorial seas. In addition, the EEZ Act is not currently supported by guiding documents such as national environmental standards, policy statements or plans that apply to the management of New Zealand's coastal marine area.… The absence of such documents or a planning framework presents a significant challenge for EEZ decision makers charged with deciding marine consent applications, and operators looking to exploit New Zealand's mineral resources" (Lamping 2016).

[16] When China cut its export quotas on rare earth minerals by 40% in 2010, prices soared, leading the US, joined by the EU and Japan, to bring a case against China in the WTO's dispute settlement body, which ruled against China. China subsequently dropped its quotas (World Trade Organization 2017).

of its rare earths' diplomacy initiative. In August 2017, Japan became the first country to successfully mine its seabed, tapping into a deposit of mineral resources 1600 m below the ocean's surface off the coast of Okinawa (McDonald 2017). In April 2018, a study published in *Scientific Reports* revealed that a deep-sea mud deposit at depths of close to 6000 m located within Japan's EEZ near Minami-Torishima Island could contain enough rare earth metals to potentially meet the world's supply "on a semi-infinite basis" (Takaya et al. 2018). Japan also holds two exploration contracts under the ISA to explore cobalt-rich ferromanganese crusts in 3000 km^2 in the Western Pacific Ocean, as well as polymetallic nodules in the CCFZ.

Japan's original Mining Act dates from 1950 and contains few provisions on environmental protection as well as no mention of the ocean or seabed minerals (Mining Act 1950). The original law also reportedly contained no regulations to check whether applicants had adequate technology, financing, track records and exploration and development plans; projects were automatically approved on a first-come first-served basis, resulting in companies receiving approval before better-qualified firms could apply (Kikkawa 2013). By the end of March 2010, the Japanese government had reportedly granted 8179 exploration rights, of which 81% remained undeveloped (Kikkawa 2013).

In 2011, Japan amended the Mining Act for the first time, responding to the need to ensure resource independence and increase domestic production. The changes, effectuated in January 2012, were primarily concerned with restricting the "first to file" arrangement in order to stimulate domestic natural resource development and make the mining system more efficient (Kikkawa 2013).

In 1982, Japan adopted the Law on Interim Measures for Deep Seabed Mining. This relatively short (~20 page) law is designed to "contribute to the promotion and extension of the public welfare through the rational development of deep seabed mineral resources" (Law on Interim Measures for Deep Seabed Mining (1982), Article 1). Permission to deep-sea mine is granted through the Minister of International Trade and Industry (Article 4), based on compliance with the standards set for mining areas by the Minister as well as sufficient financial standing and technological capability (Article 12). There is virtually nothing in the law regarding the environment or consultation, although there is a section on compensation for damages caused by "the discharge of wastewater, the accumulation of rubble or slag or the release of mineral smoke accompanying deep seabed mining in Japan" (Article 27), as well as a section establishing penalties and fines for violating provisions of the law (Chapter 6, Penal Provisions).

Although neither the Mining Act nor Japan's seabed mining law contains provisions on impacts to marine ecosystems, Japan possesses numerous domestic environmental laws that would inform its DSM regulatory regime, including the Basic Environment Law (1993), the Act on Prevention of Marine Pollution and Maritime Disaster (1970, No. 136), the Environmental Impact Assessment Law (1997, No. 81), the Law Concerning the Promotion of Business Activities with Environmental Considerations (2004, No. 77), the Act on the Exclusive Economic Zone and Continental Shelf (1996, No. 74) and the Act on Protection of Cultural Properties, as well as a Basic Act and Plan on Ocean Policy and biodiversity and climate change regimes. The Ocean Policy in particular emphasizes conservation and securing marine biodiversity (Basic Plan on Ocean Policy (2013), Chap. 2 Sect. 2). However, it also clearly calls for the promotion and development of energy and mineral resources including seabed minerals.

It is beyond the scope of this chapter to assess the entirety of Japanese laws relevant to DSM. However, the technical, sparse nature of the older mining laws raises concerns that environmental and social protections have not yet been adequately incorporated into the seabed mining regime and that a clear, integrated approach may not exist (Tatsuya 2017). Given Japan's rush to secure domestic mineral sources, there is a risk that environmental and social consequences, including transboundary harm to other states resulting from mining in Japan's waters and subsequent liability claims, could occur without adequate domestic regulatory legislation.

Domestic DSM Legislation Summary

In sum, Sects. 3 and 4 illustrate the need for domestic legislation to regulate DSM activities, even in the context of activities in the Area. As such, we find it useful to collate a non-exhaustive list of states with domestic legislation in place to govern DSM activities (taking place either within domestic jurisdiction or in the Area or both), as shown in Table 2.

5 Conclusion

This chapter has explored the national and international regulatory frameworks for DSM activities from the perspective of marine environmental protection. In particular, it has highlighted the need to standardize a precautionary, protective approach between the national and international seabed mining regimes, emphasizing that the current state of affairs – in which the international regime under the purview of the ISA is subject to the common heritage of mankind, but DSM in areas within national jurisdiction is left entirely to states – requires due attention.

Given the critical role of ocean ecosystems, in particular the seabed, in regulating climate and any number of other vital biodiversity functions, DSM activities, irrespective of where they take place, are a matter of "common concern to humankind" due to the harmful effects they are likely to cause to the marine environment. The latest advances in science validate these invaluable ecosystem services, while the deep ocean and its biodiversity are already established as common concerns of humankind. DSM activities should likewise be subjected to the same treatment (Hunter et al. 2018). As such, specific considerations under international law apply, including the due diligence obligation to regulate the activity effectively, as well as to exercise control over and, where necessary, enforce such regulations. Thus, in the case of activities in the Area, DSM contractors (i.e. the actual "polluters" that are responsible for the environmental degradation) in particular must account for the harm caused to the marine environment as it affects the community interest in favouring its protection (Sun 2018). This obligation is owed erga omnes and all

Table 2 Countries with domestic legislation pertaining to DSM activities

Country	DSM legislation
Belgium	Act on prospecting and exploration for, and exploitation of, resources of the seabed and ocean floor and subsoil thereof, beyond the limits of national jurisdiction/Loi relative à la prospection, l'exploration et l'exploitation des ressources des fonds marins et leur sous-sol au-delà des limites de la jurisdiction nationale (2013)
China	Law of the People's Republic of China on Exploration for and Exploitation of Resources in the Deep Seabed Area (2016)
Cook Islands	Seabed Minerals Act (2009)
Czech Republic	Prospecting, Exploration for and Exploitation of Mineral Resources from the Seabed beyond Limits of National Jurisdiction, Act No. 158 of 18 May 2000
Fiji	International Seabed Mineral Management Decree 2013 (Decree No. 21)
France	Law on the Exploration and Exploitation of Mineral Resources on the Deep Seabed 1981, Law No. 81-1135 of 23 December 1981
Germany	Seabed Mining Act of 6 June 1995; amended by article 74 of the Act of 8 December 2010
Italy	Regulations on the Exploration and Exploitation of the Mineral Resources of the Deep Seabed, Law No. 41 of 20 February 1985
Japan	Law on Interim Measures for Deep Seabed Mining, 1982
Nauru	International Seabed Minerals Act, No 26 of 2015
Kiribati	Seabed Minerals Act 2017
Russia	Provisional Measures to Regulate the Activity of Soviet Enterprises Relating to the Exploration and Exploitation of Mineral Resources of Seabed Areas Beyond the Limits of the Continental Shelf, 17 April 1982
Singapore	Deep Seabed Mining Act (2015)
Tonga	Tonga Seabed Minerals Act 2014
Tuvalu	Tuvalu Seabed Minerals Act 2014
UK	Deep-Sea Mining (Temporary Provisions) Act, 1981, chapter 53, 28 July 1981; Deep-Sea Mining (Exploration Licences) (Applications) Regulations 1982, No. 58; Deep-Sea Mining (Exploration Licences) Regulations 1984, No. 1230; Deep-Sea Mining Act 2014
US	Deep Seabed Hard Mineral Resources Act, 1980. Public Law 96-283, 28 June 1980, 94 Stat. 553 (30 U.S.C. 1401 et seq.), as amended 1 July 2000

*Note that multiple countries are believed to be in the midst of drafting or considering similar legislation: e.g. India, Republic of Korea, Federated States of Micronesia, Marshall Islands, Niue, Papua New Guinea, Solomon Islands and Vanuatu (ISA National Legislation Database).

states have an inherent interest in ensuring that it is effectively observed (Harrison 2017; Sun 2018). In this regard, treating international and national DSM activities as a matter of common concern would serve as a resounding call to ensure that regulations and standards adopted under both regimes are streamlined and harmonized.[17]

[17] Although states are required, pursuant to Article 208 of UNCLOS, to streamline their seabed exploitation activities and align them to standards agreed globally (or at least regionally), there is little evidence to indicate that such an endeavour is forthcoming. As a result differing standards and stringency in regulation would apply to both regimes, potentially causing efforts to protect the marine environment in one to cancel out the other. On the one hand, it is possible that stringent

Accordingly, failure by either regime to adopt and enforce necessary measures to protect the marine environment may attract responsibility under international law as well as domestic law. This could include substantial transboundary harm claims from states affected by DSM, particularly when their "rights and legitimate interests" are impacted from those activities,[18] as well as liability claims from private parties and other entities (e.g. commercial fishermen, tourism operators, indigenous groups, etc.).

Additionally, states are bound by an array of other laws applicable in the seabed mining arena, including broad and overlapping areas of environmental law stemming from the climate change regime among others, as well as international human rights obligations and special protective duties owed to indigenous peoples. All of these rights and laws may be implicated by DSM and must accordingly be considered and incorporated into national and international legislation, in order to create truly effective regulatory regimes designed to promote sustainability, prevent harm and conserve the common heritage of mankind – understood as extending beyond the pure monetary value of seabed minerals to include climate, biodiversity and other vital functions of the deep seabed.

As a parting note, we call upon states to come together pursuant to UNCLOS, in particular Article 197 (to cooperate both on a regional and global basis in formulating international rules and standards to protect the marine environment), Article 143 and Article 200 (to undertake scientific research to better understand the effects of DSM on the marine environment), Article 201 (to develop appropriate scientific criteria for the regulation of DSM) and Article 208 (to adopt standards that are no less stringent and effective than those agreed internationally). This effort would help ensure that if DSM activities take place (irrespective of whether they occur in the national or international seabed), they should be subjected to the highest environmental standards and the latest scientific knowledge; conducted in accordance with the precautionary approach and the polluter-pays principle; and remain accountable under all relevant areas of international law, with adequate monitoring and compliance mechanisms as well as appropriately enforceable legislative frameworks.

regulation and exacting standards adopted in the Area could shift contractor interest to areas within national jurisdiction where operational costs could be lower. This, in turn, could induce coastal states to engage in a race to the bottom to attract investors with weaker, less protective regulations. On the other hand, some contractors could elect to proceed with activities in the Area under the same scenario. This would entail increased operational costs, resulting in reduced revenue, and thereby potentially undermining the common heritage of mankind (as reduced revenue results in fewer financial benefits being available for equitable distribution among states, while the mineral resources of the Area continue to deplete). Consequently, streamlining environmental standards in both areas is critical.

[18] Article 142(1) of UNCLOS.

References

Advisory Opinion Responsibilities and Obligations of States Sponsoring Persons and Entities with respect to Activities in the Area. (2011). ITLOS decision No 17.

Anaya, J. (2005). Indigenous peoples' participatory rights in relation to decisions about natural resource extraction: The more fundamental issue of what rights indigenous peoples have in lands and resources. *Arizona Journal of International & Comparative Law, 22*(7), 16–17.

Anaya, J. (2015). Report of the special rapporteur on the rights of indigenous peoples on extractive industries and indigenous peoples. *Arizona Journal of International & Comparative Law, 32*(109), 118–120.

Aguon J. & Hunter J. (2019). Second Wave Due Diligence: The Case for Incorporating Free, Prior, and Informed Consent into the Deep Sea Mining Regulatory Regime. *Stanford Environmental Law Journal* 38, 3–55.

Basic Plan on Ocean Policy, March 2008, provisional translation. (2013). Accessible at: http://www.kantei.go.jp/jp/singi/kaiyou/kihonkeikaku/080318kihonkeikaku_E.pdf.

Beder, S. (2006). *Environmental principles and policies: An interdisciplinary introduction*. London: Earthscan.

Blue Ocean Law & Pacific Network on Globalisation. (2016). Resource roulette: How deep sea mining and inadequate regulatory frameworks imperil the Pacific and its peoples. Accessible at: http://nabf219anw2q7dgn1rt14bu4.wpengine.netdna-cdn.com/files/2016/06/Resource_Roulette-1.pdf.

Boschen, R. E., et al. (2013). Mining of deep-sea seafloor massive sulfides: A review of the deposits, their benthic communities, impacts from mining, regulatory frameworks and management strategies. *Ocean and Coastal Management, 84*(54), 62.

Boyle, A., & Chinkin, C. (2007). *The making of international law*. Oxford: Oxford University Press.

Brown, C. (2018). Mining at 2,500 fathoms under the sea: Thoughts on an emerging regulatory framework. *Ocean Science Journal, 53*(2), 287–300.

Chircop, A. (2011). Managing adjacency: Some legal aspects of the relationship between the extended continental shelf and the international seabed area. *Ocean Development and International Law, 42*(4), 307–316.

Christiansen, S., Ginzky, H., Singh, P., Thiele, T. (2018). The international seabed – the common heritage of mankind: Recommendations for future governance by the International Seabed Authority, IASS Policy Brief, 2018, 2. https://doi.org/10.2312/iass.2018.012.

Churchill, R., & Lowe, V. (1999). *The law of the sea* (3rd ed.). Manchester: Manchester University Press.

Commentary. (2011). Exclusive economic zone and continental shelf (environmental effects) bill: Government bill 321—2, As reported from the Local Government and Environment Committee, New Zealand Legislation. Accessible at: http://www.legislation.govt.nz/bill/government/2011/0321/23.0/whole.html.

Cook Islands Seabed Minerals Authority. (2018). Frequently asked questions. Accessible at: https://www.seabedmineralsauthority.gov.ck/faqs.

Crown Minerals Act. (1991). New Zealand legislation. Accessible at; http://www.legislation.govt.nz/bill/government/2011/0321/23.0/versions.aspx.

Decision on Marine Consent Application Chatham Rock Phosphate Limited: To mine phosphorite nodules on the Chatham Rise. (2015). Environmental protection authority. New Zealand Government. Accessible at: https://www.epa.govt.nz/assets/Uploads/Documents/Marine-Activities-EEZ/Activities/CRP-Decision-EEZ000006.pdf.

Draft Documents. (2019). Seabed Minerals Authority, Cook Islands. Accessible at: https://www.seabedmineralsauthority.gov.ck/seabed-minerals/documents_idl=10003_idt=514_id=2730_.html.

Egede, E. (2018). The area: Common heritage of mankind, sponsoring states of convenience and developing states. In M. Kotzur, N. Matz-Lück, A. Proelss, R. Verheyen, & J. Sanden (Eds.),

Sustainable ocean resource governance: Deep seabed mining, marine energy and submarine cables. Leiden: Brill.

Environment Act. (2000). No 64. Independent state of Papua New Guinea. Accessible at: http://www.mra.gov.pg/Portals/2/Publications/Enironment%20Act%202000.PDF.

Exclusive Economic Zone and Continental Shelf (Environmental Effects) Act. (2012). New Zealand legislation, reprint June 2017. Accessible at: http://www.legislation.govt.nz/bill/government/2011/0321/23.0/whole.html.

Franckx, E. (2010). The International Seabed Authority and the common heritage of mankind: The need for states to establish the outer limits of their continental shelf. *The International Journal of Marine and Coastal Law, 25*(4), 543–567.

Geddis, E. (2017). The due diligence obligation of a sponsoring state: A framework for implementation. In M. H. Nordquist, J. N. Moore, & R. Long (Eds.), *International marine economy: Law and policy*. Leiden: Brill.

Halpern, B. S., Frazier, M., Potapenko, J., Casey, K. S., Koenig, K., Longo, C., Lowndes, J. S., Rockwood, R. C., Selig, E. R., Selkoe, K. A., & Walbridge, S. (2015). Spatial and temporal changes in cumulative human impacts on the world's ocean. *Nature Communications, 6*, 7615.

Harrison, J. (2011). *Making the law of the sea: A study in the development of international law*. Cambridge, UK: Cambridge University Press.

Harrison, J. (2017). *Saving the oceans through law: The international legal framework for the protection of the marine environment*. Oxford: Oxford University Press.

Hartley, S. (2018). Chatham has plans for new mining consent. Otago Daily Times. Accessible at: https://www.odt.co.nz/business/chatham-has-plans-new-mining-consent.

Howard, R. (2018). Environmentalists plan ironsands cross-appeal. Newsroom. Accessible at: https://www.newsroom.co.nz/2018/10/05/265886/environmentalists-plan-ironsands-cross-appeal.

Hunter, J., Singh, P., & Aguon, J. (2018). Broadening common heritage: Addressing gaps in the deep sea mining regulatory regime. *Harvard Environmental Law Review* (online), 16 April 2018. Accessible at: http://harvardelr.com/2018/04/16/broadening-common-heritage/.

ISA National Legislation Database. Accessible at: https://www.isa.org.jm/national-legislation-database.

Jaeckel, A. (2017). *The International Seabed Authority and the precautionary principle: Balancing deep seabed mineral mining and marine environmental protection*. Leiden: Brill.

Jaeckel, A., Gjerde, K. M., & Ardron, J. (2017). Conserving the common heritage of humankind – Options for the deep-seabed mining regime. *Marine Policy, 78*, 150–157.

Jin, J., & Zhang, G. (2017). The forthcoming breakthrough: China's legislation on activities in the deep seabed area. In M. H. Nordquist, J. N. Moore, & R. Long (Eds.), *International marine economy: Law and policy*. Leiden: Brill.

Kaye, S. (2016). A zonal approach to maritime regulation and enforcement. In R. Warner & S. Kaye (Eds.), *Routledge handbook of maritime regulation and enforcement*. Oxon: Routledge.

Kielland, N. (2015). *Supporting aboriginal participation in resource development: The role of impact and benefit agreements*. Ottawa: Library of Parliament.

Kikkawa, T. (2013). Why has Japan revised the mining act? Historical and contemporary background. *Hitotsubashi Journal of Commerce & Management, 47*, 33–42.

Lamping, S. (2016). Seabed mining: Managing the future of mineral exploitation. Environmental Protection Authority. Accessible at: https://www.planning.org.nz/Attachment?Action=Download&Attachment_id=3780.

Law on Interim Measures for Deep Seabed Mining. (1982). Government of Japan. Accessible at: http://www.jstor.org/stable/20692545?read-now=1&loggedin=true&seq=19#page_scan_tab_contents.

Levin, L., et al. (2016). Hydrothermal vents and methane seeps: Rethinking the sphere of influence. *Frontiers Marine Policy, 3*(72), 1–23.

Lodge, M. (2012). The common heritage of mankind. *The International Journal of Marine and Coastal Law, 27*(4), 733–742.

Lodge, M., Johnson, D., Gurun, G. L., Wengler, M., Weaver, P., & Gunn, V. (2014). Seabed mining: International Seabed Authority environmental management plan for the Clarion–Clipperton zone: A partnership approach. *Marine Policy, 49*, 66–72.

Lodge, M., Segerson, K., & Squires, D. (2017). Sharing and preserving the resources in the deep sea: Challenges for the International Seabed Authority. *The International Journal of Marine and Coastal Law, 32*(3), 427–457.

Marae Moana Act. (2017). Cook Islands. Accessible at: http://www.paclii.org/ck/legis/num_act/mma201793.pdf.

Markus, T., & Singh, P. (2016). Promoting consistency in the deep seabed: Addressing regulatory dimensions in designing the International Seabed Authority's exploitation code. *Review of European Community and International Environmental Law, 25*, 347–362.

May, R. (2017). Papua New Guinea in 2016: Growing civil frustration. *Asian Survey, 57*, 194–198.

McDonald, B. (2017). Japan just mined the ocean floor and people want answers. CBC radios. Accessible at: http://www.cbc.ca/radio/quirks/october-14-2017-1.4353185/japan-just-mined-the-ocean-floor-and-people-want-answers-1.4353198.

Miller, K. A., Thompson, K. F., Johnston, P., & Santillo, D. (2018). An overview of seabed mining including the current state of development, environmental impacts, and knowledge gaps. *Frontiers in Marine Science, 4*, 418.

Mining Act 1992 Regulation, No. 20. Government of Papua New Guinea. Accessible at: http://www.mra.gov.pg/Portals/2/Publications/MINING_ACT%201992.pdf.

Mining Act, Act No. 289 of December 20, 1950. Government of Japan. Accessible at http://www.japaneselawtranslation.go.jp/law/detail_main?id=29&vm=2&re=.

New Zealand Herald. (2018). Seabed mining appeal wraps up in Wellington. Accessible at: https://www.nzherald.co.nz/nz/news/article.cfm?c_id=1&objectid=12036414.

Nielsen, S. H. H., et al. (2015). Chatham Rise nodular phosphate — modelling the prospectivity of a lag deposit (off-shore New Zealand): A critical tool for use in resource development and deep sea mining. *Oregon Geological Reviews, 71*, 545–557.

Pulu, T. B. (2013). The kingdom of Tonga and deep sea minerals. *Te Kaharoa, 6*, 50–76.

Roche, C. & Feenan, J. (2013). Drivers for the development of deep sea minerals in the Pacific. Mineral Policy Institute. Accessible at: http://www.mpi.org.au/wp-content/uploads/2014/05/Roche-and-Feenan-2013-Drivers-for-the-Development-of-Deep-Sea-Minerals-in-the-Pacific.pdf.

Rothwell, D., & Stephens, T. (2016). *The international law of the sea* (2nd ed.). London: Bloomsbury Publishing.

Seabed Minerals (Prospecting and Exploration) Regulations 2015. Cook Islands. Accessible at: http://www.seabedmineralsauthority.gov.ck/PicsHotel/SeabedMinerals/Brochure/2015Tender/Seabed%20Minerals%20_Prospecting%20and%20Exploration_%20Regulations%202015_2.pdf.

Seabed Minerals Act. 2009. Cook Islands. Accessible at: https://www.seabedmineralsauthority.gov.ck/laws-and-regulations.

Seabed Minerals Act 2014, Act 10. Tonga. Accessible at: http://extwprlegs1.fao.org/docs/pdf/ton143350.pdf.

Sun, L. (2018). *International environmental obligations and liabilities in deep seabed mining* (dissertation Leiden). Amsterdam: Ipskamp Printing.

Szablowski, D. (2011). Operationalizing free, prior, and informed consent in the extractive industry sector? Examining the challenges of a negotiated model of justice. *Canadian Journal of Development Studies, 30*, 111–130.

Singh, P., & Jaeckel, A. (2018). Future prospects of marine environmental governance. In: M. Salomon & T. Markus (eds.). Handbook on marine environment protection. Springer, Cham.

Takaya, Y., et al. (2018). The tremendous potential of deep-sea mud as a source of rare-earth elements. *Scientific Reports, 8*, 1–8.

Tanaka, Y. (2015). *The international law of the sea* (2nd ed.). Cambridge: Cambridge University Press.

Tatsuya, N. (2017). *Japanese legal system related to DSM*. Tokyo: Tokyo University of Marine Science and Technology, presentation accessible at: http://www.cosie-sip.ynu.ac.jp/wp-content/uploads/2017/12/SIP_IntSympo171107_2.2_Nakada.pdf.

UNDP. (2014). National human development report, Papua New Guinea, from wealth to wellbeing: Translating resource revenue into sustainable human development. Accessible at: http://www.pg.undp.org/content/dam/papua_new_ guinea/docs/Publications/FINAL%20PNG%20NHDR_low%20res.compressed.pdf.

Van Dover, C. L., Ardron, J. A., Escobar, E., Gianni, M., Gjerde, K. M., Jaeckel, A., Jones, D. O. B., Levin, L. A., Niner, H. J., Pendleton, L., Smith, C. R., Thiele, T., Turner, P. J., Watling, L., & Weaver, P. P. E. (2017). Biodiversity loss from deep-sea mining. *Nature Geoscience, 10*, 464–465.

Verlaan, P. (2018). Deep seabed mining. In M. Solomon & T. Markus (Eds.), *Handbook on marine environment protection: Science, impacts and sustainable management*. Cham: Springer.

Wedding, L. M., Reiter, S. M., Smith, C. R., Gjerde, K. M., Kittinger, J. N., Friedlander, J. N., Gaines, S. D., Clark, M. R., Thurnherr, A. M., Hardy, S. M., & Crowder, L. B. (2015). Managing mining of the deep seabed. *Science, 349*(6244), 144–145.

World Trade Organization. (2017). China – rare earths. Accessible at: https://www.wto.org/english/tratop_e/dispu_e/cases_e/1pagesum_e/ds431sum_e.pdf. Brief biography and photos of authors.

Pradeep Singh is a doctoral researcher at the Faculty of Law, University of Bremen and Centre for Marine Environmental Sciences (MARUM), Bremen. He is a member of the International Research Training Group INTERCOAST, an interdisciplinary collaboration between the Universities of Bremen (Germany) and Waikato (New Zealand) on coastal and marine environmental issues. His academic qualifications include an LL.B. degree from the University of Malaya (Hons.) (Dist.), an LL.M. in Global Environment and Climate Change Law from the University of Edinburgh and an LL.M. from Harvard Law School.

Julie Hunter is an attorney for Blue Ocean Law, an international law practice based in Guam specializing in indigenous and environmental law. She has produced reports and conducted fieldwork on national deep-sea mining legislation across the Pacific, including in Papua New Guinea, Tonga and Fiji. Other recent policy and academic work include research into the environmental and human impacts of extractive and large-scale development projects throughout Oceania and Southeast Asia. She holds a J.D. from Yale Law School, a M.Sc. in International Relations from the London School of Economics and Political Science and a B.A. from Harvard University.

Part V
Economic Considerations

Deep-Sea Natural Capital: Putting Deep-Sea Economic Activities into an Environmental Context

Torsten Thiele

Abstract The natural capital of the vast deep ocean is significant yet not well quantified. The ecosystem services provided by the deep sea provide a wide range of benefits to humanity. Proposed deep-sea economic activities such as fishing, deep-sea mining and bioprospecting therefore need to be assessed in this context. In addition to quantifying the economic benefits and costs of such activities on their own, their potential impact on the deep-sea natural capital also needs to be considered.

This article describes such a natural capital approach, identifies relevant ecosystem services and looks at how a range of proposed commercial activities could be assessed in this context. It suggests a methodology for such analysis and suggests an approach to a sustainable blue deep-sea economy that is consistent with environmental precaution. It will close with suggestions of how potential risks can best be handled.

The article aims to show that modern environmental economics based on natural capital can provide a useful framework for deciding future deep-sea efforts.

Keywords Deep-sea mining · Natural capital · Economic activities

1 Natural Capital and the Deep Ocean

When deep-sea mining was first proposed in the last century, little was known of the major ecosystem services that the deep ocean provides and which are crucial to life on earth (Jobstvogt et al. 2014). Today, we are aware of the important regulating and supporting services that the deep-sea environment delivers, in addition to provisioning us with seafood and other resources. We now know that deep-sea biodiversity is highly complex and rich, with unique deep-sea habitats (e.g. seamounts, submarine canyons, hydrothermal vents, cold seeps) sustaining a multitude of marine populations and endemic species (Van den Hove and Moreau 2007). Decisions to allow

T. Thiele (✉)
Institute for Advanced Sustainability Studies, Potsdam, Germany
e-mail: torsten.thiele@iass-potsdam.de

R. Sharma (ed.), *Environmental Issues of Deep-Sea Mining*,
https://doi.org/10.1007/978-3-030-12696-4_18

activities in the deep therefore need to take potential impacts on these ecosystems into account (Worm et al. 2006) and provide for the protection of habitats and the reduction and elimination of pollution as otherwise some of these benefits of the deep ocean may be lost as a consequence of such activities (Rogers et al. 2015).

Quantifying these natural capital benefits can help to address this challenge, as otherwise decision-makers only compare the costs and benefits of the proposed activity, thereby valuing the ecosystem services of nature by default at zero. Yet we have limited knowledge of the economic value of protecting this natural capital of the deep ocean. The ecosystem services approach helps bridge the gaps between natural science and economics (Gowdy et al. 2010; Braat and de Groot 2012). Economists have made progress in developing a methodology for assessing this critical form of capital (Parks and Gowdy 2013), including how to deal with remote locations and complex marine value chains (Drakou et al. 2017). The Law of the Sea Convention (LOSC 1982) considers the international seabed, the "area", as common heritage of humankind, with corresponding obligations to advance equity (Bourrel et al. 2016) and the need to address also economic considerations (EPRS 2015). This paper suggests that the emerging natural capital approach can to be applied to the deep sea.

1.1 The Concept of Natural Capital

Natural capital is defined as the stock of ecosystems and biodiversity that yield a flow of valuable ecosystem services (WBCSD 2009). Natural capital is a way to describe Earth's natural assets, including soil, air, water and living things, existing as complex ecosystems, which provide a range of services to humankind. Depleting and degrading these reserves may irreversibly reduce the availability of benefits to future generations (Science for Environmental Policy 2017). Viewing natural capital in an inclusive way allows to develop a comprehensive language for sustainability science (UNU-IHDP 2012). The total stock of capital employed in the economic system, including natural capital, determines the well-being for present and future generations.

The social worth of natural resources can be divided into three parts: use value, intrinsic value, and option value (UNU-IHDP and UNEP 2012). The value of both use and option value can be calculated directly, whereas intrinsic value requires indirect valuation approaches, such as through surveys. Viewing ecosystems as capital assets capable of producing goods and services allows the application of the standard tools and analysis developed by natural resource economics for modelling these complex systems (Helm 2015). It allows focusing on competing uses and the need to account for the value of ecosystem services in order to make efficient choices between uses (Barbier 2012). As an example, a recent study (Beaumont et al. 2008) detailed goods and services resulting from UK marine biodiversity and valued 8 of 13 services in monetary terms in order to prioritise management approaches. Provided that a society saves the rents from extraction, a sustainable increase in consumption may be possible (Hamilton and Hartwick 2017). This approach also helps to identify "perverse subsidies" which encourage ecosystem destruction such cheap fuel for deep-sea bottom trawling and take decisions that regulate such activities.

The natural capital approach provides a helpful starting point, methodology and research agenda (Atkinson et al. 2014). Mainstreaming natural capital into decisions however requires that the methodology is straightforward and routine (Guerry et al. 2016). By developing habitat quality and habitat diversity models, cost-benefit decisions can be embedded into a larger policy context (Polasky et al. 2011). Analysis of ecosystem services in quantitative financial terms, of the benefits and costs that specific parts of nature provide to humans, offers a framework to analyse the interaction between human activities and conservation (Boyd and Banzhaf 2007). It balances the use of a resource and its conservation according to how societies value consumptive and nonconsumptive goods (Perrings et al. 2010). Human activities can lead to ecosystem disservices, including loss of biodiversity, chemical contamination and sedimentation, poisoning of nontarget organisms and emissions of greenhouse gases and pollutants, that need to be quantified. The ecosystem services perspective needs however to be a part of a broader approach. Economics is best suited to describe choices made at the margin, where its language of stocks and flows can be adapted to the environment. A range of techniques such as hedonic pricing, contingent valuation and replacement cost methods have been used to calculate an overall value of nature (Costanza et al. 1997), which the TEEB report confirmed to be substantive (TEEB 2009). This thinking can then be integrated into a broader policy mandate that considers all social and societal goals and ambitions.

1.2 Natural Capital and Deep-Sea Activities

Once we have established a clear methodology, we can apply this approach to the deep sea in order to assess whether proposed commercial activities such as deep-sea mining provide positive net benefits to society (Halvar and Fujita 2009). Due to the nature of the deep ocean (the immense pressure, the hard to reach bottom, the lack of data and the distance from land), the exploration and especially the exploitation of the resources on the seabed pose immense technical and environmental challenges (Rozemeijer et al. 2017). Before embarking on this effort, a balanced assessment is ever more important in the context of a deep-sea legal regime outlined in Annex XI of the United Nations Convention on the Law of the Sea that is committed to precaution and long-term sustainability. Our knowledge of deep-ocean biodiversity only hints at thousands of undiscovered organisms and their benefits (TEEB 2012). Some threatened species, such as cold-water corals, have life spans of hundreds or even thousands of years (Barbier et al. 2014).

In 1997, an initial assessment of the overall value of ocean ecosystems came up with a tentative valuation of US$ 24tr (Costanza et al. 1997). A more recent paper has confirmed this calculation with an even higher number (Costanza et al. 2017). Whilst these are numbers of the ocean as a whole rather than for the deep sea alone, as the ocean is a single and complex, interconnected system, this provides an appropriate starting point against which to assess the impact of human activities. Thus the total value of ecosystem services is considerable and, according to another recent study (De Groot et al. 2013), ranges between US$ 490/year for the total bundle of ecosystem services that can potentially be provided by an "average" hectare of open oceans to almost US$ 350,000/year for the potential services of an "average" hectare of coral reefs.

Therefore, we need to consider how costs of any (direct and indirect) environmental impacts are treated, for purposes of weighing the net private and public benefits of activities such as deep-sea mining against the public cost of such environmental impacts (Pacific Possible and World Bank 2016). That way, potential benefits of mining can be assessed against a broad set of criteria. These also need to consider the potential benefits of alternative uses of the deep ocean, such as discovery of new substances and marine genetic resources for the development of medicine in future. Yet human impacts (Thiel 2003) are already significantly affecting the deep ocean (Ramirez-Llodra et al. 2011), with climate change being particularly relevant (Sweetman et al. 2017). As an example, anthropogenic greenhouse gas emissions are starting to seriously impact the ocean as a whole, through warming and acidification, with implications that are independent from the specific location of the emission. Global marine ecosystems are rapidly degrading as a result of overfishing, pollution and lack of regulatory protection (Glover and Smith 2003). It is therefore necessary to take into account both the cumulative and synergistic effects of human activities and the future state of the natural capital of the deep sea.

2 Deep-Sea Ecosystem Services

Progress is being made in identifying key aspects of deep-sea ecosystem services and how to value them (Armstrong et al. 2012). We now understand, for instance, that the resource use value of provisioning services "is likely to be dwarfed" by the value the ocean provides in regulating the global climate. What matters for most ecosystem services is the diversity of traits that different species possess, such as nitrogen fixers, pollinators and nutrient recyclers (Perrings et al. 2010). Using an ecosystem services framework both allows for a better understanding of the range of services the oceans provide (McLeod and Leslie 2009) and provides a way to draw in wider audiences, including other stakeholders and policymakers (Guerry et al. 2011). Marine ecosystems provide all four main categories of services identified in the Millennium Ecosystem Assessment, i.e. provisioning, regulating and supporting services and existence values. However, quantifying these in the deep is still a challenge (Atkinson et al. 2014) even as the methodologies have matured (Goulder and Kennedy 2011).

2.1 Provisioning Services

The oceans provide a wide range of provisioning services, from marine food such as fish to genetic resources discovered through bioprospecting (Arrieta et al. 2010). Yet our limited knowledge and lack of regulation have, for instance, led to unsustainable trawling practices (Pusceddu et al. 2014) and overfishing (Victorero et al. 2018). Deep-sea trawling resulted in taking of deep-sea fish for human consumption which has led to significant damage of deep-sea habitats (Puig et al. 2012), as deep-sea species are particularly vulnerable to overexploitation, due to their slow growth and late maturity (Morato et al. 2006). As a result, a precautionary approach has been recommended, limiting any extraction of marine products to activities that have been fully studied and assessed within an ecosystem-based management framework.

2.2 Regulating and Supporting Services

Regulating services provide benefits such as climate regulation, disease regulation, water regulation and purification. The ocean contributes to air quality and treating waste as well as acting as a key global carbon stock (Laffoley et al. 2014). The ocean is critical to keeping our global climate stability, and any impact on such system-wide regulatory services needs to be considered as serious, as these reinforcing regulatory services are subject to tipping points which can move the entire system to a different state. The ocean supplies crucial supporting services such as nutrient cycling, primary production via photosynthesis and oxygen production which are necessary for the production of all other ecosystem services and are linked to key planetary boundaries, a concept developed to identify the constraints our planet faces (Whiteman et al., 2012).

2.3 Intrinsic Values and Services

Non-material benefits obtained from ecosystems include spiritual and religious values, recreation and ecotourism, aesthetic, inspirational, educational, sense of place and cultural heritage. Intrinsic values and cultural services are of great significance (Chan et al. 2012). There are societies that value pristine nature because it's a connection (Cohn 2012). This important area requires further work both from a scientific and from a systemic point of view (Balmford et al. 2002). One of the proposed methodologies to deal with the lack of familiarity with deep-sea ecosystem functions is stated preference theory (Barkmann et al. 2008) and discrete choice experiments (Hoyos 2010). By applying such approaches to value specific outcomes, economists have been able to find ways to calculate specific amounts that can then be used, for instance, to award damages where specific ecosystems have been affected. Research applied choice experiment and contingent valuation methods to value the diversity of biological diversity (Christie et al. 2006). These examples suggested that society puts a high intrinsic value on a healthy ocean.

3 Human Activities in the Deep Ocean

3.1 Non-invasive

Some activities in the deep ocean, such as scientific research or the collection of genetic material in the form of eDNA, if properly organised and undertaken professionally and with the necessary caution, are likely to have minimal impact on deep-sea ecosystems (Arico and Salpin 2005). They may however deliver significant breakthroughs, such as the discovery of new substances. The evident benefits of potential medicines even whilst taking uncertainty into account could, for instance, provide additional arguments to justify protecting certain areas, such as all hydrothermal vents for their high biotechnological utility (Leary et al. 2009). The negotiations that have begun in 2018 at the United Nations for a new Marine Biodiversity Agreement aim to address such marine genetic resource issues in a way to create maximum benefit for humankind (Harden-Davies 2017).

3.2 Extractive

Extractive activities, in particular those related to non-renewable resources, are likely to cause negative impacts to the marine ecosystem, by removing the resource it can no longer be accessed by marine life, and in the process additional harm is caused (Halfar and Fujita, 2007). Environmental data plays a key role in understanding such impacts and in designing of the mining system as well as planning of the mining operation, in particular since restoration without net loss is not possible (Van Dover et al. 2017). Areas likely to be affected by deep-sea mining will range from the surface and water column due to particles discharged (accidentally or otherwise) during lifting, at-sea processing and transportation to the seafloor where the mineral will be separated from the associated substrate either due to scooping or drilling leading to resuspension and redistribution of debris in the bottom water along the path of the collector device as well as in the vicinity of the mining tracks and the land due to metal extraction and tailing disposal (Sharma 2011). The spatial extent of these efforts needs to be taken into account (Benn et al. 2010). A recent EU study of deep-sea mining impacts identified a wide range of aspects to be considered, through environmental impact assessments (Ecorys 2014).

3.3 *Harmful Activities*

Different international conventions ban harmful activities in the ocean, such as dumping, (IMO 1996) altogether. These sectoral measures have been important to limit such negative impacts on the marine environment.

4 Assessment of Proposed Commercial Activities

Natural capital accounting aims to integrate the stock of nature into national accounts; it can then show whether an activity such as mining will improve or reduce overall economic value. An important benefit of this approach is its systemic nature; it shows that only when looking at the overall impact on the system can we visualise the real impacts of an activity. It is therefore necessary to fully integrate all costs and benefits into the calculation so that price signals can provide the right incentives. This includes assessing the alternatives to mining such as recycling (Teske et al. 2016). By dividing ecosystem into provisioning, regulating and other services and applying direct and indirect valuation techniques, the externalities these services provide can be captured, allowing public policy to take them into account to craft better regulation (Goulder and Kennedy 2011). In the case of deep-sea mining activities, these need to reflect different deep-sea habitats, such as manganese nodules (Vanreusel et al. 2016), ferromanganese crusts (Probert et al. 2007), sulphides (Boschen et al. 2013) and hydrothermal vents (Gollner et al. 2011; Van Dover et al. 2018), including their differences in ecology and in economics (Martino and Parson, 2012).

5 Policy Implications

In order to effectively regulate activities, it is therefore not sufficient to address each impact independently of each other but consider their interaction to prevent the deterioration of its natural capital. As the High-Level Expert Group on Sustainable Finance noted, the risk is exploitation beyond its rate of renewal, not least due to policies that do not value it sufficiently (HLEG 2018). As a consequence, it requires a comprehensive assessment of costs and benefits of deep-sea activities (SPC 2016). There are not only technological challenges to offshore mining; it is also trapped in a vicious circle of uncertain operations, the need for high capital investments and fluctuating prices for mineral resources (Rozemeijer et al. 2017). In the European Union, for instance, the European Union Biodiversity Strategy aims to halt the loss of biodiversity and ecosystem services in the EU and to help stop global biodiversity loss by 2020. The Strategy aims to ensure "no net loss of biodiversity and ecosystem services" (Action 7, Target 2). Maximising value from the oceans requires that significant areas be set aside for marine protection (Helson et al. 2010), precaution (Mengerink et al. 2014) and adopting new management approaches, including vertical zoning (Levin et al. 2017).

Since the formation of the ISA in 1994 based on the United Nations Convention on Law of the Sea (LOSC 1982), it has served as the regulating agency for all activities related to the resources in the area (i.e. defined as the seabed and subsoil beyond the limits of national jurisdiction) (Sharma 2017). Through a series of international workshops, the ISA has also issued recommendations for assessment of possible environmental impacts from exploration of nodules and for establishment of environmental baselines and associated monitoring programme for exploration of polymetallic sulphides and cobalt crusts. Going forward, the application of the common heritage principle to the area will require an equitable assessment of any future mining activities (Bourrel et al. 2016). This is particularly relevant to fairly address the needs of developing nations (Egede 2011) as the ISA progresses in developing its regulatory framework (ISA 2015) based on the precautionary principle (Jaeckel 2017). The same applies mutatis mutandis for deep-sea mining in the EEZ, e.g. in the Pacific (Pacific Possible and Worldbank 2016). As national accounts increasingly include natural capital (Obst et al. 2016), likewise international bodies such as the ISA will need to take it into consideration, in particular in the light of UN Sustainable Development Goal 14 (Shepherd et al. 2016). As the ISA develops options for the deep-seabed mining regime (Jaeckel et al. 2017), an economic perspective that takes the deep-sea natural capital fully into account is required to deliver fully assess the opportunity ahead (Wedding et al. 2015).

References

Arico, S., & Salpin, C. (2005). *Bioprospecting of genetic resources in the deep seabed: Scientific, legal and policy aspects* (p. 72). Yokohama: United Nations University – Institute of Advanced Studies.

Armstrong, C. W., Foley, N. S., Tinch, R., & van den Hove, S. (2012). Services from the deep: Steps towards valuation of deep sea goods and services. *Ecosystem Services, 2*, 2–13.

Arrieta, J. M., Arnaud-Haond, S., & Duarte, C. M. (2010). What lies underneath: Conserving the oceans' genetic resources. *PNAS, 107*(43), 18318–18324.

Atkinson, G., Bateman, I. J., & Mourato, S. (2014). Valuing ecosystem services and biodiversity. In D. Helm & C. Hepburn (Eds.), *Nature in the balance: The economics of biodiversity* (pp. 101–134). Oxford: Oxford University Press.

Balmford, A., Bruner, A., Cooper, P., et al. (2002). Economic reasons for conserving wild nature. *Science, 297*, 950–953.

Barbier, E. B. (2012). Progress and challenges in valuing coastal and marine ecosystem services. *Review of Environmental Economics and Policy, 6*(1), 1–19.

Barbier, E. B., Moreno-Mateos, D., Rogers, A. D., Aronson, J., Pendleton, L., Danovaro, R., et al. (2014). Protect the deep sea. *Nature, 505*, 475–477.

Barkmann, J., Glenk, K., Keil, A., Leemhuis, C., Dietrich, N., Gerold, G., & Marggraf, R. (2008). Confronting unfamiliarity with ecosystem functions: The case for an ecosystem service approach to environmental valuation with stated preference methods. *Ecological Economics, 65*(1), 48–62.

Beaumont, N. J., Austen, M. C., Mangi, S. C., & Townsend, M. (2008). Economic valuation for the conservation of marine biodiversity. *Marine Pollution Bulletin, 56*(3), 386–396.

Benn, A. R., Weaver, P. P., Billet, D. S. M., Hove, S. V. D., Murdock, A., Doneghan, P., et al. (2010). Human activities on the deep seafloor in the North East Atlantic: An assessment of spatial extent. *PLoS One, 5*(9), e12730.

Boschen, R. E., Rowden, A. A., Clark, M. R., & Gardner, J. P. A. (2013). Mining of deep-sea seafloor massive sulfides: A review of the deposits, their benthic communities, impacts from mining, regulatory frameworks and management strategies. *Ocean and Coastal Management, 84*, 54–67.

Bourrel, M., Thiele, T., & Currie, D. (2016). The common of heritage of mankind as a means to assess and advance equity in deep sea mining. *Marine Policy*. https://doi.org/10.1016/j.marpol.2016.07.017.

Boyd, J., & Banzhaf, S. (2007). What are ecosystem services? The need for standardized environmental accounting units. *Ecological Economics, 63*(2–3), 616–626.

Braat, L. C., & de Groot, R. (2012). The ecosystem services agenda: Bridging the worlds of natural science and economics, conservation and development, and public and private policy. *Ecosystem Services, 1*(1), 4–15.

Chan, K. M. A., Satterfield, T., & Goldstein, J. (2012). Rethinking ecosystem services to better address and navigate cultural values. *Ecological Economics, 74*, 8–18.

Christie, M., Hanley, N., Warren, J., Murphy, K., Wright, R., & Hyde, T. (2006). Valuing the diversity of biodiversity. *Ecological Economics, 58*(2), 304–317.

Cohn, R. (2012). Putting a price on the real value of nature. *Yale Environment, 360*. http://e360.yale.edu/feature/putting_a_price_on_the_real_value_of_nature/2481/.

Costanza, R., d'Arge, R., de Groot, R., Farber, S., Grasso, M., Hannon, B., et al. (1997). The value of the world's ecosystem services and natural capital. *Nature, 387*(6630), 253–260.

Costanza, R., de Groot, R., Braat, L., Kubiszewski, I., Fioramonti, L., Sutton, P., et al. (2017). Twenty years of ecosystem services: How far have we come and how far do we still need to go? *Ecosystem Services, 28*, 1–16.

De Groot, R., Brander, L., van der Ploeg, S., Costanza, R., Bernard, F., Braat, L., et al. (2013). Global estimates of the value of ecosystems and their services in monetary units. *Ecosystem Services, 1*, 50–61.

Drakou, E. G., Pendleton, L., Effron, M., Ingram, C., & Teneva, L. (2017). When ecosystems and their services are not co-located: Oceans and coasts. *ICES Journal of Marine Science, 74*, 1531–1539. https://doi.org/10.1093/icesjms/fsx026.

Ecorys. (2014). Study to investigate the state of knowledge of deep-sea mining. Final Report under FWC MARE/2012/06 – SC E1/2013/04, 271 pp.

Egede, E. (2011). *Africa and the deep seabed regime: Politics and international law of the common heritage of mankind*. Berlin: Springer. http://www.springer.com/us/book/9783642176616.

EPRS. (2015). Deep-seabed exploitation – tackling economic, environmental and societal challenges. European parliamentary research service scientific foresight unit (Stoa) PE 547.401, 92 pp.

Glover, A. G., & Smith, C. R. (2003). The deep-sea floor ecosystem: Current status and prospects of anthropogenic change by the year 2025. *Environmental Conservation, 30*(03), 219–241.

Gollner, S., Kaiser, S., Menzel, L., Jones, D. O. B., Brown, A., Mestre, N. C., et al. (2011). Resilience of benthic deep-sea fauna to mining activities. *Marine Environmental Research, 129*, 76–101.

Goulder, L. H., & Kennedy, D. (2011). Interpreting and estimating the value of ecosystem services. In P. Kareiva, H. Tallis, T. H. Ricketts, G. C. Daily, & S. Polasky (Eds.), *Natural capital: Theory and practice of mapping ecosystem services* (pp. 15–33). Oxford: Oxford University Press.

Gowdy, J., Hall, C., Klitgaard, K., & Krall, L. (2010). What every conservation biologist should know about economic theory. *Conservation Biology, 25*, 628–630.

Guerry, A. D., Plummer, M. L., Ruckelshaus, M. H., & Harvey, C. J. (2011). Ecosystem service assessments for marine conservation. In P. Kareiva, H. Tallis, T. H. Ricketts, G. C. Daily, & S. Polasky (Eds.), *Natural capital: Theory and practice of mapping ecosystem services* (pp. 296–232). Oxford: Oxford University Press.

Guerry, A., Polasky, S., Lubchenco, J., Chaplin-Kramer, R., Daily, G. C., Griffin, R., et al. (2016). Natural capital and ecosystem services informing decisions: From promise to practice. *Proceedings of the National Academy of Sciences of the United States of America, 112*(24), 7348–7355.

Halfar, J., & Fujita, R. M. (2007). Danger of deep-sea mining. *American Association for the Advancement of Science, 316*(5827), 987.

Hamilton, K., & Hartwick, J. (2017). Wealth and sustainability. In K. Hamilton & C. Hepburn (Eds.), *National wealth* (pp. 343–363). Oxford: Oxford University Press.

Harden-Davies, H. (2017). Deep-sea genetic resources: New frontiers for science and stewardship in areas beyond national jurisdiction. *Deep Sea Research Part II: Topical Studies in Oceanography, 137*, 504–513.

Helm, D. (2015). *Natural capital*. New Haven: Yale University Press. 266 pages.

Helson, J., Leslie, S., Clement, G., Wells, R., & Wood, R. (2010). Private rights, public benefits: Industry-driven seabed protection. *Marine Policy, 34*, 557–566.

HLEG. (2018). Final report 2018 by the high-level expert group on sustainable finance. European Commission, 99 pp.

Hoyos, D. (2010). The state of the art of environmental valuation with discrete choice experiments. *Ecological Economics, 69*, 1595–1603.

IMO. (1996). Protocol on the prevention of marine pollution by dumping of wastes and other matter, London 1996, amended in 2006.

ISA. (2015). *Developing a regulatory framework for deep sea mineral exploitation in the area: Draft framework, high level issues and action plan, version II* (pp. 35–36). https://www.isa.org.jm/files/documents/EN/OffDocs/Rev_RegFramework_ActionPlan_14072015.pdf, 15 July 2015.

Jaeckel, A. (2017). *The international seabed authority and the precautionary principle*. Leiden: Brill. 362 pages.

Jaeckel, A., Gjerde, K. M., & Ardron, J. A. (2017). Conserving the common heritage of humankind – options for the deep-seabed mining regime. *Marine Policy, 78*, 150–157.

Jobstvogt, N., Hanley, N., Hynes, S., Kenter, J., & Witte, U. (2014). Twenty thousand sterling under the sea: Estimating the value of protecting deep-sea biodiversity. *Ecological Economics, 97*, 10–19.

Laffoley, D., Baxter, J. M., Thevenon, F., & Oliver, J. (2014). *The significance and management of natural carbon stores in the open ocean*. Gland: IUCN. 124 pages.

Leary, D., Vierros, M., Hamon, G., Arico, S., & Monagle, C. (2009). Marine genetic resources: A review of scientific and commercial interest. *Marine Policy, 33*(2), 183–194.

Levin, N., Kark, S., & Danovaro, R. (2017). Adding the third dimension to marine conservation. *Conservation Letters, 11*, 1–1. September 1–14.

LOSC. (1982). United Nations convention on the law of the sea, Montego Bay. http://www.un.org/depts/los/conventionagreements/conventionoverviewconvention.html.

Martino, S., & Parson, L. M. (2012). A comparison between manganese nodules and cobalt crust economics in a scenario of mutual exclusivity. *Marine Policy, 36*, 790–800.

McLeod, K., & Leslie, H. (Eds.). (2009). *A review of "Ecosystem-Based Management for the Oceans"*. Washington, DC: Island Press. 368 pp.

Mengerink, K. J., Van Dover, C. L., Ardron, J., Baker, M., Escobar-Briones, E., Gjerde, K. G., et al. (2014). A call for deep-ocean stewardship. *Science, 344*, 696–698.

Morato, T., Watson, R., Pitcher, T. J., & Pauly, D. (2006). Fishing down the deep. *Fish and Fisheries, 7*(1), 24–34.

Obst, C., Hein, L., & Edens, B. (2016). National accounting and the valuation of ecosystem assets and their services. *Environmental and Resource Economics, 64*(1), 1–23.

Pacific Possible and Worldbank. (2016). *Precautionary management of deep sea mining potential in Pacific Island countries* (p. 87). Washington, DC: Worldbank.

Parks, S., & Gowdy, J. (2013). What have economists learned about valuing nature? A review essay. *Ecosystem Services, 3*, e1–e10.

Perrings, C., Naeem, S., Ahrestani, F., Bunker, D. E., Burkill, B., Canziani, G., Elmqvist, T., et al. (2010). Ecosystem services for 2020. *Science, 330*(6002), 323–324.

Polasky, S., Caldarone, G., Duarte, T. K., Goldstein, J., Hannahs, N., Ricketts, T., et al. (2011). Putting ecosystem services models to work: Conservation, management, and trade-offs. In P. Kareiva, H. Tallis, T. H. Ricketts, G. C. Daily, & S. Polasky (Eds.), *Natural capital: Theory and practice of mapping ecosystem services* (pp. 249–263). New York: Oxford University Press.

Probert, P. K., Christiansen, S., Gjerde, K. M., Gubbay, S., & Santos, R. S. (2007). Management and conservation of seamounts (chapter 20). In T. J. Pitcher, T. Morato, P. J. B. Hart, M. R. Clark, N. Haggan, & R. S. Santos (Eds.), *Seamounts: Ecology, fisheries, and conservation* (Blackwell fisheries and aquatic resources series, 12) (pp. 442–475). Oxford: Blackwell Publishing.

Puig, P., Canals, M., Company, J. B., et al. (2012). Ploughing the deep-sea floor. *Nature, 489*, 286–289.

Pusceddu, A., Bianchelli, S., Martín, J., et al. (2014). Chronic and intensive bottom trawling impairs deep-sea biodiversity and ecosystem functioning. *Proceedings of the National Academy of Sciences, 111*, 8861–8866.

Ramirez-Llodra, E., Tyler, P. A., Baker, M. C., Bergstad, O. A., Clark, M. R., Escobar, E., et al. (2011). Man and the last great wilder-ness: Human impact on the deep sea. *PLoS One, 6*(7), e22588. https://doi.org/10.1371/journal.pone.0022588.

Rogers, A. D., Brierley, A., Croot, P., Cunha, M. R., Danovaro, R., Devey, C., et al. (2015). Delving deeper: Critical challenges for 21st century deep-sea research. In K. E. Larkin, K. Donaldson, & N. McDonough (Eds.), *Position paper 22* (p. 224). Ostend, Belgium: European Marine Board.

Rozemeijer, M. J. C., Sander, W., van den Burg, K., Jakk, R., Lallier, L. E., & van Craenenbroeck, K. (2017). Seabed mining. In K. Johnson, G. Dalton, & I. Masters (Eds.), *Building industries at sea: 'Blue Growth' and the new maritime economy*. Gistrup: River Publishers. http://hdl.handle.net/1854/LU-8545099.

Science for Environment Policy. (2017) Taking stock: Progress in natural capital accounting. In-depth report 16 produced for the European Commission, DG Environment by the Science Communication Unit, UWE, Bristol. Available at: http://ec.europa.eu/science-environment-policy.

Sharma, R. (2011). Deep-sea mining: Economic, technical, technological and environmental considerations for sustainable mining. *Marine Technology Society Journal, 45*, 28–41.

Sharma, R. (2017). *Deep-sea mining: Resource potential, technical and environmental considerations* (pp. 535). Basel: Springer International Publishing AG.

Shepherd, E., Milner-Gulland, E. J., Knight, A. T., Ling, M. A., Darrah, S., van Soesbergen, A., & Burgess, N. D. (2016). Status and trends in global ecosystem services and natural capital: Assessing progress toward Aichi Biodiversity Target 14. *Conservation Letters, 9*(6), 429–437.

SPC. (2016). An assessment of the costs and benefits of mining deep-sea minerals in the Pacific Island Region. Deep-sea mining cost-benefit analysis. Prepared by Cardno for the Secretariat of the Pacific Community. Suva, Fiji. SPC Technical Report SPC00035, ISBN: 978-982-00-0955-4.

Sweetman, A. K., Thurber, A. R., Smith, C. R., Levin, L. A., Mora, C., Wei, C.-L., et al. (2017). Major impacts of climate change on deep-sea benthic ecosystems. *Elementa: Science of the Anthropocene, 5*, 4. https://doi.org/10.1525/elementa.203.

TEEB. (2009). *TEEB—The Economics of Ecosystems and Biodiversity for National and International Policy makers—summary: Responding to the value of nature*. Wesseling: Welzel and Hardt. 31 pp.

TEEB. (2012). *Why value the oceans? A discussion paper* (p. 31). UNEP – The Economics of Ecosystems & Biodiversity.

Teske, S., Florin, N., Dominish, E., & Giurco, D. (2016). *Renewable energy and deep-sea mining: Supply, demand and scenarios* (p. 38). Sydney: Institute for Sustainable Futures, University of Technology.

Thiel, H. (2003). Anthropogenic impacts on the deep sea. In P. A. Tyler (Ed.), *Ecosystems of the world 28: Ecosystems of the deep oceans* (pp. 427–471). Amsterdam: Elsevier Science.

UNU-IHDP and UNEP. (2012). *Inclusive wealth report 2012. Measuring progress toward sustainability. Summary for decision-makers*. Bonn: UNU-IHDP.

Van den Hove, S., & Moreau, V. (2007). *Deep-sea biodiversity and ecosystems: A scoping report on their socio-economy, management and governance*. Nairobi: UNEP-WCMC. 84.

Van Dover, C. L., Ardron, J. A., Escobar, E., Gianni, M., Gjerde, K. M., Jaeckel, A., et al. (2017). Biodiversity loss from deep-sea mining. *Nature Geoscience, 10*, 464–465. https://doi.org/10.1038/ngeo2983.

Van Dover, C. L., Arnaud-Haond, S., Gianni, M., Helmreich, S., Huber, J. H., Jaeckel, A. L., et al. (2018). Scientific rationale and international obligations for protection of active T hydrothermal vent ecosystems from deep-sea mining. *Marine Policy, 90*, 20–28.

Vanreusel, A., Hilario, A., Ribeiro, P. A., Menot, L., & Martinez Arbizu, P. (2016). Threatened by mining, polymetallic nodules are required to preserve abyssal epifauna. *NPG Scientific Reports, 6*, 6.

Victorero, L., Watling, L., Palomares, M. L., & Nouvian, C. (2018). Out of sight, but within reach: A global history of bottom-trawled deep-sea fisheries from >400 m depth. *Frontiers in Marine Science*. https://doi.org/10.3389/fmars.2018.00098.

Wedding, L. M., Reiter, S. M., Smith, C. R., Gjerde, K. M., Kittinger, J. N., Friedlander, A. M., et al. (2015). Managing mining of the deep seabed. *Science, 349*, 144–145. https://doi.org/10.1126/science.aac6647.

Whiteman, G., Waler, B., & Perego, P. (2012). Planetary boundaries: Ecological foundations for corporate sustainability. *Journal of Management Studies*. https://doi.org/10.1111/j.1467-6486.2012.01073.x.

World Business Council for Sustainable Development (WBCSD). (2009). *Corporate ecosystem valuation: A scoping report*. Geneva: World Business Council for Sustainable Development. 34 pages.

Worm, B., Barbier, E. B., Beaumont, N., Duffy, J. E., Folke, C., Halpern, B. S., Jackson, et al. (2006). Impacts of biodiversity loss on ocean ecosystem services. *Science, 314*(5800), 787–790.

Torsten Thiele (torsten.thiele@iass-potsdam.de) is a Senior Research Associate at the Institute for Advanced Sustainability Studies, Potsdam, Visiting Fellow at the LSE IGA and Founder of Global Ocean Trust, promoting sustainable governance solutions for the marine space, based on technology, finance and innovation. Torsten is an expert in the financing of complex infrastructure. He worked at leading financial institutions in London for over 20 years, both at development banks such as the EBRD and at private sector firms such as JP Morgan and CIT. Till 2013 Torsten was Head of Telecoms Project Finance at Investec Bank plc, financing mobile and fixed line operators, including subsea cables.

Torsten holds graduate degrees in law and economics from Bonn University, an MPA from the Harvard Kennedy School, an MPhil in Conservation Leadership from the University of Cambridge and was a 2014 Advanced Leadership Fellow at Harvard.

Review of Mining Rates, Environmental Impacts, Metal Values, and Investments for Polymetallic Nodule Mining

Rahul Sharma, Farida Mustafina, and Georgy Cherkashov

Abstract This chapter reviews different mining rates for polymetallic nodules and evaluates the ensuing environmental impacts as well as metal production with respect to land reserves, metal prices, and projected investments. It is expected that these can be applied for different mineral deposits in other ocean basins as well.

Keywords Mining rates · Environmental impacts · Metal values · Investments · Polymetallic nodules

1 Introduction

Mero (1965) unraveled the economic potential of deep-sea manganese deposits and predicted that "deep-sea mining would start in 20 years time," a projection that was subsequently revised due to the discovery of new land-based ores and decline in metal prices (Lenoble 1990; Cronan 2000). This was revived by suggestions that these mineral deposits would be the alternative source of metals in the twenty-first century (Lenoble 2000; Kotlinski 2001), and subsequently activities related to deep-sea mining have seen a sudden spurt with the number of entities rising from just eight "Pioneer Investors" recognized under the UN Law of the Sea in the first four decades (1970–2010) to as many as 29 contractors registered currently for exploration for polymetallic nodules, polymetallic sulfides, and cobalt-rich ferromanganese crusts in the international seabed area (www.isa.org.jm).

This coupled with the development of environmental guidelines by the International Seabed Authority for the Contractors (ISA 2013; being revised in 2019) as well as reports of technology development for mining of deep-sea resources, as demonstrated by the "large-scale deep-sea mineral extraction"

R. Sharma (✉)
CSIR-National Institute of Oceanography, Dona Paula, Goa, India
e-mail: rsharma@nio.org

F. Mustafina · G. Cherkashov
Saint Petersburg State University, Saint Petersburg, Russia

R. Sharma (ed.), *Environmental Issues of Deep-Sea Mining*,
https://doi.org/10.1007/978-3-030-12696-4_19

(www.japantimes.co.jp) and "completion of submerged trials for seafloor production tools" (www.nautilusminerals.com), seems to suggest the possibility of commercial deep-sea mining becoming a reality in the near future. With this impending eventuality, this chapter reviews the estimates of mineable area, environmental impacts, and metal production for different mining rates for polymetallic nodules, as well as re-evaluates the estimated expenditures based on current metal prices and investments calculated as per inflation rate from earlier studies (Sharma 2017a, b) in order to help the industry optimize their investments and production while taking environmental impacts into consideration.

2 Proposed Mining Rates for Polymetallic Nodules

Different mining rates have been proposed in various studies over the last five decades (Table 1) wherein most of the mining rates vary between 1 and 3 Mt/year with an overall average of 2.1 Mt/year (excluding the anomalous range of 1–25 Mt/year) and 2.9 Mt/year (including all values). It is critical to note that all studies (except #12) have expressed mining rates for "dry" nodules, which represent the quantity of nodules that will eventually be available for metal extraction as per their concentrations in nodules. On the other hand, mining rates for "wet" nodules that would include the moisture content represent the rate at which eventual mining should be carried out in order to meet the required quantity of dry nodules for further processing (Sharma 2017a).

Table 1 Proposed mining rates for polymetallic nodules (Sharma 2017b)

Sr. No	Proposed by	Mining rate
1	Flipse et al. (1973)	1–5 Mt $year^{-1}$ (dry)
2	Kaufman (1974)	1 Mt $year^{-1}$ (dry)
3	Siapno (1975)	1 Mt $year^{-1}$ (dry)
4	Pearson (1975)	1–25 Mt $year^{-1}$ (dry)
5	Lenoble (1980)	2.1–3 Mt $year^{-1}$ (dry)
6	OMI (1982)	3 Mt $year^{-1}$ (dry)
7	OMA (1982)	2.1 Mt $year^{-1}$ (dry)
8	Academie des Science (1984)	3 Mt $year^{-1}$ (dry)
9	Dick (1985)	1–2 Mt $year^{-1}$ (dry)
10	UNOET (1987)	3 Mt $year^{-1}$ (dry)
11	Herrouin et al. (1991)	1.5 Mt $year^{-1}$ (dry)
12	ISA (2008)	1.5 Mt $year^{-1}$ (wet)
	Average (considering all individual values as well as median values in case of range)	2.9 Mt $year^{-1}$
	Average (considering all individual values as well as median values in case of range and excluding anomalous range of Sr. no. 4)	2.1 Mt $year^{-1}$

3 Estimation of Ore Production, Mineable Area and Environmental Impacts

3.1 Criteria and Factors for Nodule Mining

Mining of deep-sea minerals such as polymetallic nodules would be a combination of resource parameters as well as the geological and environmental setting of the resource. According to the criteria suggested by United Nations Ocean Economics and Technology Branch (UNOET 1987), the mine site should meet the following conditions:

- Cutoff grade : 1.8% Ni + Cu
- Cutoff abundance : 5 kg/m^2
- Topography : acceptable
- Duration of recovery : 20 years (life of a mine site)
- Annual recovery : 3 million tons/year or 1.5 million tons/year (ISA 2008)
- Operation per year : 300 days

Once the above criteria are fulfilled, the mineability of the resource will be influenced by features associated with it such as the following:

- Distribution characteristics of the minerals
- Association of minerals with substrates
- Seabed topography (slopes) in the mining area
- Environmental setting in the mining area

A detailed description of various environmental and geological factors that would influence the design and operation of deep-sea mining system is given by Sharma (2019, Chap. 12, this volume).

3.2 Estimation of Variables Associated with Mining

Estimation of ore production, mineable area, size of mine site, area of contact, volume and weight of sediment to be disturbed, and quantity of mine tailings can be made using formulae suggested in Sharma (2017a):

Ore Production

Ore production per day (O_p) can be calculated as:

$$O_p = \text{MR}_{(\text{dry})} / D = 5000\,\text{t} / \text{day}$$

where:

$MR_{(dry)}$ = mining rate (1.5 Mt/year),
D = no. of days of operation/year (300 days)

Total Mineable Area (*M*)

Mineable area (*M*) is calculated by subtracting the percentage of unmineable areas, due to various factors that have been estimated in different studies and include unfavorable topography (15–45%), abundance (10–25%), and grade (10–15%), from the total area (UNOET 1987). Considering conservative percentages for these factors for an area of 75,000 km^2 allotted for polymetallic nodules to a contractor, *M* can be estimated as:

$$M = A_t - \left(A_u + A_g + A_a\right) = 37{,}500\,\text{km}^2$$

where:

A_t = the total area (75,000 km^2)
A_u = area unmineable due to topography (30%)
A_g = area below cutoff grade (10%)
A_a = area below cutoff abundance (10%)

However, this could vary for each mine site as per actual area (A_t) allotted as well as the ground conditions in different mining areas.

Size of Mine Site (A_s)

Size of mine site is estimated (UNOET) as:

$$A_s = \frac{(A_r)(D)}{(A_n)(E)(M)} = \frac{1.5\times10^9\times20\ \text{years}}{5\ \text{kg/m}^2\times25\%\times37{,}500\ \text{km}^2} = 6400\ \text{km}^2$$

where:

$MR_{(dry)}$ = mining rate (1.5 Mt/year)
D = duration of mining operation (20 years)
A_n = average nodule abundance in a mineable area (5 kg/m^2)
E = overall efficiency of the mining device (25%)
M = mineable area (37,500 km^2)

Other components being constant, the actual size of the mine site would depend on two variables A_n (average nodule abundance) and E (overall efficiency of the mining device). Studies have shown that higher mean nodule abundances are expected in the first-generation mine sites in CCZ as well as CIOB than the cutoff value considered here (Singh and Sudhakar 2015), hence reducing the actual area of the mine site.

The overall efficiency (E) $= e_s \times e_d$, where e_s is the sweep efficiency and e_d is dredge (or pickup) efficiency and a conservative estimate of mean efficiency based on different studies works out to 25% (UNOET 1987). However, a better efficiency of the mining device should be possible with development in technology that would result in smaller area of the mine site leading to lower environmental impacts. As per recent research by Hong et al. (2019, Chap. 5, this volume), if e_s is around 60%, the overall efficiency (E) will be about 50%. On the other hand, in case e_s is 80%, the E of collecting system will rise up to 64%. This makes a big difference in the aspect of sustainability: reduction of 25% in the size of mine site (A_s).

Area of Contact

For a mining rate of 1.5 Mt/year (MR_{dry}) and average nodule abundance (A_n) of 5 kg/m^2, the Area of contact (A_c) on the seafloor would be:

$$A_c = MR_{(dry)} / A_n = 300{,}000{,}000\,m^2, \text{i.e.}, 300\,km^2,$$
$$\text{i.e.}, 1\,km^2/\text{day}\,(\text{for } 300\text{ days/year of operation})$$

where:

$MR_{(dry)}$ = mining rate (1.5 Mt/year)
A_n = average nodule abundance (5 kg/m^2)

So, for an annual operation time of 300 days, the contact area would be 1 km^2 per day, and for a 20 year lifetime of a mine site, it would be 6000 km^2, which is only 8% of the entire allotted area (i.e. 75,000 km^2 'Contract' area). It is also expected that the average nodule abundance would be higher in the mining area than the cut-off considered here, implying that the actual contact area (that is the area to be scraped on the seafloor) will be much smaller than that estimated here, thus reducing the environmental impact.

Volume and Weight of Sediment Disturbed

During mining of nodules, the associated sediments will also be disturbed causing serious environmental impacts. Considering a ratio of 1:9 for the proportion of nodules with respect to sediments on the seafloor (Sharma 2011) and penetration by the nodule miner to a depth of 10 cm in to the sediment to collect nodules in 300 km^2 area, the volume of sediment (V_s) to be disturbed is estimated as:

$$V_s = A_d \times D_p \times C_s / 100$$

where:

A_d = area of disturbance (300 km^2 or 3 × 10^8 m^2/year)
D_p = depth of penetration (10 cm or 0.1 m)
C_s = coverage of sediment (90%) (nodule to sediment ratio is 1:9)

Hence,

$$V_s = 3\times10^8 \times 0.1\times 90/100 = 3\times10^5 \times 90\,\mathrm{m}^3/\mathrm{year}$$
$$= 2700\times10000\,\mathrm{m}^3/\mathrm{year\ in\ 300\ days}$$
$$= 90{,}000\,\mathrm{m}^3/\mathrm{day}$$

This is equal to 103,500 t/day (wet sediment) or 20,700 t/day (dry sediment) considering 1.15 g/cm^3 as density and 80% of water content in deep-sea sediment (Khadge and Valsangkar 2008). The redistribution of this sediment is expected to cause environmental impact in the marine environment. However, as the nodule to sediment ratio considered here is for a large area that includes locations without nodules as well as low coverage, the actual sediment disturbed in a typical first-generation mine site would be lower as it would have higher nodule abundance and proportionately lesser associated sediment.

Estimation of Mine Tailings

A simplified method of estimating the quantity of mine tailings after processing of nodules can be:

$$\mathrm{MR}_{(\mathrm{dry})} - \left(\mathrm{MR}_{(\mathrm{dry})} \times C_\mathrm{m}/100\right),$$

where:

$\mathrm{MR}_{(\mathrm{dry})}$ = mining rate (1.5 Mt/year)
C_m = total concentration (%) of metals to be extracted from nodules

However, as the nodules will be processed through chemical leaching process, the exact estimation of mine tailings can only be attempted after the nature of the final discarded tails are known. For example, in a four metal recovery process, subsequent to manganese recovery, a benign quantity of discard slag is produced. Further, the environmental impact of tailings is dependent on the process route (P. K. Sen, personal communication).

4 Estimates for Different Mining Rates

Estimates of variables described above for different mining rates are compiled in Table 2.

4.1 Ore Production

Mining of nodules at different mining rates (1–3 MT/year) for a duration of 300 days/year would lead to production of 3333–10,000 t of ore/day (Table 2) irrespective of nodule abundance. This calls for not only adequate mechanism to collect the nodules from the seafloor and a lifting mechanism through >5 km of water column but also proportionate handling and storage infrastructure on the surface platform as well as transport vessels to carry ore to the shore for further processing. Given the disparity between ratios of constituent metals in the nodules and world demand (Pearson 1975), optimum production of ore with desired composition of metals would be required to maintain a balance in metal prices as overproduction could lead to lowering of prices making deep-sea mining uneconomical.

4.2 Mineable Area and Area of Contact

Estimates show that area (size) of the mine site would range between 4267 km^2 and 12,800 km^2 (Table 2) for different mining rates (1–3 MT/year) which is 5.68–17.06% of the "Contract" area (75,000 km^2) for nodule abundance of 5 kg/m^2 and 25% efficiency of the mining system. However, it can easily be expected that the first-generation mine site would have higher average nodule abundances (8–15 kg/m^2) and much better efficiency for mining due to improvement in technology in order to make it cost-effective. Studies based on statistical analysis of nodule abundance data have shown that for a typical mining area of 75,000 km^2 allotted to a contractor for polymetallic nodules, about 10% of the area will be adequate as first-generation mine site for supplying 3 million tones of dry nodules for a period of 20 years in the Clarion Clipperton Fracture Zone in the Pacific Ocean, whereas the same would be about 20% in the Central Indian Basin (Singh and Sudhakar 2015). Further, the area of contact that would be actually scraped would vary between 200 km^2/year and 600 km^2/year to mine at these rates which comes to 0.66–2.0 km^2 per day (Table 2) and would further reduce with higher nodule abundance expected in the first-generation mine site, proportionately reducing direct environmental impacts. These are confirmed by a study on production key figures for planning of nodule mining (Volkmann and Lehman 2018) which suggests that with an overall mining efficiency of 30–40% and for abundances of 13.7 kg/m^2 and 16.5 kg/m^2, an

Table 2 Estimates for mining of polymetallic nodules at different mining rates (Sharma 2017b)

Estimates for operation of 300 days $year^{-1}$	Mining rate					Remark/implication
	1.0 Mt $year^{-1}$	1.5 Mt $year^{-1}$	2.0 Mt $year^{-1}$	2.5 Mt $year^{-1}$	3.0 Mt $year^{-1}$	
Ore production/day	3333.3 t day^{-1}	5000 t day^{-1}	6666.6 t day^{-1}	8333.25 t day^{-1}	10,000 t day^{-1}	Proportionate storage and transport facility required
Area (size) of mine site[a]	4267 km^2	6400 km^2	8533 km^2	10,667 km^2	12,800 km^2	Negligible (5.68–17.06%) of the contract area
Area of contact per year[a]	200 km^2	300 km^2	400 km^2	500 km^2	600 km^2	0.66–2 km^2 day^{-1}
Volume of sediment disturbed at seafloor	60,000 m^3 day^{-1}	90,000 m^3 day^{-1}	120,000 m^3 day^{-1}	150,000 m^3 day^{-1}	180,000 m^3 day^{-1}	Major source of environmental impact
Wt. of disturbed sediment (wet) (at 1.15 g cm^{-3} density)	69,000 t day^{-1}	103,500 t day^{-1}	138,000 t day^{-1}	172,500 t day^{-1}	207,000 t day^{-1}	In slurry form that can travel with bottom currents to adjacent areas
Wt. of disturbed sediment (dry) (at 80% water content)	13,800 t day^{-1}	20,700 t day^{-1}	27,600 t day^{-1}	34,500 t day^{-1}	41,400 t day^{-1}	Dominant (50–60%) fine clays may remain suspended for longer periods

[a]For cutoff abundance of 5 kg m^{-2} (minimum required for nodule mining, UNOET 1987), Contract area of 75,000 km^2, mining for 20 years for 300 days of operation per year

area of approximately 114–182 km^2 will be mined per year or ~2300 to 3600 km^2 in a period of 20 years for production rates of 1.5–2 MT per year.

4.3 *Volume and Weight of Sediment Distributed*

Although, the actual mineable area and area of contact during mining would be small implying significant respite to the direct impact on seafloor environment and its biological communities, a significant volume of sediments associated with nodules will be disturbed (60,000–180,000 m^3/day) for different mining rates that translates into significant weight of wet sediment (69,000–207,000 t/day) and equivalent dry sediment (13,800–41,400 t/day) as shown in the estimates (Table 2). Hence, the key to minimize the ensuing impacts would be to identify high abundance areas so as reduce the amount of associated sediment that would be disturbed and also to screen the sediments close to the seafloor and minimize their lateral migration to adjacent areas so as to restrict the impact to the mining strip.

4.4 *Mine Tailings After Extraction of Metals*

In case of four metal recovery, the concentration of metals of interest (Mn + Cu + Ni + Co) is ~26%, whereas the remaining is likely to be discarded as tailings after processing, the form and quantity of which would depend on the process adopted for the purpose. As the processing of minerals is expected to be done at shore-based processing plants, the disposal of these tailings could lead to environmental issues at the disposal sites on land unless alternative uses are found. Given various chemical and physical properties of these tailings, studies have indicated several possible uses ranging from agricultural applications to being used for highway construction and for landfill purposes as well as industrial fillers in order to make deep-sea mining sustainable (Wiltshire 2017).

5 Estimation of Metal Values for Different Mining Rates

Concentrations of manganese, nickel, copper, and cobalt reported for Pacific Ocean (Morgan 2000) and Indian Ocean (Jauhari and Pattan 2000) were used for estimating metal production at mining rates between 1 and 3 Mt/year (Table 3) for each year as well as for a period of 20 years that is considered as the lifetime of a mine site (UNOET 1987). For four metals, the expected yield is ~5 to 15 million tons for different mining rates for the total period of 20 years, whereas the actual demand may vary over different years during the mining operations. However, this is assuming that 100% of the metals in the ore will be extracted, whereas in reality the percent recovery of metals depends on the process adopted and necessary corrections need to be applied. These metal production values would be critical for arriving at the decision of an optimum mining rate based on the projected demand of these metals in the world market for ensuing years from the time when mining is expected to commence. It should be noted that these values are indicative, and the actual yield for each mine site would vary with the concentrations of metals in the area. For example, higher concentrations for Mn, Ni, and Cu have been reported for Pacific Ocean by Hein and Koschinsky (2014).

Further considering the metal prices for manganese, nickel, copper, and cobalt for 2018, metal values were evaluated (in US$) for different mining rates between 1 and 3 Mt/year for known concentrations of these metals in the Pacific and Indian Oceans. Estimates show that metals worth $ 16.5–49.7 billion in Pacific Ocean and $ 15.9–48 billion in Indian Ocean would be produced over a period of 20-year lifetime of a mine site for mining rates varying between 1 and 3 million tons/year (Table 4). However, the metal values will be critical to evaluate the returns on investment with respect to different mining rates in order to optimize the same for

Table 3 Estimated metal production (Mt) for different mining rates

	Mining rate (Mt year^{-1}) of dry nodules									
	1.0		1.5		2.0		2.5		3.0	
Metal	Pacific[a]	Indian[b]	Pacific	Indian	Pacific	Indian	Pacific	Indian	Pacific	Indian
Mn	0.22	0.24	0.33	0.36	0.44	0.48	0.55	0.6	0.66	0.72
Ni	0.01	0.011	0.015	0.165	0.02	0.022	0.025	0.0275	0.03	0.033
Cu	0.0078	0.0104	0.0117	0.0156	0.0156	0.0208	0.0195	0.0260	0.0234	0.312
Co	0.0023	0.001	0.00345	0.0015	0.0046	0.002	0.00575	0.0025	0.0069	0.0030
Total/year	0.2401	0.2624	0.36015	0.3936	0.4802	0.5248	0.60025	0.6560	0.7203	0.7872
Total (in 20 years)[c]	4.802	5.248	7.203	7.872	9.604	10.496	12.005	13.120	14.406	15.744

[a]Mn = 22%, Ni = 1.0%, Cu = 0.78%, Co = 0.23% (Morgan 2000) for Pacific Ocean
[b]Mn = 24%, Ni = 1.1%, Cu = 1.04%, Co = 0.1% (Jauhari and Pattan 2000) for Indian Ocean
[c]As life of a mine site is expected to be 20 years (UNOET 1987)

Table 4 Estimated metal values (million US$)[a] for different mining rates

	Mining rate (Mt $year^{-1}$) of dry nodules									
	1.0		1.5		2.0		2.5		3.0	
Metal	Pacific	Indian	Pacific	Indian	Pacific	Indian	Pacific	Indian	Pacific	Indian
Mn	448.8	489.6	673.2	734.4	897.6	979.2	1122.0	1224.0	1346.4	1468.8
Ni	141.5	155.65	212.25	233.47	283.0	311.3	353.75	389.12	424.50	466.95
Cu	53.11	70.82	79.67	106.23	106.23	141.64	132.79	177.06	159.35	212.47
Co	185.72	80.75	278.58	121.12	371.45	161.50	464.31	201.87	557.17	242.25
Total/year	829.13	796.82	1243.7	1195.2	1658.28	1593.64	2072.85	1992.0	2487.42	2408.47
Total (in 20 years)[b]	16,582	15,936	24,874	23,904	33,165	31,873	41,457	39,840	49,748	48,169

[a]Average metal prices as of 21 June 2018: Mn = $2.04/kg, Ni = $14.15/kg, Cu = $6.81/kg, Co = $80.75/kg (www.metalprice.com, www.lme.com)
[b]As life of a mine site is expected to be 20 years (UNOET 1987)

Fig. 1 Long-term fluctuations in metal prices for manganese, nickel, copper, and cobalt (www.infomine.com/investment/metal-prices)

Fig. 1 (continued)

sustaining the mining operations for the entire period. Another study that assumed 3 million dry tons of nodules mined per year as per metal prices of March 2018 estimated 845,325 tons of Mn + Ni + Co + Cu valued at $ 2370 million (Roth 2018).

It is important to note that metal prices have been fluctuating over long periods of time (Fig. 1), and between 2011 and 2018, the prices have risen for manganese (+54%) and cobalt (+205%), whereas the prices have fallen for nickel (−38%) and copper (−28%) which have been used for estimation of resource potential (see Sect. 8). In view of the fluctuating metal prices, it is appropriate to include a time-series analysis to predict likely prices for a "would be" sea nodule plant (P. K. Sen, personal communication). Hence, the mining rates will have to be optimized based on metal prices, actual quantity of metals produced, their consumption, and the returns on investment at the time of mining.

6 Analysis of Global Production Rates of Metals and Duration of Availability of Metals from Land Reserves

Evaluation of production rates of selected metals in the last 20 years shows consistent increase (Fig. 2a–f) due to rising demands for metals for their industrial applications that include production of automobiles, electronic goods, batteries, heavy machinery, and others (www.usgs.gov).

Mean increase of production rate for 20 years was calculated using production of each metal from 1997 to 2016. This was based on estimation of increase of production rate for each year with respect to the previous year, and then a mean value of these pairs was calculated (in percentage) as follows:

$$P_{\mathrm{m}} = \frac{\sum \frac{(P_{x+1} - P_x) * 100\%}{P_x}}{19}$$

where:

P_{m} is increase of production rate of metal m (%).
P_x is production rate of current year (t).
P_{x+1} is production rate of next year (t).

Calculations showed that the mean increase of production rate of different metals in the last 20 years is as follows: Co, 8.3% per year; Cu, 3.1%: Ni, 3.7%; Mn, 4.3%; Pb, 2.6%; and Zn, 2.9%.

Based on the mean increase in production rate of metals in the last 20 years as well as availability of metal reserves on land in the world, the duration of availability of land reserves can be evaluated in order to evaluate the possible timing when deep-sea minerals may be required to be mined.

There are two ways of this estimation:

Approach 1: Having total metal reserves on land for 2016, which is considered as a base year for calculation, and the production rate of the metal for 2016, the number of years for which the reserve will last (Table 5) can be estimated as follows:

$$N_{\mathrm{m}} = R_{\mathrm{m}} / P_{\mathrm{b}}$$

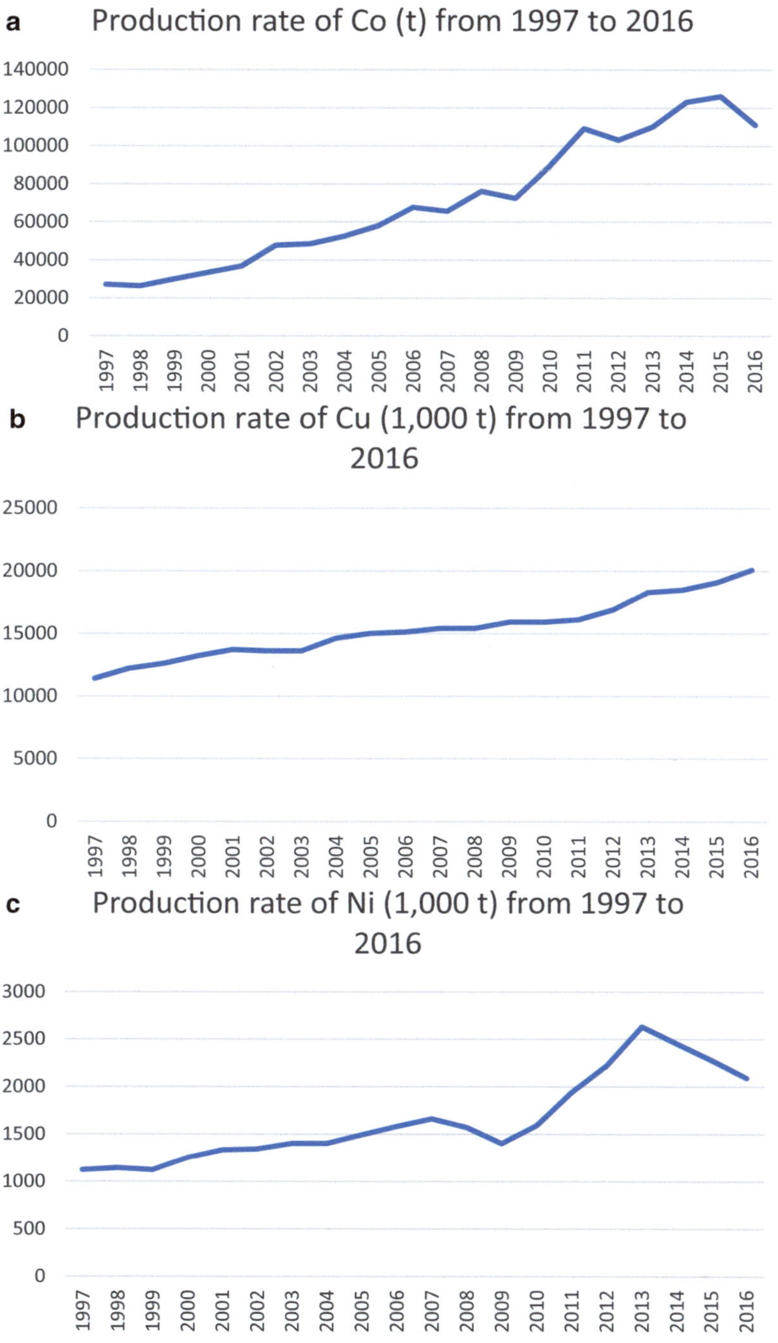

Fig. 2 (**a**) Co production rate from 1997 to 2016. (**b**) Cu production rate from 1997 to 2016. (**c**) Ni production rate from 1997 to 2016. (**d**) Mn production rate from 1997 to 2016. (**e**) Pb production rate from 1997 to 2016. (**f**) Zn production rate from 1997 to 2016

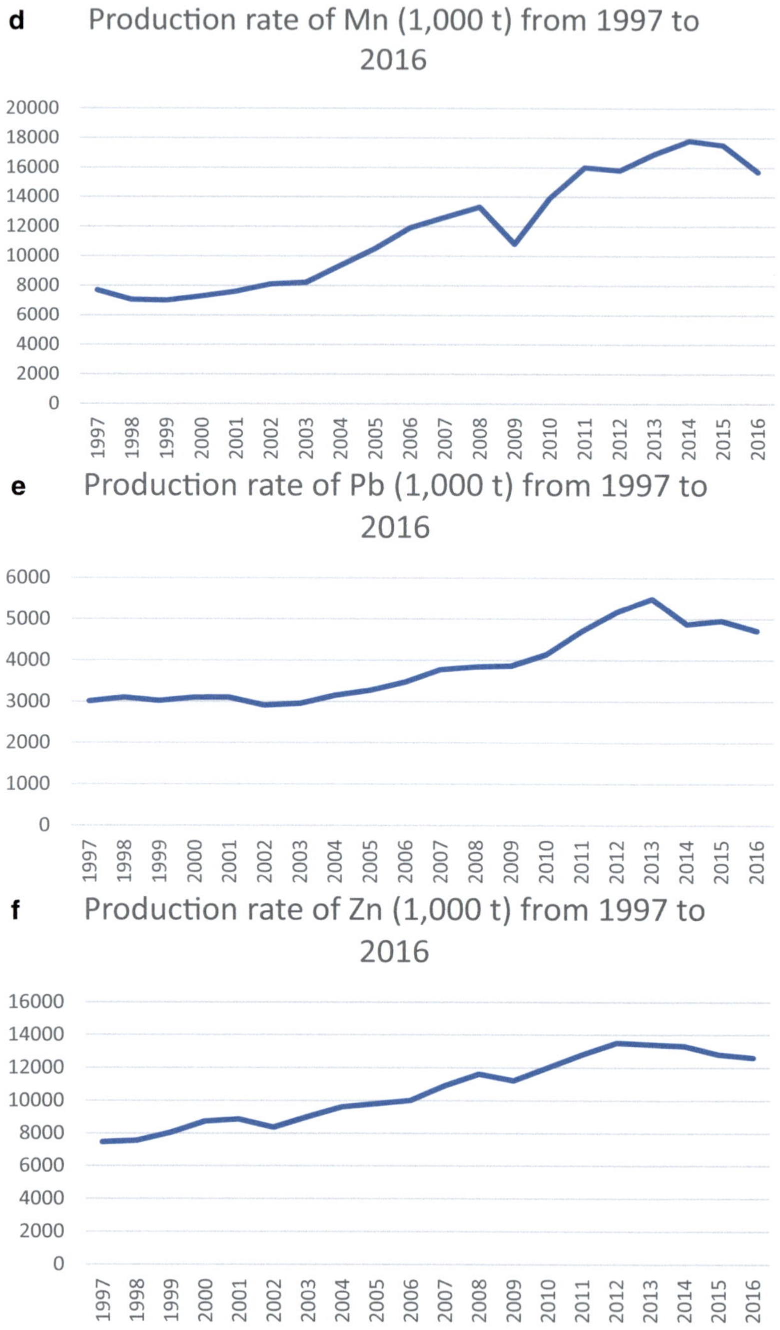

Fig. 2 (continued)

where:

N_m is number of years that last for metal m.
R_m is the latest world reserves of metal m on land in the base year (t),
P_b is production rate of the base year (t).

However, this method presumes that production rate of the base year will remain constant in the future, whereas the analysis of 20-year production of metals has shown that the production rates of metals are increasing every year. Therefore, for accurate estimation of the number of years, it is needed to consider mean increase in production over a period of time as described below.

Approach 2: The calculations of availability of metal reserves on land were performed beginning from 2016, which is the base year. Subsequently, the considering production rate for 2016 as the baseline and increase of mean production rate in %, the projected production rates for future were found with respect to increase of production rate as follows:

$$P_{x+1} = P_x + \frac{P_x \times P_m}{100\%}$$

where:

P_x is production rate of current year (t).
P_{x+1} is production rate of next year (t).
P_m is increase of production rate for metal m (%).

When this is repeated for each of the subsequent years, it is possible to estimate how many reserves remain after each year of mining with respect to the increase in production rate by:

$$R_l = R_m - P_{x+1}$$

where:

R_l is the remaining reserve (t).
R_m is the latest world reserves of metal m on land in the base year (t),

Table 5 Availability of metal reserves on land

Metal	World reserves on land in the base year (t) (www.usgs.gov)	Production rate of the base year (t) (www.usgs.gov)	Number of years for which land reserves will last
Cu	790,000,000	20,100,000	39
Co	7,100,000	111,000	64
Ni	74,000,000	2,090,000	35
Zn	230,000,000	12,600,000	18
Pb	88,000,000	4,710,000	19
Mn	680,000,000	15,700,000	43

P_{x+1} is production rate of next year (t).

Making these calculations number of times, it was found out that from 2018 Cu reserves on land will last 24 years; Co, 21 years; Ni, 21 years: Zn, 13 years: Pb, 14 years: and Mn, 23 years (Table 6).

However, these estimates are based on current figures, and the number of years can vary for the following reasons:

(a) It is possible that new technology for processing the ore will be developed so that ores with low grade and not considered as a reserve currently can be processed in the future. New techniques can encourage extraction in previously impossible conditions (Zepf et al. 2014).
(b) Recycling of metals can be considered as another reason, if new technology will be available and used extensively.
(c) During exploration new resources can be found or known resources can be converted to reserves (Zepf et al. 2014) as new geological understanding and new technologies drive exploration and reserve estimations. As can be seen from the data on metal resources in Russia (Table 7), it is clear that resources can be converted to the reserves using geological exploration of new deposits and revaluation of known resources. Sometimes total balance can be negative with regard to the production of reserves, expenditures on mining, and revaluation for the worse.

This shows that calculation of the longevity of a reserve is a complex process and needs to be carried out taking several factors into consideration as large investments will be based on these estimates.

Table 6 Availability of metal reserves on land with regard to the increase of production rate

Metal	World reserves on land in the base year (t) (www.usgs.gov)	Production rate in the base year (t) (www.usgs.gov)	Number of years for which land reserves will last
Cu	790,000,000	20,100,000	24
Co	7,100,000	111,000	21
Ni	74,000,000	2,090,000	21
Zn	230,000,000	12,600,000	13
Pb	88,000,000	4,710,000	14
Mn	680,000,000	15,700,000	23

Table 7 Changes in Russian reserve base in 2016 (Kiselev 2016)

	Estimated resources[a]			Reserves[b]		Changes with respect to 2015[c]	
Metal	P1	P2	P3	A + B + C1	C2	A + B + C1	C2
Cu	12.6 Mt	23 Mt	36.2 Mt	69.6 Mt	28.2 Mt	1 Mt	4.9 Mt
Mn	232 Mt	138 Mt	615 Mt	137,802 Kt	92,352 Kt	−5 Kt	0
Ni	1.2 Mt	5.9 Mt	5.5 Mt	Private information		−0.7%	5.1%
Pb	3 Mt	10.6 Mt	27.6 Mt	10.2 Mt	7.6 Mt	−2.1 Mt	0.4 Mt
Zn	10.7 Mt	34.9 Mt	57.4 Mt	40.8 Mt	19 Mt	−0.9 Mt	0.4 Mt
Mo	220.9 Kt	665.4 Kt	2460 Kt	1417 Kt	726.4 Kt	−73.2 Kt	127.6 Kt
Au	5497.6 t	10499.8 t	25247.9 t	8159.6 t	5657.8 t	153.4 t	530.1 t
Ag	24.8 Kt	75.4 Kt	104.2 Kt	65 Kt	53.8 Kt	−3.8 Kt	0.9 Kt
Pt	33.9 t	237.3 t	400 t	9782.4 t	5288.1 t	−82.5 t	−7 t
REE	2335.2 Mt	8379.5 Mt	359.9 Mt	17328.6 Mt	9569.9 Mt	−311.2 Kt	−1.9 Kt

[a]Estimated resource categories: P1–measured, P2–inferred, P3–estimated. Do not include reserves
[b]Reserve categories: A, reliable (detailed explored); B, ascertained (pre-explored); C1, estimated (poorly explored); A + B + C1, proved; and C2, probable
[c]Total balance of reserves is increase or decrease of reserve amount with respect to the geological exploration, revaluation, production, and mining expenditures

7 Estimation of Resource Potential for Different Abundances of Polymetallic Nodules

In order to address the question whether deep-sea mining will be profitable or not, polymetallic nodules were taken as an example, and a model of calculations was made that could be applied for other types of marine minerals.

For the exploration of nodules, each contractor has signed a Contract for an area of 75,000 km^2 with the International Seabed Authority, within which several mine sites will be located. The abundance of nodules in these areas is known to generally vary from less than 5 to 20 kg/m^2 (Sharma 2017a). As 5 kg/m^2 is considered as the cutoff (UNOET 1987), estimations of total resources for abundances of 5 kg/m^2, 10 kg/m^2, or 15 kg/m^2 were calculated as follows:

$$N_w = A \times n$$

where:

N_w is a total resource for abundance n – i.e., wet nodules (t).
A is a contract area of 75,000 km^2.
n is an abundance of nodules in the area – 5 kg/m^2, 10 kg/m^2, or 15 kg/m^2.

Estimated total resources are wet nodules, as taken directly from seabed and contain water, whereas the true total resource will be amount of these nodules after they are dried. It is considered that during drying nodules will lose about 25% of their mass (Mero 1977):

Table 8 Total resource potential of nodules for mining area of 75,000 km^2

Mean abundance on the area (kg/m^2)	5	10	15
Mean abundance on the area (t/km^2)	5000	10,000	15,000
Total resource potential of the area, wet nodules (t)	375,000,000	750,000,000	1,125,000,000
Total resource potential of the area, dry nodules (t)	281,250,000 (281.25 Mt)	562,500,000 (562.50 Mt)	843,750,000 (843.75 Mt)
Duration for which these will last at 1.5 Mt/year (y)	187.5	375	562.5

$$N_{\mathrm{d}} = N_{\mathrm{w}} \times 75\%$$

where:

N_{d} is a total resource for abundance n – i.e., dry nodules (t).

Hence, the estimation of period for which nodules for different mean abundances at a mining rate of 1.5 million tons/year leads to several hundred years from a single contract area (Table 8).

8 Value of Metals Versus Investments for Mining

Considering a typical Contract area of 75,000 km^2 for polymetallic nodules and metal concentrations for a 5 kg/m^2 of nodule abundance and average metal production rates of 1.5 and 3.0 MT/year, an exercise conducted with metal prices of January 2011 yielded gross metal value of \$18.7 billion and \$37.4 billion, respectively, over a period of 20 years (Sharma 2017b). Evaluating the same for 3-month average metal prices for June 2018, and considering conservative metal values, yields gross metals worth \$21.725 billion and \$43.450 billion, respectively, in view of the sharp increase in the price of cobalt as well as manganese, while the other metals show a downward trend (Table 9). Similarly, estimates of capital and operating expenditures for different components of mining system (mining system, ore transfer, and processing plant) based on figures provided by different contractors for 1.5 MT/year of mining (ISA 2008) had worked out to total investment of \$11.90 billion over a period of 20 years (Sharma 2017) which, based on cumulative inflation rate of 16.90% (www.usinflationcalculator.com), works out to \$13.90 billion in June 2018 (Table 10).

Comparing the estimates of total metal values (Table 9) and total expenditures (Table 10) for 1.5 MT/year of nodule mining, it appears that for an investment of \$13.90 billion, metals worth \$ 21.725 billion can be extracted as per current prices. It should be noted that figures in Table 10 are based on those suggested by the contractors for different activities as per the technology and material cost as of

Table 9 Resource potential and metal production estimates

Nodule/ metal	Mean concentration[a]	Resource potential t (Mt)[b]	Metal production per year t (Mt) at 1.5 Mt/year, at 3 Mt/ year		Price of metal ($/kg)	Gross in-place value of metal $/year at 1.5 Mt/year, at 3 Mt/ year		Gross in-place value of metal $/20 years at 1.5 Mt/year, at 3 Mt/year	
Wet nodules	–	375,000,000 (375)	–	–	–	–	–	–	–
Dry nodules	25% of wet nodules[c]	281,250,000 (281.25)	–	–	–	–	–	–	–
Manganese	22/24% of dry nodules	61,800,000 (61.8)	330,000 (0.33)	660,000 (0.66)	1.32[d]	435,600,000	871,200,000	8.712 billion	17.424 billion
					2.04[e]	673,200,000	1346,400,000	13.464 billion	26.928 billion
Nickel	1.0/1.1% of dry nodules	2,810,750 (2.81)	15,000 (0.015)	30,000 (0.030)	23.00[d]	345,000,000	690,000,000	6.90 billion	13.80 billion
					14.15[e]	212,250,000	424,500,000	4.245 billion	8.490 billion
Copper	0.78/1.04% of dry nodules	2,190,000 (2.19)	11,700 (0.0117)	23,400 (0.0234)	8.30[d]	97,110,000	194,220,000	1.9422 billion	3.8844 billion
					6.81[e]	79,677,000	159, 354,000	1.593 billion	3.187 billion
Cobalt	0.23/0.1% of dry nodules	281,250 (0.281)	1500 (0.0015)	3000 (0.003)	39.20[d]	58,800,000	117,600,000	1.176 billion	2.352 billion
					80.75[e]	121,125,000	242,250,000	2.4225 billion	4.845 billion
Total (metals)	24.01/26.24%	67,081,000 (67.081)	358,200 (0.3582)	716,400 (0.7164)	–	936,510,000	187,302,000	18.7302 billion	37.460 billion
						1,086,252,000	2,172,504,000	21.725 billion	43.450 billion

Modified from Sharma (2017b)

[a]Morgan (2000) for Clarion Clipperton Zone in Pacific Ocean/Jauhari and Pattan (2000) for Central Indian Ocean

[b]At 5 kg/m^2 for 75,000 km^2 (75 × 10^9 m^2) considering the lower value of concentration of metals between the Pacific and Indian Ocean (in col. 2)

[c]Mero (1977)

[d]Average metal prices as of June 2011 (source: www.metalprices.com, www.lme.com) and corresponding gross-in-place values (col.7–10)

[e]Average metal prices as of January 2018 (source: www.metalprices.com) and corresponding gross-in-place values (col. 7–10)

Table 10 Estimated capital and operating expenditures for polymetallic nodule mining (for 1.5 mi t/year)

Item	Capital expenditures	Operating expenditures	Total
Mining system	\$ 550 million[a] (\$ 372–562 million)[b]	\$ 100 million/year[a] (\$ 69–96 million)[b] ×20 years = \$ 2.0 billion	\$ 2.55 billion
Ore transfer	\$ 600 million[a] (\$ 495–600 million)[b]	\$ 150 million/year[a] (\$ 93–132 million/year)[b] ×20 years = \$ 3.0 billion	\$ 3.60 billion
Processing plant	\$750 million	\$250 million/year ×20 years = \$5.0 billion	\$ 5.75 billion
Total	\$ 1.90 billion	\$ 10.0 billion	\$ 11.90 billion (\$ 13.90 billion)[c]

Modified from Sharma (2017b)
[a]Rounded off to nearest 50 of the highest value
[b]Figures in brackets show the range for different systems (Source: ISA 2008)
[c]Estimated total expenditure considering cumulative inflation rate at 16.90% for June 2018 (www.usinflationcalculator.com)

2008. A recent techno-economic assessment of nodule mining (Van Nijen et al. 2018) shows that for 3 million (dry) tons of nodules and a net exploitation period of 25 years with two independent mining vessels, the total capital expenditure for mining and processing would be about \$ 4 billion, with an operating expenditure of \$ 1.014 billion (\$ 0.325 billion for mining and \$ 0.689 billion for processing) per year. As the mining rate considered by Van Nijen et al. (2018) is two times than that considered for calculations in Table 10, the capital expenditure and the operating costs are also twice as much. However, it is important to note that in both the calculations, the estimated expenditure for processing is 50–60% of the total cost for exploiting the nodules.

It is suggested that the assumptions made regarding costs of processing can be revisited because process plant estimates (capital and operating) will markedly depend on the process route chosen. Following some reports (e.g., Nyhert et al. 1983), process plant experience has become enriched, and focused estimates are available. It has also been well established that earlier-assumed recoveries cannot be attained for poorer nodule grade (P. K. Sen, IIT Kharagpur – personal communication). Hence, detailed techno-economic assessment based on fluctuations in metal prices and demand as well as cost estimates due to changes in technology and materials at regular intervals would be critical to determine the timing for commencement of mining operations.

9 Conclusion

Estimates for ore production, area of mine site, contact area, and weight and volume of sediment to be disturbed for different mining rates from 1 to 3 Mt/year have been presented in the chapter. Based on production rates over 20 years, estimations have also been made as to the duration for which current reserves, available for different metals on land, will last. However, it must be recognized that these are conservative estimates based on cutoff abundance values and indicative metal concentrations in order to demonstrate the methodology of arriving at these variables, as also for a range of mining rates that need to be considered for arriving at the optimum mining rate. The actual estimates for each mine site will have to be made separately based on "ground" conditions.

In view of several variables for optimization of mining rates, the following factors need to be taken into consideration:

- Concentration and distribution characteristics of the mineral deposits on the seafloor
- Capability/efficiency of the mining system
- Metal prices and demand (consumption) in the world market
- Investment versus returns from mining infrastructure
- Environmental impact due to mining at sea and on land

Estimates have shown that production of ore between 3333 and 10,000 t/day with 24–26% of combined metal concentrations for different mining rates would require development of appropriate infrastructure for lifting, handling, storage, and transport on the mining platform. From the environment point of view, the actual mine site and the area of contact on the seafloor could be an extremely small proportion (<10%) of the entire area allotted to the contractors. However, the volume of sediment expected to be disturbed due to mining calls for serious consideration of restricting the sediments to the seafloor and reducing the penetration of the miner into the sediments. Volkmann and Lehman (2018) have suggested that strip mining along 20 m wide strips (which is the collecting width of the nodule collector) laid parallel in the mining area leaving 5 m wide unmined strip between them would provide connectivity of habitat for native species to recolonize the mined strips. Similarly, alternative use of tailings on land (Wiltshire 2017) requires immediate attention for sustainable development of deep-sea mining.

Estimates of total metal values that can be recovered versus total expenditures for 1.5 MT of nodules mined per year have shown that for an investment of $13.90 billion, metals worth $21.725 billion can be extracted as of current metal prices. However, regular techno-economic assessments based on metal prices and their demand as well as cost estimates due to changes in technology, materials and processes would be critical together with development of environment-friendly mining techniques. These could be done by applying different economic models (Van Nijen et al. 2018) as well as developing ecosystem-based management (Cormier et al. 2017). International Seabed Authority (ISA) is also considering various approaches in order to meet the goals of promoting deep-sea mining on one hand as well as protection of the marine environment on the other (Lodge et al. 2017). Nevertheless, it is a collective responsibility of all stakeholders including the contractors, industry, and environmental groups as well as the ISA to find technological solutions as well as develop regulations, so as to make deep-sea mining sustainable (Morgan et al. 1999) in order to fulfill the growing demand for metals in the world as also conserving the environment.

Detailed environmental impact assessment is becoming a key requirement in planning and execution of upcoming projects and obtaining regulatory approvals for which robust EIA process is necessary (Durden et al. 2018). At the same time, comprehensive techno-economic assessment of deep-sea mining aimed towards responsible utilization of mineral resources of the deep sea requires that exploitation technologies and methodologies as well as tools to plan for sustainable exploitation must be developed (Volkmann et al. 2018). Hence, sustainable deep-sea mining is a combination of a suitable EIA process, supported by adequate technological innovations in the mining system design so as to mitigate the potential serious harm to the deep-sea ecosystem and to ensure confidence among all stakeholders.

Acknowledgment The author acknowledges the inputs by Dr. TRP Singh, Ex-Deputy General Manager, Engineers India Limited, New Delhi, and Prof. P. K. Sen, Indian Institute of Technology, Kharagpur, during the preparation of this paper.

References

Academie des Sciences. (1984). *Les nodules polymetalliques. Faut-il exploiter les mines oceaniques?* Paris: Gauthier-Villiars.

Cormier, R., Kelble, C. R., Anderson, M. R., Allen, J. I., Grehan, A., & Gregersen, O. (2017). Moving from ecosystem-based policy objectives to operational implementation of ecosystem-based management measures. *ICES Journal of Marine Science, 74*(1), 406–413. https://doi.org/10.1093/icesjms/fsw181.

Cronan, D. S. (Ed.). (2000). *Marine mineral deposits handbook*. Boca Raton: CRC Press.

Dick, R. (1985). Deep-sea mining versus land based mining. In J. B. Donges (Ed.), *The economics of deep-sea mining* (pp. 2–60). Berlin: Springer-Verlag.

Durden, J. M., Lallier, L. E., Myrphy, K., Jaeckel, A., Gjerde, C., & Jones, D. O. B. (2018). Environmental impact assessment process for deep-sea mining in the 'Area'. *Marine Policy, 87*, 194–202. https://doi.org/10.1016/j.marpol.2017.10.013.

Flipse, J. E., Dubs, M. A., & Greenwald, R. J. (1973). Pre-production manganese nodules activities and requirements. Mineral resources of the deep seabed, hearings of the sub-committee on minerals, materials and fuels of the US Senate Committee on Interior and Insular affairs (15 March) (pp. 602–700).

Hein, J. R., & Koschinsky, A. (2014). Deep-ocean ferromanganese crusts and nodules, Chapter 11. In H. D. Holland & K. K. Turekian (Eds.), *Treatise on geochemistry* (Vol. 13, 2nd ed., pp. 273–291). Amsterdam: Elsevier.

Herrouin, G., Lenoble, J. P., Charles, C., Mauviel, F., Bernard, J., & Taine, B. (1991). French study indicates profit potential for industrial manganese nodule venture. *Transactions of Society for Mining, Metallurgy, and Exploration, 288*, 1893–1899.

Hong, S., Kim, H.-W., Yeu, T., Choi, J.-S., Lee, T. H., Bae, D. S., & Lee, J.-G. (2019). Technologies for safe and sustainable mining of deep-seabed minerals. In R. Sharma (Ed.), *Deep-sea mining and environmental issues*. Cham: Springer International Publishers.

ISA. (2008). *Report on the International Seabed Authority's workshop on Polymetallic nodule mining technology: Current status and challenges ahead* (p. 4). Jamaica. ISBA/14/LTC/3: International Seabed Authority.

ISA. (2013). *Recommendations for the guidance of contractors for the assessment of the possible environmental impacts arising from exploration for marine minerals in the Area* (p. 32). International Seabed Authority ISBA/19/LTC/8.

Jauhari, P. J., & Pattan, J. N. (2000). Ferromanganese nodules from the Central Indian Ocean Basin. In D. S. Cronan (Ed.), *Handbook of marine mineral deposits* (pp. 171–195). Boca Raton: CRC Press.

Kaufman, R. (1974). The selection and sizing of tracts comprising a manganese nodule ore body. In *Pre-prints offshore technology conference,* vol. II Houston, USA (pp. 283–289).

Khadge, N. H., & Valsangkar, A. B. (2008). Geotechnical characteristics of siliceous sediments from the Central Indian Basin. *Current Science, 94*, 1570–1573.

Kiselev, Y. (Ed.) (2016). Государственный доклад «О состоянии и использовании минерально-сырьевых ресурсов РФ в 2015 году». Министерство природных ресурсов и экологии РФ, Москва, Россия (pp. 101–266).

Kotlinski, R. (2001). Mineral resources of the world's ocean – their importance for global economy in the 21st century. In *Proceedings of ISOPE ocean mining symposium*, Szczecin, Poland (pp. 1–7).

Lenoble, J. P. (1980). Polymetallic nodule resources and reserves in the North Pacific from the data collected by AFERNOD. *Oceanology International, 80*, 11–18.

Lenoble, J. P. (1990). *Future deep sea bed mining of polymetallic nodules*. Issy-les-Moulineaux Cedex: IFREMER.

Lenoble, J. P. (2000). A comparison of possible economic returns from mining deep-sea polymetallic nodules, polymetallic massive sulphides and cobalt-rich ferromanganese crusts. In *Proceedings of the workshop on mineral resources* (pp. 1–22). Jamaica: International Seabed Authority.

Lodge, M. W., Segerson, K., & Squires, D. (2017). Sharing and preserving the resources in the deep-sea: Challenges for the international Seabed Authority. *International Journal of Marine and Coastal Law, 32*, 427–457.

Mero J. L. (1965). The mineral resources of the sea. Amsterdam, The Netherlands: Elsevier. 312 pp.

Mero, J. L. (1977). Economic aspects of nodule mining. In G. P. Glasby (Ed.), *Marine manganese deposits* (pp. 327–355). Amsterdam: Elsevier.

Morgan, C. L. (2000). Resource estimates of the Clarion-Clipperton manganese nodule deposits. In D. S. Cronan (Ed.), *A handbook of marine mineral deposits* (pp. 145–170). Boca Raton: CRC Press.

Morgan, C. L., Odunton, N., & Jones, A. T. (1999). Synthesis of environmental impacts of deep seabed mining. *Marine Georesources & Geotechnology, 17*, 307–357.

Nyhart, J. D., Triantafyllou, M. S., Averback, J., Bliek, A., Sklar, B., Gillia, M., Kirkpatrick, D., Muggerridge, J., Nakagawa, R., Newman, J., & Will, A. (1983). *A pioneer deep ocean mining venture* (p. 274). Cambridge, MA., Sea Grant Program, June 1983: Massachusetts Institute of Technology.

OMA. (1982). Application by Ocean Mining Associates for an exploration license, filed with NOAA, US Department of Commerce (18-2-1982).

OMI. (1982). Application for and notice of claim to exclusive exploration rights for manganese nodule deposits in the NE Equatorial Pacific Ocean, filed with NOAA, US Deparment of Commerce (19-2-1982).

Pearson, J. S. (1975). *Ocean floor mining*. Park Ridge: Noyes Data Corporation.

Roth, R. (2018). Understanding the economics of seabed mining for polymetallic nodules. Presentation at International Seabed Authority Council Meeting, Jamaica, 6 March 2018.

Sharma, R. (2011). Deep-sea mining: Economic, technical, technological and environmental consideration for sustainable development. *Marine Technology Society Journal, 45*, 28–41.

Sharma, R. (2017a). Assessment of distribution characteristics of polymetallic nodules and their implications on deep-sea mining. In R. Sharma (Ed.), *Deep-sea mining – resource potential, technical and environmental considerations* (pp. 229–258). Cham: Springer International Publishing AG.
Sharma, R. (2017b). Deep-sea mining: Current status and future considerations. In R. Sharma (Ed.), *Deep-sea mining – resource potential, technical and environmental considerations* (pp. 3–21). Cham: Springer International Publishing AG.
Sharma, R. (2019). Environmental factors for design and operation of deep-sea mining system-based on case studies. In *Environmental issues of deep-sea mining – impacts, consequences and management*. Cham: Springer International Publishing AG.
Siapno, W. D. (1975). *Exploration technology and ocean mining parameters* (p. 24). San Francisco: American Mining Congress Convention.
Singh, T. R. P., & Sudhakar, M. (2015). Estimating potential of additional mine-sites for polymetallic nodules in Pacific and Indian Oceans. *International Journal of Earth Sciences and Engineering, 8*, 1938–1941.
UNOET. (1987). *Delineation of mine sites and potential in different sea areas*. London: UN Ocean Economics and Technology Branch and Graham & Trotman Limited.
Van Nijen, K., Passel, S. V., & Squires, D. (2018). A stochastic techno-economic assessment of seabed mining of polymetallic nodules in the Clarion Clipperton Fracture Zone. *Marine Policy, 95*, 133–141. https://doi.org/10.1016/j.marpol.2018.02.027.
Volkmann, S. E., & Lehnen, F. (2018). Production key figures for planning the mining of manganese nodules. *Marine Georesources and Geotechnology, 36*, 360–375. https://doi.org/10.1080/1064119X.2017.1319448. DOI:10.1080/1064119X.2017.1319448.
Volkmann, S. E., Kuhn, T., & Lehnen, F. (2018). A comprehensive approach for a techno-economic assessment of nodule mining in the deep sea. *Mineral Economics, 31*, 319–336. https://doi.org/10.1007/s13563-018-0143-1.
www.isa.org.jm. (2018). Authority has entered into 15-year contracts with twenty-seven contractors for exploration for polymetallic nodules, polymetallic sulphides and cobalt-rich ferromanganese crusts in the international seabed area.
www.japantimes.com. Japan successfully undertakes large-scale deep-sea mineral extraction (26 September 2017).
www.lme.com. Metal prices for Mn, Ni, Cu, Co (June, 2018).
www.metalprices.com. Metal prices for Mn, Ni, Cu, Co (January, 2011).
www.nautilusminerals.com. Nautilus completes successful trials in PNG (12 February 2018).
www.usgs.gov. U.S. Geological Survey (referred on September 2018).
www.usinflationcalculator.com. To calculate inflation over time (June, 2018).
Wiltshire, J. C. (2017). Sustainable development and its application to mine tailings of deep sea minerals. In R. Sharma (Ed.), *Deep-sea mining: Resource potential, technical and environmental considerations* (pp. 423–444). Cham: Springer International Publishing AG.
Zepf, V., Simmons, J., Reller, A., Ashfield, M., & Rennie, C. (2014). *Materials critical to the energy industry. An introduction* (2nd ed., p. 94). London: British Petroleum.

Dr. Rahul Sharma (rsharma@nio.org, rsharmagoa@gmail.com) retired as Chief Scientist from the CSIR-National Institute of Oceanography in Goa, India, with a career spanning 35 years in the field of exploration and exploitation of marine minerals. He has led a multidisciplinary group on "environmental studies for marine mining." He has a master's degree in Geology and a doctorate in Marine Science. His professional interests include application of exploration and environmental data to deep-sea mining. He has edited three special issues of journals, published 37 scientific papers, authored 22 articles and 41 technical reports, and presented more than 50 papers at national and international conferences. He has also edited a book on *Deep-Sea Mining* published by Springer International Publishers that has chapters contributed by experts from around the world.

His international assignments include Visiting Scientist to Japan; Visiting Professor to Saudi Arabia; member of the UNIDO mission "to assess the status of deep-sea mining technologies" in Europe, the USA, and Japan; and invited speaker and consultant for the International Seabed Authority, Jamaica. He has contributed to the "World Ocean Assessment report I" of the United Nations and has also been invited to contribute a chapter on "Potential impacts of deep-sea mining on marine ecosystem" for the *Oxford Encyclopedia* for environmental science. In addition to his research career, he has been involved with several activities relating to science communication and outreach, as well as training programs for international participants, professionals, and students.

Farida Mustafina is a student of "Polar and Marine Research" Master program, which is collaboration between St. Petersburg State University, Russia, and Hamburg University, Germany. Currently she is undergoing a master's course in "Integrated Climate System Science" at Hamburg University, and her topic of research is economics of marine mineral resources. Previously, she has done an internship in National Institute of Oceanography, Goa, India, and worked on the Economic assessment of deep-sea mining.

Dr. Georgy Cherkashov is Deputy Director of the Institute for Geology and Mineral Resources of the Ocean (VNIIOkeangeologia, St. Petersburg, Russia) of the Ministry of Natural Resources (since 1996). He holds a Dr. Sci. for research of seafloor massive sulfide (SMS) deposits of the Mid-Atlantic Ridge. He has been the Chief Scientist of 13 ocean-going Russian and international expeditions for prospecting of SMS deposits in the Pacific, Atlantic, and Indian Oceans (1983–2007). He was also the President of International Marine Minerals Society (2011–2012) and member of the Legal and Technical Commission of the International Seabed Authority (since 2012) and is Professor at St. Petersburg State University (Marine Geology), part time (since 2005).

Techno-economic Perspective on Processing of Polymetallic Ocean Nodules

Navin Mittal and Shashi Anand

Abstract This chapter looks at the techno-economic feasibility of processing of polymetallic nodules obtained from Pacific as well as Indian Ocean with an emphasis on cobalt market. Numerous processing routes have been developed for recovery of three (Cu + Ni + Co) as well as four (Cu + Ni + Co + Mn) metals. Some of the processes tested on scales varying from tens to a few hundreds of kilograms per batch indicate that there are no major gaps in the processing technologies. With the passage of time, recovery of molybdenum and rare earth elements (Mo and REE) from this resource has also gained importance. Various feasibility studies for extraction of metals from polymetallic nodules are presented in this chapter with respect to operating and capital investments for a 1.5/3.0 million tonne plant. It appears that enhanced requirement of cobalt and nickel mainly for the battery industry may drive the deep-sea mining in the future. The study concludes that it would be advisable to target 4+ metal recoveries (including Mo, REE, etc.) and design a flexible flow sheet for optimal product mix to get maximum value addition.

Keywords Techno-economic analysis · Resource classification · Metallurgical processing · Ocean nodules

1 Introduction

The existence of polymetallic nodules (PMN) commonly known as manganese nodules has been known since the late 1800s since their discovery during the Challenger expedition of 1873–1876 (Hein 2016). On the 7th March 1873, the dredge hauled up on its deck contained several peculiar black oval bodies which were composed of manganese oxide. The total amount of polymetallic nodules lying on the sea floor was estimated at more than 1.5 trillion tonnes by Mero (1965). Later the estimate was reduced to 500 billion tonnes by Archer of the London Geological Museum in1981 (Wikipedia 2018). Polymetallic nodules are spread over most of the oceans

N. Mittal (✉) · S. Anand
Hydromet Technology Solutions Pvt. Ltd, Vadodara, India
e-mail: navin@hydrometsolutions.com

R. Sharma (ed.), *Environmental Issues of Deep-Sea Mining*,
https://doi.org/10.1007/978-3-030-12696-4_20

in the world and are abundantly available. This has led to global efforts to develop processes for extraction of metals like copper (Cu), nickel (Ni), cobalt (Co), and manganese (Mn) from polymetallic nodules. Recently recovery of rare earth elements (REE) and molybdenum (Mo) from PMN has also received much interest (Mohwinkel et al. 2014; Parhi et al. 2011, 2013, 2015).

Initially the process development was confined to recovery of three metals (Cu, Ni, and Co), but later Mn was also included because of economic reasons. From time to time, several authors have discussed and reviewed the progress of extraction technologies for three as well as four metals from manganese nodules (Huberd 1980; Monhemius 1980; Fuerstenau and Han 1983; Haynes et al. 1985; Han and Fuerstenau 1986; Han 1997; Premchand and Jena 1999; Sen 2010; Das and Anand 2017). With the inclusion of REE and Mo, the flow sheets would look techno-commercially more attractive. Some of the recent studies related to rare earths and Mo extraction from manganese nodule have been published (Mohwinkel et al. 2014; Parhi et al. 2011, 2013, 2015).

Several studies have been reported to highlight the techno-economic feasibility of mining and processing of manganese nodule (Agarwal et al. 1976, 2012; Clark et al. 1995; Flentje et al. 2012; Volkmann et al. 2018). However, with the changed market dynamics, the flow sheets are continually being modified to address market requirements. This chapter focuses on techno-economic perspective of processing of polymetallic nodules and also looks at environmental consequences of nodule processing.

2 Polymetallic Nodules Characterization

Mn nodules are concretions that occur on or near the surface of the sediment that covers abyssal plains throughout all the oceans, where sedimentation rates are low, less than 10 centimeters (cm) per thousand years. Nodules are generally golf ball sized, most commonly 1–12 cm, but can vary in diameter from millimeter to as large as 20 cm (Hein 2016). The formation of Mn nodules takes place at depths of about 3500–6500 m. It is now known that the various regions where the highest concentrations of metal-rich nodules occur are (i) in the NE Pacific Clarion and Clipperton fracture zones (CCZ), (ii) in the SE Pacific Peru Basin, (iii) in the central South Pacific Penrhyn Basin, and (iv) in the Central Indian Ocean Basin (Hein 2016). The chemical composition of nodules from selected areas of the global ocean is provided in Table 1 (Hein and Koschinsky 2014).

The CCZ is the area of greatest economic interest and where about 15 exploration contracts for polymetallic nodules have been signed or are pending with International Seabed Authority (ISBA) (Hein 2016); and there is also one nodule contract in the Central Indian Ocean Basin signed with India.

Table 1 Chemical composition of nodules from selected areas of the global ocean (Hein and Koschinsky 2014)

Element	UOM	CCZ nodules	Peru Basin nodules	Indian Ocean nodules	Cook Islands nodules
Cu	wt %	1.07	0.60	1.04	0.23
Ni	wt %	1.30	1.30	1.10	0.38
Co	wt %	0.21	0.05	0.11	0.41
Mn	wt %	28.4	34.2	24.4	15.9
Fe	wt %	6.16	6.12	7.14	16.1
Al	wt %	2.36	1.5	2.92	3.25
Si	wt %	6.55	4.82	10.02	8.24
Ti	wt %	0.32	0.16	0.42	1.15
Bi	ppm	8.8	3.3	–	11
Li	ppm	131	311	110	61
Mo	ppm	590	547	600	262
Nb	ppm	22	13	98	92
Pb	ppm	338	121	731	2000
Pt	ppm	0.13	0.04	–	0.21
Te	ppm	3.6	1.7	4	23
Th	ppm	15	6.9	76	34
Tl	ppm	199	129	347	138
V	ppm	445	431	497	920
W	ppm	62	75	92	59
Zn	ppm	1366	1845	1207	516
Zr	ppm	307	325	752	468
TREE	ppm	813	403	1039	1707

Dash (–) no data, *TREE* total rare earth elements including yttrium

3 Resource Consideration

The evaluation of the nodule resources, as identified through field exploration, includes laboratory analysis to estimate its grade (Ni + Cu + Co %), abundance (kg/m^2), and reserve (million metric tonnes). A mine site was defined as a portion of the seabed where a commercial operation could be maintained for 20–25 years with a production of 1.5–4 million tonnes per year of "good nodules" (Mukhopadhyay et al. 2018). Good nodules were defined as those containing 1.25–1.5% Ni and 1–1.4% Cu, as well as 27–30% Mn and 0.2–0.25% Co (Glasby 1977; Cronan 1980). The average abundance of polymetallic nodules at four major locations is illustrated in Fig. 1 (Hein and Petersen 2013).

A conservative estimate for only CCZ nodules is about 21,100 million dry tonnes. This tonnage of nodules may provide more than the entire land-based resources of Mn, twice that of nickel, thrice that of cobalt, and about fifth of copper available in land-based resources (Hein et al. 2013). The newly calibrated estimate places the total resource in the 75,000 km^2 Pioneer Area, awarded to India by

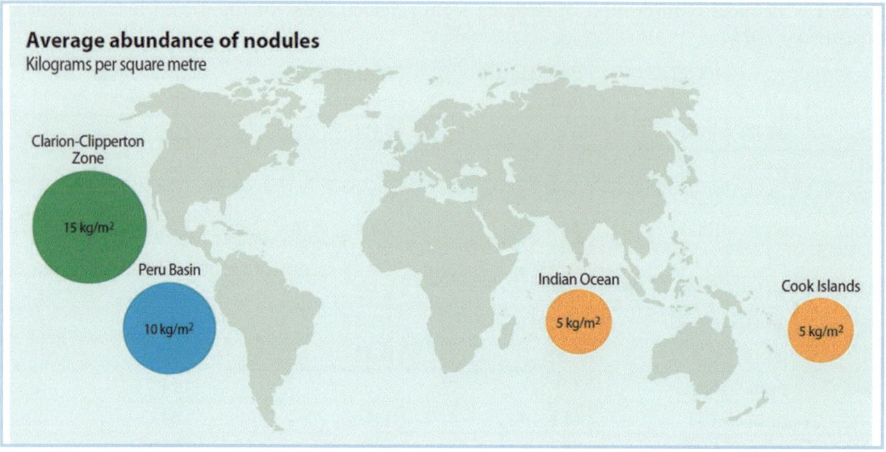

Fig. 1 Average abundance of nodules at four major locations (Hein and Petersen 2013). (Source: James R. Hein, U.S. Geological Survey)

UNCLOS, at 365 million tonnes, with Mn metal taking the lion's share of 95.17 million tonnes, Ni 4.508 million tonnes, Cu 4.455 million tonnes, and Co 0.418 million tonnes on wet nodule basis (Mukhopadhyay et al. 2018).

Besides the above mentioned elements, it is reported that the concentration of heavy REEs (HREEs) is higher in seabed deposits than in the largest land-based REE mines, for example, the largest REE mine, Bayan Obo (China), and the second largest, Mountain Pass (USA) (Hein 2012; Hein et al. 2013). It may be mentioned here that economically heavy rare earths, namely, Tb, Dy, Ho, Er, Tm, Yb, and Lu, along with the medium ones (Sm, Eu, Gd) are of more economic interest. The above mentioned land-based deposits contain less than 1% HREEs (percentage of total REE content), whereas the CCZ nodules have a relative content of 26% of HREEs (Hein 2012; Hein et al. 2013). According to the unofficial reports from the Tonga offshore Mining Ltd. (Nautilus Minerals), polymetallic nodules from the CCZ of the northeast Pacific Ocean contain total REE concentrations of 800 g/t (0.08%) on a dry basis or 600 g/t on as received basis. This translates to a rare earth oxide (REO) grade of 960 g/t or (0.096%) on a dry basis. The split of rare earths is 73% lights (Ce, La, Nd, Pr), 7% medium (Sm, Eu, Gd), and 10% heavies (Tb, Dy, Ho, Er, Tm, Yb, Lu) and 8% yttrium (Y) (Apavasileiou 2014). Further, it takes numerous steps to process a land-based ore to isolate one of the HREEs. In contrast, in nodules because the REEs are adsorbed onto the main iron and manganese phases, perhaps these can be dissolved using relatively simple procedures in a more cost-efficient way.

4 Metallurgical Processing of Nodules

The nodule processing techniques are broadly divided into three categories:

- Pyrometallurgical treatment followed by hydrometallurgical processing
- Hydrometallurgical processing (only)
- Hydrometallurgical processing followed by pyrometallurgical (for recovery of Mn from leach residue) treatment

It is well known that the metals of primary interest (Cu + Ni + Co) are in oxide forms and are associated in the lattices of iron and manganese minerals. For extraction of these metals, the lattices are to be broken either by reductive pyrometallurgy or through hydrometallurgical reduction. The initial work on polymetallic nodules process flow sheet development was started in the 1970s on ammoniacal leaching known as Cuprion process (Agarwal and Wilder 1974, 1975, Agarwal et al., 1978, 1979; Sazbo 1976) and by nodules smelting known as INCO process (Sridhar 1974; Sridhar et al. 1976, 1977). Hydrochloric acid-based processes were developed both by Deep Sea Ventures and Métallurgie Hoboken-Overpelt (Monhemius 1980). Efforts were also made to develop high-pressure sulfuric acid leaching for extraction of three metals (Cu + Ni + Co) (Hanieg and Meixner 1974; Neuschutz et al. (1977) in line with treatment of nickel laterites.

Subsequently numerous processing routes have been developed for recovery of three as well as four metals (Das and Anand 2017). However, only some of the processes have been tested on a scale varying from tens of kilograms to hundreds of kilograms per batch. Some of the processes which are actively being followed and tested in ≥100 kg scale were discussed in the ISBA workshop (ISBA 2008). While there does not appear to be any major gap in the processing technologies, the results available may/may not be fully adequate for upscaling and use in the feasibility estimates. As a result there has been continual prospecting, exploration, metallurgical testing, and economic analysis to assess the potential of polymetallic nodules as an exploitable mineral resource. More comprehensive economic viability of polymetallic nodules has been developed. Some of the studies are those of Andrews et al. (1983), Hillman and Gosling (1985), Charles et al. (1990), and Soreide (2001). These processes are briefly described below.

4.1 Reduction Hydrochloric Acid Leach Process

This process is based on dissolution of all major metals including manganese in hydrochloric acid. It uses solvent extraction technique in chloride media to separate key metals. High metal recoveries have been reported for this process; however, the corrosive nature of chloride is the main constraint for this process. During leaching the insoluble manganese dioxide reduces with hydrogen chloride and releases the copper, nickel, and cobalt for dissolution as chlorides in leach solution (Andrews et al. 1983). The schematic flow sheet of the process is illustrated in Fig. 2.

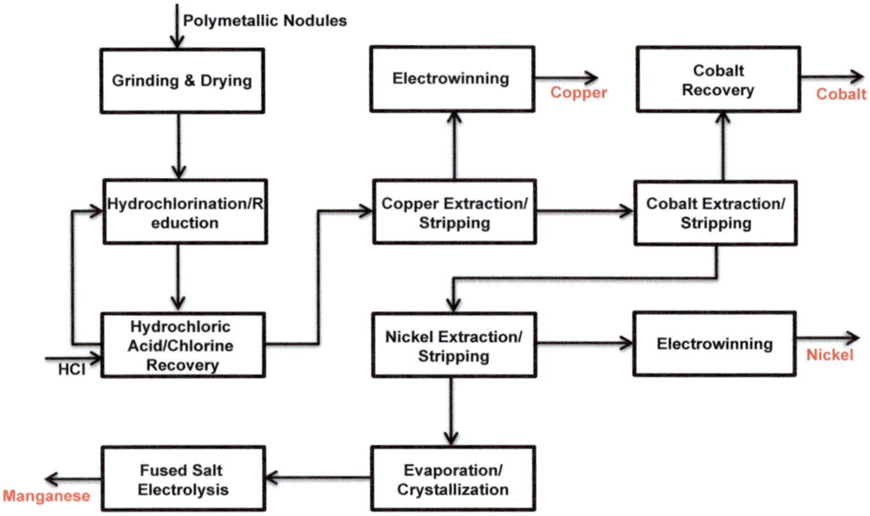

Fig. 2 Schematic flow sheet of reduction hydrochloric acid leach process (Andrew et al. 1983)

4.2 *The Cuprion Process*

Reductive leaching process was pioneered by Kennecott Copper by using carbon monoxide (CO) as a reducing agent under ammoniacal conditions (Agarwal et al. 1979). The process which Kennecott has developed for the metallurgy of manganese nodules is unique in concept. For the first time, a mineral ore was reduced, and its metal values were freed through the action of a carbonaceous gas on water slurry at ambient temperature and pressure. The first step in this process is a low-temperature (50 °C) hydrometallurgical reduction of manganese dioxide, by an aqueous ammoniacal solution containing cuprous ion (Agarwal et al. 1979). The process has been successfully demonstrated in a continuous pilot plant processing 350 kg/day of nodule ore. The process flow sheet is illustrated in Fig. 3 (Agarwal et al. 1979).

4.3 *High-Temperature and High-Pressure Sulfuric Acid (HPAL) Leach Process*

This process is quite similar to laterite processes used in Canada, Cuba, and Australia. The disadvantage of the process is that it emphasizes on three metal recoveries while keeping the fourth metal manganese in residue. The process involves high-pressure leaching using sulfuric acid (Soreide 2001; Flentje et al. 2012). The schematic of the process is illustrated in Fig. 4.

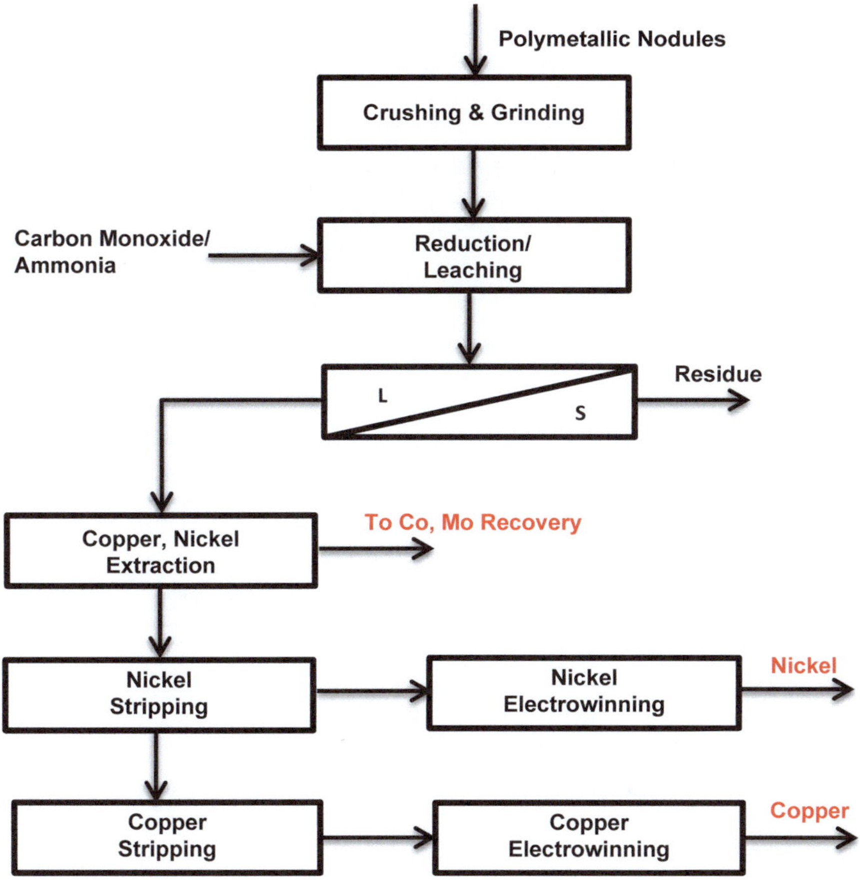

Fig. 3 Schematic flow sheet of Cuprion ammoniacal leach process (Agarwal et al. 1979)

5 Economic Evaluation of Nodules

Because of the relatively high content of manganese, copper, cobalt, and nickel, the manganese nodules are considered as potential source of multimetals. Various researchers are also focusing on extraction of other metals like molybdenum and rare earths from polymetallic nodules. The revenue potential from one tonne of polymetallic nodules at current metal prices from CCZ and Indian Ocean is provided in Table 2. It can be seen that the value of metals in CCZ is almost 20% higher than that in the Indian Ocean.

More than 50% of revenue comes from manganese which clearly indicates that economic viability of commercial plant flow sheet based on three metal recoveries may not be achieved. It is important to note that though manganese does seem to be

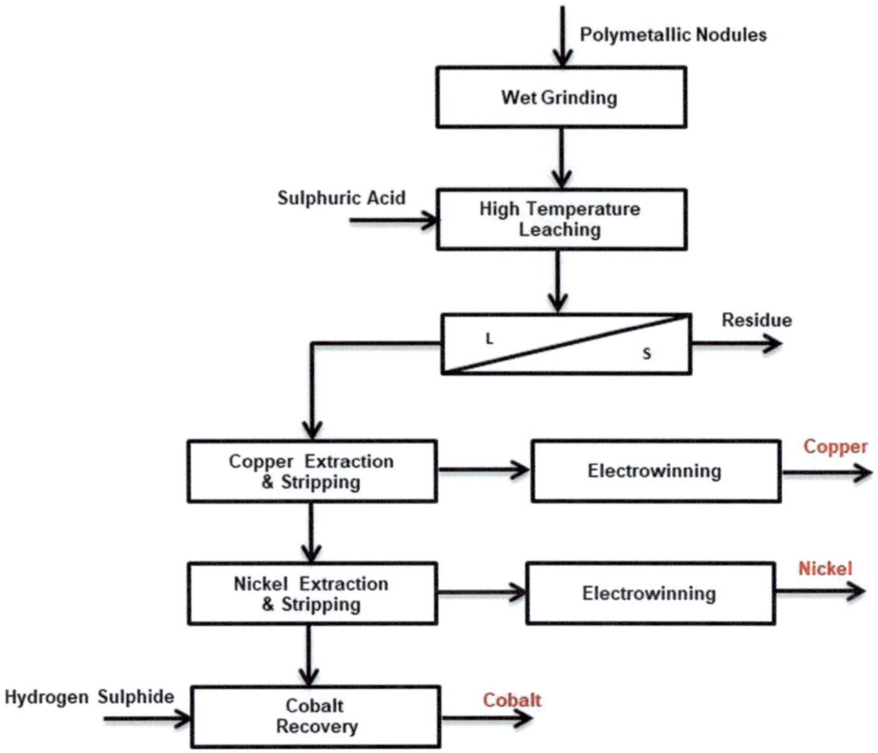

Fig. 4 Schematic flow sheet of high-pressure acid leaching process (Soreide 2001)

Table 2 Revenue potential from one tonne of polymetallic nodules based on April 2018, metal prices

	Price	Pacific Ocean nodules (CCZ)			Indian Ocean nodules		
		Grade	Value	Contribution	Grade	Value	Contribution
Pay metal	USD/ kg	%	USD/ tonne of nodule	%	%	USD/ tonne of nodule	%
Copper	6.8	1.07	$ 72.8	7.0%	1.04	$ 70.7	8.4%
Cobalt	91.0	0.21	$ 191.1	18.3%	0.11	$ 100.1	11.8%
Nickel	14.0	1.31	$ 183.4	17.5%	1.10	$ 154.0	18.2%
Manganese	2.0	28.40	$ 568.0	54.3%	24.40	$ 488.0	57.6%
Zinc	3.0	0.14	$ 4.1	0.4%	0.12	$ 3.6	0.4%
Molybdenum	26.0	0.06	$ 15.3	1.5%	0.06	$ 15.6	1.8%
Rare earth oxide	14.0	0.08	$ 11.4	1.1%	0.10	$ 14.5	1.7%
Total value of metals per tonne of nodules			**$ 1046**			**$ 847**	

the most important revenue source, the figure is misleading. The recovery of manganese depends upon the upstream process. In some cases, the Mn is recovered from the leach residue, while in other processes it is recovered from leach solution. It can be recovered as electrolytic manganese dioxide, Mn metal, or ferromanganese. Considering the application of manganese, a major part is used in the production of steel, whereas only a small part is used as electrolytic manganese metal (EMM) or electrolytic manganese dioxide (EMD) (Charles River Associates 1980). The manganese metal/ferromanganese output from the commercial polymetallic nodule plant needs to be carefully evaluated.

6 Feasibility Studies

As mentioned earlier, a number of feasibility studies have been conducted since 1970 for manganese nodules. Comparative data on these studies is presented in this chapter outlining various cost aspects in each area of operation, viz., mining, transportation, and processing.

6.1 Cost Estimates for Reduction and Hydrochloric Acid Leach Process (Four Metal Recovery)

This study was undertaken by Andrews et al. in 1983 and is based on reduction and hydrochloric acid leach process. In this process four metals are recovered. The capital expenditure (CAPEX) and operating expenditure (OPEX) estimate are illustrated in Table 3.

Table 3 Cost estimates for reduction and hydrochloric acid leach process (Andrews et al. 1983)

Description	UOM	Mining (Wet)	Transport (Dry)	Processing (Dry)	Total
Production	MMTPA	2.3	1.5	1.5	
Capital cost	MM $	180.0	176.0	513.0	869.0
	$/tonne of dry nodules	120.0	117.3	342.0	579.3
Capital cost ratio		20.7%	20.3%	59.0%	
Operating cost	MM $	45.0	25.0	165.0	235.0
	$/tonne of dry nodules	30.0	16.7	110.0	156.7
Operating cost ratio		19.1%	10.6%	70.2%	

6.2 *Cost Estimates for Cuprion Ammoniacal Leach Process (Three Metal Recovery)*

Hillman and Gosling (1985) undertook this study wherein the process is based on ammoniacal leach process. In this process, three metals, i.e., copper, nickel, and cobalt, are recovered. The CAPEX and OPEX estimates are illustrated in Table 4.

6.3 *Cost Estimates for Reduction and Hydrochloric Acid Leach Process (Four Metal Recovery)*

This study was done by Charles et al. (1990) and is based on reduction and hydrochloric acid leach process. The CAPEX and OPEX estimates are given in Table 5.

Table 4 Cost estimates for Cuprion ammoniacal leach process (Hillman and Gosling 1985)

Description	Unit	Mining (Wet)	Transport (Dry)	Processing (Dry)	Total
Production	MMTPA	4.2	3.0	3.0	
Capital cost	MM $	590.0	310.0	727.0	1627.0
	$/tonne of dry nodules	196.7	103.3	242.3	542.3
Capital cost ratio		36.3%	19.1%	44.7%	
Operating cost	MM $	77.0	37.0	111.0	225.0
	$/tonne of dry nodules	25.7	12.3	37.0	75.0
Operating cost ratio		34.2%	16.4%	49.3%	

Table 5 Cost estimates for reduction and hydrochloric acid leach process (Charles et al. 1990)

Description	Unit	Mining (Wet)	Transport (Dry)	Processing (Dry)	Total
Production	MMTPA	2.3	1.5	1.5	
Capital cost	MM $	282.0	188.0	470.0	940.0
	$/tonne of dry nodules	188.0	125.3	313.3	626.7
Capital cost ratio		30.0%	20.0%	50.0%	
Operating cost	MM $	48.0	36.0	156.0	240.0
	$/tonne of dry nodules	32.0	24.0	104.0	160.0
Operating cost ratio		20.0%	15.0%	65.0%	

6.4 Cost Estimates for High-Temperature and High-Pressure Sulfuric Acid Leach Process (Three Metal Recovery)

The study was reported by Soreide (2001), and it is the first such study based on high-pressure and high-temperature sulfuric acid leach process which is similar to the process adopted for nickel laterites. The CAPEX and OPEX estimates are illustrated in Table 6.

6.5 Cost Estimates at Current Level

The cost estimates presented above are based on studies conducted in different years. In order to get comparative numbers, CPI inflation calculator by the Bureau of Labor Statistics, US Department of Labor, is used. The base month in all cases has been taken as April 2018 for calculation purposes, and the cost estimates are illustrated in Table 7.

Since the plant capacities varied from 0.7 MMTPA to 3.0 MMTPA, a rule of thumb developed over the years known as the rule of six-tenths was applied to esti-

Table 6 Cost estimates for high-temperature and high-pressure sulfuric acid leach process (Soreide 2001)

Description	Unit	Mining (Wet)	Transport (Dry)	Processing (Dry)	Total
Production	MMTPA	1.1	0.7	0.7	
Capital cost	MM $	127.0	93.0	271.0	491.0
	$/tonne of dry nodules	181.4	132.9	387.1	701.4
Capital cost ratio		25.9%	18.9%	55.2%	
Operating cost	MM $	21.8	13.5	22.9	58.2
	$/tonne of dry nodules	31.1	19.3	32.7	83.1
Operating cost ratio		37.5%	23.2%	39.3%	

Table 7 Comparative cost estimates (calculated) as on April, 2018

Process	Year	Pay metals	Capacity MMTPA	Capital cost $/tonne of dry nodules	Operating cost $/tonne of dry nodules
Reduction and hydrochloric acid leach process	1983	4	1.50	1472.02	398.18
Cuprion ammoniacal leach process	1985	3	3.00	1271.01	175.78
Reduction and hydrochloric acid leach process	1990	4	1.50	1218.13	311.00
High-temperature and high-pressure sulfuric acid leach process	2001	3	0.70	993.40	117.70

mate the comparative capital cost for a 3.0 MMTPA plant for all the four cases as illustrated in Fig. 5. The following equation expresses the rule of six-tenths (Chilton 1950):

$$C_b = C_a \left({}^{S_b}\!/\!{}_{S_a} \right) 0.6$$

where:

C_b is the approximate cost of plant of size S_b.
C_a is the known cost of plant having corresponding size S_a.
S_b/S_a is the ratio known as size factor, dimensionless.

Figure 5 reveals that for three metal recoveries, capital cost is considerably lower for Cuprion process than that of high-pressure sulfuric acid leaching process. In fact, for better economics, Cuprion process was modified for producing manganese concentrate from leach residue. However, not much information is available for this aspect. In the future, economic rent for commercial exploitation of nodules from seabed may also come into picture (Cornwell 1974). It may affect the overall economics of manganese nodule processing.

In the above sections, only a few processes have been taken into consideration. Das and Anand (2017) have described in detail the present scenario with respect to development of recent processes by the major organizations/investors/contractors of ISBA (100–500 kg/day). Most of these processes are modifications of Cuprion process: Caron's process (Caron 1924; Jana and Akrekar 1989; Srikanth et al. 1997; Jana et al. 1999a, b), INCO process (Xiang et al. 1999; Kotlinski et al. 2008), and HPAL process (Rodriguez et al. 2013). However, a few new flow sheets have also

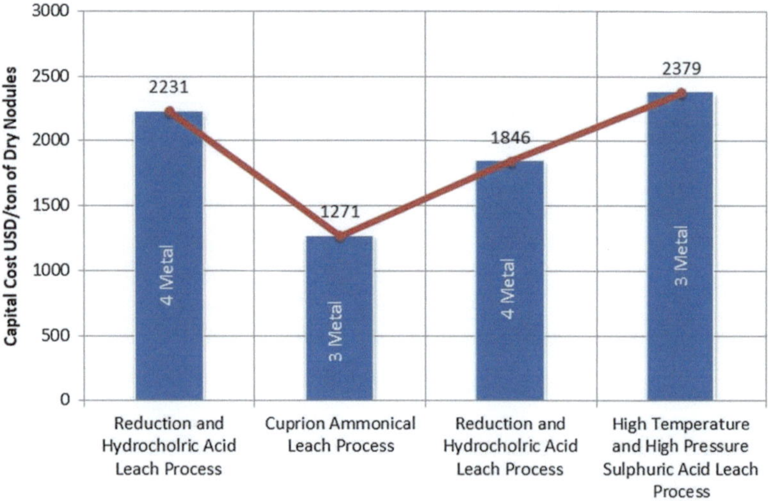

Fig. 5 Comparative estimated capital cost for a 3.0 MMTPA Plant as on April, 2018

been developed based on use of SO_2 in ammonia (Das 2001; Mittal and Sen 2003) or in acid medium (Kotlinski et al. 2008). All the present efforts are toward recovery of four metals either during leaching or from the leach residues. Most of these processes claim ≥90% recovery of four metals in addition to recovery of molybdenum and rare earths. As of now, the complete picture of these processes is not available; therefore, it is difficult to make a cost estimate of these processes.

When the manganese nodules are compared with nickel laterites, the potential values of metals are higher for manganese nodules as besides nickel and cobalt, additional potential of Cu and Mn recovery comes into picture. In the case of nickel laterites, since both Caron's processes (reduction roast ammoniacal leaching) and HPAL are the commercial processes used all over the world, the processing of nodules should also be commercially viable provided mining cost becomes comparable to land-based mining. The two main issues of concern for nodules are (i) the mining cost of nodules and (ii) its inherent water content. With the continuing efforts for making mining cost-effective and extraction processes more efficient combined with emerging markets for nickel and cobalt, the likelihood of processing manganese nodules may head toward a commercial reality.

6.6 The Environmental Considerations for Processing Nodules

Several processes have been developed and/or are being developed for efficient recovery of metal values from manganese nodules, but not much information is available on disposal of solids/liquid effluents generated during these processes. With time the environmental regulations are becoming more and more stringent, and it is mandatory for the industries to comply with these regulations. In fact, presently the emphasis is on developing processes with zero waste. Taking an example of ammonia leaching of nodules, the only soluble metal ions are Cu, Ni, Co, and to some extent Mo. These metals form only ~2.5% of total weight of nodules. Further treatment to recover manganese from leach residue by hydrometallurgical process will recover another 20–25% of Mn, and still the remaining solid waste will be about ~70% by weight of the original nodule. If we consider the moisture content of wet cake of residue to be as low as 30%, the wet residue cake that need to be dumped will almost match to the weight of initial nodules. Further, depending on the lixiviant used for leaching the residue, the treatment/washing of the residue has to be done to make it safe for disposal. Water balance needs to be carefully monitored for various process steps so as to produce minimum liquid effluent. Though SX-EW works in closed circuit, a bleed step is invariably incorporated to manage the alkali concentration. Both solid and liquid effluents need treatment before discarding. These steps will result in additional costs. Generally in ammonia leaching processes, efficient ammonia recovery circuits are included, but still ammonia losses do occur polluting the environment. These issues need to be addressed while finalizing the flow sheets. In the case of hydro-pyrometallurgical processes, after extracting, Cu, Ni, and Co, the leach residues are treated at high temperature to produce an alloy of ferromanganese or silico manganese. The slag produced in these processes is mostly stable and may be stored or used for land filling and other applications. Its use as a construction material or cement needs to be investigated.

In the case of pyro-hydrometallurgical processes, the first step is to produce a white metal of Cu-Ni-Co or a matte, and the slag produced may be stable. The white metal/matte does have some Fe and Mn as impurities, but the amount of residue generated after leaching of white metal/matte is small. The slag obtained can be treated again for the production of ferromanganese/silico manganese. The intensive energy requirements for pyrometallurgical treatment add to processing costs. More research and development efforts are required to tie the loose ends of metallurgical processes keeping in view the recent environmental regulations for disposal of effluents generated on processing the nodules. The idea of dumping effluents back to sea needs careful assessment of its impact on polluting the seawater.

7 Evolving Market

Metal market has undergone tremendous change since the time the above-discussed feasibility reports were prepared. As millions of people in emerging economies adopt a modern lifestyle, the demand for critical metals is soaring. However, the increasing demand causes the crisis of their supply because of either simple deficiency in the Earth's crust or geopolitical constraints which might create political issues for their supply. Consumer electronics (CE) and electric vehicles (EV) worldwide have been dramatically changing the human life landscape, which will or have demonstrated a wide application of lithium-ion battery (LIB). There are different kinds of lithium-ion batteries as illustrated in Table 8 (Reaugh 2018) available in the

Table 8 Lithium-ion battery chemistry (Reaugh 2018)

Type	Use	Chemistry	Li (%)	Co (%)	Ni (%)	Mn (%)	P (%)	Fe (%)
LCO	Portable electronics	$LiCoO_2$	11	89				
NCA	Tesla model S (EV)	$LiNiCoAlO_2$	11	14	73			
LMO	Nissan leaf (EV)	$LiMn_2O_4$	6	–		94		
LFP	Popular in China but switching to NMC	$LiFePO_4$	7	–			33	60
NMC (111)	Most popular in EV	$LiNiMnCoO_2$	12	30	30	28		
NMC (433)	Most popular in EV	$LiNiMnCoO_2$	11	27	36	26		
NMC (532)	Most popular in EV	$LiNiMnCoO_2$	11	18	45	26		
NMC (622)	Most popular in EV	$LiNiMnCoO_2$	11	18	54	17		
NMC (811)	Most popular in EV	$LiNiMnCoO_2$	11	9	72	8		

market depending upon the usage and cost. However, the trade-off is on account of the following factors:

- Specific energy
- Specific power
- Safety
- Cost
- Performance

As demand for electric cars increases, the need for cobalt, nickel, and manganese is also expected to soar, as these metals are the key components in EV batteries. In total, surging demand for electric vehicles is expected to push demand for lithium-ion batteries above 400 GWh by 2025 as per Benchmark Mineral Intelligence.

Lithium-ion batteries for EV contain about 11 kg of cobalt each. The battery demand for cobalt is expected to go from 46,000 tonnes in 2016 to 76,000 tonnes by the end of 2020. The Cobalt Institute (CI) estimates that total refined cobalt supply in 2017 from the main sources reporting their production was 116,937 tonnes, which is 24.5% higher than in the previous year. It was reported to the CI that Chinese refined production for 2017 was 69,600 tonnes which is an increase of 24,554 tonnes (or 54.5%) over that produced in calendar year 2016.

We are witnessing yet another revolution with the advent of electric vehicles (EVs) in Europe, North America, and Asia. The World Energy Council study published recently argues that 1 in 6 cars to be sold in 2020 will be electric to meet carbon dioxide emission standards. Global electric vehicle sales by region are illustrated in Fig. 6 (Colin 2017).

Nickel, which is primarily used for the production of stainless steel, is already one of the world's most important metal markets at over $20 billion in size. For this

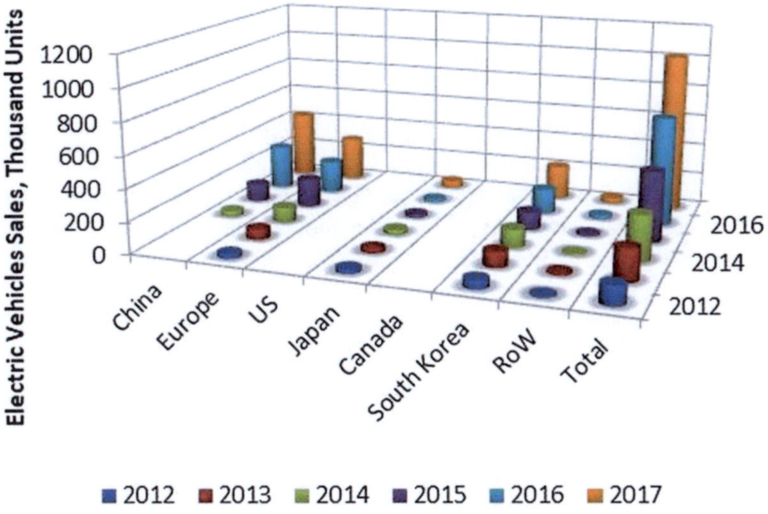

Fig. 6 Global electric vehicles sales by region (Colin 2017)

reason, how much the nickel market is affected by battery demand depends largely on EV penetration. EVs currently constitute about 1% of auto demand – this translates to 70,000 tonnes of nickel demand, about 3% of the total market. However, as EV penetration goes up, nickel demand will increase rapidly as well. A shift of just 10% of the global car fleet to EVs would create demand for 400,000 tonnes of nickel, in a 2 million tonne market (Nickel Institute, 2018).

While a number of different technologies are being explored for use in electric vehicle batteries, two in particular are considered the front runners, because of greater efficiency and ease of manufacturing. Both of these are nickel-based, comprising nickel/manganese/cobalt (NMC) or nickel/cobalt/aluminum (NCA). The cathode in the NMC-type battery is comprised of about 30% nickel and that in the NCA battery is about 80% nickel. In fact, while the two battery technologies are both called lithium-ion, nickel is by far the largest element in each of them. Demand for nickel for electric vehicles is forecast to climb to around 500,000 tonnes by 2030 according to a study by metals market analyst Roskill (https://sciencebusiness.net).

Electrolytic manganese dioxide (EMD) for the battery industry is expected to be the fastest-growing segment of the manganese market. The hydrometallurgical process developed by American Manganese Inc. along with Kemetco Research Inc. is capable of producing high-purity electrolytic manganese dioxide (EMD) and chemical manganese dioxide (CMD) critical for improving performance and safety of high-tech lithium-ion batteries. The company has received patent from the US Patent and Trademark Office for the company's manganese recovery process (American Manganese Inc. 2013). Now considering Mn, Ni, and Co of the manganese nodules, besides conventional applications of these metals, the battery applications for electronic vehicles may evolve as one of the most important usages. This may lead to a much better return on investments for manganese nodule plants.

8 Conclusions

Various feasibility studies have been conducted to estimate the CAPEX and OPEX of a commercial nodule processing plant. The technology for nodule processing is well developed for four metal recoveries. However, to make the flow sheet commercially more attractive, recovery of other metals like molybdenum, rare earths, etc. is currently being explored. Technology for nodule mining still seems to be a challenge, and more efforts are required to make it cost-effective.

The nodule processing flow sheet should be flexible in terms of product mix as metals of relevance in polymetallic nodules are finding newer applications in industries like infrastructure, building and construction, electrical and electronics, batteries, etc. Lithium-ion batteries have pushed the demand for cobalt, nickel, and manganese in recent times. Traditional markets like Europe and the USA are still important but have been overtaken by China during the last decade. India is also becoming an important player in this sector being one of the fastest-growing economies in the world.

In view of these emerging markets, sustainable exploitation of deep-sea minerals seems to be the answer in order to meet the growing demands of the world industry.

References

Agarwal, J. C., & Wilder, T. C. (1974). Recovery of metal values from manganese nodules. US Patent 3,788,841, Jan 29, 1974.

Agarwal, J. C., & Wilder, T. C. (1975). Recovery of metal values from manganese nodules. Canadian Patent 980,130, Dec 23, 1975.

Agarwal, J. C., Beecher, N., Davies, D. S., Hubred, G. L., Kakaria, V. K., & Kust, R. N. (1976). Processing of ocean nodules: A technical and economic review. *JOM, 28*(4), 24–31.

Agarwal, J. C., Barner, H. E., Beecher, N. et al. (1978). Kennecott process for recovery of copper, nickel, cobalt and molybdenum from ocean nodules. *Paper presented at AIME annual meeting*. Denver, CO, Feb, SME preprint 788–789.

Agarwal, J. C., Barner, H. E., Beecher, N., et al. (1979). Kennecott process for recovery of copper, nickel, cobalt and molybdenum from ocean nodules. *Mining and Engineering, 31*(12), 1704–1709.

Agarwal, B., Hu, P., Placidi, M., Santo, H., & Zhou, J. J. (2012). Feasibility study on manganese nodules, recovery in the Clarion-Clipperton Zone. In R. A. Shenoi, P. A. Wilson, & S. S. Bennett (Eds.), *Seabed exploitation* (The LRET (The Lloyd's Register Educational Trust) Collegium 2012 Series) (Vol. 2). Highfield, Southampton: The Print Centre, University of Southampton. isbn:978-0-854-32950-2.

American Manganese Inc. (2013). Manganese extraction process. US patent No, 8460681, June 2013.

Andrews, B. V., Flipse, J. E., Brown, F. C. (1983). The economic viability of a four metal pioneer deep sea venture, Texas A and M University, Sea Grant College program, TAMU-SG-84-201, pp. 201.

Apavasileiou Konstantinos. (2014). Near future REE resources for Europe – the new frontier of marine exploration, mining and processing ERES2014: 1st European rare earth resources conference, Milos, 04–07/09/2014.

Caron, M. H. (1924). Process of recovering values from nickel and cobalt-nickel ores. US Patent 1,487,145, Mar 18, 1924.

Charles River Associates, Inc. (1980). Energy requirements for metals production: Comparison between ocean nodules and land-based resources-; Prepared for National Oceanic and Atmospheric Administration, Washington, DC, Sept 1980

Charles, C., Herouin, G., Mouviel, F., & Bernard, J. (1990). Views on future nodule technology based on INFREMER GEMONOD studies. *Materials and Society, 14*(3–4), 299–326.

Chilton, C. H. (1950). Six-tenths factor applies to complete plant costs. *Chemical Engineering, 57*, 112–114.

Clark, A. L., Lum, J. A., Li, C., Icay, W., Morgan, C. Yoshiaki, I. (1995). Economic and development potential of manganese nodules within the Cook Island Economic Zone (EEZ), Pacific Islands Development Program, and Program on Resources, Energy and Minerals (https://www.google.co.in/search?q=Economic+and+development+potential+of+ manganese+nodules+within+the+CooK+Island+economic+zone+(EEZ)&oq=Economic+and+development+potential+of+manganese+nodules+within+the+CooK+Island+economic+zone+(EEZ)&aqs=chrome..69i57.1641j0j7&sourceid=chrome&ie=UTF-8).

Cornwell, N. W. (1974). Manganese nodule mining and economic rent. *Natural Resources Journal, 14*(4), 519–531.

Cronan, D. S. (1980). *Underwater minerals* (p. 362). London: Academic.
Das, R. P. (2001). India's demonstration metallurgical plant to treat ocean nodule. *Proceedings of 4th ocean mining symposium*, ISOPE, Szczecin, Poland, Sept 23–27, 163–167.
Das, R. P., & Anand, S. (2017). Metallurgical processing of polymetallic nodules – chapter 12. In R. Sharma (Ed.), *Deep sea mining* (pp. 365–394). Cham: Springer.
Flentje, W., Lee, S. E., Virnovskaia, A., Wang, S., & Zabeen, S. (2012). In R. A. Shenoi, P. A. Wilson, & S. S. Bennett (Eds.), *Polymetallic nodule mining: Innovative concepts for commercialization* (The LRET (The Lloyd's Register Educational Trust) Collegium 2012 Series) (Vol. 5). Highfield, Southampton: The Print Centre University of Southampton. isbn:978-0-854-32953-3.
Fuerstenau, D. W., & Han, K. N. (1983). Metallurgy and processing of marine manganese nodules. *Mineral Processing Technology Revolutionary, 1*, 1–83.
Glasby, G. P. (Ed.). (1977). *Marine manganese deposits* (p. 523). Amsterdam: Elsevier.
Han, K. N. (1997). Strategies for processing of ocean floor manganese nodules. *Transactions of the Indian Institute of Metals, 51*(1), 41–54.
Han, K. N., & Fuerstenau, D. W. (1986). Extraction behaviour of metallic elements from deep sea manganese nodules in reducing medium. *Marine Mining, 2*, 155–169.
Hanieg, G., & Meixner, M. J. (1974). Pressure leaching of manganese nodule with sulphuric acid. *Erzmetall, 27*(7–8), 335.
Haynes, B. W., Law, S. L., Barron, D. C., et al (1985). Pacific manganese nodules: Characterization and processing. Bulletin 679, US Bureau of Mines.
Hein, J. R. (2012). Prospects for rare earth elements from marine minerals, International Sea Bed Authority Briefing Paper 02/12.
Hein, J. R. (2016). *Manganese nodules*. Santa Cruz: U.S. Geological Survey.
Hein, J. R., & Koschinsky, A. (2014). Deep-ocean ferromanganese crusts and nodules, Chapter 11. In H. D. Holland & K. K. Turekian (Eds.), *Treatise on geochemistry* (Vol. 13, 2nd ed., pp. 273–291). Amsterdam: Elsevier.
Hein, J. R., & Petersen, S. (2013). The geology of manganese nodules, Chapter 1. In E. Baker, Y. Beaudoin (Eds.), *Deep sea minerals: Manganese nodules; a physical, biological, environmental, and technical review* (pp. 7–18). Secretariat of the Pacific Community.
Hein, J. R., Mizell, K., Koschinsky, A., & Coinrad, T. A. (2013). Deep-ocean mineral deposits as a source of critical metals for high and green technology applications: Comparison with land based resource. *Ore Geology Reviews, 51*, 1–14.
Hillman, T., & Gosling, B. B. (1985). *Mining deep ocean manganese nodules: Description and economic analysis of a potential venture*. Washington, DC: U.S. Department of the Interior. https://sciencebusiness.net/news/how-nickel-makes-electric-vehicles-go-green. Accessed Jan 18, 2018.
Hubred, G. L. (1980). Manganese nodule extractive metallurgy: A review 1973–1978. *Marine Mining, 2*, 191–212.
ISBA. (2008). *Workshop on polymetallic nodule mining technology, status and challenges ahead*. Chennai: NIOT. http://www.isa.org.jm/files/documents/EN/Pubs/Chennai.pdf. Accessed 18–22 Feb 2008.
Jana, R. K., & Akerkar, D. D. (1989). Studies of the metal–ammonia–carbon dioxide–water system in extraction metallurgy of poly metallic sea nodules. *Hydrometallurgy, 22*, 363–378.
Jana, R. K., Srikanth, S., Pandey, B. D., et al. (1999a). Processing of deep sea manganese nodules at NML for recovery of copper, nickel and cobalt. *Metals and Materials Processing, 11*, 133–144.
Jana, R. K., Pandey, B. D., & Premchand. (1999b). Ammoniacal leaching of roast reduced deep-sea manganese nodules. *Hydrometallurgy, 53*, 45–56.
Kotlinski, R., Stoyanova, H. V., & Avramov, H. A. (2008). An overview of the interoceanmetal (IOM) deep sea technology development (mining and processing) http://www.isa.org.jm/files/documents/EN/Workshops/Feb2008/IOM-Abst.pdf

Mero, J. L. (1965). *The mineral resources of the sea* (Vol. 1, pp. iii–iix). Amsterdam: Elsevier, 1–312.

Mittal, N. K., & Sen, P. K. (2003). India's first medium scale demonstration plant for treating polymetallic nodules. *Minerals Engineering, 6*, 865–868.

Mohwinkel, D., Kleint, C., & Koschinsky, A. (2014). Phase associations and potential selective extraction methods for selected high-tech metals from ferromanganese nodules and crusts with siderophores. *Applied Geochemistry, 43*, 13–21.

Monhemius, A. J. (1980). The extractive metallurgy of deep sea manganese nodule. In R. Burkin (Ed.), *Topics in non ferrous extractive metallurgy* (pp. 42–69). London: Society of chemical industry.

Mukhopadhyay, R., Ghosh Anil, K., & Iyer Sridhar, D. (2018). Resource assessment, Chapter 8. In *The Indian Ocean nodule field: Geology and resource potential* (2nd ed.). Amsterdam: Elsevier.

Neuschutz, D., Scheffler, U., & Junghans, H. (1977). Verfahren ZurAufarbeitung von Manganknollen Durch Schwefelsaure Druck laugung. (Method for the processing of manganese nodules by sulphuric acid pressure leaching). *Erzmetall, 30*(2), 61–67.

Nickel Institute. (2018). Science/business, www.sciencebusiness.net/news/how-nickel-makes-electric-vehicles-go-green.

Parhi, P. K., Park, K. H., Kim, H. I., et al. (2011). Recovery of molybdenum from the sea nodule leach liquor by solvent extraction using Alamine 304-I. *Hydrometallurgy, 105*, 195–200.

Parhi, P. K., Park, K. H., Nam, C. W., Park, J. T., et al. (2013). Extraction of rare earth metals from deep sea nodule using H_2SO_4 solution. *International Journal of Mineral Processing, 119*, 89.

Parhi, P. K., Park, K. H., Nam, C. W., et al. (2015). Liquid-liquid extraction and separation of total rare earth (RE) metals from polymetallic manganese nodule leaching solution. *Journal of Rare Earths, 3*(2), 207–213.

Premchand, Jena R. K. (1999). Processing of polymetallic sea nodules: an overview. *Proceedings of 3rd ocean mining symposium*, ISOPE, Goa, India, Nov 8–10, 237–245.

Reaugh, L. W. (2018). American Manganese Inc., www.americanmanganeseinc.com

Rodriguez, M. P., Aja, R., Miyares, R. C. (2013). Optimization of the existing methods for recovery of basic metals from polymetallic nodules. *Proceedings of the 10th ISOPE ocean mining and gas hydrates symposium*. Szczecin, Poland, Sept 22–26, 2013, p 173.

Sazbo, L. J. (1976). Recovery of metal values from manganese deep sea nodules using ammoniacal cuprous leach solutions US patent 3,983,017, Sept 28, 1976.

Sen, P. K. (2010). Metals and materials from deep sea nodules: An outlook for the future. *International Materials Reviews, 55*(6), 364–391.

Soreide, F. (2001). Deep ocean mining reconsidered: A study of the manganese nodule deposits. In *Cook Island*. fourth ISOPE ocean mining symposium, Sept 23–27, Szczecin, Poland, ISOPE-M-01-015, International Society of Offshore and Polar Engineers.

Sridhar, R. (1974). Thermal upgrading of sea nodules. *Journal of Metals, 26*(12), 18–22.

Sridhar, R., Jones, W. E., & Warner, J. S. (1976). Extraction of copper, nickel, cobalt from sea nodules. *Journal of Metals, 28*(4), 32–37.

Sridhar, R., Warner, J. S., Bell, M. C. E. (1977). Non-ferrous metal recovery from deep sea nodules. US Patent 4,049,438, September 20, 1977.

Srikanth, S., Alex, T. C., Agrawal, A. et al. (1997). Reduction roasting of deep-sea manganese nodules using liquid and gaseous reductants. *Proceedings of 2nd ocean mining symposium*, Seoul, South Korea, Nov 24–26, 1997, (pp. 177–184).

Volkmann, S. E., Kuhn, T., & Lehnen, F. (2018). A comprehensive approach for a techno-economic assessment of nodule mining in the deep sea. *Mineral Economics*. https://doi.org/10.1007/s13563-018-0143-1.

Wikipedia. (2018). https://ipfs.io/ipfs/QmXoypizjW3WknFiJnKLwHCnL72vedxjQkDDP1mXWo6uco /wiki/Manganese_nodule.html.

Xiang, Z., Zequan, H., Yujun, S. et al. (1999). The smelting –rusting-solvent extraction processes to recover valuable metals from polymetallic nodules. *Proceedings of 3rd ocean mining symposium*, ISOPE, Goa, India, Nov 8–10, 1999, (pp. 227–231).

Navin Mittal obtained his Bachelor of Engineering degree in Chemical Engineering from National Institute of Technology, Srinagar, India. He has worked in the area of extractive metallurgy for almost 32 years. Navin led a team of engineers in design of 500 kg/day polymetallic nodules pilot plant in India. He has also worked on Cuprion ammoniacal leach process during his association with SGS in Canada. He has designed India's first column flotation plant for zinc cleaning duty. Navin was also associated with the design of 10,000 tonnes/annum zeolite-A plant. He has to his credit design of commercial solvent extraction plants for copper, nickel, and cobalt. He along with co-workers published several technical papers in various journals and conferences. Currently, Navin runs his own company and provides consultancy services for hydrometallurgical processes with special emphasis on recovery of metals from lean and secondary resources.

Dr. Shashi Anand obtained her MSc and PhD degrees in chemistry from Indian Institute of Technology, Delhi. She worked in hydro-electrometallurgy Department of Institute of Mineral and Materials Technology, IMMT (earlier RRL) for almost 32 years. Dr. Anand was associated on processing of Indian Ocean polymetallic nodules for three decades. She along with her co-workers published several papers and obtained a number of patents related to processing of Manganese nodules. Her main achievements include development of low-temperature leaching of polymetallic nodule and impurity control during leaching. Dr. Anand successfully led a team of researchers from IMMT side, who operated India's first 500 kg/d nodule processing plant. Currently, she provides consultancy for hydrometallurgical processes and nanomaterial synthesis.

Index

R. Sharma (ed.), *Environmental Issues of Deep-Sea Mining*,
https://doi.org/10.1007/978-3-030-12696-4

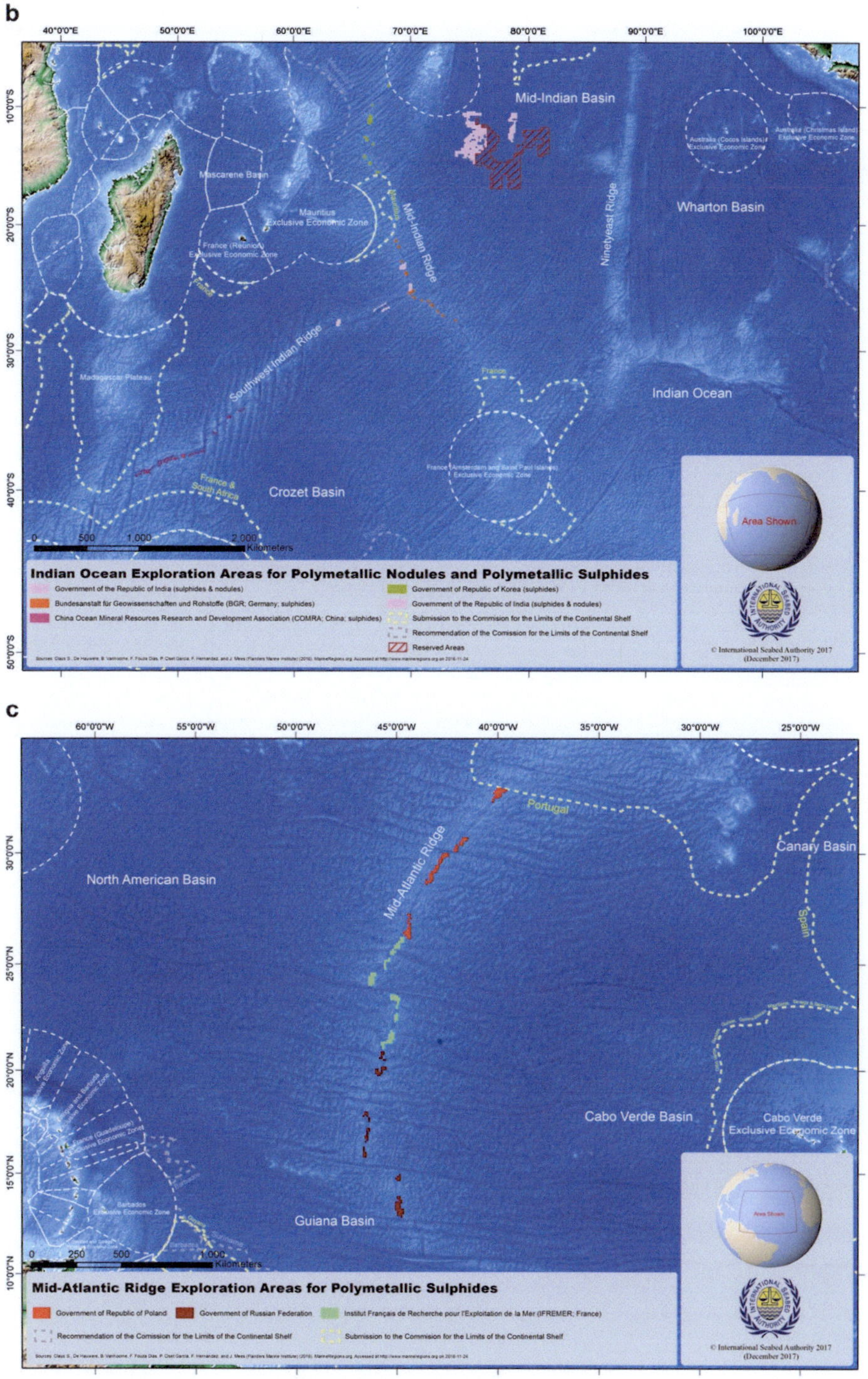

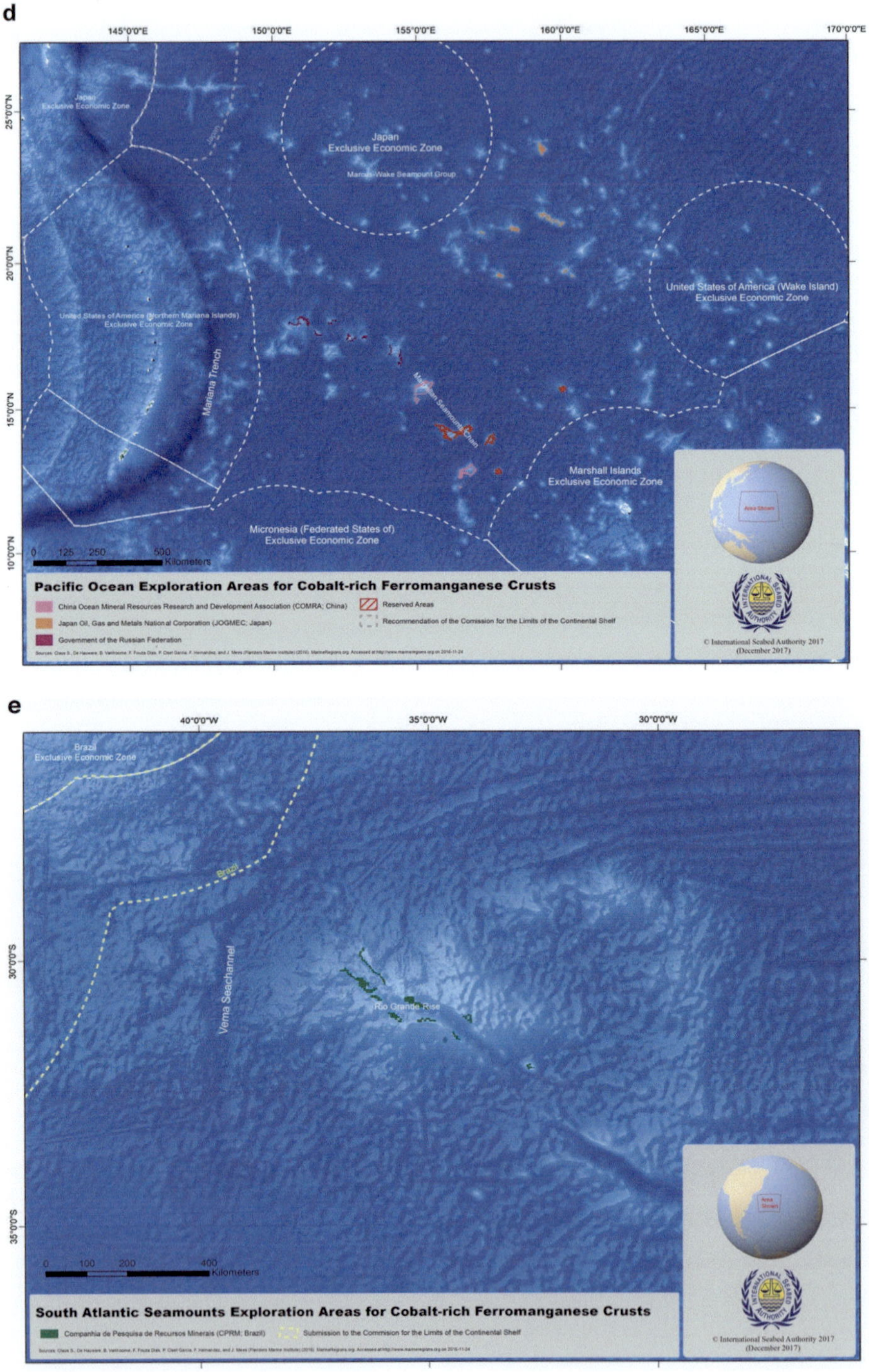

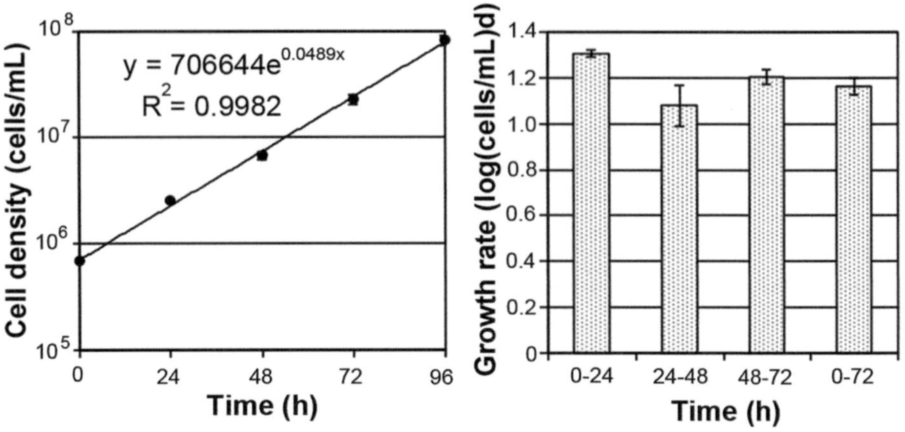

Fig. 2 Growth rates of *Cyanobium* sp. (NIES-981) for 96 h after culture under control conditions (left) and section-by-section-specific growth rates (0–24, 24–48, and 48–72 h) (right). Data are means of three independent experiments. Error bars show standard deviation

Table 2 Comparisons of EC_{50} and low-toxic-effect values (EC_{10}, EC_5, and NOEC) between *Cyanobium* sp. and *Pseudokirchneriella subcapitata*

	Cyanobium sp.				*P. subcapitata*		
	72-h EC_{50} (95% confidence limit)	72-h EC_{10} (95% confidence limit)	72-h EC_5 (95% confidence limit)	NOEC	72-h EC_{50}	72-h EC_{10}	NOEC
3,5-DCP (mg/L)	1.71 (1.60–1.82)	0.940 (0.934–0.947)	0.793 (0.534–0.896)	0.625	1.8–2.3[a, b]	0.91[b]	0.75[a]
CAT (μg/L)	105 (94.7–118)	26.8 (24.2–28.4)	18.2 (15.3–20.3)	12.5	100–297[c, d, e]	–	10[e]
DFF (μg/L)	4.10 (3.70–4.62)	1.26 (1.23–1.28)	0.902 (0.733–1.25)	0.750	0.27–1.23[f, g]	–	–
Cu^{2+} (mg/L)	0.550 (0.475–0.632)	0.139 (0.0793–0.203)	0.0947 (0.045–0.135)	0.113	0.0075–0.047[h, i]	0.012[i]	0.0018–0.01[h, i]
Cr(VI) (mg/L)	4.61 (4.24–4.96)	2.19 (2.06–2.23)	1.77 (1.27–2.13)	1.76	0.30–0.488[a, b]	0.092[b]	0.064[a]

EC effective concentration, *NOEC* no observed effect concentration, *3,5-DCP* 3,5-dichlorophenol, *CAT* simazine, *DFF* diflufenican
[a]Comber et al. (1995)
[b]Mayer et al. (1998)
[c]Okamura et al. (2000)
[d]Pérez et al. (2011)
[e]Sbrilli et al. (2005)
[f]Katsumata et al. (2009)
[g]Weyman et al. (2012)
[h]Franklin et al. (2002)
[i]Radix et al. (2000)

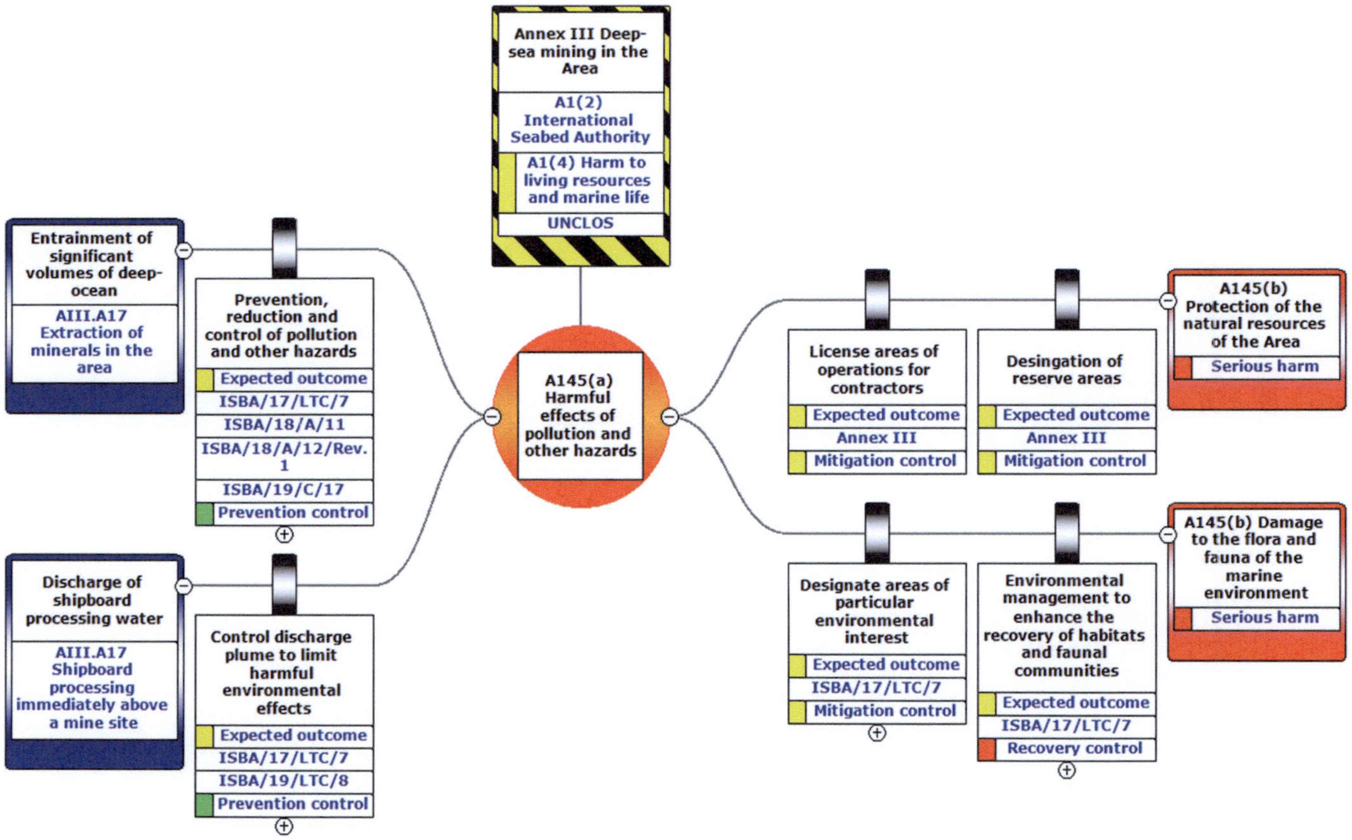

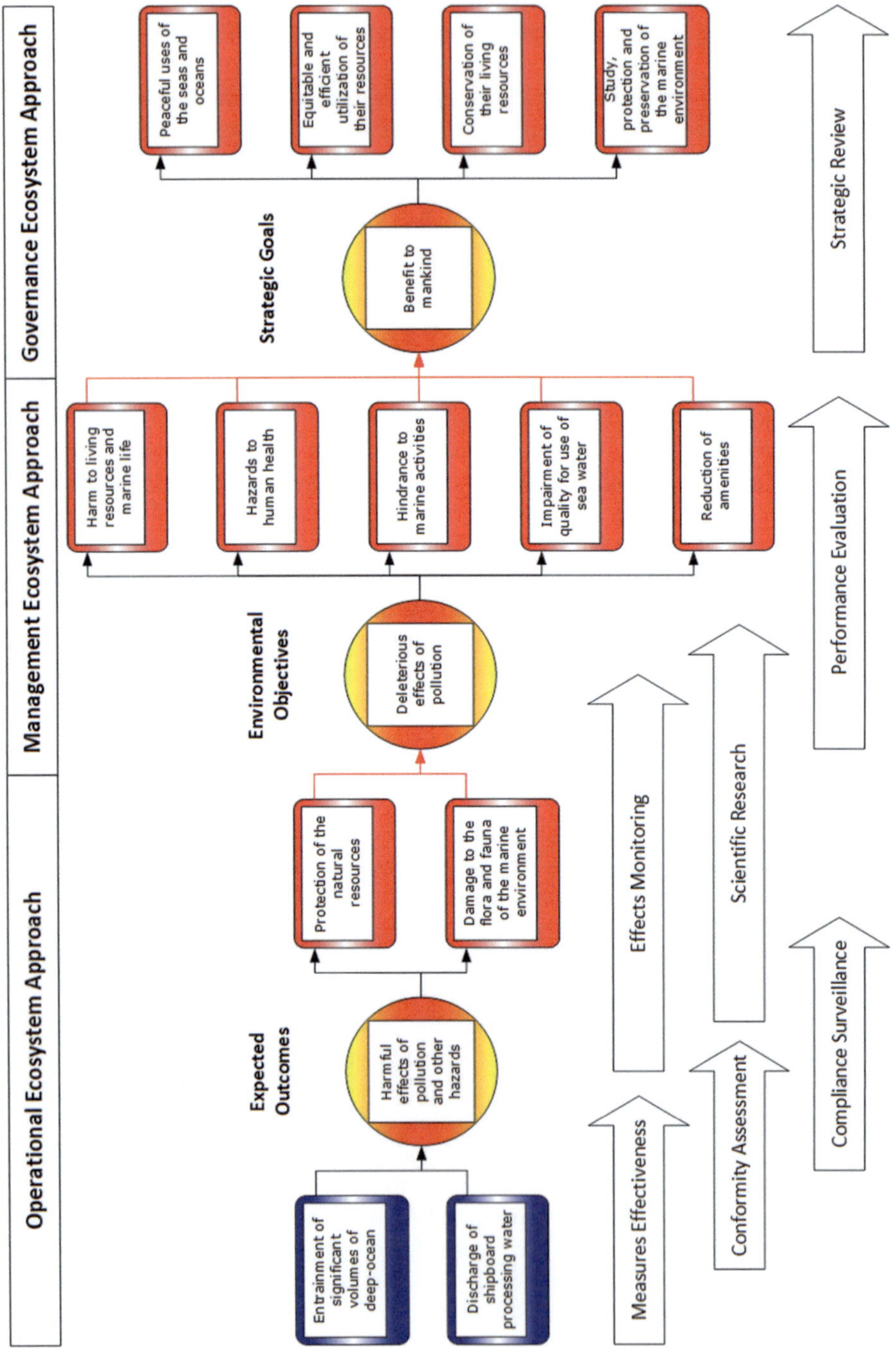

MIX
Papier aus verantwortungsvollen Quellen
Paper from responsible sources
FSC® C105338

If you have any concerns about our products,
you can contact us on
ProductSafety@springernature.com

In case Publisher is established outside the EU,
the EU authorized representative is:
Springer Nature Customer Service Center GmbH
Europaplatz 3, 69115 Heidelberg, Germany

Printed by Libri Plureos GmbH
in Hamburg, Germany